Lecture Notes in Networks and Systems

Volume 76

The series "Lecture Notes in Networks and Systems" publishes the latest developments in Networks and Systems—quickly, informally and with high quality. Original research reported in proceedings and post-proceedings represents the core of LNNS.

Volumes published in LNNS embrace all aspects and subfields of, as well as new challenges in, Networks and Systems.

The series contains proceedings and edited volumes in systems and networks, spanning the areas of Cyber-Physical Systems, Autonomous Systems, Sensor Networks, Control Systems, Energy Systems, Automotive Systems, Biological Systems, Vehicular Networking and Connected Vehicles, Aerospace Systems, Automation, Manufacturing, Smart Grids, Nonlinear Systems, Power Systems, Robotics, Social Systems, Economic Systems and other. Of particular value to both the contributors and the readership are the short publication timeframe and the world-wide distribution and exposure which enable both a wide and rapid dissemination of research output.

The series covers the theory, applications, and perspectives on the state of the art and future developments relevant to systems and networks, decision making, control, complex processes and related areas, as embedded in the fields of interdisciplinary and applied sciences, engineering, computer science, physics, economics, social, and life sciences, as well as the paradigms and methodologies behind them.

**** Indexing: The books of this series are submitted to ISI Proceedings, SCOPUS, Google Scholar and Springerlink ****

More information about this series at http://www.springer.com/series/15179

Isak Karabegović
Editor

New Technologies, Development and Application II

Editor
Isak Karabegović
Technical Faculty Bihać
University of Bihać
Bihać, Bosnia and Herzegovina

ISSN 2367-3370 ISSN 2367-3389 (electronic)
Lecture Notes in Networks and Systems
ISBN 978-3-030-18071-3 ISBN 978-3-030-18072-0 (eBook)
https://doi.org/10.1007/978-3-030-18072-0

This Springer imprint is published by the registered company Springer Nature Switzerland AG
The registered company address is: Gewerbestrasse 11, 6330 Cham, Switzerland

Interdisciplinary Research of New Technologies, Their Development and Application

The content of this book is very interesting and important as it covers a wide range of new technologies and technical disciplines including complex systems such as: cyber-physical systems, robotics, mechatronics systems, automation, manufacturing, autonomous systems, sensor, networks, control systems, energy systems, automotive systems, biological systems, vehicular networking and connected vehicles, effectiveness and logistics systems, smart grids, nonlinear systems, power systems, social systems, economic systems and other. The papers included in this content have been presented at the International Conference New Technologies, Development and Application, held in Sarajevo, Bosnia and Herzegovina, on 27–29 June 2019. Majority of organized conferences are usually focusing on a narrow part of the issues within a certain discipline while conferences such these are rare. There is a need to hold such conferences. The value of this conference is that various researchers, programmers, engineers and practitioners come to the same place where ideas and latest technology achievements are exchanged. Such events lead to the creation of new ideas, solutions and applications in the manufacturing processes of various technologies. New coexistence is emerging, horizons are expanding, and unexpected changes and analogies arise. Best solutions and applications in technologies are critically evaluated.

The first chapter starts with the implementation of Industry 4.0, cyber-physical systems, robots, robots systems, mechatronic systems, automation of production processes and advanced production. The first article provides an analysis and implementation of Industry 4.0 and robotic systems. The next article presents the new concept of restructuring system in the manufacturing processes of companies. The following article analyses the adaptation of intelligent mobile robots in a changing environment. One of the papers provides a series of approaches that serve as the basis for development and application of the principles and the Kaizen concept of Lean Six Sigma (LSS) in the wood processing sector. The last article offers the definition of a mass unit, kilogram, in two experiments.

The second chapter is devoted to innovative and interdisciplinary applications of advanced technologies (IATs). The first article is about Smart Campus Project which addresses necessary analysis and recommendations for the implementation of a smart and sustainable campus. The second article gives a brief overview of the swarm intelligence algorithm used in software engineering. The third article provides an analysis of the performance of the HDLC protocol for data transmission, detection and correction of traffic error, which is used in data exchange for one physical channel. The last work in this chapter offers an analysis of the economic efficiency and its components of B&H and groups of comparative countries.

The third chapter is devoted to transport systems, logistics and intelligent transport systems. This chapter starts with intelligent mobility. The second article provides a computer simulation of the collision phase of two vehicles with two approaches: one approach is based on energy, and the other approach is based on the impulse. Another article is TSCLab (laboratory for controlling traffic), MATLAB's tool for the evolution of the adaptive traffic control system (ATCS). One of the articles offers an ergonomic driver's seat for the created virtual environment of mid-range cars.

The fourth chapter is devoted to electric power systems with a different spectrum of topics. The first article is about development in solar power micro gas turbines and waste heat recovery. The second is about laboratory research of the influence of pulsating flow of flue gases on heat transfer. Next are the development of a system for the drainage of oil vapour from the housing of large hydraulic aggregates and then the possibility of using coal with biomass in a thermal power plant to the impact of using electric cars on the environment.

Chapter Five is devoted to new methods in the agricultural culture of a wide range of topics: removing heavy metals from water such as Pb (II) and Zn (II), improving the quality of mineral water using a selective ion-exchange column, then establishing the level of heavy metals (Pb, Cu, Zn) and health risks assessment based on the contents of heavy metals in urban soil, assessment of the water quality of the Una River in the Una National Park on the basis of selected microbiological parameters, as well as research of the biological characteristics of the distribution of chestnut gall wasp (Dryocosmus kuriphilus Yasumatsu) in Bosnia and Herzegovina.

The sixth chapter focuses on the field of geodesy, construction, new materials and sustainable innovation and others. The first article focuses on the detection of changes in the river bed, and the identification of boundary changes using topographic data from different epochs. The following article analyses the building thermal insulation material based on sheep wool, the advantages and disadvantages of the production of explosive materials "IN-SITU" ("NALIM" technology). One article gives new approaches and techniques of motivation for construction industry engineers in B&H, then assessing the active efficiency of active charcoal with filter in the process of water purification. One of the articles describes sustainable innovations in architectural design.

The whole content of this book is intended to a wide range of technical systems, different technical disciplines in order to apply the latest solutions and achievements in technologies and to improve manufacturing processes in all disciplines where systemic thinking have a very important role in the successful understanding and building of human, natural and social systems. This content is the second in a series of publications that are intended to the development and implementation of new technologies in all industries.

Isak Karabegović

Contents

Computer Science, Information and Communication Technologies, e-Business

Intelligent Transport Systems, Logistics, Traffic Control

List of Contributors

Oday I. Abdullah Department of Energy Engineering, College of Engineering, University of Baghdad, Baghdad, Iraq; Hamburg University of Technology, Hamburg, Germany

Naida Ademović Faculty of Civil Engineering, University of Sarajevo, Sarajevo, Bosnia and Herzegovina

Avdul Adrović Faculty of Natural Sciences and Mathematics, University of Tuzla, Tuzla, Bosnia and Herzegovina

Mohammed Alazeezi Emirates Advanced Research & Technology Holding LLC, Abu Dhabi, UAE

Asmir Aldžić Faculty of Health Studies, University of Bihac, Bihać, Bosnia and Herzegovina

Aldina Aldžić-Baltić Faculty of Health Studies, University of Bihac, Bihać, Bosnia and Herzegovina

Aida Šahinović Alešević Faculty of Science, University of Sarajevo, Sarajevo, Bosnia and Herzegovina

Ismar Alagić TRA Tešanj Development Agency, Tešanj, Bosnia and Herzegovina; Faculty of Mechanical Engineering, University of Zenica, Zenica, Bosnia and Herzegovina; Faculty of Engineering and Natural Sciences, International University of Sarajevo, Ilidža, Sarajevo, Bosnia and Herzegovina

Jafar Alzaili Centre for Compressor Technology, City University of London, London, UK

Pontus Anderson Novacast Systems AB, Ronneby, Sweden

Nedim Babic Faculty of Electrical Engineering, University of Tuzla, Tuzla, Bosnia and Herzegovina

Alen Bajrić Faculty of Natural Sciences and Mathematics, University of Tuzla, Tuzla, Bosnia and Herzegovina

Senaid Bajrić Zagrebinspekt d.o.o., Mostar, Bosnia and Herzegovina

Ivan Balashev TU Gabrovo, Gabrovo, Bulgaria

Lejla Banjanovic-Mehmedovic Faculty of Electrical Engineering, University of Tuzla, Tuzla, Bosnia and Herzegovina

Krsto Batinić Faculty of Mechanical Engineering, University of East Sarajevo, Istočno Sarajevo, Bosnia and Herzegovina

Hande Bayazıt Middle East Technical University, Ankara, Turkey

Sabina Begić Faculty of Technology Tuzla, University of Tuzla, Tuzla, Bosnia and Herzegovina

N. V. Bekk Department of Industrial Design, Novosibirsk State University of Architecture, Design and Arts, Novosibirsk, Russia

Svetlana K. Belošević Faculty of Technical Science, University of Priština, Kosovska Mitrovica, Serbia

Mirha Bičo Ćar School of Economics and Business, University of Sarajevo, Sarajevo, Bosnia and Herzegovina

Rejhana Blazevic Faculty of Mechanical Engineering, University of Sarajevo, Sarajevo, Bosnia and Herzegovina

Savković Borislav Faculty of Technical Science, University of Novi Sad, Novi Sad, Serbia

Liubov Bovnegra Odessa National Polytechnic University (ONPU), Odessa, Ukraine

Irina Andreevna Boychenko Department of Industrial Design, Novosibirsk State University of Architecture, Design and Arts, Novosibirsk, Russia

Jožef Božo Technical College of Applied Sciences in Zrenjanin, Zrenjanin, Republic of Serbia

Amra Bratovcic Faculty of Technology, University of Tuzla, Tuzla, Bosnia and Herzegovina

Lucija Brezočnik Faculty of Electrical Engineering and Computer Science, University of Maribor, Maribor, Slovenia

Adis Bubalo JP Electric Power Industry BiH d.d. Sarajevo, Jablanica, Bosnia and Herzegovina

Adis Bubalo Faculty of Mechanical Engineering, University “Džemal Bijedić” of Mostar, Mostar, Bosnia and Herzegovina

Aleksandr I. Burya Ukrainian Technological Academy (UTA), Kiev, Ukraine; Dniprovsk State Technical University, Kamianske, Ukraine

Malik Čabaravdić University of Zenica, Zenica, Bosnia and Herzegovina

Fuad Ćatović Faculty of Civil Engineering, "Džemal Bijedić" University of Mostar, Mostar, Bosnia and Herzegovina

Zoran Cekic Faculty of Civil Engineering and Management, University Union Nikola Tesla, Belgrade, Serbia

Aida Čolo Engineering Faculty, University of Sarajevo, Sarajevo, Bosnia and Herzegovina

Francesca Colucci Meid4 S.r.l., Fisciano, Salerno, Italy

Ramzija Cvrk Faculty of Technology Tuzla, University of Tuzla, Tuzla, Bosnia and Herzegovina

Predrag Dašić High Technical Mechanical School of Professional Studies, Trstenik, Serbia; SaTCIP Publisher Ltd., Vrnjačka Banja, Serbia; Faculty of Strategic and Operational Management (FSOM), Novi Beograd, Serbia

Remzo Dedić University of Mostar, Mostar, Bosnia and Herzegovina

Samira Dedić Faculty of Health Studies, University of Bihać, Bihać, Bosnia and Herzegovina

Luka Dedic HR-10000, Treskavicka, Zagreb, Croatia

Zemira Delalić Biotechnical Faculty, University of Bihac, Bihac, Bosnia and Herzegovina

Alen Delić TTU energetik d.o.o., Tuzla, Bosnia and Herzegovina

Senid Delic Faculty of Mechanical Engineering, University of Sarajevo, Sarajevo, Bosnia and Herzegovina

Daut Denjo Faculty of Mechanical Engineering, "Džemal Bijedić" University of Mostar, Mostar, Bosnia and Herzegovina

Fatih Destović Faculty of Educational Sciences, University of Sarajevo, Sarajevo, Bosnia and Herzegovina

Evgueni A. Deulin MT-11, BMSTU, Moscow, Russia

Mensud Đidelija JP Elektroprivreda BiH d.d. Sarajevo, Jablanica, Bosnia and Herzegovina

Maida Djapo Faculty of Education, University "Dzemal Bijedic" of Mostar, Mostar, Bosnia and Herzegovina

Ivana Domljan University of Mostar, Mostar, Bosnia and Herzegovina

Vjekoslav Domljan SSST Sarajevo, Sarajevo, Bosnia and Herzegovina

Sarjanovic Dražen Sara-Mont Doo, Beograd, Serbia

Darko Drev Faculty of Civil and Geodetic Engineering, University of Ljubljana, Ljubljana, Slovenia

Himzo Đukić Faculty of Mechanical Engineering, University "DžemalBijedić" of Mostar, Mostar, Bosnia and Herzegovina

Ješić Dušan International Technology Management Academy, Novi Sad, Serbia

Aida Džaferović Faculty of Biotechnical, University of Bihac, Bihać, Bosnia and Herzegovina

Edin Džiho Faculty of Mechanical Engineering, "DžemalBijedić" University of Mostar, Mostar, Bosnia and Herzegovina

Ahmed El Sayed Faculty of Engineering and Natural Sciences, International Burch University, Sarajevo, Bosnia and Herzegovina

Predrag Elek Faculty of Mechanical Engineering, University of Belgrade, Belgrade, Serbia

Şeyda Ertekin Middle East Technical University, Ankara, Turkey

Iztok Fister Jr. Faculty of Electrical Engineering and Computer Science, University of Maribor, Maribor, Slovenia

Vodopivec Franc Institute of Metals and Technology, Ljubljana, Slovenia

Mehdi Ganji Willdan Energy Solutions, Anaheim, USA

Lucian Gal Aurel Vlaicu University, Arad, Romania

Melisa Gazdić International Burch University, Sarajevo, Bosnia and Herzegovina; Green Council, Sarajevo, Bosnia and Herzegovina

Ivan Grujic Faculty of Engineering, University of Kragujevac, Kragujevac, Serbia

Jasna Glisovic Faculty of Engineering, University of Kragujevac, Kragujevac, Serbia

Murat Göl Middle East Technical University, Ankara, Turkey

Dušan Golubović Faculty of Mechanical Engineering, University of East Sarajevo, Istočno Sarajevo, Bosnia and Herzegovina; University of East Sarajevo, Lukavica, Bosnia and Herzegovina

Martin Gregurić Faculty of Traffic and Transport Sciences, University of Zagreb, Zagreb, Croatia

Domenico Guida Department of Industrial Engineering, University of Salerno, Fisciano, Salerno, Italy

Rašid Hadžović “DzemalBijedic” University of Mostar, Mostar, Bosnia and Herzegovina; Faculty of Civil Engineering, University “DžemalBijedić”, USRC “Mithat Hujdur Hujka”, Mostar, Bosnia and Herzegovina

Emina Hadžić Faculty of Civil Engineering, University of Sarajevo, Sarajevo, Bosnia and Herzegovina

Edina Hajdarević Faculty of Natural Sciences and Mathematics, University of Tuzla, Tuzla, Bosnia and Herzegovina

Ismar Hajro Engineering Faculty, University of Sarajevo, Sarajevo, Bosnia and Herzegovina

Jasmin Halilović University of Tuzla, Tuzla, Bosnia and Herzegovina

Saud Hamidović Faculty of Agriculture and Food Sciences, University of Sarajevo, Sarajevo, Bosnia and Herzegovina

Enes Hatibović University of Sarajevo School Science and Technology, Sarajevo, Bosnia and Herzegovina

Nihad Hodzic Faculty of Mechanical Engineering, University of Sarajevo, Sarajevo, Bosnia and Herzegovina

Emir Horozić Faculty of Technology, University of Tuzla, Tuzla, Bosnia and Herzegovina

Elvis Hozdić Faculty of Mechanical Engineering, Department of Control and Manufacturing Systems, University of Ljubljana, Ljubljana, Slovenia

Ermin Husak Technical Faculty Bihać, University of Bihać, Bihać, Bosnia and Herzegovina

E. I. Ikonnikova MT-11, BMSTU, Moscow, Russia

Safet Isić Faculty of Mechanical Engineering, University Campus, University “Džemal Bijedić” of Mostar, Mostar, Bosnia and Herzegovina

Jandrlić Ivan Faculty of Metallurgy, University of Zagreb, Sisak, Croatia

Edouard Ivanjko Department of Intelligent Transportation Systems, Faculty of Transport and Traffic Sciences, University of Zagreb, Zagreb, Republic of Croatia

Nudžejma Jamaković Faculty of Agriculture and Food Science, University of Sarajevo, Sarajevo, Bosnia and Herzegovina

Gordan Jančan Chemilab d.o.o., Ljubljana, Slovenia

Zlata Jelačić Faculty of Mechanical Engineering, University of Sarajevo, Sarajevo, Bosnia and Herzegovina

Jovana Jovanovic Faculty of Civil Engineering and Management, University Union Nikola Tesla, Belgrade, Serbia

Mihailo Jovanović Faculty of Management Herceg Novi, Herceg Novi, Montenegro

Huska Jukić Faculty of Health Studies, Faculty of Biotechnical, University of Bihac, Bihać, Bosnia and Herzegovina

Zlatan Jukic Faculty of Electronics and Computer Engineering, HTL Rankweil, Rankweil, Austria; TU Wien, Vienna Uniersity of Technology, Wien, Austria

Halid Junuzović Faculty of Technology Tuzla, University of Tuzla, Tuzla, Bosnia and Herzegovina

Suvada Jusić Faculty of Civil Engineering, University of Sarajevo, Sarajevo, Bosnia and Herzegovina

Zlatko Jusufhodžić Public Institution "Veterinary Institute", Bihać, Bosnia and Herzegovina

Josip Kacmarcik Faculty of Mechanical Engineering, University of Zenica, Zenica, Bosnia and Herzegovina

Tatjana Kandikjan University Ss. Cyril and Methodius Skopje, Skopje, Republic of Macedonia

Kiriakos P. Kapetis Vehicles Laboratory, School of Mechanical Engineering, National Technical University of Athens, Athens, Greece

Petar Karabadjakov ET "Ingeborg Demirova", Gabrovo, Bulgaria

Edina Karabegović Technical Faculty Bihać, University of Bihać, Bihać, Bosnia and Herzegovina

Isak Karabegović Technical Faculty Bihać, University of Bihać, Bihać, Bosnia and Herzegovina

Enver Karahmet Faculty of Agriculture and Food Science, University of Sarajevo, Sarajevo, Bosnia and Herzegovina

Emir Krivić Mechanical Engineering, Tobacco Factory Sarajevo, Sarajevo, Bosnia and Herzegovina

Amira Kasumović Faculty of Civil Engineering, "Džemal Bijedić" University of Mostar, Mostar, Bosnia and Herzegovina

Anes Kazagic JP Electric Power Industry B&H d.d., Sarajevo, Bosnia and Herzegovina

Maja Kazazic Faculty of Education, University "Dzemal Bijedic" of Mostar, Mostar, Bosnia and Herzegovina

Anera Kazlagić Faculty of Agriculture and Food Sciences, University of Sarajevo, Sarajevo, Bosnia and Herzegovina

Ozan Keysan Middle East Technical University, Ankara, Turkey

Emir Klarić Green Council, Sarajevo, Bosnia and Herzegovina

Sanela Klarić International Burch University, Sarajevo, Bosnia and Herzegovina

Slobodanka Ključanin Technical Faculty Bihać, University of Bihać, Bihać, Bosnia and Herzegovina

Suzana Koprivica Faculty of Civil Engineering and Management, University Union Nikola Tesla, Belgrade, Serbia

V. V. Kostyleva The Kosygin State University, Moscow, Russia

Dimitris V. Koulocheris Vehicles Laboratory, School of Mechanical Engineering, National Technical University of Athens, Athens, Greece

Džemal Kovačević University of Tuzla, Tuzla, Bosnia and Herzegovina

Olena Kovalevska Donbas State Engineering Academy (DSEA), Kramatorsk, Ukraine

Sergiy Kovalevskyy Donbas State Engineering Academy (DSEA), Kramatorsk, Ukraine

Bakir Krajinović Federal Hydrometeorological Institute, Sarajevo, Bosnia and Herzegovina

Mario Krzyk Faculty of Civil and Geodetic Engineering, University of Ljubljana, Ljubljana, Slovenia

Andrej Kump Novacast Systems AB, Ronneby, Sweden

Azra Kurtović Faculty of Civil Engineering, University of Sarajevo, Sarajevo, Bosnia and Herzegovina

Omer Kurtanović Faculty of Economics, University of Bihac, Bihać, Bosnia and Herzegovina

T. S. Lapina Novosibirsk Institute of Technology of The Kosygin State University, Novosibirsk, Russia

Ana M. Lazarevska University Ss. Cyril and Methodius Skopje, Skopje, Republic of Macedonia

Lamija Lemeš University of Sarajevo, Sarajevo, Bosnia and Herzegovina

Samir Lemeš Polytechnic Faculty, University of Zenica, Zenica, Bosnia and Herzegovina

Samir Lemes Faculty of Mechanical Engineering, University of Zenica, Zenica, Bosnia and Herzegovina

Alma Leto "Džemal Bijedić" University of Mostar, Mostar, Bosnia and Herzegovina

Amar Leto Faculty of Mechanical Engineering, University Campus, University "Džemal Bijedić" of Mostar, Mostar, Bosnia and Herzegovina

Jovanka Lukić Faculty of Engineering, University of Kragujevac, Kragujevac, Serbia

Tetiana Lysenko Odessa National Polytechnic University (ONPU), Odessa, Ukraine

Slavica Mačužić Faculty of Engineering, University of Kragujevac, Kragujevac, Serbia

Mehmed Mahmić Technical Faculty Bihać, University of Bihać, Bihać, Bosnia and Herzegovina

Emad Mamoua Faculty of Engineering and Natural Sciences, International Burch University, Sarajevo, Bosnia and Herzegovina

Sadko Mandžuka Faculty of Traffic and Transport Sciences, University of Zagreb, Zagreb, Croatia

Mersida Manjgo University "DžemalBijedić" of Mostar, Mostar, Bosnia and Herzegovina

Andrea Marr Willdan Energy Solutions, Anaheim, USA

Semir Mehremić Faculty of Mechanical Engineering, University "Džemal Bijedić" of Mostar, Mostar, Bosnia and Herzegovina

Sadjit Metovic Faculty of Mechanical Engineering, University of Sarajevo, Sarajevo, Bosnia and Herzegovina

Alma Mičijević "Džemal Bijedić" University of Mostar, Mostar, Bosnia and Herzegovina; Faculty of Agromediteranean, University "Džemal Bijedić", Mostar, Bosnia and Herzegovina

Davor Milić Faculty of Mechanical Engineering, University of East Sarajevo, Istočno Sarajevo, Bosnia and Herzegovina

Hata Milišić Faculty of Civil Engineering, University of Sarajevo, Sarajevo, Bosnia and Herzegovina

Zoran Miljković Faculty of Mechanical Engineering, University of Belgrade, Belgrade, Serbia

Milan M. Milosavljavić Faculty of Technical Science, University of Priština, Kosovska Mitrovica, Serbia

Milutin M. Milosavljević Faculty of Technical Science, University of Priština, Kosovska Mitrovica, Serbia

Dragoljub Mirjanić Academy of Sciences and Arts of RS, Sarajevo, Bosnia and Herzegovina

Slađana Mirjanić University of Banja Luka, Banja Luka, Bosnia and Herzegovina

Ivana Radonjić Mitić Faculty of Science, University of Niš, Nis, Serbia

Daniela Mladenovska University Mother Teresa, Skopje, Republic of Macedonia

Doina Mortoiu Aurel Vlaicu University, Arad, Romania

Alen Mujkic DCCS GmbH, Tuzla, Bosnia and Herzegovina

Valentin Muller Aurel Vlaicu University, Arad, Romania

Edis Nasić University of Tuzla, Tuzla, Bosnia and Herzegovina

Daniela Koltovska Nechoska Faculty of Technical Sciences, St. Kliment Ohridski University, Bitola, Republic of Macedonia

Emir Nezirić Faculty of Mechanical Engineering, "DžemalBijedić" University of Mostar, Mostar, Bosnia and Herzegovina

Adnan Novalić Faculty of Engineering and Natural Sciences, International Burch University, Sarajevo, Bosnia and Herzegovina

Mirna Nožić Faculty of Mechanical Engineering, University "DžemalBijedić" of Mostar, Mostar, Bosnia and Herzegovina

Murco Obučina Mechanical Faculty, University Sarajevo, Sarajevo, Bosnia and Herzegovina

Salah-Eldien Omer SAG CONSULTING d.o.o., Zagreb, Croatia
Technical Faculty Bihać, University of Bihać, Bihać, Bosnia and Herzegovina

Enisa Omanović-Mikličanin Faculty of Agriculture and Food Sciences, University of Sarajevo, Sarajevo, Bosnia and Herzegovina

Goran Orašanin Faculty of Mechanical Engineering, University of East Sarajevo, Istočno Sarajevo, Bosnia and Herzegovina

Mirsada Oruč University of Zenica, Zenica, Bosnia and Herzegovina

Sıla Özkavaf EPRA Elektrik Enerji, Ankara, Turkey

Ena Pantic Faculty of Education, University "Dzemal Bijedic" of Mostar, Mostar, Bosnia and Herzegovina

Carmine M. Pappalardo Department of Industrial Engineering, University of Salerno, Fisciano, Salerno, Italy

Kovač Pavel Faculty of Technical Science, University of Novi Sad, Novi Sad, Serbia

Daniel Pavleski Traffic Department of the City of Skopje, Skopje, Republic of Macedonia

Tomislav Pavlović Faculty of Science, University of Niš, Nis, Serbia

Ekrem Pehlić Faculty of Health Studies, University of Bihac, Bihać, Bosnia and Herzegovina

Milica Petrović Faculty of Mechanical Engineering, University of Belgrade, Belgrade, Serbia

Milenko Petrović Faculty of Technical Science, University of Priština, Kosovska Mitrovica, Serbia

Senada Pobrić Faculty of Mechanical Engineering, "Džemal Bijedić" University of Mostar, Mostar, Bosnia and Herzegovina

Vili Podgorelec Faculty of Electrical Engineering and Computer Science, University of Maribor, Maribor, Slovenia

Alina Bianca Pop SC TECHNOCAD SA, Baia Mare, Romania

Pravdić Predrag High Technical Mechanical School of Professional Studies, Trstenik, Serbia

Senad Rahimic University "DžemalBijedić" of Mostar, Mostar, Bosnia and Herzegovina; Faculty of Mechanical Engineering, University "DžemalBijedić" of Mostar, Mostar, Bosnia and Herzegovina

Vahid Redžić Polytechnic Faculty, University of Zenica, Zenica, Bosnia and Herzegovina

Milenko Rimac University of Zenica, Zenica, Bosnia and Herzegovina

Miloš Rodić Public Institution "Veterinary Institute", Bihać, Bosnia and Herzegovina

Olga Rybak Odessa National Polytechnic University (ONPU), Odessa, Ukraine

Elvir Šahić School of Economics and Business, University of Sarajevo, Sarajevo, Bosnia and Herzegovina

Merima Šahinagić-Isović Faculty of Civil Engineering, "Džemal Bijedić" University of Mostar, Mostar, Bosnia and Herzegovina

Mirsada Salihović Faculty of Pharmacy, University of Sarajevo, Sarajevo, Bosnia and Herzegovina

Kemal Salkić Agricultural Institute of the Una-Sana Canton, Sarajevo, Bosnia and Herzegovina

Aida Šapčanin Faculty of Pharmacy, University of Sarajevo, Sarajevo, Bosnia and Herzegovina

Edita Saric Federal Institute for Agriculture in Sarajevo, Ilidza, Bosnia and Herzegovina

Abdulnaser Sayma Centre for Compressor Technology, City University of London, London, UK

Edin Šemić Faculty of Mechanical Engineering, University “Džemal Bijedić” of Mostar, Mostar, Bosnia and Herzegovina

Jakub Secic MBT BH, Tuzla, Bosnia and Herzegovina

Indji Selim International School of Architecture and Design, University American College Skopje, Skopje, Republic of Macedonia

Amra Selimović Faculty of Technology Tuzla, University of Tuzla, Tuzla, Bosnia and Herzegovina

Tetiana Sidelnykova Odessa National Medical University (ONMU), Odessa, Ukraine

Sofija Sidorenko University Ss. Cyril and Methodius Skopje, Skopje, Republic of Macedonia

Stojan Simić Faculty of Mechanical Engineering, University of East Sarajevo, Istočno Sarajevo, Bosnia and Herzegovina

Marco Claudio De Simone Department of Industrial Engineering, University of Salerno, Fisciano, Salerno, Italy

Isat Skenderović Faculty of Natural Sciences and Mathematics, University of Tuzla, Tuzla, Bosnia and Herzegovina

Pero Škorput Faculty of Traffic and Transport Sciences, Zagreb, Croatia

Izet Smajevic University of Sarajevo, Sarajevo, Bosnia and Herzegovina

Alisa Smajović Faculty of Pharmacy, University of Sarajevo, Sarajevo, Bosnia and Herzegovina

Mirza Softić Faculty of Technology, University of Tuzla, Tuzla, Bosnia and Herzegovina

Azra Špago Faculty of Civil Engineering, “Džemal Bijedić” University of Mostar, Mostar, Bosnia and Herzegovina

Suad Špago University “DžemalBijedić”, Mostar, Bosnia and Herzegovina

Denijal Sprečić University of Tuzla, Tuzla, Bosnia and Herzegovina

Mirna Habuda Stanić Faculty of Food Technology, University of Osijek, Osijek, Croatia

Andjelina Marić Stanković Faculty of Science, University of Niš, Nis, Serbia

Svetlana Stevović Innovation Center of the Faculty of Mechanical Engineering in Belgrade, Belgrade, Serbia

Ivan Stoianov “Podem Gabrovo” Ltd., Gabrovo, Bulgaria

Rešković Stoja Faculty of Metallurgy, University of Zagreb, Sisak, Croatia

Nadica Stojanovic Faculty of Engineering, University of Kragujevac, Kragujevac, Serbia

Savo Stupar School of Economics and Business, University of Sarajevo, Sarajevo, Bosnia and Herzegovina

Aida Šukalić "Džemal Bijedić" University of Mostar, Mostar, Bosnia and Herzegovina

Xiaoqin Sun University of Science and Technology, Changsha, China

Edin Šunje Faculty of Mechanical Engineering, University Campus, University "Džemal Bijedić" of Mostar, Mostar, Bosnia and Herzegovina

Suad Sućeska Sarajevo, Bosnia and Herzegovina

Jasmin Suljagić Faculty of Technology, University of Tuzla, Tuzla, Bosnia and Herzegovina

Mihail Aurel Țîțu Lucian Blaga University of Sibiu, Sibiu, Romania
The Academy of Romanian Scientists, Bucharest, Romania

Aurelia Tanasoiu Aurel Vlaicu University, Arad, Romania

Bogdan Tanasoiu Aurel Vlaicu University, Arad, Romania

Petar Tasić Engineering Faculty, University of Sarajevo, Sarajevo, Bosnia and Herzegovina

M. V. Taube Department of Industrial Design, Novosibirsk State University of Architecture, Design and Arts, Novosibirsk, Russia

Saeed Teimourzadeh EPRA Elektrik Enerji, Ankara, Turkey

Brlić Tin Faculty of Metallurgy, University of Zagreb, Sisak, Croatia

Osman Bülent Tör EPRA Elektrik Enerji, Ankara, Turkey

Muris Torlak University of Sarajevo, Sarajevo, Bosnia and Herzegovina

Dzenana Tomasevic Faculty of Mechanical Engineering, University of Zenica, Zenica, Bosnia and Herzegovina

Vladimir Tonkonogyi Odessa National Polytechnic University (ONPU), Odessa, Ukraine

Raul Turmanidze Donbas State Engineering Academy (DSEA), Kramatorsk, Ukraine; Department of Production Technologies of Mechanical Engineering, Georgian Technical University (GTU), Tbilisi, Georgia

Branka Varešić Faculty of Agriculture and Food Science, University of Sarajevo, Sarajevo, Bosnia and Herzegovina

Sasa Vasiljevic High Technical School of Professional Studies, Kragujevac, Serbia

Krešimir Vidović Ericsson Nikola Tesla, Zagreb, Croatia

Samir Vojić University of Bihač, Bihać, Bosnia and Herzegovina

V. I. Volokh Dniprovsk State Technical University, Kamianske, Ukraine

Clio G. Vossou Vehicles Laboratory, School of Mechanical Engineering, National Technical University of Athens, Athens, Greece

Maja Tonec Vrančić Faculty of Traffic and Transport Sciences, Zagreb, Croatia

Miroslav Vujic Faculty of Traffic and Transport Sciences, University of Zagreb, Zagreb, Croatia

Martin White Centre for Compressor Technology, City University of London, London, UK

Alexey Yakimov Odessa National Polytechnic University (ONPU), Odessa, Ukraine

Ye. A. Yeriomina Dniprovsk State Technical University, Kamianske, Ukraine

Tuna Yıldız Middle East Technical University, Ankara, Turkey

Minela Žapčević Faculty of Health Studies, University of Bihac, Bihać, Bosnia and Herzegovina

Muhidin Zametica Green Council, Sarajevo, Bosnia and Herzegovina

Nermina Zaimovic-Uzunovic Faculty of Mechanical Engineering, University of Zenica, Zenica, Bosnia and Herzegovina

Milutin Živković High Technical Mechanical School of Professional Studies, Trstenik, Serbia

New Technologies in Mechanical Engineering, Metallurgy, Mechatronics, Robotics and Embedded Systems

Implementation of Industry 4.0 and Industrial Robots in the Manufacturing Processes

Isak Karabegović(✉), Edina Karabegović, Mehmed Mahmić, and Ermin Husak

Technical Faculty Bihać, University of Bihać, Ul.Irfana Ljubujankića bb, 77000 Bihać, Bosnia and Herzegovina
isak1910@hotmail.com

Abstract. The fourth industrial revolution or Industry 4.0 is already present around us, but the concept itself is not widespread. The implementation of Industry 4.0 will improve many aspects of human life in all segments. The implementation will initiate changes in business paradigms and manufacturing models, which will be reflected on all levels of manufacturing processes, as well as supply chains, including all workers in the manufacturing process, managers, cyber-physical system designers, and end-users. We are witness to big changes in all industry branches, with new business methods, product system transformation, consumption, delivery and transportation emerging, owing to the implementation of new technological discoveries that include: robotics & automation, Internet of Things (IoT), 3D printers, smart sensors, Radio Frequency Identification (RFID,) etc. The implementation strategy of Industry 4.0 is to enable the adjustment of industrial production to complete intelligent automation, which means introducing self-automation, self-configuration, self-diagnosis and problem-solving, knowledge and intelligent decision-making. On the other hand, there are many challenges in the coming period, such as issues of changing business paradigms, legal issues, resource planning, security issues, standardization issues, and many other. We must point out that the success or failure to implement Industry 4.0 lies in the hands of all participants in the production chain, from the manufacturer to the end-users.

Keywords: Manufacturing process · Industry 4.0 · Robot · Automation · Internet of Things (IoT) · Smart factory

1 Introduction

The industry is presently in the process of implementing the fourth industrial revolution. The WEF World Economic Forum (held in Davos in 2016) called the changes, that took place on the world industrial and digital scene, the fourth industrial revolution. The term "Industry 4.0" appeared for the first time in 2011 at the fair in Hanover, Germany. Industry 4.0 presents the vision of the advanced industrial production that is already being partly implemented as new technology in the automation of manufacturing processes, data exchange and data processing. Within the fourth industrial revolution a new value chain is formed which primarily relies on the Cyber-Physical Systems, which is also the second name for the Internet of Things, and the associated

I. Karabegović (Ed.): NT 2019, LNNS 76, pp. 3–14, 2020.
https://doi.org/10.1007/978-3-030-18072-0_1

service most often realized in the cloud (Cloud Computing). The analysis and discussion about the fourth industrial revolution that is taking place today aims to explain its multiple influence, the speed of technological revolution, the comprehensiveness, and the increased awareness with the purpose of its implementation. The book by Klaus Schwab "The Fourth Industrial Revolution", published by the World Economic Forum, Geneva Switzerland, in 2016 has the same objective. It is necessary to create a framework for reflection on "Industry 4.0" which describes key issues and highlights possible answers. In other words, it is necessary to provide a platform for encouraging public and private cooperation and partnerships on issues related to the technological revolution. In all communities, including industry, interaction and cooperation are needed to create a positive shift in the implementation of "Industry 4.0" in order to enable individuals and groups from all over the world to participate in implementation and benefit from current transformations.

2 The Concept of the Fourth Industrial Revolution – "Industry 4.0"

The following revolutions have so far been recorded in the world: the first industrial revolution that occurred with the invention of steam engine in 1784, the second industrial revolution with the discovery of electricity in 1870, and the third industrial revolution with the application of the IT system and the Internet in 1969. We are presently in the fourth industrial revolution [1–12]. The difference between the two last industrial revolutions is shown in Fig. 1.

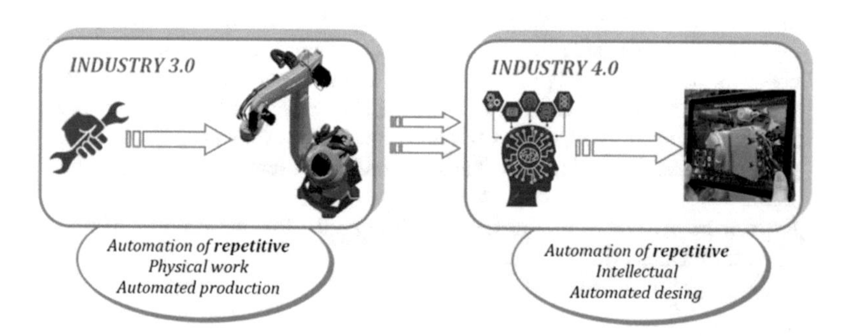

Fig. 1. The difference between the third and the fourth industrial revolution in the manufacturing process

Based on Fig. 1, it is possible to conclude that the third industrial revolution uses physical work and automated production for the automation of repetition, whereas the fourth industrial revolution employs intellectual work and automatic design for the automation of repetition. Digital technologies that have computer hardware, software and network, with core that is not new but rather in the midst of the third industrial revolution, are becoming increasingly sophisticated and integrated, which results in the transformation of society and the global economy. Like any industrial revolution, it is created by increasing knowledge in the world through innovation that can be evolutionary and revolutionary, by using advanced software, as shown in Fig. 2.

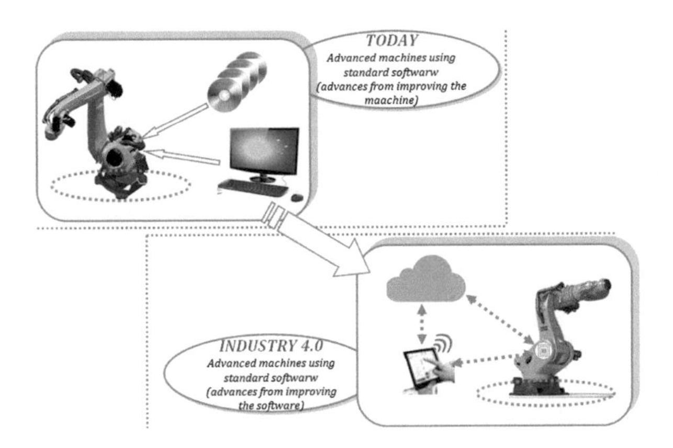

Fig. 2. Industry 4.0 uses advanced software

Almost all industrial branches are using advanced machines with standard software, while the fourth industrial revolution uses advanced software for typical machines, as shown in Fig. 2. Evolutionary innovations are causing many of the gradual advances in technology and processes, and can be continuous or dynamic. Revolutionary innovations are called discontinuous innovations, and are often new. Revolutionary innovations dominate the time of industrial revolution, while evolutionary innovations dominate in the period between the industrial revolutions. Revolutionary innovations Internet of Things, Big Data and Autonomous Intelligent Systems are present today in all segments of society: new education, new ecology, connectivity, individuality, new health paradigm, urbanization, mobility and globalization. Both evolutionary and revolutionary innovations are present in the fourth industrial revolution. Industry 4.0, as shown in Fig. 3, shows a new paradigm of industrial production, indicating the existence of the revolution of distributed artificial intelligence, where technology and manufacturing systems in all industrial branches are rapidly changing, i.e. changes are made from the line system to the closed loop system. The fourth industrial revolution creates a world in which virtual and physical production systems cooperate globally with each other in a flexible way in the digital environment. The fourth industrial revolution "Industry 4.0" relies on the Cyber-Physical System and is based on digital technology, as shown in Fig. 3.

CPS – Cyber-Physical System is nothing more than a system of physical objects and corresponding virtual objects connected to communicate to all present information networks. Digital technologies enable us to collect data about production and working conditions, and perform 3D simulation anywhere and anytime. Cyber service based on the dynamic integration of service providers and services through information exchange is the real algorithm based on which work process documents are completed, including 3D modeling and topology, which enable us to influence intelligent and conscious physical objects from any place and any time. This is no longer a science

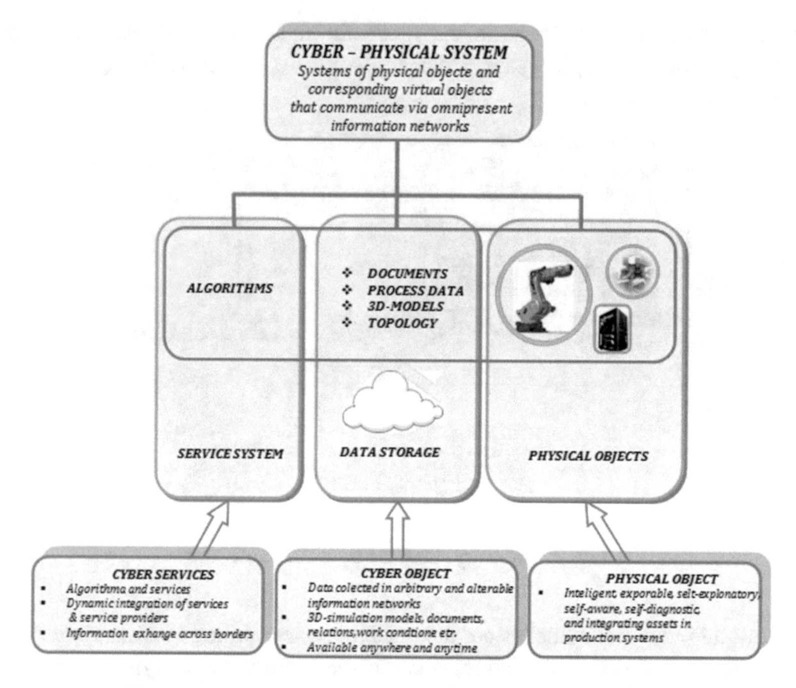

Fig. 3. Schematic representation of Cyber-Physical System

fiction, it already exists among us. Every industrial revolution has its foundations. The fourth industrial revolution has its foundations in the following technologies: cloud computing, robotics & automation, intelligent sensors, 3D printers and radio frequency identification - RFID, as shown in Fig. 4.

Fig. 4. The foundation of the fourth industrial revolution – "Industry 4.0"

The above technologies represent the foundation of the fourth industrial revolution "Industry 4.0", which best illustrates the trend in the number of reported innovations from cloud computing, robotics & automation, intelligent sensors, 3D printers and radio frequency identification - RFID in the following five countries: China, USA, Japan, Korea Germany, as shown in Table 1 [4.5].

Table 1. Number of innovations in "Industry 4.0" in 2016

	CHINA	*USA*	*JAPAN*	*Rep.of Korea*	*GERMANY*	*TOTAL ΣΣ*
Cloud Computing	*1.144*	*1.559*	*132*	*314*	*70*	*3.219*
Robotics	*2.709*	*164*	*638*	*142*	*349*	***4.002***
Smart Sensors	*2.586*	*126*	*83*	*88*	*132*	*3.015*
3D Printing	*504*	*536*	*307*	*52*	*0*	***1.399***
RFID	*384*	*1.055*	*1.243*	*1.039*	*98*	*3.819*
TOTAK ΣΣ	***7.327***	***3.440***	***2.403***	***1.635***	***649***	***15.454***

The analysis of the trend of innovation in 2016 from the technologies that create the foundation of "Industry 4.0" in five most technologically developed countries in the world, brings us to the conclusion that in the last five years China ranked first in the world with 7.327 patents in 2016, while Germany ranked fifth with 649 patents in 2016, although it announced its strategy as "Industry 4.0" in 2011 at the Hannover Fair. In regard to technology, the first place is held by robotics & automation with 4002 innovations applied globally in 2016, followed by RFID, cloud computing, intelligent sensors and the latest 3D printers [4, 5]. Many researchers around the world are working on the development and implementation of various implementation scenarios of "Industry 4.0". Professor M. Herman with colleagues from the University of Dortmund identified the principles underlying the scenario for the implementation of "Industry 4.0". Six project principles have been identified:

- Interoperability,
- Decentralization,
- Virtualization,
- Real-time work ability,
- Service Orientation, and
- Modularity.

The basic implementation of "Industry 4.0" requires the ability to develop connection and communication between Cyber-Physical Systems (CPS), as shown in Fig. 3, people and production processes, as well as the successful linking of CPSs of different manufacturers. It is necessary to enable CPS to monitor physical systems by linking data obtained through intelligent sensors with virtual and simulation production models, i.e. to create a virtual copy of the physical manufacturing process. Powerful computers will enable CPS to make decisions within the "smart factory" manufacturing process. In other words, they will enable decentralization process, without the need for central planning and monitoring. By using intelligent sensors, they collect and process data, thus constantly monitoring the analysis of the production process and all the machines in it, and in case of failure or stop, the production is automatically directed to the other machine. The internet includes all the participants within the manufacturing process and most of them outside of the process.

Modularity (standardized hardware and software) allows flexible adaptation of "smart factory" to changing requirements, as well as replacement and extension of particular models. The fourth industrial revolution is nothing more than closed circle that is made up of customers, devices and tools and applications of new technologies such as IoT, ICT, CPS, etc. Regardless of the principles underlying the scenario, the implementation of "Industry 4.0" requires the establishment of a value chain, as shown in Fig. 5. Based on this we can conclude that number one is the application of software innovation and hardware structures, followed by education and training programs for the wider population and integration with the new above-mentioned technologies. Industry 4.0 is nothing more than a real-time ability to link production processes and IT systems to dynamic control of all complex systems, or in other words, the creation of innovative solutions that will allow intelligent systems to do what is overwhelming and tedious for human beings.

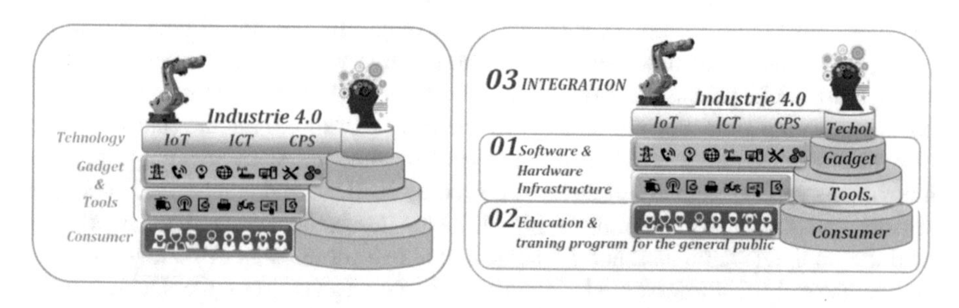

Fig. 5. Value chain of "Industry 4.0"

3 The Possibility of Application of "Industry 4.0" in Manufacturing Processes

It is a well-known fact that "Industry 4.0" has been increasingly present in all industrial branches and manufacturing processes for more than a decade, including transportation, and supplying customers with finished products. "Industry 4.0" is present in all segments of society [1–4, 8–11]. In the fourth industrial revolution, new technologies are based on innovations that are much faster and wider implemented in production processes than in previous industrial branches, thus changing manufacturing processes, as shown in Fig. 6.

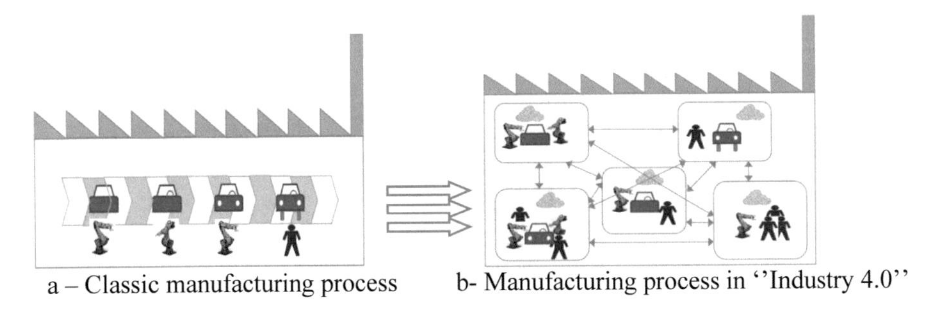

Fig. 6. Difference between the classic manufacturing process and the manufacturing process in "Industry 4.0"

The classic manufacturing process used line production (Fig. 7a), and first-generation industrial robots in the automation process, which were separated from the workers to prevent injuries. In order to change the product, they had to be reprogramed, which required a lot of time for reorganization, i.e. that was rigid automation. Industry 4.0 changed the production process to a working cell or a closed loop with a closed connection. The rapid rise in the number of patents from "Industry 4.0", and especially from the following technologies: cloud computing, robotics & automation, intelligent sensors, 3D printers and radio frequency identification - RFID, as mentioned in the introduction, and their application in the manufacturing process, increases their implementation in "Industry 4.0". Many companies in all industries implement the above-mentioned technologies to increase productivity, quality, introduce innovations, increase creativity, and better design to make them competitive on the global market. New technologies enable them to maintain the company's management headquarters in their own country, and manufacturing facilities in other countries. The development of new technologies and application of new materials in the industry, as well as placement of finished products and growing demand for alternative energy sources (solar cell production) will increase the use of new technologies, both in manufacturing and sale of finished products across all industries. The variety of applications of new technologies is on the rise, requiring flexible automation in all industries, as well as reduction in the manufacturing time and sale of continuously high-quality products. It is expected that the introduction of the innovative technologies in industry will lead to greater flexibility, increased productivity, increased product quality, and customer co-operation in order to increase sales of finished products. The development and implementation of "Industry 4.0" in the industry is leading to a new kind of smart manufacturing processes. The characteristics of the smart manufacturing process in "Industry 4.0" are defined in Fig. 7 [18–20].

The characteristics of a smart manufacturing process using "Industry 4.0" are intelligence, awareness, connectivity and controllability. The manufacturing process is equipped with advanced sensors that allow access to information on the state of production and the environment, and is equipped with powerful computers that allow for autonomous decision-making and algorithm-based self-sufficiency process. An example of an automated digital/virtual production cell is shown in Fig. 8 for machining the

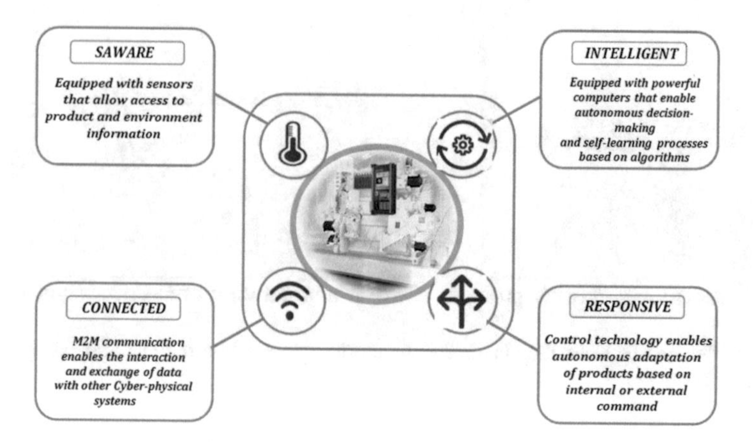

Fig. 7. Definition of the characteristics of the manufacturing process in "Industry 4.0" [13–15]

car rim. The automated and digital/virtual production cell for car rim machining requires safety, CNC lathe, industrial robot, transport station, transport robot, RFID chip conveyor, control center for simulation, and the mobile device with the applications to enter the parameters, as shown in Fig. 8.

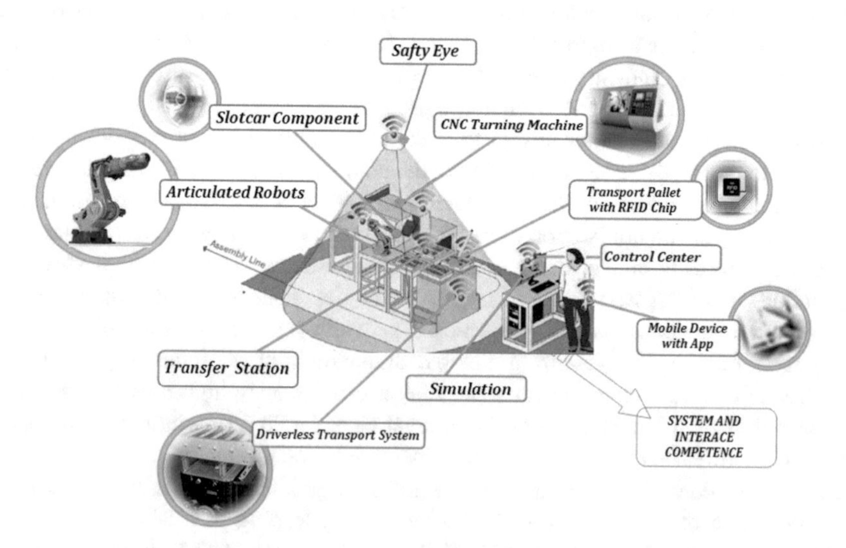

Fig. 8. Physical automated and digital/virtual production cell

All devices in the production process are interconnected and have communication with M2M, interaction and data exchange with other cyber-physical systems. The control technology enables autonomous adaptation of a product based on the internet or an external command. By networking the production cells, we get a smart manufacturing process, or a smart factory. Figure 9 shows an exemplary scheme view of the "Industry 4.0" application in the manufacturing process in the industry.

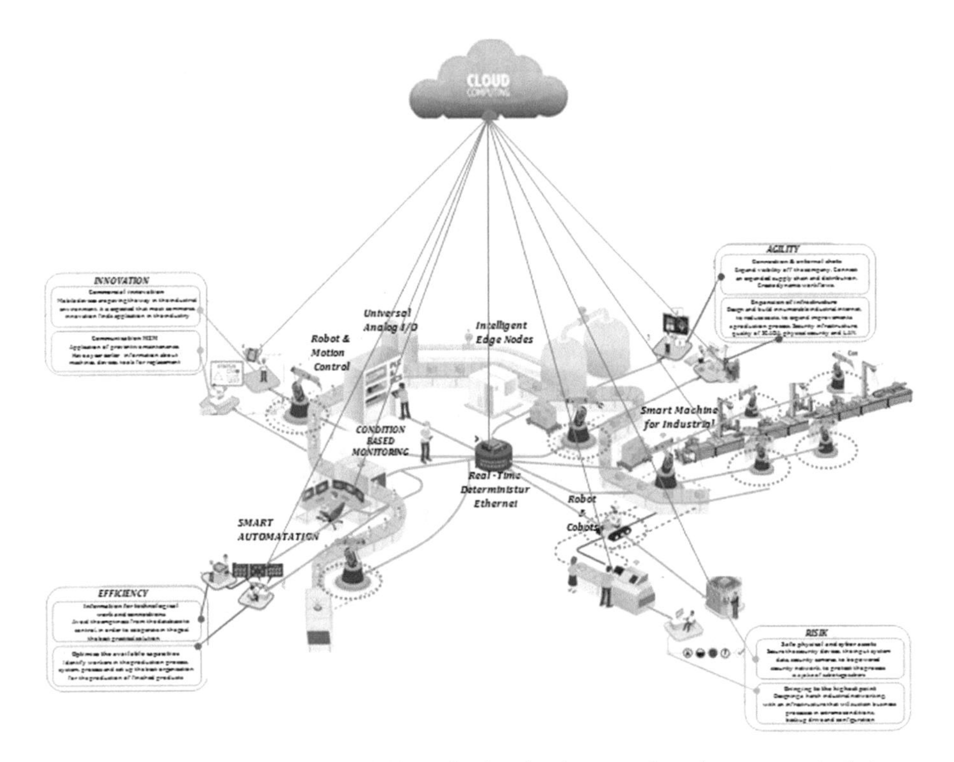

Fig. 9. Scheme of the "Industry 4.0" application in the manufacturing process in industry [12–17]

Based on Fig. 9, we see that in every production process, regardless of the industry, the most important components in the manufacturing process are innovation, efficiency, agility and risk. In regard to commercial innovations, where it is necessary to mobilize employees and supervisors to access knowledge wherever they are, which is enabled by the mobile devices that are already in an industrial environment, most of the commercial innovations will be applied in the manufacturing processes of the industry. M2M communication and the environment aims to use preventive maintenance of the plant, thus obtaining information on machines and equipment for replacement or repair a year earlier. It is an efficient way to have information on technological work and connections, to avoid blanks from base to control, and work together to achieve the best practical solution. It is important to optimize available capacities: workers, manufacturing processes, systems, and adjust the best organization to produce finished products. As far as agility is concerned, it is necessary to have connectivity and external cooperation, to have information about production and connections, and build an internal industrial internet to make modular expansion of the infrastructure. In addition, there is a risk, so it must be ensured that the physical and cyber assets are safe. By making decisions based on data and insight, "Industry 4.0" opens up the world of possibilities for the manufacturing process. Sensors help detect defects or abnormalities in the manufacturing process, enabling quick adaptation of settings and changing parameters to prevent production shortfalls in the upcoming period. The permanent maintenance of the machine enables companies to plan stopping times more effectively and prevent breaks or failures. Based on the data and insights gained from the production process to the delivery process, the entire supply chain is managed much more efficiently. Health and safety of workers are improved as data and incident analysis helps in identifying and preventing dangerous situations. Production line changes are simulated practically before they are physically implemented to adequately assess the impact and reduce the chance of errors. Manufacturers can provide a variety of new customer services – including predictive maintenance and quality control – using sensor data. Energy consumption can be optimized using advanced analytics. Based on data and insights derived from production to delivery, the entire supply chain is managed much more efficiently. Health and safety of workers are improved as data and incident analysis helps identify and prevent dangerous situations. For "Industry 4.0" implementation in all industry branches, it is necessary to have SPS developers, IT specialists, software engineers, electrical and mechanical engineers, and other engineers who have multidisciplinary skills, which requires the complete change of education system. Considering the fact that, as we have already mentioned, the world knowledge is doubled in one or two years, and having in mind the big changes in all segments of society, it is necessary to change the education system that needs to "prepare young people for jobs that do not yet exist in order to use technologies that have not yet been discovered, in the era of global competitiveness".

4 Conclusion

The fourth industrial revolution is already present in all industrial branches, from the production to the sale of finished products. By introducing the technologies that are the foundation of the fourth industrial revolution, such as Cloud Computing, Robotics & Automation, Intelligent Sensors, 3D printers, and RFID, we witness changes in processes and technologies as well as organization of manufacturing and sales. The application of "Industry 4.0" in the manufacturing processes of all industrial branches transforms the line manufacturing process into a network manufacturing process, i.e. with above-mentioned technologies it is transformed in the closed loop production process, which provide complete information on product design at any time. By using a large number of smart sensors in the production process, we have information on production plants and machines at any given moment, based on which decisions are made when replacing the machine, i.e. we have permanent maintenance. The implementation strategy of "Industry 4.0" is to make the adjustment of industrial production to complete intelligent automation, which means introducing self-automation, self-configuration, self-diagnosis and problem-solving, understanding and intelligent decision-making. Using this technology, we can monitor demands on the global market. Figure 9 shows a schematic representation of a production process in which "Industry 4.0" is implemented. The complete "Industry 4.0" application leads us in the direction of the “smart” “factories”, which is the goal that allows the companies to remain competitive on the global market. Smart factories are not isolated from other social changes such as the development of businesses, the development of science and education, which require changes in all segments because world knowledge is doubled in one or two years. On the other hand, there are many challenges in the coming period, such as questions of changing business paradigms, legal issues, resource planning, security issues, the issue of standardization, and many other. We must point out that the success or failure to implement “Industry 4.0” lies in the hands of all participants in the chain, from the manufacturer to the end user.

References

1. Klaus, S.: The Fourth Industrial Revolution, World Economic Forum, Geneva, Switzerland (2016). https://luminariaz.files.wordpress.com/2017/11/the-fourth-industrial-revolution-2016-21.pdf
2. Isak, K.: The role of industrial and service robots in fourth industrial revolution with focus on China. J. Eng. Archit. **5**(2), December 2017. Published by American Research Institute for Policy Development, USA, ISSN 2334-2986 (Print), 2334-2994 (Online), pp. 110–117. http://jea-net.com/journals/jea/Vol_6_No_1_June_2018/7.pdf, https://doi.org/10.15640/jea.v5n2a9
3. Crnjac, M., Veža, I., Banduka, N.: From concept to the introduction of industry 4.0. Int. J. Ind. Eng. Manag. (IJIEM), **8**(1), 21–30 (2017). https://www.researchgate.net/publication/319007861_From_concept_to_the_introduction_of_industry_40
4. Isak, K., Karabegović, E., Mahmić, M., Husak, E.: Contribution of fourth industrial revolution to production processes in China. In: 1st International Conference “Engineering and Entrepreneurship”, ICEE-2017, Tirana, Albania, 17–18 November 2017, Proceedinga Book, pp. 295–301 (2017). ISBN 987-9928-146-47-2
5. Isak, K., Karabegović, E., Mahmić, M., Husak, H.: Applicationof new technologies in final product sales for clothing industry. In: 8th Scientific-Professional Symposium Textile Science and Economy, Zagreb, Croatia, pp. 128–131, 26 January 2015 (2015). ISSN 1847-2877
6. Nikolić, G., Katalinić, B., Rogale, D., Jerbić, B., Čubrić, G.: Roboti & primjena u industriji tekstila i odjeće. Tekstilno-Tehnološki fakultet Zagreb, Zagreb (2008)
7. Gojko, N.: Industrija 4.0 – pravac razvoja tekstilne i odjevne industrije. Tekstil **66,** (3–4), Tekstilo-tehnološki fakultet Zagreb, Hrvatska, pp. 65–73. ISSN 0492-5882. file:///C:/Users/isak/Downloads/Nikolic_Rogale_Tekstil_3_4_2017%20(4).pdf
8. Vlatko, Doleček: Karabegović Isak. Roboti u industriji, Tehnički fakultet Bihać, Bihać (2008)
9. Knez, B., Miličić, J., Rudan, R., Szirovieza, L., Taboršak, D.: Primjenjena antropometrija za antropologiju, biomedicine, ergonomiju i standardizaziju, Priručnik za terensko istraživanje Republike Hrvatske, Ministarstvo odbrane, Zagreb (1995)
10. d’Aquin, M.: Data analytics beyond data processing and how it affects Industry 4.0, Insight Centre for Data Analytic, Dublin, Irska (2017). https://www.slideshare.net/mdaquin/data-analytics-beyond-data-processing-and-how-it-affects-industry-40

11. Kusmin, K.-L.: Industry 4.0 Analytical Article, IFI8101 - Information Society Approaches and ICT Processes, School of Digital Technologies, Tallinn University, Estonia (2016). http://www.tlu.ee/~pnormak/ISA/Analytical%20articles/2-Industry%204.0%20-%20Kusmin.pdf
12. Bechtold, J., Lauenstein, C., Kern, A., Bernhofer, L.: Executive Summary: The Capgemini Consulting Industry 4.0 Framework, Capgemini Consulting, Pariz (2014). https://www.capgemini.com/consulting/wp-content/uploads/sites/30/2017/07/capgemini-consulting-industrie-4.0_0_0.pdf
13. Davies, R.: Industry 4.0: Digitalisation for productivity and growth. http://www.europarl.europa.eu/RegData/etudes/BRIE/2015/568337/EPRS_BRI(2015)568337_EN.pdf
14. Bunse, B., Kagermann, H., Wahlster, W.: Smart Manufacturing for the Future, Germany Trade & Invest, Berlin, Germany (2017). http://belarus.ahk.de/fileadmin/ahk_belarus/Publikationen/GTAI_industrie4.0-smart-manufacturing-for-the-future-en__1_.pdf
15. Must, I., Anton, M., Kruusmaa, M., Aabloo, A.: Linear modeling of elongated bending EAP actuator at large deformations. In: Proceedings of the SPIE, vol. 7287, 72870V (2009). https://www.researchgate.net/publication/236115802_Linear_modeling_of_elongated_bending_EAP_actuator_at_large_deformations
16. Isak, K., Ujević, D., Hodžić, D.: Intelligent system for manufacturing of clothes on distance. In: 86th Textile Institute World Conference, Fashion and Textiles: Heading Towards New Horizons, Hong Kong, 18–21 November 2008, pp. 928–934 (2008)
17. Isak, K., Kadić, S., Ujević, D.: Application of modular robotization line and intelligent textiles in clothing production. In: 2nd DAAAM International Conference on Advanced Technologies for Developing Countries - ATDC 2003, Tuzla, BiH, 25–27 juni 2003, pp. 207–202 (2003)
18. Sihn, W., Jäger, A., Hummel, V., Ranz, F.: Implications for Learning Factories from Industry 4.0, ESB Business School, Reutlingen University, Reutlingen (2016). https://www.kth.se/polopoly_fs/1.481977!/industry%204.0%20-%20full.pdf
19. Chand, S.: The Smart Factory – Risk Management Perspectives, CRO Forum, Amsterdam, Netherlands (2015). file:///C:/Users/isak/Downloads/cro-the-smart-factory%20(2).pdf
20. Ermel, U., Hülsebusch, M.: Industrial IoT Risk Assessment of Smart Factories, PLUS-2016, Munchen, Deutschland (2016). http://www.fair-module-backend.int.live.fsnmm.de

Socio-Cyber-Physical Systems Alternative for Traditional Manufacturing Structures

Elvis Hozdić(✉)

Faculty of Mechanical Engineering, Department of Control and Manufacturing Systems, University of Ljubljana, Aškerčeva 6, 1000 Ljubljana, Slovenia
ehozdic@yahoo.com

Abstract. This work presents a new concept for the restructuring of systemic and organization manufacturing structures in manufacturing enterprises. In the proposed concept is the role of man improved and the role of manager will be given to man, in real time. It is developed the basic concept of socio-cyber-physical manufacturing systems (SCPMS) that represents a building blocks for the new conception of an advanced manufacturing systems in a spirit of socio-cyber-physical systems (SCPS). The proposed concepts enable cybernetization of the functional and managerial competences of the manufacturing structures. This approach aims to improve the performance of manufacturing systems by increasing their productivity, availability, responsiveness and agility, all of which increase competitiveness of manufacturing companies.

Keywords: Cyber-physical systems · Industry 4.0 · Manufacturing systems · Socio-cyber-physical manufacturing systems

1 Introduction

The beginning of the new millennium, from the point of view of industrial production, is characterized by the rapid development of information and communication technologies and the Internet on one side, as well as globalization, uncertainty and ever-increasing demands of the contemporary market on the other [1].

A new production philosophy, which emerges recently under the name *Industry 4.0*, opens space for novel approaches to industrial production. Industry 4.0 stands for a new way of organization and control of complete value-adding systems [2]. The key objective is to fulfill individual customer needs at the cost of mass production. The foundation for the new opportunities is the digitalization of production with help of cyber-physical manufacturing systems (CPMS). Therefore all involved resources like workers, products, resources and systems have to be integrated as smart, self-organized, cross-corporate, real-time and autonomously optimized instances [3].

In the last two or three decades, several influencing attempts to reshaping industrial production have been published, which can be recognized as cornerstones of Industry 4.0 [4]. With the effort of adapting manufacturing and its systems to the time in which they exist, new manufacturing approaches integrating new manufacturing systems with new business models and architectures have emerged [5]. Among these approaches, which all aim of restructuring the manufacturing systems in a way to become more

I. Karabegović (Ed.): NT 2019, LNNS 76, pp. 15–24, 2020.
https://doi.org/10.1007/978-3-030-18072-0_2

flexible, efficient and effective in responding to the changing requirements of the markets, the following approaches have to be mentioned due to their impact on further research and development: intelligent production systems (IPS), biological manufacturing systems (BMS), holonic manufacturing systems (HMS), fractal factories, complex adaptive manufacturing systems (CAMS), reconfigurable production systems (RMS), production networks (PN), distributed manufacturing systems (DMS), complex production systems, ubiquitous manufacturing systems (UMS), and cloud manufacturing systems (CMfgS). A review of the listed approaches and concepts of manufacturing systems are presented in the paper [2].

Recently, a new approach is emerging in terms of a cyber-physical production system (CPPS) [3], which upgrades in a way the mentioned attempts. It originates from the concept of cyber–physical systems (CPS) [6], and links physical and virtual components of a production system into a coherent whole.

All the mentioned approaches including the latest CPPS somehow underestimate, neglect or ignore the role of people in manufacturing. But, according to [7] human *Subjects* are a vital element of any manufacturing system, also in a highly automated or robotized one. However, the humans' role in manufacturing has been significantly changed during the time. It has evolved from a plain workforce, over machinists and supervisors of computer controlled machinery, toward value creators, knowledge workers and decision makers in today's manufacturing systems. They turned from rigid replicators of simple tasks in mass production and became the most intelligent, flexible and capable manufacturing resource. Therefore, people must be incorporated into the new manufacturing philosophy in order to establish a new type of manufacturing systems integrating physical and cyber subsystems with the social one into a socio-cyber-physical system [8].

The aim of this paper is conceptualization of a socio–cyber–physical manufacturing system (SCPMS) and definition of its building blocks.

2 Description of Manufacturing Systems

The term *"Manufacturing system"*, for the first time in its paper [9], introduced M.E. Merchant of the sixties of the last century. Merchant introduced into the manufacturing philosophy the systems approach as an important conceptual tool in modeling and the control of the objects in manufacturing on the macro scale.

The modeling of the manufacturing system on micro scale followed two directions: Peklenik proposed in 1966 a closed loop manufacturing system, consisting of a manufacturing process and a machine tool. It is based on the input/output relation, considering the manufacturing process as a linear transfer system [10]. A similar approach was proposed also by Sata [11] for the process identification. The second approach was initiated by Spur [12] and is conceptually based on the relation manufacturing system – machine tool.

Today's manufacturing systems are composed of many elements, which are interrelated and trying to achieve mutual communication and interaction in order to perform production activities. Due to increased uncertainty on markets as well as intensive communication interactions between the elements, the behavior of manufacturing systems turns out to be more complex and unpredictable and thus managing becomes more difficult. Besides, one has to consider that today the elements of manufacturing systems are not only real, physical elements, such as machinery, processes, work pieces, tools, hardware and similar; there exist also many digital and virtual elements (in the paper, the word cyber is used to denote these elements), such as computer programs, information systems, software packages, data bases, files, world wide web, a variety of communication protocols, and many others. Besides, there are humans involved in the system, who perform physical work, operate machinery, engineer design, plan tasks, maintain equipment, supervise and manage resources, etc. Such systems represent the generation of advanced manufacturing systems. According to [13], advanced manufacturing systems flourish manufacturing systems in service systems, processes, plants, automation, robots, measure systems, advanced process information, process signals in the manufacturing of high-rise information systems in communication systems. The basic structure of advanced manufacturing systems is shown in Fig. 1.

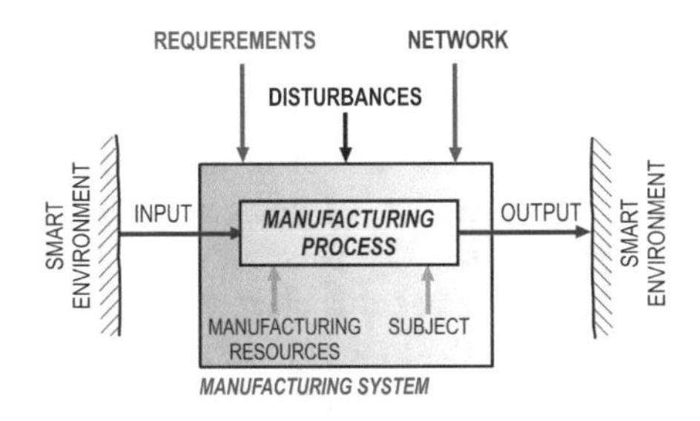

Fig. 1. Structure of advanced manufacturing systems

Prevailing decentralization is an essential feature of advanced manufacturing systems. Especially this characteristic allows us to structure a advanced manufacturing systems from bottom – up, i.e. from the level of technological processes and work systems. Therefore, it is of upmost importance to define the basic building blocks of future manufacturing systems.

Today's production is a networked process, in which more production units are included. Therefore, a comprehensive investigation of the manufacturing system is necessary. To focus on different problems, the whole manufacturing system is divided into hierarchical levels [14]. In the paper [14], authors are six production levels identified and scaled in a top-down fashion: production network, production location, production system, machine and process.

Network production systems (NPS) represent the network-related production systems that wish to take advantage of linking and presentation in the form of larger wholes. The importance of networks is that its building blocks retain autonomous decision-making, and the coordination of the network achieves a competitive advantage. NPS is based on an application from the environment and form by highly competent partners, based on the need for a rapid response to any demand of the global market. With this kind of a network, production systems are viewed as a new complex system [15]. These types of network organizations, open a new opportunities for competitiveness, innovation, agility and adaptability in a production environment, which is based on communication between partners, and exchange of information, knowledge, resources, competence, based on mutual understanding, participation and collaboration [16].

Manufacturing network, clusters and virtual enterprises, which are based on cooperation, are being developed [17]. A manufacturing network provides a basis for competitiveness, innovativeness, agility and adaptiveness by enabling interconnected partners to (1) form long-term business coalitions, (2)develop mutual understanding and trust, (3) jointly react to business opportunities, (4) gain synergetic effects by cooperation and (5) share information, knowledge, resources, competencies and risks [18].

Today's the manufacturing system is a complex socio-technical-economic system, which integrates production processes, equipment to perform these processes and people, and with the intention of converting inputs of raw materials into products and their use or sale.

Under the pressure of globalization and the development of new ICT, highly complex manufacturing systems are emerging. In such systems, the human factor plays a key role.

In the analysis, structuring, modeling, simulation and optimization of advanced complex manufacturing systems, it is inevitable to respect the psycho-sociological indicators of human potential. This sort of event requires the need for structuring advanced manufacturing systems from the perspective of social factors.

Based on the described observations it is clear that in an advanced manufacturing system three systems are confronted, i.e. a physical, a cyber and a social one. Therefore, the emerging cyber-physical manufacturing systems are a logical and viable step in evolution of manufacturing systems. The vision of convergence of physical, cyber and social elements in the next generation manufacturing systems calls for a new concept, which will be based not only on advanced technologies, but will also recognize the dominance a human *Subject* in this context and will take into account his new role in manufacturing systems. In the following section, a concept of a socio-cyber-physical system (SCPMS) is presented.

3 Concept of Socio-Cyber-Physical Manufacturing Systems

3.1 Background and Objectives

As previously mentioned in introduction, the factory is complex adaptive manufacture system structured by three levels. At the operation level there are the functional units such as elementary work system, logistics systems, service systems, etc. The elementary work systems which are horizontal included in the tings network. At the coordination level there are the functional units such as autonomous work systems (AWS) [18] and service units (SU) [19]. AWS and SU are horizontal included in the manufacturing-oriented service network. At the business level there is production-business system (PBS). PBS enables vertical integration of manufacturing structures and horizontal included of factory in production network. The traditional manufacturing structure are based on the traditional automation pyramid (field level, control level, process control level, plant management level and enterprise resource planning level).

As previously mentioned, cyber-physical manufacturing systems are becoming a viable alternative to traditional factory architectures. CPMS partly break with the traditional automation pyramid. Vogel-Heuser et al. [20] described how the automation pyramid, which used to be the *"common sense"* for industrial and automation IT architecture, is evolving into a new kind of architecture. A similar approach is described in the paper [21] and introduced a unified 5-level architecture as a guideline for implementation of cyber-physical systems. Those approaches didn't define the role of *Subject* in the CPMS. The role of *Subject* will change significantly in the future generation of manufacturing systems [22].

The concept of a SCPMS is introduced here. This approach proposes restructuring manufacturing structures in a spirit of CPS. In next, reference model of SCPS is presented.

3.2 Reference Model of SCPS

The reference model of socio-cyber-physical systems (SCPS), which consists of three constitutive elements, i.e. the *Subject* as a social element, the cyber system (CS) as a cyber element and the physical work system (PWS) as a physical element is presented at Fig. 2.

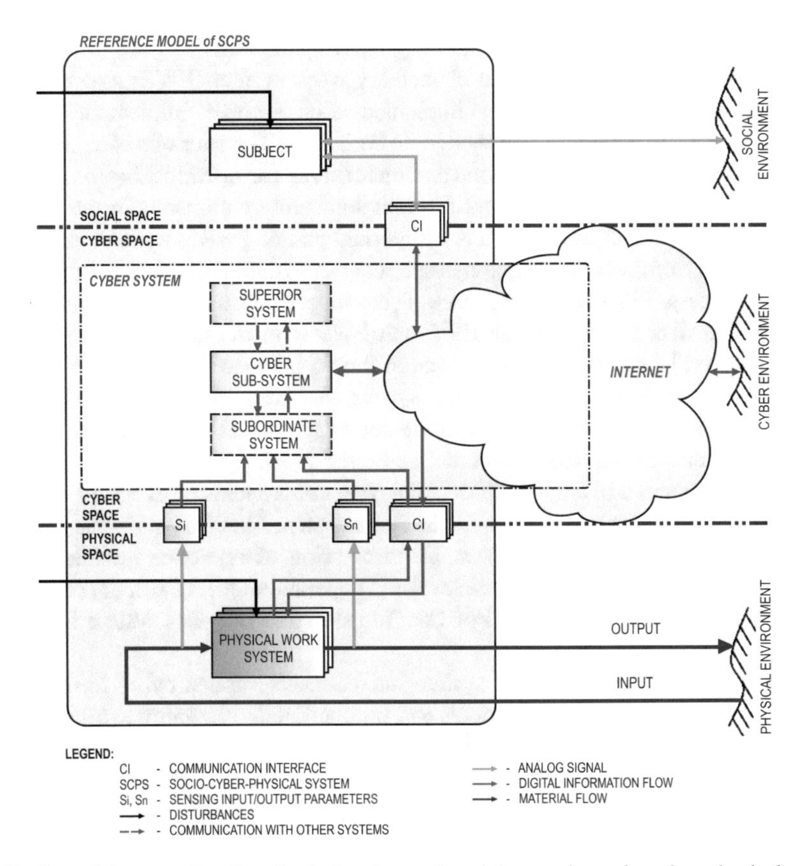

Fig. 2. A model connecting the physical, caber and social space in socio cyber-physical system

The elements are interconnected among each other with information flows which run over a network. For being connected on the network, the *Subject* and the physical work system require adequate communication interfaces. Each element has corresponding relations with its specific environment; the *Subject* with the business and social environment, CS with the cyber environment and PWS with the physical environment. Besides, the system as a whole is placed in a surrounding, which is called a smart environment, as shown in Fig. 2. Thus, the constituting elements are linked with this environment as well.

Now let us look how an SCPS is connected to the outside world, which are shown in Fig. 2. The constitutive elements are acting in their specific spaces, i.e. a *Subject* in a social space, PWS in a physical space, and CS in a cyberspace. The *Subject* belongs to one or several social systems (e.g. organization, enterprise, etc.) and builds collaborative relationships with other *Subjects* of social and business processes by networking with the aim of unfolding business processes. The *Subject* communicates with other human *Subjects* either directly and/or via the cyberspace over social networks and/or particular collaborative environments.

On the other side, PWS is connected to one or several logistic systems, through which input and output material flows are regulated. In the physical space there are also other service systems, such as maintenance, tool management, quality control, etc., which provide physical services supporting PWS operations.

The PWS structure is based on elementary work system (EWS) structure [7]. Its core represents the process, which is implemented on a process implementation device (e.g. machine-tool) and controlled by a logic controller. The later one is connected on a network via a communication interface, which serves for downloading program code (e.g. NC code) on the controller and for communicating commands in the downward direction and statuses as a feedback. An important part of PWS is a monitoring system. It collects data from sensors, which measure characteristic input and output parameters related to process, PWS resources, work pieces, operations and physical environment. The important distinction between EWS and PWS is in the fact that in the latter the *Subject* is moved from the physical space to the social space of SCPS.

Social, logistic and other supporting systems are well known from existing manufacturing systems. The new system in the context of social-cyber-physical system is the cyber system, which needs to be defined more precisely.

CS is incorporated in a cyber space. His cyber sub-systems or cyber manufacturing structures, such as cyber system of EWS, cyber system of AWS, and cyber system of production-business system (PBS) enable structuring of reference models of SCPS. Cyber sub-systems are acting in three levels: (1) operation level of CS, (2) coordination level of CS, and (3) business level of CS. The structure of cyber system is shown in Fig. 3.

As shown in Fig. 3, the cyber system interconnects various cyber forms, such as Internet of Things, Internet of Services, and production and social networks. Hence, it provides an open platform for performing various processes related production, which are based on interaction of elements through communication, exchange and processing of information, access to data and knowledge sources, provision of services, cloud computing, etc.

CS enables vertical connectivity between cyber manufacturing structures and horizontal connectivity cyber manufacturing structures in cyber network structures, such as things network (TN), service network (SN), and production network (PN).

Cyber manufacturing structures communicate with the *Subjects* via the cyber system. The communication CS with elements of PWS enables cyber system of EWS structure.

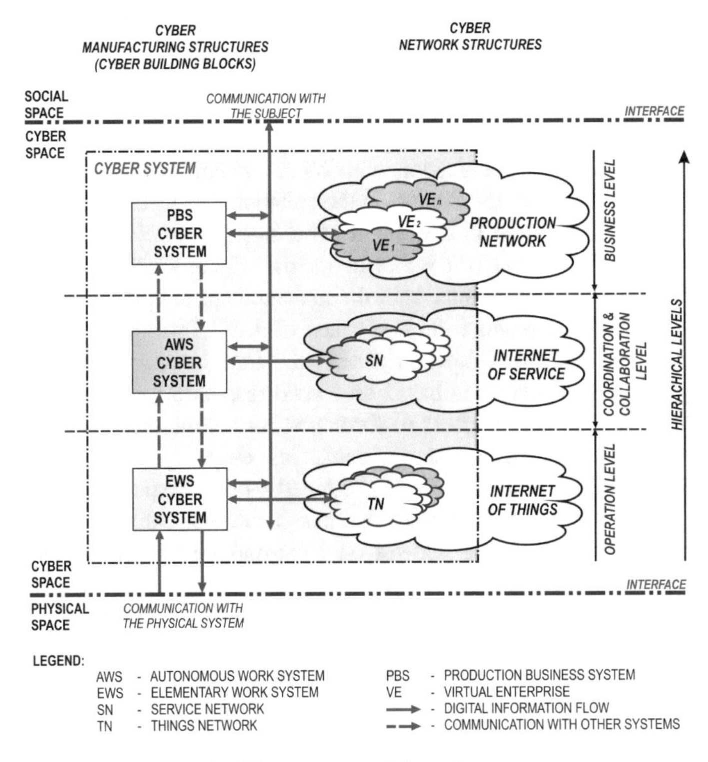

Fig. 3. The structure of the cyber system

The connectivity CS and PWS enables multi-agents technology. The method of connectivity is shown in the paper [23].

3.2.1 Cyber System of EWS

The basic structure of cyber system of EWS is structured with three basic parts (1) the part of data acquisition, (2) the part of self-learning, and (3) the part of monitoring and identification of state. An important part of cyber system of EWS is a data acquisition. It collects data from sensors, which measure characteristic input and output parameters related to process, physical system resources, work pieces, operations and physical environment. The collected data are stored in a data base where they serve for visualization, interpretation, performance evaluation, diagnostics, prognostics, data mining, and other data analytic purposes. The second part of cyber system of EWS is s self-learning. The function of self-learning is founded on the data base where data collected in the real-time from the function of data acquisition. Data from the data base represent the input for the function self-learning. The function of self-learning is structured with sub-function data mining and sub-function knowledge elicitation. The result of self-learning is the newly discovered knowledge. The newly discovered knowledge is stored in the form of knowledge models in the knowledge base. The third part of cyber system of EWS is monitoring and identification of the state of the PWS of the EWS.

3.2.2 Cyber System of AWS

The basic part of the cyber system of AWS is management & control of AWS. The management and control enable communication cyber system of AWS with cyber systems of EWSs and communication with other systems in cyber system. Also, it enables communication with the Subject in the collaborative layer of cyber system. The data connection cyber system of AWS and cyber systems of EWSs is based on connectivity individual D&K base of EWSs with the data & knowledge management in the cyber system of AWS. The outputs from the management & control (actual data from cyber system of EWSs, the data from D&K base of AWS, information from Subject or other systems) represent input for the cybernetized functionality of AWS. The cybernetized functionality of AWS are based on the corresponding management functionality that is described in [18] (such as performance evaluation, diagnostics, prognostics, planning and scheduling, management of resources, etc.).

The support for realization functions presented models respectively partial models for corresponding function from knowledge base of AWS. Outputs of a cybernitized function are information that needed for decision-making Subject, management and control EWSs, and participation AWS in the service networks. The cybernetized functions in cyber system of AWS enable collaborative participation *Subject* and cyber elements (such as agents and multi-agents technology, expert systems, genetic algorithms, neural networks, web service, etc.) in realized certain tasks.

3.2.3 Cyber System of PBS

As previously mentioned, a factory is structured in three levels (the corporate, the managerial, and the manufacturing [7]. The function of corporate level and managerial level such as the business policy, the management, the research & development, the sales and commercial activities, the management of resources, the marketing & analysis, etc., are the part of a production-business system (PBS). The concept SCPS is introduced the cyber system of PBS here, see Fig. 3. The cyber system of PBS is cyber manufacturing structure on the business level of cyber system. The management of the cyber system of PBS is an important part that enabling communication between the cyber system of PBS and the cyber system of AWS or SU. Also, the communication between the cyber system of PBS and the other systems is enabled through the management (connectivity in the production network, in the virtual enterprises, etc.).

The *Subject*, also, communicates with the cyber system of PBS via the function of management. He is a manager and controller in the realization of certain functions on the business level of cyber system. Digitalization of inputs and outputs of the functions, in cyber system of PBS, enables a more advanced way of doing business. Key advantages about by digitalization will be reflected through: (1) minimizing operating costs, (2) logical linking between different functions so that the output of one function is always useful input of other functions, (3) real-time view of the entire company's production-business system, (4) minimizing the space for human error in the realization of certain functions.

4 Conclusion

This paper presents the development of the concept of socio-cyber-physical manufacturing systems (SCPMS) with emphasis on the social element as an indispensable component of all manufacturing systems. The concept was developed cyber structure of elementary work system, cyber structure of autonomous work system, and cyber structure of production-business system. Based on the developed concept enabled the integration of physical and cyber, and social elements of the manufacturing system.

The implementation of the concepts SCPS in real industrial environment is a challenge for further research in the domain of the concept of cyber-physical structure of manufacturing systems, thus contributing to research that defines the consultation document - *"Factory of the Future"* of the European Commission.

References

1. Hozdić, E., Jurković, Z.: Cybernetization of industrial product-service systems in network environment. In: New Technologies, Development and Application, pp. 262–270. Springer (2019)
2. Bauernhansl, T.: Industry 4.0: challenges and opportunities for the automation industry. In: 7th EFAC Assembly Technology Conference 2013 (2013)
3. Monostori, L., Kádár, B., Bauernhansl, T., Kondoh, S., Kumara, S., Reinhart, G., Sauer, O., Schuh, G., Sihn, W., Ueda, K.: Cyber-physical systems in manufacturing. CIRP Ann. Manuf. Technol. **65**, 621–641 (2016)
4. Hozdić, E.: Smart factory for Industry 4.0: a review. Int. J. Modern Manuf. Technol. **VII** (2015)
5. Jovane, F., Koren, Y., Boer, C.R.: Present and future of flexible automation: towards new paradigms. CIRP Ann. Manuf. Technol. **52**, 543–560 (2003)
6. Gill, H.: NSF perspective and status on cyber-physical systems. In: National Workshop on Cyber-Physical Systems (2006)
7. Peklenik, J.: Fertigungskybernetik, Eine Neue Wissenschaftliche Dusziplin Fur Die Produktionstechnik (Festvortrag anlasslich der Verleihung des Georg - Schlesinger Preises 1988 des Landes Berlin, 1988) (1988)
8. Morosini, E., Hartmann, J., Makuschewitz, T., Scholz-Reiter, B.: Towards socio-cyber-physical systems in production networks. Procedia CIRP **7**, 49–54 (2013)
9. Merchant, M.E.: The manufacturing system concept in production engineering research. CIRP An. X **2**, 77 (1962)
10. Peklenik, J.: Contribution to a correlation theory for the grinding process. In: ASME Production Engineering Conference; Journal of Engineering for Industry, vol. 86, No. 2 (1964). (1963)

11. Sata, T.: New identification methods for manufacturing processes. In: Advances in Manufacturing Systems, Research and Development, pp. 11–22. Pergamon Press, Oxford (1971)
12. Spur, G.: Betrachtungen zur Optimierung des Fertigungssystems, Werkzeugmaschine. Werkstattstechnik **57**, 411–417 (1967)
13. High Level Group: High Level Group on Key Enabling Technologies (2010)
14. Wiendahl, H.P., Nyhuis, P., Hartmann, W.: Should CIRP develop a production theory? motivaton - development path – framework. In: Sustainable Production and Logistics in Global Networks, Conference Proceedings of 43rd CIRP International Conference on Manufacturing Systems, pp. 3–18 (2010)
15. Westkämper, E., Hummel, V.: The stuttgart enterprise - integrated engineering of strategic & operational functions. In: 38th CIRP ISMS (2005)
16. Zaletelj, V., Sluga, A., Butala, P.: The B2MN approach to manufacturing network modeling. In: Proceedings of the 6th International Workshop on Emergent Synthesis IWES 2006, pp. 9–16 (2006)
17. Peklenik, J.: A new structure of an adaptable manufacturing system based on elementary work units and network integration. In: 7th International Conference AMST 2005, pp. 27–40. Springer (2005)
18. Butala, P., Sluga, A.: Autonomous work systems in manufacturing networks. CIRP Ann. Manuf. Technol. **55**, 521–524 (2006)
19. Zupančič, R., Sluga, A., Butala, P.: A service network for the support of manufacturing operations. Int. J. Comput. Integr. Manuf. **25**, 790–803 (2012)
20. Vogel-Heuser, B., Kegel, G., Wucherer, K.: Global information Architecture for Industrial Automation. atp Ed. - Sutomatisierungstechnische Prax. 51 (2009)
21. Lee, J., Bagheri, B., Kao, H.-A.: A cyber-physical systems architecture for Industry 4.0-based manufacturing systems. Manuf. Lett. **3**, 18–23 (2015)
22. Hozdić, E.: Cybernetization of manufacturing systems. In: 13th International Scientific Conference - MMA2018 Flexible Technologies, Novi Sad, Serbia (2018)
23. Leitão, P., Colombo, A.W., Karnouskos, S.: Industrial automation based on cyber-physical systems technologies: prototype implementations and challenges. Comput. Ind. **81**, 11–25 (2016)

Blockchain in Distributed CAD Environments

Samir Lemeš[1(✉)] and Lamija Lemeš[2]

[1] Polytechnic Faculty, University of Zenica, Fakultetska 1, 72000 Zenica, Bosnia and Herzegovina
slemes@unze.ba
[2] University of Sarajevo, 71000 Sarajevo, Bosnia and Herzegovina

Abstract. Distributed and collaborative CAD (Computer Aided Design) environments gained wide popularity recently, in engineering fields such as BIM (Building Information Modelling) and GIS (Geographical Information Systems) in Civil Engineering, or PDM/PLM (Product Data Management/ Product Lifecycle Management) in Mechanical Engineering. One of key issues in these applications is data integrity and confidence in information stored in information systems. Blockchain technology, initially invented to support the cryptocurrency trusted authority, opened its way to other areas, and it could be the solution for the data integrity issue in distributed CAD environments.

Keywords: Blockchain · CAD · BIM · PDM/PLM · Data integrity

1 Introduction

The use of cloud computing in CAD is still at an early stage in the development and application. CAE applications that are demanding for computer resources make use of cloud computing to overcome that limitation of a single PC, but regular CAD modelling and design applications still have limitations for changing the paradigm and using this technology [1]. On the other hand, collaborative CAD environments such as BIM or PDM/PLM, already embraced the advantages of cloud computing, which brought another issue - the problem of information security. Blockchain could be one of the solutions to overcome this problem.

Blockchain is an emerging technology that is likely to affect almost every business process that requires a trusted digital environment. It was conceptualized in 2008 by a person or a group using alias “Satoshi Nakamoto” to serve as the public transaction ledger of the cryptocurrency bitcoin, and it has ever since inspired other applications. Blockchain is a simple distributed database containing information, such as transactions or agreements, that are stored chronologically across a network of computers, decentralised and is not usually managed by a central authority.

One of the greatest features in every computer software is the “undo” function, which is used to correct accidental mistakes. Blockchain actually killed the “undo” feature, because adding a new entry into the blockchain is irreversible, which is then used to provide the data integrity. The most important characteristic of blockchain is that once it is published, the information on the blockchain cannot be changed. Since every block in the blockchain contains information about its predecessor block,

I. Karabegović (Ed.): NT 2019, LNNS 76, pp. 25–32, 2020.
https://doi.org/10.1007/978-3-030-18072-0_3

providing a mechanism to verify data integrity of the entire blockchain, an extremely resilient trusted record is created. This could be the major driving force in introducing this technology in distributed CAD environments.

Valero opened a set of questions about BIM interoperability in [2]. He claims that blockchain is the solution that could allow the exchange of information between participants through completely secure and irreversible coding, which is completely traceable. The first integrated BIM and blockchain application aimed at its productive use on site is under development in Spanish technological research and development project DELFOS [2].

A comprehensive analysis covering the viable application of blockchain and real use cases developed by startups, has recently been published by Institution of Civil Engineers (ICE) [3]. The report claims that we are at a very early stage of blockchain's inception, and we should begin thinking about its implementation in our systems, business and processes. The report is distinguished into three main parts with regards to the three potential applications: payment and project management; procurement and supply chain management; and BIM and Smart asset management. Blockchain enabled transparency, traceability and collaboration for the construction industry. In addition to advantages, the report also highlighted some barriers to blockchain adoption identified by industry leaders: early challenges are regulatory uncertainty and lack of trust among users, and possible obstacles in the next 2–3 years are cost, lack of knowledge and lack of governance.

A French tech startup is developing a blockchain-based solution to integrate its distributed ledger technology into BIM processes, tools, and data. The project is backed by Autodesk and the French BIM Task Group, and a market-ready product was launch as an early beta-version in January 2019. They aim to create a new collaborative process that bridges the gap between 3D digital modelling and the formal and legally binding paper-based processes related to project administration, building control, insurance and payment [4].

Mathews et al. in [5] claim that "What the Internet did for society and industry 20 years ago is what blockchain can do for the next 20 years, that is to provide a platform for application development to drive efficiencies and effect a digital transformation in the targeted industry."

Dounas and Lombardi in [6] remained in the CAD environment, trying to implement blockchain into CAD system by an integration strategy between CAD and blockchain technology in an algorithmic platform using Ethereum.

Li et al. in [7] discussed how blockchain can be used in a distributed CAM (Computer Aided Manufacturing) system. They demonstrated how to successfully employ the blockchain technology to facilitate the cloud manufacturing and establish a new type of trustable platform, aiming to develop peer to peer and decentralized network infrastructure for cloud manufacturing. The major gain was increase in system security.

2 Distributed CAD Environments

With the increasing use of BIM, architectural, engineering and construction professionals are experiencing radical changes to working practices. BIM as a relatively new technology and approach reflects many of the changes, challenges and opportunities triggered by the introduction of PLM in the aerospace and automotive sectors almost two decades ago [8]. BIM and PLM also share a number of similarities relative to their approach to data sharing, project management, organisation of teams around deliverables and timelines, and object-based visualisation activities. The PLM concept emerged from PDM to primarily manage design files created by CAD tools. PLM services have expanded to cover not only product definition and design phases but also manufacturing and operations. PLM systems actas a hub connecting intangible asset information (i.e. virtual products of design and analysis activities) to physical assets information managed by systems like Enterprise Resource Planning (ERP) and Customer Relationship Management (CRM).

2.1 Building Information Model (BIM)

BIM is a digital representation of physical and functional characteristics of a facility. As such it serves as a shared knowledge resource for information about a facility forming a reliable basis for decisions during its life-cycle from inception onward. A basic premise of BIM is collaboration by different stakeholders at different phases of the life cycle of a facility to insert, extract, update or modify information in the BIM process to support and reflect the roles of that stakeholder [9].

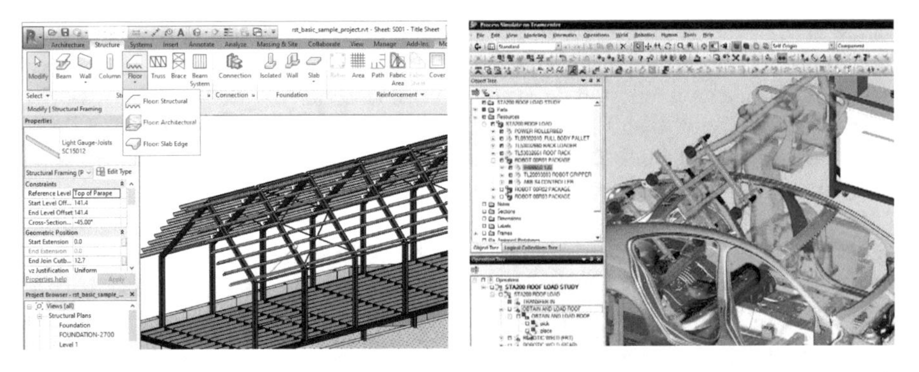

Fig. 1. Autodesk Revit (www.autodesk.com)

Fig. 2. Siemens Teamcenter (www.plm.automation.siemens.com)

This technology is still under development and new definitions are continuously adopted. The baseline for BIM 3D is the 3D CAD geometry. BIM 4D adds a component of time, BIM 5D adds information about costs, BIM 6D adds sustainability through an intelligent connection of individual 3D objects with all aspects of product life cycle information, and BIM 7D deals with the performance of assets. Figure 1 is a screenshot of the most famous BIM software Autodesk Revit.

2.2 Product Data Management/Product Lifecycle Management (PDM/PLM)

What BIM is for Construction and Civil Engineering, PDM/PLM is for Mechanical Engineering. PDM is a specialized information system used to manage product data. It represents the extension of 3D CAD models to a specialized design environment that manages a set of CAD files in hierarchically distributed files. PLM is a wider term used to manage the product life cycle in a networked environment, in a way that multiple users access CAD models from the database, rather than from individual files. In a PLM system, CAD files represent only a single attribute in a set of attributes describing a machine element, assembly, or construction.

Although most CAD software vendors develop and offer PDM/PLM solutions, such as: Dassault Systems Enovia, Autodesk Fusion Lifecycle, PTC Windchill, Siemens Teamcenter (Fig. 2), SAP, Aras Innovator, some surveys surprisingly reveal that only one third of CAD users actually uses PDM/PLM to create a Bill of Materials for their CAD drawings, while most of them still relies on spreadsheet software. Therefore, this is still an emerging technology, not mature enough.

3 Blockchain

The literature usually defines a blockchain as follows [10]: "A blockchain is a ledger of facts, replicated across several computers assembled in a peer-to-peer network. Facts can be anything from monetary transactions to content signature. Members of the network are anonymous individuals called nodes. All communication inside the network takes advantage of cryptography to securely identify the sender and the receiver. When a node wants to add a fact to the ledger, a consensus forms in the network to determine where this fact should appear in the ledger; this consensus is called a block." Technically, blockchain relies on three well-known IT concepts: peer-to-peer networks, public-key cryptography, and distributed consensus based on the resolution of a random mathematical challenge. The combination of these concepts allows a breakthrough in computing.

Facts stored in the blockchain can't be lost. They are there forever, replicated as many times as there are nodes. Even more, the blockchain doesn't simply store a final state, it stores the history of all passed states, so that everyone can check the validity of the final state by replaying the facts from the beginning. In order to do that, blockchain uses cryptographic hashfunctions to generate a digital signature of the previous information. A hash function is a mathematical algorithm that maps data of arbitrary size to a bit string of a fixed size (a hash) and is designed to be a one-way function, that is, a function which is infeasible to invert. Hash functions are in common use for decades, and all share the same four properties: they should be computationally efficient (based on a fast algorithm), deterministic (always giving a unique result for every distinct input), pre-image resistant (their output must not reveal any information about the input), and collision resistant (it is practically impossible to find two different inputs producing the same output).

The most commonly used classes of hash functions include: Secure Hashing Algorithm (SHA-2, SHA-3), RACE Integrity Primitives Evaluation Message Digest (RIPEMD), Message Digest Algorithm 5 (MD5), BLAKE2 etc. Each of these classes may contain several different algorithms. All these hash functions are similar, but they differ slightly in the way the algorithm creates a digest, or output, from a given input. They also differ in the fixed length of the digest they produce. Figure 3 illustrates how changing only one letter in an information gives entirely different hash digest.

Blockchain concept uses hash functions to create fixed size blocks of information, which are then added to an array of blocks (hence the term block-chain). The new block is again irreversibly encrypted by a hash function to get the fixed size output, and the chain contains the complete history of changes which cannot be falsified, because a change in any step of data transfer will give completely different final output.

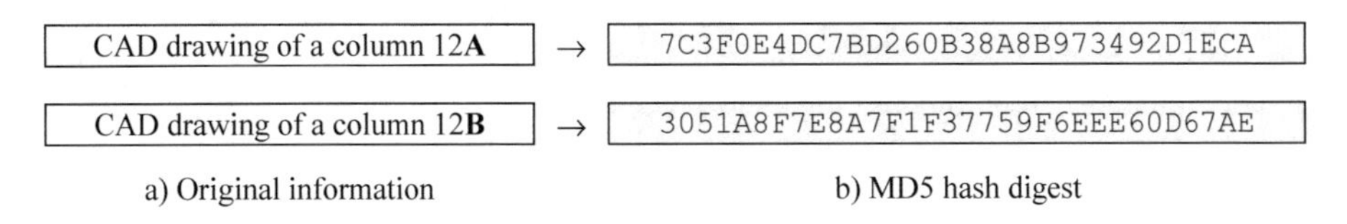

Fig. 3. Using MD5 hash function to irreversibly encrypt data

How does blockchain technology disable the "undo" function? The answer to that question lies in the fact that any change in any step of the process changes all blocks following that step, creating a completely new branch. Since hash functions used to generate blocks are irreversible, there is no way to find out which change could be "undone" to recreate the original blockchain, making the "undo" function impossible, or at least unusable. The entropy of the entire chain increases, but not indefinitely, since the size of the hash function result is finite, making this technology efficient and fast.

4 Blockchain in Distributed CAD Environments

Figure 4 illustrates how blockchain is implemented in a CAD environment used to perform computer simulation. Block "A" is created by hashing the initial CAD model "1" supplemented by a random string "0" (which in this case serves as a cryptographic public key). Block "B" is made from the CAE model "2", accompanied with hashed signature of the block "A". The simulation result "3" with block "B" from the previous step is hashed to create the block "C". If data is corrupted (e.g. an unauthorised intruder changes a parameter in CAE model - force, constraint, material), both blocks "B" and "C" are changed, pointing out that data corruption occurred in the CAE model "2". Since there was no change in the block "A", one can conclude that there were no changes in 3D CAD data, i.e. in the geometry. Therefore, blockchain provides both data integrity (by detecting changes in the final digest "C") and traceability (pointing out who is responsible for data corruption – in this case, CAE model "2").

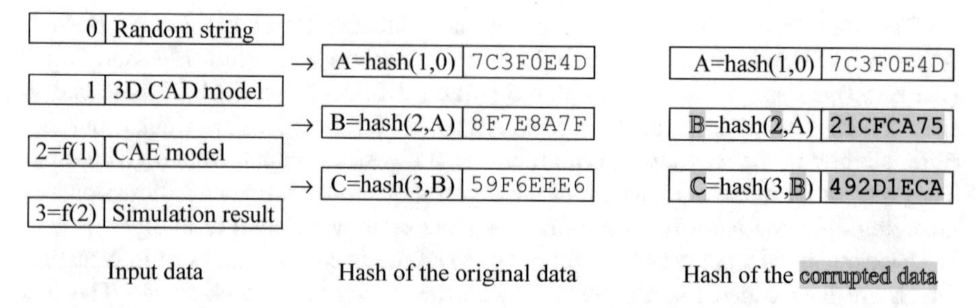

Fig. 4. Any change in a single blockchain step is traceable in a final digest

Large engineering projects demand a lot of teamwork and collaboration between different team members. In such a distributed environment, it is essential to provide the reliability of data, and to enable fast method to identify the responsibility of team members in every step of the engineering project. All stakeholders have their own digital signature, an automated payment process can be established by using so called "Smart Contracts", and complete supply chain can be kept transparent.

Blockchain has no centralized authority, which could make information vulnerable to external attacks, making that technology extremely secure and reliable. Wherever there is need for a secure, reliable and traceable information, there is apotential to use blockchain.

However, the relative immaturity of this model will probably be an obstacle to wider use in distributed CAD environments. The report [3] points out some specific challenges of blockchain implementation in construction industry: regulatory implications (sometimes related to the differences between national/regional regulations), vested interests (technological potential to eliminate intermediaries could influence the business model), a need for the new skills, potential of cost optimization to increase investments into R&D, securing intellectual property rights through digital ownership, and overcoming fragmentation in legacy systems and practices.

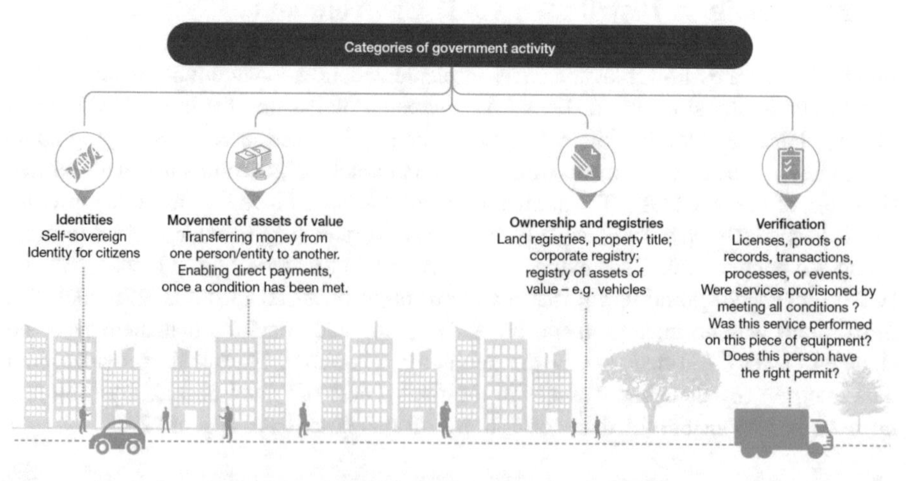

Fig. 5. Impact of blockchain technology on smart cities [11]

Blockchain technology is one the hottest topics in modern industry, and expert groups, business initiatives, research institutes and startups keep searching for new applications of blockchain in a modern world. The challenges arisen from the climate change and resource scarcity, rapid urbanization, demographic and social changes and shift in global economic power, lead to new initiatives, such as the one presented in PwC report [11]. The purpose of this report was to understand and analyse how blockchain can be utilised for making cities smarter. It identifies areas of smart cities where blockchain innovation can be used to enhance them and provide for better liveability and economic development.

BIM is likely to expand into the smart cities initiatives, since it provides digital baseline for interconnecting the buildings, infrastructure (traffic, utilities, communications), people and natural resources into a complex interconnected system. Fostering technological improvements is a challenge because of the complexity, vulnerability and reliability of these systems. Blockchain can be used to eliminate risks connected to widespread use of technology in an everyday life and human activities (Fig. 5).

It is noticeable that the major vendors in PDM/PLM/BIM market (SAP, Siemens, Autodesk) are still careful in implementing blockchain technology into their products. It looks like they rely on startups which are much more prominent in this field, waiting for their solutions, which are likely to be acquired by the big players.

The efforts of Siemens are focused on Industry 4.0 and additive manufacturing, while SAP collaborates with companies such as Capgemini and Deloitte, aimed at leveraging the SAP blockchain framework and IoT, beginning from the inception of products through the manufacturing and logistics phases. The presentations at the SAP TechEd conference in September 2017 announced that a number of companies will collaborate with SAP to determine how to better leverage the SAP blockchain framework and IoT technologies to manage business processes, beginning from the inception of products through the manufacturing and logistics phases that span multiple industries [12]. Autodesk, on the other hand, is sceptical about this technology, speculating "Since there are traditional technologies that raise traditional database reliability and security to an acceptable level, it is not clear that blockchain would add much value to core PLM capabilities" [13]. This statement should be challenged, and will surely be the topic of further research.

5 Conclusion

We can conclude that, despite the great potential of blockchain technology in a distributed CAD environment, we shall wait for a few more years to see the advantages of this technology to be used in BIM/PDM/PLM. The major disadvantages of blockchain are the transparency of data and the slowness of storing data in the blockchain, as it requires a distributed consensus. It seems like there is still not much interest within the major software vendors to leverage blockchain in their BIM/PDM/PLM solutions, and they are still waiting what small startups will have to offer in a near future. On the other hand, increasing demand by the Industry 4.0, IoT, Smart Cities, and other initiatives, could foster the change in their approach, in order to use the best what blockchain has to offer: data integrity, reliability, and traceability.

References

1. Lemeš, S.: Računarstvo u oblaku kao alat za razmjenu CAD podataka. In: Brdarević, S., Jašarević, S. (eds.) 10th International conference "Quality 2017", pp. 39–46, Neum, Bosnia and Herzegovina, 17–20 May 2017 (2017). ISSN 1512-9268
2. Valero, F.: BIM and Blockchain, Zigurat Global Institute of Technology (2018). https://www.e-zigurat.com/blog/en/bim-and-blockchain/. Accessed 19 Jan 2019
3. Penzes, B.: Blockchain technology in the construction industry: Digital Transformation for High Productivity, ICE (Institution of Civil Engineers) report (2018). https://www.ice.org.uk/ICEDevelopmentWebPortal/media/Documents/News/Blog/Blockchain-technology-in-Construction-2018-12-17.pdf. Accessed 19 Jan 2019
4. Cousins, S.: French start-up develops Blockchain solution for BIM, BIM + Task Group UK (2018). http://www.bimplus.co.uk/news/french-start-develops-blockchain-solution-bim/. Accessed 19 Jan 2019
5. Mathews, M., Robles, D., Bowe, B.: BIM+Blockchain: A Solution to the Trust Problem in Collaboration?, CITA BIM Gathering 2017, 23–24 Nov 2017 (2017). https://doi.org/10.21427/D73N5K, http://arrow.dit.ie/bescharcon/26. Accessed 19 Jan 2019
6. Dounas, T., Lombardi, D.: A CAD-Blockchain Integration Strategy for Distributed Validated Digital Design, BIM Applications - Volume 1 - eCAADe 36, pp. 223–230 (2018). http://papers.cumincad.org/data/works/att/ecaade2018_226.pdf. Accessed 19 Jan 2019
7. Li, Z., Barenji, A.V., Huang, G.Q.: Toward a blockchain cloud manufacturing system as a peer to peer distributed network platform. Robot. Comput. Integr. Manuf. **54**, 133–144 (2018). https://doi.org/10.1016/j.rcim.2018.05.011
8. Jupp, J., Nepal, M.: BIM and PLM: comparing and learning from changes to professional practice across sectors. In: IFIP Advances in Information and Communication Technology, AICT-442, pp 41–50. Product Lifecycle Management for a Global Market (2014). https://doi.org/10.1007/978-3-662-45937-9
9. Cramer, M., Hunt, S.: BIM Basics - Past, Present, Future, Partners in progress conference (2010). https://www.pinp.org/wp-content/uploads/2016/03/Cramer-Hunt-presentation.pdf. Accessed 19 Jan 2019
10. Zaninotto, F.: The Blockchain Explained to Web Developers (2016). https://marmelab.com/blog/2016/04/28/blockchain-for-web-developers-the-theory.html. Accessed 20 Jan 2019
11. Baru, S.: Blockchain: the next innovation to make our cities smarter, FICCI- PwC (Federation of Indian Chambers of Commerce and Industry – Price waterhouse Coopers) report at the Smart Cities Summit, New Delhi, India (2018). https://www.pwc.in/assets/pdfs/publications/2018/blockchain-the-next-innovation-to-make-our-cities-smarter.pdf. Accessed 20 Jan 2019
12. Bjorlin, C.: SAP blockchain, IoT effort takes vertical dive into PLM, asset management (2017). https://www.iotworldtoday.com/2017/09/27/sap-blockchain-iot-effort-takes-vertical-dive-plm-asset-management/. Accessed 20 Jan 2019
13. Smith, F.: Blockchain Technologies and PLM. https://www.autodeskfusionlifecycle.com/en/blog/blockchain-technologies-plm/. Accessed 20 Jan 2019

Single Mobile Robot Scheduling Problem: A Survey of Current Biologically Inspired Algorithms, Research Challenges and Real-World Applications

Zoran Miljković(✉) and Milica Petrović

Faculty of Mechanical Engineering, University of Belgrade, Kraljice Marije 16, 11120 Belgrade, Serbia
{zmiljkovic,mmpetrovic}@mas.bg.ac.rs

Abstract. Intelligent mobile robots belong to advanced material handling systems that are finding increasing applications in modern manufacturing environments. Due to their mobility, mobile robots can adapt to changing environments in manufacturing systems and carry out various tasks such as transportation, inspection, exploration, or manipulation. On the other hand, features such as autonomy, intelligence, flexibility, and the capability to learn allow mobile robots to be widely used for many tasks, including material handling, material transporting, or part feeding tasks. Motion planning and scheduling of an intelligent mobile robot are one of the most vital issues in the field of robotics since these factors are essential for contributing to the efficiency of the overall manufacturing system. The robot scheduling problem belongs to the class of NP-hard problems and numerous efforts have been made to develop methodologies for obtaining optimal solutions to the problem. Therefore, this paper presents a review of the literature sources and gives a comparative analysis of biologically inspired optimization algorithms used to solve this problem. Four different optimization algorithms, namely genetic algorithms (GA), particle swarm optimization algorithm (PSO), chaotic particle swarm optimization algorithm (cPSO), and whale optimization algorithm (WOA) are proposed and implemented in Matlab software package. The experimental verification is carried out by using real-world benchmark examples. The experimental results indicate that all aforementioned algorithms can be successfully used for optimization of single mobile robot scheduling problem.

Keywords: Single mobile robot scheduling · Biologically inspired algorithms · Intelligent manufacturing system

1 Introduction

Implementation of intelligent mobile robots in manufacturing systems is one of the major requirements for creating an intelligent manufacturing environment. Due to their mobility, they can operate on the shop floor as a component of the material transport system and perform various tasks such as transportation, inspection, or manipulation [1]. In such conditions, transportation tasks need to be carefully planned and optimally

I. Karabegović (Ed.): NT 2019, LNNS 76, pp. 33–41, 2020.
https://doi.org/10.1007/978-3-030-18072-0_4

scheduled due to its impact on the efficiency of the overall manufacturing system [2]. Having these facts in mind, this paper analyses integration and scheduling of mobile robot in the manufacturing environment.

Single mobile robot scheduling problem belongs to the class of NP-hard optimization problems, which has attracted the interest of numerous researchers in recent decades. GA-based methodology for creating an optimal path and location of the workpiece with the objective to minimize the time required for a robot to transport a workpiece was developed in [3]. In reference [4], the authors presented tabu search and probabilistic tabu search to solve the single-vehicle pickup and delivery problem with time windows. The aim was to minimize the total distance traveled by the vehicle. A single-machine scheduling problem in a job-shop environment where the jobs have to be transported between the machines by a single transport robot was analyzed in [5]. They regarded the robot scheduling problem as a generalization of the traveling salesman problem with time windows and used a local tabu search algorithm to solve it. The objective was to determine a sequence of all nodes and corresponding starting times in the given time windows in such a way that all generalized precedence relations were respected and the sum of all traveling and waiting times was minimized. Furthermore, the same authors extended their work and proposed two different approaches to integrate a transportation stage into procedures which schedule the machines [6]. Besides transportation times for the jobs, the authors also analyzed empty moving times for the robot. The objective was to determine a schedule with minimal makespan. The problem of finding an optimal feeding sequence in a manufacturing cell with feeders fed by a single mobile robot with manipulation arm was presented in [7–9]. The mathematical formulation for mobile robot scheduling problem in a manufacturing cell was given in [7]. A genetic algorithm-based heuristic developed to find the near optimal solution for the robotic cell scheduling problem was presented in [8]. The focus in reference [9] was on modeling and scheduling of autonomous mobile robot for a real-world industrial application. The system and concept have been evaluated in a real-world case study at an impeller production line with the mobile robot prototype “Little Helper”.

2 Problem Formulation and Mathematical Modeling

In this paper, we analyze the manufacturing environment where three parts have to be transported between 8 machine tools by using a single mobile robot (SMR). Therefore, the problem of SMR scheduling is incorporated into integrated process planning and scheduling problem (IPPS). In this integration, the sequences of operations for all parts as well as information about manufacturing resources (machines, tools, TADs) are defined with flexible AND/OR process plan networks (see details in [10, 11]). Schedule of mobile robot activities such as retrieving a part from storage, transporting of the part or part returning to storage is defined by the scheduling plan [10]. Additionally, the following transportation times are taken into consideration: (i) the time needed for the SMR to arrive at the machine where the part is manufactured (empty trip), and (ii) the time needed to transport the part for the next operation (loaded trip). Furthermore, the following two assumptions about SMR are considered: (i) is on the machine one at the start of the process and (ii) it can transport only one part at a time.

In order to achieve optimal scheduling plan, we propose the mathematical models for three fitness functions. The first fitness function used in this paper considers the time needed to manufacture all parts in the intelligent manufacturing environment. Fitness function that minimizes makespan is defined with Eq. (1):

$$f_1 = \min\left(\max\left(c_{ijl}^k\right)\right), \tag{1}$$

where $\max(c_{ijl}^k)$ represents the completion time of the last operation in the scheduling plan, i.e. completion time of the j-th operation in an alternative process plan l of the i-th job that is manufactured on the k-th machine tool.

The second fitness function is the robot finishing time. In large manufacturing environments, where there are numerous machine tools, one mobile robot is hardly sufficient to satisfy all needed transportation tasks. In the aforementioned case, the fitness function representing robot finishing time should be optimized as in Eq. (2):

$$f_2 = \min\left(\max\left(tr_{ijl}\right)\right), \tag{2}$$

where tr_{ijl} represents robot finishing time for j-th operation in the alternative process plan l of the i-th job.

The last fitness function named total job and robot waiting time is advised when there a need to eliminate most of the wastes in the manufacturing environment. This fitness function is minimized by using Eq. (3):

$$f_3 = \min\left(\sum_{i=1}^{N}\sum_{j=1}^{Pil} Jwt_{ijl} + Rwt_{ijl}\right) \tag{3}$$

where Jwt_{ijl} and Rwt_{ijl} are job and robot waiting time during current operation. Robot waiting time represents the time robot waits before it can start to transport part for the next operation. When current operation is completed, a job waiting time represents the period job waits for the robot to transport it to the next machine tool.

3 Biologically Inspired Optimization Algorithms

3.1 Gentic Algorithms

The genetic algorithms (GA) are one of the most widely used metaheuristics, which is inspired by the biological phenomenon of evolution. The main idea of the algorithm is to apply GA operators such as selection, crossover, and mutation in order to modify individuals in a population over generations and to obtain an optimal solution. In order to represent one possible solution of the SMR scheduling problem, each individual in the population is coded as a scheduling plan string. Pseudocode of the GA approach for single mobile robot scheduling problem is described by Algorithm 1.

Algorithm 1. Pseudo code of GA algorithm	
1:	**Initialize** the parameters of the GA (size of population, probability of crossover, probability of mutation, maximum number of generations)
2:	**Initialize randomly the** population **of individuals**
3:	Evaluate the fitness of the individuals (equations (1), (2), or (3))
4:	**Repeat:**
5:	Select the best individuals according to roulette selection
6:	Apply crossover and mutation operators [10]
7:	Evaluate the fitness of the individuals (equations (1), (2), or (3))
8:	**Until** the maximum of generation is not met
9:	**Output** the optimal scheduling plan

3.2 Particle Swarm Optimization Algorithm

Particle swarm optimization (PSO) algorithm is a population-based optimization method inspired by the movement and intelligence of the organisms in a swarm (e.g., a flock of birds migrating). In order to search for food, each member of the swarm (aka "particle") determines its velocity based on their personal experience as well as information gained through interaction with other members of the swarm. The equations of the traditional PSO algorithm are originally proposed by [12] and pseudocode for implementation on the SMR scheduling problem is given by Algorithm 2.

Algorithm 2. Pseudo code of PSO algorithm	
1:	**Initialize** the parameters of the PSO (swarm size, maximum number of generation, inertia weight W, acceleration constants C_1 and C_2)
2:	**Initialize** a swarm of particles with velocities and positions $V_{id}^{t+1} = W \cdot V_{id}^{t} + C_1 \cdot rnd() \cdot (P_{ld}^{t} - X_{id}^{t}) + C_2 \cdot Rnd() \cdot (P_{gd}^{t} - X_{id}^{t})$ $X_{id}^{t+1} = X_{id}^{t} + V_{id}^{t+1}$
3:	**Evaluate** the fitness of each particle by using equations (1), (2), or (3)
4:	find the global best position P_{gd} and local best position P_{ld}
5:	**Repeat:**
6:	generate next swarm by updating the velocities V_{id} and positions X_{id} of the particles
7:	round off the real number of positions for machine, tool, and TAD to the nearest integer number from machine, tool and TAD sets
8:	evaluate swarm
9:	compute each particle's fitness by using equations (1), (2), or (3)
10:	find new global best position and local best position
11:	update the global best position of the swarm and the local best position of each particle
12:	**Until** the maximum of generation is not met
13:	**Output** the optimal scheduling plan

3.3 Chaotic Particle Swarm Optimization Algorithm

The major advantages of traditional PSO are easy implementation and efficient computation. However, on the other side, the major drawback of the algorithm is its premature convergence, especially while handling problems with more local optima. One of the ways to overcome this problem is to hybridize the PSO algorithm with chaos. Chaotic particle swarm optimization algorithm (cPSO) applies traditional PSO to perform global exploration and chaotic local search to perform local search on the solutions produced in the global exploration phase. In this paper, we employ ten non-invertible, one-dimensional chaotic maps (see e.g. [10] and [11]) and replace the values of *rand()* and *Rand()* with chaotic number generators *chaos()* and *Chaos()*, respectively. Pseudocode of the cPSO approach is given as follows:

Algorithm 3. Pseudo code of cPSO algorithm	
1:	**Initialize** the parameters of the cPSO for the process plans optimization (swarm size, maximum number of generation, inertia weight W, acceleration constants C_1 and C_2)
2:	**Initialize** a swarm of particles with velocities and positions $V_{id}^{t+1} = W \cdot V_{id}^{t} + C_1 \cdot chaos() \cdot (P_{ld}^{t} - X_{id}^{t}) + C_2 \cdot Chaos() \cdot (P_{gd}^{t} - X_{id}^{t})$ $X_{id}^{t+1} = X_{id}^{t} + V_{id}^{t+1}$
3:	**Evaluate** the fitness of each particle by using equations (1), (2), or (3)
4:	find the global best position P_{gd} and local best position P_{ld}
5:	**Repeat:**
6:	generate next swarm by updating the velocities V_{id} and positions X_{id} of the particles
7:	round off the real number of positions for machine, tool, and TAD to the nearest integer number from machine, tool and TAD sets
8:	evaluate swarm
9:	compute each particle's fitness by using equations (1), (2), or (3)
10:	find new global best position and local best position
11:	update the global best position of the swarm and the local best position of each particle
12:	**Until** the maximum of generation for process planning is not met
13:	**Output** the optimal process plan

3.4 The Whale Optimization Algorithm

The whale optimization algorithm (WOA) belongs to a family of newly developed population-based metaheuristic algorithms. The main idea of this algorithm is to mathematically model the way humpback whales hut their prey. Hunting is represented with two phases, and one of them is selected for each variable in the whale according to random parameters A and p. Each agent (whale) in the population represents one possible scheduling plan. Encoding and decoding method for each scheduling plan is given in [10]. According to reference [13], pseudocode of the WOA approach for single mobile robot scheduling problem is described as follows:

Algorithm 4. Pseudo code of WOA	
1:	**Initialize** the parameters of the WOA (size of population, number of iterations, defining range for variables a, C, and l)
2:	Initialize randomly the population of agents
3:	Evaluate the fitness of the agents (equations (1), (2), or (3))
4:	**Repeat:**
5:	Leader determination and assignment of parameters A and p
6:	Modification of agents' position
7:	Evaluate the fitness of the agents (equations (1), (2), or (3))
8:	**Until** the maximum of iteration is not met
9:	**Output** the optimal process plan

4 Experimental Results

In order to analyze the performance of selected algorithms on the real-world problem, the authors employ the experiment with three parts; details about the parts can be found in reference [10]. The considered manufacturing environment consists of 8 machine tools, 12 cutting tool, 6 TADs and a single mobile robot used to perform the transportation tasks. The robot can carry only one part at a time. At the beginning of the process, the mobile robot is at machine one and all parts are on the machine of their first operation. Parameters of the algorithms are set according to the recommendations made in [10]. After some preliminary experiments and parameter tuning, the following parameters are chosen for the experiment. The number of iterations is set to 50 and the size of the population is 80 for all considered algorithms. For both PSO and cPSO the acceleration constants (C_1 and C_2) are set to 2, and the inertial weight is linearly decreasing from 1.4 to 0.4. For the cPSO algorithm, the Tent map is used to replace the random number generator. The parameter a in WOA algorithm is set to 3 and it linearly decreases through the iterations to 1. Finally, for GA the crossover and mutation probability are set to 0.9 and 0.2, respectively. All the algorithms are run 10 times for

Table 1. Comparison of optimization results

Fitness function		WOA	PSO	GA	cPSO
F1	Best	116.5	116.5	129.6	**112.1**
	Average	116.9	122.13	146.45	115.15
	Std	0.51	4.68	8.65	6.12
F2	Best	97	**86.3**	100.4	90.3
	Average	102.15	90.59	118.41	94.37
	Std	5.02	9.51	7.77	5.93
F3	Best	20.9	**17.7**	55.7	37.1
	Average	32.25	32.83	72.37	54.41
	Std	8.40	9.83	13.36	11.86

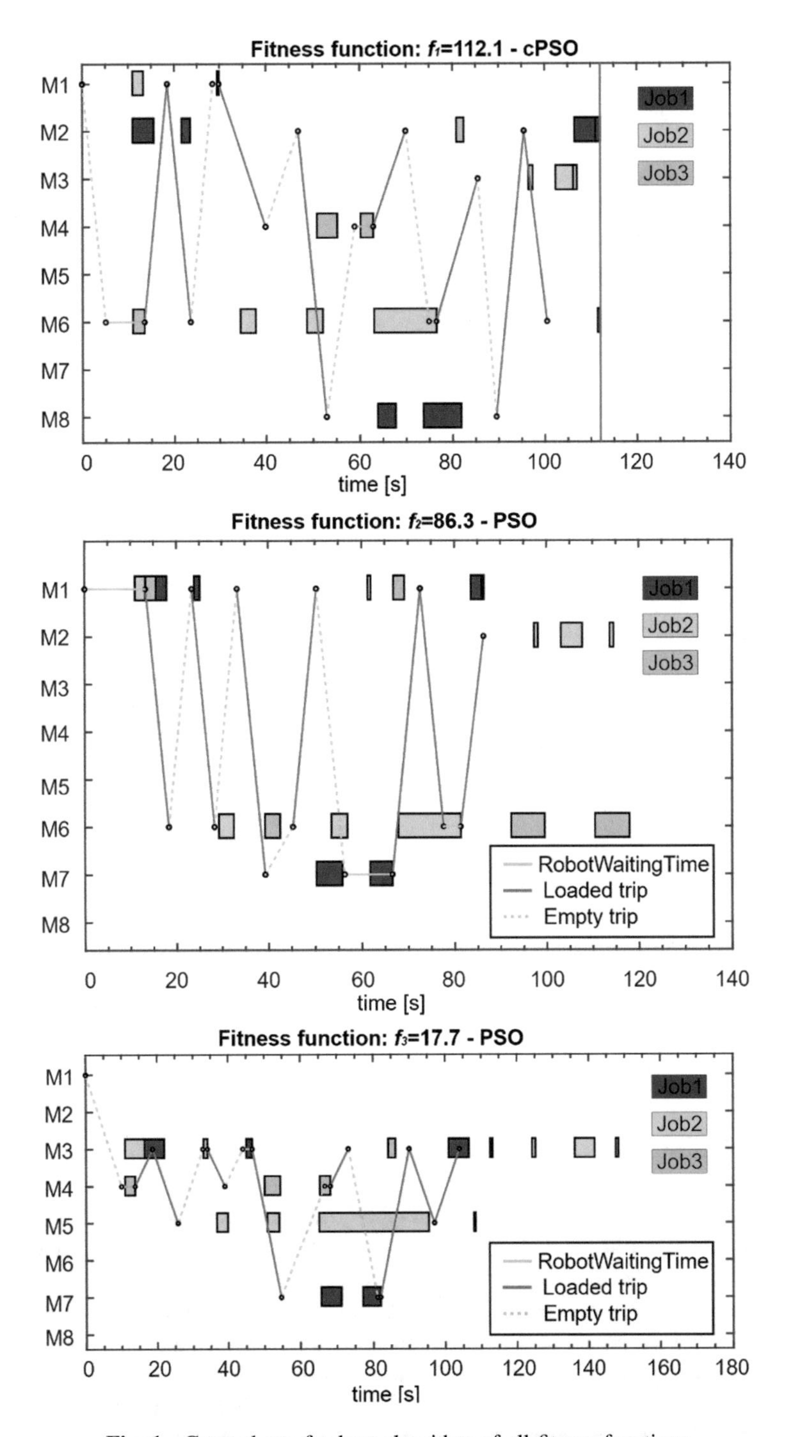

Fig. 1. Gantt charts for best algorithm of all fitness functions

each of the fitness functions and the statistical results are shown in Table 1. As it can be seen, cPSO achieves the best result for fitness function f_1, while traditional PSO gets best results for f_2 and f_3. While these algorithms achieve the best results, the best average result for f_3 is achieved by WOA. Scheduling results are presented in Fig. 1.

The best scheduling plan for all fitness functions is illustrated with a Gantt chart in Fig. 1. The Gantt chart represents the order of operations, machine tools on which all operations are manufactured and the route robot took in order to perform the transportation tasks. Convergence curves of the best run for all algorithms are represented with Fig. 2. It can be seen that algorithms have different convergence rates for each of the fitness functions.

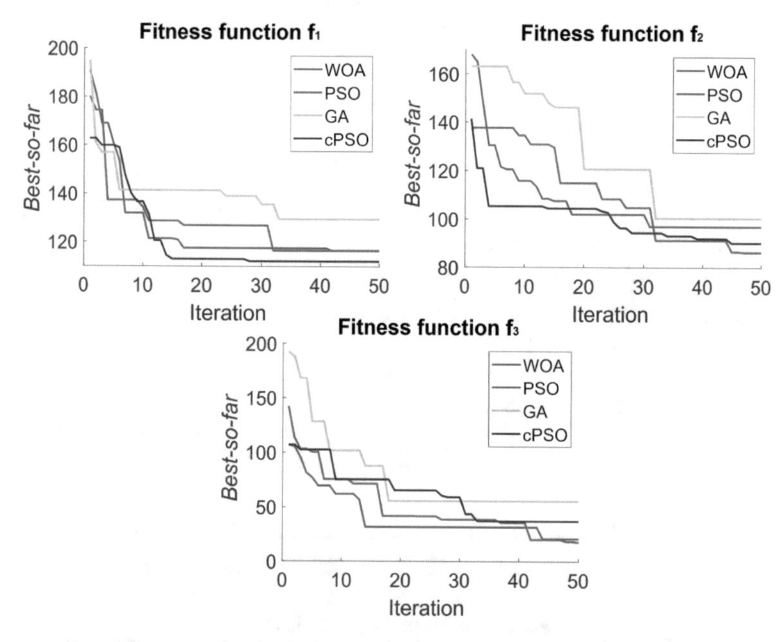

Fig. 2. Convergence curves for best run of each algorithm

5 Conclusion

In recent years, the application of biologically-inspired methods has been gaining popularity in research related to engineering optimization problems. One of the problems crucial for the development of intelligent manufacturing systems is the implementation of an intelligent mobile robot into a manufacturing environment and its scheduling. In this paper, the authors propose four biologically-inspired methods, namely genetic algorithms (GA), particle swarm optimization algorithm (PSO), chaotic particle swarm optimization algorithm (cPSO), and whale optimization algorithm (WOA) to optimize combinatorial NP-hard single mobile robot scheduling problem. Solutions of the scheduling problem are encoded into a scheduling plan in order to intelligently search for the optimal sequence of mobile robot transportation tasks. Optimal scheduling sequence is found in accordance with three fitness functions (i) makespan (ii) robot finishing time and (iii) total robot and job waiting time. All the algorithms are developed and implemented in the Matlab software environment. The experimental results show that biologically inspired optimization algorithms can be used to generate an optimal motion path of an intelligent mobile robot according to the requirements of the manufacturing process and defined scheduling functions. One of the future research directions is oriented towards analysis of biologically inspired optimization algorithms in solving the multi-objective single mobile robot scheduling problem.

References

1. Miljković, Z., Vuković, N., Mitić, M., Babić, B.: New hybrid vision-based control approach for automated guided vehicles. Int. J. Adv. Manuf. Technol. **66**(1–4), 231–249 (2013)
2. Mitić, M., Vuković, N., Petrović, M., Miljković, Z.: Chaotic metaheuristic algorithms for learning and reproduction of robot motion trajectories. Neural Comput. Appl. **30**(4), 1065–1083 (2018)
3. Chen, C.J., Tseng, C.S.: The path and location planning of workpieces by genetic algorithms. J. Intell. Manuf. **7**(1), 69–76 (1996)
4. Landrieu, A., Mati, Y., Binder, Z.: A tabu search heuristic for the single vehicle pickup and delivery problem with time windows. J. Intell. Manuf. **12**(5–6), 497–508 (2001)
5. Hurink, J., Knust, S.: Tabu search algorithms for job-shop problems with a single transport robot. Eur. J. Oper. Res. **162**(1), 99–111 (2005)
6. Hurink, J., Knust, S.: A tabu search algorithm for scheduling a single robot in a job-shop environment. Discrete Appl. Math. **119**(1), 181–203 (2002)
7. Dang, Q.V., Nielsen, I., Steger-Jensen, K.: Mathematical formulation for mobile robot scheduling problem in a manufacturing cell. In: IFIP International Conference on Advances in Production Management Systems, pp. 37–44. Springer, Heidelberg (2011)
8. Dang, Q.V., Nielsen, I.E., Bocewicz, G.: A genetic algorithm-based heuristic for part-feeding mobile robot scheduling problem. In: Trends in Practical Applications of Agents and Multiagent Systems, pp. 85–92. Springer, Heidelberg (2012)
9. Nielsen, I., Dang, Q.V., Bocewicz, G., Banaszak, Z.: A methodology for implementation of mobile robot in adaptive manufacturing environments. J. Intell. Manuf. **28**(5), 1171–1188 (2015)
10. Petrović, M., Vuković, N., Mitić, M., Miljković, Z.: Integration of process planning and scheduling using chaotic particle swarm optimization algorithm. Expert Syst. Appl. **64**, 569–588 (2016)
11. Petrović, M., Mitić, M., Vuković, N., Miljković, Z.: Chaotic particle swarm optimization algorithm for flexible process planning. Int. J. Adv. Manuf. Technol. **85**(9–12), 2535–2555 (2016)
12. Kennedy, J., Eberhart, R.C.: Particle swarm optimization. In: Proceedings of the IEEE International Conference on Neural Network, Perth, Australia, pp. 1942–1948 (1995)
13. Petrović, M., Jokić, A., Miljković, Z.: Single mobile robot scheduling: a mathematical modeling of the problem with real-world implementation. In: Proceedings of the 13th International Scientific Conference MMA 2018 - Flexible Technologies, Novi Sad, Serbia, pp. 175–178 (2018). ISBN 978-86-6022-094-5

Hexapod Robot Navigation Using FPGA Based Controller

Lejla Banjanovic-Mehmedovic[1](✉), Alen Mujkic[2], Nedim Babic[1], and Jakub Secic[3]

[1] Faculty of Electrical Engineering, University of Tuzla, 75000 Tuzla, Bosnia and Herzegovina
lejla.mehmedovic@untz.ba
[2] DCCS GmbH, Tuzla, Bosnia and Herzegovina
[3] MBT BH, Tuzla, Bosnia and Herzegovina

Abstract. In order to improve efficiency and achieve higher performance, motor control mechanism on a robotic platform realized by microcontroller-based system last time is changing with the reconfigurable hardware platforms. This paper presents the field programmable gate array (FPGA) implementation of the hexapod robot navigation using the tripod gate sequence. The servo motor controller is implemented in the Cyclone IV FPGA chip by Altera using Verilog as Hardware Description Language (HDL). The implementation of the servomotor controller in FPGA has several advantages as circuit design flexibility and parallel command executions when compared to the conventional microcontroller-based system. Particular advances introduced in this field have impact on motor control design of multiple-output requirements as well as parallel co-work of multiple robotic platforms in different applications in scope of the Industry 4.0.

Keywords: FPGA · Hardware Description Language (HDL) · Hexapod robot navigation · Verilog

1 Introduction

A biomimetic system is a system that mimics biology in a way that it adopts mechanisms found in natural organisms and applies them in engineering [1]. Biomimetic robots do not copy the complete behavior of the animals, only the most useful abilities are extracted and modified for more practically design [2].

The mobility of animals, especially many insects is often inspiration for robot locomotion. Insects have six legs, which gives them clear stability advantages over four legged animals [3]. The hexapod walker using Raspberry Pi is presented in this paper [4]. A control strategy for achieving stable walking in complex and unknown environments is proposed in [5]. In paper [6], the implementation of a hexapod mobile robot with a fuzzy controller navigating in unknown environments is presented. The number of investigations of six-legged robots has been performed including studies on dynamic walking on irregular terrain [7].

I. Karabegović (Ed.): NT 2019, LNNS 76, pp. 42–51, 2020.
https://doi.org/10.1007/978-3-030-18072-0_5

The primary concerns in the last decade in automated real-time systems have been the design of Systems-on-Chip hardware architectures as well as development of algorithms feasible for execution on hardware platforms such as a Field Programmable Gate Array (FPGA). The FPGA has allowed designers to create large designs, test them and make modifications easily and quickly [8].

The common hexapod robot is controlled by a discrete conventional microcontroller whereby the configuration is fixed and included in the hexapod robot which would require new chip replacement due to multiple-output requirements (for example, 18 motors instead 3). The sequential nature of program execution in the microcontroller would significantly affect the synchronization of the hexapod movement [9]. The implementation of the servomotor controller of robots platform in FPGA has several advantages such as circuit design flexibility and parallel command executions when compared to the conventional microcontroller-based system. The particular advances introduced in this field have impact on the motor control design of multi-legged robotic platform and their parallel co-work.

In [10], the hexapod robot has legs inspired by the spider, which it uses to manoeuvre across horizontal surfaces. FPGA is used as a controller, with one soft processor controlling each leg, adding additional modularity. The paper [11] presents the robot, which was programmed with a device NI myRIO-1900 embedded hardware. The robot was developed to be adapted to any type of surface, considering the angle at which the device is located. Data acquisition and handling of the servomotors that the hexapod has and variations in angles were performed in the FPGA Xilinix Z-7010.

Our paper proposes the FPGA implementation of the hexapod robot navigation. It focuses on the design and implementation of a servo motor controller in an FPGA platform. Our target goal is development of FPGA-based systems used for multi-robot navigation where FPGA based solutions have a great privilege with real-time simultaneous command executions.

The contribution is organized as follows: after an introductory part describing the state-of-art in FPGA-motor control algorithms of the hexapod robot, Sect. 2 presents the basic key aspects about FPGA and HDL. In Sects. 3, the robot control architectures are introduced. Section 4 proposes the description of the hexapod robot navigation using Verilog implementation of finite state machine (FSM) based behaviour with the experimental results. The last section concludes the paper with insights forextensions.

2 FPGA as Controller Solution

FPGAs are a types of reconfigurable devices and can be programmed to do any type of digital function. Configurable microcontrollers present a valuable solution in real-time applications and they can be used in the design of a variety of system-on chip designs [12]. The advantages of the FPGA over a microprocessor chip are [8]:

- the FPGA has the ability to be reprogrammed on the fly, which means that there is no need for any down time for the controller;
- the new FPGAs that are on the market will support hardware that is upwards of 1 million gates;
- the FPGA will operate faster than a microprocessor chip.

If a microprocessor needs to be reprogrammed then the entire system must be taken down and the microprocessor will be reprogrammed and then the system can be brought back up on line. The FPGA's can be programmed while they are running, because they have reprogram times of the order of microseconds. This short time means that the system will not even know that the chip was reprogrammed and there may be a short waiting period and the system will not have to be shut down. The fact that the FPGA is a programmable chip means that the controller will be running as an Application Specific Integrated Circuit (ASIC). When a piece of hardware is custom made for an application, the design will be able to run much faster than a general purpose microprocessor that is running from software that has been downloaded on it.

Hardware Description Languages (HDL) have been developed to describe how hardware behaves. There are two main differences between traditional programming languages and HDL [8]:

- traditional languages are a sequential process, whereas HDL is a parallel process;
- HDL runs forever whereas traditional programming languages will only run forever if directed.

Two common ways used in the development of FPGA systems are:

- direct VHDL (Very High Speed Integrated Circuit Hardware Description Language) or Verilog coding and
- usage of particular HDL Code Generators.

Although existing HDL Code Generator (for example from MATLAB software tool) simplifies the procedure since the deep knowledge about VHDL/Verilog programming language is not required, direct VHDL/Verilog coding provides much higher degree of freedom in terms of directly accessible resources of FPGA. From this reason, Verilog code is used for the hexapod navigation implementation in our research.

3 Hexapod Robot Control Approaches

The main part of a walking robot is a hexapod controller, which determines the order and range of the leg movements. One approach is an exact mathematical model, which determines the movement and contact with the environment of each leg. This can be done using mathematical formulations or inverse kinematics models. However, this can be difficult, because complete modeling of all aspects of the robot and the environment and other influences is very complex task [13].

Another approach consists of different methods, which would generate the best gait according to a given situation. The robot control architectures are related to sensing, monitoring and acting actions of the robot. Different kinds of robot controllers can be distinguished [13]:

- reactive control architectures
- behavior-based control architectures
- deliberative controllers (hierarchical) or sense-plan-act control architectures
- hybrid control architectures

The reactive control architecture is stimulus-response based. Although the reactive control architecture response speed is rather high, which is an advantage when operating in the real world where the response time is very important, the reactive architecture is not suitable for predictive planned outcomes.

The behavior based is an alternative to the reactive system architecture. It is based on priority behaviors organized into layers. The problem of this control architecture is the right order of the layers.

The deliberative control architectures are based on the Sense-Plan-Act principle and for their optimal functioning they usually need full knowledge about the environment. The advantage of this architecture is that the goal can be easily achieved owing to the goal oriented architecture. On the other hand, this architecture is rather slow and it is not suitable for purposes where a quick reaction is needed. The limitations seen by reactive and the deliberative architectures can be solved by combining both approaches into a hybrid architecture. The advantage of hybrid architecture is the goal oriented architecture represented by deliberative layer and at the same time the reactive layer can execute low level actions [7].

We propose the finite state machine (FSM) based behavior control approach, which is implemented on FPGA so it could be reprogrammed and tested on site.

4 Hexapod Robot Navigation and Results

The designing and controlling of six legged robots (hexapods) are inspired by insects. The legs of a hexapod robot with a rectangular body are usually symmetrically distributed into two groups. Each group is located on the opposite side of the body. Another type is the hexapod robot with the circle body, which has the legs evenly distributed around the body. Some robots use mechanisms to couple their joints for the purposes of reducing the number of actuators, because actuators are typically heavy and reducing their number can increase robot range [3].

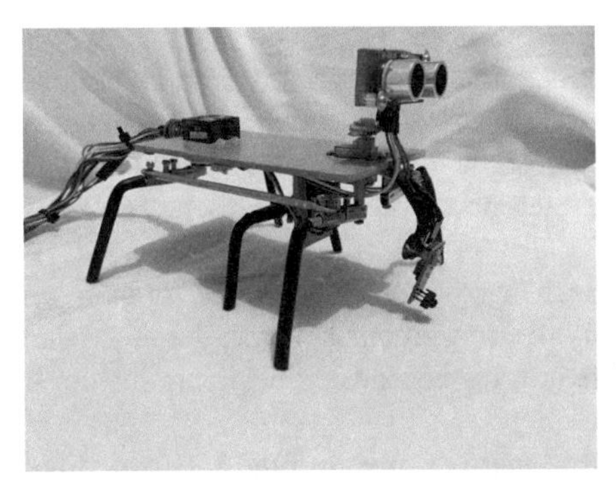

Fig. 1. Hexapod robot construction design

In this paper we propose the hexapod robot construction which has a rectangular body and legs symmetrically distributed into two groups. This construction design is presented in Fig. 1. It has four servo motors, one for turning the head with an ultrasonic sensor and three servo motors for navigation in environment: forward, backward, turn right, and turn left. In our control approach we used one motor to control two legs. To detect the ground, the hexapod used the IR sensor, so the movement on rough terrain is possible.

The Altera Cyclone IV Board was used for generating control pulses for the hexapod movement, Fig. 2. Using a FPGA decrease the power consumption of the controller and the required chip space.

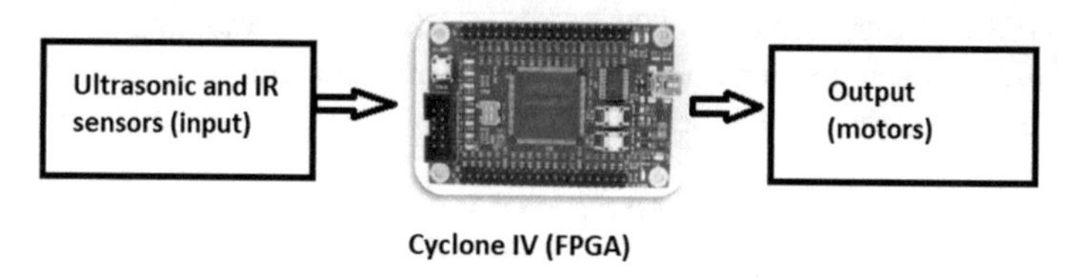

Fig. 2. Control block diagram using Altera Cyclone IV FPGA

4.1 Hexapod Robot Sequence Diagram

The reliability of the exchange of supporting legs is essential for stability of walking. A gait refers to the locomotion achieved through the movement of robot legs. There are several basic gaits, observed in insects: tripod, wave or ripple [7].

The tripod gait is based on two groups of legs. During each step the first group of the legs is lifted and is rotated forward and is laid on the ground. Then the other group is lifted. Both groups are moving, the first group backward, the second group forward and finally the second group is laid on the ground. It is obvious that both groups perform the same movement, but they are shifted by half a period. The tripod gait is very fast, but also very unstable.

Another gait is wave (also known as a metachronal wave gait), which is the most stable gait, but also the slowest. The wave gait consists of a sequential adjustment of the robot's legs forward. Once all the legs are set to the new positions, the step is completed. Maximally one leg is lifted up in each phase of a step. This leads to high stability of this gait.

The ripple gait is inspired by insects. Each leg performs the same move – up, forward, down, backward. Leg moves partially overlap. In other words, the time when the first foot is lifted and begins to move forward, the second leg begins to lift up. In this way the robot cycles through all legs.

The hexapod sequence diagram of the tripod gait is shown in Fig. 3. We modified the sequence diagram presented in [4], where the hexapod robot was controlled by discrete conventional microcontroller. Our hexapod robot controller is programmed in Verilog on Altera Cyclone IV FPGA chip, using Quartus environment.

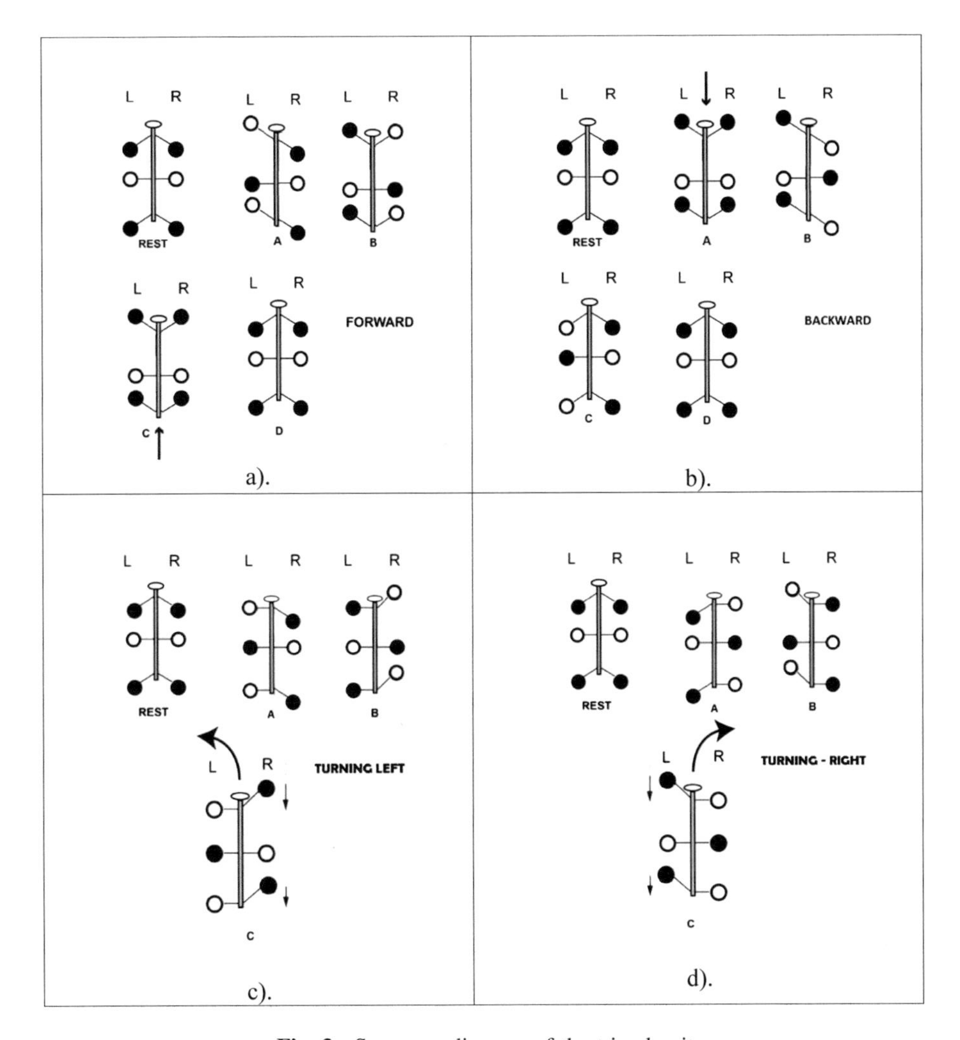

Fig. 3. Sequence diagram of the tripod gait

The middle legs are moved by a single servomotor, while the front and rear legs on the left side are controlled together and moved by second servomotor. Similar case is for the right side legs. Each leg is represented by a circle. The black circle indicated that the weight of the robot is on that leg, while the white one indicates that the leg is in the air. The REST position indicates the initial state, while the other states (A, B, C, D) are the sequence of movements of the corresponding legs. Except these three servomotors, the fourth servomotor is used to turn the head with the ultrasonic sensor.

The sequences tripod gate for the hexapod robot navigation using FPGA (forward, backward, turning right and turning left) with certain motor values for each case are presented in Table 1.

Table 1. The hexapod robot sequence diagram of the tripod gait with servo motors values

	Forward			Backward		
	Left motor	Middle motor	Right motor	Left motor	Middle motor	Right motor
1.	Bottom (150)	Middle (300)	Bottom (255)	Bottom (150)	Middle (300)	Bottom (255)
2.	Top (255)	Left (450)	Bottom (255)	Top (255)	Middle (300)	Top (150)
3.	Top (255)	Right (150)	Top (150)	Top (255)	Right (150)	Bottom (255)
4.	Top (255)	Middle (300)	Top (150)	Bottom (150)	Left (450)	Bottom (255)
5.	Bottom (150)	Middle (300)	Bottom (255)	Bottom (150)	Middle (300)	Bottom (255)
	Turning left			Turning right		
	Left motor	Middle motor	Right motor	Left motor	Middle motor	Right motor
1.	Bottom (150)	Middle (300)	Bottom (255)	Bottom (150)	Middle (300)	Bottom (255)
2.	Middle (200)	Left (450)	Bottom (255)	Bottom (150)	Right (150)	Middle (200)
3.	Middle (200)	Right (150)	Top (150)	Top (255)	Left (450)	Middle (200)
4.	Middle (200)	Left (450)	Top (150)	Top (255)	Right (150)	Middle (200)

4.2 FSM Based Robot Navigation and Results

The navigation approach of hexapod robot is based on finite state machine (FSM) modelling. The FSM of manoeuvres in the environment is presented in Fig. 4. It consists of 8 states: S0 (Standby state), S1 (Normal walking), S2 (Backward walking), S3 (Turn head to the right), S4 (Turn right), S5 (Turn head to the left), S6 (Turn left) and S7 (Rough terrain). The transitions between the states are T1–T11.

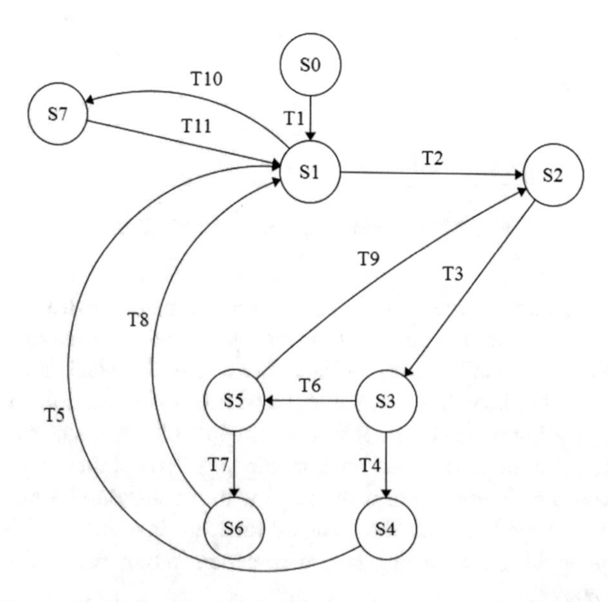

Fig. 4. Finite State Machine (FSM) of hexapod robot navigation

The default state is S0, where the hexapod waits on pushbutton activation. After this event, S0 has transition toward state S1 (Normal walking). From this state, it is possible to make transition in case of obstacle detection to the state S2 (Backward walking), or in case of rough terrain detection to state S7 (Rough terrain).

In state S2, the hexapod goes three steps reverse and switches to S3 (turns head to the right). The hexapod remains in the state S7 until it gets a signal for rough terrain detection. From the state *S3*, the two transitions with new events are possible toward other states: in case of no obstacle detection toward the state *S4 (Right Turn)* and in case of obstacle detection toward the state *S5 (Turn head left)*. From the state S4, with finishing turning, the transition leads toward the state S1. From the state S5, if the obstacle is not detected, the transition leads to the *state S6 (Turn left)* and with finishing this action toward state S1. All states and transitions descriptions are presented in Table 2.

Table 2. Finite state machine (FSM) with states description for hexapod robot navigation

States	Description
S0	Standby state
S1	Normal walking
S2	Backward walking
S3	Turn head to the right
S4	Turn right
S5	Turn head to the left
S6	Turn left
S7	Over a rough terrain
S0	Standby state
S1	Normal walking
S2	Backward walking
S3	Turn head to the right
Transitions	Description
T1	Turn the main switch on; Switch to **Normal walking** mode
T2	Obstacle detected; Switch to **Backward** mode
T3	Go three step in reverse; Switch to **Turn head right** mode
T4	Obstacle is not detected; Switch to **Turn right** mode
T5	Turning right is finished; Switch to **Normal walking** mode
T6	Obstacle is detected; Switch to **Turn head left** mode
T7	Obstacle is not detected; Switch to **Turn left** mode
T8	Turning left is finished; Switch to **Normal walking** mode
T9	Obstacle is detected; Switch to **Backward** mode
T10	Rough terrain is detected; Switch to **Over a rough terrain** mode
T11	Rough terrain is not detected; Switch to **Normal walking** mode
*	From each state we can switch to state S0 if the main switch is off

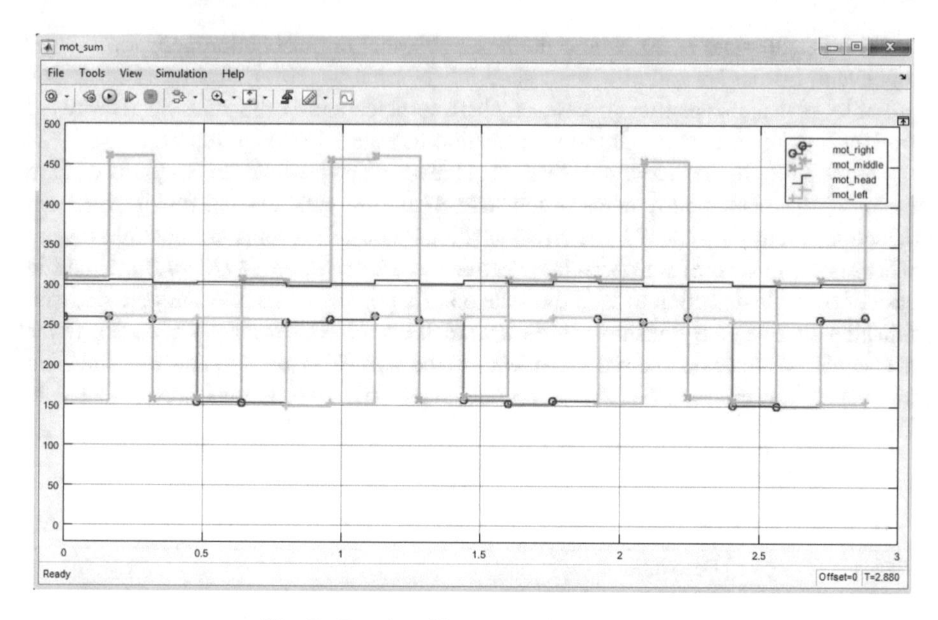

Fig. 5. Results of hexapod robot navigation

The results of hexapod robot navigation during one experimental scenario are presented in Fig. 5. If we track values in sequence for left, middle and right motors from Table 1, we can see that the hexapod robot goes forward in a few steps.

5 Conclusion

The development of FPGA-based systems used for multi-robot navigation has to include both real-time navigation of each robot platform as well as real-time cooperation algorithm. The implementation of the servomotor controller in FPGA has several advantages such as circuit design flexibility and parallel command executions when compared to the conventional microcontroller-based system. In this initial stage of development, the navigation approach for the particular hexapod robot is implemented on real-time FPGA-based platform using Verilog in order to synchronize the position and direction of the servo motor at every state of movement. In the future we will first focus on the design of the hexapod robot controller using adaptive control approaches as well as the fuzzy based behaviour solution. Multiple robots can be used to do manufactory operation in order to finish the operation quicker, therefore our next step in future research would be FPGA based implementation of parallel coordination algorithm of hexapod robots as well. This proves that the new technological era in scope of the Industry 4.0 determined by increased use and sudden expansion of systems based on real-time platforms is yet to come.

References

1. Na, S.Y., Shin, D., Kim, J.Y., Baek, S.-J., Lee, B.-H.: Pipelines monitoring system using bio-mimetic robots. Int. J. Comput. Inf. Eng. **3**(1), 23–29 (2009)
2. Ranjan, V.: Biomimetic Robotics: Mechanisms and Control. Cambridge University Press, New York (2009)
3. Quinn, R.D., Nelson, G.M., Bachmann, R.J., Kingsley, D.A., Offi, J.T., Ritzmann, R.E.: Insect design for improves robot mobility. In: Berns, K., Dillmann, R. (eds.) From Biology to Industrial Applications, pp. 69–76. Professional Engineering Publishing Limited, UK (2001)
4. Watson, R.: Raspberry Pi powers this walking robot that can either run autonomously or be controlled by a smartphone (2017). https://maker.pro/raspberry-pi/projects/hexapod-walker-raspberry-piA
5. Guo, Y.: An Abstract of Hexapod gait planning and Obstacle avoidance algorithm, Master of Science Degree in Mechanical Engineering, The University of Toledo (2016)
6. Kern, M., Woo, P.-Y.: Implementation of a hexapod mobile robot with a fuzzy controller. Robotica **23**, 681–688 (2005)
7. Zak, M., Rozman, J., Zboril, F.: Overview of bio-inspired control mechanisms for hexapod robot. Int. J. Comput. Inf. Syst. Industr. Manag. Appl. **8**, 125–134 (2016). ISSN 2150-7988
8. McKenna, M., Wilamowski, B.M.: Implementing a fuzzy system on a field programmable gate array. In: International Joint Conference on Neural Networks, IJCNN 2001, Washington, DC, vol. 1, pp. 189–194 (2001)
9. Soon, C.-Y.: FPGA-based hexapod robot controller, Bachelor Degree of Electronic Engineering (Computer Engineering), Universiti Teknikal Malaysia, Melaka (2016)
10. Ahmed, A., Henrey, M., Bloch, P.: A miniature legged hexapod robot controlled by a FPGA. Int. J. Mech. Eng. Mechatron. **1**(2), 1–6 (2013)
11. Aguilar, L.M., Tores, J.P., Jimenes, C.R., Cabrera, D.R.: Balance of a hexapod in real time using a FPGA. In: 2015 CHILEAN Conference on Electrical, Electronics Engineering, Information and Communication Technologies (CHILECON) (2015)
12. Reichenbach, M., Schmidt, M., Pfundt, B., Fey, D.: A new virtual hardware laboratory for remote FPGA experiments on real hardware. In: Proceedings of the 2011 International Conference on E-learning, E-business, Enterprise Information System E-Government (EEE), pp. 17–23 (2011)
13. Jakimovski, B.: Biologically Inspired Approaches for Locomotion, Anomaly Detection and Reconfiguration for Walking Robots. Springer, Berlin (2011)

Vision Guided Robot KUKA KR 16-2 for a Pick and Place Application

Samir Vojić(✉)

University of Bihač, 77000 Bihać, Bosnia and Herzegovina
samir.vojic@unbi.ba

Abstract. In this paper, we describe a robotic vision system for picking and placing objects using industrial robot KUKA KR 16-2. The proposed system recognizes the object and estimates the position and orientation of the object. It then sends the result to the robot control system. The system was installed in a lab for robotics at the Technical faculty in Bihać.

Keywords: Robotics · Vision system · VisionTech · Kuka · Image processing

1 Introduction

A vision guided robot system is basically a robot fitted with one or more cameras used as sensors to provide a secondary feedback signal to the robot controller to more accurately move to a variable target position [1]. Vision guided robot is rapidly transforming production processes by enabling robots to be highly adaptable and more easily implemented, while dramatically reducing the cost and complexity of fixed tooling previously associated with the design and set up of robotic cells, whether for material handling, automated assembly and more.

Pick and place robots have become commonplace in today's manufacturing environment. Typically relegated to simple, repetitive and monotonous tasks that robots naturally excel at, pick and place robots bring a number of benefits for manufacturers. Pick and place robots are usually mounted on a stable stand, strategically positioned to reach their entire work envelope. Advanced vision systems enable them to grasp and move objects on a conveyor belt, which can be used in a variety of different ways [3].

Pick and place robots are used in many ways, depending on the product being handled and the manufacturer's need for automation. There are four main ways that pick and place robots are used [5]:

Assembly: pick and place robots, during assembly processes, grab an incoming part from a conveyor belt and then place this part onto another work piece, which is then typically carried away by another conveyor belt.

Packaging: similar to assembly processes, a pick and place robot grabs a part on an incoming conveyor belt and, rather than assemble the part, the robot places it in a packaging container at a high speed.

Bin Picking: pick and place robots equipped with advanced vision systems can grab a part out of a bin, sometimes even when parts are randomly mixed together in a bin, and place this part on a conveyor for production.

I. Karabegović (Ed.): NT 2019, LNNS 76, pp. 52–58, 2020.
https://doi.org/10.1007/978-3-030-18072-0_6

Inspection: vision systems can monitor products moving on an incoming conveyor belt and detect defective products, and then a pick and place robot can remove the defective product before it reaches the final phases of production.

While pick and place robots are used in a number of different ways, the four types of applications listed above are some of the most common in today's manufacturing facilities.

We use KUKA. VisionTech to determine and correct the position of a robot relative to the position of a component with the aid of one or more cameras. Up to 3 cameras may be operated simultaneously with the system.

2 Vision System

VisionTech is an add-on technology package and consists of an image processing package and a plug-in for the KUKA smartHMI. The acquisition and processing of images is used to calculate a base correction. The base correction can be used to correct the position of the robot relative to the position of a component. Areas of application are detecting the position of components, deracking and depalletizing. The robot controller communicates with one or more GigE cameras via the image processing system. The connection between the kernel system and the image processing system is established via the Ethernet KRL interface [1].

Vision system components are (Fig. 1):

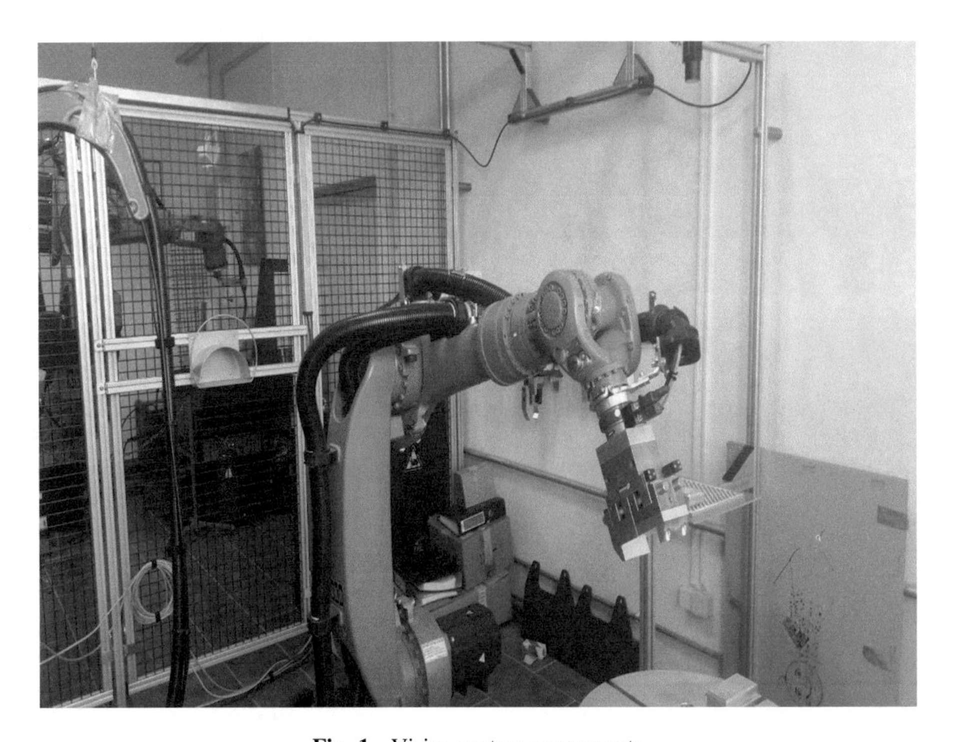

Fig. 1. Vision system components

- Robot KUKA KR 16-2
- KR C4
- KUKA GigE switch
- KUKA MXG20 camera
- KUKA lens
- Calibration plate

2.1 Configuring the Vision System

To establish a connection between the kernel system and the image processing system, the number of the flag that is to trigger execution of the interrupt program must be entered in the configuration file for the Ethernet KRL interface. The interrupt program monitors the result of the image processing. It is triggered once the calculation of the image processing is completed [1].

The camera network must not be connected to another network. During productive operation, the network configuration must not be changed.

VisionTech can be used with both stationary and moving cameras. A stationary camera is fixed in its position, e.g. mounted on a stand or on the ceiling. A moving camera is mounted on the robot flange. In this system we use a stationary camera (Fig. 2).

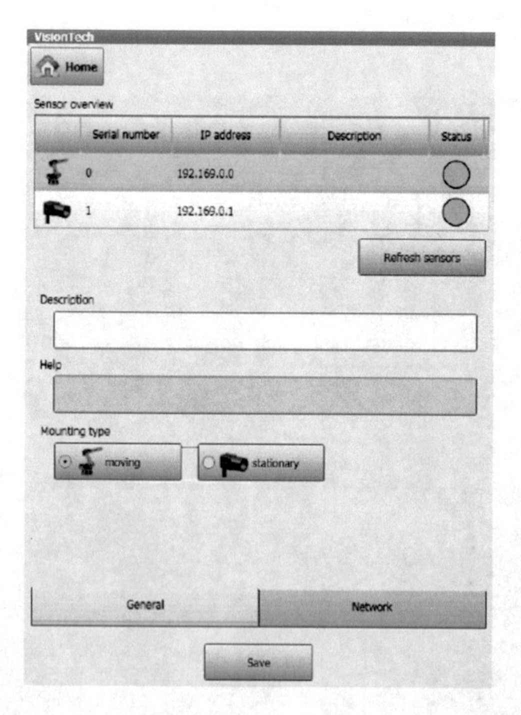

Fig. 2. Sensor overview

For the calibration we use calibration wizard and calibration plate (Fig. 3).

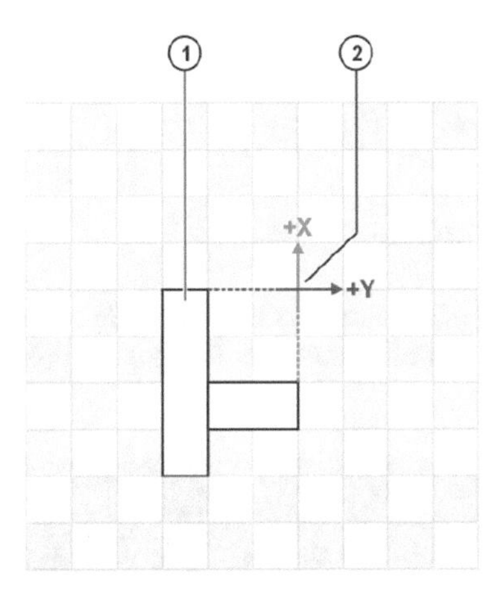

Fig. 3. Calibration plate with fiducial mark (1-fiducial mark, 2- origin)

During verification of the calibration, a measurement is used to detect whether the position of a reference feature has changed relative to a reference position that must be created once. This offset is compared with limit values defined by the user. If the offset is greater than the defined limit values, this may indicate a deterioration of the calibration. A deterioration of the calibration also results in a deterioration of the image processing results [2].

2.2 Configuring a 2D Task

During configuration of a 2D task, a tool block is assigned to the camera. A tool block contains image processing tasks and has the file extension VPP (Fig. 4).

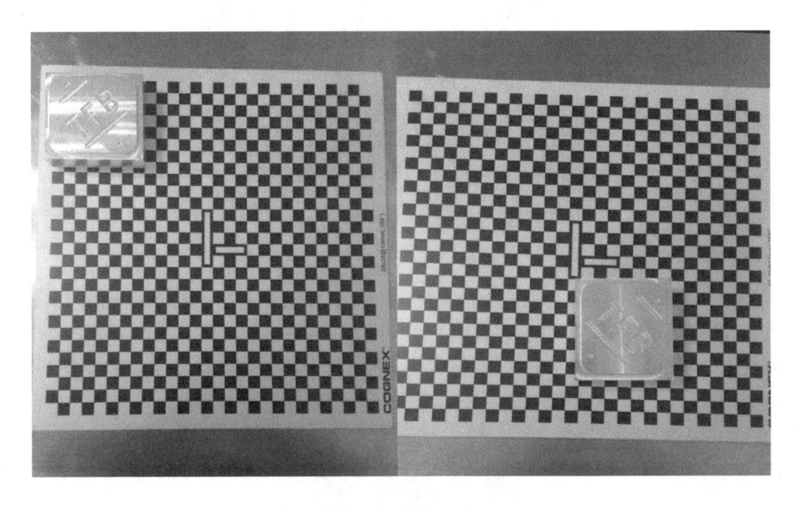

Fig. 4. 2D task

For 2D tasks, only 1 camera is needed; the calibration is carried out at the workpiece level. Once the model has been generated, a result window with images and a table is displayed. The result is the position of the component in the calibration base. Areas detected by the cameras are indicated in green in the images. Areas that have not been detected are marked in red [4].

3 Example Programming

An experimental example of a vision system for guide robot KUKA KR 16-2 in pick and place application is given below. Object for pick and place is shown in the Fig. 5.

Fig. 5. Vision guided robot

The proposed system recognizes the object and estimates its position and orientation.

A programming code used in vision guiding of the robot in the example of picking and placing and object which is randomly placed on the picking surface is shown in Fig. 6.

When using the vision system it is only necessary to program the pick and place of the object once, all other pick and place actions, regardless of the position and orientation of the object, are performed automatically.

```
VT_Clear("VisionTechConfig")
for Part = 1 to PartCounter
if((Results[Part].score <>-1) and
(Results[Part].score >= 0.5))then
      CorrFrame =
VT_GetCorrectionFrame(Results[Part])
      Base_Data[16] = Base_Data[10] : CorrFrame
halt; teach the gripping points
PTP P5 CONT Vel=100 % PDAT1 Tool[5]:vision Base[0]
PTP P6 CONT Vel=100 % PDAT2 Tool[5]:vision Base[0]
LIN P4 Vel=0.1 m/s CPDAT4 Tool[5]:vision Base[16]
LIN P2 Vel=0.05 m/s CPDAT2 Tool[5]:vision Base[16]
LIN P3 Vel=0.1 m/s CPDAT3 Tool[5]:vision Base[16]
endif
endfor
PTP p6 CONT Vel=100 % PDAT3 Tool[5]:vision Base[0]
PTP p5 Vel=100 % PDAT4 Tool[5]:vision Base[0]
PTP HOME  Vel= 100 % DEFAULT
endloop
END
```

Fig. 6. Programing code

4 Conclusion

In industrial pick and place applications, vision guided robots are typically robotic arms with integrated machine vision systems. The machine vision system helps the robot discover the location of an object in order to guide the robot to a desired point for pick and place.

Robots have become a core element of Industry 4.0 and flexibility can be incorporated to them by vision systems and other sensor technologies in order to achieve the requirements and functionalities of the new applications. New tasks are becoming more or more complex and it is necessary to improve the accuracy and to work collaborative with humans, which means making decisions in real-time and triggering actions. For these goals, visual feedback is the key issue, and this is in fact what vision systems provide to robots.

References

1. KST VisionTech 3.0 V3.1, KUKA Roboter GmbH (2014)
2. Vojić, S.: Kalibracija kamere u sistemima robotske vizije. In: 8th International Scientific Conference on Production Engineering "Development and Modernization of Production" RIM 2011, Velika Kladuša, 29.09.-01.10 (2011)

3. Kang, S., Kim, K., Lee, J., Kim, J.: Robotic vision system for random bin picking with dual-arm robots. In: MATEC Web of Conferences, vol. 75, ICMIE (2016)
4. Nisha, D.K., Sekar, Indira: Vision assisted pick and place robotic Arm. Adv. Vis. Comput. Int. J. (AVC) **2**(3), September 2015
5. https://www.robotics.org/blog-article.cfm/Pick-and-Place-Robots-What-Are-They-Used-For-and-How-Do-They-Benefit-Manufacturers/88, November 2018

TLD Design and Development for Vibration Mitigation in Structures

Francesca Colucci[1], Marco Claudio De Simone[2(✉)], and Domenico Guida[2]

[1] Meid4 S.r.l., 84084 Fisciano, Salerno, Italy
[2] Department of Industrial Engineering, University of Salerno, via Giovanni Paolo II, 84084 Fisciano, Salerno, Italy
mdesimone@unisa.it

Abstract. Steel structures are widely used all over the world. Steel interprets the most current synthesis between engineering and architecture, creating constructions that translate into investments that are advantageous over time. Thanks to the strength of its expressiveness and its known characteristics of elasticity and malleability, the architectural work and the structural one become the interpreter of the other, enhancing the project and its peculiarities. The variability of constructive solutions is significantly increased by the ease with which steel is combined with other materials. Steel is able to intelligently exploit the performance of other construction materials such as glass, where natural lighting allows fascinating transparencies. The design of steel structures, however, involves considerable skills in the field of structure dynamics and vibration mitigation. For this reason, it was decided to develop an experimental apparatus for vibration tests for structures in order to analyse the dynamic behaviour for several factors, including symmetrical and asymmetrical configurations. The experimental apparatus is also designed to test active and passive control systems for vibration control and mitigation. To demonstrate the flexibility of the implemented apparatus, this article reports the study and design of a TLD, Tuned Liquid Damper, for passive vibration mitigation for a two-dimensional structure.

Keywords: Test rig · Vibration mitigation · Tuned liquid dampers · Passive control · Structural design

1 Introduction

Vibrations are a desired phenomenon in many cases as for musical instruments in which vibrations are essential for the generation of sound. As for machines and structures, these vibrations are often unwanted. In the case of vehicles, vibrations can disperse energy and create unwanted noises as in the case of disc brake squeal phenomena in which there is an amplification of the vibrations due to the presence of friction. In the case of structures, however, it can be very dangerous for the safety of the structures and those who use them. In fact, it is famous in literature, the case of the earthquake that involved Mexico City in 1985 of magnitude 8.1 that caused the collapse of numerous structures. After the event, the damage analysis revealed that

I. Karabegović (Ed.): NT 2019, LNNS 76, pp. 59–72, 2020.
https://doi.org/10.1007/978-3-030-18072-0_7

buildings that had reported serious structural damage to collapse all had a height between six and fiftieth floors [1–4]. As for the buildings up to five floors and those with more than sixteen floors resisted the earthquake. From this experience we can deduce the key factor for the mitigation of vibrations in structures: the natural period of a structure [5]. Vibrations are unwanted for many domains, and different methods have been developed for mitigating them. Vibrations propagate via mechanical waves and specific mechanical linkages conduct vibrations more efficiently than others [6]. Passive vibration isolation makes use of materials and mechanical linkages that absorb and damp these undesirable mechanical waves. Every object has a fundamental natural frequency [7]. When vibration is applied, energy is transferred most efficiently at the natural frequency, somewhat efficiently below the natural frequency, and with decreasing efficiency above the natural frequency. This can be seen in the transmissibility curve, which is a plot of transmissibility vs. frequency [8–11].

Transmissibility is the ratio of vibration of the isolated surface respect to the source. Vibrations are never completely eliminated, but they can be abundantly reduced. The transmissibility curve shows the typical performance of a passive, negative-stiffness isolation system. Below the natural frequency, transmissibility hovers near 1. A value of 1 means that vibration is going through the system without being amplified or reduced. At the resonant frequency, energy is transmitted efficiently, and the incoming vibration is amplified [12–15]. Damping in the system limits the level of amplification. Above the resonant frequency, little energy can be transmitted, and the curve rolls off to a low value [16–19]. A passive isolator can be seen as a mechanical low-pass filter for vibrations. Among the various alternatives used to reduce the vibrations on the structures a successful technique are the TLD (Tuned Liquid Dampers) [20–23]. One of the first applications of TLDs to ground structures was proposed by Modi and Welt [24]. Due to their low maintenance and operating costs, the ease of installation and design, and relatively good performance, TLDs have been a more favoured damping system for vibration mitigation of structures. A TLD is water confined in a tank that uses the sloshing energy of the water to reduce the dynamic response of the system when the system is subjected to excitation [25, 26]. The applied external energy is dissipated by TLDs through the intrinsic friction of the liquid, wave breakage, and liquid boundary layer friction. Additionally, when subjected to dynamic forces, the tuned liquid inside the water tank gives an anti-phase inertial force to the structure, resulting in a reduced vibration [27, 28]. In the previous decades many modifications have been applied to conventional TLDs in order to make them more efficient in reducing structural vibrations. Zhao and Fujino [29] employed metal screens and they showed that them attenuated the nonlinearity of liquid motion and augmented the damping of water sloshing [30–33]. Cassolato et al. [34] proposed usage of inclined slat screens for adjusting the damping ratio of TLDs. They observed that an enhance in the screen angle decreased the damping ratio of the TLD [35, 36]. It was also found that over a range of excitation amplitudes the new TLD system could maintain a constant damping ratio [37, 38]. Zahrai et al. [39] proposed a TLD system with rotatable baffles and analysed its efficiency experimentally for vibration control of a 5-story scaled-down steel structure [40]. At the end of free vibration tests, it was noticed that respect the case where no baffle was installed inside the TLD, the presence of baffles reduced displacement and acceleration responses by 2.5% and 3.9%, respectively [41–43]. In another study, Ruiz et al. [44] proposed a new type of TLD with a floating roof for controlling structural vibration. The presence of the floating roof impeded the wave breaking in the sloshing water and led to a linear response even under a high excitation amplitude. They proposed a numerical model for the invented TLD and validated it by scaled experimental tests. Samantha and Banerji [45] suggested a modified configuration for TLDs by adding a flexible rotational spring to the bottom of a TLD tank. They remarked that for the optimal rotational stiffness of the spring the modified TLD was more effective and robust for structural control under imposed movement, base excitations, when compared with the standard type of TLDs [46]. In an attempt to increase the effectiveness of conventional TLDs, Fujino and Sun [47] studied the use of multiple tuned liquid dampers [MTLDs]. The MTLDs consisted of a number of TLDs whose sloshing frequency was distributed over a certain range around the natural frequency of a structure. It was shown that, for a small amplitude of excitations, MTLDs were more effective in suppressing structural vibrations when compared with the single TLD. After all, when the sloshing amplitude was large the MTLDs had almost the same effectiveness as a single TLD [48–52]. For many decades, conventional TLDs have been employed as far as possible for mitigation of vibrations in real full-scale structures [53–56]. Next, TLDs were successfully employed for vibration mitigation of many types of structures, including air traffic control towers, jacket offshore platforms, tall buildings, elevated water tanks and bridges [57–68]. In the previous applications, the sloshing frequency of TLDs has been mostly tuned to the first natural frequency of structures, because in regular structures the first mode of vibration has the lofty contribution to the dynamic responses. But, in irregular structures, in addition to the first mode shape, higher mode shapes often play a significant role in the dynamic behaviour [69]. Ergo, tuning the sloshing frequency of a TLD to the first natural frequency of an irregular structure may not effectively abate structural dynamic responses [70, 71]. It should be also noted that under lateral loads, in regular structures, TLDs are often installed at the highest level, where displacement responses are larger than on other floors [72]. However, considering the significant effects of higher mode shapes on the dynamic responses of irregular structures, the highest level may not be the optimal location for installation of TLDs in irregular structures [73–80]. This paper investigates the efficiency of conventional TLDs for vibration mitigation of vertically irregular structures. In this study, a three-story one-bay scaled steel structure was designed to represent a vertically irregular structure. It was then submitted to the free and forced vibration tests with and without the presence of TLDs. In this study, two different arrangements were considered for the installation of conventional TLDs on the test structure. The displacement and acceleration responses of each floor were measured and compared. In the next sections (test rig, experiment and conclusion) after an accurate description of the structure to be analysed, an application of TLD system for the mitigation of the vibrations will be shown on it, to give substance to the concepts so far exposed.

2 Test-Rig

For the design of the vibrating table it was necessary to study first of all the class of structures that one intends to study. To do this, it was necessary to realize a CAD model of the structure to be studied followed by a modal analysis to study the natural frequencies of the system so as to be able to design the vibrating table in safety. For the 3D design, the Solid works modelling software was used, while for the modal analysis the ANSYS finite element modelling software.

In Fig. 1(a) we have reported the rendering of the shaking table while in Fig. 1(b) the real device during assembly is reported.

Fig. 1. 3D CAD design of the test-rig (a) and real shaking table device (b).

2.1 Instrumental Apparatus

The instruments that will be used for the experimental tests are all present in the Mechanics Applied to Machines of the Department of Industrial Engineering of the University of Salerno and are listed below:

- Shaker (Brüel&Kjær Modal Exciter – Type 4824);
- Power Amplifier (Brüel&Kjær – Type 2732);
- Wave generator (Textronix Arbitrary Function Generator AFG320);
- Oscilloscope (Textronix TDS3014B);
- Accelerometers Type 4371 della Brüel&Kjær

The structure, under investigation, represents a two-dimensional model of a three-storey building. The model was assembled by using Bosch aluminium profiles with a characteristic section of 45 $\times$ 45 mm^2, characterized by a good stiffness, compared to the application in question. The Bosh profiles will be used to model the floors of the structure. The various floors will be connected by harmonic steel bars, a material with remarkable elastic properties that in no way compromise the total rigidity of the component, thus guaranteeing a good flexibility of the structure. In the structure, bars of equal length will be used between the base and the first floor and between the second and third floors as shown in Fig. 2a. Between the first and second floor, instead, the bars have a greater length in order to make the structure uneven. For the mobile platform it was decided to use an aluminium frame weighing about 128 kg. On the casing, two Hiwin guides are installed on which two carriages each slide. The purpose of this vibrating table is to also test structures capable of reproducing the behaviour of bridges. For this reason, 2.5 m long guides were used. The vibration test platform was mounted on a support with anti-vibration feet to ensure good passive isolation.

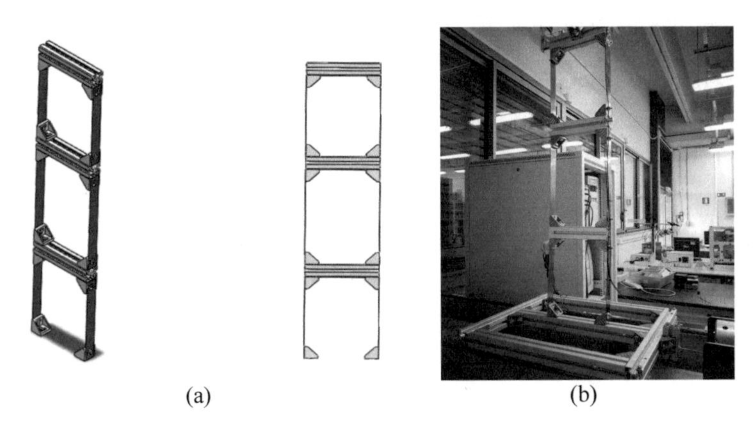

(a) (b)

Fig. 2. 3D CAD design of the structure (a) and real structure during experimental activities (b).

In Fig. 2b, is reported the structure assembled on the shaking table during a resonance test.

3 Mathematical Model

Before testing the effectiveness of the test-rig we created a 3dof lumped-mass model in order to analyse the dynamic behaviour of the system and compare the system's response with the experimental data.

By using Lagrange equation, it is possible to write the equations of motion of the system reported in Fig. 3, by evaluating the kinetic energy and potential energy as reported in the following equations:

$$E_k = \frac{1}{2}\begin{bmatrix} \dot{x}_1(t) & \dot{x}_2(t) & \dot{x}_3(t) \end{bmatrix}\begin{bmatrix} m_1 & 0 & 0 \\ 0 & m_2 & 0 \\ 0 & 0 & m_3 \end{bmatrix}\begin{bmatrix} \dot{x}_1(t) \\ \dot{x}_2(t) \\ \dot{x}_3(t) \end{bmatrix} = \frac{1}{2}\dot{x}^T(t)M_s\dot{\mathbf{x}}(t) \quad (1)$$

$$E_p = \frac{1}{2}\begin{bmatrix} x_1(t) & x_2(t) & x_3(t) \end{bmatrix}\begin{bmatrix} k_1+k_2 & -k_2 & 0 \\ -k_2 & k_2+k_3 & -k_3 \\ 0 & -k_3 & k_3 \end{bmatrix}\begin{bmatrix} x_1(t) \\ x_2(t) \\ x_3(t) \end{bmatrix} = \frac{1}{2}\mathbf{x}^T(t)K_s\mathbf{x}(t) \quad (2)$$

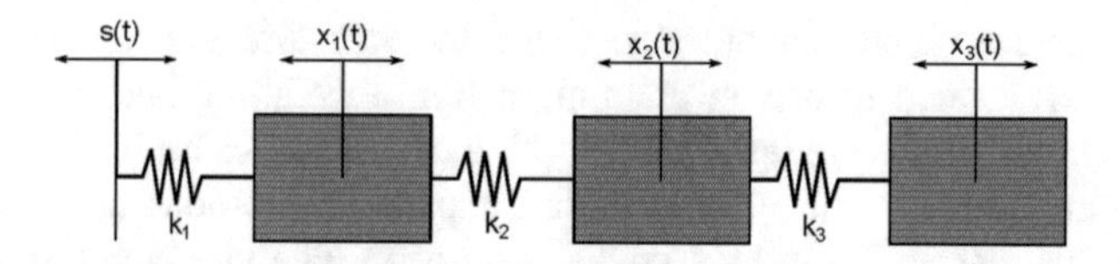

Fig. 3. Lumped-mass model

with m_1, m_2 and m_3 are respectively the mass of the first floor, second floor and third floor while k_1, k_2 and k_3 are the stiffness values for the six harmonic steel pillars. For evaluating the values of such stiffness we used the following relation:

$$k_i = 12\ EI \backslash L^3\ \forall i = 1:3 \tag{3}$$

with the length of the pillars L_1 and L_3 = 0.205 m while L_2 = 0.255 m, I is the moment of inertia of the rectangular section of the pillars equal to $2.917 * 10^{-12}$ m^4 and Young's modulus for the harmonic steel is $207 * 10^9$ m^4.

In Eqs. 4 and 5 are reported the mass matrix and the stiffness matrix evaluated for the system.

$$M_s = \begin{bmatrix} 0.890 & 0 & 0 \\ 0 & 0.951 & 0 \\ 0 & 0 & 1.115 \end{bmatrix} \tag{4}$$

$$K_s = \begin{bmatrix} 2556 & -874 & 0 \\ -874 & 2556 & -1682 \\ 0 & -1682 & 1682 \end{bmatrix} \tag{5}$$

Once the mass and stiffness matrices were calculated, it was possible to calculate the natural frequencies of the analysed system. To validate the frequencies evaluated by using the lumped-mass model calculated, it was decided to perform a modal analysis thanks to the ANSYS finite element analysis software. In Table 1 are reported the natural frequencies evaluated for the lumped-mass model and the modal analysis conducted in Ansys.

Table 1. Natural frequencies of the 3DOF structure

$f_{n,i}$	Lumped-mass model	Modal analysis ANSYS software
1	2.05	1.95
2	6.23	5.97
3	8.23	8.03

In order to evaluate the goodness of the frequencies reported in Table 1, we used one of the most popular method for evaluating the natural frequencies of a structure: the hammer impact test. For the test we used the instrumented hammer to generate our impulse force and used accelerometers for measuring the response. The accelerometers are positioned at every floor of the structure. In this way the test is very time efficient as it requires a single impact and data capture. In Fig. 4 are reported fft signal evaluated for the third floor. As it is possible to notice from the graph, the three frequencies validate the frequencies evaluated on the two models reported previously.

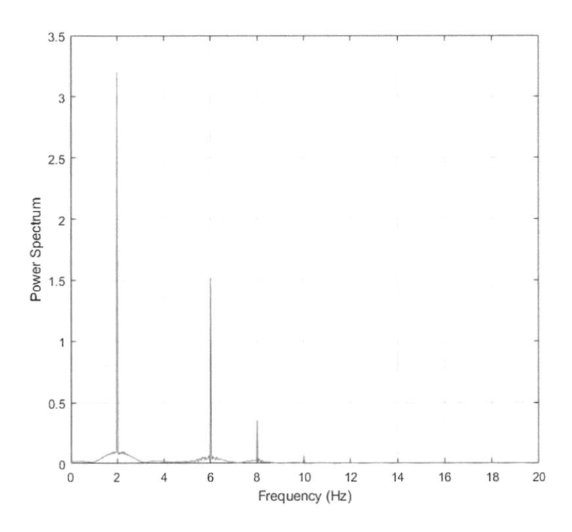

Fig. 4. FFT result for the recorded time history of acceleration on the third floor.

4 Experimental Activity

The goal is to test the goodness of the use of TLDs in the mitigation of vibrations in structures. For this reason, an experimental campaign was conducted on the structure by varying the frequency of the input signal given to the shaker. The experimental apparatus is reported in Fig. 5. The input signal is generated by using the Tektronix

Fig. 5. Experimental activity

FG320 wave generator and sent to the power amplifier. The input signal and the output signals are acquired through the acquisition board Bruel&Kjr-PULSE Labshop v. 4.2 in which it is possible to set the sensitivity of the inputs to the signal analyser, the acquisition time of 32 s, the sampling frequency of 50 Hz and with the offset function it is possible to adjust the measured signal making sure that the average value of noise is zero.

4.1 Design of the Tuned Liquid Damper

In this paper we investigated the interaction between the structure and the tuned liquid damper. The TLD which was tuned to the first natural frequency of the bare. During the experimental investigation, we considered two cases with the damper installed on the first and second floor respectively.

The mass ratio (i.e. mass of liquid respect to the mass of the structure) for the two cases was at 33.3%. In literature, previous studies conducted on TLDs propose a mass ratio range of 1–4%. In our study, such ratio is higher, in order to increase the effectiveness of TLDs.

The sloshing frequencies of TLDs were determined using the equation proposed by Housner and reported in (6).

$$f_{n,liquid} = \frac{1}{2}\sqrt{\frac{3.16g}{L}\tanh\left(\frac{3.16H}{L}\right)} \tag{6}$$

In Fig. 6 is reported the final dimensions of the TLD used for this activity dimensioned taking into account the hypothesis of shallow waters ($H/L \leq 0.15$).

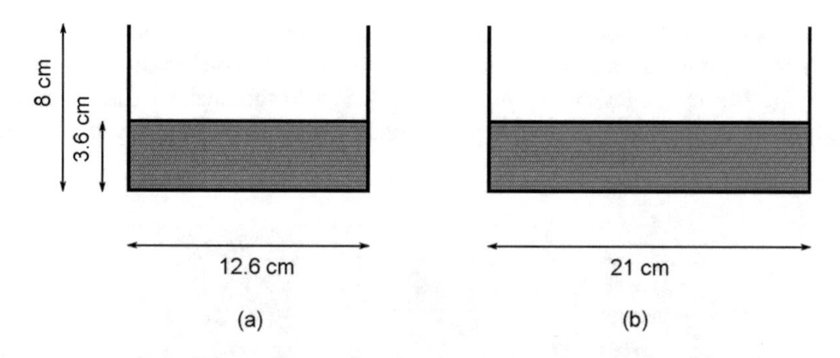

Fig. 6. Dimensions of the constructed TLD (a) side view of TLD (b) front view of TLD

In Fig. 7 are reported the peak values of acceleration recorded by each floor for several values of frequencies of the input signal of the structure without (see Fig. 7a) and with the use of TLD positioned on each floor.

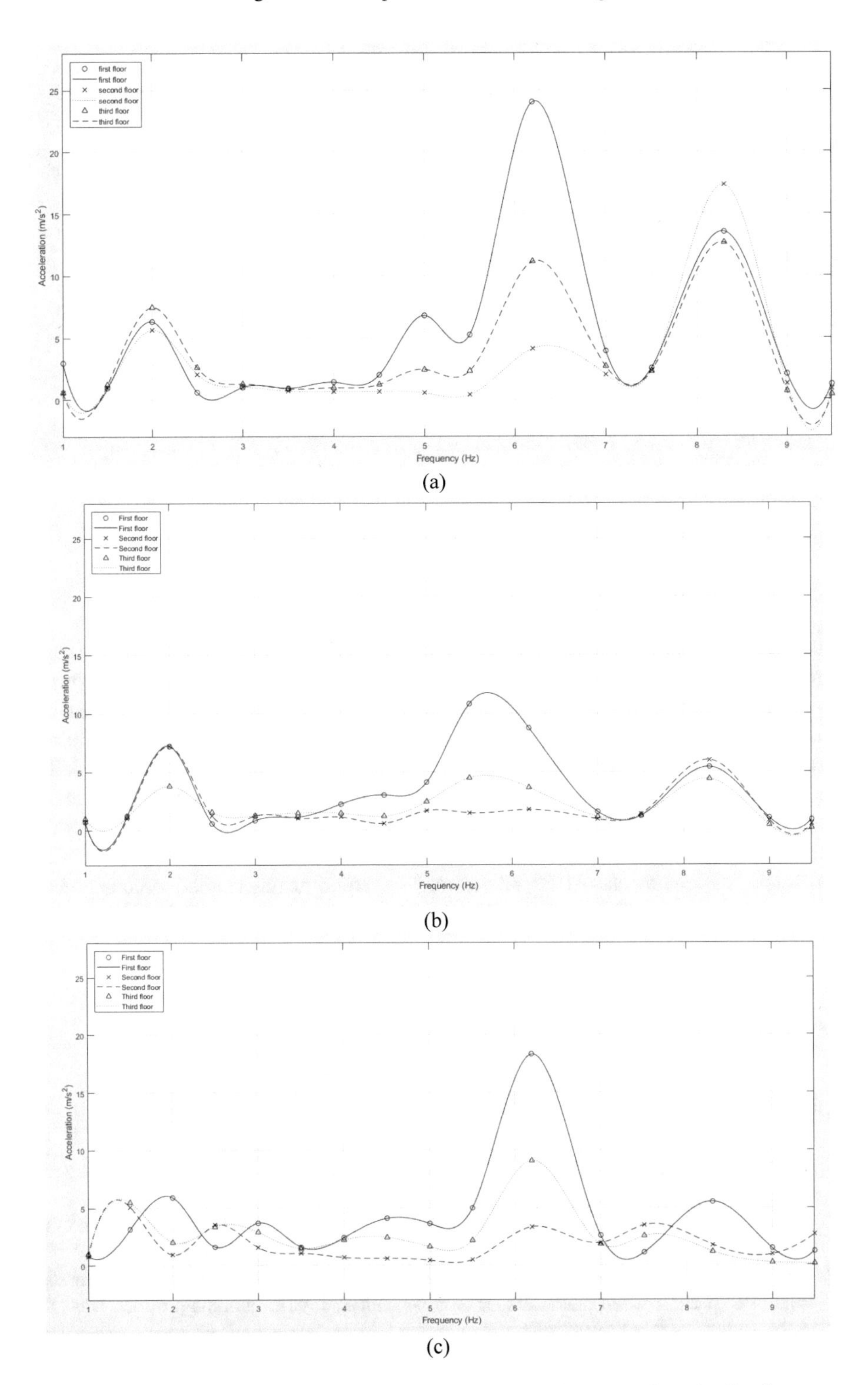

Fig. 7. Peak acceleration responses of the structure obtained by forced vibration tests: (a) structure without absorber (b) TLD mounted on the first floor (c) TLD mounted on the second floor (d) TLD mounted on the third floor.

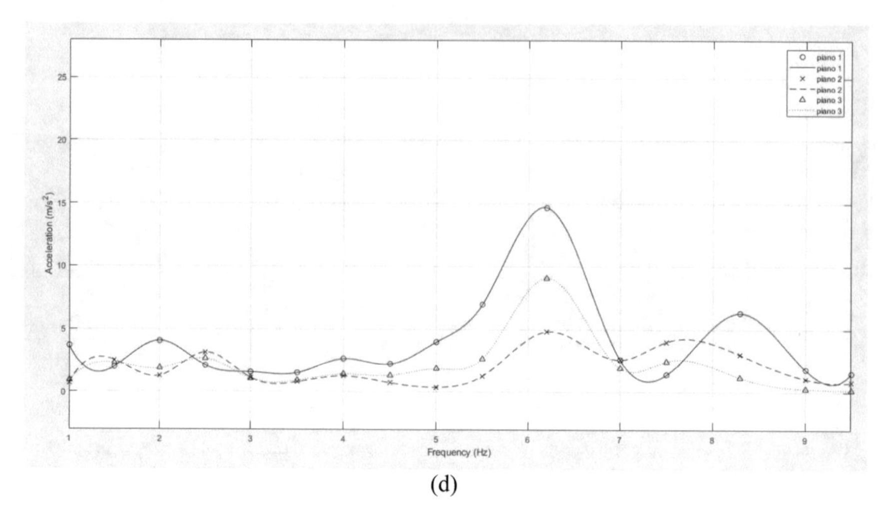

(d)

Fig. 7. (*continued*)

5 Conclusion

This study investigated experimentally the effectiveness of conventional Tuned Liquid Dampers (TLDs) for vibration mitigation of a vertically irregular structure. A three-story steel structure with the total weight of 2.7 kg was specifically designed and constructed in order to represent a vertically irregular structure. The test structure was subjected to free and forced vibration tests and natural frequencies were extracted. The test structure was equipped with TLD that were tuned to its first resonance frequencies. The mass ratio used for the TLD was equal to 33.3%. The response of the structure was studied by placing the TLD on each floor separately and the dynamic responses of the structure-TLD was measured for all floors. The experimental results demonstrate the excellent ability of a TLD to mitigate vibrations in resonance conditions if properly dimensioned.

References

Journal Papers:

1. Ozaki, M., Adachi, Y., Iwahori, Y., Ishii, N.: Application of fuzzy theory to writer recognition of Chinese characters. IOSR J. Eng. **2**(2), 112–116 (2012)
2. Debnath, N., Deb, S.K., Dutta, A.: Multi-modal vibration control of truss bridges with tuned mass dampers under general loading. J. Vib. Control **22**(20), 4121–4140 (2016)
3. Lu, Z., Wang, D., Zhou, Y.: Experimental parametric study on wind-induced vibration control of particle tuned mass damper on a benchmark high-risebuilding. Struct. Des. Tall Spec. Build. 26(8), e1359 (2017)

4. Guida, D., Nilvetti, F., Pappalardo, C.M.: Optimal control design by adjoint-based optimization for active mass damper with dry friction. In: Programme and Proceedings of (COMPDYN 2013) 4th International Conference on Computational Methods in Structural Dynamics and Earthquake Engineering, Kos Island, Greece, 12–14 June, pp. 1–19 (2013)
5. Xu, X., Lai, F., Li, G., Zhu, X., Zhu, L.: A novel vibration suppression device for floating offshore wind generator. J. Energy Resour. Technol. **141**(6) (2019). Transactions of the ASME, art. no. 061201, https://doi.org/10.1115/1.4042404
6. Lu, X., Zhang, Q., Weng, D., Zhou, Z., Wang, S., Mahin, S.A., Qian, F.: Improving performance of a super tall building using a new eddy-current tuned mass damper. Struct. Control Health Monit. **24**(3), e1882 (2017)
7. De Simone, M.C., Rivera, Z.B., Guida, D.: Obstacle avoidance system for unmanned ground vehicles by using ultrasonic sensors. Machines **6**(2) (2018). Art. no. 18, https://doi.org/10.3390/machines6020018
8. Pappalardo, C.M., Guida, D.: A time-domain system identification numerical procedure for obtaining linear dynamical models of multibody mechanical systems. Arch. Appl. Mech. **88**(8), 1325–1347 (2018)
9. Love, J.S., Lee, C.S.: Nonlinear series-type tuned mass damper-tuned sloshing damper for improved structural control. J. Vib. Acoust. **141**(2) (2019). Transactions of the ASME, art. no. 0210061. https://doi.org/10.1115/1.4041513
10. Marano, G.C., Greco, R.: Robust optimum design of tuned mass dampers for high-rise buildings under moderate earthquakes. Struct. Des. Tall Spec. Build. **18**(8), 823–838 (2009)
11. De Simone, M.C., Guida, D.: Modal coupling in presence of dry friction. Machines **6**(1) (2018). Art. no. 8. https://doi.org/10.3390/machines6010008
12. Greco, R., Marano, G.C., Fiore, A.: Performance–cost optimization of tuned mass damper under low-moderate seismic actions. Struct. Des. Tall Spec. Build. **25**(18), 1103–1122 (2016)
13. Furtmüller, T., Di Matteo, A., Adam, C., Pirrotta, A.: Base-isolated structure equipped with tuned liquid column damper: an experimental study. Mech. Syst. Signal Process. **116**, 816–831 (2019). https://doi.org/10.1016/j.ymssp.2018.06.048
14. Zhang, Z., Basu, B., Nielsen, S.R.K.: Real-time hybrid aeroelastic simulation of wind turbines with various types of full-scale tuned liquid dampers. Wind Energy **22**(2), 239–256 (2019). https://doi.org/10.1002/we.2281
15. Altunişik, A.C., Yetisşken, A., Kahya, V.: Experimental study on control performance of tuned liquid column dampers considering different excitation directions. Mech. Syst. Sig. Process. **102**, 59–71 (2018)
16. De Simone, M.C., Russo, S., Rivera, Z.B., Guida, D.: Multibody model of a UAV in presence of wind fields. In: 2018 Proceedings - 2017 International Conference on Control, Artificial Intelligence, Robotics and Optimization, ICCAIRO 2017, pp. 83–88, January 2018. https://doi.org/10.1109/iccairo.2017.26
17. Hemmati, A., Oterkus, E., Khorasanchi, M.: Vibration suppression of offshore wind turbine foundations using tuned liquid column dampers and tuned mass dampers. Ocean Eng. **172**, 286–295 (2019). https://doi.org/10.1016/j.oceaneng.2018.11.055
18. Pappalardo, C.M., Guida, D.: On the computational methods for the dynamic analysis of rigid multibody mechanical systems. Machines **6**(2), 20 (2018)
19. Rozas, L., Boroschek, R.L., Tamburrino, A., Rojas, M.: A bidirectional tuned liquid column damper for reducing the seismic response of buildings. Struct. Control Health Monit. **23**(4), 621–640 (2016)
20. Pappalardo, C.M., Guida, D.: System identification algorithm for computing the modal parameters of linear mechanical systems. Machines **6**(2), 12 (2018)

21. Pabarja, A., Vafaei, M.C., Alih, S., MdYatim, M.Y., Osman, S.A.: Experimental study on the efficiency of tuned liquid dampers for vibration mitigation of a vertically irregular structure. Mech. Syst. Signal Process. **114**, 84–105 (2019). https://doi.org/10.1016/j.ymssp.2018.05.008
22. Pappalardo, C.M., Guida, D.: System identification and experimental modal analysis of a frame structure. Eng. Lett. **26**(1), 56–68 (2018)
23. Ashasi-Sorkhabi, A., Malekghasemi, H., Ghaemmaghami, A., Mercan, O.: Experimental investigations of tuned liquid damper-structure interactions in resonance considering multiple parameters. J. Sound Vib. **388**, 141–153 (2017)
24. Modi, V.J., Welt, F.: Vibration control using nutation dampers. In: Proceedings of the International Conference on Flow Induced Vibration, London, pp. 369–376 (1987)
25. Chaiviriyawong, P., Panedpojaman, P., Limkatanyu, S., Pinkeaw, T.: Simulation of control characteristics of liquid column vibration absorber using a quasi-elliptic flow path estimation method. Eng. Struct. **177**, 785–794 (2018). https://doi.org/10.1016/j.engstruct.2018.09.088
26. De Simone, M.C., Guida, D.: Identification and control of a unmanned ground vehicle by using arduino. UPB Sci. Bull. Ser. D Mech. Eng. **80**(1), 141–154 (2018)
27. Pappalardo, C.M., Guida, D.: Dynamic analysis of planar rigid multibody systems modelled using natural absolute coordinates. Appl. Comput. Mech. **12**, 73–110 (2018)
28. Kareem, A., Kijewski, T., Tamura, Y.: Mitigation of motions of tall buildings with specific examples of recent applications. Wind Struct. **2**(3), 201–251 (1999)
29. Nakamura, S.I., Fujino, Y.: Lateral vibration on a pedestrian cable-stayed bridge. Struct. Eng. Int. **12**(4), 295–300 (2002)
30. De Simone, M.C., Guida, D.: Control design for an under-actuated UAV model. FME Trans. **46**(4), 443–452 (2018). https://doi.org/10.5937/fmet1804443D
31. Zhao, Z., Fujino, Y.: Numerical simulation and experimental study of deeper-water TLD in the presence of screens. J. Struct. Eng. **39**, 699–711 (1993)
32. Pappalardo, C.M., Guida, D.: Use of the adjoint method in the optimal control problem for the mechanical vibrations of nonlinear systems. Machines **6**(2), 19 (2018)
33. De Simone, M.C., Rivera, Z.B., Guida, D.: Finite element analysis on squeal-noise in railway applications. FME Trans. **46**(1), 93–100 (2018). https://doi.org/10.5937/fmet1801093D
34. Cassolato, M.R., Love, J.S., Tait, M.J.: Modelling of a tuned liquid damper with inclined damping screens. Struct. Control Health Monit. **18**(6), 674–681 (2011)
35. Pappalardo, C.M., Guida, D.: Forward and inverse dynamics of a unicycle-like mobile robot. Machines **7**(1), 5 (2019)
36. La, V.D., Adam, C.: General on-off damping controller for semi-active tuned liquid column damper. JVC/J. Vib. Control **24**(23), 5487–5501 (2018). https://doi.org/10.1177/1077546316648080
37. Shad, H., Adnan, A., Behbahani, H.P., Vafaei, M.: Efficiency of TLDs with bottom-mounted baffles in suppression of structural response when subjected to harmonic excitations. Struct. Eng. Mech. **60**(1), 131–148 (2016)
38. De Simone, M.C., Guida, D.: On the development of a low-cost device for retrofitting tracked vehicles for autonomous navigation. In: AIMETA 2017 - Proceedings of the 23rd Conference of the Italian Association of Theoretical and Applied Mechanics, 4, pp. 71–82 (2017)
39. Zahrai, S.M., Abbasi, S., Samali, B., Vrcelj, Z.: Experimental investigation of utilizing TLD with baffles in a scaled down 5-story benchmark building. J. Fluids Struct. **28**, 194–210 (2012)
40. Bhattacharyya, S., Ghosh, A.D., Basu, B.: Design of an active compliant liquid column damper by LQR and wavelet linear quadratic regulator control strategies. Struct. Control Health Monit. **25**(12) (2018). Art. no. e2265, https://doi.org/10.1002/stc.2265

41. Pappalardo, C.M., Guida, D.: On the use of two-dimensional euler parameters for the dynamic simulation of planar rigid multibody systems. Arch. Appl. Mech. **87**(10), 1647–1665 (2017)
42. Altay, O., Klinkel, S.: A semi-active tuned liquid column damper for lateral vibration control of high-rise structures: theory and experimental verification. Struct. Control Health Monit. **25**(12) (2018). Art. no. e2270, https://doi.org/10.1002/stc.2270
43. Pappalardo, C.M., Guida, D.: Adjoint-based optimization procedure for active vibration control of nonlinear mechanical systems. ASME J. Dyn. Syst. Meas. Control **139**(8), 1–11 (2017), 081010
44. Ruiz, R.O., Lopez-Garcia, D., Taflanidis, A.A.: Modeling and experimental validation of a new type of tuned liquid damper. Acta Mech. **227**(11), 3275–3294 (2016)
45. Samanta, A., Banerji, P.: Structural vibration control using modified tuned liquid dampers, IES. J. Part A: Civil Struct. Eng. **3**(1), 14–27 (2010)
46. Akbarpoor, S., Dehghan, S.M., Hadianfard, M.A.: Seismic performance evaluation of steel frame structures equipped with tuned liquid dampers. Asian J. Civil Eng. **19**(8), 1037–1053 (2018). https://doi.org/10.1007/s42107-018-0082-8
47. Fujino, Y., Sun, L.M.: Vibration control by multiple tuned liquid dampers (MTLDs). J. Struct. Eng. **119**(12), 3482–3502 (1993)
48. Pappalardo, C.M., Guida, D.: On the Lagrange multipliers of the intrinsic constraint equations of rigid multibody mechanical systems. Arch. Appl. Mech. **88**(3), 419–451 (2018)
49. Concilio, A., De Simone, M.C., Rivera, Z.B., Guida, D.: A new semi-active suspension system for racing vehicles. FME Trans. **45**(4), 578–584 (2017). https://doi.org/10.5937/fmet1704578C
50. Younes, M.F.: Effect of different design parameters on damping capacity of liquid column vibration absorber. J. Eng. Appl. Sci. **65**(6), 447–467 (2018)
51. Tsao, W.H., Hwang, W.-S.: Tuned liquid dampers with porous media. Ocean Eng. **167**, 55–64 (2018). https://doi.org/10.1016/j.oceaneng.2018.08.034
52. Park, W., Park, K.S., Koh, H.M., Ha, D.H.: Wind-induced response control and serviceability improvement of an air traffic control tower. Eng. Struct. **28**(7), 1060–1070 (2006)
53. Pappalardo, C.M., Guida, D.: Control of nonlinear vibrations using the adjoint method. Meccanica **52**(11–12), 2503–2526 (2017)
54. Ruggiero, A., Affatato, S., Merola, M., De Simone, M.C.: FEM analysis of metal on UHMWPE total hip prosthesis during normal walking cycle. In: AIMETA 2017 - Proceedings of the 23rd Conference of the Italian Association of Theoretical and Applied Mechanics, 2, pp. 1885–1892 (2017)
55. Pappalardo, C.M.: A natural absolute coordinate formulation for the kinematic and dynamic analysis of rigid multibody systems. Nonlinear Dyn. **81**(4), 1841–1869 (2015)
56. Love, J.S., Tait, M.J.: Multiple tuned liquid dampers for efficient and robust structural control. J. Struct. Eng. **141**(12), 04015045 (2015)
57. Tamura, Y., Kohsaka, R., Nakamura, O., Miyashita, K.I., Modi, V.J.: Wind-induced responses of an airport tower—efficiency of tuned liquid damper. J. WindEng. Ind. Aerodyn. **65**(1), 121–131 (1996)
58. Chu, C.-R., Wu, Y.-R., Wu, T.-R., Wang, C.-Y.: Slosh-induced hydrodynamic force in a water tank with multiple baffles. Ocean Eng. **167**, 282–292 (2018). https://doi.org/10.1016/j.oceaneng.2018.08.049
59. Quatrano, A., De Simone, M.C., Rivera, Z.B., Guida, D.: Development and implementation of a control system for a retrofitted CNC machine by using Arduino. FME Trans. **45**(4), 565–571 (2017). https://doi.org/10.5937/fmet1704565Q
60. Jin, Q., Li, X., Sun, N., Zhou, J., Guan, J.: Experimental and numerical study on tuned liquid dampers for controlling earthquake response of jacket offshore platform. Mar. struct. **20**(4), 238–254 (2007)

61. Ruggiero, A., De Simone, M.C., Russo, D., Guida, D.: Sound pressure measurement of orchestral instruments in the concert hall of a public school. Int. J. Circ. Syst. Sig. Process. **10**, 75–812 (2016)
62. Housner, G.W.: The dynamic behavior of water tanks. Bull. Seismol. Soc. Am. **53**(2), 381–387 (1963)
63. Guida, D., Pappalardo, C.M.: Control design of an active suspension system for a quarter-car model with hysteresis. J. Vibr. Eng. Technol. **3**(3), 277–299 (2015)
64. Sun, L., Kikuchi, T., Goto, Y., Hayashi, M.: Tuned Liquid Damper (TLD) using heavy mud. WIT Trans. Built Environ. **38**, 87–96 (1998)
65. De Simone, M.C., Guida, D.: Dry friction influence on structure dynamics. In: COMPDYN 2015 - 5th ECCOMAS Thematic Conference on Computational Methods in Structural Dynamics and Earthquake Engineering, pp. 4483–4491 (2015)
66. Behbahani, H.P., Bin Adnan, A., Vafaei, M., Pheng, O.P., Shad, H.: Effects of TLCD with maneuverable flaps on vibration control of a SDOF structure. Meccanica **52**(6), 1247–1256 (2017)
67. Alkmim, M.H., Fabro, A.T., de Morais, M.V.G.: Optimization of a tuned liquid column damper subject to an arbitrary stochastic wind. J. Braz. Soc. Mech. Sci. Eng. **40**(11) (2018). Art. no. 551, https://doi.org/10.1007/s40430-018-1471-3
68. Behbahani, H.P., Bin Adnan, A., Vafaei, M., Shad, H., Pheng, O.P.: Vibration mitigation of structures through TLCD with embedded baffles. Exp. Tech. **41**(2), 139–151 (2017)
69. Pappalardo, C.M., Guida, D.: A Comparative Study of the Principal Methods for the Analytical Formulation and the Numerical Solution of the Equations of Motion of Rigid Multibody Systems". Arch. Appl. Mech. **88**(12), 2153–2177 (2018)
70. Iannone, V., De Simone, M.C.: Modelling of a DC Gear Motor for Feed-Forward Control Law Design for Unmanned Ground Vehicles (2019) Actuators, Submitted
71. Fujino, Y., Pacheco, B.M., Chaiseri, P., Sun, L.M.: Parametric studies on tuned liquid damper (TLD) using circular containers by free-oscillation experiments. Doboku Gakkai Ronbunshu **1988**(398), 177–187 (1988)
72. Tamura, Y., Kousaka, R., Modi, V.J.: Practical application of nutation damper for suppressing wind-induced vibrations of airport towers. J. Wind Eng. Ind. Aerodyn. **43**(1–3), 1919–1930 (1992)
73. Roy, A., Ghosh, A.D., Chatterjee, S.: Influence of tuning of passive TLD on the seismic vibration control of elevated water tanks under various tank-full conditions. Struct. Control Health Monit. **24**(6), e1924 (2017)
74. Butterworth, J., Lee, J.H., Davidson, B.: Experimental determination of modal damping from full scale testing. In: 13th World Conference on Earthquake Engineering, Vancouver, B.C., Canada, 1–6 August 2004, Paper no. 310
75. Tait, M.J., Isyumov, N., El Damatty, A.A.: Performance of tuned liquid dampers. J. Eng. Mech. **134**(5), 417–427 (2008)
76. Graham, E.W., Rodriguez, A.M.: The characteristics of fuel motion which affect airplane dynamics. J. Appl. Mech. **19**(3), 381–388 (1952)
77. Rivera, Z.B., De Simone, M.C., Guida, D.: Modelling of Mobile Robots in ROS-based Environments (2019) Robotics, Submitted
78. Sun, L.M., Fujino, Y., Chaiseri, P., Pacheco, B.M.: The properties of tuned liquid dampers using a TMD analogy. Earthquake Eng. Struct. Dyn. **24**(7), 967–976 (1995)
79. Sun, L.M.: Semi-analytical modelling of tuned liquid damper (TLD) with emphasis on damping of liquid sloshing (1991). Doctoral thesis, the University of Tokyo, Tokyo, Japan, Section 5.1.3, pp. 61–62
80. Rivera, Z.B., De Simone, M.C., Guida, D.: Waipoint Navigation for Wheeled Mobile Robots in Ros-based Environments (2019). Machines, Submitted

Vibration Analysis of Rotating Machinery as Excitation of Concrete Structure Vibrations

Emir Nezirić(✉), Edin Džiho, and Edin Šunje

Faculty of Mechanical Engineering, "DžemalBijedić" University of Mostar, Sjeverni logor bb., 88000 Mostar, Bosnia and Herzegovina
emir.neziric@unmo.ba

Abstract. Concrete structures are exposed to ground-borne vibrations caused by transient or continuous excitation. Increased vibration levels and long-time exposure to vibrations could cause different levels of structural damages. As the continuous source of vibration, excitation is common to have rotational machinery. Rotational machinery faults usually increase the machinery vibrations, which are one of the excitation sources. In this paper, it would be presented an application of rotational machinery vibration analysis in determining the cause of increased concrete structure vibrations.

Keywords: Vibration analysis · Excitation · Concrete structure vibrations · Rotational machinery vibrations

1 Introduction

Vibration sources such as machinery vibrations, traffic or blasting are inducing vibrations of nearby structures. Vibrations could be conducted through soil or through direct contact between the source of vibration and the reciever. Increased level of structure vibrations could damage structures, especially if structures are exposed to excitation for longer periods of time. Measuring and analyzing vibrations of the structures is an important part of structure reliability tracking [1, 2].

Determining the natural frequency of the structure is the first step in the analysis of the structure vibrations. The most usual method for determining natural frequencies of the structure is finite element modelling [3–8]. This method has developed procedures for natural frequency calculations and is very well documented. Machinery faults cause vibrations in machines and machine vibration excites the supporting structure. If excitation force frequency and natural frequency of structure have the same or approximate values, resonance or beating could be a possible problem in the structure.

In most of the available literature, vibration excitation caused by rotational machinery vibrations is mostly considered as machine unbalance. In practical applications, it could be handy that vibration analyst could decide which machinery fault is producing an excitation force since all faults do not have the same frequency range and layout.

This paper shows the procedure of the vibration analysis of the reinforced concrete structure excited by rotational machinery with unknown fault. It also shows a short review of the current literature, basic assumptions on natural frequencies and resonance and beating phenomena, and an analysis of vibration of the real structure excited by rotational machinery.

I. Karabegović (Ed.): NT 2019, LNNS 76, pp. 73–81, 2020.
https://doi.org/10.1007/978-3-030-18072-0_8

2 Vibration Analysis of Machinery Excited Structures

It is important to find out how rotational machinery would affect its supporting structure. Displacements of some of the parts of the structure could reveal the structure members weakness. As a possible step in the solution on the machinery caused vibrations in structures could be discovering the fault which is causing the rotational machinery vibrations at first place.

Gaznavi et al. [9] have presented a literature review on a dynamic analysis of machine structure with short descriptions of methods of the research and most important conclusions. Documented researches are developing different approaches in the machine vibration impact on pedestal vibration response. Elvin et al. in [3] have developed 1DOF frame structure model where the unbalanced machine is used as excitation force. The focus of the research was how the startup and shut down of the machine impacts on the resonance of the structure with different damping values. Silva et al. in [4] have presented an analysis of the steel oil platform based on real construction. After determining the natural frequency of the structure, levels of the maximum displacements generated by rotating machinery vibrations are analyzed. Authors have concluded that levels of vibration (displacement, velocity) are very high and that design of structure should be reevaluated. Displacement levels of steel frame structures are also investigated by Ferro in [5]. Levels of frame structure displacements caused by machinery vibrations were the key factor in determining the type of connections between frame members. As the main conclusion mentioned by authors is that a fully rigid frame is the best solution for rotational machinery carrying structure. A similar methodology was used by Patel et al. in [6], where authors have developed two different types of concrete foundations for blower machines. Analysis has shown that hollow block foundation with piles is a better choice as a foundation than a table mounted block foundation according to displacement levels. Leso et al. in [7] have presented an analysis of the foundation effect on the vibration response of the fan. Two different mounting situations are presented and analyzed, where are presented frequency responses of both cases. In [8], Tamariz et al. have presented an analysis of the damage caused by long exposure of the concrete pedestal to rotational machinery vibrations. As the main conclusion of the research is that this rotational vibration excitation could create fatigue damages for 50 years.

All mentioned papers have a similar idea on rotational machinery vibration impact on structures: they should be considered and monitored during the exploitation of structure.

3 Natural Frequency and Resonance

In this section would be presented a short introduction to resonance and beating phenomena. General equation of motion could be written as

$$m\ddot{x} + c\dot{x} + kx = F(t) \tag{1}$$

where m is mass, c is damping, k is stiffness, F(t) is excitation force as function of time, and x as displacement [10]. Equation (1) could be rearranged as follows

$$\ddot{x} + \frac{c}{m}\dot{x} + \frac{k}{m}x = \frac{F(t)}{m} \tag{2}$$

In Eq. (2), k/m represents squared circular frequency of the motion. Harmonic excitation force as function of time most common representation is

$$F(t) = A\sin(\Omega t + \beta) \tag{3}$$

where A is force amplitude, Ω i circular frequency of the excitation force and β is phase angle. Circular frequencies ω and Ω could be used to calculate frequencies of the oscilations.

$$\begin{aligned} f_n &= \frac{\omega}{2\pi} = \frac{1}{2\pi}\sqrt{\frac{k}{m}} \\ f_e &= \frac{\Omega}{2\pi} \end{aligned} \tag{4}$$

where f_n is natural frequency of the system and f_e is excitation force frequency. Resonance occurs when these two frequencies are equal. Beating phenomenon occurs when these two frequencies do not have exactly the same value, but the difference is very small. Characteristic change of amplitude caused by these phenomena is shown in Fig. 1.

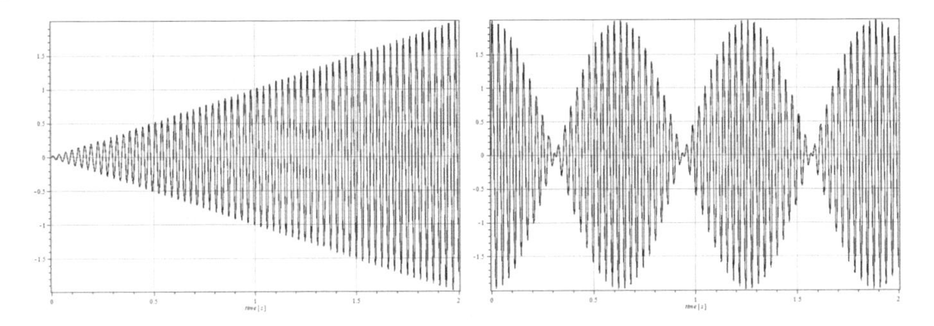

Fig. 1. Resonance (left) and beating (right) amplitude change

Since the real structures have multiple degrees of freedom, they would also have multiple eigenmodes with multiple frequencies. If one of those frequencies are matching with excitation force frequency, it could initiate the presence of resonance or beating phenomenon.

4 Vibration Analysis of Pumping Plant as Excitation Force of Concrete Structure

The analyzed system in this investigation is used for city water supply and it consists of 9 pumps located in the reinforced concrete structure. The piping system of this pumping plant is located in the basement part of the structure. Problems with the increased vibrations of the structure are reported after the reconstruction of the concrete pedestal for three pumps with the horizontal axis (Fig. 2).

Fig. 2. Investigated pumping plant

Characteristics of the pumps are given in Table 1.

Table 1. Characteristics of the pumps

Pump name	Axis orientation	Flow [m^3/h]	Power [kW]	RPM
Kobilja glava 1	Horizontal	360	250	1489
Kobilja glava 2	Horizontal	324	200	1490
Kobilja glava 3	Horizontal	324	200	1490
Kobilja glava 4	Vertical	360	200	2975
Buća potok 1	Vertical	360–486	132	2975
Buća potok 2	Vertical	360–486	132	2975
Centar 1	Vertical	540	160	1485
Centar 2	Vertical	540	160	1485
Centar 3	Vertical	648	160	1490

4.1 Description of the Measurement Procedure

As a preparation for a detailed analysis of the pumping plant vibrations, the measuring spots are chosen to give enough data to observe all possible vibration causes. The following locations are chosen for these measurements:

A. Opening frame for pipe from Kobilja glava 2 pump to basement pipe system (location with subjectively highest vibrations noticed).
B. Support pole for pipe from Kobilja glava 2 pump.
C. Abutment bottom (in the level of the pump carrying concrete slab) – according to BS 7385-2 standard [2].
D. Bearing on the motor of the Kobilja glava 2 pump.
E. Bearing on the motor of the Kobilja glava 3 pump.

For location C are chosen both of the abutments on each side of the concrete slab where pedesals of the pumps are reconstructed. For location D and E are chosen bearings of the neigboring pumps of the most excited structure part. Locations of the measuring points are shown in Fig. 3.

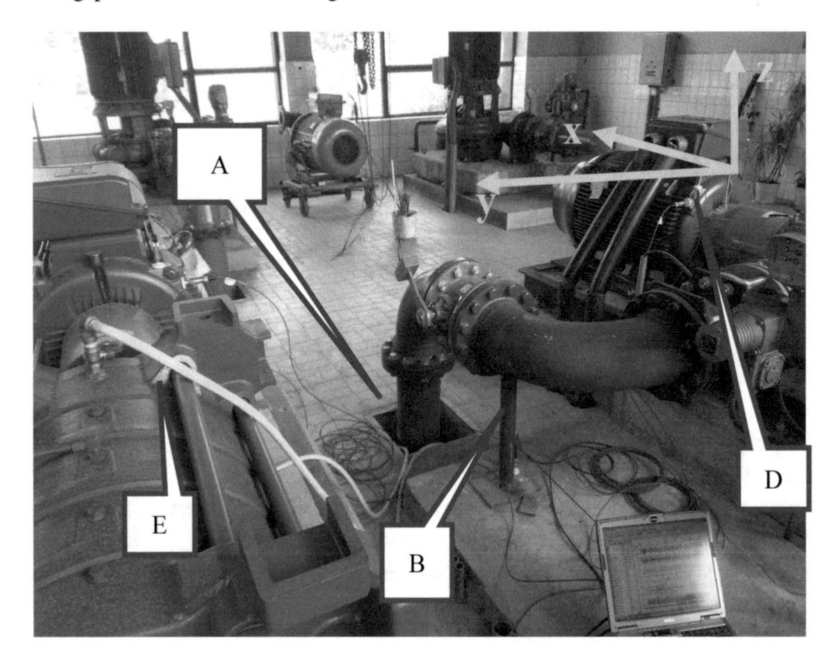

Fig. 3. Locations of the measuring points

Conditions with the water supply levels could not allow that all pumps are active at the same time. It could be possible to combine only some of the pumps active, where the neigbouring pumps of the increased vibration level location are active in all measurements. Combinations of the active pumps in the different measurements are shown in Table 2.

Table 2. Combinations of the active pumps in the two measurements.

Measurement	Active pumps								
	Kobilja glava 1	Kobilja glava 2	Kobilja glava 3	Kobilja glava 4	Buća potok 1	Buća potok 2	Centar 1	Centar 2	Centar 3
M1	X	X	X			X		X	X
M2	X	X	X	X	X			X	X

4.2 Analysis of the Results

As the first step in this analysis, with the handheld device are measured RMS of vibrations on all pumps. Any extremely increased vibrations of pumps could imply on structure vibrations. Results of the measurements are shown in Table 3.

Table 3. Velocity RMS of the pumps

Pump[a]	Bearing 1			Bearing 2			Bearing 3			Bearing 4[b]		
	H	V	A	H	V	A	H	V	A	H	V	A
KG1	0,4	0,5	0,2	0,5	0,6	0,3	0,3	0,4	0,4	1,0	1,0	0,6
KG2	1,3	0,8	0,8	1,6	1,4	0,8	0,8	0,4	0,4	1,2	0,8	0,6
KG3	1,6	1,7	1,1	1,9	2,3	1,0	0,5	0,3	0,6	1,0	1,0	0,6
KG4	3,7	3,2	2,1	1,2	1,1	1,8	2,5	2,0	2,2	–	–	–
BP1	1,3	2,1	0,9	1,0	2,1	2,1	1,1	2,7	1,4	–	–	–
BP2	2,0	1,2	1,2	1,2	1,4	1,3	1,1	1,3	1,1	–	–	–
C1	–	–	–	–	–	–	–	–	–	–	–	–
C2	1,5	1,7	1,2	0,9	0,9	0,6	1,3	0,8	0,8	–	–	–
C3	1,5	1,0	0,7	0,5	0,6	0,5	0,4	0,5	0,6	–	–	–

[a] *Abbreviations of pump name: KG-Kobilja glava, BP-Buća potok i C-Centar*
[b] *Vertical pumps have only 3 bearings*

Vibrations are measured in the horizontal, vertical and axial direction on bearings for horizontal pumps. For vertical pumps H stands for water outlet direction, V stands for the perpendicular direction of water outlet direction and A stands for the axial direction of the bearing.

Velocity RMS measured on all pumps have low values. All pumps have satisfying vibrations for a new equipment category, according to standard ISO 10816-3 proposed limits. Those small vibrations should only excite natural frequency oscillations of structure, and not be a problem by itself.

Since it is hard to calculate natural frequencies of the structure because of the reconstruction of the large concrete pedestals, analyzing the excitation force frequency and structure oscillating frequency could be done by analyzing the measurements. The first step was to confirm the rotational frequency of the pumps KG2 and KG3. Frequency spectrums are shown in Fig. 4.

On the frequency spectrum it could be seen the peaks of the frequency 24.84 Hz and it 4X and 7X orders. Other orders are also noticeable, but with much lower amplitude. 24.84 Hz frequency corresponds to the rotational frequency of the pump. 4X and 7X could be a frequency which depends on pump number of the impellers, which is unknown. Since those frequencies are high, it is hard to expect them to be close enough to the natural frequency of the structure.

For the possible cause of the increased structure vibrations, it could be looked in frequency spectrum measured on the location of the noticed increased vibrations. Frequency spectrums measured on the frame between KG2 and KG3 (location A) are shown in Fig. 5.

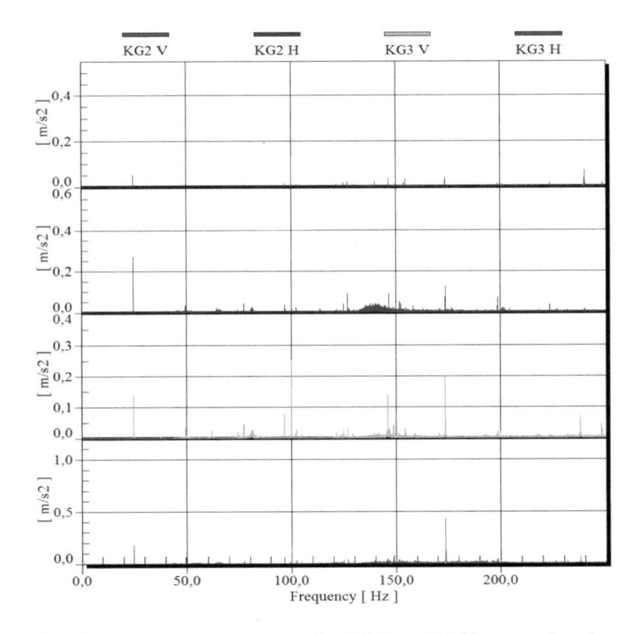

Fig. 4. Frequency spectrums for KG2 and KG3 motor bearing

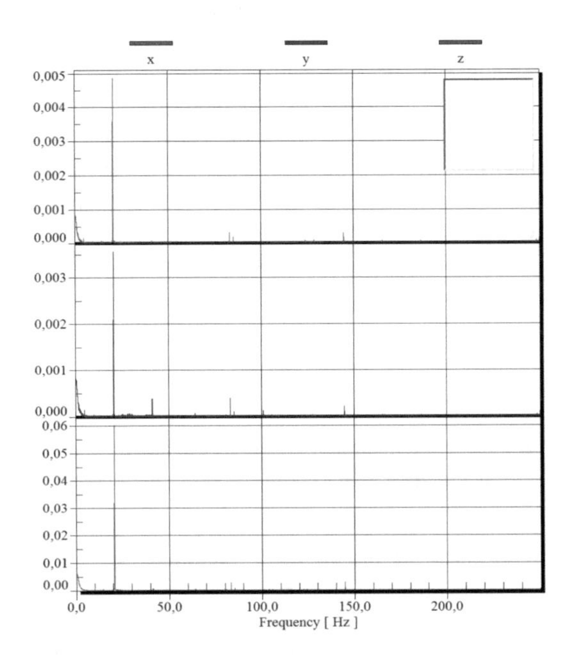

Fig. 5. Frequency spectrums of vibrations measured on location A

Most dominant frequency on the frequency spectrum measured on the location A is 20.69 Hz. Difference between the excitation force frequency and this natural frequency of the structure is small enough to cause the beating phenomenon. Confirmation of this is shown on timewave of displacement in z-direction measured on location A (concrete slab) (Fig. 6).

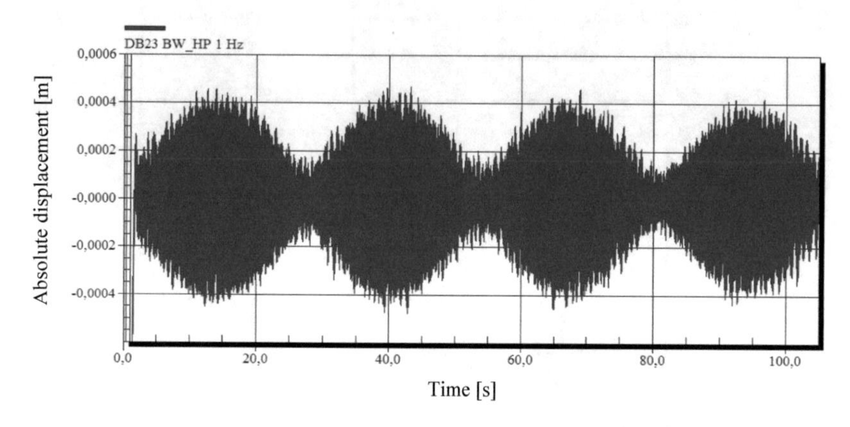

Fig. 6. Timewave of displacement measured in z direction (location A)

Clear proof of the presence of the beating phenomena is visible on the timewave measured on the location A in the z-direction. Machinery rotation frequency is close enough to excite the natural frequency of the concrete slab, which excited mode has dominant vertical movement in the middle of the concrete slab.

Since the increased vibrations are not noticed before the pedestal reconstruction, the additional mass of concrete pedestal and stiffness had changed one of the natural frequencies. That change of frequency impacted on the appearance of the beating phenomenon by making the excitation and natural frequency difference small enough for introducing this phenomenon.

5 Conclusion

After the presented theoretical appearance of the resonance and the beating phenomena, a short review of the current researches on this topic and case study in this paper, the following conclusions could be made:

- The most common cause of the structure vibrations are resonance and beating phenomena.
- It is possible to determine resonance or beating even if the natural frequencies of the structure are unknown. Unknown natural frequencies are common with structures which have been reconstructed.
- Standard tools of the rotational machinery vibration analysis (waveform analysis, frequency spectrum analysis) are key tools in discovering the real cause of the increased vibrations of the carrying structure.

References

1. ISO 4866:2010, Mechanical vibration and shock – Vibration of fixed structures – Guidelines for the measurement of vibrations and evaluation of their effects on structures, International Organization for Standardization (2010)
2. BS 7385-2, Evaluation and measurement for vibration i buildings – Guide to damage levels from groundborne vibration, British standard (1993)
3. Elvin, A.A., Elvin, N.G.: Structures subjected to startup and shutdown of rotating machinery. J. S. Afr. Inst. Civil Eng. **57**(1), 38–46 (2015)
4. da Silva, J.G., Sieira, A.C., da Silva, L.S., Rimola, B.: Dynamic analysis of steel platforms when subjected to mechanical equipment-induced vibrations. J. Civil Eng. Archit. **10**(2016), 1103–1113 (2016)
5. Ferro, R.M., Ferreira, W.G., Calenzani, A.F.G.: Dynamic analysis of support frame structures of rotating machinery. Global J. Res. Eng. Civil Struct. Eng. **XIV**(5), 27–31 (2014)
6. Patel, U., Mangukiya, S., Miyani, A., Patel, H., Vora, S., Sevelia, J.: Dynamic analysis of foundation supporting rotary machine. Int. J. Eng. Res. Appl. **5**(8), 34–45 (2015)
7. Leso, N., Puttonen, J., Porkka, E.: The effect of foundation on fan vibration response. Rakendeiden Mekaniika (J. Struct. Mech.) **44**(1), 1–20 (2011)
8. Tamariz, E., Lopez, A., Gonzalez, D.: Damage assessment in a reinforced concrete pedestal based on rotating machinery vibration analysis. In: Third International Conference on Mechanical Models in Structural Engineering, University of Sevile, 24–26 June 2015
9. Gaznavi, M., Kumar, A., Satheesh, J., Madhusudhan, T.: A review on dynamic analysis of machine structure. Int. Res. J. Eng. Technol. **4**(6), 1120–1123 (2017)
10. Isic., S., Doleček, V., Voloder, A.: An analysis of postbuckling frequency change of beam structures using finite elements method. Sci. J. Univ. Rev. **2**(3), 13–18 (2008). Journal of Alexander Dubček University of Trenčin

On the Use of the Udwadia-Kalaba Equations for the Nonlinear Control of a Generalized Van Der Pol-Duffing Oscillator

Carmine M. Pappalardo(✉) and Domenico Guida

Department of Industrial Engineering, University of Salerno,
Via Giovanni Paolo II, 132, 84084 Fisciano, Salerno, Italy
cpappalardo@unisa.it

Abstract. In this paper, a new method for controlling nonlinear mechanical systems is proposed. The methodology developed in this work is based on the use of the Udwadia-Kalaba equations in conjunction with the modern techniques of optimal control. The Udwadia-Kalaba equations represent an effective method for solving forward and inverse dynamics problems in the same analytical framework. On the other hand, the optimal control method is used in this work in combination with the inverse dynamic approach based on the Udwadia-Kalaba equations in order to obtain a nonlinear tracking controller. The mechanical system considered in this paper for performing numerical experiments is a nonlinear oscillator which includes in a generalized form the Van der Pol model for the system damping and the Duffing model for the system stiffness. The numerical results presented in this paper demonstrate the effectiveness of the method developed in this investigation.

Keywords: Nonlinear dynamics · Optimal control · Udwadia-Kalaba equations · Van Der Pol oscillator · Duffing oscillator

1 Introduction

In mechanical and industrial engineering applications, the study of the dynamical behavior of mechanical systems in response to applied forces and described by a nonlinear structure represents an important issue [1–10]. This problem is particularly challenging when the construction of a mechanism or of a machine is based on an approximated design solution affected by a certain degree of uncertainty [11–15]. In nonlinear control problems, the main goal is to construct the control laws necessary for guiding the dynamic behavior of a mechanical system in order to obtain the desired performance. To this end, appropriate analytical methods and computational procedures are necessary [16–25]. In the design of new control strategies, it is of fundamental importance to realize the fact that the classical approaches based on the linearization of the dynamic equations are not suitable for solving control problems associated with nonlinear mechanical systems. Unlike the conventional methods used in the literature, in this investigation, a new approach for the solution of nonlinear control problems is devised considering an inverse dynamic approach combined with the optimal control techniques.

I. Karabegović (Ed.): NT 2019, LNNS 76, pp. 82–95, 2020.
https://doi.org/10.1007/978-3-030-18072-0_9

The remaining part of this paper is organized as follows. In Sect. 2, a concise literature review on the analytical and computational techniques for the forward and inverse dynamics of nonlinear mechanical systems is reported. Section 3 describes in detail the research methodology employed in this paper for the development of an effective nonlinear control algorithm based on the combined use of the Udwadia-Kalaba equations with the optimal control theory. In Sect. 4, the numerical results obtained for the generalized Van Der Pol-Duffing oscillator used as a benchmark example are reported and a discussion on the performance of the proposed control algorithm is provided. Section 5 contains the summary of the manuscript and the conclusions drawn in this research work.

2 Literature Review

In this section, a short literature survey is provided. A general and reliable method for solving nonlinear control problems exploits the use of advanced nonlinear optimization techniques based on the adjoint method [26, 27]. However, the analytical formulation as well as the computer implementation of these techniques are complex also in the case of dynamical systems described by a simple set of nonlinear equations of motion [28–30]. Conversely, the nonlinear control technique devised in this investigation is based on an inverse dynamic method and employs the Udwadia-Kalaba equations [31–33]. The Udwadia-Kalaba equations, called with the names of their discoverers, originates from a combination of the fundamental principles of analytical dynamics and the modern methods of numerical linear algebra [34, 35]. In the literature, several interesting research works based on the use of the Udwadia-Kalaba equations can be found [36, 37]. For example, Koganti and Udwadia developed a methodology for the dynamics and control of tumbling multibody systems [38, 39]. Schutte and Udwadia derived a new approach for modeling complex multibody dynamical systems [40]. Cho et al. devised a continuous sliding mode controllers for multi-input-multi-output nonlinear systems [41]. Mylapilli and Udwadia developed a control method for three-dimensional incompressible hyperelastic beams modelled using the Absolute Nodal Coordinate (ANCF) method [42]. Wanichanon et al. used the Udwadia-Kalaba equations for the satellite formation-keeping [43–45]. Udwadia and Phohomsiri investigated the mathematical forms assumed by the fundamental equations of constrained motion in presence of redundant algebraic constraints and when the mass matrix is singular [46–50]. On the other hand, several studies on the application of the methods of the optimal control theory are available in the literature. Heydari and Balakrishnan proposed a finite-horizon control-constrained nonlinear optimal control using single network adaptive critics [51]. Lin et al. proposed a survey on the control parameterization methods used in the nonlinear optimal control [52]. Dierks and Jagannathan used the optimal control theory in the case of affine nonlinear continuous-time systems [53]. Xin and Pan employed the nonlinear optimal control method for a spacecraft approaching a tumbling target [54]. Liu and Wei used a finite-approximation-error-based optimal control approach for discrete-time nonlinear systems [55]. Wang et al. considered an adaptive dynamic programming approach for the optimal control of unknown non-affine nonlinear discrete-time systems [56]. Bryson et al. investigated the performance of a set of numerical methods for the computer implementation of the optimal control techniques [57–60].

3 Research Methodology

In this section, the fundamental aspects related to the methodology developed in this paper are described. To this end, the Udwadia-Kalaba equations and the optimal control theory are analyzed. Considering a mechanical system having n generalized coordinates and subjected to m constraint equations, the equations of motion are given by:

$$\begin{cases} \mathbf{M}\ddot{\mathbf{q}} = \mathbf{Q} + \mathbf{Q}_c \\ \mathbf{C} = \mathbf{0} \end{cases} \tag{1}$$

where $\mathbf{q}$ is the vector of generalized coordinates, $\mathbf{M}$ is the mass matrix, $\mathbf{Q}$ is the total vector of generalized forces, $\mathbf{Q}_c$ is the vector of constraint forces, and $\mathbf{C}$ is the vector of algebraic constraints. The vector of algebraic constraints $\mathbf{C}$ can be a mathematical representation of the mechanical joints or can be associated with the desired dynamic behavior of the mechanical system. In the former case, the generalized constraint force vector $\mathbf{Q}_c$ identifies the reaction forces and moments of the kinematic joints (forward dynamics), while in the latter case this vector denotes the generalized force vector relative to the control actions applied on the mechanical systems (inverse dynamics). In both cases, the Udwadia-Kalaba equations can be readily used for the explicit determination of the constraint generalized force vector $\mathbf{Q}_c$. Without loss of generality, these equations can be cast in the following compact form:

$$\begin{cases} \mathbf{a} = \mathbf{M}^{-1}\mathbf{Q}, \quad \mathbf{e} = \mathbf{Q}_d - \mathbf{C}_{\mathbf{q}}\mathbf{a}, \quad \mathbf{K} = \mathbf{C}_{\mathbf{q}}\mathbf{M}^{-1}\mathbf{C}_{\mathbf{q}}^T \\ \mathbf{F} = \mathbf{K}^{+}, \quad \boldsymbol{\lambda} = -\mathbf{F}\mathbf{e}, \quad \mathbf{Q}_c = -\mathbf{C}_{\mathbf{q}}^T\boldsymbol{\lambda} \end{cases} \tag{2}$$

where $\mathbf{C}_{\mathbf{q}}$ is the constraint Jacobian matrix, $\mathbf{Q}_d$ is the constraint quadratic velocity vector, $\mathbf{a}$ is the acceleration of the mechanical system released by the algebraic constraints, $\mathbf{e}$ is the error vector obtained by introducing the unconstrained acceleration vector $\mathbf{a}$ into the constrained equations defined at the acceleration level, $\mathbf{K}$ is the kinetic matrix of the constrained system, $\mathbf{F}$ is the constraint feedback matrix, $\boldsymbol{\lambda}$ is the vector of Lagrange multipliers, and $\mathbf{Q}_c$ is the constraint generalized force vector. Denoting with $\mathbf{q}_r$ the reference trajectory for the mechanical system under consideration, one can define an error vector $\boldsymbol{\varepsilon}$ given by:

$$\boldsymbol{\varepsilon} = \mathbf{q} - \mathbf{q}_r \tag{3}$$

By doing so, the simplest constraint equation associated with inverse dynamic problems can be written as follows:

$$\mathbf{C} = \boldsymbol{\varepsilon} = \mathbf{0} \tag{4}$$

which leads to:

$$\dot{\mathbf{C}} = \dot{\boldsymbol{\varepsilon}} = \dot{\mathbf{q}} - \dot{\mathbf{q}}_r = \mathbf{0} \quad \Rightarrow \quad \ddot{\mathbf{C}} = \ddot{\boldsymbol{\varepsilon}} = \ddot{\mathbf{q}} - \ddot{\mathbf{q}}_r = \mathbf{C}_{\mathbf{q}}\ddot{\mathbf{q}} - \mathbf{Q}_d = \mathbf{0} \tag{5}$$

where:

$$\mathbf{C}_{\mathbf{q}} = \mathbf{I}, \quad \mathbf{Q}_d = \ddot{\mathbf{q}}_r \tag{6}$$

Considering the previous definition of the algebraic constraint equations, one can readily use the Udwadia-Kalaba equations for obtaining the generalized force vector that solves the tracking control problem formulated in terms of an inverse dynamic problem. Furthermore, in order to improve the effectiveness of the control action found by means of the Udwadia-Kalaba equations, one can interpreter the constraint equations defined at the acceleration level as a set of linear equations of motion defined in terms of the error vector $\boldsymbol{\varepsilon}$ as follows:

$$\ddot{\boldsymbol{\varepsilon}} = \mathbf{0} \quad \Rightarrow \quad \ddot{\boldsymbol{\varepsilon}} = \mathbf{u}_b \tag{7}$$

where $\mathbf{u}_b$ is an additional vector of control inputs which improves the performance of the nonlinear controller calculated by using the Udwadia-Kalaba equations. The vector of control inputs $\mathbf{u}_b$ can be computed by using the infinite-horizon Linear-Quadratic Regulator (LQR) method. To this end, define the following state space equations:

$$\dot{\mathbf{z}}_b = \mathbf{A}_b\mathbf{z}_b + \mathbf{B}_b\mathbf{u}_b \tag{8}$$

where:

$$\mathbf{z}_b = \begin{bmatrix} \boldsymbol{\varepsilon} \\ \dot{\boldsymbol{\varepsilon}} \end{bmatrix}, \quad \mathbf{A}_b = \begin{bmatrix} \mathbf{O} & \mathbf{I} \\ \mathbf{O} & \mathbf{O} \end{bmatrix}, \quad \mathbf{B}_b = \begin{bmatrix} \mathbf{O} \\ \mathbf{I} \end{bmatrix} \tag{9}$$

where $\mathbf{z}_b$ is the state vector, $\mathbf{A}_b$ is the state matrix, and $\mathbf{B}_b$ is the input influence matrix. In order to compute the feedback control action $\mathbf{u}_b$ by using the LQR method, consider the following infinite-horizon cost function:

$$J_b = \int_0^\infty \frac{1}{2}\mathbf{z}_b^T\mathbf{Q}_z\mathbf{z}_b + \frac{1}{2}\mathbf{u}_b^T\mathbf{Q}_u\mathbf{u}_b dt \tag{10}$$

where $\mathbf{Q}_z$ is the weight matrix associated with the state vector and $\mathbf{Q}_u$ is the weight matrix associated with the control vector. Employing this algorithm one obtains an optimal feedback gain denoted with $\mathbf{F}_b$ and given by:

$$\mathbf{u}_b = -\mathbf{F}_b\mathbf{z}_b = -\begin{bmatrix} \mathbf{K}_b & \mathbf{R}_b \end{bmatrix}\begin{bmatrix} \boldsymbol{\varepsilon} \\ \dot{\boldsymbol{\varepsilon}} \end{bmatrix} = -\mathbf{K}_b\boldsymbol{\varepsilon} - \mathbf{R}_b\dot{\boldsymbol{\varepsilon}} \tag{11}$$

where $\mathbf{K}_b$ and $\mathbf{R}_b$ respectively represent the equivalent stiffness and damping matrices arising from the feedback control action. Therefore, one can write:

$$\ddot{\boldsymbol{\varepsilon}} = \mathbf{u}_b \quad \Leftrightarrow \quad \ddot{\boldsymbol{\varepsilon}} = -\mathbf{K}_b\boldsymbol{\varepsilon} - \mathbf{R}_b\dot{\boldsymbol{\varepsilon}} \quad \Leftrightarrow \quad \ddot{\boldsymbol{\varepsilon}} + \mathbf{R}_b\dot{\boldsymbol{\varepsilon}} + \mathbf{K}_b\boldsymbol{\varepsilon} = \mathbf{0} \tag{12}$$

which represents a linear set of equations of motion that is statically and dynamically stable provided that the matrices $\mathbf{K}_b$ and $\mathbf{R}_b$ are symmetric positive definite matrices. It is, therefore, apparent that the original nonlinear controller can be effectively improved considering the following definitions:

$$\mathbf{C}_\mathbf{q} = \mathbf{I}, \quad \mathbf{Q}_d = \ddot{\mathbf{q}}_r - \mathbf{R}_b(\dot{\mathbf{q}} - \dot{\mathbf{q}}_r) - \mathbf{K}_b(\mathbf{q} - \mathbf{q}_r) \tag{13}$$

where $\mathbf{C}_\mathbf{q}$ is the modified constraint Jacobian matrix and $\mathbf{Q}_d$ is the modified constraint quadratic velocity vector.

4 Numerical Results and Discussion

In this section, a simple numerical example is proposed in order to demonstrate the effectiveness of the nonlinear control method developed in this investigation. For this purpose, the performance of the proposed method is analyzed in the case of the vibration control problem of a nonlinear mechanical oscillator. A simple mechanical oscillator subjected to a nonlinear force field is considered as the case study and is shown in Fig. 1.

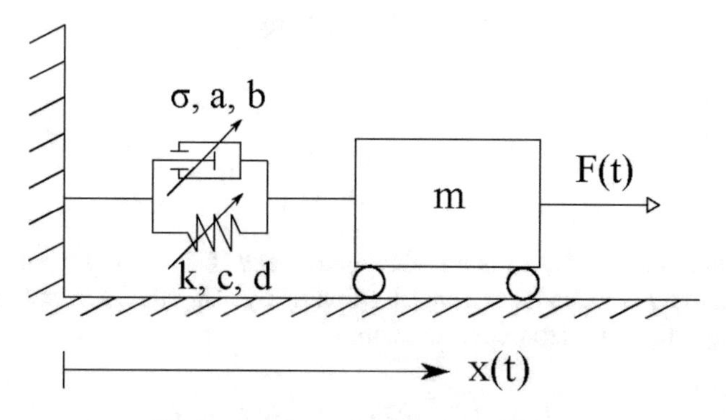

Fig. 1. Nonlinear mechanical oscillator.

The mechanical system shown in Fig. 1 is a nonlinear oscillator with one degree of freedom. The numerical data used for performing dynamical simulations of the mechanical oscillator are reported in Table 1.

Table 1. System data.

Descriptions	Symbols	Data (units)
Oscillator Mass	m	4.0 (kg)
Damping Coefficient	σ	1.75 (N×s/m)
Van Der Pol First Coefficient	a	−2.0 (−)
Van Der Pol Second Coefficient	b	$(1/\mathrm{m}^2)$
Stiffness Coefficient	k	1.5 (N/m)
Duffing First Coefficient	c	1.0 (−)
Duffing Second Coefficient	d	$(1/\mathrm{m}^2)$
Initial Displacement	x_0	5.0 (m)
Initial Velocity	v_0	5.0 (m/s)

The configuration of the nonlinear oscillator is identified by the parameter x which represents the displacement of the point mass m. The concentrated mass is subjected to a nonlinear force field produced by the combined action of a nonlinear dashpot with damping σ and a nonlinear spring with stiffness k. The force law of the nonlinear dashpot is characterized by a generalized Van der Pol damping model having constant coefficients a and b, while the force law of the nonlinear spring is characterized by a generalized Duffing elastic model having constant coefficients c and d. Considering a control action u equal to the external force F applied on the system, the equation of motion of the nonlinear oscillator can be explicitly written as follows:

$$m\ddot{x} + \sigma\dot{x}\left(a + bx^2\right) + kx\left(c + dx^2\right) = u \tag{14}$$

The equation of motion can be readily casted in the following configuration-space form:

$$\mathbf{M}\ddot{\mathbf{q}} = \mathbf{Q} + \mathbf{Q}_c \tag{15}$$

where:

$$\mathbf{q} = x, \quad \mathbf{M} = m, \quad \mathbf{Q} = -kx\left(c + dx^2\right) - \sigma\dot{x}\left(a + bx^2\right), \quad \mathbf{Q}_c = u \tag{16}$$

The reference trajectory assumed for the mechanical oscillator is denoted with the function g and is given by:

$$g = e^{-\xi\omega_n t}\left(Ae^{\omega_a t} + Be^{-\omega_a t}\right) \tag{17}$$

where:

$$A = \frac{1}{2\omega_a}((\omega_a + \xi\omega_n)x_0 + v_0), \quad B = \frac{1}{2\omega_a}((\omega_a - \xi\omega_n)x_0 - v_0), \quad \omega_n = 2\pi f_n, \\ \omega_a = \omega_n\sqrt{\xi^2 - 1} \tag{18}$$

where the parameters ξ, f_n, ω_n, and ω_a are defined in Table 2.

Table 2. Trajectory parameters.

Descriptions	Symbols	Data (units)
Damping Ratio	ξ	1.1 (–)
Natural Frequency	f_n	0.4 (Hz)
Angular Frequency	ω_n	2.513 (rad/s)
Aperiodic Coefficient	ω_a	1.152 (1/s)

For the mechanical oscillator, the trajectory error is given by:

$$\boldsymbol{\varepsilon} = x - g \tag{19}$$

which leads to:

$$\mathbf{C_q} = 1, \quad \mathbf{Q}_d = \ddot{g} \tag{20}$$

The control law based on the inverse dynamic method can be enhanced considering the following vector and matrix quantities:

$$\mathbf{C_q} = 1, \quad \mathbf{Q}_d = \ddot{g} - k_c(x - g) - \sigma_c(\dot{x} - \dot{g}) = \ddot{g} - \beta_c^2(x - g) - 2\alpha_c\beta_c(\dot{x} - \dot{g}) \tag{21}$$

where the feedback parameters k_c, σ_c, α_c, and β_c can be readily computed by using the LQR method considering the following matrices:

$$\mathbf{A}_b = \begin{bmatrix} 0 & 1 \\ 0 & 0 \end{bmatrix}, \quad \mathbf{B}_b = \begin{bmatrix} 0 \\ 1 \end{bmatrix}, \quad \mathbf{Q}_z = \begin{bmatrix} 10^3 & 0 \\ 0 & 10^3 \end{bmatrix}, \quad \mathbf{Q}_u = 1 \tag{22}$$

It follows that:

$$\alpha_c = \frac{\sigma_c}{2\sqrt{k_c}}, \quad \beta_c = \sqrt{k_c} \tag{23}$$

The optimal parameters for the compensation controller obtained by using the LQR method are reported in Table 3.

Table 3. Compensator parameters.

Descriptions	Symbols	Data (units)
Controller Stiffness Coefficient	k_c	31.623 (N/m)
Controller Damping Coefficient	σ_c	32.607 (N×s/m)
Controller Damping Ratio	α_c	2.899 (–)
Controller Angular Frequency	β_c	5.623 (rad/s)

The nonlinear control law for the mechanical oscillator obtained using the Udwadia-Kalaba equations in conjunction with the LQR method can be obtained as:

$$\begin{cases} \mathbf{a} = -\frac{k}{m}x(c+dx^2) - \frac{\sigma}{m}\dot{x}(a+bx^2) \\ \mathbf{e} = \ddot{g} - \beta_c^2(x-g) - 2\alpha_c\beta_c(\dot{x}-\dot{g}) + \frac{k}{m}x(c+dx^2) + \frac{\sigma}{m}\dot{x}(a+bx^2) \\ \mathbf{K} = \frac{1}{m}, \quad \mathbf{F} = m \\ \boldsymbol{\lambda} = -m\ddot{g} + m\beta_c^2(x-g) + m2\alpha_c\beta_c(\dot{x}-\dot{g}) - kx(c+dx^2) - \sigma\dot{x}(a+bx^2) \\ \mathbf{Q}_c = m\ddot{g} - m\beta_c^2(x-g) - m2\alpha_c\beta_c(\dot{x}-\dot{g}) + kx(c+dx^2) + \sigma\dot{x}(a+bx^2) \end{cases} \tag{24}$$

Considering a small error in the initial conditions, dynamical simulations are performed in order to test the performance of the nonlinear control law described above. To this end, a numerical integration scheme based on the fourth-order Adams-Bashforth method is used employing a time step $\Delta t = 10^{-3}$ (s) and a time span $T = 30$ (s). The nonlinear control action obtained by employing the proposed approach is represented in Fig. 2.

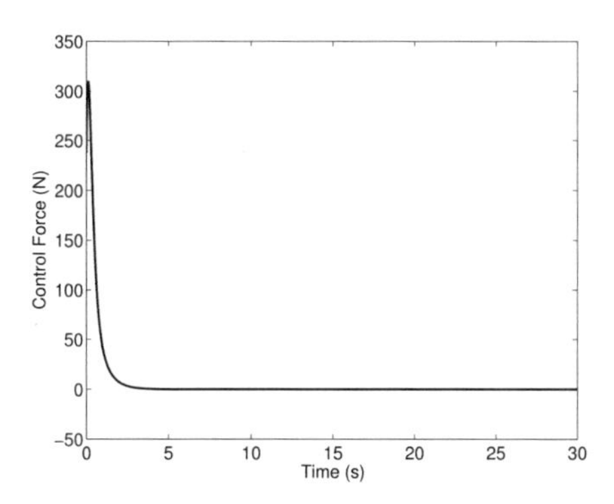

Fig. 2. Nonlinear control force.

The resulting displacement, velocity, and state-space trajectory of the mechanical oscillator are respectively represented in Fig. 3(a), (b) and (c).

In these figures, the solid lines refer to the dynamical behaviors obtained without the action of the nonlinear control force while the dashed lines represent the state trajectories arising from the introduction of the nonlinear control force. As shown in Figs. 3(a), (b), (c), the nonlinear control force designed by employing the proposed method makes the mechanical oscillator follow the prescribed state trajectory. The performance of the nonlinear control force can be readily assessed by comparing the displacement and velocity root-mean-square deviations obtained with and without the use of the nonlinear controller. For the mechanical oscillator, the displacement and velocity root-mean-square deviations are respectively defined as:

$$d_{RMSD} = \sqrt{\frac{1}{N}\sum_{k=1}^{N} x^2(k)}, \quad v_{RMSD} = \sqrt{\frac{1}{N}\sum_{k=1}^{N} \dot{x}^2(k)} \tag{25}$$

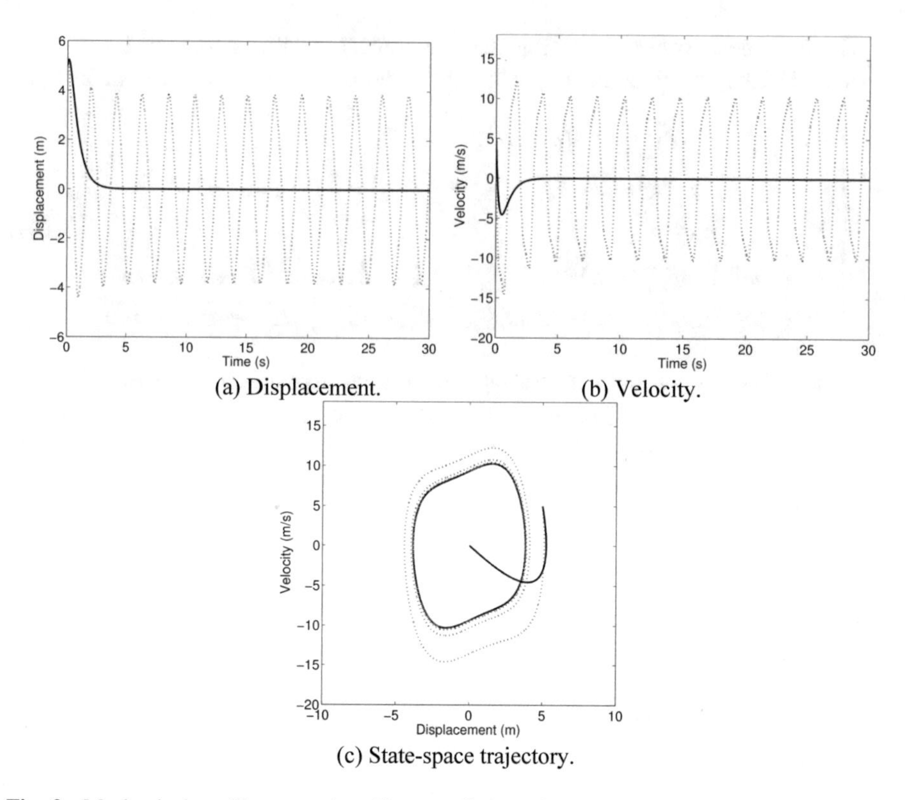

(a) Displacement.

(b) Velocity.

(c) State-space trajectory.

Fig. 3. Mechanical oscillator motion. Uncontrolled motion (dashed lines), controlled motion (solid lines).

Table 4 shows the displacement and velocity root-mean-square deviations with and without the presence of the nonlinear control force calculated by using the proposed method.

Table 4. Displacement and velocity root-mean-square deviations.

	d_{RMSD}	$v_{RMSD,1}$
Uncontrolled motion	2.641 (m)	7.762 (m/s)
Controlled motion	7.644×10^{-1} (m)	7.428×10^{-1} (m/s)
Percentage reduction	71.051 (%)	90.430 (%)

The numerical results reported in this section demonstrates the effectiveness of the approach developed in the paper.

5 Summary and Conclusions

The authors are mainly focused on three research fields, namely multibody dynamics, nonlinear control, and system identification [61–75]. This investigation fits this research framework with particular attention to the control algorithms suitable for nonlinear dynamical systems.

In this investigation, a new algorithm for the nonlinear control of mechanical systems is developed. The fundamental aspects of this research work can be summarized as follows. For a given set of holonomic and/or nonholonomic constraint equations, the Udwadia-Kalaba equations allow for computing in a closed form the generalized constraint force vector associated with the kinematic constraints. If the constraint equations are virtual, namely they do not correspond to mechanical joints but represent the desired behavior for the configuration vector of the mechanical systems under study, the Udwadia-Kalaba equations can be effectively used for calculating the generalized control force vector necessary for obtaining a prescribed state trajectory. The vector of control inputs can be subsequently recovered from the generalized control force vector obtained by doing so. The simplest system of virtual constraint equations sets equal to zero the error between the actual configuration vector of the mechanical system and the desired dynamic behavior. At the acceleration level, this set of constraint equations is equivalent to a linear set of equations of motion of a multiple degrees of freedom mechanical system. Considering the set of constraint equations mentioned before, one can readily compute the constraint Jacobian matrix and the constraint quadratic velocity vector necessary for the analytical implementation of the Udwadia-Kalaba equations.

An improvement of the performance of the control action obtained from the virtual set of constraint equations can be achieved considering a feedback compensator applied to the virtual set of linear equations of motion, namely on the virtual set of constraint equations defined at the acceleration level. To this end, one only needs to modify the constraint quadratic velocity vector necessary for the application of the Udwadia-Kalaba equations in order to include the effect of the feedback compensator. In order to optimize the selection of the control gain used for the feedback compensator, an algorithm based on the continuous-time infinite-horizon Linear-Quadratic-Regulator (LQR) technique can be employed. For this purpose, the selection of the weight matrices that appear in the definition of the performance index facilitates the computation of the control gain. Finally, one can recover from the resulting feedback gain the equivalent stiffness and damping matrices employed in the definition of the optimal compensation control action. The numerical experiments performed in the paper confirmed the efficacy of the proposed control method.

References

1. Villecco, F.: On the evaluation of errors in the virtual design of mechanical systems. Machines **6**, 36 (2018)
2. Sena, P., Attianese, P., Pappalardo, M., Villecco, F.: FIDELITY: fuzzy inferential diagnostic engine for on-line support to physicians. In: Proceedings of the 4th International Conference on the Development of Biomedical Engineering in Vietnam, Ho Chi Minh City, Vietnam, 8–10 January 2012, pp. 396–400 (2012)
3. Ghomshei, M., Villecco, F., Porkhial, S., Pappalardo, M.: Complexity in energy policy: a fuzzy logic methodology. In: Proceedings of the 6th International Conference on Fuzzy Systems and Knowledge Discovery, Tianjin, China, 14–16 August 2009, vol. 7, pp. 128–131. IEEE, Los Alamitos (2009)
4. Zhai, Y., Liu, L., Lu, W., Li, Y., Yang, S., Villecco, F.: The application of disturbance observer to propulsion control of sub-mini underwater robot. In: Proceedings of the ICCSA 2010 International Conference on Computational Science and Its Applications, Fukuoka, Japan, 23–26 March 2010, pp. 590–598
5. Sena, P., D'Amore, M., Pappalardo, M., Pellegrino, A., Fiorentino, A., Villecco, F.: Studying the influence of cognitive load on driver's performances by a fuzzy analysis of lane keeping in a drive simulation. IFAC Proc. **46**, 151–156 (2013)
6. Ghomshei, M., Villecco, F.: Energy metrics and sustainability. In: Proceedings of the International Conference on Computational Science and Its Applications, Seoul, Korea, 29 June–2 July 2009, pp. 693–698 (2009)
7. Sena, P., Attianese, P., Carbone, F., Pellegrino, A., Pinto, A., Villecco, F.: A fuzzy model to interpret data of drive performances from patients with sleep deprivation. Comput. Math. Methods Med. **2012**, 5 (2012). 868410
8. Zhang, Y., Li, Z., Gao, J., Hong, J., Villecco, F., Li, Y.: A method for designing assembly tolerance networks of mechanical assemblies. Math. Probl. Eng. **2012**, 26 (2012). 513958
9. Villecco, F., Pellegrino, A.: Evaluation of uncertainties in the design process of complex mechanical systems. Entropy **19**, 475 (2017)
10. Pellegrino, A., Villecco, F.: Design optimization of a natural gas substation with intensification of the energy cycle. Math. Probl. Eng. **2010**, 10 (2010). 294102
11. Barbagallo, R., Sequenzia, G., Cammarata, A., Oliveri, S.M., Fatuzzo, G.: Redesign and multibody simulation of a motorcycle rear suspension with eccentric mechanism. Int. J. Int. Des. Man. **12**, 517–524 (2018)
12. Barbagallo, R., Sequenzia, G., Oliveri, S.M., Cammarata, A.: Dynamics of a high-performance motorcycle by an advanced multibody/control co-simulation. Proc. Inst. Mech. Eng. Part K J. Eng. **230**, 207–221 (2016)
13. Oliveri, S.M., Sequenzia, G., Calí, M.: Flexible multibody model of desmodromic timing system. Mech. Based Des. Struct. **37**, 15–30 (2009)
14. Barbagallo, R., Sequenzia, G., Cammarata, A., Oliveri, S.M.: An integrated approach to design an innovative motorcycle rear suspension with eccentric mechanism. Advances on Mechanics, Design Engineering and Manufacturing, pp. 609–619. Springer, Cham (2017)
15. Calí, M., Oliveri, S.M., Sequenzia, G.: Geometric modeling and modal stress formulation for flexible multi-body dynamic analysis of crankshaft. In: Proceedings of the 25th Conference and Exposition on Structural Dynamics 2007, Orlando, FL, USA, 19–22 February 2007, pp. 1–9 (2007)
16. Cammarata, A.: A novel method to determine position and orientation errors in clearance-affected overconstrained mechanisms. Mech. Mach. Theory **118**, 247–264 (2017)

17. Cammarata, A., Calió, I., Greco, A., Lacagnina, M., Fichera, G.: Dynamic stiffness model of spherical parallel robots. J. Sound Vib. **384**, 312–324 (2016)
18. Cammarata, A., Lacagnina, M., Sequenzia, G.: Alternative elliptic integral solution to the beam deflection equations for the design of compliant mechanisms. Int. J. Interact. Des. Manuf. (IJIDeM), 1–7 (2018). https://doi.org/10.1007/s12008-018-0512-6
19. Cammarata, A., Sinatra, R., Maddio, P.D.: A two-step algorithm for the dynamic reduction of flexible mechanisms. In: Mechanism Design for Robotics, pp. 25–32. Springer, Cham (2018)
20. Muscat, M., Cammarata, A., Maddio, P.D., Sinatra, R.: Design and development of a towfish to monitor marine pollution. Euro-Mediterr. J. Environ. Integr. **3**, 11 (2018)
21. Cammarata, A., Sinatra, R.: On the elastostatics of spherical parallel machines with curved links. In: Recent Advances in Mechanism Design for Robotics, pp. 347–356. Springer, Cham (2015)
22. Cammarata, A., Lacagnina, M., Sinatra, R.: Dynamic simulations of an airplane-shaped underwater towed vehicle marine. In: Proceedings of the 5th International Conference on Computational Methods in Marine Engineering, Hamburg, Germany, 29–31 May 2013, pp. 830–841, Code 101673 (2013). ISBN 978-849414074-7
23. Cammarata, A., Angeles, J., Sinatra, R.: Kinetostatic and inertial conditioning of the McGill Schonflies-motion generator. Adv. Mech. Eng. **2**, 186203 (2010)
24. Cammarata, A.: Unified formulation for the stiffness analysis of spatial mechanisms. Mech. Mach. Theory **105**, 272–284 (2016)
25. Cammarata, A.: Optimized design of a large-workspace 2-DOF parallel robot for solar tracking systems. Mech. Mach. Theory **83**, 175–186 (2015)
26. Kirk, D.E.: Optimal Control Theory: An Introduction. Springer, New York (1970)
27. Lewis, F.L., Vrabie, D., Syrmos, V.L.: Optimal Control. Wiley, Chichester (2012)
28. Khalil, H.K.: Nonlinear Control. Pearson, New York (2015)
29. Udwadia, F.E., Kalaba, R.E.: Analytical Dynamics: A New Approach. Cambridge University Press, Cambridge (2007)
30. Udwadia, F.E., Weber, H.I., Leitmann, G.: Dynamical Systems and Control. CRC Press, Boca Raton (2016)
31. Udwadia, F.E.: Equations of motion for constrained multibody systems and their control. J. Optim. Theory Appl. **127**, 627–638 (2005)
32. Udwadia, F.E.: Inverse problem of Lagrangian mechanics for classically damped linear multi-degrees-of-freedom systems. J. Appl. Mech. **83**(10), 104501 (2016)
33. Udwadia, F.E.: Optimal tracking control of nonlinear dynamical systems. Proc. R. Soc. Lond. A Math. Phys. Eng. Sci. **464**, 2341–2363 (2008)
34. Udwadia, F.E., Kalaba, R.E.: A new perspective on constrained motion. Proc. Math. Phys. Sci. **1992**, 407–410 (1992)
35. Udwadia, F.E., Kalaba, R.E.: On the foundations of analytical dynamics. Int. J. Non-Linear Mech. **37**(6), 1079–1090 (2002)
36. Udwadia, F.E., Koganti, P.B.: Optimal stable control for nonlinear dynamical systems: an analytical dynamics based approach. Nonlinear Dyn. **82**(1–2), 547–562 (2015)
37. Koganti, P.B., Udwadia, F.E.: Unified approach to modeling and control of rigid multibody systems. J. Guidance Control Dyn. **2016**, 2683–2698 (2016)
38. Koganti, P.B., Udwadia, F.E.: Dynamics and precision control of tumbling multibody systems. J. Guid. Control Dyn. **40**(3), 584–602 (2017)
39. Koganti, P.B., Udwadia, F.E.: Dynamics and precision control of uncertain tumbling multibody systems. J. Guidance Control Dyn. **40**(5), 1176–1190 (2017)
40. Schutte, A., Udwadia, F.: New approach to the modeling of complex multibody dynamical systems. J. Appl. Mech. **78**, 021018 (2011)

41. Cho, H., Wanichanon, T., Udwadia, F.E.: Continuous sliding mode controllers for multi-input multi-output systems. Nonlinear Dyn. **94**(4), 2727–2747 (2018)
42. Mylapilli, H., Udwadia, F.E.: Control of three-dimensional incompressible hyperelastic beams. Nonlinear Dyn. **90**(1), 115–135 (2017)
43. Wanichanon, T., Cho, H., Udwadia, F.: Satellite formation-keeping using the fundamental equation in the presence of uncertainties in the system. In: AIAA SPACE 2011 Conference and Exposition 2011, vol. 7210 (2011)
44. Wanichanon, T., Udwadia, F.E., Cho, H.: Formation-keeping of uncertain satellites using nonlinear damping control. J. Res. Appl. Mech. Eng. **2**(1), 20–33 (2014)
45. Wanichanon, T., Udwadia, F.E.: Nonlinear damping control for uncertain nonlinear multi-body mechanical systems. J. Res. Appl. Mech. Eng. **2**(1), 7–19 (2014)
46. Udwadia, F.E., Phohomsiri, P.: Generalized LM-inverse of a matrix augmented by a column vector. Appl. Math. Comput. **190**(2), 999–1006 (2007)
47. Udwadia, F.E., Phohomsiri, P.: Recursive formulas for the generalized LM-inverse of a matrix. J. Optim. Theory Appl. **131**(1), 1–16 (2006)
48. Udwadia, F.E., Phohomsiri, P.: Explicit equations of motion for constrained mechanical systems with singular mass matrices and applications to multi-body dynamics. In: Proceedings of the Royal Society of London A: Mathematical, Physical and Engineering Sciences, vol. 462, No. 2071, pp. 2097–2117, July 2006
49. Udwadia, F.E., Phohomsiri, P.: Recursive determination of the generalized Moore-Penrose M-inverse of a matrix. J. Optim. Theory Appl. **127**(3), 639–663 (2005)
50. Udwadia, F.E., Kalaba, R.E., Phohomsiri, P.: Mechanical systems with nonideal constraints: explicit equations without the use of generalized inverses. J. Appl. Mech. **71**(5), 615–621 (2004)
51. Heydari, A., Balakrishnan, S.N.: Finite-horizon control-constrained nonlinear optimal control using single network adaptive critics. IEEE Trans. Neural Networks Learn. Syst. **24**(1), 145–157 (2013)
52. Lin, Q., Loxton, R., Teo, K.L.: The control parameterization method for nonlinear optimal control: a survey. J. Ind. Manag. Optim. **10**(1), 275–309 (2014)
53. Dierks, T., Jagannathan, S.: Optimal control of affine nonlinear continuous-time systems. In: IEEE American Control Conference (ACC), pp. 1568–1573, June 2010
54. Xin, M., Pan, H.: Nonlinear optimal control of spacecraft approaching a tumbling target. Aerosp. Sci. Technol. **15**(2), 79–89 (2011)
55. Liu, D., Wei, Q.: Finite-approximation-error-based optimal control approach for discrete-time nonlinear systems. IEEE Trans. Cybern. **43**(2), 779–789 (2013)
56. Wang, D., Liu, D., Wei, Q., Zhao, D., Jin, N.: Optimal control of unknown non affine nonlinear discrete-time systems based on adaptive dynamic programming. Automatica **48**(8), 1825–1832 (2012)
57. Bryson, A.E.: Applied optimal control: optimization, estimation and control. Routledge (2018)
58. Bryson, A., Ho, Y.C.: Applied Optimal Control: Optimization, Estimation, and Control (Revised Edition). Taylor and Francis, Pennsylvania (1975)
59. Bryson, A.E.: Optimal control-1950 to 1985. IEEE Control Syst. **16**(3), 26–33 (1996)
60. Weinreb, A., Bryson, A.: Optimal control of systems with hard control bounds. IEEE Trans. Autom. Control **30**(11), 1135–1138 (1985)
61. De Simone, M.C., Rivera, Z.B., Guida, D.: Obstacle avoidance system for unmanned ground vehicles by using ultrasonic sensors. Machines **6**, 18 (2018)

62. De Simone, M.C., Russo, S., Rivera, Z.B., Guida, D.: Multibody model of a UAV in presence of wind fields. In: Proceedings of the 2017 International Conference on Control, Artificial Intelligence, Robotics and Optimization (ICCAIRO), Prague, Czech Republic, 20–22 May 2017, pp. 83–88 (2017)
63. De Simone, M.C., Guida, D.: Identification and control of a unmanned ground vehicle by using arduino. UPB Sci. Bull. Ser. D **80**, 141–154 (2018)
64. De Simone, M.C., Guida, D.: On the development of a low-cost device for retrofitting tracked vehicles for autonomous navigation. In: Proceedings of the 23rd Conference of the Italian Association of Theoretical and Applied Mechanics, Salerno, Italy, 4–7 September 2017, vol. 4, pp. 71–82 (2017)
65. De Simone, M.C., Guida, D.: Control design for an under-actuated UAV model. FME Trans. **46**, 443–452 (2018)
66. De Simone, M.C., Guida, D.: Modal coupling in presence of dry friction. Machines **6**, 8 (2018)
67. De Simone, M.C., Rivera, Z.B., Guida, D.: Finite element analysis on squeal-noise in railway applications. FME Trans. **46**, 93–100 (2018)
68. De Simone, M.C., Guida, D.: Object Recognition by Using Neural Networks For Robotics Precision Agriculture Application. Eng. Lett. (2019, in press)
69. Concilio, A., De Simone, M.C., Rivera, Z.B., Guida, D.: A new semi-active suspension system for racing vehicles. FME Trans. **45**, 578–584 (2017)
70. Quatrano, A., De Simone, M.C., Rivera, Z.B., Guida, D.: Development and implementation of a control system for a retrofitted CNC machine by using Arduino. FME Trans. **45**, 565–571 (2017)
71. Ruggiero, A., Affatato, S., Merola, M., De Simone, M.C.: FEM analysis of metal on UHMWPE total hip prosthesis during normal walking cycle. In: Proceedings of the 23rd Conference of the Italian Association of Theoretical and Applied Mechanics, Salerno, Italy, 4–7 September 2017, vol. 2, pp. 1885–1892 (2017)
72. Ruggiero, A., De Simone, M.C., Russo, D., Guida, D.: Sound pressure measurement of orchestral instruments in the concert hall of a public school. Int. J. Circuits Syst. Signal Process **10**, 75–81 (2016)
73. De Simone, M.C., Guida, D.: Dry friction influence on structure dynamics. In: Proceedings of the COMPDYN 2015 - 5th ECCOMAS Thematic Conference on Computational Methods in Structural Dynamics and Earthquake Engineering, Crete Island, Greece, pp. 4483–4491 (2015)
74. Iannone, V., De Simone, M.C., Modelling of a DC gear motor for feed-forward control law design for unmanned ground vehicles. Actuators (2019, Submitted)
75. Rivera, Z.B., De Simone, M.C., Guida, D.: Modelling of mobile robots in ROS-based environments. Robotics (2019, Submitted)

Systems for Passive and Active Vibration Damping

Safet Isić[1(✉)], Semir Mehremić[1], Isak Karabegović[2], and Ermin Husak[2]

[1] Faculty of Mechanical Engineering, University "Džemal Bijedić" of Mostar, University Campus, 88000 Mostar, Bosnia and Herzegovina
safet.isic@unmo.ba

[2] Technical Faculty Bihać, University of Bihać, 77000 Bihać, Bosnia and Herzegovina

Abstract. Reduction of the vibration amplitude is often problem in mechanical and civil engineering. Damping of the oscillation could be done by changing the mass and stiffness. This approach often requires a change in the constructive system solution. The alternative is active vibration damping systems, which, depending on the vibration states of the system they are part of, can induce the vibrations that by superimposition lead to their reduction. This paper presents an overview the vibration damping system (active and passive) and the principle of the mechatronic system for active vibration damping.

Keywords: Vibrations · Active damping · Mechatronics

1 Introduction

Vibrations are a mechanical phenomenon that accompanies the work of most systems with moving elements. Increased vibration can negatively affect the performance of these systems (e.g. the accuracy of the cutting machine) or, in some cases, endanger the strength and stability of the system (e.g. oscillation of bridges under the influence of wind, oscillation of buildings in the earthquake, etc.). Reduction of the amplitude of the oscillation could be done by changing the mass arrangement, changing the stiffness, and installing special supports. This approach often requires a change in the constructive system solution, what is in many cases difficult or impossible to do. The alternative is active vibration damping systems, which, depending on the vibration states of the system they are part of, can induce the vibrations that by superposition lead to their reduction. With the development of mechatronics, the possibilities of such systems are becoming ever greater. This paper reviews the vibration damping system (active and passive) and presents the principle of the mechatronic system for active vibration damping.

I. Karabegović (Ed.): NT 2019, LNNS 76, pp. 96–104, 2020.
https://doi.org/10.1007/978-3-030-18072-0_10

2 Passive Vibration Damping

One of the most important passive vibration damper is dynamic vibration absorber or tuned mass damper system (Fig. 1). Adding additional mass m_1 over spring k_1 to the mass m_0 which is in resonance state induced by forcing load $F_o \cos \Omega t$, change system property and hold both amplitudes of the initial mass m and additional mass finite. In the spatial case of the initial condition of the additional mass (e.g. setting initial displacement of the additional mass to produce force in sprig k_1 which is opposite of the perturbation force), amplitude of the initial mass could be setted to zero, remaining amplitude of the additional mass finite. This effect is based on the changing of eigenfrequency of the initial 1 DOF system transforming it in the 2 DOF system.

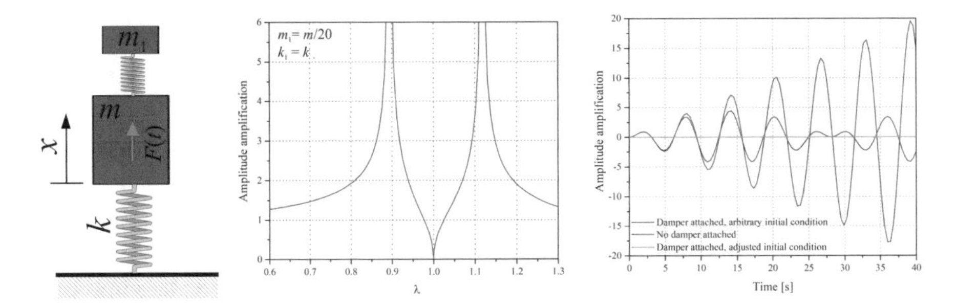

Fig. 1. Dynamical vibration dumper of a originally undamped 2 DOF system.

Damping effect and parameters adjustments could be analysed from the following equations

$$\begin{aligned} m\ddot{x} + (k + k_1)x - k_1 x_1 &= F_O \cos \Omega t \\ m_1 \ddot{x}_1 + k_1 x_1 - k_1 x &= 0 \end{aligned} \quad (1)$$

One of practical realisation of dynamic vibration absorber with additional mass is shown in Fig. 2.

Fig. 2. Dynamic vibration absorber.

Dynamic vibration absorber is frequently used as seismic damper and mounted in large structures to reduce the amplitude of mechanical vibrations. One of the most famous application of this damper is in Taipei 101, the highest building in Asia in the period 2004–2010 (Fig. 3). They are frequently used in power transmission, automobiles and buildings.

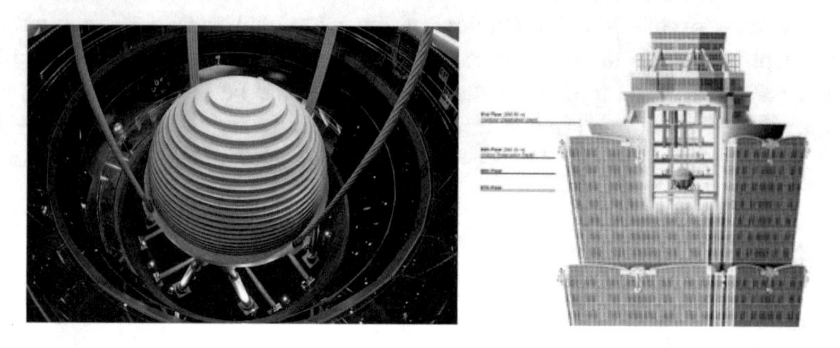

Fig. 3. The largest tuned mass damper at top Taipei 101 and its location on Taipei 101

One another example of passive vibration control by dynamic vibration damper is shown in Fig. 4. High-tension wires of overhead electrical power lines often have small barbell-shaped Stockbridge dampers hanging from the wires. This reduce the high-frequency, low-amplitude oscillation (flutter).

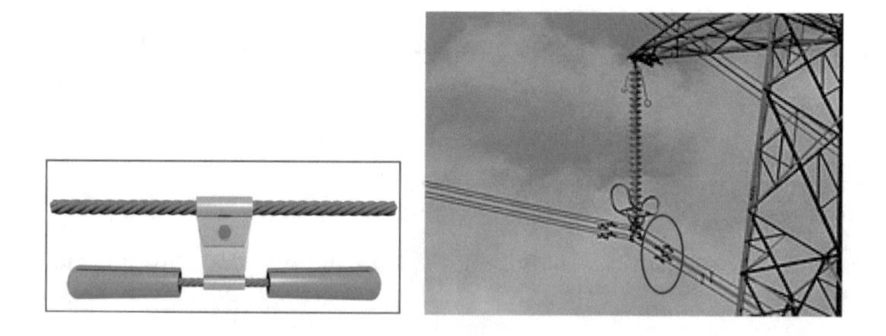

Fig. 4. Stockbridge dampers on a 400 kV power line near Castle Combe, England.

Passive vibration control could be obtained adding damping elements (viscous or frictional) to increase structural damping. The examples of this approach to vibration control in a large scale bearing constructions and metal machining are shown in Figs. 5 and 6. In both presented cases, vibration control of the system with initial mass [**m**], stiffness [**k**] and damping [**c**] could be analysed from the following equation

$$[\mathbf{m}]\{\ddot{\mathbf{x}}\} + ([\mathbf{k}] + [\Delta\mathbf{k}])\{\mathbf{x}\} + ([\mathbf{c}] + [\Delta\mathbf{k}])\{\mathbf{x}\} = \{\mathbf{F}(t)\} \qquad (2)$$

where [Δ**k**] and [Δ**c**] represents influence of added damping elements.

Fig. 5. A viscous damper for vibration reduction in a bearing constructions.

Fig. 6. Passive vibration control in metal machining equipment.

3 Active Vibration Damping

Insufficiency of passive damping systems are resolved by development and implementation of active damping systems. There are almost two approaches to active vibration damping: active changing system damping according to induced vibration state and enhancement of classic dynamic vibration absorber by adding appropriate actuators to induce its forced vibrations.

Active vibration control by active damping is used now in suspended bridge construction to increase the structural damping in case of increased vibration (Fig. 7). Bridge damping is changed by increasing tension of some suspension cables. Cable tension increases bridge stiffness and its damping. The active damping of bridges is also a solution to vibration problems which could occurre during the construction process, when the various parts are not yet connected resulting in increased flexibility and increased possibility of damage under increased vibrations.

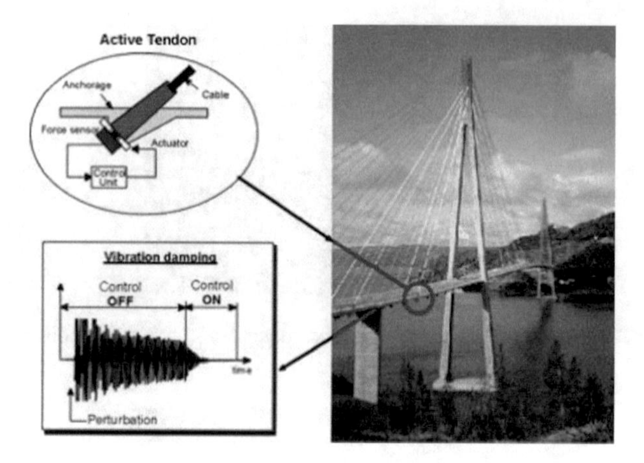

Fig. 7. Active tendon control reduces bridge vibration

Implementation of dynamic damper absorber to produce semi-active vibration damping system is used in Danube City Tower in Vienna (Fig. 8). A 300 ton mass pendulum balances the vibrations. The semi-active dampers continuously adapt to the occurring load cases and respond with an adapted force.

Fig. 8. (a) Danube City Tower in Vienna, Austria; (b) MR-SVA of Danube City Tower with two controlled MR dampers and (c) two independent real-time controllers.

Semi-active vibration damping is used in some vehicles. Typically engines use passive rubber isolators that support the engine, limiting engine movement and isolating the engine from the chassis. An example of this is the Porsche 911 Dynamic Engine Mount System. In this system, engine mounts employ a magnetizable fluid and an electrically generated magnetic field that provide variable stiffness. The internal excitation of the gear transmission system is periodic. The vibration energy generated is mainly concentrated on the gear mesh frequency and its harmonic frequencies. As a new vibration control method is developed in recent years, the vibration active control which is based on the vibration signal obtained by the sensor generates a vibration signal to neutralize the harmful vibration (Fig. 9).

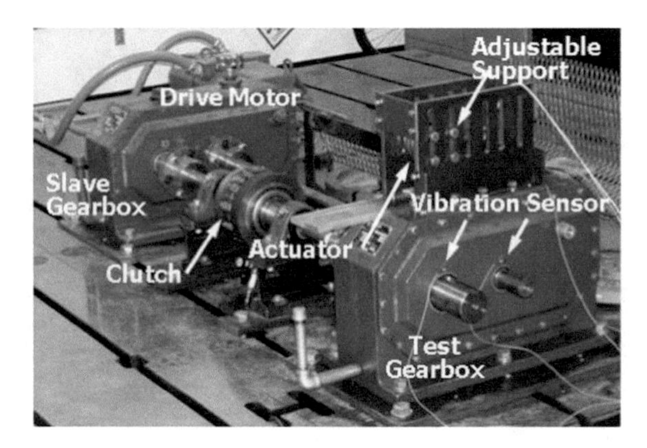

Fig. 9. Experimental setup of active vibration control of the gearbox system

One of the most significant structural dynamics problems of large gantry cranes are elastic vibrations in trolley travel direction. They put additional mechanical stresses on a crane construction and reduce crane operation performance. As these vibrations are mostly excited by trolley acceleration forces, they can be taken into account in the trolley motion control system. Figure 10 presents a robust control-based approach for active damping of the gantry crane elastic structural vibrations.

Fig. 10. Active vibration control in gantry cranes.

High vibration often occurs at a top of wind turbine where nacelle with heavy equipment is placed. In a classic configuration of wind turbine damping system, an auxiliary mass hung below the nacelle of a wind turbine supported by dampers or friction plates. Modern solution includes sensors, actuators, amplifier and controller, what enables to detect characteristics of increased vibration and generate vibrations of damping system which will reduce vibration of a wind turbine in a safe margins (Fig. 11).

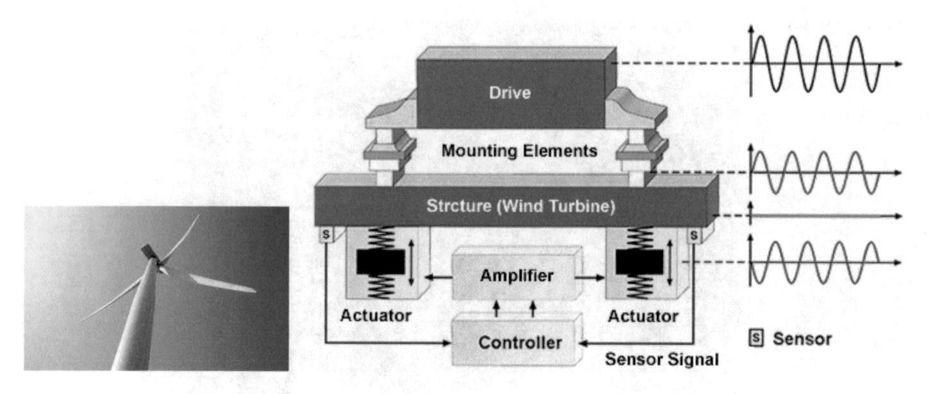

Fig. 11. The principle of active vibration damping in wind turbines.

In general, active vibration control (damping) system is a mechatronic system. It is in practice closed loop feedback system consisting of sensors, controllers and actuators (Fig. 12). In most cases, the sensor is a piezoelectric accelerometer that senses the excitation of the passive engine mount. The acceleration signal is then processed by the controller. The controller generates a canceling signal that is fed to a power amplifier. The amplifier converts the controller's low-voltage signal to an actuator current. The actuators in most cases are electromagnetic transducers. The force generated by the actuator cancels the primary disturbance signal resulting in near zero vibration of the system.

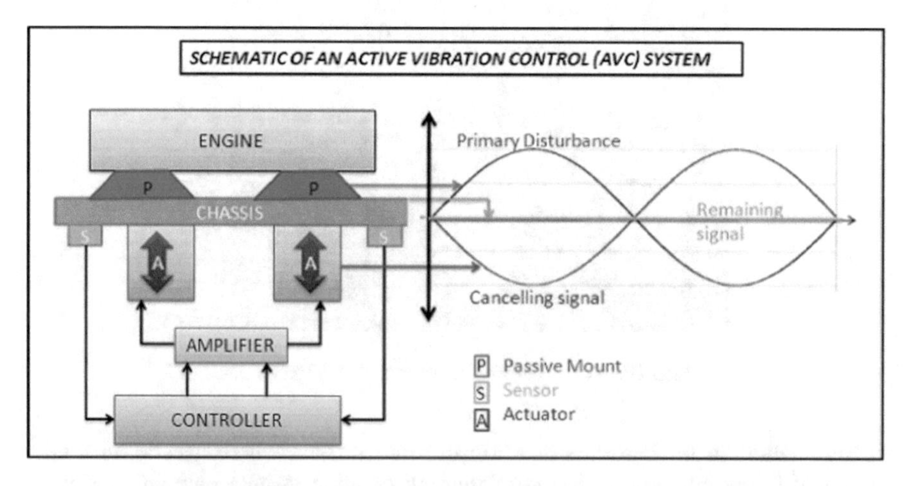

Fig. 12. Working principal of a active vibration control system.

Controller in a damping system almost generate canceling sygnal which induce actuators to oscillate in a counter-phase to the primary disturbance. Governing equation used for active vibration control could be:

$$[\mathbf{m}]\{\ddot{\mathbf{x}}\} + ([\mathbf{k}] + [\Delta\mathbf{k}(t)])\{\mathbf{x}\} + ([\mathbf{c}] + [\Delta\mathbf{c}(t)])\{\mathbf{x}\} = \{\mathbf{F}(t)\} + \{\mathbf{F}_{actuator}(t)\} \quad (3)$$

where $[\Delta\mathbf{k}(t)]$, $[\Delta\mathbf{c}(t)]$ and F_{actuator} represents time response determined by the controller of the active damping system to the detected of increasing vibration and primary disturbance.

4 Conclusion

A vibrations follows work of a many machines and structures. They presence in some cases may lead to undesirable condition works or even to damage of the systems. The classical approach to vibration reduction and control are dynamic vibration absorber and/or classical damping elements of constant characteristics. This approach could resolve problems in a limited frequency range. Modern approach to vibration reduction and control includes senzors for vibration stage detectors and additional mass with actuators to produce vibrations which reduce vibration of the main system or reduce vibration by changing internal damping characteristics. Many of these control system produce only axial vibration and reduce vibration in only one direction. In a more complex systems, as mechatronics systems, different random excitation and vibration in a different directions are possible. Because of that for modern vibration control should be putted attention on the development of devices which could combine both vibration reduction by forced motion and by changing internal damping characteristics simultaneously.

References

1. Beards, C.F.: Structural Vibration Analysis and Damping, New York-Toronto (1996)
2. Doleček, V., Voloder, A., Isić, S.: Vibracije (2009)
3. Hermanrud, O.C.: Active and passive damping systems for vibration control of metal machining equipment, Norwegian University of Science and Technology. Master thesis (2017)
4. Wang, H., Zhang, F., Li, H., Sun, W., Luo, S.: Experimental Analysis of an Active Vibration Frequency Control in Gearbox, College of Mechanical Engineering and Automation, Huaqiao University, Xiamen, Fujian 361021, China
5. Golovin, I., Palis, S.: Robust control for active damping of elastic gantry crane vibrations. Mech. Syst. Signal Process. **121**(15), 264–278 (2019)
6. Heysami, A.: Types of Dampers and their Seismic Performance During an Earthquake. Int. Res. J. Environ. Sci. **10**(1), 1002–1015. ISSN: 0973-4929
7. Weber, F., Distl, H., Fischer, S., Braun, C.: MR damper controlled vibration absorber for enhanced mitigation of harmonic vibrations. J. Actuators **5**(4) (2016). ISSN 2076-0825

8. https://cecas.clemson.edu/cvel/auto/systems/active_vibration_control.html
9. https://en.wikipedia.org/wiki/Taipei_101
10. http://www.esa.int/spaceinimages/Images/2001/04/Active_tendon_control_reduces_bridge_vibration

Testing of Tribological Properties of ADI

Kovač Pavel[1(✉)], Savković Borislav[1], Ješić Dušan[2], and Sarjanovic Dražen[3]

[1] Faculty of Technical Science, University of Novi Sad, Trg D Obradovica 6, 21000 Novi Sad, Serbia
pkovac@uns.ac.rs
[2] International Technology Management Academy, 21000 Novi Sad, Serbia
[3] Sara-Mont Doo, 110000 Beograd, Serbia

Abstract. This experimental work consists of testing of isothermally improved specimens for wear process without lubrication. In the first part investigated was tribological properties on tribometer pin on disc made produced from ADI. Tested was loss of length and loss of mass of pin for different increasing number of revolutions. Both wear parameters were increasing with number of revolutions. In the second part this was accomplished by conducting face milling tests for a range of machining conditions. More specifically, the effects of cutting speed on tool life was tested. In the study was found that the machinability of ADI, in terms of tool life, degrades as the cutting speed increases. Tool wear progression in time cubic polynomial fit equations for workpiece ADI material. The Tool life equation developed by Taylor to quantify the relationship between cutting speed and tool life is determined as well.

Keywords: Tribology tests · Pin on disc · Wear · Tool life · ADI

1 Introduction

Ductile iron with austenitic-ferritic matrix (ADI), due to the specific heat treatment, shows an advantageous combination of engineering properties. The superior wear resistance and strength-to-weight ratio, together with high stress resistance and good ductility, make such material suitable for cost- and weight-efficient automotive structural applications. In automotive applications, these weight savings can be transferred to consumers in the form of lighter vehicles and better fuel efficiency, which reduces CO_2 emissions. The uses of ADI have expanded to a large range of applications replacing metals such as steel, aluminum, and other light weight alloys in the agricultural and heavy vehicle industries. Today ADI has expanded to applications in agricultural equipment, construction equipment, gears or powertrain, heavy truck or trailers, light vehicles and buses, mining or forestry equipment, railway equipment, farm and oilfield machinery, conveyor and tooling equipment, defense, energy generation, and even sporting goods [1].

However, the difficulties experienced when machining ADI remain one of the major restricting factors in the growth of the market for ADI. In this paper, results of milling experimental trials carried out on ADI are discussed. Tests were performed

I. Karabegović (Ed.): NT 2019, LNNS 76, pp. 105–114, 2020.
https://doi.org/10.1007/978-3-030-18072-0_11

using uncoated and titanium aluminum nitride or titanium carbo-nitride coated cutting inserts obtained from both recycled and traditionally prepared tungsten carbide cobalt (WC-Co)-based materials [1].

In the work [2], a series of test were conducted and the effects of austempering temperature and times onto the material properties was investigated. During the experiments cutting forces, flank wear and surface roughness values were measured throughout the tool life and the machining performance of ADI having different structures were compared.

Austempered Ductile Iron is characterised by improved mechanical properties but low machinability compared to conventional ductile iron materials and steels of similar strengths. Paper [3] describes material and machining investigations as well as cutting simulations to reveal the wear mechanisms being responsible for the low machinability of ADI.

The main advantages of ADI are seen in its excellent machinability, increased processing speed, increased vibration damping, near net shape casting or product weight reduction. Mechanical properties of ADI depend strongly on the morphology of graphite particles [4]. The Performance of PVD Coated Grade in Milling of ADI 800 tested in [5]. The strength of ADI material is up to two times that of the nodular iron of standard quality, with the same value of toughness and ductility. Hardness and wear resistance can be increased by additions of titanium and boron in the hardfacings [6].

Previous studies have shown that alloying elements such as Cu and Ni influence the austempering temperature, initiation time and completion of the austempering reaction, which provide a larger processing window and ease control of the reaction [7]. The size of the heat treatment processing window is one of the many factors of ADI that are affected by carbon and different alloying elements as well as the segregation of those elements.

Dry and minimum quantity lubrication drilling of novel austempered ductile iron (ADI) for automotive industry applications is in study [8].

Austempering temperature and time, which have the largest effect on the final microstructure, are chosen based on the desired mechanical properties. At lower austempering temperatures, the diffusion rate of carbon is decreased and the driving force for the formation of ferrite is increased [4]. At higher austempering temperatures, the increased rate of diffusion of carbon creates more stabilized austenite. Therefore, higher austempering temperatures (350 to 450 °C) lead to lower strength and hardness but higher elongation, toughness, and in many cases better fatigue characteristics. While, lower austempering temperature (250 to 350 °C) lead to higher strength, hardness and abrasion resistance but lower elongation and toughness [4]. The austempering time is critical because it determines where in the processing window the casting is being produced. Shorter austempering times drive the operations toward a martensitic reaction, when cooled to room temperature, thus resulting in an increase in strength and hardness but a significant decrease in ductility. Very short austempering time can create a hard but brittle casting with poor mechanical properties. Longer austempering times drive the operation towards an upper bainitic reaction, which results in an increase in ductility and a decrease in strength [4].

The influence of austempering temperature on tribological properties of austem-peredpearlitic ductile iron (ADI) was investigated. The austempered samples were subjected to abrasion and adhesion wear resistance tests using steel and ductile iron plates [9].

2 Experimental Work

The isothermal treatment of the ductile iron was carried out according to Fig. 1, and consists of austenitization at 900 °C for 90 min, rapid cooling to the temperature of the isothermal transformation (300 °C), holding at that temperature (120 min), and then cooling at room temperature.

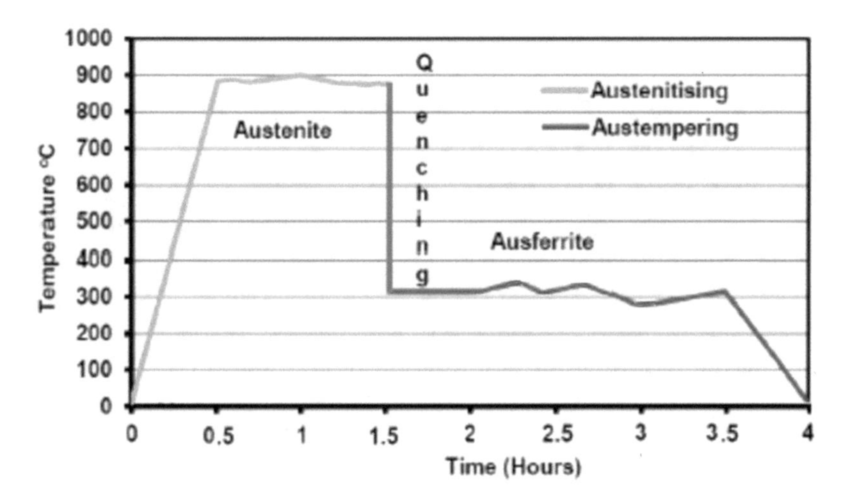

Fig. 1. Ductile iron heat treatment process [4]

Experimental work consists of testing of isothermally improved specimens for wear process without lubrication. ADI microstructure consists of an ausferrite microstructure that includes ferrite needles within an austenite background with spheroidal graphite nodules.

The test is carried out on a perlite ductile cast iron (EN 700) EN-GJS-700-2. It can be seen from Fig. 2 that the material has an extremely perlite structure. In the first part of the test, the wear testing was carried out in the friction pair plate of ADI and pin of ADI. Plate dimensions was Ø 100 × 5 mm, grinded surface (Table 1).

Table 1. The chemical composition of as-cast ductile iron (wt%)

C	Si	Mn	Cu	Mg	P	S
3.76	2.35	0.51	1.43	0.066	0.02	0.004

The mechanical properties of as-cast material are given in Table 2. The pearlitic matrix had the most significant influence on mechanical properties resulting in higher strength of ductile iron.

Table 2. Mechanical properties of as-cast material

Rm (MPa)	$Rp_{0.2\ \%}$ (MPa)	A_5 (%)	K_0 (J)	HV10 EN	EN 1563:1997
768	516	4.8	21.3	274	EN-GJS-700-2

The microstructure consists mainly of pearlite (78 vol.%), ferrite (10 vol.%) and graphite particles (12 vol.%). The applied heat treatments of the ductile iron samples lead to the spheroidization of the majority of graphite particles (more than 85%) with a nodule size ranging from 45 to 55 μm.

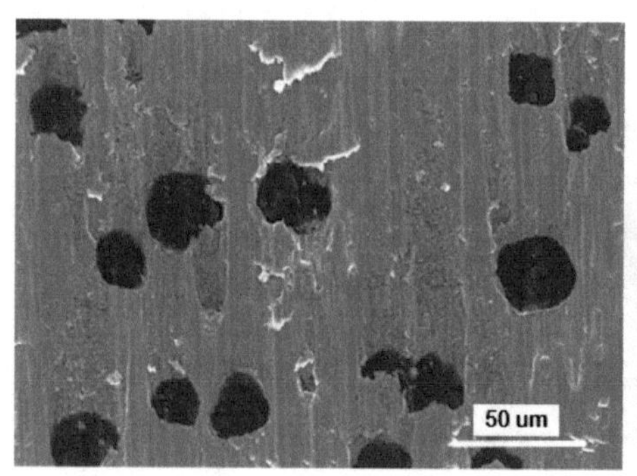

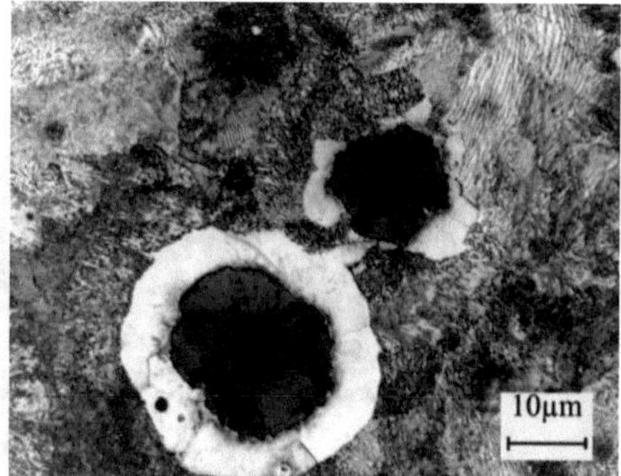

Fig. 2. Microstructure of ductile iron EN-GJS-700-2, before heat treatment

The adhesion and abrasive wear tests were carried out according to ASTN-D 3389 using Taber Abrazer model 503 equipment.

During the testing on the device, the speed was changed from 1000 rpm to 5000 rpm in five degrees of change duration 10 min and with a pressing force 10 N, Fig. 3. After tribology testing, specimen recordings of the trace wear on the pin were carried out.

The measurement of the length of the pin test specimens due to wear was performed using a micrometer. Testing the mass of the pin specimens was done using an analytical balance.

The main objective of this study was to assess the machinability of ADI, in order to establish cost-effective processing conditions for the machining of ADI. This was accomplished by conducting face milling studies for a range of machining conditions. More specifically, the effects of cutting speed on tool life. The face milling operations

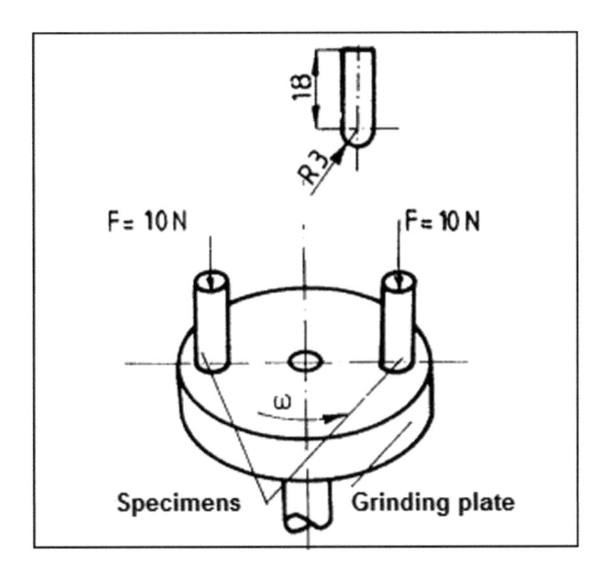

Fig. 3. Schematic presentation of wear tests

were carried out on 80 mm thick plates using the following cutting conditions with a tooth, 80 mm face milling tool: constant chip load of 0.1 mm/tooth and depth of cut of 1 mm, with varying cutting speed, 235, 290 and 355 m/min. A Taylor tool life model was also developed, using the tree cutting speeds wear progression curves.

The cutting tool used was a single insert in an indexable holder. The holder was an indexable face milling tool holder (125 mm diameter) and the insert was a TiN coated carbide insert, 0° rake, 75° entrance angle.

Tool life is widely considered the most important machinability metric. Tool life was defined as maximum flank wear penetration (VB_{max}) by uniform wear of 0.3 mm (average wear) or localized wear of 0.5 mm (on any individual tooth). Average tool wear was measured in cutting time intervals. In some cases, tool life was limited by the occurrence of catastrophic tool failure, rather than progressive tool wear.

3 Results and Discussion

The results of the tribology tests are presented as a graph display depending on the of wear trace, are expressed in terms of the decrease in the pin length Δl loss of the pin mass Δm and the number of revolutions during treatment.

In Figs. 4 and 5 is shown comparative diagrams of the wear process of the contact pairs of austempered ductile iron pin on ADI plate.

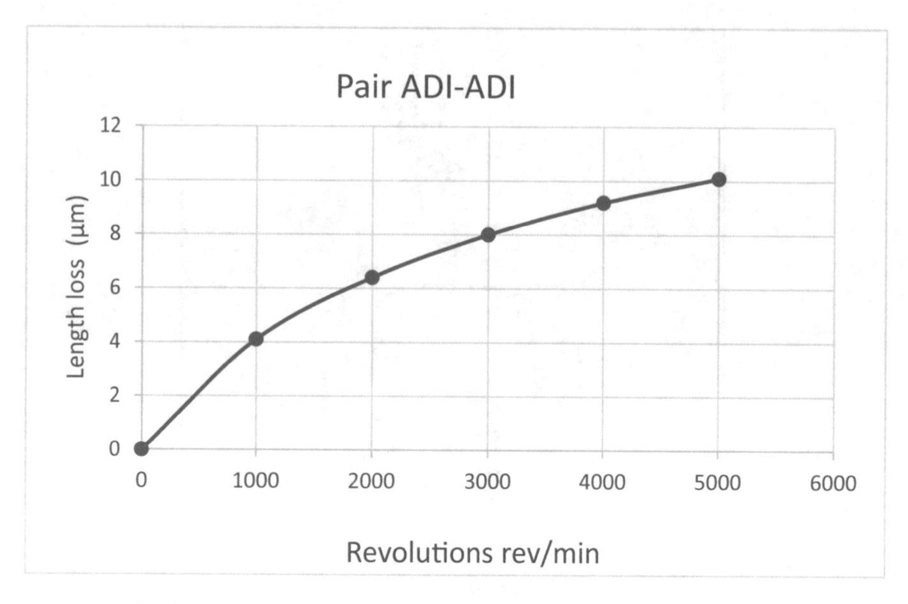

Fig. 4. Graph of wear process expressed as a loss of length Δl in μm

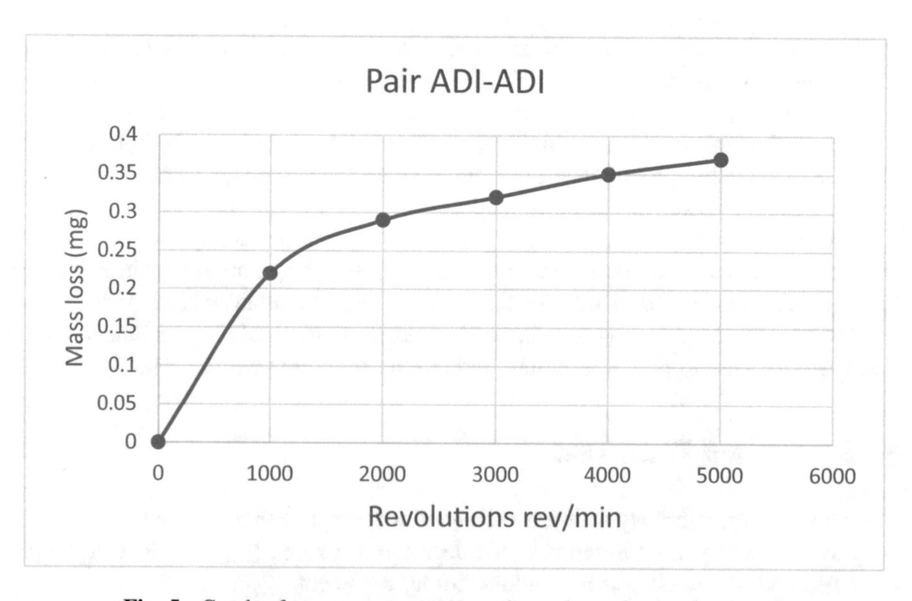

Fig. 5. Graph of wear process expressed as a loss of mass Δm in mg

It is clear from these figures that the increasing number of revolutions leads to the increase of length loss (Fig. 4) and mass loss (Fig. 5). The length and mass loss are stabilised at 5000 rev/min.

After tribology testing of the specimens, recording of the trace of the wear was carried out, Fig. 6 shows the traces of abrasion wear process on the ADI specimen surface.

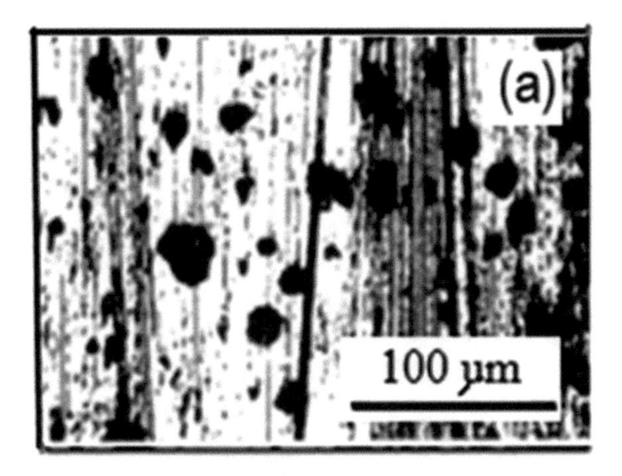

Fig. 6. Traces of wear on the ADI surface.

In metal machining, there are numerous types of tool wear, such as notch wear, nose radius wear, thermal and mechanical cracking, edge buildup, plastic deformation, edge chipping or frittering, chip hammering and tool fracture or breakage. The two most prevalent types of wear are flank and rake wear (forming a crater). The average wear land width is known as VB and the maximum wear land width is VB_{max}. The flank wear criterion specified for the width of the flank wear, which typically corresponds to VB_{max}. A commonly used flank wear criterion value is 0.3 mm for uniform wear. The typical tool life progression curve reported, which consist of three stages in Fig. 7. Is shown tool wear progressions curves for ADI at tree various cutting speeds, maximum wear values (VB_{max}) as a function of effective cutting time ($a = 1$ mm, $f = 0.1$ mm/tooth are constant).

In the study was found that the machinability of ADI, in terms of tool life, degrades as the cutting speed increases. Increasing the cutting speed leads to decreasing tool life by rapid wear development. The rapid wear development, can be attributed to the effect that cutting speed has on the temperature at the cutting interface. Large tool wear rates, thus short tool life and high cutting forces, indicate high temperatures at the cutting interface.

The tool wear curve for cutting speed followed the expected behavior for the most cost-effective machining operations in that the profile consists of a rapid increase in the wear rate at the beginning of cutting that is followed by a linear wear rate until rapid tool wear beyond a threshold cutting time.

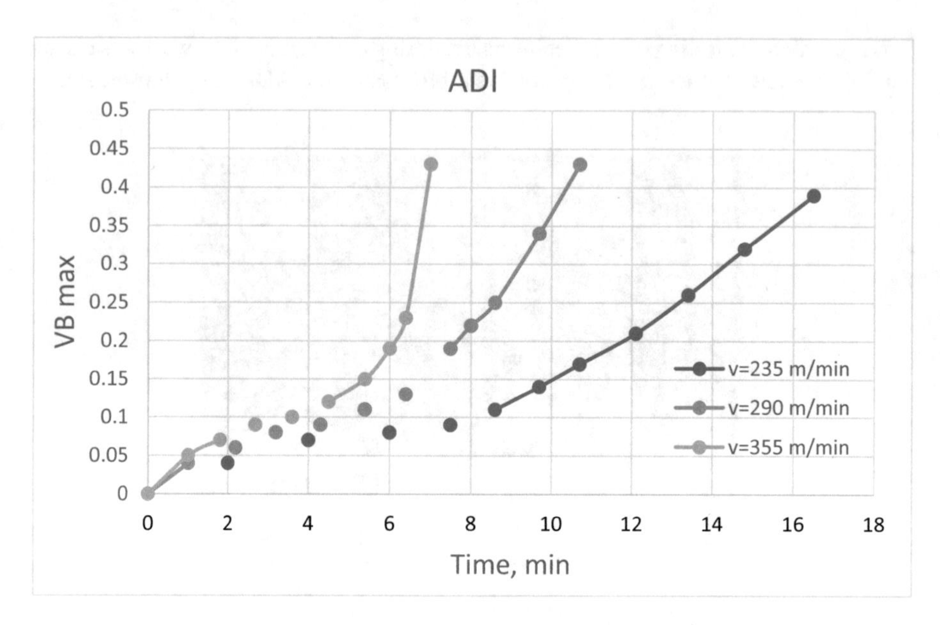

Fig. 7. Tool wear curves for 3 cutting speeds

Tool wear progression cubic polynomial fit equations for workpiece ADImaterialEN-GJS-700-2, at various cutting speeds, maximum wear values (VB_{max}) as a function of effective cutting time (a = 1 mm, f = 0.1 mm/tooth) are:

$$v = 355\,\text{m/min} \quad y = 0.0049x^3 - 0.0411x^2 + 0.1074x - 0.009 \quad R^2 = 0.9836 \quad (1)$$

$$v = 290\,\text{m/min} \quad y = 9\text{E-}05x^3 - 0.001x^2 + 0.0154x + 0.0065 \quad R^2 = 0.9956 \quad (2)$$

$$v = 235\,\text{m/min} \quad y = 0.0007x^3 - 0.0076x^2 + 0.0421x + 0.0034 \quad R^2 = 0.9966 \quad (3)$$

Cutting speed expectedly had a major effect on tool wear rate, with higher cutting speeds yielding faster wear rates. A cubic polynomial fit was applied to the tool wear data to represent the tool wear progression. The fitted curves are accurate representations of the tool wear progression for cutting speeds 235, 290, and 355 m/min due to the high correlation coefficients R^2 values, as shown above.

The equation developed by Taylor to quantify the relationship between cutting speed and tool life was determined for the ADI. The tool life (T) tests above were used to determine the empirical constants, C and exponent in equation. In order to derive the constants, the relationship between tool life and cutting speed were plotted on a log-log graph, shown in Fig. 8, and Eqs. (4) and (5) was developed. From the linear fit of the data the constant n was extracted from the slope and the constant C was extracted from

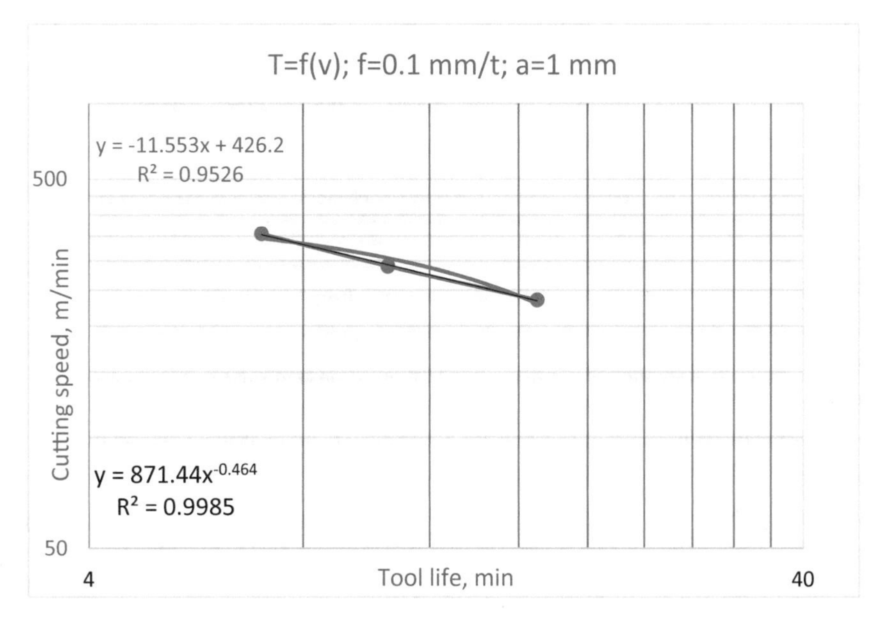

Fig. 8. Tool life versus cutting speed

the y-intercept, of the linear regression equation or with higher correlation coefficient R^2 in the power form:

$$T = -11.553v + 426.2 \qquad R^2 = 0.9526 \tag{4}$$

$$T = 871.44v^{-0.464} \qquad R^2 = 0.9985 \tag{5}$$

The results of the fit are provided in equation the constant C in equation it is simply an empirical constant relating the cutting speed and tool life (T) for a particular work and tool material combination. The constant on speed is the slope of the generally linear curve.

4 Conclusion

Determination of tribological properties for the austemperedpearlitic ductile iron has been studied. The achieved results can be summarized as follows.

Formation of spheroidized graphite nodules during austempering treatments decreases the wear loss due to lubrication effect of graphite.

Tested was loss of length and loss of mass of pin for different increasing number of revolutions. Both wear parameters were increasing with number of revolutions and constant increasing was achieved foe 5000 no of revolutions.

Tool wear progression cubic polynomial fit equations for work ADI material for tree cutting speeds.

The tool life tests above were used to determine the empirical constants in Taylors equation.

Acknowledgements. This paper presents a part of the research of the Research project number TR 35015.

References

1. Priarone, P.C., Robiglio, M., Settiner, L.: Milling of austempered ductile iron (ADI) with recycled carbide tools. Int. J. Adv. Manuf. Technol. **82**(1–4), 501–507 (2016)
2. Cakir, C., Bayram, A., Isik, Y., Salar, B.: The effects of austempering temperature and time onto the machinability of austempered ductile iron. Mater. Sci. Eng. A **407**(1–2), 147–153 (2015)
3. Klocke, F., Klöpper, C., Lung, D., Essig, C.: Fundamental wear mechanisms when machining austempered ductile iron (ADI). CIRP Ann. **56**(1), 73–76 (2007)
4. Sadik, I., Myrtveit, T.: The performance of PVD coated grade in milling of ADI 800. Mater. Metallur. Eng. **3**(5), 246–249 (2009)
5. Golubovic, D., Kovač, P.P., Savkovic, B., Jesić, D., Gostimirović, M.: Testing the tribological characteristics of nodular iron austempered by conventional and an isothermal procedure. Mater. Technol. **48**(2), 293–298 (2014)
6. Kaptanoglu, M., Eroglu, M.: Microstructure and wear of iron-based hard facings reinforced with in-situ synthesized TiB2 particles. Kovove Mater. Metallic Materials **55**, 123–131 (2017). https://doi.org/10.4149/km.2.123
7. Rajnovic, D., Eric, O., Sidjanin, L.: The standard processing window of alloyed ADI materials. Kovove Mater. Metallic Materials 50, 199–208 (2012). https://doi.org/10.4149/km_2012_3_199
8. Meena, A., El Mansori, M.: Study of dry and minimum quantity lubrication drilling of novel austempered ductile iron (ADI) for automotive applications. Wear **271**(9–10), 2412–2416 (2011)
9. Ješic, D., Kovač, P., Plavšic, M., Šooš, Ľ., Sarjanovic, D.: Wear resistance of austempered pearlitic ductile iron. Kovove Mater. **56**, 415–418 (2018). https://doi.org/10.4149/km_6_415
10. Harding, R.A.: The production, properties and automotive applications of austempered ductile iron. Kovove Mater. **45**, 1–16 (2007)

Influence of Strip Cooling Rate on Lüders Bands Appearance During Subsequent Cold Deformation

Rešković Stoja[1(✉)], Brlić Tin[1], Jandrlić Ivan[1], and Vodopivec Franc[2]

[1] Faculty of Metallurgy, University of Zagreb, Aleja narodnih heroja 3, 44000 Sisak, Croatia
reskovic@simet.hr
[2] Institute of Metals and Technology, 1000 Ljubljana, Slovenia

Abstract. This paper presents the results of study of Lüders bands appearance on hot rolled strip during cold deformation. The research was carried out on niobium microalloyed steel. After the thermomechanical treatment the strip was cooled at different rates. Samples taken from hot rolled strip were tested by stretching to fracture with simultaneous application of the methods thermography and digital image correlation. The analysis of measurement results was performed with the software packages IRBIS 3 professional and MatchID. Significant differences were found in the samples tested at different cooling rates. In the samples cooled at a lower cooling rate, the appearance of Lüders bands in the elastoplastic area was determined. The appearance of Lüders bands was not observed in the samples cooled at a higher rates.

Keywords: Lüders band · Microalloyed steel · Hot rolled strip · Cooling rate

1 Introduction

In the last few years the occurrence and propagation of the Lüders bands [1, 2] have been intensively investigated. Lüders bands were observed and studied in the fifties of the last century and connected with retarding and accumulation of dislocations on obstacles [3, 4]. At that time it could not fully explain the mechanisms of Lüders bands formation and propagation. This phenomenon is again intensively investigated by the development of modern methods such as thermography and visioplasticity with digital image correlation (DIC) [5, 6]. The appearance of Lüders bands, Lüders bands propagation during deformation, and the stresses and deformations in, on and beyond the Lüders band front can be determined using thermography and digital image correlation simultaneously with a static tensile test [7].

Numerous studies of the Lüders bands occurrence and propagation have been carried out on various metal materials under different test conditions [8–10]. The influence of strain rate [9, 10], microstructure and dislocation density [3, 8] is studied. However, there is still no generally accepted conclusion of the cause of the Lüders bands formation and which factors influencing Lüders bands formation and propagation through the deformation zone.

I. Karabegović (Ed.): NT 2019, LNNS 76, pp. 115–121, 2020.
https://doi.org/10.1007/978-3-030-18072-0_12

In this paper are presented the results of research of Lüders bands appearance using thermography and digital image correlation (DIC) on niobium microalloyed steel. The studies were carried out on hot-rolled strip which is cooled at a different cooling rates after thermomechanical treatment.

2 Experimental

Studies were carried out on microalloyed steel with 0.035% Nb. Chemical composition of steel is given in Table 1. Strip with thickness of 3 mm is rolled from steel. Rolling parameters of strip have shown in Table 2. After the thermomechanical treatment, Table 2, the strip was air-cooled (A) and water-cooled up to 510 °C and then air-cooled (B). The measurements were performed by contact platinum-rhodium pyrometry. Measured values have shown in Table 3.

Table 1. Chemical composition of steels [wt.%]

	C	Mn	Si	Al	S	P	Nb	N
%	0.09	0.75	0.05	0.020	0.014	0.018	0.035	0.0081

Table 2. Rolling parameters of strip 50 × 3.0 mm

Rolling parameters	Number of rolls						
	1	2	3	4	5	6	7
T deformation, °C	990	968	947	942	912	858	820
h_0, mm	20.20	17.11	13.28	10.10	7.05	5.28	3.70
h_x, mm	17.11	13.28	10.10	7.05	5.28	3.70	3.04
Reduction in the roll Δh, mm	3.09	3.63	3.18	3.05	1.77	1.58	0.66
Total reduction ε, %	15.30	34.26	50.00	65.10	73.39	81.86	84.95

Table 3. Cooling of strip

Cooling time, s		0	10	25	35	50
Temperature, °C	A	820		690	650	520
	B	820	680	510	–	–

After cooling, samples were taken for structural testing. The samples were prepared for metallography and tested on the Olympus DP70 optical microscope.

Research of the Lüders bands occurrence and propagation was carried out by stretching to the fracture on the static tensile machine EU 40mod 400 kN, at stretching rate of 20 mm/min (strain rate 0.007 s^{-1}). The tests were performed simultaneously with thermography and digital image correlation (DIC). Samples were prepared by grinding the oxide layer and applying a black matte coating with emissivity factor of

0.95. White speckle patterns were applied on the black background for digital image correlation (DIC). Samples were recorded with infrared camera VarioCAM M82910 with temperature sensitivity of 80 mK and digital camera Panasonic HDC-SD9 with 2.1 MP resolution. The results of thermographic measurements were analyzed by Irbis professional software package and digital image correlation measurements were analyzed with MatchID software package.

3 Results and Discussion

The results of the metallographic tests are shown in Fig. 1. The both samples from strip have a fine-grained ferrite-pearlite microstructure. The strip cooled at a lower average cooling rate, sample A (average 6 °C/s), Fig. 2A, has a homogeneous fine-grained microstructure. Grain size is below 10 μm. Sample B, the strip cooled at a higher average cooling rate (average 12.4 °C/s), has a microstructure with a slightly higher percentage of pearlite. Grain size is 10–20 μm.

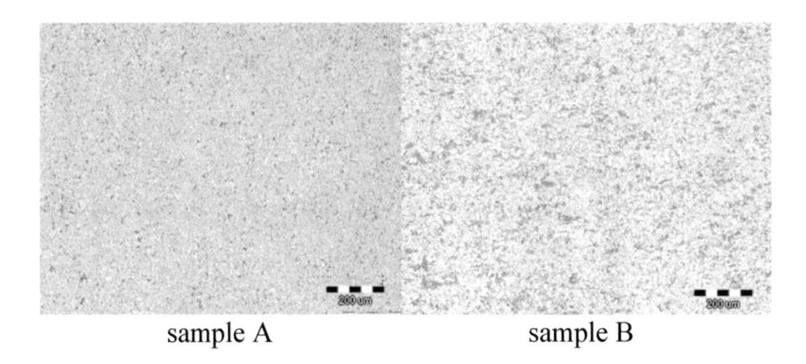

Fig. 1. The microstructure of strip with: (A) average cooling rate 6 °C/s, (B) average cooling rate 12.4 °C/s

It is well-known that niobium microalloyed steels have a fine-grained ferrite-pearlite microstructure which is a consequence of the performed thermomechanical treatment. In the final phase of thermomechanical treatment, niobium precipitates are excreted, which have a strong influence on recrystallization. After the phase transformation a fine-grained microstructure is obtained as in Fig. 1 because of the high density of dislocation. The cooling rate has a significant effect on the grain size. At this stage of thermomechanical treatment $\gamma \rightarrow \alpha$ phase transformation takes place. As the cooling rate is higher, the greater proportion of carbide phases is in the structure.

The mechanical properties of the strip were examined. The obtained results are shown in Table 4. As expected, sample B has a higher tensile strength and hardness but lower elongation. It was found during mechanical properties testing that in homogeneous, fine-grained microstructure of sample A, inhomogeneous deformations occur at the start of plastic material flow, Fig. 3. Sample B do not have pronounced yield strength.

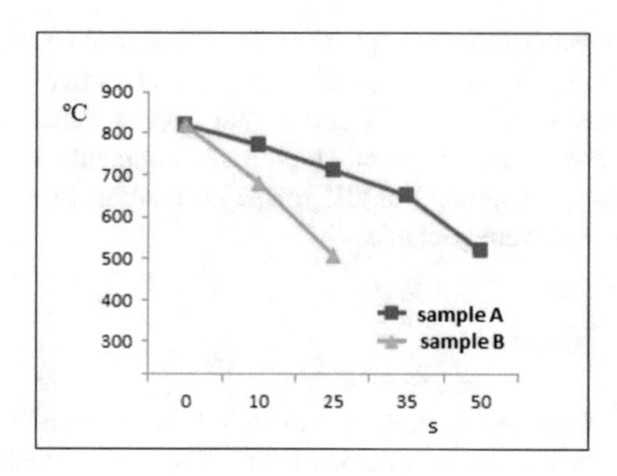

Fig. 2. Cooling rate of strip

Table 4. Mechanical properties of the strip

	R_p, MPa	R_m, MPa	A_5, %	HB
A	436	505	32.2	160
B	457	621	18.9	218

Inhomogeneous deformations in the sample A are associated with the occurrence of the Lüders bands [6], which was confirmed by thermography and digital image correlation on Figs. 4 and 5.

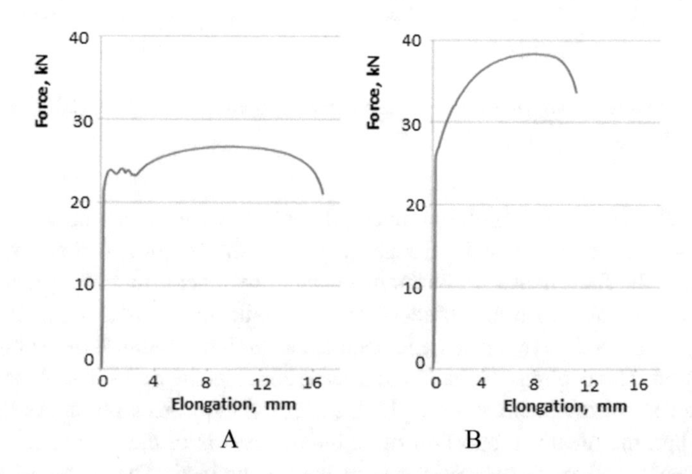

Fig. 3. Force-elongation diagrams

The captured thermograms, Fig. 4, show that in the sample A with inhomogeneous deformation, Fig. 3, Lüders band occurs. After reaching R_p, Lüders band formation begins at one end of the sample. From the point R_p to the start of inhomogeneous deformations, the Lüders band is formed at an angle approximately equal to the angle 45°.

After that, the Lüders band propagates through the deformation zone to the other end of the deformation zone. In this area inhomogeneous deformations are visible on the force-elongation diagram. There are no temperature changes in front of the Lüders band front. In this field elastic deformation takes place. Behind the Lüders front, there is a certain temperature increase that increases as Lüders band propagates through the deformation zone. This is associated with redistribution of dislocations. In sample B, with no pronounced yield strength, there are no Lüders bands. Stresses are concentrated in the middle of the sample and spreads through the deformation zone.

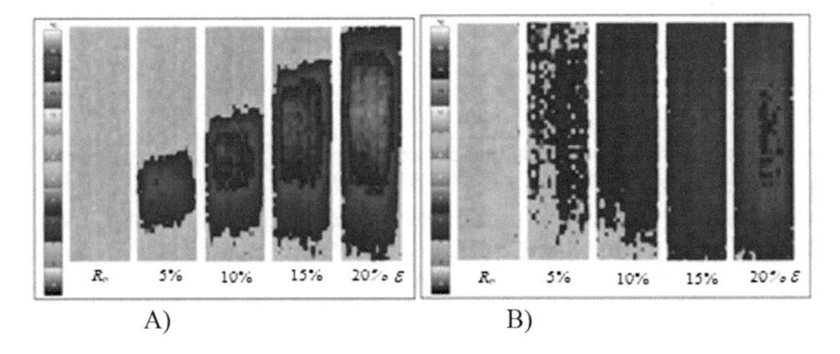

Fig. 4. Thermograms at the start of plastic flow of material

The results obtained by thermographic analysis were also confirmed with digital image correlation (DIC), Fig. 5. On the sample A, the occurrence and propagation of the Lüders band was observed while on the sample B, Lüders band was not observed. By comparing the recorded maps, Fig. 5, it can be seen that increasing of total deformation increases the amount of deformation behind the Lüders band front. Sample B shows considerably lower deformation amounts than sample A.

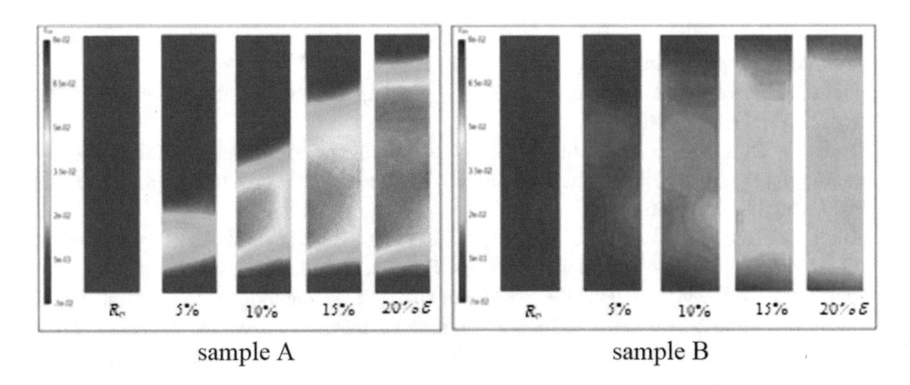

Fig. 5. Deformation maps at the start of plastic flow of material

The results of the recent research [7] of niobium micro-alloyed steel behaviours during cold deformation show that Lüders bands do not occur in inhomogeneous ferrite-pearlite microstructures with a significant difference in grain size. Comparing the microstructure of Fig. 1 with the research in [11], it results that the microstructure is not a reason for the different behaviour of samples A and B at the start of the plastic material flow. It can be concluded from Table 4 that the internal stresses in sample B are higher due to higher cooling rate. This indicates higher density of less mobile dislocations. In niobium micro-alloyed steels at the end of thermomechanical treatment are excreted deformation induced precipitates that interact with dislocations [2]. It is a realistic assumption that the different behaviour of samples A and B at the start of the plastic flow is related to the dislocations that interact with the precipitates, i.e. with dislocations that are less mobile.

4 Conclusion

The conducted research has shown that the cooling rate of strip after the hot rolling has a significant effect on the start of the plastic material flow during subsequent cold deformation. At the start of the plastic flow during cold deformation, Lüders bands appear at cooling rate of 6 °C/s. At higher cooling rate 12.4 °C/s, Lüders bands do not appear. It was not found differences in microstructure that could be associated with differences in behaviour of differently cooled samples. The appearance of the Lüders bands can be related to the differences in the density of dislocations and their interaction with the precipitates. These influences will be studied in detail by further research using transmission electron microscopy.

Acknowledgment. This work has been fully supported by Croatian Science Foundation under the project IP-2016-06-1270.

References

1. Schwab, R., Ruff, V.: On the nature of the yield point phenomenon. Acta Mater. **61**(5), 1798–1808 (2013)
2. Rešković, S., Jandrlić, I.: Influence of niobium on the beginning of the plastic flow of material during cold deformation. Sci. World J. **3**(5), 5 (2013)
3. Cottrell, H., Bilby, B.A.: Dislocation theory of yielding and strain ageing of iron. Proc. Phys. Soc. A **62**, 49–62 (1949)
4. Brlić, T., Rešković, S., Jandrlić, I.: Study of inhomogeneous plastic deformation using thermography. In: 16th International Scientific Conference on Production Engineering - CIM 2017, Zagreb, pp. 65–68 (2017)
5. Kutin, M., Ristić, S., Burzić, Z., Puharić, M.: Testing the tensile features of steel specimens by thermography and conventional methods. Sci. Tech. Rev. **60**(1), 66–70 (2010)
6. Tretyakova, T.V., Wildemann, V.E.: Study of spatial-time inhomogeneity of serrated plastic flow Al-Mg alloy: using DIC-technique. Frattura ed Integrità Strutturale **27**, 83–97 (2014)
7. Jandrlić, I., Rešković, S., Vodopivec, F.: Determining the amount of Lüders band in niobium microalloyed steel. Metalurgija **55**(4), 631–634D (2016)

8. Johnson, H., Edwards, M.R., Chard-Tuckey, P.: Microstructural effects on the magnitude of Luders strains in a low alloy steel. Mater. Sci. Eng. A **625**, 36–45 (2015)
9. Tsuchida, N., Tomota, Y., Nagai, K., Fukaura, K.: A simple relationship between Lüders elongation and work-hardening rate at lower yield stress. Scripta Mater. **54**, 57–60 (2006)
10. Rešković, S., Jandrlić, I., Vodopivec, F.: Influence of testing rate on Lüders band propagation in niobium microalloyed steel. Metalurgija **55**(2), 157–160 (2016)
11. Rešković, S., Jandrlić, I., Brlić, T.: The influence of niobium content and initial microstructure of steel. In: 5th International Conference on Recent Trends in Structural Materials, Materials Science and Engineering, vol. 461 (2019)

The Impact of the Deformation Redistribution on the Special Narrowing Force

Mirna Nožić(✉) and Himzo Đukić

Faculty of Mechanical Engineering, University "DžemalBijedić" of Mostar, Univerzitetskikampus, 88000 Mostar, Bosnia and Herzegovina
mirna.nozic@unmo.ba

Abstract. The paper presents the results of experimental investigations of the special narrowing process with two variants of tool construction in the production of the same workpiece. Both tools had three narrowing operations. Deformation analysis was obtained through logarithmic deformation and narrowing ratio, and was performed after each narrowing operation. In laboratory conditions, a deformation force was recorded on a hydraulic test machine with custom manufacturing tools. For the experiments, cylindrical preparations, obtained by deep drawing with the reduction of wall thickness from steel and brass, were used.

Keywords: Special narrowing · Die · Narrowing press · Deformation redistribution · Special narrowing force

1 Introduction

A classic narrowing reduces the cross-section of the pipe or ends of cylindrical vessels. The forms of narrowing can be different, and the most common are: moving from a larger to a smaller cylinder or moving from a cylindrical shape to a conical shape. The special narrowing is the narrowing of the bottom of the cylindrical vessel, which was previously made by deep drawing with the reduction of wall thickness, so that the thickness of the bottom s_0 is significantly greater than the thickness of the wall. From the bottom of the container, a conical shape with a minimum diameter of the cone (d) at the top is made by tapering (Fig. 1a).

The cone geometry changes from the cylindrical part (ØD) to the tip of the variable radius (R), in order to reduce the air resistance to a minimum. The special narrowing tool consists of press and matrix. The number of narrowing operations depends on the overall degree of deformation and the type of material. In the last narrowing operation, the dimensions of the matrix correspond to the outer dimensions of the conical shape, and the dimensions of the printers correspond to the internal dimensions of the workpiece. The 3D display of the tool in the last special compression operation is given in Fig. 1b.

Tool construction can be done in two ways. The first variant in all narrowing operations uses the same matrix, the dimensions of which are determined by the dimensions of the last operation. The dimensions of the printers according to this

I. Karabegović (Ed.): NT 2019, LNNS 76, pp. 122–129, 2020.
https://doi.org/10.1007/978-3-030-18072-0_13

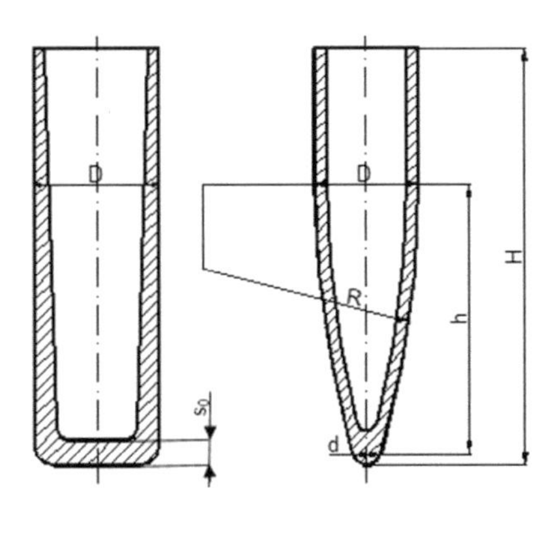

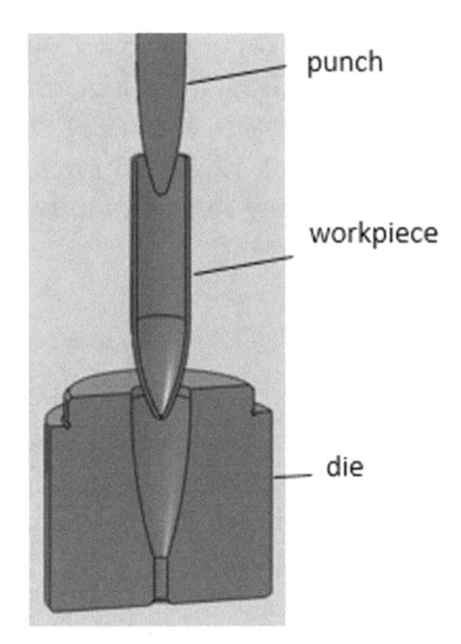

Fig. 1. a. Preparation and work item after narrowing. b. 3D display of the tool in the last operation of special narrowing

variant are adapted to the dimensions of the matrix in such a way that the height of the conical part increases gradually.

According to the second variant in all operations, matrices of different dimensions are used, and the dimensions of the printers are adapted to the dimensions of the constructed matrices. The common feature of both variants is that in the last narrowing operation, matrices of the same dimensions are used.

2 Deformational Analysis of Work Courses Constructed by Different Variants of Special Narrowing

Different variants of tool construction lead to different redistribution of deformation from first to last narrowing operation. Deformation in narrowing can be expressed through various parameters, and in this paper the following are used: total logarithmic deformation and narrowing ratio. The total logarithmic deformation is calculated by the expression:

$$\varphi_u = \ln(D/d) \tag{1}$$

where: D - diameter of the preparation, d - diameter of the tip of the cone (Fig. 1a).

The ratio of narrowing is calculated by the expression:

$$m_s = s_i/s_0 \tag{2}$$

where: s_0 - the thickness of the bottom of the preparation, and the wall thickness of the narrowed cone section.

By measuring the outer dimensions of the conical shape in all narrowing operations, the values of the logarithmic degree of deformation for both versions of the tool construction were calculated. A common deformation diagram for the first, second and third special narrowing operation according to variant 1 is shown in Fig. 2. It is evident from the figure that in all three narrowing operations, the logarithmic deformation has the same current.

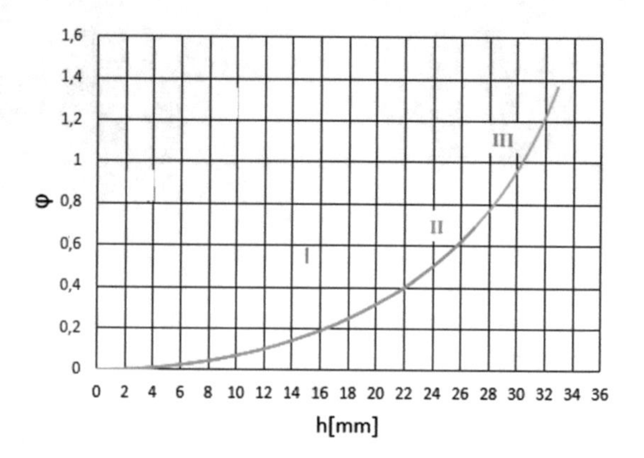

Fig. 2. Joint deformation diagram for first, second and third special narrowing operation according to variant 1

A common logarithmic deformation diagram for the first, second and third special narrowing operation according to variant 2 is shown in Fig. 3.

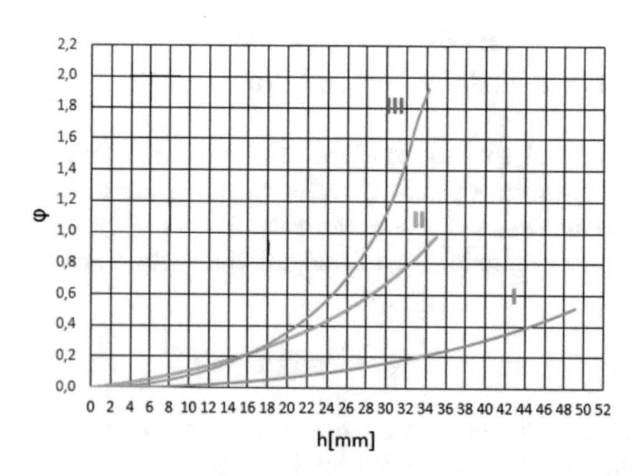

Fig. 3. Joint deformation diagram for the first, second and third special compression operation according to variant 2

From Fig. 3, it is evident that the logarithmic deformations differ significantly in the first, second and third operations, that is, each operation has its own logarithmic deformation.

In the tool constructed with three different dies, the logarithmic deformations distributed by operation are: on the first die $\varphi_1 = 0.51$, on the other die $\varphi_2 = 0.465$, and on the third die $\varphi_3 = 0.948$ (Fig. 3). The total logarithmic deformation for all three contraction operations in this tool is $\varphi_u = 1.923$. In the tool where one logarithmic deformation die is distributed in operations in all operations: $\varphi_1 = 0.43$, $\varphi_2 = 0.258$ and $\varphi_3 = 0.679$ (Fig. 2). The total logarithmic deformation for tools with equal dies is $\varphi_u = 1.367$.

The values of the narrowing ratio in all three operations performed on a tool with one die are shown in Fig. 4. With m_s, the ratio of narrowing on the conical part of the workpiece are denoted, and with m_{sv} we use the narrowing of the operations at the top of the cone.

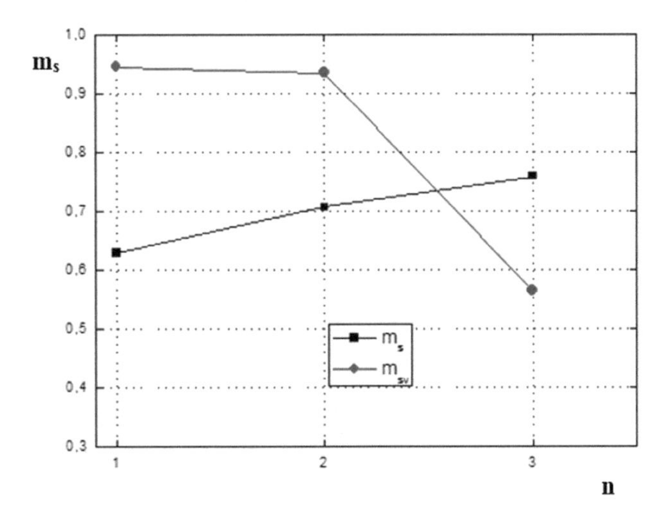

Fig. 4. The narrowing ratio for a tool with the same die

The influence of the geometry of the punch on the redistribution of the degree of deformation was analyzed through the ratio of the height of the conic section of the punch and the depth of the die (h_i/h_m). The height ratio for both tool variants is shown in the diagram in Fig. 5.

3 Experimental Research

Experimental investigations were carried out in production conditions at a special press. Two tools were developed for the same workpiece, made according to variant 1 and variants 2. Both tools had three special narrowing operations. Deformation analysis was performed after each narrowing operation. The recording of the narrowing force diagram in all operations was performed in laboratory conditions on a hydraulic test machine with a rated force of up to 400 kN, on production tools adapted to a hydraulic test machine.

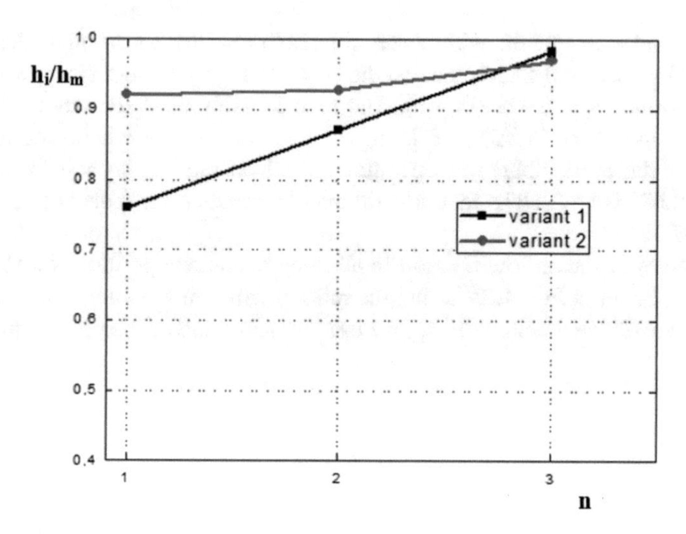

Fig. 5. The relationship between the height of the punch and the depth of the die

For the research, cylindrical preparations of diameter 14 mm and height 50 mm were used, obtained by deep drawing with the reduction of the wall thickness of steel Č.0148 and brass CuZn10. The mechanical properties of these materials are given in Table 1.

Table 1. Mechanical properties of CuZn10 and Č.0148

Mechanical properties of materials			
Materials	σ_m[N/mm^2]	δ_{10}[%]	HB
CuZn10	Min. 245	Min. 45	~50
Č.0148	294–362	Min. 27	85–110

The narrowing force diagrams were recorded on both tools for all three narrowing operations with two different materials. An example of a common diagram of the narrowing force for the third operation on two different tools is shown in Fig. 6.

The influence of the material on the narrowing force is shown in the diagrams shown in Figs. 7 and 8. In Fig. 7, diagrams for the first, second and third slide operation of the workpieces made of steel Č.0148 are given, and in Fig. 8, the diagrams of the narrowing force for the first, second and third operation of narrowing the workpieces made of brass CuZn10.

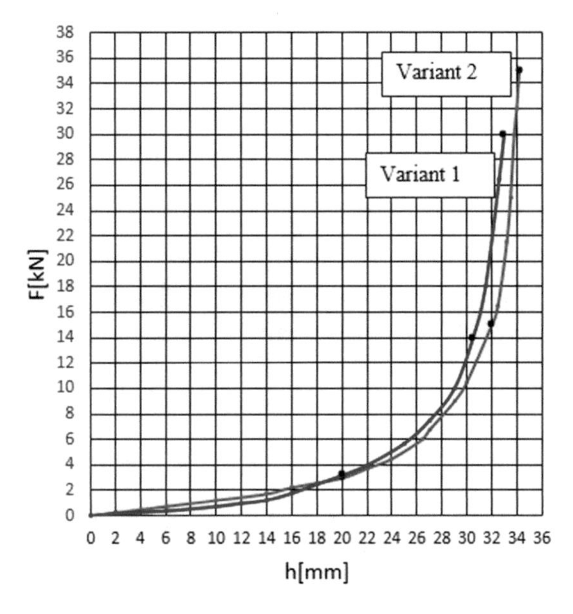

Fig. 6. Common diagram of the narrowing force for the third operation on tools 1 and 2

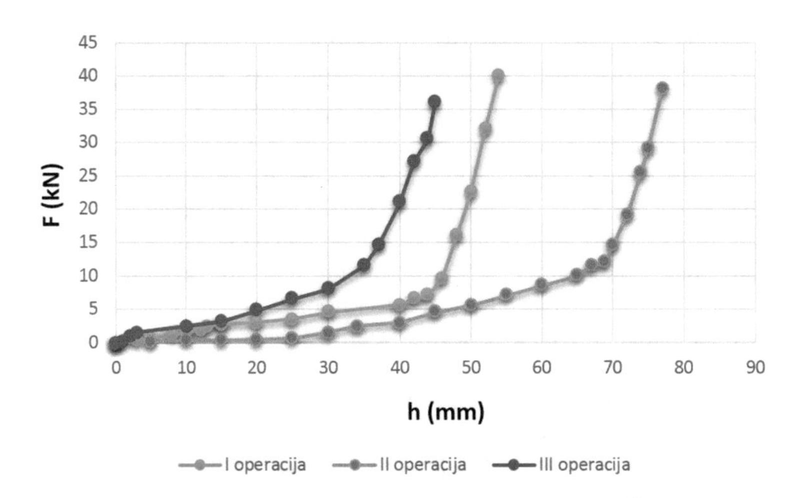

Fig. 7. Common diagrams of the narrowing force for Č.0148

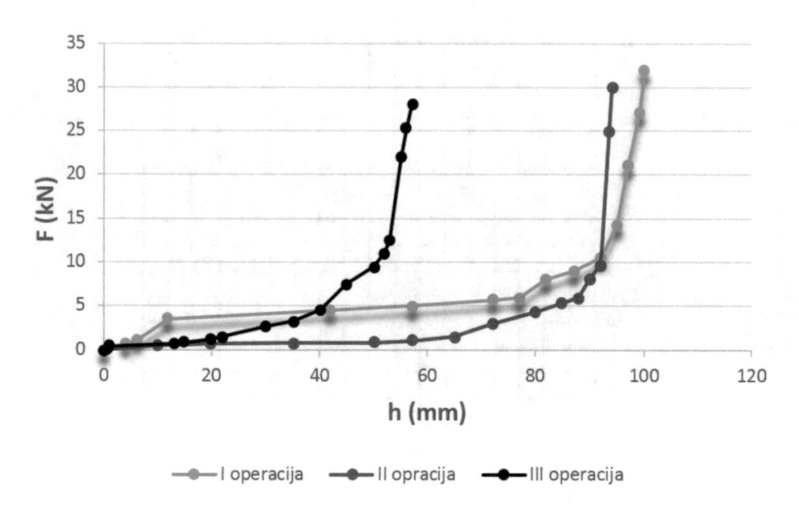

Fig. 8. Common diagrams of the narrowing force for CuZn10

4 Conclusion

Based on the experimental investigations of the effect of redistribution of the deformation on the force of special narrowing, caused by the different construction of the tools for making the same workpiece, the following conclusions can be drawn:

- For tools constructed according to variant 1, where all three dies are the same, the total logarithmic deformation is $\varphi_u = 1.367$, and the total logarithmic deformation for tools constructed according to variant 2 is $\varphi_u = 1.923$. Since this is the same work item, the total logarithmic deformation on the tool made according to variant 2 is higher by 40.6%. In all deformation processes, the theoretical deformation force depends largely on the type of material and the logarithmic degree of deformation;
- By analyzing the ratio of the height of the punch and the depth of the die in both tools (Fig. 5), it can be concluded that for the tool according to variant 1, the height of the punch gradually approaches the total depth of the die, and with the tool according to variant 2, the height of the punch is very close in all operations depth of the die. The ratio of the height of the punch and the depth of the die directly influences the necessary deformation work of narrowing;
- By analyzing the narrowing relationship on both tools, it is evident that the deformation degree in the peak design is greatest in the last narrowing operation, where the calibration of the tip is simultaneously performed (Fig. 4);
- The narrowing force diagrams for both tools shown in Fig. 6 confirm the deformation analysis according to which the narrowing forces are greater on the tool made according to variant 2;
- The influence of the type of material on the narrowing force is visible from the diagrams shown in Figs. 7 and 8. The steel narrowing forces are 33% larger than the braking force, which is proportional to the mechanical properties of these two materials, given in Table 1.

Based on the investigations carried out, it can be concluded that in special narrowing, the tool construction should be made according to variant 1, where the final form of the die is used in all narrowing operations.

References

1. Đukić, H., Nožić, M.: Limit values of maximal logarithmics strain in multi-stage cold forming operations. J. Technol. Plast. **40**(1) (2015)
2. Nožić, M., Đukić, H.: Eksperimentalno određivanje graničnih stupnjeva deformacije u procesima proširivanja i sužavanja. In: International Conference on MATRIB 2018, Vela Luka, Hrvatska (2018)
3. Gagula, E.: Istraživanje uticajnih parametara izrade košuljice zrna različitih kalibara, magistarski rad. Mašinski fakultet Mostar (2017)
4. Audoly, B., Hutchinson, J.W.: Analysis of necking based on a one-dimensional model. J. Mech. Phys. Solids **97**, 68–91 (2016)
5. Audoly, B., Hutchinson, J.W.: One-dimensional modeling of necking in rate-dependent materials. J. Mech. Phys. Solids **123**, 149–171 (2019). ISSN 0022–5096
6. Zhang, Z.L., Hauge, M., Odegard, J.O., Thaulow, C.: Determining material true stress-strain curve from tensile specimens with rectangular cross-section. Int. J. Solids Struct. **36**, 3497–3516 (1999)
7. Kim, H.S., Kim, S.H.,. Ryu, W.-S.: Finite element analysis of the onset of necking and the post-necking behaviour during uniaxial tensile testing. Mater. Trans. **46**(10), 2159–2163 (2005)
8. van den Boogaard, A.H., Huetink, H.: Prediction of sheet necking with shell finite element models. In: 6th International ESAFORM Conference on Material Forming, Salerno, Italy (2003)
9. Đukić, H., Nožić, M.: Obrada deformisanjem, drugo izdanje. Univerzitet "Džemal Bijedić", Mašinski fakultet Mostar (2018)
10. Nožić, M., Đukić, H.: Projektovanje tehnologija obrade deformisanjem. Univerzitet "Džemal Bijedić" Mašinski fakultet Mostar (2016)

Investigation of Primer Influence on Strength of Aluminium Specimens Bonded by VHB Tape

Aida Čolo, Petar Tasić(✉), and Ismar Hajro

Engineering Faculty, University of Sarajevo,
Vilsonovo šetalište 9, 71 000 Sarajevo, Bosnia and Herzegovina
tasic@mef.unsa.ba

Abstract. Very High Bond tape is a type of adhesive often used for fast and efficient bonding of repair work as well as assemblies. One of many types of VHB tape is the acrylic based 4611F, which according to the product specifications has good properties at slightly elevated temperatures and is used for bonding parts of roof constructions of busses and trucks, glazing systems and traffic signs. This paper presents the results of lap sheer testing of VHB bonded aluminium specimens. The tests were carried out on both primed and unprimed specimens at the following temperatures: 20, 50, 100 and 150 °C. The results obtained show that benefits of priming exist only at room temperatures.

Keywords: VHB · Aluminium · Primer · Lap sheer test · Elevated temperature

1 Introduction

Manufacturing requires simple, repeatable and economically acceptable processes. In other words, they must be efficient. Therefore, new technologies and techniques are developed, while existing are enhanced. One of very popular joining technologies is adhesive bonding. Nowadays, it is present in almost all areas of manufacturing, from shoemaking to aerospace industry, latter using epoxy adhesives modified with rubber to bond structural components made of aluminium alloys. Reason for such popularity lies in fact that adhesive can create very strong joints. In particular, this is important for materials that can be hardly welded, or even not possible to be welded at all. Such materials are wood, stone, ceramics and glass. Adhesive bonding enables joining very thin sheets of material, as well as distribution of load over entire contact surface. Aesthetical requirements are easy achievable, especially because its application can be done by various mechanisms, or even robots. It is important to mention that this is joining technology applicable under almost any conditions, since it usually does not require any special surface preparation [1].

Significant advantage of adhesive bonding over welding, soldering or brazing is visible in cases when heat input would lead to deterioration of mechanical properties. This is of great importance in aerospace and space industry, where components must withstand great loads, while having high rigidity and minimal weight. This led to usage of light-metal alloys, in particular those of aluminium and titan. Development of

I. Karabegović (Ed.): NT 2019, LNNS 76, pp. 130–135, 2020.
https://doi.org/10.1007/978-3-030-18072-0_14

adhesive bonding suppressed riveting, technology dominant in joining of light-metal alloys. Adhesive bonds can also absorb vibrations, and, in general, they have lower price compared with other joining technologies.

Principal disadvantages of adhesive bonds are limited operating temperature (from –50 to +200 °C), brittle behaviour at lower temperatures and decrease of dynamic strength with increased adhesive layer thickness. Sometimes, it also can be prolonged cure time. In most cases, surface must be excellently prepared, and beside mechanical preparation, it can include chemical preparation as well [1, 2]. Even tough adhesive joints have relatively low strength, there is number of ways it can be improved. One of them, application of primer, is considered in this paper.

Primer is material applied to surfaces to be joined, increasing surface energy prior to joining. Generally, low surface energy is problematic for adhesive bonding, disabling appropriate wetting of surface. Low surface energy materials are usually polymers, such as polypropylene and polyethylene, but generally can be any other material in case surface is not adequately prepared. Table 1 gives an overview of surface energies of some common materials in case of well prepared surface. Listed values are approximate, since they can vary significantly (e.g. for steel and aluminium that range depends on chemical composition and surface condition).

Table 1. Overview of surface energies of some common materials [3, 4]

Material	Surface energy (mN/m)
Copper	1,100
Steel	1,000
Aluminium	750
Glass	500
PVC	39
Acryl	38
PTFE	18

Strength of adhesive joint increases as the contact surface between material and adhesive increases. This is case for both liquid and tape adhesives. Waviness or shape distortion can weaken adhesive joint, but roughness can improve it [4, 5]. Grease, dust and other kinds of surface impurities can interfere with adhesive joining process, so surfaces must be cleaned in accordance with adhesive manufacturer's recommendations. This is particularly important for joining with adhesive tapes.

2 VHB Tapes

Pressure Sensitive Adhesive tape is adhesive that consists from adhesive layer applied to tape made of paper, polymer, rubber or other materials [4].

Unique characteristic of this adhesive is possibility to create joint by simply applying pressure, what can be made even by hand. There is no chemical reaction nor phase change during cure time. PSA tapes maintain excellent adhesive properties mainly at low temperatures (e.g. room temperature). Such tapes are used for joining negligibly loaded structural components, signs and labels [6]. It is possible to modify PSA tapes by adding various enhancers, and use acrylic sponge as carrier. Such adhesive tape can be alternative to joining by rivets, nuts and bolts, and even spot resistance welding. That is VHB tape, and it can be used for various purposes, including bonding roof bows on trucks and buses bonding glazing bars or muntin bars positioned internally and signage applications [7].

VHB tapes have many advantages, among other possibility to bond dissimilar materials without galvanic corrosion, damping vibrations and noise, as well as sealing. It also requires no fixture or pressure during solidification.

There are few disadvantages, though. Among most important ones is prolonged solidification time (up to 72 h), although it can be significantly reduced by increasing temperature during solidification. These tapes are also prone to creep under prolonged static load.

3 Investigation

For this investigation, VHB tape 3M 4611F has been used. This is double-sided general purpose VHB tape, commonly used for joining of metals. It is suitable for using at slightly elevated temperatures. It maintains properties at up to 120 °C, while it can withstand temperatures up to 230 °C for short periods of time [8].

As a test material aluminium alloy AlMg3 has been used. This material does not require any specific surface preparation except usual cleaning. Cleaning has been done by applying trichloroethylene and rinsing with distilled water. Tape manufacturer recommend use of 3M 94 primer for better adhesion [8, 9].

Two sets of specimens were made, so the influence of primer application could be examined. One set was with primer applied prior to bonding (primed specimens) and the other without (not primed specimens). Tests were made at different temperatures, i.e. room temperature, 50, 100 and 150 °C. For each set and temperature five specimens have been tested. For primed set, primer has been applied after cleaning and only at one side of specimen. Geometry and dimensions of specimens are given in Fig. 1, where (1) denotes aluminium sheet and (2) VHB tape.

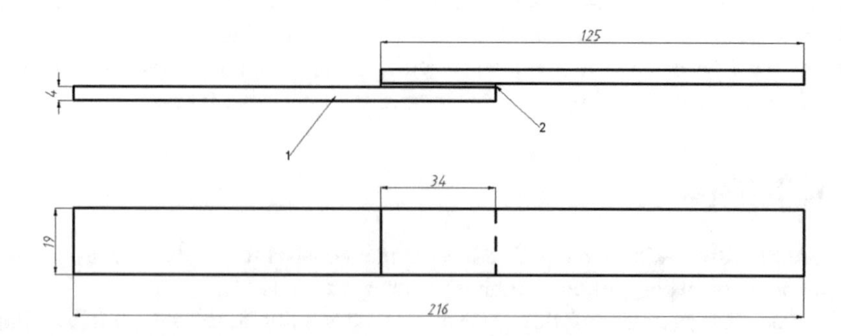

Fig. 1. Geometry and dimensions of specimens

Adhesive has been activated by applying pressure of approximately 0.2 MPa, and after that, specimens were left without fixation or additional pressure for 72 h at room temperature.

A quasi-static tensile test has been conducted (Fig. 2), with constant elongation rate at all temperatures (13 mm/min). Temperature has been monitored during testing, directly at furnace and by thermocouples mounted inside of it.

Fig. 2. Samples in jaws at testing device (left) and furnace for specimens (right)

4 Results and Discussion

Lap shear strength has been obtained by dividing maximal force measured during testing with lap surface calculated after measuring specimen dimensions. This has been done for both primed and not primed specimens set, and for all temperatures. Figure 3 shows dependence of average lap shear strength as a function of temperature for primed specimens, and Fig. 4 for not primed ones. Additional lines represent minimal and maximal values obtained during experiment.

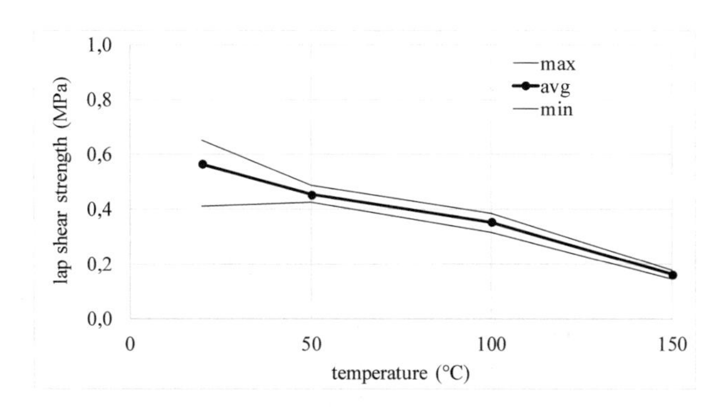

Fig. 3. Results obtained for primed specimens

As possible to see from Figs. 3 and 4, some scattering is present at room temperature, i.e. there is relatively wide range of values obtained during tests. This is

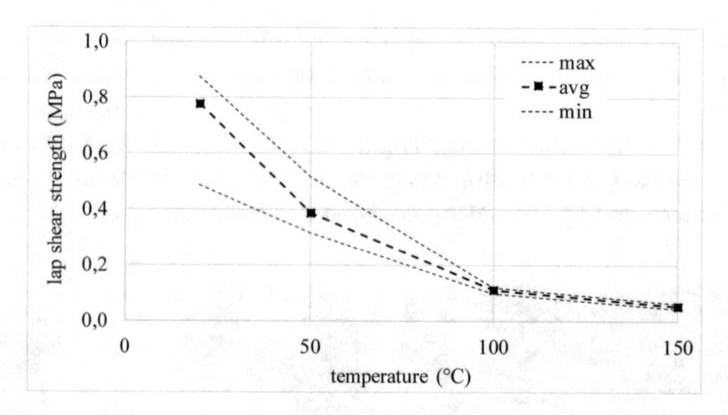

Fig. 4. Results obtained for not primed specimens

slightly more visible for primed specimens. As temperature increases, scattering is lower, for both primed and not primed specimens.

Figure 5 gives summary of tests to enable comparison between results for primed and not primed specimens. Average values of lap shear strength are displayed for all temperatures. As possible to see, not primed joints show relatively high strength up to 100 °C. Afterwards comes slight and uniform decrease.

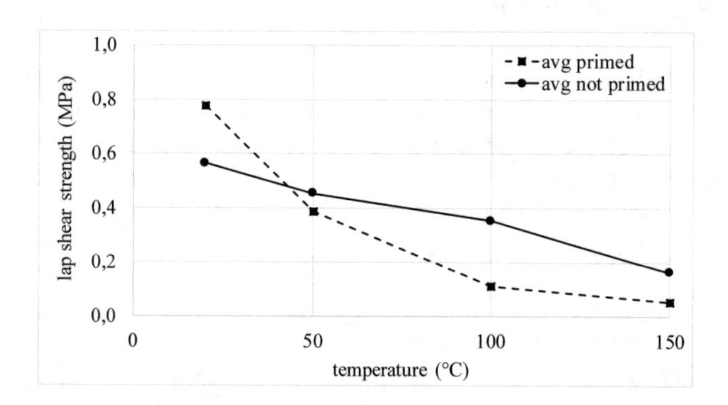

Fig. 5. Comparison of average lap shear strength values for primed and not primed specimens

Strength of primed joint is higher than not primed one, but only a room temperature. There is sharp drop in strength at elevated temperatures. The reason for this could be that primer is not resistant to higher temperatures.

According to Fig. 5, at temperatures above approximately 40 °C primed specimens have lower strength than not primed ones. Hence, it can be concluded that application of primer for joints intended to be used at such temperatures is not only unnecessary, but also undesirable.

Applying primer should increase strength, since it increases adhesive forces. Therefore, it is interesting to compare strength values obtained with primed and not primed specimens. Even though Fig. 5 clearly depicts difference between strength of primed and not primed specimens at room temperature, it is necessary to confirm this. Statistical analyse has been done only for results at room temperature, because at higher temperatures primed specimens shown worse result than not primed ones.

T-test is done by using Microsoft Excel. Different values of variance have been assumed for two different sets of specimens. It has been shown that there is statistically significant difference between mean values of two sets, with probability of 4.4% (it is acceptable up to 5.0%). Hypothesis that two sets have same mean value has been rejected. Therefore, it is possible to say that two strength values, for primed and not primed set, are statistically significantly different.

5 Conclusion

VHB tapes poses many useful properties, making them usable for joining various materials, including polymers, metals and glass. Even though they cannot achieve high joint strength, typical for structural adhesives (e.g. epoxy resins), they have relatively wide use in automotive and aerospace industry, as well as in architecture.

One of common methods to increase joint strength is application of primer. Purpose of this paper was to analyse whether application of primer to aluminium alloy specimens leads to increased strength. This is interesting to analyse because aluminium itself has relatively high value of surface energy, and under normal conditions it is not absolutely required to apply primer. Two sets of specimens were made, and primer has been applied to one set, while second set was just normally cleaned.

After experiments, it has been concluded that primed specimens exhibit significant drop in strength at elevated temperatures. On the other side, not primed specimens have just slight drop. Interestingly, at elevated temperatures, not primed specimens display higher strength than primed ones. Hence, using primer prior to joining aluminium with VHB tape is not recommended. It has been shown that primer increases strength, but only at room temperature.

References

1. Kuczmaszewski, J.: Fundamentals of Metal-Metal Adhesive Joint Design. Polish Academy of Sciences, Lublin (2006)
2. Pizzi, A., Mittal, K.L.: Handbook of Adhesive Technology. Marcel Dekker, New York (2003)
3. Petrie, E.M.: Handbook of Adhesives and Sealants. McGraw Hill, New York (1999)
4. Lind, M., Petersson, D., Petersson, E.: Double-Sided Pressure Sensitive Adhesive Tape. Blekinge Institute of Technology, Karlskrona (2014)
5. Obućina, M.: Lijepljenje. Mašinski fakultet Sarajevo, Sarajevo (2011)
6. Creton, C.: Pressure-sensitive adhesives: an introductory course. MRS Bull. **28**, 434–439 (2003)
7. https://www.can-dotape.com/adhesive-tape-consultant/pressure-sensitive-adhesive-tape/. pristupljeno 25 Dec 2018
8. 3M, VHB Tapes: Tech Data (2011)
9. 3M, Innovations in Bonding to Low Surface Energy Surfaces (2015)

Improving the Performance Properties of Abrasive Tools at the Stage of Their Operation

Vladimir Tonkonogyi[1], Tetiana Sidelnykova[2], Predrag Dašić[3(✉)], Alexey Yakimov[1], and Liubov Bovnegra[1]

[1] Odessa National Polytechnic University (ONPU), Odessa, Ukraine
[2] Odessa National Medical University (ONMU), Odessa, Ukraine
[3] High Technical Mechanical School of Professional Studies, Str. RadojaKrstića 19, 37240 Trstenik, Serbia
dasicp58@gmail.com

Abstract. Grinding performance largely depends on the durability of the abrasive tool. Developed and tested solid lubricant for impregnating the working surface of the grinding wheel, which increases its cutting ability and, as a result, reduces the number of edits. Reducing the need for frequent revisions of the abrasive tool reduces dust generation in the workshop and helps to prevent the occupational disease of the grinder - pneumoconiosis. It is established that the depth of cut when grinding according to an elastic scheme depends not only on the coefficient of grinding and the coefficient of friction, but on the magnitude of their difference.

Keywords: Higher · Fatty carboxylic acids · Friction coefficient · Grinding wheel · Solid lubricant

1 Introduction

Intensification of the industry leads to severe environmental pollution. The point of abrasive dust generated during grinding wheel grinding affects the respiratory organs.

Lightweight possess a very important property. They are constantly cleared of dust using phagocytes (a special type of white blood cells) but with a high content of abrasive dust in the air, the protective effect of the body weakens [1].

Dust accumulating in the lungs affects them, leading to a dash of pneumoconiosis. This disease is characterized by a slow transformation of the lung tissue from elastic ones that can significantly stretch and increase the area of air exchange when inhaling into a tissue with many scars (fibrosis) [2].

There are many types of pneumoconiosis. The most common and dangerous is silicosis, which is the result of a large amount of dust containing free silica entering the lungs Al_2O_3) And carborundum (silicon carbide Sio) And contains a small amount of silica SiO_2Z - report content in a bunch of quartz (sand). Due to this content of silicon dioxide in abrasive dust, formed directly in the process of grinding, it is insignificant, but its content increases dramatically when editing the grinding wheel, used to restore

I. Karabegović (Ed.): NT 2019, LNNS 76, pp. 136–145, 2020.
https://doi.org/10.1007/978-3-030-18072-0_15

its cutting properties. To prevent occupational disease of the grinder, caused by the impact on the bronchopulmonary apparatus of abrasive dust, it is necessary to reduce its content in the air of the working area. To do this, it is necessary to make technological decisions to reduce dust generation at the workplace of the grinder. One of such solutions is an increase in time of the cutting ability of grinding wheels, i.e. an increase in the time interval between two changes of the abrasive tool. The following methods are known for improving the performance properties of abrasive tools at the stage of technological preparation of production: creation of grinding wheels with a discontinuous working surface [4], Impregnation of wheels with special impregnating compounds [5–15]. For the same purpose at the stage of operation of abrasive tools use: special cutting fluids [19] and solid lubricants [20–22]. Of the above methods for improving the performance properties of an abrasive tool, the most easily accessible and economical way is to impregnate the working surface of the grinding wheel with a solid lubricant directly during its operation.

2 Materials and Methods

Steel samples 12 × 2H4A width 3 mm and length 150 mm processed on the machine 3Г71M by wheel ПП200 × 20 × 76 24A 40 C2 5 K6 mortise grinding mode: V_{kr} = 30 m/s, V_{zag} = 6 m/min with constant clamping force of the sample to the circle F_y = *60 H*, without coolant.

Constant force F_y. It was provided with the help of a special device, which is a body inside which the slider freely moved with a fixed sample. The sample was pressed to the circle with a lever, one end of which was connected to the slider, and the load was suspended at the other end.

3 Results and Discussion

It is known that good lubricating properties with a sufficient long period of grinding are preserved in the case of using the binder component with high thermal stability. What are the components of higher fatty carboxylic acids and their esters with monatomic higher alcohols (waxes), at elevated temperatures, fatty acids react with metals to form salts (soaps). Thermal and lubricating properties of metallic soaps (for example, salts of higher fatty carboxylic acids and iron) are higher than those of the original derivatives of hydrocarbons.

$$\underset{\text{(stearic acid)}}{2C_{17}H_{35} - COOH} + Fe \rightarrow \underset{\text{(iron stearate)}}{Fe(C_{17}H_{33}COO)_2}$$

In the case of grinding steel, such reactions take place when the grain contacts the metal being processed on the surface of the cutting grains. There is a decrease in the friction coefficient as a result of the formation of metallic soaps on the submicro-profile of the cutting grains. The consequence of this process is to reduce the abrasive ability of the grain. For grinding processes that are performed at small lengths of the contact arc of a circle and parts, such changes in the conditions of micro-cutting are beneficial as the number of cutting grains increases. Depth of penetration of active grains is reduced, which allows you to load previously overlapped screened grains.

It was proposed to increase the heat capacity and lubricity of the binder component at the stage of manufacturing solid lubricants to carry out partial saponification of the used fatty acids with caustic alkalis (for example, potassium hydroxide KOH). The presence of alkali metal soap in a solid lubricant along with free higher fatty carboxylic acids makes it possible to increase the heat capacity of the lubricant and its anti-friction properties and also provides the possibility of chemical interaction with the material being processed. In this paper, to optimize the grinding process, we propose the following composition of solid lubricant (Table 1). In the last row of the table is the composition of solid lubricant does not contain potassium hydroxide [4.21].

Table 1. Chemical compositions of solid lubricants

No composition of imprenator	Stearic acid $C_{17}H_{35}COOH\%$	Oleic acid $C_{17}H_{33}COOH\%$	Acetamide $CH_3CONH_2\%$	Potassium hydroxide $KOH\%$
1	70	15	14	1
2	67	17	14	2
3	65	19	13	3
4	63	21	12	4
5	60	23	12	5
6 (prototype)	63	23	14	–

Oleic acid is an unsaturated higher fatty acid, which, under severe conditions, is oxidized with breaking of carbon-carbon bonds with the formation of the corresponding acids according to the scheme:

$$CH_3-(CH_2)_7-CH=CH-(CH_2)_7-COOH \xrightarrow{t^\circ} CH_3(CH_2)_7COOH + HOOC-(CH_2)_7-COOH \quad (1)$$

With an increase in the number of carboxyl groups, reactivity in terms of interaction with the material being processed increases.

In an alkaline medium, acetamide is hydrolyzed to form the salt of carboxylic acid and ammonia.

$$\underset{\text{(Acetamide)}}{CH_3-C\begin{matrix}\diagup\!\!\!\!/ O \\ \diagdown NH_2\end{matrix}} + KOH \overset{(H_2O)}{\rightleftarrows} CH_3-C\begin{matrix}\diagup\!\!\!\!/ O \\ \diagdown OK\end{matrix} + NH_3 \quad (2)$$

In turn, when heated, ammonia decomposes into hydrogen and nitrogen, using iron as a catalyst of the process:

$$2NH_3 \overset{Fe}{\rightleftarrows} 3H_3 + N_2 \ (1200 - 1300)^{\circ}C \tag{3}$$

Hydrogen with air oxygen at $t = 550$ °C burns to form water vapor:

$$2H_2 + O_2 = 2H_2O \tag{4}$$

The formation of water vapor contributes to the chemical interaction of the lubricant with the treated surface.

Manufacturing techniques of pencils solid lubrication and testing.

Stearic acid is poured into the tank, heated to melt (90 – 100) °C, oleic acid, potassium hydroxide and acetamide are added, stirred until complete dissolution in stearin and poured into a cylindrical shape. After hardening and cooling, the solid lubricant pencils are ready for use. Solid lubricant is applied to the surface of the abrasive wheel at a working speed of rotation. Lubrication is considered complete when the working surface of the abrasive tool changes its color.

When the content of stearic acid is more than 67%, the pencils of solid lubricant become brittle and inconvenient to use. In terms of their technological properties, they approach the action of pure stearin, that is, the effectiveness of the lubricant decreases when the grinding mode becomes more stringent. When the amount of stearic acid is less than 60%, the consistency of the composition changes, the melt hardens worse when using solid lubricants in the form of pencils.

According to the technology described above, 5 lubricant compositions were prepared (Table 1): 4 within the limits of the proposed concentrations of stearic acid (60–67%), and 1 composition with extreme concentration (70%). The composition of the exorbitant concentration of stearic acid less than 60% was not prepared, since samples with such a component content are not well cured.

The compositions of solid lubricants 1–5 tested on surface grinding machine model 3Г71 wheel ПП200 × 200 × 76 24A 40 C2 5 K6 on steel samples 65Г. This steel is viscous and when it is ground there is a rapid salinization of the working surface of the abrasive wheel. Grinding was carried out on the modes: the depth of grinding t = 0.01; 0.02; 0.03; 0.04 $\frac{mm}{propulsionstem}$ (the magnitude of the wheel on the limb, the speed of rotation of the circle $V_{кр} = 30$ M/c, table moving speed $V_{заг} = 10\,\mathrm{m/min}$. Surface area $5 \times 150\,\mathrm{mm}^2$.

Before applying on the working surface of the Circle of solid lubricant was carried out editing. Experiments with masks having the compositions № = № = 1–5, repeated 5 times. Moreover, the Circle was edited and its working surface was impregnated only before the first experiment. Grinding with the use of a prototype lubricant (composition No. 6 in Table 1) and grinding without using a lubricant. Repeated in the flesh before the Circle was used, a sign of which was a sharp increase in power and the appearance of burns on the surfaces to be treated. Table 2 shows the results of power measurements during grinding “dry” by using lubricants of different compositions, and Table 3 shows photographs of the chips formed as a result. From Table 2 it can be seen that the appearance of burn-throughs on the surfaces being processed and the salting of the working surface of the abrasive circle occur during grinding using lubricants No. 1–5. After 5 experiments, using lubricant No. 6 (of the prototype) after four experiments, and when grinding dry - after the third experiment.

Table 2. Grinding power measurement results

Grease composition	Grinding depth, mm	The number of repetitions of experiments				
		1	2	3	4	5
		Grinding power, W				
1	0,01	120	140	140	140	150
	0,02	225	240	240	245	250
	0,03	345	345	345	350	360
	0,04	500	530	505	520	530
2	0,01	120	100	100	110	110
	0,02	210	210	220	220	220
	0,03	350	425	310	340	340
	0,04	530	460	440	460	470
3	0,01	140	145	145	150	150
	0,02	240	250	255	280	285
	0,03	340	345	370	410	410
	0,04	485	485	505	510	520
4	0,01	135	140	140	145	145
	0,02	235	240	245	260	265
	0,03	335	240	350	380	400
	0,04	470	480	490	500	510
5	0,01	130	135	135	140	140
	0,02	230	235	235	255	260
	0,03	330	335	345	375	395
	0,04	465	475	485	495	500
6 (prototype)	0,01	160	165	170	175	
	0,02	260	265	270	280	
	0,03	360	365	370	380	
	0,04	480	485	490	burned	
Dry grinding	0,01	290	300	310		
	0,02	400	410	415		
	0,03	500	510	burned		
	0,04	610	630			

From this it follows that the proposed solid lubricants (compositions No. 1–5) Significantly increase the period of operation of the abrasive tool between revisions. The need for frequent revisions in connection with the salting of m Circle and, as a result, the appearance of burn-throughs on the treated surfaces when grinding “dry” is one of the significant reasons that reduces the productivity of abrasive machining.

Released during the decomposition of ammonia into hydrogen and the formation of water vapor during grinding with a solid lubricant (formulations nos. 1–6) Intensifies the oxidation process of iron shavings. Iron oxidation is accompanied by the release of

Table 3. Chips formed during grinding with circles using lubricants having a different ratio of chemical elements

№ impregnator composition	Grinding chips
1	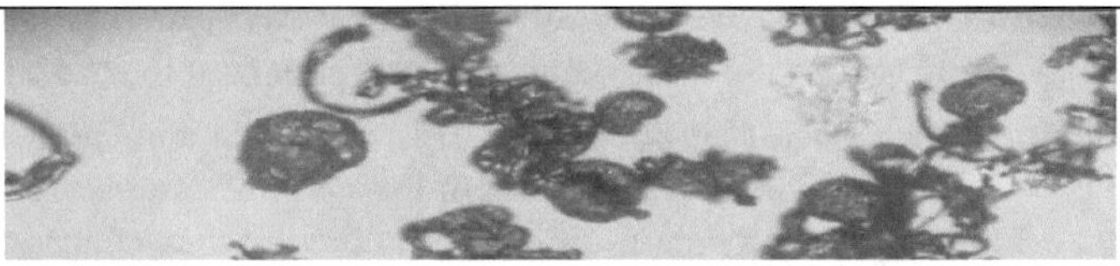
2	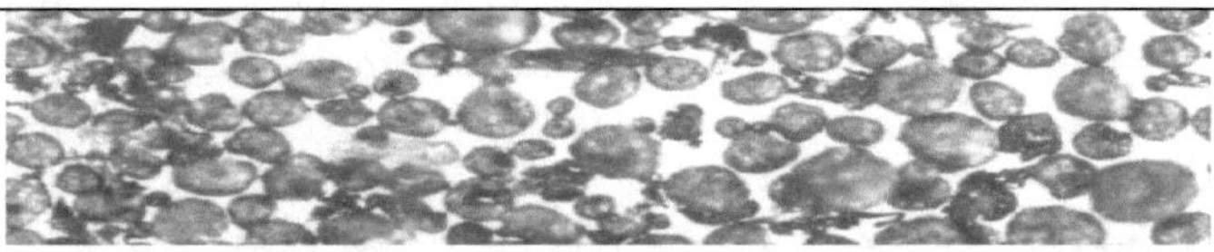
3	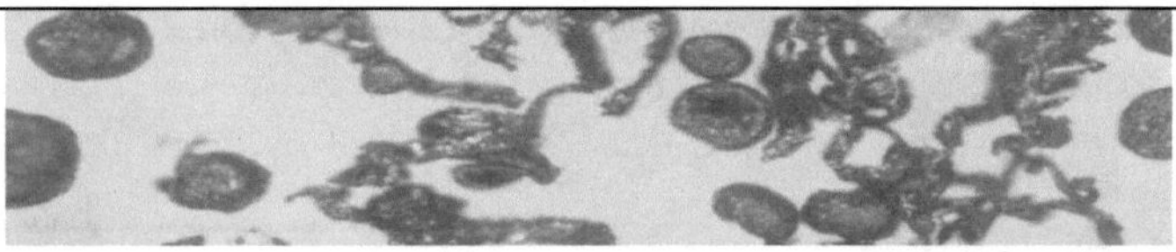
4	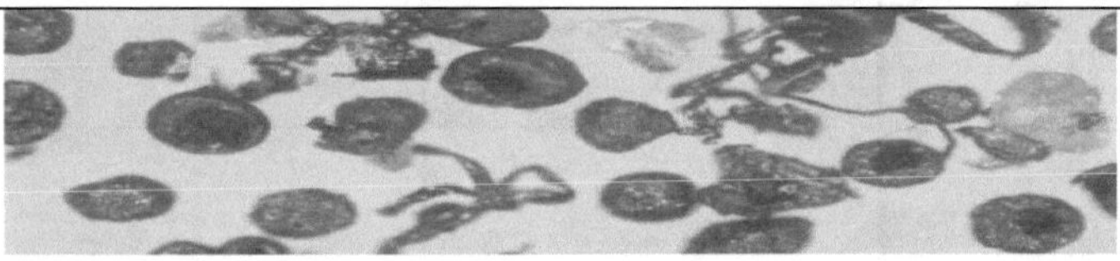
5	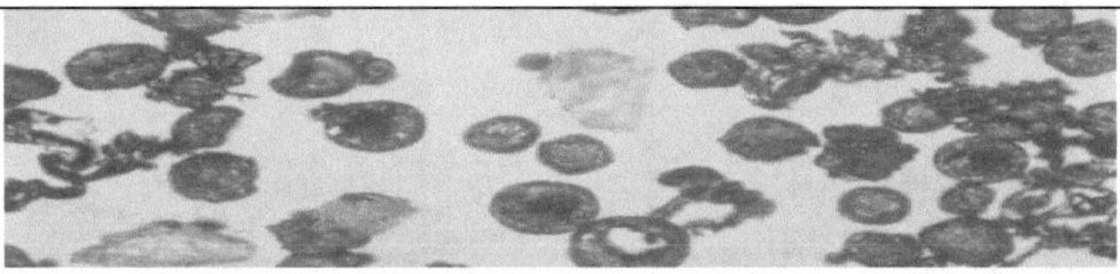
6 prototype	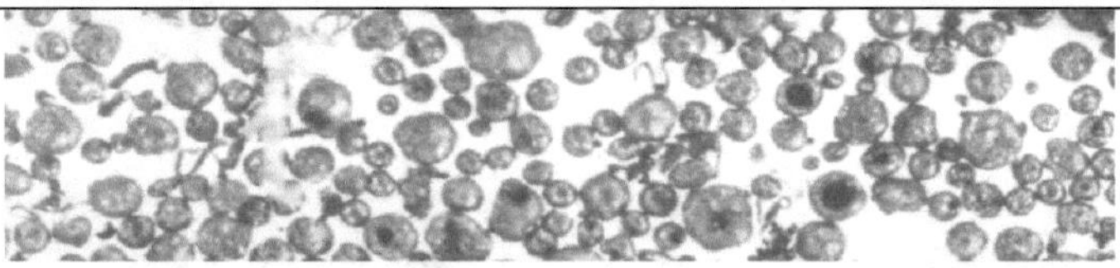
Dry grinding	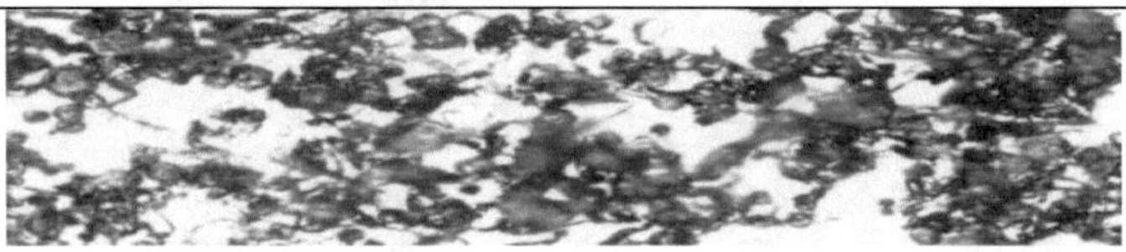

a significant amount of heat, which leads to control chips. This is confirmed by the presence of spherical chips after grinding with an impregnated wheel (Table 3). The formation of a fragile oxide film on chips facilitates removal from the surface of the Circle and improves the course of the grinding process.

Figure 1 shows the nature of the change in the grinding coefficient over time. $K_s = \frac{Fz}{Fy}$, Equal to the ratio of the tangential component of the cutting force to its normal component when machining steel $12 \times 2H4A$ normal and impregnated grinding wheels $\prod\prod 200 \times 20 \times 7624A\,40\,C2\,5K6$. From the graph it can be seen that in the interval of the 12-min processing period, a decrease in the grinding coefficient is observed both for the impregnated and for the usual circles. Moreover, the grinding coefficient for a conventional circle is greater than for an impregnated one. In the time interval $12\,\text{min} \leq T \leq 25\,\text{min}$. The grinding ratio for a conventional Circle continues to decline, while for the impregnated one it is increasing. Reduced grinding ratio in the time interval $1\,\text{min} \leq T \leq 25\,\text{min}$ when machining, the circle impregnated is explained by a decrease in the tangential component of the cutting force Fz due to the formation of lubricating films on cutting grains.

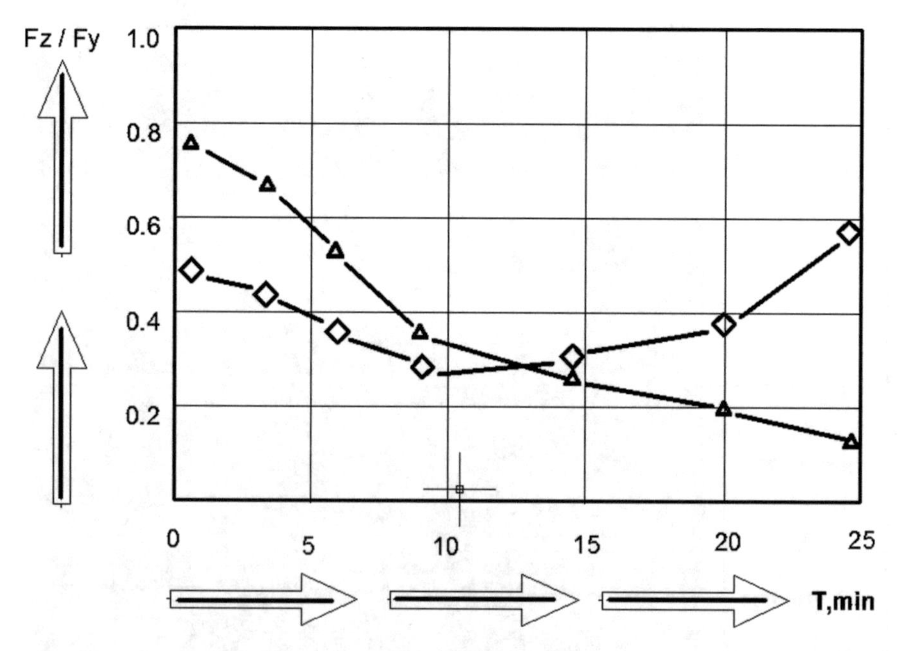

Fig. 1. Changing the coefficient of grinding in time in the process of working with ordinary (curve with triangular points) and impregnated (curve with square points) circles

It is established that when grinding by an elastic scheme, the depth of cut T depends not only on the grinding coefficient K_s or coefficient of friction f but of the magnitude of the difference $(K_s - f)$:

$$t = \frac{1}{2} \cdot \left[\frac{(3-n) \cdot (6-n) \cdot k_s \cdot F_y}{1,32 \cdot AB} \right]^{\frac{6}{(6-n)}} \cdot \left[\frac{V_{kr}}{6,9 \cdot V_{zag} \cdot (1 + l_2/l_1)} \right]^{\frac{2 \cdot (3-n)}{(6-n)}}$$
$$\times \left[\frac{29,369 \cdot \bar{x}^3 \cdot \sqrt{R} \cdot (k_s - f)^2}{tg\gamma \cdot m \cdot [0,37 \cdot (k_s - f) + 1,49 - k_s \cdot 1,016] \cdot [0,37 \cdot (k_s - f) + 2,98 - k_s \cdot 2,032]} \right]^{\frac{2n}{(6-n)}} \quad (5)$$

where R is the radius of the grinding wheel, m; B - grinding width, m; F_y - grinding wheel rotation speed, m/s; V_{zag} - longitudinal moving speed, m/s; l_1, l_2 - the length of the depression and the length of the cutting protrusion on the intermittent abrasive wheel, respectively, m; m - volume concentration of grains of a circle, %; $\underline{x}$ - circle grain, m; A - parameter characterizing the strength properties of the material being processed; n - parameter determined experimentally.

For the usual abrasive wheel, which does not have cavities on its working surface, the parameter $(1 + l_2/l_1) = 1$, because $l_2 = 0$.

From formula (5) it is possible to isolate the parameter M, which characterizes the cutting ability of the grinding wheel during the time of grinding:

$$M = k_s^{\frac{6}{6-n}} \left[\frac{(k_s - f)^2}{[0{,}37(k_s - f) + 1{,}49 - k_s \cdot 1{,}016] \cdot [0{,}37 \cdot (k_s - f) - 2{,}98 - k_s \cdot 2{,}032]} \right]^{\frac{2n}{(6-n)}} \quad (6)$$

Ratio $\frac{M_{pr}}{M_{sp}}$ shows how many times the cutting ability of a discontinuous circle is greater than the cutting ability of a continuous circle.

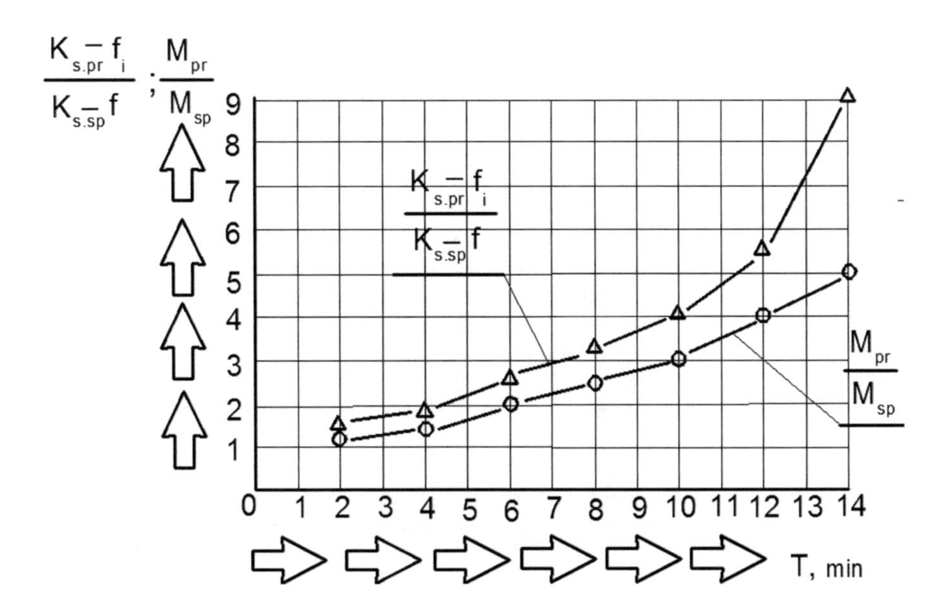

Fig. 2. Experimental data showing how the cutting ability of solid circles decreases over time compared to intermittent and intermittent impregnated circles

From Fig. 2, it can be seen that after the 14-min periods, the periods of grinding using an elastic scheme began to cut a continuous circle 5 times worse than intermittent ones. In Fig. 2 shows the dependence $(k_{s.pr} - fi)/(k_{s.sp} - f) = f(t)$, showing how the cutting ability of a continuous circle deteriorates over time compared to the cutting ability of a discontinuous impregnated circle of the same characteristic. From the graph it can be seen that after a 14-min grinding period, the cutting ability of the solid wheel decreased by 9 times compared with the cutting properties of the intermittent impregnated wheel.

4 Conclusion

Created and tested solid lubricant for abrasive machining of steels 12 × 2 H4A and 65Γ, in which acetamide and potassium hydroxide are introduced along with stearic and oleic acid. This provides improved lubricating and cooling properties of the lubricant, thereby increasing the time between grinding wheel changes, which increases productivity and reduces the release of abrasive dust in the workplace of the polisher, which reduces the likelihood of occupational lung disease.

References

1. Izmerov, N.F., Shifman, B.B., Artamonova, V.G., et al.: A new study on field lung disease caused by silicate dust medical and environmental problems of workers. Bull. Sci. Counc. 1, 26 (2003)
2. Poteryaeva, E.L., Ivashchenko, I.E., Logvinenko, I.N., Nesina, I.A., Erzin, D.A., Permina, I. I.: Evaluation of the effectiveness of aerosol therapy in the rehabilitation programs for patients with occupational diseases of the broncho-pulmonary apparatus. In: Occupational Medicine and Industrial Ecology, Moscow, no. 4, pp. 25–27 (2007)
3. Ivashchenko, I.E., Poteryaeva, E.L., Nesina, N.A.: Assessment of the quality of life of patients with occupational pathology of the bronchopulmonary apparatus in the dynamics of treatment and rehabilitation programs in therapy. In: Modern Aspects of Restorative Medicine. Materials II Scientific Conference of Doctors, Novosibirsk, p. 27 (2007)
4. Yakimov, A.A.: Technological basis for ensuring the quality of the surface layer when grinding gear wheels, p. 456. Odessa, Astroprint (2003)
5. Zhukov, N.P., Maynikova, N.F., Chudrilin, A.V.: Increasing the life of the abrasive tool by modifying the impregnating composition. In: Problems of Energy and Resource Saving: Collection of Scientific, Proceedings, pp. 219–227. SSTU, Saratov (2010)
6. Chirikov, G.V.: Effect of grinding wheel impregnation on the quality of processing, no. 2, pp. 22–23. Mechanical Engineering Technology (2007)
7. Krumnev, G.P., Naddachin, V.B., Sokolov, V.F.: Expansion of technological capabilities of grinding wheels by impregnating their surface or volume. Inf. Technol. Educ. Product. **3**(8), 160–165 (2015)
8. Yusupov, G.Kh., Koleiv, S.A.: The influence of physico-chemical phenomena on the relationship of abrasive grains with the processed material during the cutting process. Intellectual Systems in Production, no. 1, pp. 206–209 (2010)

9. Morozov, A.V., Mitrofanov, A.P.: Impact of impregnation on the durability of abrasive tools and the quality of the machined surface when grinding bearing rings. In: Problems of Tukhniki and Technology (Technology 2011): Colloquium of XIV International Scientific and Technical Conference (Orel, 5–7 okm. 2011), pp. 40–43. Technological Institute, N.N. Polikarpova Federal State Budgetary Educational Institution of Higher Professional Education “State University UNPK”, Orel (2011)
10. Stepanov, E.V.: Impregnation of abrasive wheels. Young Sci. **10**, 206–209 (2013)
11. Feng, B.F., Su, H.L., Zhang, Q.Z., Zheng, L., Gai, Q.F., Cai, G.Q.: Grinding and energy in high speed. Grinding for quenched steel. In: Key Engineering Materials, vol. 416, pp. 504–508 (2009)
12. Xiv, S.C., Geng, Z.J., Xiu, P.B.: Experimental study on point grinding technical parameters affecting coolant jet parameters. In: Key Engineering Materials, vol. 416, pp. 13–17 (2009)
13. Nikitin, A.V.: Grinding of difficult-to-work materials with impregnated circles as a way to increase their cutting properties. Tools Technol. **2**, 52–58 (2010)
14. Nosenko, V.A., Mitrofanov, A.A.: Research and application of azodicarbonamide for impregnating abrasive tools. In: Latysheva, V.N. (ed.) Physics, Chemistry and Mechanics of Tribosystems: Interuniversity Collection of Scientific Papers, no. 9, pp. 145–150. Ivan. State University, Ivanovo (2010)
15. Nosenko, V.A., Mitrofanov, A.P., Butov, G.M.: Study of the use of impregnators from the class of porophores for the impregnation of abrasive tools, no. 8, pp. 35–40. STIN (2011)
16. Stepanov, A.V., Vetkasov, N.N.: Simulation of the thermal stress of flat grinding with the use of a solid lubricant with fillers of nanomaterials and highly dispersed materials, no. 5 (184), pp. 85–90. News of the Volgograd State Technical University (2016)
17. Vettsov, N.N., Stepanov, A.V., Sapunov, V.: Improving the quality of the surface layer of polished blanks by applying solid lubricant pencils with fillers of ultrafine natural and nanomaterials. In: Nauklemkie Technologies in Mechanical Engineering and Aeronautical Engineering: Proceedings of the IV International Scientific and Technical Conference in 2 Parts, RGATU them. P.A. Solovyov, Rybinsk, 4. II, pp. 59–62 (2012)
18. Nosenko, V.F., Mitrofanv, A.P., Butov, G.M.: Impregnation of abrasive tools with foaming agents. Russ. Eng. Res. **31**(11), 1160–1163 (2011)
19. Zelentsova, L.V., Denisov, A.V.: To the question about the influence of the SOTS on the process of defect formation in round external grinding. In: Proceedings of the IV All-Russian Conference-Seminar “Scientific and Technical Creativity: Problems and Prospects”, Syzran, 22 May 2009, pp. 13–18. Sam Publishing House, Samara, State Tech. un-that. (2009)
20. Vekatov, N.N., Stepanov, A.V.: Calculation of surface roughness, polished using solid lubricants, no. 3–1 (33-1), pp. 36–41. Vector of Science at TSU (2015)
21. Yakimov, A.A., Vinnikova, V.I.: Lubricant for mechanical processing of materials. Patent 9586 Ukraine, C10M133; No. 93111448; declare 31.13.92, Bul# 1. – 4p (1999)
22. Vektasov, N.N., Stepanov, A.V., Sapunov, V.V.: Improving the efficiency of flat grinding by applying pencils of solid lubricant with nanomaterial fillers. In: Drozorov, A.V. (ed.) Youth and Science of the XXI Century: Proceedings of the III International Scientific Practical Conference, 23–26 November 2010, pp. 128–131. State Agricultural Academy, Ulyanovsk (2010)

Simulation Centrifugal Casting of the Heat Resistant Austenitic Steel HK 30 Modified by Niobium

Alen Delić[1(✉)], Mirsada Oruč[2], Milenko Rimac[2], Andrej Kump[3], and Pontus Anderson[3]

[1] TTU energetik d.o.o., 75000 Tuzla, Bosnia and Herzegovina
alen.delic@ttuenergetik.ba
[2] University of Zenica, 72000 Zenica, Bosnia and Herzegovina
[3] Novacast Systems AB, 372 38 Ronneby, Sweden

Abstract. Simulation casting of has become a powerful tool to visualize mould filling, solidification and cooling, and to predict the location of internal defects such as shrinkage porosity, sand inclusions and cold shuts. It can be used for existing castings, and for developing new castings. This process in casting technology is an innovative casting process simulation tool that basically simulates mold filling and solidification, and provides the possibility of simulating casting production. The simulation process in any case reduces production cost and optimizes the technological process of casting. It's possible simulate most commercial casting methods, as well as the process of centrifugal casting of heat resistant steel and alloys. Simulations visualize the consequences of a specific design of gating systems and moulds. Optimising the design of the gating and venting system can be avoided casting defect, such as oxide inclusions due to excessive turbulence, cold-shuts, shrinkage and porosity.

In this paper presents the application of the 3D CAD model in technology centrifugal casting tube of HK 30 Nb steel.

Keywords: 3D model · Simulation · Casting · Centrifugal casting

1 Introduction

The main input to a simulation program is the 3D CAD model of the casting. The CAD model are mainly delivered ba the customer, and this is followed by the design and modeling of mould, cores, feeders, feed aids and gating system. The program introduced the specification of materials that include cast metal, mould, core, feedaids, and process parameters, type of mould, heat transfer coefficient of metal and pouring temperature.

The main outputs of simulation programs include animated visualisation of mould filling, casting solidification, and further cooling to room temperature. Mould filling simulation helps in predicting the total filling time, mould erosion, incomplete filling and air etrapment. The simulation casting solidification shows the temperatures, gradients and cooling rates inside the casting, which are used for predicting the location of shrinkage porosity based on Niyama and other criteria. Further cooling to room

I. Karabegović (Ed.): NT 2019, LNNS 76, pp. 146–151, 2020.
https://doi.org/10.1007/978-3-030-18072-0_16

temperature can also be simulated, which is useful for predicting microstructure, mechanical properties, residual stresses and distortion.

2 Simulation Casting

There are three main applications of casting simulation programs:

- solving the problems for existing castings,
- method optimization, and
- improvement design part.

Solving the problems for existing castings is needed for existing castings that have unusually high or unexpectedly level of internal defects or poor yield.

Thus, method optimization is useful for both existing castings, and those under development for the first time, by eliminating trials. Figure 1 shows the comparison between experimental and virtual casting. The method design (casting orientation, mould layout, feeders, feedaids, and gating) is modified on computer, and simulated to check for defects, if any. Several iterations are carried out until the desired quality and yield are achieved. Even minor improvements in existing castings that are produced in large numbers, can lead to significant improvements in utilization of material, energy, equipment and labour resources.

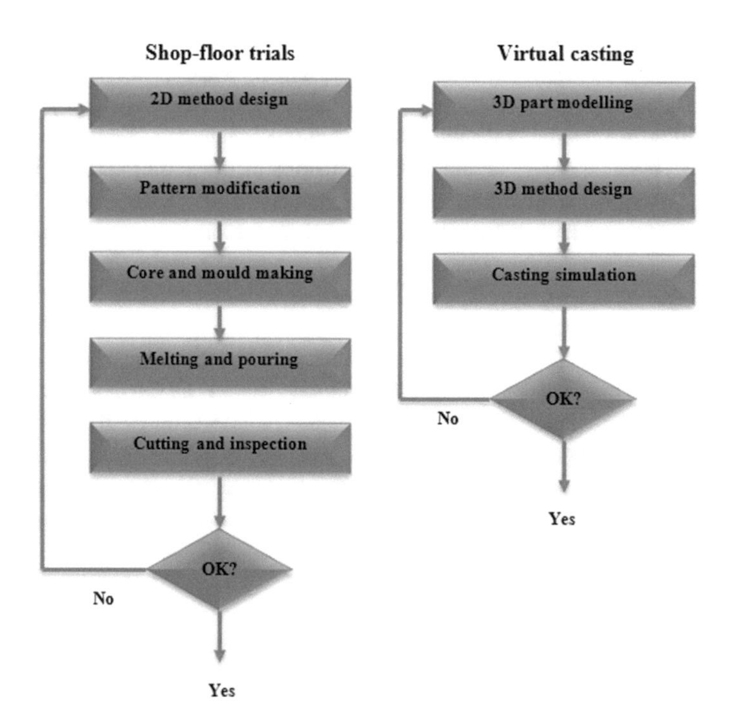

Fig. 1. Comparison of manual and computer-aided method optimization

Minor changes in part design can significantly improve its production without affecting its functionality. However, this decision can only be made by the designer of parts, in consultation with the foundry, with the help of simulation programs.

The simulation project has five levels as shown in Fig. 2.

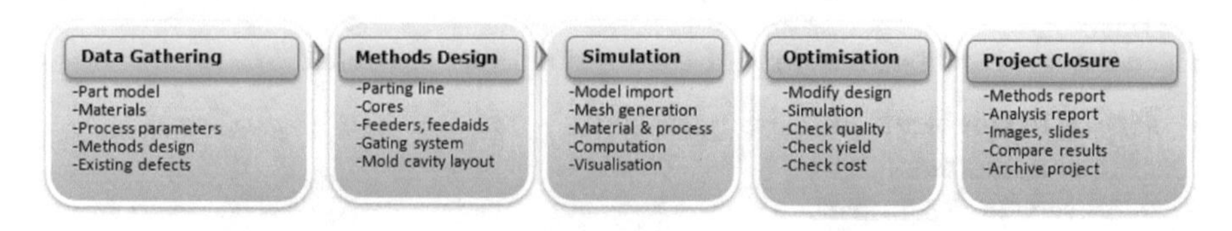

Fig. 2. Major stages in casting simulation and optimization

In the simulation process, the most important stage of the process is data collection, as inaccurate and incomplete data leads to inaccurate simulations and wrong conclusions. This is the basis for defining a problem that chooses the type of simulation. The casting model should be 3D, should be saved in an exchange standard like STL (abbreviation of "stereolithography").

The properties of the cast metal including density, thermal conductivity, specific heat, latent heat, volumetric shrinkage, viscosity and surface tension of the cast metal as a function of temperature are available in the simulation program database. The database also contains data on mold properties and process parameters related to the temperature, thermal conductivity and specific heat of the mold material, as well as the required process parameters of the temperature and the time of casting.

The data required for design refers to quality assurance and yield improved of existing casting in relation to design used in foundry, including details of parts of the model, cores, feeders, cavity layout and dimensions of the gating system. Photograph of the full casting or radiographs of internal defects is the starting point for solving them. In the case of new casting, the suggest design must be provided as a starting point.

At this level of design, the 3D model of castings is converted into a 3D mold model with feeders, gating channels, core and feed-aids. If design is not available the engineer must design based on the mentioned elements and his experience and the foundry practice. This requires knowledge of both methods as well as CAD skills. Also, after each round of the simulation the design can be changed based on the result, so the mold model can be recreated again.

The main stages in the process is the simulation of casting. The first step is to form the correct FEM (Finite Element Method) mesh. The size of the elements must be optimal, the mesh must cover the interior of the model without any gaps. The second key input factor and the adjustment are different boundary conditions, such as the reciprocal transfer coefficient of heat.

The three main types of relationships in the simulation are:

– metal-mold,
– mold-environment and
– metal-environment.

Optimization includes design enhancement to eliminate defects and improve yield. For existing castings, the simulation results are compared with the observed defects to determine the cause of the defect. Then the design is modified, the model is introduced into the simulation program and after the necessary pre-processing the casting is simulated again.

2.1 Simulation of Tube Casting in the Centrifugal Casting Process

The scheme of the complete horizontal centrifugal casting system used in the simulation of casting tube of steel HK 30 Nb according standard, is given in Fig. 3, which is also indicated on the coat coating, which was also considered in the simulation of casting.

The input for the centrifugal casting simulation program was the 3D CAD drawing of the casting, molding, front and rear shutter models, as shown in Fig. 4.

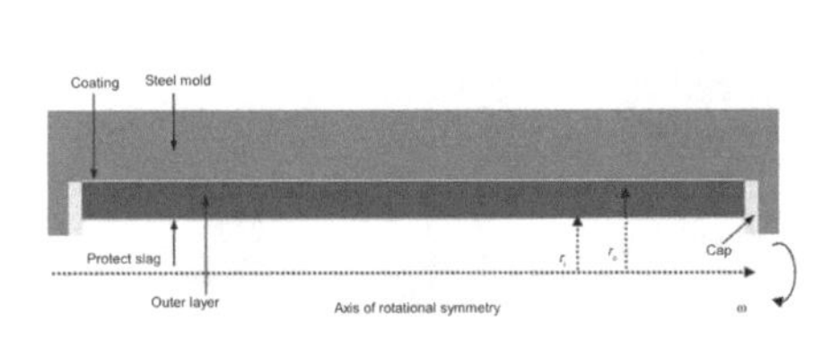

Fig. 3. The scheme of the complete horizontal centrifugal casting

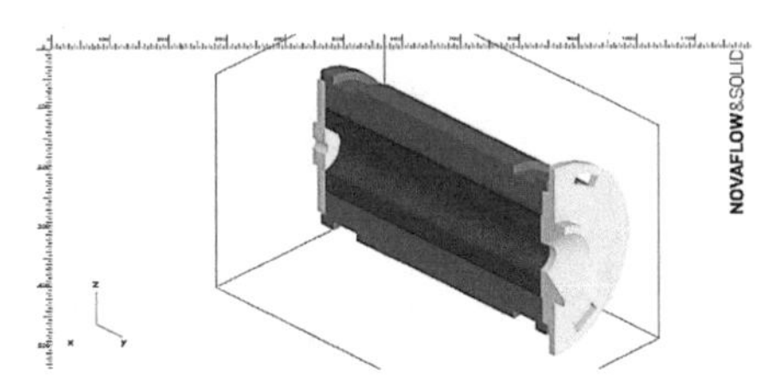

Fig. 4. The input 3D modeling program for simulation

The orientation of the parts in the model was not compatible, so the rotation function was used, after which a cast model was made, or the orientation of the existing input 3D model was changed. In essence, instead of using the least squares method, the least volume method was used. By adjusting the parameters of the network the model is divided into small hexagons (cubes), combined with edge cells, which represents a mathematical approximation that fully corresponds to the original model. In this case, the cell size is no longer so critical, so larger cells could be used. By changing the cell size, the cell number changed, after which it was determined how many cells were generated by this adjustment. For setting the simulation parameters, the first type of casting or centrifugal casting was chosen first. Thereafter, the charging parameters and rotation speeds set out in Table 1 are set.

The next step was to select the choice of mould material and casting material, which was selected in the database. In the event that there was no adequate material in the database menu, which is the cast in this simulation, it is possible to create completely new materials with an adequate chemical composition. Together with the choice of materials, the initial temperature of the casting was made, and the same procedure was carried out for other materials, as shown on the monitor, Fig. 5.

Table 1. The process parameters centrifugal casting which used in simulation

The process parameters centrifugal casting	
Rotation speed of the mould	1600 o/min
Rotation speed before pouring	900 o/min
Pouring time	16 s (4,5 kg/s)
Acceleration time	20 s
Deceleration time	30 s
Time before cooling	27 s
Cooling time	120 s
Mold preheated temperature	220 °C

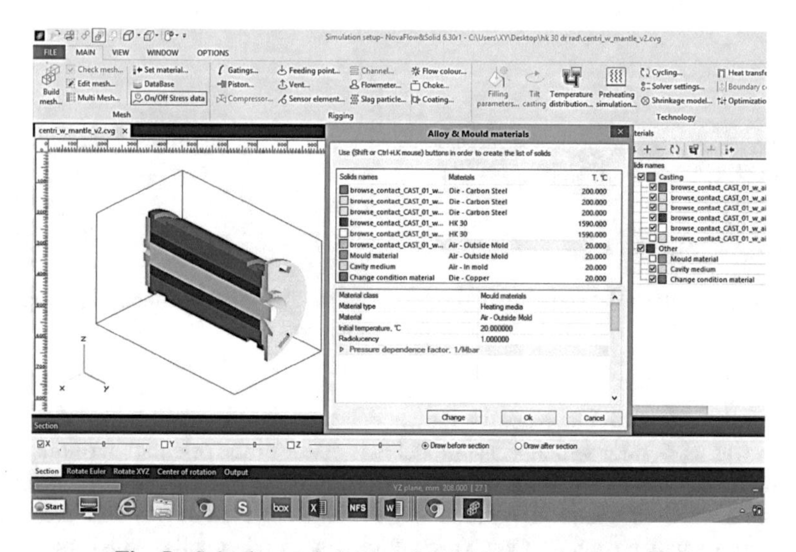

Fig. 5. Selection and change of mold and casting materials

For the simulation of the filling of the casting, an input point is defined as a part of the selective area on the front of the mold. At this stage, a flowmeter and sensors for measuring temperature, velocity and pressure are created. By measuring the efficiency of the power supply system, the amount of metal flowing in kg/s, l/s and m/s was measured. Temperature, velocity and pressure sensors are placed in the casting. The choice of insulation coil, front and rear shutter automatically generated the heat transfer value in W/mK.

In the simulation procedure, the influence of gravity on shrinkage depends on the type of casting material and is listed in Table 2. In this option (Fig. 5) the shrinkage can be excluded or modified to correlate with the experimental data of the destruction test. It should be noted that the gravity of steel casting has a great influence on the prediction of shrinkage. In the case of less intensive power supply, it is necessary to disable the influence of gravity.

Table 2. General recommendations gravitation impact on shrinkage prediction

Type of casting	Gravity influence
Steel	High
Iron, Grey	Low
Iron, Ductile	Medium
Non Ferrous	High

Heat transfer model was selected for surface heat transfer, after which simulation parameters were recorded. When a simulation module is opened, a previously prepared model appears in the window. The end point of the simulation is controlled by self-decomposition parameters. The simulation ends when the automatic stop criterion has been reached.

Results such as temperature, liquid phase, shrinkage, speed and other necessary data can be seen in the simulation program browser.

The results can be presented in the following forms:

- Powerful browsing and slicing in x, y and z directions,
- Built-in animation functions presenting results,
- Creation of AVI and real time AVI movie files as well as WMV,
- Two or more simulations can be viewed simultaneously in the browser,
- Possibility to save simulations in BMP or JPEG formats in each module,
- Automatic report generator in doc-format.

3 Conclusion

The application of the simulation of the centrifugal casting process in the tube, the safety in the selection of technological parameters, the experimental casting itself, the exclusion of certain technological procedures, and the casting achieved a satisfactory level of achieved quality, mechanical and other exploitation properties. The application of software of steel HK 30 Nb centrifugal casting technology has shown significant results in quality of production, and certainly opens up opportunities for conquering and manufacturing new types of such and similar castings.

Multi Roller Cyclo Reducing Gear

Petar Karabadjakov[1], Ivan Balashev[2], and Ivan Stoianov[3](✉)

[1] ET "Ingeborg Demirova", Gabrovo, Bulgaria
pkarabat@gmail.com
[2] TU Gabrovo, Gabrovo, Bulgaria
balashev@tugab.bg
[3] "Podem Gabrovo" Ltd., Gabrovo, Bulgaria
creative_studio1979@abv.bg

Abstract. On the basis of the patents of one of the authors, constructions of a range of multi roller cyclo reducing gear have been developed, having the advantage of achieving a high gear ratio of one stage. The main element of the cyclic gearbox is a generator delivering a translational movement created by input eccentric shaft rollers. The transmission of the torque from the generator is accomplished by engaging a large number of rollers in a ring gear to redistribute the load of the cyclo reducing gear. It is compact, with small dimensions. In the elaboration, dependencies are derived for determination of the transmission ratio and the geometric dimensions of the elements of the cyclo reducing gear.

Keywords: Multi roller cyclo drive

1 Introduction

In the drive systems mechanical transmissions are used to match the parameters (angular speed, rotary torque, etc.) of the motor and the operated machine. In this paper, basing on patents of one of the authors, constructions of a range of multi-roller cyclo reducers have been developed, having the advantage of realizing great transmission ratio with one stage.

The wave reducers meet these requirements excluding the technology of the elastic ring gear with external teeth which is subjected to the greatest loading and must endure a large number of deformations. To avoid this disadvantage, Eng. Petar Karabadjakov's patents [1, 2] consider replacement of the elastic ring gear with external teeth by a standard bush-roller chain, or by calibrated rollers only, placed on the external cylindrical surface of the generator at even steps. Reducers type K-H-V are known [3, 4, 6] which, in case of shank engagement, can be made with difference between the number of teeth of the ring gear with internal teeth (rollers) and gear with external teeth (cycloid) equal to one. Working out epicycloid or hypocycloid requires special technological equipment. In the paper [5, 7] as adjusted cyclo transmissions the gearings with internal shank epicycloid and shank hypocycloid engagement are considered,

I. Karabegović (Ed.): NT 2019, LNNS 76, pp. 152–157, 2020.
https://doi.org/10.1007/978-3-030-18072-0_17

with difference between the number of teeth of the two gears equal to one. The main element of the cyclo reducer is a generator with a great number of rollers performing transmission motion from the input eccentric shaft. The transmission of the generator torque is effected by engaging a great number of rollers in a ring gear with internal teeth, thus re-distributing the load of the cyclo reducer elements. The cyclo reducer is a compact and small sized unit.

2 Construction of a Multi-roller Cyclo Reducer

The multi-roller cyclo reducer can be realized under two construction schemes:

Scheme 1 (Fig. 1): the ring gear 2 is connected with the output shaft 3 of a travel wheel of an electric hoist monorail trolley and the separator 4 (Fig. 3) is fixed; or

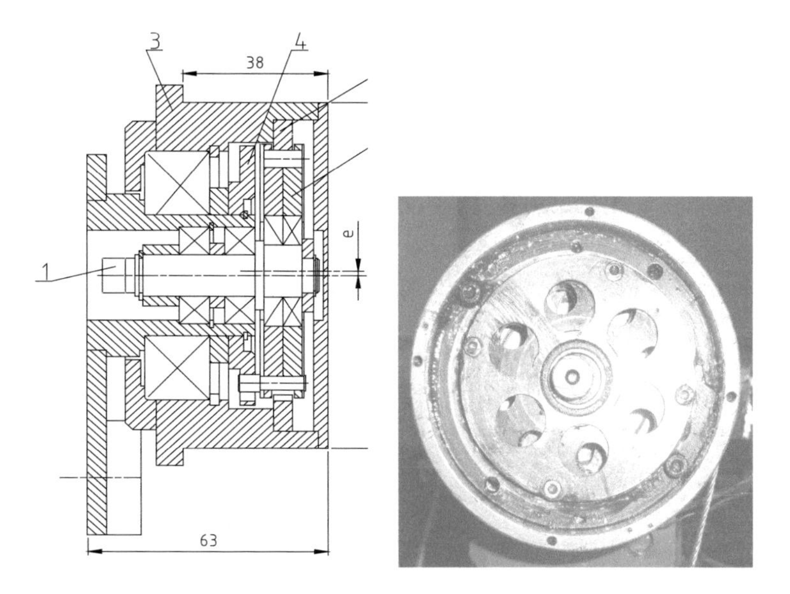

Fig. 1. Constructive scheme 1

Scheme 2 (Fig. 2) – (coaxial cyclo reducer): the separator 4 is connected with the output shaft 1, and the ring gear 2 (Fig. 4) is fixed.

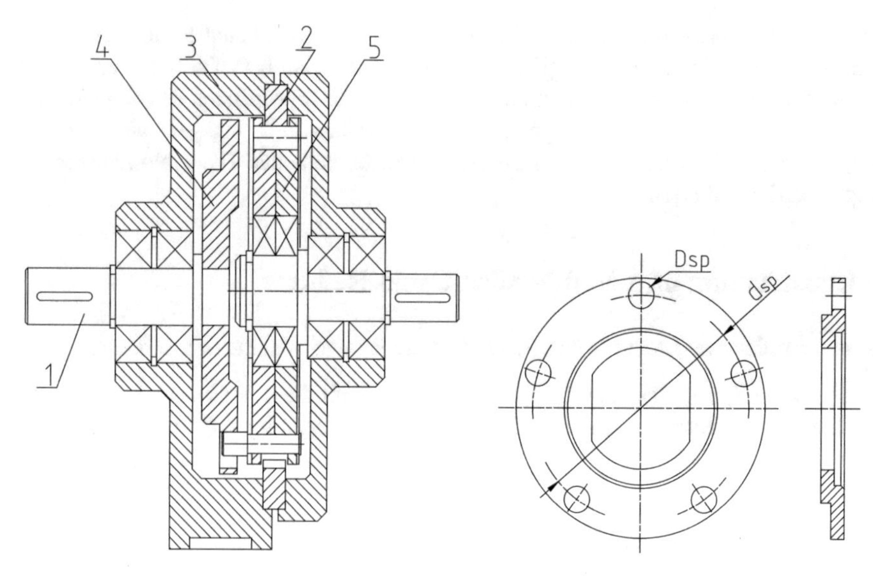

Fig. 2. Constructive scheme 2

Fig. 3. Separator

2.1 Kinematic and Geometric Dependences

Kinematic Dependences:

The cyclo reducer transmission ratio is defined as follows:

For Scheme 1 (Fig. 1):

$$i = \frac{z_v}{z_v - z_g}, \tag{1}$$

For Scheme 2 (Fig. 2):

$$i = \frac{z_g}{z_v - z_g}, \tag{2}$$

where z_v is the number of ring gear teeth, and z_g - the number of generator rollers and $z_v > z_g;\ z_v - z_g = 1$.

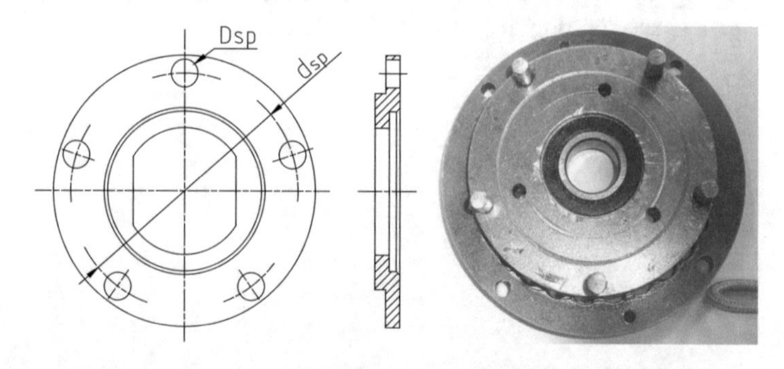

Fig. 4. Generator and a set with ring gear

Geometric Dependences:
As a basic parameter a standard module m of the transmission is selected, and roller diameter $d = 2.m$ is set; then, input shaft eccentricity e = 0,5.m (Fig. 5) is selected.

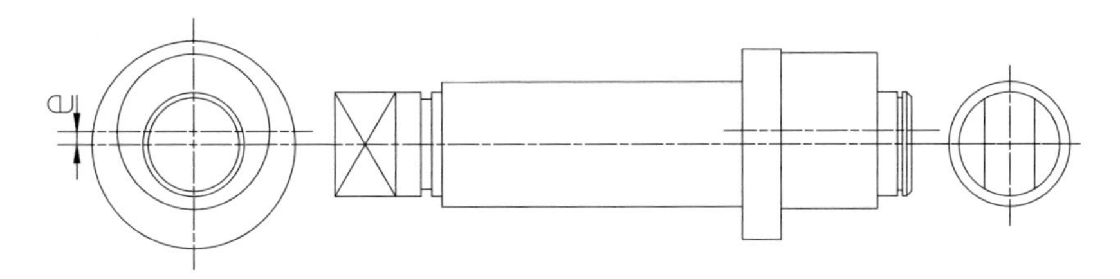

Fig. 5. Input shaft

The generator sizes (Fig. 4) are defined as follows:

$$d_g = m.z_g \ d_{ga} = d_g + d \ d_{gf} = d_g - d \ R_g = d/2 \tag{3}$$

where d_g is the reference diameter, d_{gf} - the internal diameter, d_{ga} - the external diameter of the generator.

For the engagement between the external generator diameter d_{ga} and the external ring gear diameter d_{va}, the gap $w = d_{va} - d_{ga}$ = 0,2.m (Fig. 4) is set.

The ring gear sizes (Fig. 6) are defined as follows:

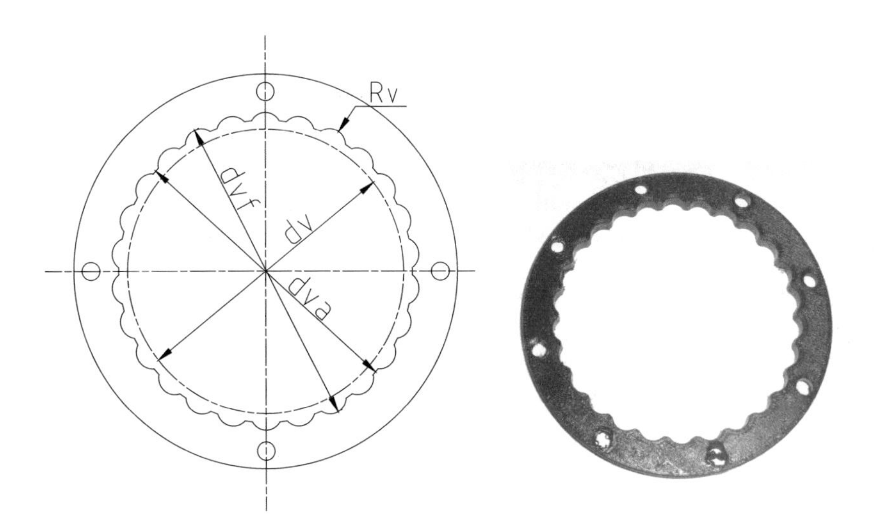

Fig. 6. Ring gear

$$\begin{aligned} d_v &= m.z_v \; d_{vf} = z_g.m + d + 2.e + w = m.(z_g + 3,3) = m.(z_v + 2,2) \\ d_{va} &= z_g.m + d - e + w = m.(z_g + 1,2) \\ d_{va} &= z_g.m + d - e + w = m.(z_g + 1,2) = z_g.m + d - 2.e + w \end{aligned} \quad (4)$$

where d_v is the reference diameter, d_{vf} - the internal diameter, d_{va} - the external ring gear diameter, $Rv = 2.m$ - the radius of the arcs forming the tooth.

The separator sizes (Fig. 3) are defined as follows:

The separator reference diameter - $d_{sp} = m.z_g$; the diameter of separator holes - $D_{sp} = 2.e + d = 3.m$; number of separator holes - $N = 5$.

Experimental model of a multi-roller cyclo-reducer under construction scheme 1 (Fig. 1) was made, with parameters set as follows: standard module m = 2,5 [mm] and input shaft eccentricity e = 1,25 [mm].

The geometric sizes are shown in Table 1.

Table 1. Data of the multi-roller cyclo-reducer experimental model

Sizes, mm									
Generator					Ring gear				
$z_g = 23$ $m = 2,5$	d_g	d_{ga}	d_{gf}	R_g	$z_v = 24$ $m = 2,5$	d_v	d_{va}	d_{vf}	R_v
	57,5	62,5	52,5	2,5		60	62	65	5
Separator	$D_{sp} = m.z_g = 57,5$			$d_{sp} = 3.m = 7,5$					

3 Conclusions

1. Methodology has been elaborated for geometric sizing of the main elements of a multi-roller cyclo-reducer;
2. Options for implementation of multi-roller cyclo-reducer in the material handling equipment have been presented;
3. The geometry of the multi-roller cyclo-reducer considered achieve great transmission ratios, great number of rollers engaged with most favourable contact surfaces and high efficiency coefficient realized in relatively small sizes;
4. Experimental model of a multi-roller cyclo-reducer has been elaborated, built in the wheel of an electric hoist monorail trolley.

References

1. Karabadjakov, P.: Patent "lifting mechanism" no. 111793 (2014)
2. Karabadjakov, P.: Patent "lifting mechanism with multi-stage reducer" no. 11829 (2014)
3. Shannikov, V.M.: Theory and construction of reducers with eccentric cycloid. Mashgiz (1959)

4. Kudrjavzev, V.N.: Planetary transmissions. Machinostroene, Moscow (1966)
5. Dolchinkov, R.: Chain-cycloid transmissions, BSU, OTT. In: Scientific Conference with International Participation, June 2013
6. Arnaudov, K., Karaivanov, D.: Planetary transmissions. Bulgarian Academy of Science Publishers Prof. Marin Drinov
7. CYCLO-Getriebe. Prospekt der Firma CYCLO-Getriebebau, Lorenz Brareu KG/BRD

Analysis of the Comparative Advantages of Gear Pumps Indicators

Milutin Živković[1], Predrag Dašić[1,2(✉)], and Pravdić Predrag[1]

[1] High Technical Mechanical School of Professional Studies, Str. Radoja Krstića 19, 37240 Trstenik, Serbia
predragdasic@gmail.com, dasicp58@gmail.com
[2] SaTCIP Publisher Ltd., Str. Tržni Centar Pijaca no. 101, 36210 Vrnjačka Banja, Serbia

Abstract. Hydraulic pumps belong to the basic devices of each hydraulic system, because they convert mechanical energy to the pressure of the working fluid. As a significant number of drive solutions are realized by gear pumps, special attention will be dedicated to them in this paper. Their basic comparative advantage is in the simplicity of conservative solutions as well as relatively simple production. This implies, for the buyer, an acceptable price and not a negligible or favorable mass ratio and the achieved strength. For this reason, the authors of this paper will define all the indicators of this conclusion. In support of this assertion, we will analyze the gears of production of the famous Serbian manufacturer, PPT-HIDRAULIKA ad from Trstenik. The same will show tabs and diagrams.

Keywords: Gear pumps · Weight ratio · Pump power · Comparative comparisons

1 Introduction

There is no hydraulic system without pump/s, the basic device of each hydraulic system. These are, it is known, devices that convert (input) mechanical energy to the pressurized energy of the working liquid. Their basic comparative advantages are confirmed in practical application, as they are: simplicity of constructivity of the solution, low production price, safe and reliable exploitation, relatively favorable volume efficiency, in comparison to other types of drives, Fig. 1 [1, 2].

For this reason, the authors of this paper will show by parallel analysis of drive parameters that they represent the key to solving many drive requirements. They relate to the choice of the optimum ratio of the purchase price and the efficiency of the hydraulic system. The subject of the analysis is the gear pumps of production of PPT Hidraulika ad from Trstenik (Serbia), which are produced in Trstenik since 1958. They were once produced in significant quantities, but in recent years, interest in placing them has grown. It is assumed that the production assortment covers all flow ranges, connection modes and working pressures. They produce them in quality, with technical characteristics, in the ranking of the world's leading manufacturers, constant monitoring of the newspaper and tendencies in design, production and testing methods, which is guarantee of placement and reliability of the supplier.

I. Karabegović (Ed.): NT 2019, LNNS 76, pp. 158–164, 2020.
https://doi.org/10.1007/978-3-030-18072-0_18

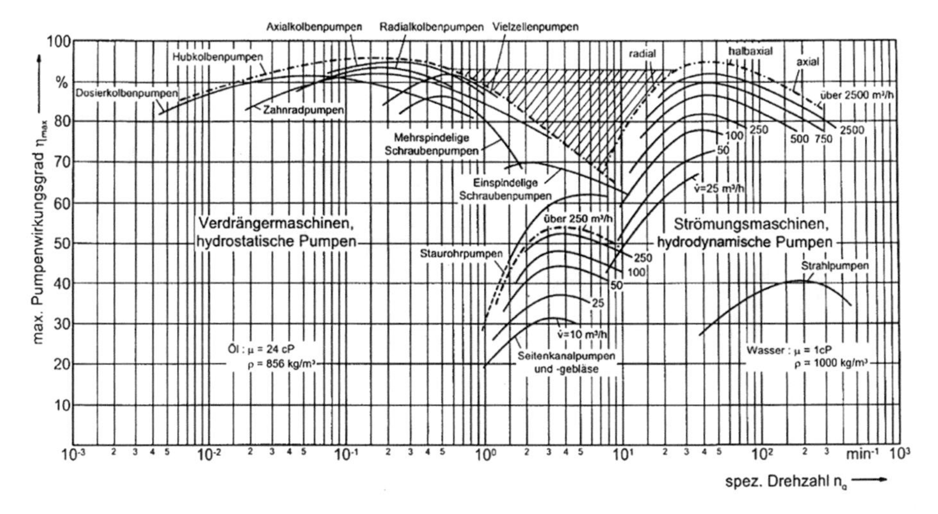

Fig. 1. Extended "Cordier" diagram

2 Gear Pumps (Tendencies in Design)

Own experience and literature data show that almost 80% of drives solutions are achieved by these drive devices. These are systems of general purpose, which gives them the universality of application. They are usually made with an external evolutionary tooth and with the same number of teeth of both gears. Most manufacturers produce them with 12 teeth, on both gears. Due to the specificity of such structures, in terms of working conditions, the devices must be constructively precisely defined, calculated, constructed and tested, in order to guarantee a high level of reliability of the required performance in all conditions of exploitation. The structural parts of the pump, such as the housing, must pay special attention to the choice of geometry, materials and mass distribution. The answer is to meet the criteria of resistance (satisfactory rigidity, desired dynamic behavior, and permissible stresses), material savings (reduction of mass), reduction noise levels (always requirement for hydraulic and pneumatic devices), etc. Product pricing must be taken into account, so the performance of the device is achieved as a compromise of geometric, material and economic criteria. The principle of work is based on changing the volume of working chambers, which allows the suction and pushing of liquids into the pushing pipeline. In the phase of increasing the volume of the working chamber, a sub-print is performed and the chamber is filled, i.e. the suction phase is carried out.

When the working element (gears) decreases the chamber volume in the chambers, a pressure phase, i.e. compression. The pressure of the working fluid depends on the drive torque and is not determined by the structural characteristics of the pump, but is in the function of internal and external resistance in the system, the components can be seen in Fig. 2, [3]. Where is:

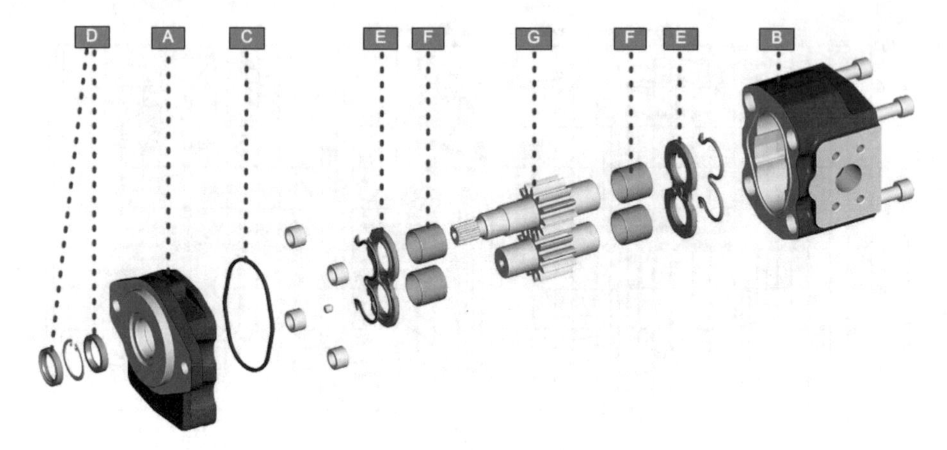

Fig. 2. Axonometric view of a gear pump with an external toothing

A - Cast iron front cover: the standard frontcover design can be fitted to two different pump interfaces,
B - Cast iron main pump body: a wide range of displacements available with two different basic bodies both with back cover integrated. Rear ports on request,
C - HNBR seal material instead of NBR,
D - Double HNBR shaft seals,
E - Strong pressure balance plate instead of aluminum. Balancing area and intermediate notches optimized,
F - Large-diameter bearings, fitted both in front cover and body,
G - A large number of teeth, a tooth profile optimized a larger shaft diameter.

Gear pumps are used as lubrication pumps for head and auxiliary devices as well as fuel pumps for lower pressures. In the high pressure area, they serve as a source of hydraulic energy in open hydraulic work circuits, in which the external loads are constant, which enables application in both stationary and mobile hydraulic systems and systems. Gear pumps, unlike other types of pumps, with their constructive solutions allow relatively high numbers of rings and pressures. New solutions allow for a pressure which until recently was characteristic for piston pumps, even 330 bar. With the possibility of flow regulation, their fields of application have been significantly expanded, which is solved by the sensory control of the frequency and momentum in the working range [4, 5]. There are also solutions with built-in non-return valve, safety valve, constant flow valve and pressure regulation in most of these solutions. Anti-cavitation valves are also installed [3, 6]. In recent years, a flow control solution has also appeared at the set pressure, the design given in Fig. 3 [7].

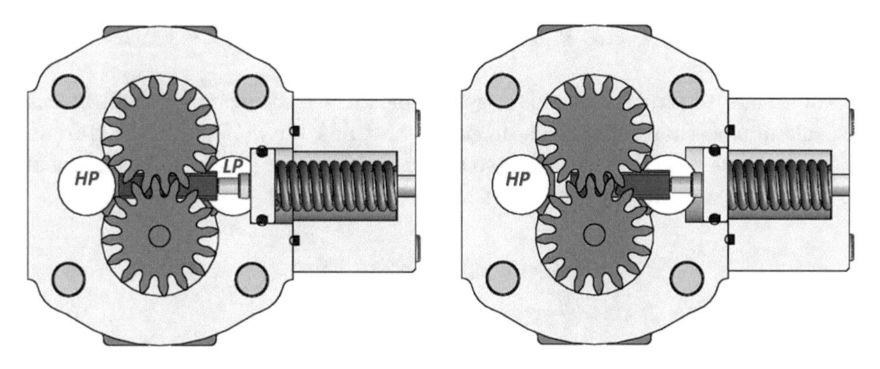

Fig. 3. Slider display for max. and min. flow of gear pumps

The basic problem to be solved in its development and production is the determination of the technical and economic characteristics of each pump. Variant solutions feature a set of positive properties (a high degree of useful effect and high reliability in work, a small specific weight, etc.) and a set of negative properties (increased forest, increase in nominal dimensions, high prices, ..). The basic task of the constructor is to choose the solution that allows to minimize the negative properties of the construction and increase the desired one. The appearance of the disassembled gear pump is shown in Fig. 3 and the analysis of the properties of contemporary solutions is given in Fig. 4 [3].

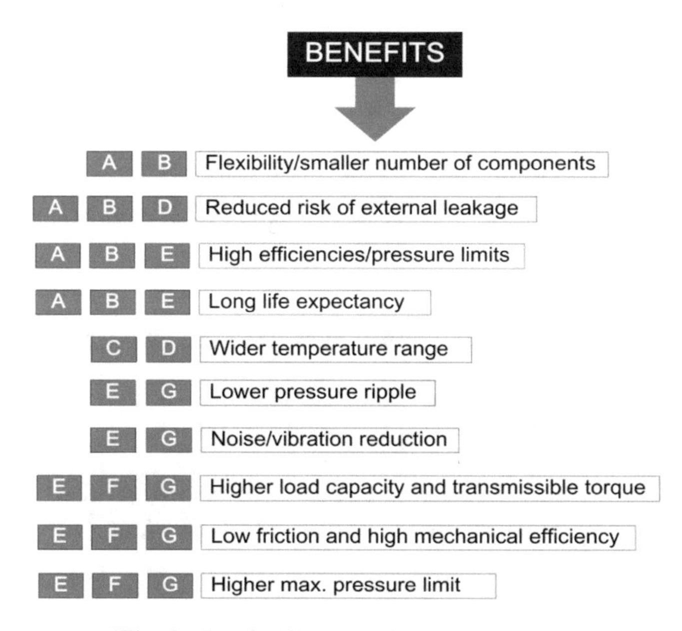

Fig. 4. Benefit of new solutions of gear pumps

3 Production of Gear Pumps PPT Hidraulika Ad - Trstenik

PPT Hidraulika ad produces standard gear pumps (for pressures up to 200) at least in six standard sizes (housings, gears, front and rear lids, etc.). Using the data from the catalog, the mass ratio [kg] according to the achieved power [kW] for the entire family 3115 ** ** ** is shown in parallel, as shown in Fig. 5 [3].

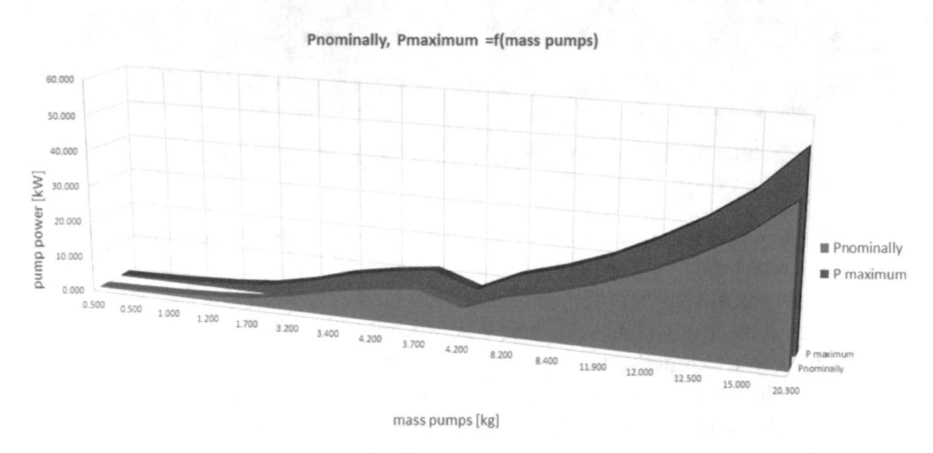

Fig. 5. Ratio of weight and power of gear pumps from group 3115 ** ** **

Due to the needs of the market and the observation of world trends, PPT Hidraulika has also developed series products of gear pumps for more pressures (280 to 320) bar, factory markings ZPB. Table 1 gives their basic characteristics shown in Fig. 6 [3].

Table 1. Technical characteristics of gear pumps from group ZPB

Specific pump flow; q [cm^3/o]		4,00	5,50	8,00	11,00	14,00	16,00	19,00	22,5
Pump speed; n [o/min]		1500							
Pressure [bar-a]	Nominal	280						250	
	Maximum	320						280	
Theoretical flow of the pump $Q = q \cdot n \cdot 10^{-3}$ [l/min]		6,00	8,25	12,0	16,5	21,0	24,0	28,5	33,75
The power of the pump; $P = \frac{p \cdot Q}{600}$ [kW]	Nominal	2,8	3,85	5,6	7,7	9,8	11,2	11,875	14,10
	Maximum	3,2	4,4	6,67	8,8	11,2	12,8	13,3	15,75
Pump weight [kg]		2,8	2,8	2,9	3,0	3,2	3,4	3,6	3,8
Weight relationship; $\eta = \frac{m}{P}$ [g/kW]	Nominal	1,0	0,73	0,52	0,39	0,33	0,30	0,30	0,27
	Maximum	0,875	0,64	0,435	0,34	0,286	0,27	0,27	0,24

It should be emphasized that PPT Hidraulika produces, in addition to other devices, also gears with an already installed non-return valve, a constant flow valve and a safety valve [3, 8, 9].

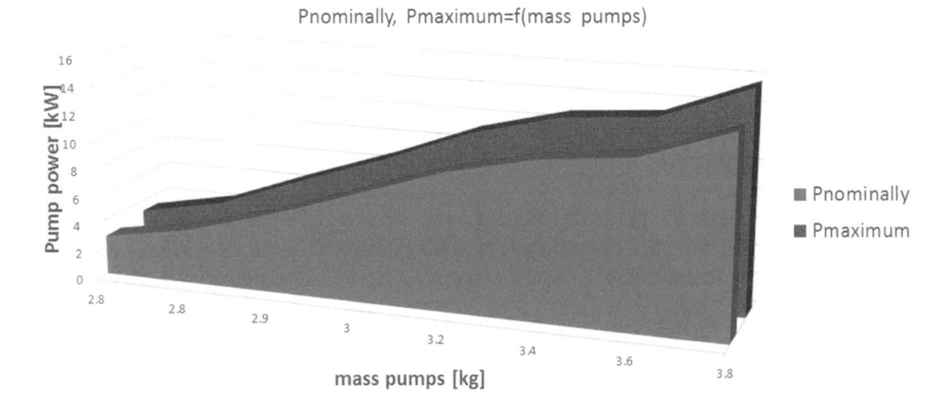

Fig. 6. The ratio of the weight and the achieved power of the gear pumps from the ZPB group

4 Conclusion

PPT - Hidraulika follows global trends in the development of these devices, which initiates the improvement of existing ones and the development of new construction and technology solutions. All manufactured devices, year after year, confirm the level of quality at home and in the increasingly demanding world market [3, 8, 9]. The development of gear pumps in the foregoing directions has led to a slight improvement in all its parameters, as well as to a significant reduction in noise, as well as an increase in utility, which is a characteristic of the achieved technical level. There is no less significant reduction of the weight ratio and the achieved strength. These are all parameters for evaluating a reliable supplier. The conclusion is that:

- Figures 5 and 6 provide the opportunity to fully monitor the benefits of ZPB pumps, which is confirmed by the data from Table 1, especially at higher specific flows;
- PPT Hidraulika manufactures and specially designed pumps, with built-in valves (non-returnable, safe and constant flow valve).

References

1. Gordon Mohn, G., Nafz, T.: Swash plate pumps – the key to the future. In: 10th International Fluid Power Conference, Dresden, Group 3-Pumps - Paper 3-1, pp. 139–150 (2016)
2. Grabow, G.: Optimalbereiche von Fluidenergiemaschinen – Pumpen und Verdichter. In: Forschung im Ingenieurwesen, vol. 67, pp. 100–106. Springer (2002). https://doi.org/10.1007/s10010-002-0084-1
3. PPT - HIDRAULIKA AD. Komercijalno tehnička dokumentacija za pumpe iz grupe 3115 i ZPB (viši pritisci) (2018)
4. Rexroth – Bosch Group: Sytronix – variable-speed pump drives (Energy-efficient – Intelligent-Cost-effective, R999000332 (2015-08)©Bosch Rexroth AG (2015)
5. ABB DRIVES: Technical guide No. 4 Guide to variable speed drives, Specifications subject to change without notice. 3AFE61389211 REVC12.5.2011 (2011)

6. CASAPPA: BUILIT-IN VALVES, for hydraulicgear pumps and motors, V01TA
7. Vacca, A., Devendran, R.S.: A flow control system for a novel concept of variable delivery external gear pump. In: 10th International Fluid Power Conference - Dresden 2016, pp. 263–274 (2016)
8. Živković, M., Todorović, D., Vasić, S.: Gear pumps in the past and present. časopis (EMIT)–Economics Management Information Technology, vol. 3, no. 4, pp. 187–195 (2015). ISSN 2217-9011
9. Živković, M., Dašić, P., Petrović, Z.: Influence of new technologies on higher energy efficiency of hydrostatic devices and systems. In: Karabegović, I. (ed.) NT 2018. LNNS, vol. 42, pp. 386–396. Springer (2019). https://doi.org/10.1007/978-3-319-90893-9_46

Flatness Measurement on a Coordinate Measuring Machine

Nermina Zaimovic-Uzunovic(✉), Samir Lemes, Dzenana Tomasevic, and Josip Kacmarcik

Faculty of Mechanical Engineering, University of Zenica, Fakultetska 1, 72000 Zenica, Bosnia and Herzegovina
nzaimovic@mf.unze.ba

Abstract. Different aspects of surface measurement strategies on Coordinate Measuring Machine (CMM) were varied and their influence on the flatness deviation result was investigated. The CMM measurements were conducted using single point and continuous scanning probing. The measurements were performed with five different point densities in rectangular grid sampling strategies and three different probe styli. The results showed a very significant influence of a sampling size on a flatness deviation measurement result.

Keywords: Coordinate Measuring Machine (CMM) · Flatness · Sampling · Measurement strategy

1 Introduction

Coordinate Measuring Machines (CMM) are widely being used to test the geometric tolerances for different industrial parts [1]. Without the CMM, it is very hard to control all the geometric requirements in modern industry. The measurement accuracy and uncertainty of CMM results depends on the different features of the measurement strategies: sample point number and distribution (sampling strategy), probe and styli selection, probing technique and approach speed, data assessment method etc. There are no exact rules for defining measurement strategies in coordinate metrology. An appropriate strategy has to be defined for every individual measurement task; requiring a reliable estimate of uncertainties of measurement results. Some standards (ASME/ANSI, ISO, BS) define various methods for different measurements of geometry. However, these standards do not define the accuracy of measurement results. NPL collected the end user and CMM manufacturer experience on such measurements, and published them recently in NPL's Guide of best practices [2]. The Guide covers some, but not all the deviations that could occur due to poor selection of measuring strategy.

Flatness deviation measurement or tolerance control is common task in coordinate metrology. The different aspects of flatness measurement strategy were widely investigated. Raghunandan and Rao [3] investigated influence of surface roughness on accuracy of flatness inspection and concluded that roughness can be parameter for definition of sample size. The same authors in [4] presented a method to determine an

I. Karabegović (Ed.): NT 2019, LNNS 76, pp. 165–172, 2020.
https://doi.org/10.1007/978-3-030-18072-0_19

optimum sample size for industrial flatness measurements. The key influence of sample size on uncertainty and accuracy of flatness measurement was discussed and showed in many papers, i.e. in [5, 6]. In [7, 8] the knowledge on the manufacturing signature was used as a tool for definition of an appropriate sampling strategy. The sampling strategies were defined using models identified with modelling geometrical features by Discrete Fourier Transform (DFT). The implementation of statistical analysis for sampling strategy design is considered in [9–11]. Zaimovic-Uzunovic and Lemes in [12] investigated measurement strategy in CMM cylindricity measurement, where beside a sampling strategy, the influence of a stylus tip size on CMM results was shown. The highest deviation results were obtained with smallest tip styli, which was explained with the mechanical filtering effect [13], dependent on stylus tip size.

Different characteristics of flatness measuring strategies were varied in this research and their influences on a final flatness deviation result were investigated. The two probing methods (single point and continuous scanning), three stylus tip sizes and number of sampling points were varied in this research.

2 Flatness Definition and Measurement Strategies on CMM

The flatness is defined in ISO 1101 [14] as a distance between two parallel planes, confining the surface, oriented in such manner that the distance between the planes is a minimum possible value, Fig. 1.

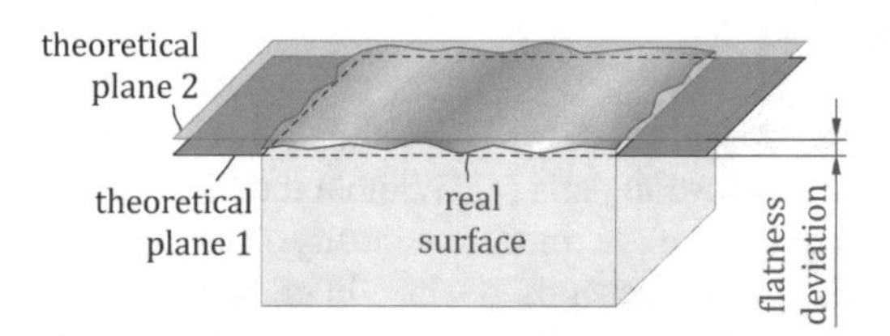

Fig. 1. Flatness deviation

In order to perform the flatness measurement, it is necessary to define a finite number of points and their position (distribution) on the surface to be extracted by CMM measurement. The sampling strategy is one of the key parameters in flatness measurement uncertainty budget, and it is necessary to trade between time (cost) and accuracy (uncertainty). There are two approaches in defining number and distribution of points [2]: application of predefined rules and detail analysis of the part to be measured in order to define the form deviation distribution over the part and hence establish an optimal sampling strategy. The first approach is recommended in different standards and aims to achieve a nearly uniform distribution of probing points on the surface. The second approach is more expensive and time consuming, but it can be justified for parts measured routinely in industrial production. The predefined sampling strategies for measuring flatness deviation were investigated here, namely the grid method defined in ISO 12781-2 [15], Fig. 2. After the points defined by sampling strategy have been acquired, a flatness deviation was calculated by CMM data processing software. There are two criteria defined in different standards and used in CMM software for evaluation of the flatness error: the least squares criterion (LSC) and the minimum zone criterion (MZC) [16]. The least squares criterion is superior regarding simplicity and calculation time [17]. It is widely used in industry and also in this paper.

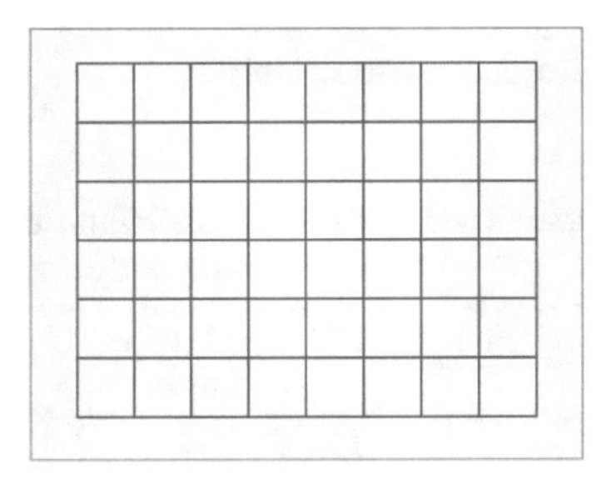

Fig. 2. Rectangular grid extraction strategy (ISO 12781-2)

3 Experimental Setup

An investigation was performed in order to analyse the effect of different probe styli (tip size) and different sampling strategies on the results of flatness measurement with two different probing methods: single point measurement and continuous scanning measurement. The experimental measurements were carried out on the top surface of the part used for CMM training, made of stainless steel, produced by classical machining technologies, Fig. 3. The gross dimensions of the part top surface are 78 × 67 mm. The experiment was performed on the CMM Zeiss Contura G2 (MPE_E = (1.8 + L/300 μm, MPE_P = 1.8 μm), equipped with ZEISS VAST XT scanning probe and CALYPSO 4.8 measurement software [18], Fig. 4.

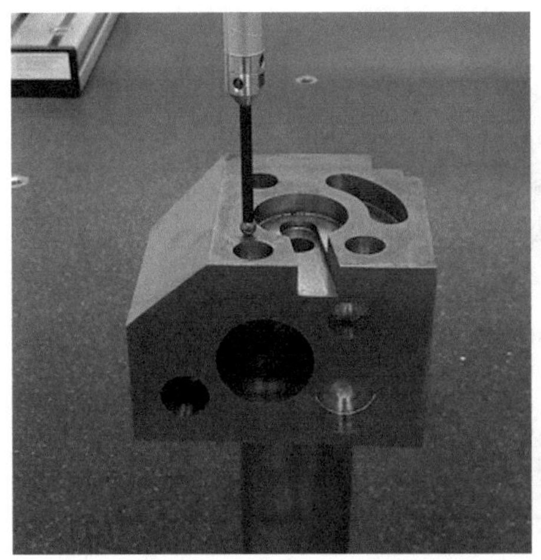

Fig. 3. Measurement of the top surface.

Fig. 4. CMM Zeiss Contura G2.

Rectangular grid sampling strategies were implemented for the measurements, and five different sample sizes were implemented for each probing method. The rectangular grid strategy definition was used and automated in CMM software. The sample size was defined by the number of lines in two directions of a grid for single points, and by the number of parallel lines (variation of rectangular grid) for scanning probing. Densities (sizes) of all sampling strategies used in this investigation, together with a number of actual sampling points are given in Tables 1 and 2, for single point and continuous scanning measurement, respectively.

Table 1. Definition of strategies used for single point measurement.

Name	Grid size	Number of points
Points 1	4 × 4	13
Points 2	6 × 6	18
Points 3	10 × 10	47
Points 4	15 × 15	96
Points 5	20 × 20	170

Table 2. Definition of strategies used for continuous scanning measurement.

Name	Number of parallel lines	Number of points
Lines 1	6	155
Lines 2	10	423
Lines 3	15	666
Lines 4	20	888
Lines 5	25	1147

The graphical representation for two selected single point strategies and for two selected continuous scanning strategies are given in Figs. 5 and 6, respectively. The scanning measurements were carried out with arbitrarily slow scanning speed of 5 mm/s in order to avoid dynamic effects on results. Hence, no filter or outlier elimination algorithm options for data filtering [16] available in CMM software were applied. Number of points extracted from the scanning lines is defined with the step width of 0.5 mm.

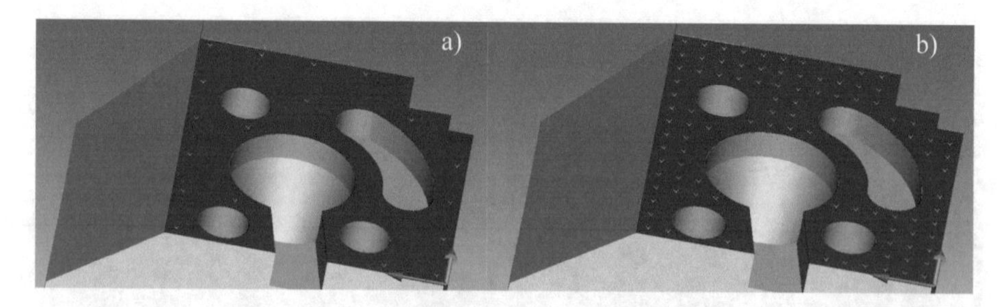

Fig. 5. Single point measurement sampling strategies grids: (a) 6 × 6, (b) 15 × 15

Measurements for every strategy were performed using the three different probes with different styli and stylus tip diameters, Table 3.

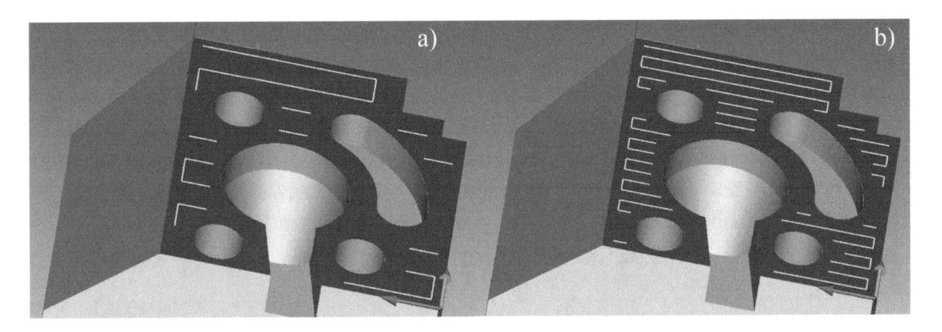

Fig. 6. Continuous scanning measurement sampling strategies grids: (a) 10 × 10, (b) 20 × 20

4 Results

Figures 7 and 8 show the graphical representation of the results. The results reveal the great influence of sampling size on flatness measurement results, similar to the findings in [8, 9]. This is especially noticeable because the part used in measurement was not produced with high quality of finishing i.e. the flatness deviations had relatively high values. All results were in range from 101.6 μm to 162.2 μm. The deviations from maximum result obtained with identical probing type and stylus, depending on the sample size, were from 24.0% to 34.6%. The maximum flatness deviation result, and maximum values for different styli, were obtained with scanning probing, which is expected because more points were extracted with measurements, analogue to the results in [8]. It is noticeable that the continuous scanning measurement results converge, and the single point measurement results diverge. However, the maximum

Table 3. Styli used for measurements [19].

Feature	Stylus A	Stylus B	Stylus C
DK	2 mm	5 mm	8 mm
L	44 mm	50 mm	114 mm
DG	11 mm	11 mm	11 mm
ML	19 mm	40 mm	101 mm
DS	1.5 mm	3.5 mm	6 mm
Weight	5 g	4g	10 g

results from single point measurement strategy are close to maximum scanning results, therefore it is believed that the further increase of sample size would not vary the single point probing results significantly, i.e. the obtained maximum results are in a converged range. One can notice that strategies defined with rectangular grid of 15 × 15 size for single point (Points 4) and strategies with 15 parallel lines for continuous scanning measurements result (Lines 3) are in range of approximately 10% deviation from the maximum result obtained with identical probing type and stylus.

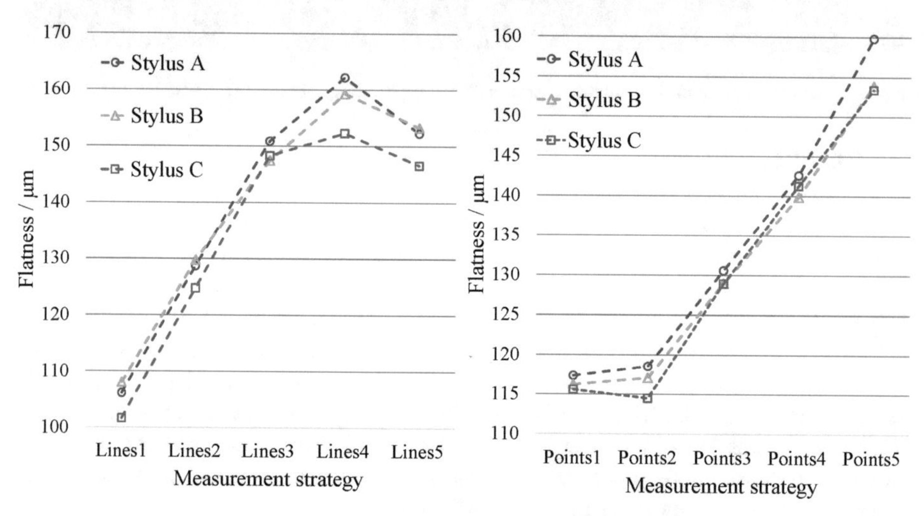

Fig. 7. Graphical representation of single point flatness measurement results.

Fig. 8. Graphical representation of continuous scanning measurement results.

The result deviation values due to different styli (tip size) are comparable with the ones in [15]. The maximum deviations observed are under 10 μm. As expected, the highest results are mostly obtained with the stylus A with the smallest tip size. Surprisingly, there are three results obtained with scanning measurement where the maximum flatness was measured with medium size stylus B. The deviations are relatively small here and no clear explanation could be provided, besides some random and unconsidered influences on the measurement process.

5 Conclusions

The results showed and confirmed the great influence of sampling size on accuracy of CMM flatness measurement. For the measured part, the rectangular grid sampling strategy with size of 15 lines provided results with small deviation from the referent, maximal values, hence this strategy can be recommended for the common measurement tasks.

Continues scanning and single point probing provided similar results and due to scanning advantages regarding speed and number of acquired points, its expected superiority for common measurement tasks was confirmed.

The experiment also showed the influence of stylus tip size on the measurement result (mechanical filtering), but for the flatness deviation values of the inspected part these result deviations were much smaller and less significant, comparing to the ones due to sample sizes.

References

1. Pereira, P.H.: Cartesian coordinate measuring machines. In: Coordinate Measuring Machines and Systems, 2nd edn., pp. 57–79. CRC Press, Boca Raton (2012)
2. Flack, D.: CMM Measurement Strategies, Measurement Good Practice Guide 41. National Physical Laboratory (2014)
3. Raghunandan, R., Rao, P.V.: Selection of sampling points for accurate evaluation of flatness error using coordinate measuring machine. J. Mater. Process. Technol. **202**(1–3), 240–245 (2008)
4. Raghunandan, R., Rao, P.V.: Selection of an optimum sample size for flatness error estimation while using coordinate measuring machine. Int. J. Mach. Tools Manuf. **47**(3–4), 477–482 (2007)
5. Lakota, S., Görög, A.: Flatness measurement by multi-point methods and by scanning methods. AD ALTA J. Interdiscip. Res. **1**(1), 124–127 (2011)
6. Jalid, A., Hariri, S., Laghzale, N.E.: Influence of sample size on flatness estimation and uncertainty in three-dimensional measurement. Int. J. Metrol. Qual. Eng. **6**(1), 102 (2015)
7. Capello, E., Semeraro, Q.: The harmonic fitting method for the assessment of the substitute geometry estimate error. Part I: 2D and 3D theory. Int. J. Mach. Tools Manuf. **41**(8), 1071–1102 (2001)
8. Capello, E., Semeraro, Q.: The harmonic fitting method for the assessment of the substitute geometry estimate error. Part II: statistical approach, machining process analysis and inspection plan optimization. Int. J. Mach. Tools Manuf. **41**(8), 1103–1129 (2001)
9. Henke, R.P., Summerhays, K.D., Baldwin, J.M., Cassou, R.M., Brown, C.W.: Methods for evaluation of systematic geometric deviations in machined parts and their relationships to process variables. Precis. Eng. **23**(4), 273–292 (1999)
10. Summerhays, K.D., Henke, R.P., Baldwin, J.M., Cassou, R.M., Brown, C.W.: Optimizing discrete point sample patterns and measurement data analysis on internal cylindrical surfaces with systematic form deviations. Precis. Eng. **26**(1), 105–121 (2002)
11. Colosimo, B.M., Moya, E.G., Moroni, G., Petro, S.: Statistical sampling strategies for geometric tolerance inspection by CMM. Econ. Qual. Control **23**(1), 109–121 (2008)
12. Zaimovic-Uzunovic, N., Lemes, S.: Cylindricity Measurement on a Coordinate Measuring Machine. Advances in Manufacturing, pp. 825–835. Springer, Cham (2018)
13. Weckenmann, A., Estler, T., Peggs, G., McMurtry, D.: Probing systems in dimensional metrology. CIRP Ann. Manuf. Technol. **53**(2), 657–684 (2004)
14. International Organization for Standardization, ISO 1101:2017: Geometrical product specifications (GPS) - Geometrical tolerancing - Tolerances of form, orientation, location and run-out (2017)
15. International Organization for Standardization, ISO 12781-2:2011: Geometrical product specifications (GPS) - Flatness - Part 2: Specification operators (2011)

16. Shakarji, C.M.: Coordinate measuring systems algorithms and filters. In: Coordinate Measuring Machines and Systems, 2nd edn., pp. 153–182. CRC Press, Boca Raton (2012)
17. Strbac, B., Hadzistevic, M., Klobucar, R., Acko, B.: The effect of sampling strategy on the equation of a reference plane measured on a CMM. In: Proceedings of 10th Research/Expert Conference with International Participation, QUALITY 2017, Neum, Bosnia and Herzegovina (2017)
18. ZEISS, ZEISS CONTURA: The Reference Machine in the Compact Class. https://www.zeiss.com/metrology/products/systems/bridge-type-cmms/contura.html. Accessed 27 Oct 2018
19. ZEISS, STYLI. https://world.probes.zeiss.com/en/Styli/category-12.html. Accessed 20 Jan 2019

Considerations on Optimizing Technological Process for Production of Low Voltage Automotive Cables

Lucian Gal, Doina Mortoiu, Bogdan Tanasoiu(✉), Aurelia Tanasoiu, and Valentin Muller

Aurel Vlaicu University, Arad, Romania
bogdan.tanasoiu@gmail.com

Abstract. The newest trends in the automotive industry are focused on reducing material usage, building more efficient engines, using aluminum and composite materials.

In accordance with the new requirements imposed by the market, related to diversity of production, adaptability, delivery times, mass production, costs, optimization of the production processes and equipment used becomes mandatory.

In the aluminum cable production process, obtaining the finite product is done by using a continuously optimized technology, whose implementation leads to the lean improvement of costs. The paper presents a number of optimization operations performed on the automotive aluminum wire production process.

Keywords: Wiring · Aluminum · Pulling · Twisting · Extrusion

1 Introduction

Automotive cables with aluminum as the base conductor, present significant advantages in reducing automobile weight, considering that new safety and comfort improving equipment (airbags, ABS, soundproofing, new communication devices etc.) tent to add to the weight of modern automobiles.

The added weight also required an increase of engine power, which implies higher fuel consumption. The only viable solution for reducing fuel consumption, after the design of more fuel efficient engines, continues to be the reduction of automobile weight.

Under these conditions, aluminum presents a series of qualities that automotive specialists could not ignore: low density, good corrosion resistance and high thermal conductivity. Specialists state that every ton of aluminum used on automobiles reduces emissions by 20 tons throughout the lifetime of the automobile.

From the additional advantages of using aluminum we have to consider also the facts that it is a metal that is 100% recyclable, the aluminum recovery rate from scrapped automobiles being over 95% currently in industrialized countries. Furthermore, using recycled aluminum in automobile production allows an electrical energy saving of 90–94%. As designers and researchers improve the fatigue, thermal shock

I. Karabegović (Ed.): NT 2019, LNNS 76, pp. 173–180, 2020.
https://doi.org/10.1007/978-3-030-18072-0_20

and mechanical behavior of aluminum alloys, and will develop new technologies for casting and working parts made of these materials, more and more producers will move to producing aluminum automobile chassis.

Future automobiles will contain increasingly more aluminum and less iron and steel. Consequently, the advantages of using aluminum in automobile construction are two fold:

- Technical and constructive
- Related to useability after leaving the factory

The constructive advantages are related to the following properties of aluminum alloys: they have a resistance to mechanical loads that can be compared to that of regular steels, but with a significantly lower specific gravity, and they posses a thermal conductivity which is superior to that of ferrous materials. Considering usage, the advantages are related to driving and fuel economy. The lighter an automobile is, the faster it can accelerate and brake, giving improved performance and safety.

Aluminum conductor cables reduce product weight and cost. This is the reason why aluminum will replace copper as a conductor for multiple applications.

Aluminum cables are designed to match high standards of the ultrasonic welding equipment producer and have a very high performance in connection obtained through ultrasonic welding.

2 Fabrication Technology

The fabrication technology is in accordance with the “Just in time” principle, which is a production process management method where the products are executed when the customer requirement is created. The fabrication steps are presented in Fig. 1.

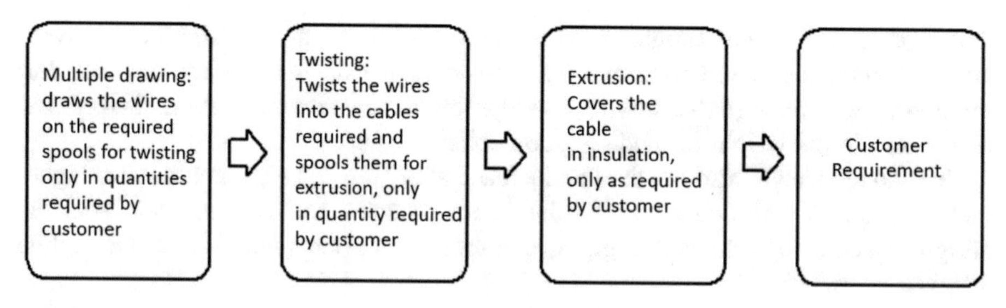

Fig. 1.

Wire drawing is the process through which wires of different diameters are produced, given specific tolerances. The process reduces the diameter of the wires by passing them through a series of dies, each having a smaller diameter orifice than the previous one. The final diameter is obtained by passing the aluminum through the final row of dies. The value of partial cross section reductions during wiredrawing is between 10–45%, depending on the nature and dimension of the raw material, the total cross section reduction being able to reach a value of 95% (Fig. 2).

Fig. 2. Multiple wire drawing

In the multiple drawing process considered, the raw material is 2.5 mm cable, and the finished product is represented by the spools of wires of the same diameter, between 0.300 and 0.700 mm. The process comprises the following operations: (a) raw material preparation (b) wire drawing (c) thermal treatment (d) completion operations.

Lengthening of the wire is obtained through its drawing through the succession of dies. Due to friction, a significant rise in wire temperature is obtained, which can damage both the wire and the dies. In order to obtain a smooth drawing and to reduce this temperature, lubricants are used. The phenomena occuring during wire lengthening vary according to the position of the wire in the drawing process, as follows:

During drawing, the change in the raw material cross section occurs mainly under the action of transverse forces exerted by the walls of die. In the beginning only the grains whose slip planes correspond with the direction of the forces undergo deformation. As the material advances through the die, the entire mass is plastically deformed under the action of compressive forces that arise from the increase of traction forces.

At the exit from the die, the material is hardened, and to avoid its fracture in case of a new reduction of cross section, it undergoes an annealing heat treatment. The deformations of the crystals during the drawing process are numerous, such that other than slipping, twisting frequently occurs.

The influencing factors of drawing: quality of metal surfaces, drawing velocity, material tension during drawing, die material, shape of die, lubricant used.

Dehardening is the thermal treatment through which the physical and chemical properties of the material are altered in order to increase ductility, such that a more malleable material is obtained. This process entails the heating of the aluminum up to the recrystallization temperature, and cooling after the recrystallization is achieved. The dehardening is a very important process before spooling which prevents the breaking of the wires. The cooling of the wires is done using an emulsion with a very low concentration of oil. After the wires come out of the thermal treatment oven, they are grouped and spooled onto rotating spools.

Process control for wire drawing must include:

- Periodic measurements of tensile properties.
- Ultrasonic welding property checks by producing a joint and testing the peeling strength.

Twisting is the process through which spools of twisted wires are produced from the spools of simple drawn wires, using equipment shown in Fig. 3. The process entails the passing of the wires through distribution panes. The following requirements are considered: cross section, cable type, composition – number of wires and diameter, feeding type of aluminum spools (drawn of pretwisted), resistance, quantity/spool, twist stepping, maximum speed setting, spooling tension, spool pressure, type of spool, guiding die, spooling step, twisting direction. Twisting dies guide and help the uniform cylindrical forming of the twisted wires.

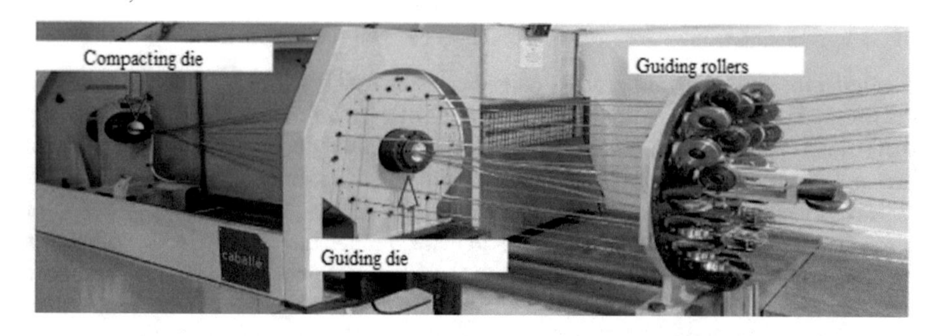

Fig. 3.

These dies do not serve in lengthening the wires or changing the diameter, their only role is to help compact the wires.

Extrusion consists of melting the plastic material and applying one or multiple layers on the aluminum conductor, in order to prevent electrical current leakage. The extrusion equipment diagram is shown in Fig. 4.

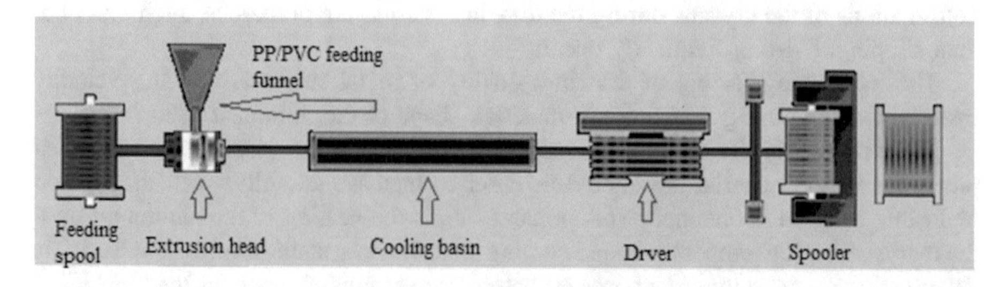

Fig. 4. Extrusion process equipment.

PP and PVC grains are fed through the funnels of the extrusion machine. The insulating material can be one of multiple types of PVC, depending on the wire diameter: (1) for cross sections smaller than: 0.35–0.50 PVC 2035 is used; 0.75–6.00 PVC 2005 is used; (2) for cross sections greater than: 8.00–20.00 - PVC 2010; 25.00–120.000 - PVC 2011. Applying coloring dye is done in accordance with customer requirements for easier identification of different circuits or electrical systems. The temperature used for melting these materials is between 160° and 205°. When the cable exits the extrusion head, the conductor passes through the diameter measurement device, then the wire is marked, upon which it enters the cooling system, which is composed of two troughs and a bath which help in cooling the insulation. Following this, the cable passes through a cold water bath, in order to solidify the plastic material, and then through a drying unit, in accordance with Fig. 5.

Fig. 5.

3 Process Optimization for Wire Drawing

The first process improvement we will discuss, is the one brought to aluminum spools which feed the multiple drawing machine, consisting of the introduction of a teflon ball on the wire, as shown in Fig. 6. This prevents the bending of the wire before it enters

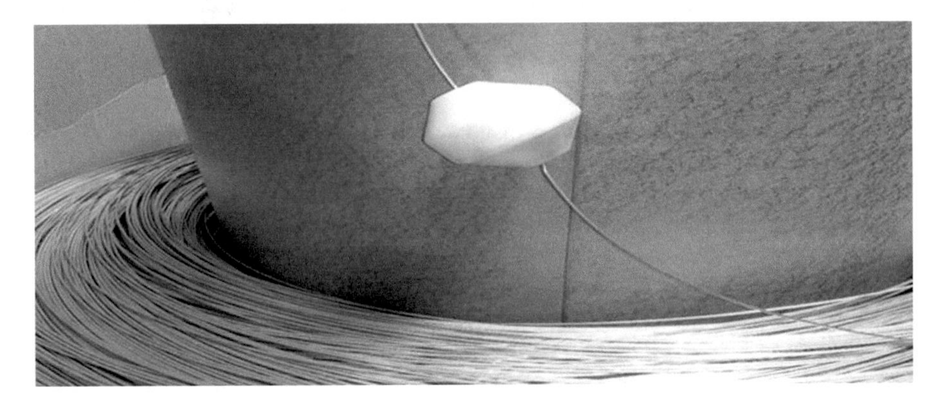

Fig. 6.

the guiding rollers of the drawing machine. This solution, which involves reduced costs, helps to make the drawing process more fluid, without accidental wire breakages, and the time allotted to solving such problems has been reduced, as well as the quantity of scrap or damaged material.

Another problem that was solved during the multiple drawing phase is related to the low temperature of the emulsion during the colder seasons. The temperature of the emulsion drops by approximately 10–15 °C, the optimal temperature being 30–42°. Consequently, the equipment was fitted with a timer for the emulsion heating system (approximately 2 h before start of production), during which time the lubricant reaches the optimal temperature.

For twisting, the equipment used to produce copper conductors were adapted with new devices and systems designed for the new process of aluminum wire processing. During copper conductor twisting, the wires are twisted using a lire with teflon guides. Through those guides, the wire is led on the roller system and crank, and then spooled onto metallic spools. In aluminum conductor production, this ceramic guiding system could not be used to maximum capacity due to the fact that aluminum wires would break and any friction with a harder material would leave marks on the conductor, affecting the aspect and dimensional parameters of the wires. The next solution was to replace the rollers with ceramic rollers, which has proven to be the method to make the process more fluid and allow aluminum conductor production to a high standard (Fig. 7).

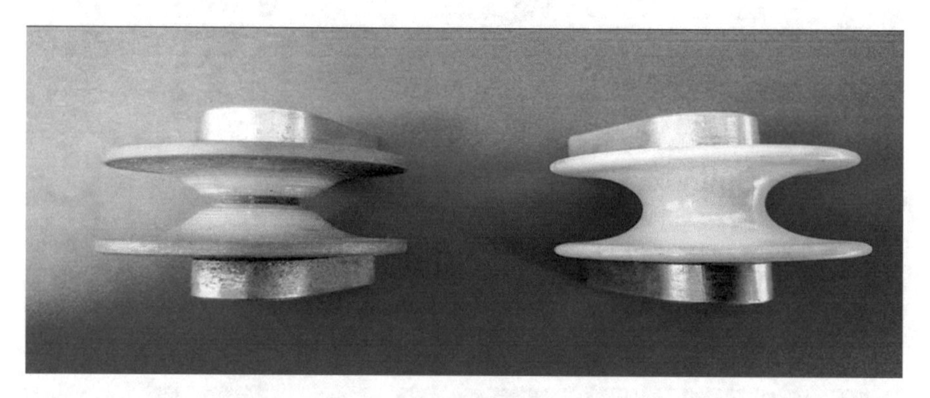

Fig. 7. Teflon (left) vs. Ceramic roller (Right)

For the higher diameter twisting of conductors (85/95/120 mm), problems were encountered with the metallic rollers as well, which caused numerous problems, both aesthetic and during the extrusion phase, where multiple errors were encountered. The cause of these problems was removed be replacing the metallic rollers with an interior V-shaped section, with ones with a larger interior opening, with a U-shaped section (Fig. 8).

Fig. 8. V-section roller (left) vs. U-section roller (right)

This type of rollers allowed the wire to more easily traverse the path in the twisting machine, without it being affected in respect to aspect or physical diameter, while at the same time eliminating the problems created on the extrusion line (Fig. 9).

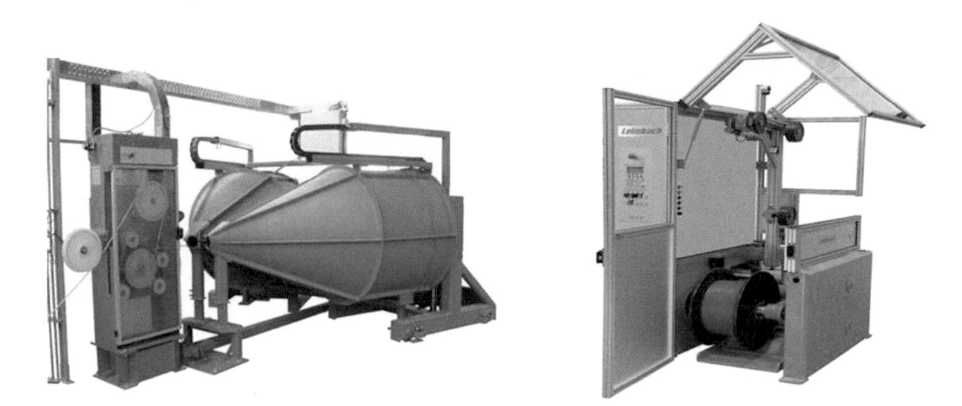

Fig. 9.

During extrusion, dynamic pay-off was introduced, replacing the classical static pay-off. The optimization consisted of eliminating multiple wire breakages during the feeding of the extrusion equipment phase, and the acceleration of the line cycle time.

4 Conclusions

Using all the qualities of aluminum, by comparison to other conductors, during actual and innovative testing in order to new and high performance products, which to ensure materials for international automotive companies, presents a new possibility and challenge in research and development.

The production process comprises three phases: multiple drawing, twisting and extrusion (possibly respooling in case of observed defects) with the presented and implemented optimizations, led to the production of aluminum cables up to the standards of the automotive industry and also to the reduction of scrap and an increase in quality indicators. The continuous optimization of the production process with the necessary implementations on the equipment used in the multiple drawing, twisting and extrusion phases, ensure production flexibility and quality improvement in accordance with customer requirements.

References

1. Cerințe tehnice – Torsadare, E.D.M., Coficab E.E. Arad
2. Condiții standard de laborator, Q.D.M., Coficab E.E. Arad
3. Control Plan - Masterbatch for Incoming Inspection, Q.D.M., Coficab E.E. Arad
4. EM – 4016 Wire Elongation Tester Utilization and Verification, Q.D.M., Coficab E.E. Arad
5. General Presentation of Coficab Eastern Europe, H.R.D.M, Coficab E.E. Arad
6. Identificarea defectelor de echipament din raportul de producție, P.D.M., Coficab E.E. Arad
7. Instrucțiune de alimentare a plasticului cu colorant, P.I.M., Coficab E.E. Arad

Benefit of Using Robots in the Production of Three-Layer Parquet

Salah-Eldien Omer[1,2](✉)

[1] SAG CONSULTING d.o.o., Vramčeva 17, 10000 Zagreb, Croatia
sagzagreb@hi.htnet.hr
[2] Technical Faculty Bihać, University of Bihać,
Ul. Irfana Ljubujankića bb., 77 000 Bihać, Bosnia and Herzegovina

Abstract. The usage of robots in timber industry processing is starting to bring more benefits. The production of three-layer parquet is advancing in the sense of quality and variety. The production of lamella in this process is very important for the quality of the parquet. Usage of robots in certain phases of production process upgrades the quality and multiplies the quantity. Robots for the wood processing must be ordered according to the need of the process with cooperation with the producer of robots to assure the planed benefit.

Keywords: Robots · Benefits of robots · Three-layer parquets · Upgrading reconstruction

1 Introduction

The history of using Robots in wood industry processing is going back for a long period. Since the wood processing in many ways is a multi-disciplinary process, the usage of Robots is followed after the high levels of automation. When the price of Robots started to be acceptable for wood industry processing, investors started to buy them when the technical teams of the factories started to ask for them. In certain areas the Robots shows very high efficiency and bring profit to companies, but as I mention based on our work in different areas of wood processing.

Five Reasons for Using Robotics in Manufacturing Process:

1. Robots used in manufacturing process create efficiencies all the way from raw material handling to finished product packing.
2. Robots can be programmed to operate 24/7 in lights-out situations for continuous production.
3. Robotic equipment is highly flexible and can be customized to perform even complex functions.
4. With robotics in greater use today than ever, manufacturers increasingly need to embrace automation to stay competitive.
5. Automation can be highly cost-effective for nearly every size of company, including small industry.

I. Karabegović (Ed.): NT 2019, LNNS 76, pp. 181–188, 2020.
https://doi.org/10.1007/978-3-030-18072-0_21

Five Ways Robots don't Eliminate Manufacturing Jobs:

1. When companies can't compete, mainly jobs are sent labour markets.
2. Robots in manufacturing help to create jobs by recreating more manufacturing work.
3. Robots protect workers from repetitive, mundane and dangerous tasks, while also creating more desirable jobs, like engineering, programming, management and equipment maintenance.
4. Robots free up manpower to let companies maximize workers' skills in other areas of the business.
5. Today's labour market generally includes fewer skilled manufacturing workers due to decades of off shoring, and robots eliminate the shortfall.

Five Ways Robotics Make Manufacturers Globally Competitive:

1. Automation allows domestic companies to be price competitive with offshore companies.
2. Robotics in manufacturing achieve higher throughput, so companies can vie for larger contracts.
3. Robots achieve ROI quickly, often within two years, offsetting their upfront cost.
4. Department of Labour statistics indicate that workers are maximizing their output capacity, and robots help manufacturers break that ceiling.
5. In a world where the importance of green manufacturing is growing, robots save on utilities since they don't require climate control or lighting, and they create cleaner spaces.

Today's fast and flexible robots work in industries ranging from wood and plastic processing to semiconductor manufacturing and research. While still a mainstay of high-volume production, robots are finding more roles in small to medium-sized operations. As we know what are robots doing for processing industry it is obvious, they always bring benefit to it.

Any repetitive task is a candidate for robotic manufacturing, especially if it's difficult or dangerous for a human, or takes place in a hostile environment. What's more, adding force sensing and vision systems lets a robot adapt to changes in part position or orientation, increasing flexibility and versatility. Good jobs for robots include:

- Machine Tool Tending
- Material Removal
- Palletization and De-Palletizing
- Material Handling
- Welding, Gas Metal Arc Welding (Submerged Arc and Resistance)
- Assembly

Moving quickly and accurately, robots handle parts too small for human eyes and fingers and never make mistakes. That's one reason growing numbers of products are designed for robotic assembly from the outset.

Vision technology is fast becoming standard, reducing the need for expensive fixturing and tooling, and force sensing lets a robot adapt when an assembly problem is encountered.

Every manufacturer can benefit from putting robots to work. So, we found out what manufacturing robots can do for wood processing business in several different productions. Here by we are going to concentrate on the production process of three-layer parquet, where Roots are very rarely used.

2 Process of Producing Three Layer Parquet

According to the European norm EN13489 a multi-layer parquet element comprises 'a combination of wood and/or wood materials that are glued together'. Each layer is made of wood. Put simply: multi-layer parquet is a solid wood floorboard with a multi-layered structure. Each layer is always glued at a 90° fibre direction to the previous layer. This achieves both higher form stability and less movement in the wood. Contraction and expansion, when the floorboards shrink or expand when humidity levels change, is substantially reduced. There are even more advantages of a multi-layered parquet flooring against either solid parquet or laminates or vinyl floors or carpets and tiles.

There are different possibilities how this kind of multi-layer parquet can be structured. Usually they are two floorboard structures range: a two-layer and a three-layer structure. Most of the floorboards have traditional tongue and groove joints and are supplied with a ready-treated surface. Thus, the floorboards are delivered ready for use as soon as they are laid and then just need normal care like any most flooring.

The three-layer parquet is constructed very simply and with high quality product elements. They mainly made using three layers of wood: a top layer, a central layer and a backing layer. All three layers are glued together in a press. This creates high quality, three-layer wooden floorboards for sophisticated construction final product.

The top layer, so-called wear layer, is always made of a layer of high-quality fine wood, which is usually made of sawn veneer with a thickness of about 5.0 mm. Usually they produce the majority of top layer flooring in their own sawmill. Every wear layer is first dried at their own drying facility to the technically correct initial moisture content before it is processed.

The central layer is made of solid soft wood or panel elements that run transversely to the fibre direction of the top layer. The backing layer, on the other hand, is made of veneer with the same fibre direction as the top layer. This creates a stable sandwich structure as a basis for the three-layer wooden parquet. The high-quality wood that the producer use for this substructure comes from sustainable forestry and is 100% PEFC certified as recommended.

Solid wood floors have some major drawbacks. Humidity fluctuations can cause the wood to warp, uneven subfloors require time-consuming sanding prior to installation, etc. In addition, a large amount of precious hardwood is needed to produce just a few square metres of flooring. Multilayer parquet from quality production process easily solves these issues by using a balanced 3-layer construction with grains running in different directions. This reduces the natural swelling and shrinking of the wood by up to 70%. and compared to solid wood floors, multilayer parquet is also the better choice in terms of sustainability, eco-friendliness, and price-performance ratio.

The superior engineering of each of high-quality parquet products guarantees that you always get the best and most special products. Only carefully selected precious wood species are used for the top layer, which is between approx. 2.5 mm and up to approx. 3.5 mm thick depending on the type of the product. They are cross glued to optimally distribute the innate tensions acting within the individual layers. The bottom layer comprising the softwood backing veneer ensures long lasting dimensional stability. As in the following Fig. 1 for example, without this complex multilayer construction Parquet would not be so perfectly suited for a floating installation.

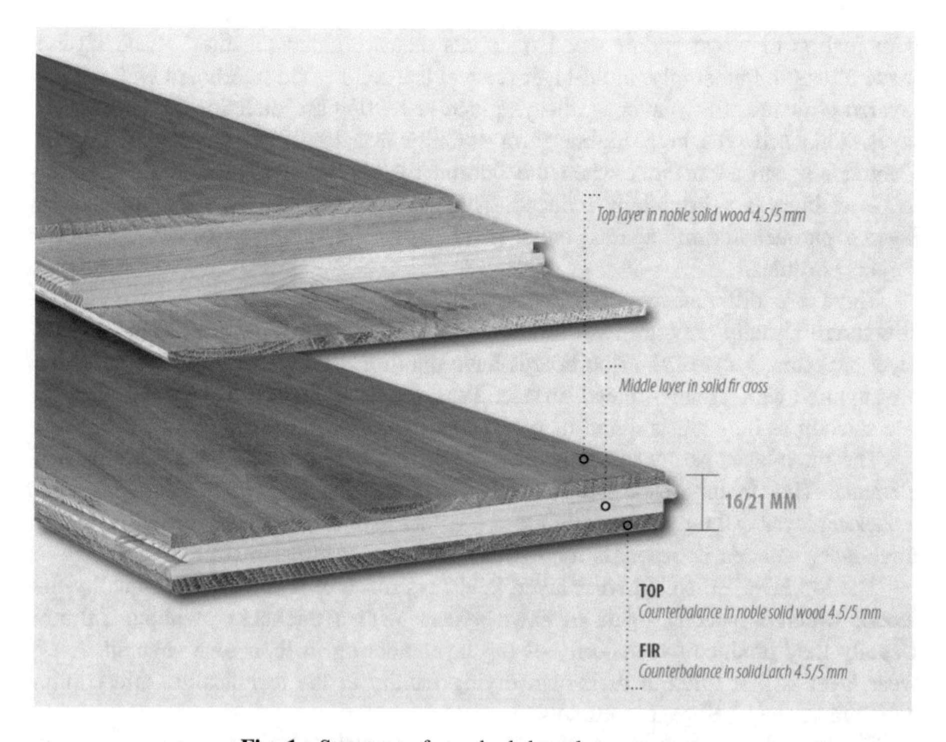

Fig. 1. Structure of standard three layer parquet.

3 Usage of Robot in the Production of Three Layer Parquet

High quality of 3-layer parquet is the result of advanced technology and constant control of production process. The perfect usage and durability parameters, the appropriate thickness, the number of varnish layers as well as the resistance all determine the uniqueness of the 3-layer parquet. The look of the parquet generally is very important to the user. Most of the producers of parquet pay tension to the look of the wood flooring parquet, because it is generally selling it.

Grades of wood flooring related to the look of the top are as follows:

1. Select grade – allows sound knots up to 8 mm and dark knots op to 2 mm in diameter (no clusters). It allows minimum differences in coloration and grain pattern without any sapwood.
2. Natural grade – a richer floor with more variation. The size of the knots is not limited, unless they affect the stability of the wood. Possible cracked knots can be filled with wood putty. This grade allows greater colour and grain pattern variations as well as thinner strips of sapwood. Stains are not allowed.
3. Rustic grade – a very vibrant selection full of character with obvious variations in colour and grain pattern. The size of knots is unlimited, unless they compromise the stability of boards. Possible cracked knots can be filled with wood putty. This grade allows healthy sapwood.

To assure the quality of parquet or the wood flooring the producer most pay big tension to the quality of input wood elements in all phases of production.

Such very sensitive selective production needs very precise detection of quality plus sorting and assembling of the final product. In the following figures we represent a standard production line of three-layer parquet. In most modern such production lines it is a combination of high automatic phases with specific sorting and control areas for the semi half products (Fig. 2).

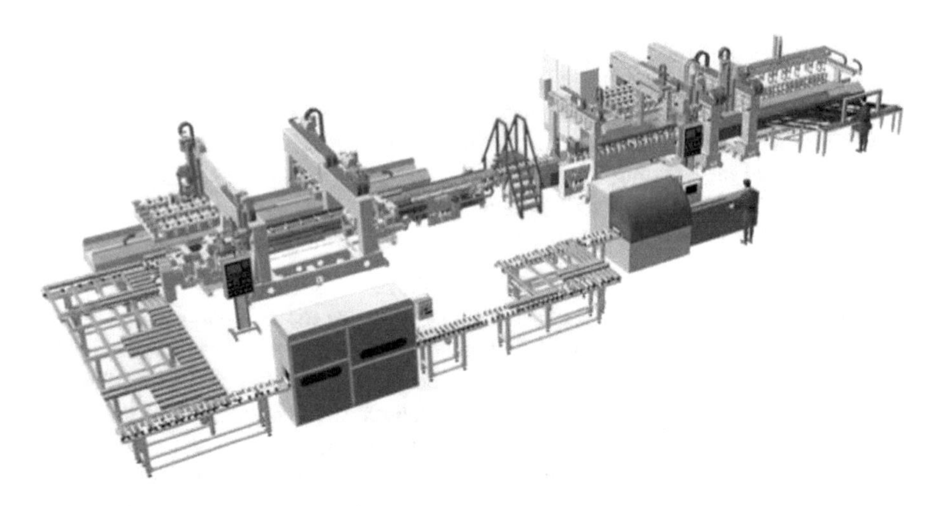

Fig. 2. Modern production line for three-layer parquet.

To insure the quality of processing of basic elements for layers of the parquet we usually recommended in the cooperation of machinery producers is to define and order Robotic units per defined phases of operations. We define to the producer phases needed like the surface elements, the middle board production and the back elements of the parquet. As an example, is the following Fig. 3.

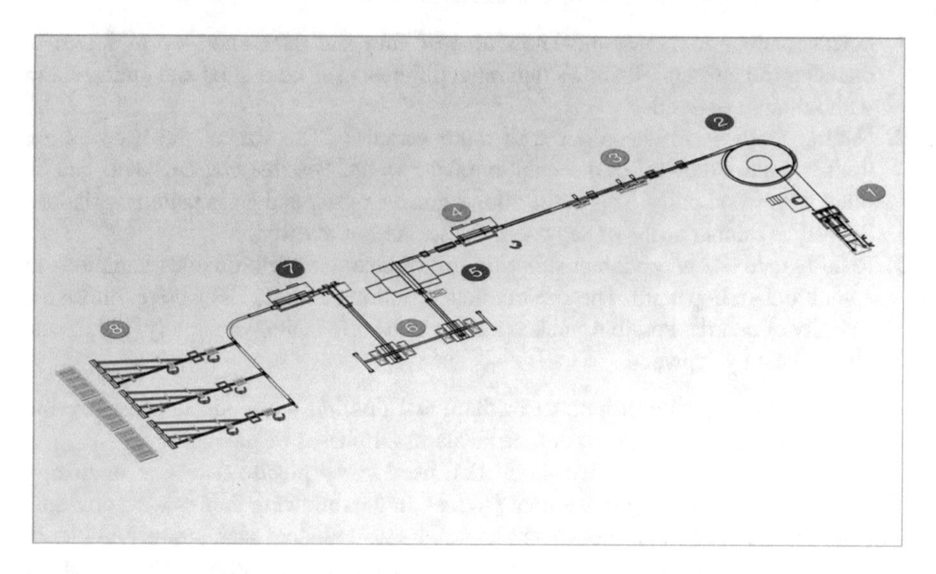

Fig. 3. An example of top layer lamellae production - using WEINIG Concept system solutions.

1. Sorting of the blanks in any stacking pattern,
2. Singularizing the blanks,
3. Measurement of dimensions and moisture content,
4. Pre-planning and precision planning in a single operation,
5. Highly precise, right angled cross-cutting of the blanks,
6. Stacking and unshackling of the blanks,
7. Splitting of blanks into lamella packs,
8. Sorting and stacking of graded lamellae

In the above scheme is the input of the solid raw material in the process. In two areas like the quality control of the boards and the quality sorting boards and as well as lamellae is recommended to install Robot units where they with high quality and precision will do the job. This will assure the quality of the needed lamellae and elements with high precision at the positions in the scheme above. In Fig. 4 we show a suggestion of some of the producers of Robots in Europe.

The second important phases are the gluing of the layers of three parquet. Also, some of Italian producers of machine for wood industry processing are offering Robot units for such phases.

Recently ITALIAPRESS has been a notable rise in interest in cold pressing, due to the reduced gluing time with the advent of a new generation of cold glues and an increase in the dimensions requested (especially of 3-ply parquet flooring), which require a more stable gluing. To respond to these new needs of increased dimensions, greater flexibility and elevated productivity with contained costs (consider just the

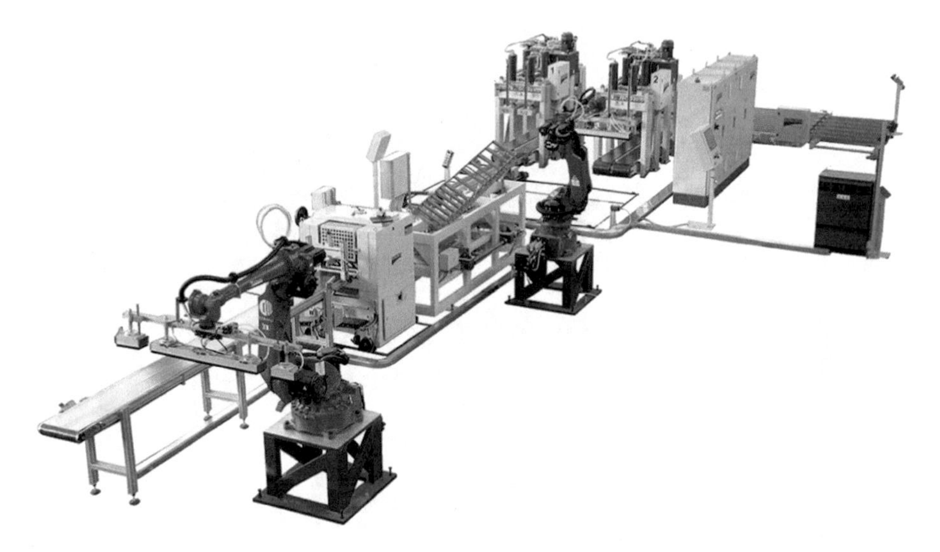

Fig. 4. Robots in certain phases production line of three layer parquet.

energy savings from cold pressing compared with hot pressing), Italpresse has developed a concept that is revolutionary in the wood flooring sector: the use of robots in the composition and joining of different layers that form the plank.

In the case of 3-ply, the robots perform an addition task of joining the intermediate layer of suitably spaced transverse slices, and precisely joining them to the other two layers.

Pressing of single stacks and high precision joining, united with particular devices which the press is equipped with, guarantee the highest quality gluing and a truly surprising stability and dimensional repeatability.

The only request for the Robot producers from the user of Robots in the three-layer parquets production is program the Robot according to the process phase to be use in.

Maximizing the productivity and profitability of the process in many cases they usually using programmed robot they use Robomaster.

Robotmaster CAD/CAM programming is completely off-line, decoupling planning and set-up from production. Its powerful workspace simulation tools assure error-free consideration of workspace and part orientation and constraints. Robotmaster delivers unmatched, full-function off-line programming capability for robots - easy programming of precise robot motion control and quick generation of path trajectories of any size, with minimal programmer intervention.

In the different models and programs of Robots produced specially for certain phases in wood processing in this case in the production of three layer parquet we find one of the company Homag which is indented for such process for the area which needed precision are represented with high success is shown in Fig. 5.

Fig. 5. Homag robot unit for the sorting and monobilation of products elements.

4 Conclusion

The production of the three-layer parquet with lamellae in the face is a multi-phases process which in growing rapidly in usage, sales and the competition in international is also growing.

Generally, in the wood industry processing the usage of Robots is growing and that with the cooperation with the producer of Robots where the users order Robots with certain description and characteristics for certain operations in the production process.

After our analysis for existing production lines and process for the modern three-layer parquet we defined several areas in this process where Robots will bring an advancement as improving productivity and quality.

The programs and software's which is also developed for such Robots are highly advanced, easy to program and proceed to production as CAD/CAM programming.

Some of the well-known producers of specialized machines for wood industry processing are already developing the needed Robots based to certain orders and also developing new types for specialized processing which bring higher benefits.

References

1. International Federation of Robotics (IFR) Statistical Department. World Robotics - Industrial Robots 2016 (2017). http://www.worldrobotics.org
2. Landscheidta, S.: Opportunities for robotic automation in wood product industries: the supplier and system integrators' perspective. In: 27th International Conference on Flexible Automation and Intelligent Manufacturing, FAIM 2017, 27–30 June 2017, Modena, Italy (2017)
3. Robot technology for wood processing (Wood, Unlimited brochures 2011)
4. Omer, S.-E.: Justified usage of robots in wood industry, Saint Petersburg (2016)

Kaizen in Practice-Case Study of Application of Lean Six Sigma Method in Working Condition of Wood-Processing Firm

Ismar Alagić[1,2,3](✉)

[1] TRA Tešanj Development Agency, Trg Alije Izetbegovića 1, 74260 Tešanj, Bosnia and Herzegovina
ismar.alagic@gmail.com

[2] Faculty of Mechanical Engineering, University of Zenica, Fakultetska 1, 72000 Zenica, Bosnia and Herzegovina

[3] Faculty of Engineering and Natural Sciences, International University of Sarajevo, Hrasnička cesta 15, 71210 Ilidža, Sarajevo, Bosnia and Herzegovina

Abstract. This article is the result of several years of author's work in the field of quality management especially in the use of methods and tools for quality management. I have paid special attention to the research of a unique and understandable concept of continuous progress - Kaizen.

What is Kaizen? Kaizen is a combination of two Japanese words KAI and ZEN. Compound KAIZEN (Japanese 改善) which means "always good" or "continuous improvement". This article provides proposal a set of approaches that are the basis for the development and application of the principles and Kaizen of Lean Six Sigma concepts. A special focus is given to the method of applying Lean Six Sigma concepts in specific working conditions of domestic company from wood-processing sector.

Keywords: Lean Six Sigma (LSS) · Kaizen · Wood-processing · Improvement

1 Introduction

Bosnia-Herzegovina's industrial production is no longer as it was once in the European market, partly because of political opportunities and partly because of the unfulfilled conditions that this market demands. Quality management and introduction of the concept of Lean Manufacturing - Six Sigma is in the hands of domestic companies and has not reached the level that is present primarily in the world industrial production [1].

This article is the result of several years of author's work in the field of quality management especially in the use of methods and tools for quality management. My attempt is reflected in the fact that through the subject article of the presentation of management philosophy, theories and management tools that have been developed and used in the highly developed countries of the world, with a special focus on so-called Japanese philosophy of quality. I have paid special attention to the research of a unique and understandable concept of continuous progress – Kaizen [2].

I. Karabegović (Ed.): NT 2019, LNNS 76, pp. 189–198, 2020.
https://doi.org/10.1007/978-3-030-18072-0_22

The last thirty years "Lean Production - Six Sigma" concept has become the dominant approach in world production, which leads to improvement in productivity and cost reduction in work processes [3]. Powerful step of the automotive industry in Japan, compared to European and American competition was based on the principle of continuous improvement as one of the basic elements of improving production. The term "Kaizen" whose application management granted by Japanese company is an element of overall control of quality related to long-term continuous access to updates, with respect to human needs and quality [4]. This article provides a set of approaches that are the basis for the development and application of the principles, methods and tools of Lean Six Sigma concepts. A special focus is given to the Kaizen of applying Lean Six Sigma concepts in specific working conditions of domestic wood-processing firm, taking into account international experience in the study of this field [2].

2 Kaizen

The Kaizen is reflected in the fact that each company develop employees who are motivated and trained to achieve the effect, to direct all aspects of business processes and to improve the competitive position of the company on the market [10].

What is Kaizen?
Kaizen is a combination of two Japanese words KAI and ZEN. Compound KAIZEN (Japanese 改善) which means "always good", "continuous improvement" or "application for the better." This business concept, is an iterative process that is repeated over PTA events and is a frequent occurrence. To illustrate it KAIZEN represents the journey or the transition from event to cause, from results to process and/or the implementation of changes to the motivation for change. In interpreting the meaning of the term Kaizen is very important to note that it makes differences in the interpretation of the term "continuous" (permanent) and "continually" (uninterrupted). Both terms represent words twins who have descended from Latin, the root of the word "continuare". For example, "continuous" means starting and stopping, and "continually" means never ending [9]. For example, "permanent" means starting and stopping, and "uninterrupted" means never ending. Kaizen is the best example of where are clashed concepts of Western developed economies and Japan. The western approach favours big ideas and significant contributions of individuals, while the Japanese approach favours teamwork, and small and everyday improvements. In the Western approach, great ideas and improvements come from the representatives of management and engineers, while the Japanese approach embodied in Kaizen means that the big improvements come from constant small improvements made by workers from department. Improvement in terms of Kaizen is just a new beginning for change. From all this it is clear that Kaizen represents a philosophy where the collective wisdom of the people brings endless results. The secret lies in the joint and teamwork of employees.

The recommendations that should be taken when applying Kaizen are [10]:

- Start with your problems, not with others;
- Start with small improvements;

- Start with the simplest things;
- Improvement is part of the daily routine;
- Never dismiss the idea before you try it;
- Highlight problems and do not hide them, and;
- Never accept the "status quo" or be motivated to change.

Achieving the previously mentioned quality management-TQM is impossible without the use of Japanese philosophy - KAIZEN. KAIZEN means continuous improvement involving all employees: Top management, managers and workers. The very notion is related to people, processes and products, and is considered the most important concepts of Japanese management. KAIZEN requires continual improvement, no matter how good are considered the product, service or process at a given time. Most of the currently known tools for quality management has evolved from KAIZEN philosophy, according to the philosophy of "good is never good enough" [5].

3 Kaizen in Practice

We will now give an example from a local company that manufactures furniture and has developed a system of Lean improvements through the use of tools: 5S, VSM, Kaizen i SMED [7]. Mentioned firm is during the last 20 years collaborating with one of the largest furniture manufacturers in the world-Swedish IKEA. The company has 360 employees and produces five product groups with 57 different items. The main focus is put on the production of composite veneered furniture as well as furniture made from mixed materials. It should be noted that mentioned company has introduced the quality management system according to the ISO 9001 standard. This quality management system was implemented in 2015. In order to provide product that meets customer requirements and relevant laws and regulations and aims to enhance customer satisfaction with effective application of the system, including continuous improvement system. In addition to the standard ISO 9001, that company has introduced ISQS (IKEA supplier quality standard) Quality GO/NOGO Standard (Standard quality GO/NOGO). This standard has to apply each company that wants to become a supplier of IKEA, a control is carried out every two years. GO/NOGO requirements include the following [1]:

IKEA product documentation;

- Connection system (Connect system);
- Process control;
- Final control;
- Conflicting products.

Measuring the quality of products and services based on the following elements: COPQ (Cost Of Poor Quality The cost of poor quality); Logistic KPI (Key Performance Indicator Key indicators characteristics); Daily assembling products by administrative workers; Measures control by pattern [6].

One of the leading furniture manufacturers in the world, a global corporation IKEA, based in Sweden has special requirements for its suppliers that are included in IWAY

Standard (Minimum requirements for environmental, social and working conditions in sourcing products, goods and services), and that in segment of environmental and human health have to apply all their suppliers, among which is the analysed company. Basic principles when working with environmental, social and working conditions within the IWAY standard are:

- What is in the best interest of the child?
- What is in the best interest of the worker?
- What is in the best interest of the environment?

Through the above principles IKEA supports business and cooperation with suppliers described in the sustainability strategy of IKEA Group: "People & Planet Positive". IKEA's way of purchasing products, materials and services (IWAY) is a code of conduct for IKEA Group that applies to suppliers. It consists of the minimum IKEA demands relating to environmental, social and working conditions (including child labour). IWAY is based on eight fundamental conventions defined in "Basic Principles of the right to work," the declaration of the International Labour Organization from June 1998 and on the 10 principles of the UN Global contract in 2000. IKEA recognizes the fundamental principles of human rights, as defined by the Universal Declaration of Human Rights (UN 1948), and abide by the UN sanctions list and the EU list of restrictive measures. IKEA Supplier shall always be in compliance with the strictest requirements, regardless of whether they are relevant to applicable laws or IKEA IWAY specific requirements. If IKEA requirements are in conflict with local laws and regulations, the law will always prevail and must be respected. In such cases, the Supplier shall immediately notify the IKEA company [2].

Successful implementation of the IWAY Standard is based on cooperation, mutual trust and respect between the Supplier and IKEA company. IKEA, its employees, as well as any third person chosen by IKEA, must with confidence to relate to all observations, discussions and written information received from the Supplier. Trust, integrity and honesty are core values of IWAY Standard and are essential for its sustainable use. On these grounds IKEA has based relations and these relations continue to deepen the respect of these values. It is important that all IKEA employees and external business partners understand the attitude of the IKEA company on corruption and the prevention of the it. This attitude is based on the rules and standards of the IKEA Group for the Prevention of Corruption and it is communicated to the business partners, including letters to service providers which will be signed [1].

IWAY standard structure is the following: IWAY binding requirements (IWAY MUST requirements); General conditions; Business ethics; Environment; Chemicals; Waste; Emergencies and fire prevention; Health and security of workers; Employment, work hours, salaries and beneficence; Accommodation; Child labour and younger workers; Discrimination; Workers' participation; Harassment, mistreatment and disciplinary measures. Binding "IWAY MUST" requirements are as follows: Child labour; Forced and obligatory work; Business ethics; Serious environmental pollution; Serious danger to safety and health; Opening hours; Earnings and; Ensuring workers from accidents [9]. Figure 1 provides some of the products of analyzed company.

Mentioned company in February 2015 began with activities to introduce its own Lean system improvements and Kaizen. The analysed company is introducing Kaizen

Fig. 1. Product range of analysed company [1].

concept through the use of numerous tools and with improvement program it achieved the following results [11]:

- Cleaner and better organized work places;
- Improved safety of employees;
- Improved life expectancy of machinery and equipment;
- Greater utilization of working surfaces;
- Reduced rate of injuries at work;
- Improved quality of their products;
- Reduced costs and losses;
- Improved employee morale;
- Improved productivity;
- Satisfying customer requirements and;
- Promoted team work.

One of the specific problems in the production process of analysed company is fraying at edges of ILSENG product, shown in Fig. 2.

Fig. 2. Example of frittering edge on products Ilseng [2].

Namely, in the analysed company at the beginning they followed only one of three products that are produced in the area of PC Mechanical. They followed the following parameters:

1. The time elapsed from the start of pressing to the side of the processing;
2. Board where it was pressed (three presses);
3. The operator that pressed them (8 operators);
4. The elements that can be tailored by the width;
5. The elements that can be tailored by the length;
6. The total scrap (not for further use).

Through the program of Kaizen improvements was established a whole mechanism of data collection and analysis in the period 18.05–23.05.2015 with monitoring of scrap appearance. Table 1 shows an overview of scrap monitoring [2].

The Fig. 3 gives a graph showing the finishing of the product ILSENG in the period 18.05–23.05.2015.

Then was continued monitoring of waste and for the next period (25–30.05.2015). The Table 2 shows the comparative review of monitoring scrap for two comparative periods (18.05–23.05) and (25–30.05.2015).

Applying Kaizen approach enabled the establishment of a monitoring system of waste through a period of time so as to identify all forms of waste and frequency of their occurrence [8], which is graphically shown in Fig. 4.

After the data collection and identification of the above mentioned types of waste and frequency of their occurrence experiencing a certain period of time, were formed the following two working teams:

(a) Team 1 that followed the technological process of pressing and;
(b) Team 2 that followed technological process of side processing on the line of IMA manufacturer.

Table 1. Data on scrap monitoring in analysed period.

KAIZEN – scrap monitoring (18/05–23/05)			
PC 2 - MACHINE			
WORKPLACE: Preparation of varnishing elements – finishing ILSENG			
Number of monitored elements: 1538			
Responsible person: xx			
Error type	Number of pieces	Percentage	Remarks
Number of usable pieces	646	42%	
Tearing	466	30%	
Pressed	68	4.4%	
Joint	182	11.8%	
Sharp edges	1	0.6%	
Other errors			
Reshaping by length	51	3.3%	
Reshaping by width	87	5.6%	
Total scrap	37	2.4%	

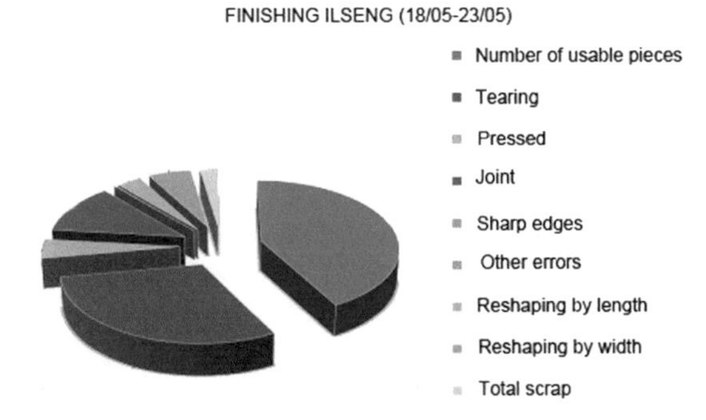

Fig. 3. Graph showing the finishing of the product ILSENG in the period 18.05–23.05.2015 [1].

Based on several weekly observation and investigation of the reasons, it was concluded that chipping occurs on the machine on which the processing is performed at the sides of ILSENG i.e. on line of IMA manufacturer. After determining the cause and place of origin of the problem, the drawing up of technological solutions which consisted in setting of two units with abrasive belts that are supposed to replace the cutter, which was done under urgent procedure. Figure 5 gives an overview of that intervention of introduction of new technological solutions.

Figure 6 gives an overview of processed elements without having chipping after the removing of the problem and the resulting improvements.

Table 2. Comparative overview of scrap monitoring for two periods [2].

KAIZEN – scrap monitoring (25/05–30/05)				KAIZEN – scrap monitoring (18/05–23/05)			
PC 2 - MACHINE				PC 2 - MACHINE			
WORKPLACE: Preparation of varnishing elements – finishing ILSENG				WORKPLACE: Preparation of varnishing elements – finishing ILSENG			
Number of monitored elements: 1279				Number of monitored elements: 1538			
Responsible person: xx				Responsible person: xx			
Error type	Number of pieces	Percentage	Remarks	Error type	Number of pieces	Percentage	Remarks
Number of usable pieces	565	44%		Number of usable pieces	646	42%	
Tearing	391	31%		Tearing	466	30%	
Pressed	100	7.8%		Pressed	68	4.4%	
Joint	128	10%		Joint	182	11.8%	
Sharp edges	3	0.2%		Sharp edges	1	0.6%	
Other errors				Other errors			
Reshaping by length	77	6%		Reshaping by length	51	3.3%	
Reshaping by width	0	0%		Reshaping by width	87	5.6%	
Total scrap	15	1.2%		Total scrap	37	2.4%	

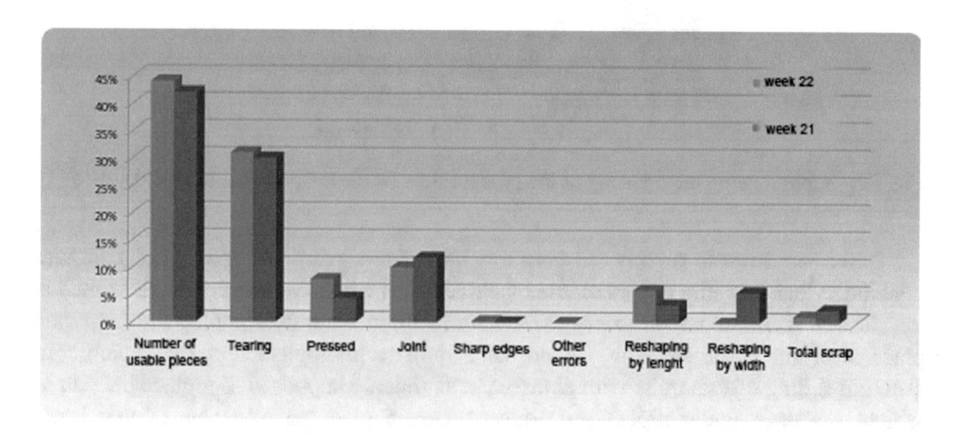

Fig. 4. Graphic types and the frequency of occurrence of waste [1].

Fig. 5. Technological solution on the machine for side processing in order to eliminate chipping [2].

Fig. 6. Processed elements "error-free" after the Kaizen improvements [2].

4 Conclusion

The paper presents the application of Kaizen and 5S as a LSS tool in the concrete case study of domestic wood processing firm.

Applied technological solution proved to be very effective, and achieved the following measurable effects [2]:

- Chipping is reduced by 30%;
- Ejected the technological operation of manual finishing of elements;
- There has been a surplus of 4 workers;
- In this way, it was reduced the technological process of product manufacturing;
- Achieved financial savings in the net work on a monthly basis by an average of: 1,100 BAM;
- Decreased technological waste from 3% to 2.5%, which brought financial savings on a monthly basis by an average of 2,500 BAM.

References

1. Alagić, I.: Upravljanje kvalitetom: Lean proizvodnja - Six Sigma, CIP - Katalogizacija u publikaciji, Nacionalna i univerzitetska biblioteka, 629.331:[658.5:005.6(075.8), COBISS. BH-ID 24216326. Štamparija-S, Tešanj (2017). ISBN 978-9958-074-08-0
2. Alagić, I.: Industrial engineering & maintenance: lean production - six sigma with application of tools and methods in specific working conditions, CIP - Katalogizacija u publikaciji, Nacionalna i univerzitetska biblioteka, 629.331:[658.5:005.6(075.8), COBISS. BH-ID 24232454, Štamparija-S, Tešanj (2017). ISBN 978-9958-074-09-7
3. Alagić, I., Božičković, R., Višekruna, V., Brkić, A.: Primjena Lean Six Sigma alata u konkretnim radnim uslovima firme iz automobilske industrije. Univerzitet u Istočnom Sarajevu, Mašinski fakultet, Festival kvaliteta 2017, Jahorina, BiH, 26–28 Oktobar 2017
4. Alagić, I.: Razvoj modela sistema upravljanja kvalitetom u upravljanju lancem dobavljača automobilske industrije i primjena Lean Six Sigma alata u radnim uslovima firme iz autoindustrije. In: XIX Nacionalni, V Međunarodni naučno-stručni skup Sistem Kvaliteta Uslov Za Uspešno Poslovanje I Konkurentnost, Hotel Kraljevi Čardaci SPA Kopaonik, Srbija, 29 November–1 December 2017
5. Alagić, I.: Application of lean six sigma tools in order to eliminate bottlenecks in working conditions firm from B&H. In: 19 Međunarodni Simpozij o Kvaliteti, Plitvice, Republika Hrvatska, 21–23 March 2018
6. Staudter, C., Meran, R.: Design for Six Sigma and Lean Toolset: Implementing Innovations Successfully. Springer, Heidelberg (2009)
7. El-Haik, B., Roy, D.M.: Service Design for Six Sigma: A Roadmap for Excellence. John Wiley & Sons, Hoboken (2005)
8. Levine, D.M.: Statistics for Six Sigma Green Belts with Minitab and JMP. Pearson Education, Upper Saddle River (2006)
9. Buehlmann, U.: Value Stream Mapping, Leadership and Skills Development Training. Bern University of Applied Sciences, Biel, Switzerland (2016)
10. Marić, B., Božičković, R.: Održavanje tehničkih sistema & Lean koncept. Univerzitet u Istočnom Sarajevu, Mašinski fakultet Istočno Sarajevo, Istočno Sarajevo (2014). ISBN 978-99976-623-09
11. Gregory, H.: Watson: Six Sigma for Business Leaders-A Guide to Implementation. GOAL/QPC, Methuen (2004). ISBN 1-57681-049-6

Application of the Modified Genetic Algorithm for Optimization of Plasma Coatings Grinding Process

Vladimir Tonkonogyi[1], Predrag Dašić[2(✉)], Olga Rybak[1], and Tetiana Lysenko[1]

[1] Odessa National Polytechnic University (ONPU), Odessa, Ukraine
[2] High Technical Mechanical School of Professional Studies, Str. Radoja Krstića 19, 37240 Trstenik, Serbia
dasicp58@gmail.com

Abstract. The problem of defining optimal conditions for grinding plasma coatings may be considered as multi-objective optimization problem with a system of bounding inequalities that contain surface roughness, temperature, local and residual stresses as well as intrinsic defects size. This approach contributes to applying evolutionary algorithms such as genetic algorithm to solve the stated problem. Taking into account special characteristics of technological process, modification of the classical genetic algorithm has been carried out in the presented research. The combined method of selection based on the mitosis and meiosis operators makes it possible to increase fitness of a population ensuring its diversity during the following iterations. It is particularly important to maintain population diversity in genetic algorithm. The reason for that is preventing premature convergence which causes the obtained solution to be far from optimal. Another way to ensure population diversity is applying the developed mutation domain model that allows to alter random genes in chromosomes with the lowest value of the fitness function. The presented algorithm is based on both the combined method of selection and the mutation domain model. In order to compare the results of solving the problem of optimization of plasma coatings grinding process using modified genetic algorithm with other evolutionary algorithms, solutions performed by the classical genetic algorithm, ant colony optimization, particle swarm optimization and scatter search algorithm are presented. It was found that applying modified genetic algorithm provides high efficiency of solving process and reliability of the obtained results.

Keywords: Genetic algorithm · Multi-objective optimization · Selection operator · Mutation operator · Surface grinding · Plasma coatings

1 Introduction

Spraying plasma coatings onto working surfaces of tools is becoming progressively widespread in modern mechanical engineering, automobile, aerospace and other branches of industry. Beyond additional opportunities provided by surface engineering to increase functionality of the finished products, economic usefulness of spraying

I. Karabegović (Ed.): NT 2019, LNNS 76, pp. 199–211, 2020.
https://doi.org/10.1007/978-3-030-18072-0_23

plasma coatings plays an important role in industry. Physical and chemical properties of tools during their service are largely determined by the surface layer, which makes it possible to avoid production of the solid tool from expensive and scarce materials.

In the process of grinding plasma coatings, different properties of a substrate and a coating layer cause the risk of their debonding due to the destruction of adhesive contacts between them. Furthermore, there is still a possibility of surface and intrinsic defects occurrence, typical for grinding workpieces with solid structure – such as grinding cracks, burns, scratches etc. These defects make the processed tools inappropriate for further application, which, in turn, causes economic losses. Along with the conditions preventing defects generation and propagation, ensuring quality of the processed surface includes requirements for needed precision and roughness of the finished workpieces. Therefore, setting technological parameters while grinding plasma coatings depends on the complexity of factors.

Applying evolutionary algorithms is an advanced approach to solve complex optimization problems with a large number of parameters and processing constraints. With the help of evolutionary methods it is possible to determine optimal operating modes for each grinding regime, size and abrasive material of the grinding wheel and other conditions. An advantage of these methods, such as genetic algorithm, is that the result obtained at the beginning of the solving process is eventually improved on the basis of the analysis of further solutions. This scheme of optimization problem solving provides an opportunity to find multiple sets of possible solutions contained in an established database of designing technological regimes.

Evolutionary algorithm modifications carried out by setting new methods of selection, crossover principles and mutation strategy allow to increase efficiency of solving process and reliability of the solution. The genetic algorithm for grinding operations CAD system is developed in this report in order to determine optimal grinding parameters during plasma coatings processing.

2 Literature Review

The problem of grinding process optimization is generally based on the conditions of cost minimizing, achieving maximum productivity and best quality of the processed surface. Attempts to define optimal grinding parameters using linear and geometric programming, gradient method and the method of Lagrange multipliers are described in the works [1–4]. However, these methods appeared to be inefficient for defining optimal processing regimes because of non-linearity of the objective functions and large number of constraints regarding the process of grinding.

Various innovative metaheuristic algorithms have advanced and become increasingly popular recently for solving complex multi-objective problems of optimization technological system parameters. In the work [5] grinding parameters selection problem is stated and solved in terms of quadratic programming. Grinding wheel speed, workpiece speed and dressing parameters are considered variables, and total production cost, production rate and roughness of the processed surface are selected as objective functions of the created model.

To find a set of Pareto optimal solutions of the grinding parameters search problem authors of [6] use multi-objective evolutionary algorithm. Obtained results appear to be much better in comparison to the previously applied methods.

An attempt of adapting genetic algorithm to grinding process optimization is made in paper [7] on the basis of mathematical model presented in paper [5]. Despite a number of advantages, this method has a high probability of convergence to the local extremum, which impedes solving the stated problem.

Authors of [8] carried out optimal grinding parameters search using ant colony optimization algorithm based on modeling the behavior of simulated ants located in the vertices of the imaginary graph and moving with some probability of one or the other route selection. Results of the algorithm application are rather effective.

In the researches [9–11] authors propose algorithms based on Taguchi method to analyze the relationship between the grinding parameters and the quality of the processed surface. Taguchi method is a robust statistical method of technological process optimization which is often used to design manufacturing systems and to control the resulting products quality.

Since the grinding process optimization is based on nonlinear objective functions of several variables, some authors [12, 13] prefer to apply differential evolution algorithm and its modifications. Differential evolution algorithm has much in common with the genetic algorithm strategy, but it does not require to represent the variables in binary code.

Modification of the scatter search method for grinding process optimization problem is presented in the work [14]. The main feature of scatter search algorithm is that the solutions from initial population should be sufficiently scattered over the feasible set region.

When reviewing advanced evolutionary methods, particle swarm optimization should also be mentioned as it can be used to search for the global extremum if calculating the gradient of an objective function raise some difficulties. Particle swarm optimization algorithm is based on the collective animal behavior model and characterized by speed adjustment which is defining for the convergence of an algorithm. An example of applying particle swarm optimization method to solve the problem of grinding process parameters optimization is presented in paper [15]. Authors of [16] suggest an enhanced particle swarm optimization algorithm to improve the results of solving the stated problem. Paper [17] is focused on the development of hybrid particle swarm optimization method which combines the dynamic neighborhood topology for particle swarm optimization with the mutation approach based on the conditions of surface grinding process. In paper [18] chosen regression model of the surface roughness in finishing turning of hardened steel with mixed ceramic cutting tools.

Analysis of the scientific literature referred to in this review has revealed a number of unresolved problems regarding optimization of the parameters of plasma coatings grinding process and the algorithm of the stated problem solving:

- previous research of the grinding regime parameters was carried out for the workpieces that have the solid structure. Structural steel was typically used as a material of the processed tool. Taking into account that structural steel is not very expensive, loss of the material during the process of grinding has no effect on the

finished products cost. However, plasma coatings materials are often characterized by the high price and low availability, so they are important factors of price formation that will influence on the optimality criterion formulating;
- while grinding plasma coatings, different properties of the coating layer and the substrate material should be taken into account as well as possibility of destruction of adhesive contacts and debonding of a coating layer from a substrate;
- genetic algorithm is one of the main evolutionary methods for solving optimal technological parameters search problem. An advantage of the classical genetic algorithm is its fast convergence. But the fast convergence may become an important disadvantage in case of convergence to the local extremum, because then it would not be possible to find optimal parameters values for the global extremum;
- the mutation probability remains almost the same for all of the solutions in the population with very little reference to the fitness function value. But the results of the mutation of chromosomes within the neighbourhood of the global maximum might be less fit than their parents. It has negative consequences for problem solving process. Hence, the mutation mechanism should particularly contribute to altering among the solutions with the least value of the fitness function.

3 The Aim and Objectives of the Research

The aim of the presented research is to develop modified genetic algorithm in order to find optimal parameters for grinding plasma coatings.

To achieve this objective, the following tasks should be accomplished:

- to define special aspects of grinding parameters selection problem statement for processing workpieces with plasma coatings, taking into account additional conditions and constraints imposed on mathematical model of grinding workpieces with solid structure;
- to develop method of selection that would allow to avoid decreasing fitness of the solutions from the global maximum neighbourhood as a result of recombination with less fit chromosomes, to maintain the highest population diversity and to prevent premature convergence of the algorithm;
- to propose mutation mechanism that would allow the solutions with the least fitness value to alter faster without decreasing population size of the system.

4 Formalization of the Optimal Technological Parameters Search Problem for Plasma Coatings Grinding Process

Optimal technological parameters search problem for the process of grinding plasma coatings studied in the work [19] is presented as multi-objective optimization problem. Productivity of the coated workpieces processing and allowance loss of coating material are considered optimum criteria which can be expressed as:

$$P = \frac{1000 \cdot a \cdot b}{L \cdot k \cdot (B + b + \Delta)} \cdot v_l^j \cdot v_{tr}^s \cdot \sum_{i=1}^{N} h_i\ , \quad j = \overline{1, M}; \quad s = \overline{1, Q} \tag{1}$$

$$\delta = \sum_{i=1}^{N} h_i \cdot n_i \tag{2}$$

where a is the length of the workpiece; b is its width; L is table length; k is precision coefficient; B is width of the grinding wheel; Δ is empty width of the table; v_l^j is longitudinal feed; v_{tr}^s is transverse feed; h_i is depth of cut for the ith pass; n_i is number of passes with the set depth of cut.

Thus, problem statement consists in searching technological parameters of plasma coatings grinding process that provide maximum grinding productivity [20] and minimum loss of coating material on allowance with the required surface quality [21]:

$$P \rightarrow \max; \tag{3}$$

$$\delta \rightarrow \min; \tag{4}$$

$$\begin{cases} R_a \leq R_a^*; \\ T(x) \leq T_{\max}; \\ \sigma_{\max}(x, \tau) \leq \sigma_{\lim}; \\ \sigma_y < \theta_{adh}; \\ \ell_0 < \ell_0^*, \end{cases} \tag{5}$$

where R_a and R_a^* – obtained and required roughness of the processed surface; T and $T_{\max}$ – grinding temperature and maximum temperature that allows to avoid grinding burns on the surface of plasma coating; $\sigma_{\max}$ – local stresses maximum value; $\sigma_{\lim}$ – stresses that trigger grinding cracks generation; θ_{adh} – adhesive strength between the coating layer and the substrate; σ_y – tangent stresses that destroy adhesion; ℓ_0 and ℓ_0^* – linear dimension of the structural defect and size of the defect that causes local fracture.

In order to solve multi-objective optimization problem, an additive function is built as a sum of the criteria $Z_1 = P$ and $Z_2 = \delta$, weighted according to their importance. Hence, complex optimality criterion can be expressed as:

$$Z = w_1 \cdot Z_1 - w_2 \cdot Z_2\ . \tag{6}$$

Depending on the priority of one or another criterion during the process of spraying and grinding plasma coatings, weight coefficients are selected so that their values are within [0; 1] and their sum is $w_1 + w_2 = 1$.

Vector of technological parameters which can be set at the grinding process design stage includes wheel speed, longitudinal and transverse feed, grinding depth and number of passes of the grinding wheel. Other parameters are constant for each case of circumstances for the stated problem. Therefore, searching for Pareto optimal solutions is carried out within the set of possible values of the vector $x = \left(v_w, v_l^j, v_{tr}^s, h_1, n_1, \ldots, h_i, n_i\right)$.

5 Strategy of the Optimization Problem Solving

5.1 Genetic Algorithm Application

General idea of the genetic algorithm consists in following methods of optimization that are usual for living organisms: genetic inheritance and natural selection. In this case biological terminology is used to explain operations of the algorithm. The main principle of natural selection is based on a statement, that it is the fittest individuals who survive and reproduce during the process of evolution. According to the law of genetic inheritance, their offsprings inherit the main characteristics of parents. Besides, being exposed to the random mutations gain them a number of new features. If the changes contribute to organism adaptation, they are inherited in the next generations. Therefore, average fitness of the population is eventually increasing.

Genetic algorithm relies on fundamental operators common to all evolutionary algorithms of optimization: phenotype building, selection, crossover, mutation etc. Genetic algorithm running consists in systematic performing operations presented in Fig. 1.

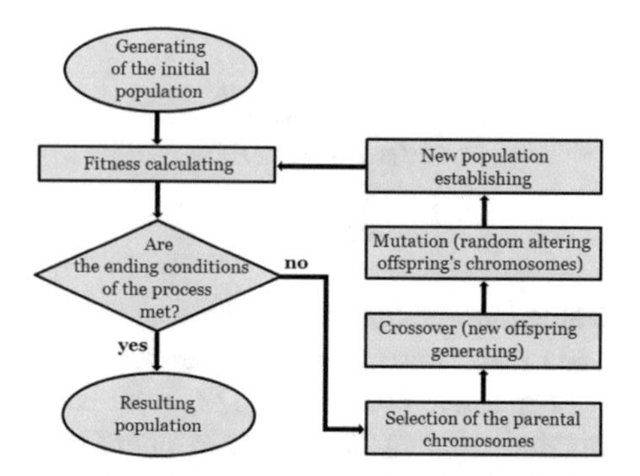

Fig. 1. General scheme of the classical genetic algorithm

5.2 Development of the Modified Genetic Algorithm for Solving Grinding Parameters Search Problem

Sufficient size of the population and maintaining its diversity are necessary conditions for genetic algorithm running. Premature stopping of an algorithm before the global extremum is reached may obstruct its reliability meaning the obtained result is not Pareto optimal. The reason of premature convergence is natural selection principle that forms the basis of the genetic algorithm. Selection of the parental solutions is carried out with regard to their fitness: the higher it is the greater is the probability that the chromosome will take part in further crossover. Hence, fitter solutions have an advantage and begin to replace other variants of the code from the population. In case the population size is insufficient for diversity maintaining and the solution with the highest value of the fitness function is far from the global extremum, algorithm prematurely converge to the local maximum or minimum. And even if after a certain number of iterations a much fitter solution is generated, it has no influence on the final result, as it remains in the minority among the large number of former leaders' offsprings. Probability of reaching the global or the local extremum of the fitness function also greatly depends on its type. Premature convergence is more likely to occur if local extremums of the function are far from one another within its definition domain [22].

In Fig. 2 distribution of chromosomes within the population during the process of solving optimal grinding parameters search problem using the classical genetic algorithm is presented.

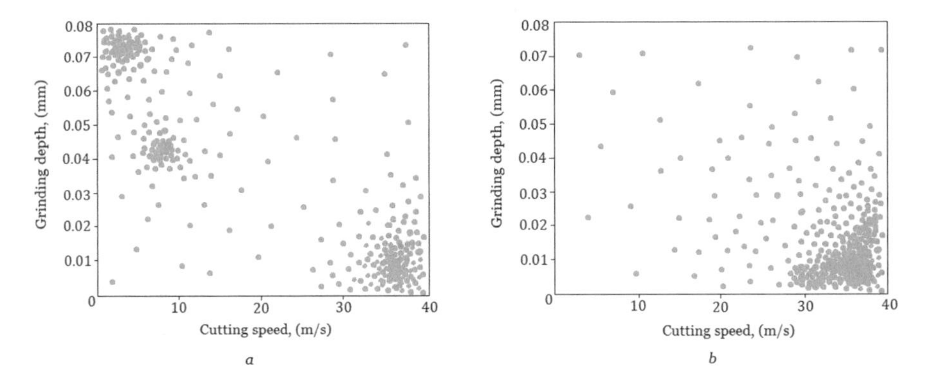

Fig. 2. Distribution of chromosomes for the classical genetic algorithm: a – grouping around several extremums; b – dominating of one of the extremums and driving out other solutions from the population

At the early stages of the genetic algorithm running fitness of the solutions vary significantly from the global extremum. During the further crossover sets of the solutions eventually group around the local and the global extremums (Fig. 2a). After that solutions concentrate in one of the areas which randomly started to dominate by the number of solutions (Fig. 2b). Chromosomes concentrated in other areas within the problem solution space are gradually driven out from the population. Therefore, in order to prevent premature convergence and avoid the loss of the useful genetic material, it is necessary to concentrate the set of the problem solutions step by step towards the global extremum area.

The fundamental idea of the classical genetic algorithm reflects the reproductive division mechanism which is called meiosis in the natural sciences. Along with meiosis, there is one more reproduction method typical for living organisms – mitosis. When applying the mitosis operator, chromosomes are exposed to replication of their genetic code. It means that the considered solution at some stage is rewritten and moved to the next iteration unchanged. Applying the mitosis operator to the best solutions in the population enables to increase its general fitness and maintain population diversity. If chromosomes with the highest values of the fitness function at the current stage of problem solving reproduced with the help of mitosis operator, it would be possible to prevent decreasing their values as a result of recombination with less fit solutions and to avoid driving out other variants of the genetic code. Proportion of elite solutions within the ith population may be defined by relative fitness function values:

$$\xi = \frac{f_k^i}{\max Z(x^i)}, \tag{7}$$

where f_k^i is fitness function value for the kth solution of the ith population; $\max Z(x^i)$ is the highest value of the fitness function within the ith population.

Hence, having determined relative fitness ξ, lower limit $\underline{\xi}$ of fitness for solutions that are moved to the next generation unchanged can be found out. Solutions whose relative fitness is within $\xi \in [\underline{\xi};\ 1]$ characterize dispersion of the elite chromosomes to the highest value of the fitness function at each stage of the algorithm running. If the elite solutions loosed their leading positions afterwards, they would be involved in crossover and new generation establishing. Thus, at the beginning of the problem solving process convergence would be slowed down, whereas there would be increasing fitness of the population with wide diversity of possible solutions. In that case, the size of the population remains the same during the whole problem solving process.

Chromosomes with less values of the fitness function form a set of solutions for further selection and crossover. Selecting chromosomes is carried out on the basis of the tournament selection. Crossover is taking place until there is the required number of solutions in the next generation.

For the least fit solutions in the population it would be useful to provide a mechanism that would allow them to alter faster under the conditions of the stated problem. For this purpose the mutation operator should be applied, which makes it possible to turn parts of the binary code into the opposite values. Knowing relative fitness of the solutions and value $\overline{\xi}$ defining upper limit of the solutions with the least fitness, chromosomes for the further mutation are easily found. Relative fitness function of these solutions lies within $(0;\ \overline{\xi}]$ values. Principle of generating each new population in the modified genetic algorithm is shown in Fig. 3.

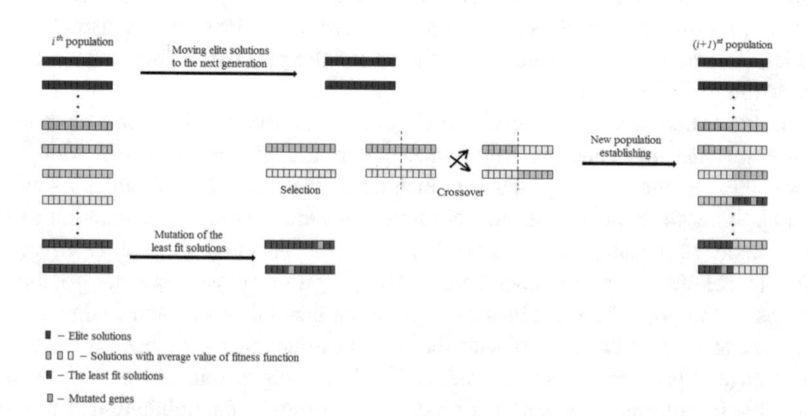

Fig. 3. Scheme of the new population establishment for the modified genetic algorithm

While solving the problem of optimization of plasma coatings grinding process, along with the conditions of diversity maintaining and premature convergence preventing, constraints ensuring quality of the processed surface (5) have a particular importance. These constraints are determined by characteristics of the processed coating, grinding wheel parameters and the workpiece geometry. Besides, grinding machine characteristics play an important role in generating the initial population and the set of possible solutions, as they define limitations of the processing speed and other key parameters of surface grinding. Information required for calculations during the stated problem solving is stored in corresponding databases of the plasma coatings grinding CAD system.

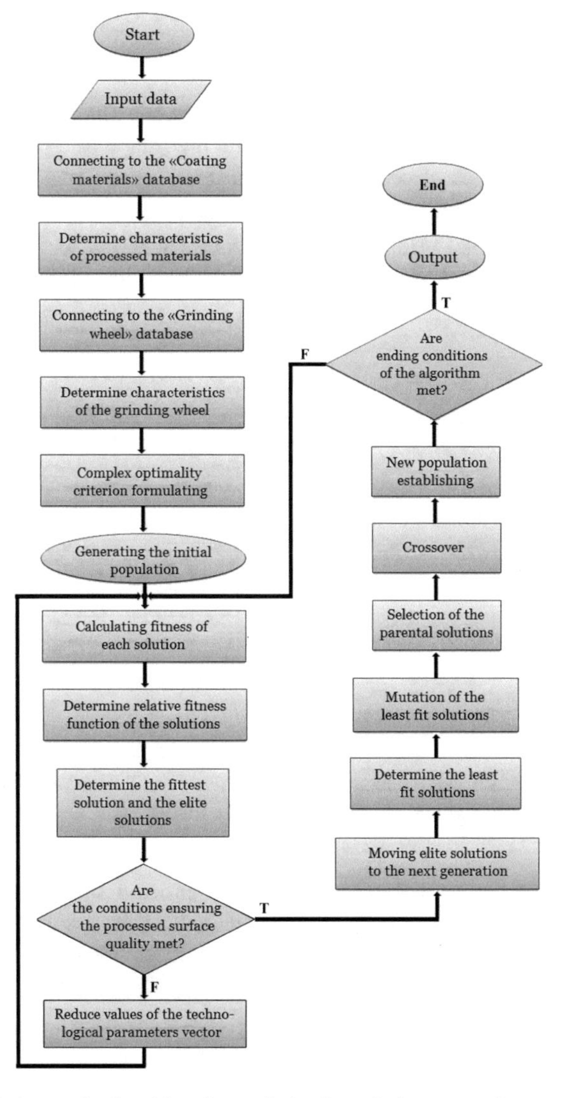

Fig. 4. Modified genetic algorithm for optimization of plasma coatings grinding process

In case the conditions ensuring surface quality are not met at some stage of the solving process, values of the technological parameters vector $x_k = \left(v_w, v_l^j, v_{tr}^s, h_1, n_1, \ldots, h_i, n_i\right)$ should be reduced by the minimum value Δ possible for the grinding machine. Therefore, reduction of values $\left(v_w - \Delta v_w,\ v_l^j - \Delta v_l^j, v_{tr}^s - \Delta v_{tr}^s, h_1 - \Delta h_1, n_1, \ldots, h_i - \Delta h_i, n_i\right)$ will take place until the conditions that ensure surface quality are met. The modified genetic algorithm for optimization of plasma coatings grinding process within the CAD system is presented in Fig. 4. An ending condition of the algorithm is achieving such level of convergence when there is no better result for several iterations.

6 Discussion of the Results of the Modified Genetic Algorithm Performance Analysis

In order to compare the efficiency of the modified genetic algorithm with the results of the classical genetic algorithm, the corresponding graphics are plotted in Fig. 5. Objective function values for each iteration are determined regarding the maximum value obtained as a result of solving optimization problem. Presented graphics demonstrate that applying the mitosis and meiosis operators as well as mutation domain model makes it possible to increase the efficiency of the optimization process for grinding plasma coatings.

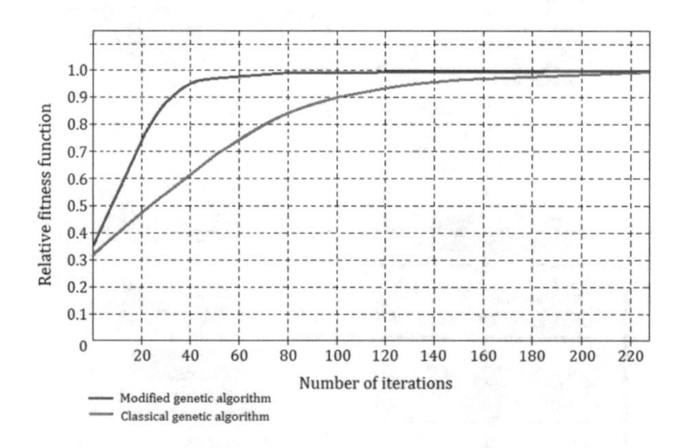

Fig. 5. Objective function value dependence on the number of iterations during the process of optimization for the modified and the classical genetic algorithm

Figure 6 shows the results of technological process optimization for grinding plasma coatings applying other evolutionary methods such as ant colony optimization algorithm, particle swarm optimization algorithm and scatter search method.

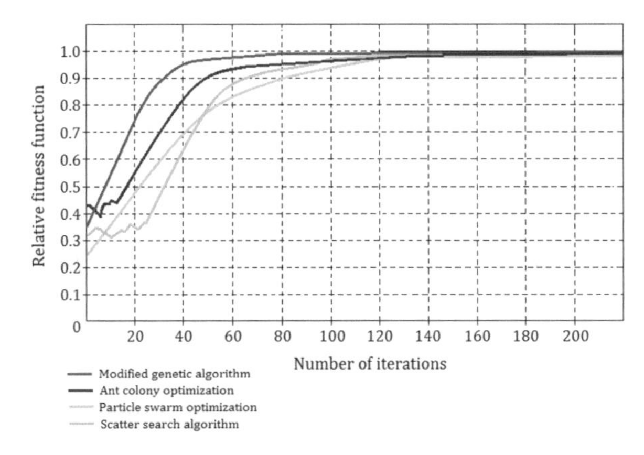

Fig. 6. Comparison of convergence of the modified genetic algorithm, ant colony optimization algorithm, particle swarm optimization method and scatter search method for the process of optimization of grinding plasma coatings

7 Conclusion

Analysis of the conditions and constraints that ensure required quality of the processed surface is carried out in the presented research. Mathematical model of the optimal technological parameters search problem is developed for the process of grinding plasma coatings. In this case, the porosity of coating layer as well as an adhesive strength between the coating layer and the substrate are considered for the first time. The resulting system of constraints provide zero-defect grinding of a workpiece and the required surface roughness, while optimality criteria include key factors of price formation for the finished products.

On the basis of the mitosis operator the combined method of selection is developed. It assumes replication of the elite chromosomes into the next generation and allows to prevent driving out less fit solutions from the population, to increase general fitness of the population and ensure its maximum diversity. This approach helps to slow down convergence of the algorithm at the beginning of problem solving process and makes it possible to avoid premature convergence to the local extrema.

Mutation operator that allows the least fit chromosomes in the population to alter faster is presented and investigated. Thus it becomes possible to maintain size of the population system and to involve the solutions with the least fitness value in the next generation establishing that has a positive effect on population diversity.

The results of the modified genetic algorithm performance are compared with other evolutionary methods of optimization. This comparison has proved an advantage of the modified genetic algorithm for the optimal technological parameters search during the process of grinding plasma coatings.

References

1. Iwata, K., Murotsu, Y., Iwatsubo, T., Fujii, S.: A probabilistic approach to the determination of the optimum cutting conditions. J. Eng. Ind. **94**(4), 1099–1107 (1972)
2. Malkin, S.: Selection of operating parameters in surface grinding of steels. ASME J. Eng. Ind. **98**(1), 56–62 (1976)
3. Gopalakrishnan, B., Faiz, A.K.: Machine parameter selection for turning with constraints: an analytical approach based on geometric programming. Int. J. Prod. Res. **29**(9), 1897–1908 (1991)
4. Agapiou, J.S.: The optimization of machining operations based on a combined criterion. Part 1: the use of combined objectives in single pass operations. ASME J. Eng. Ind. **114**(4), 500–507 (1992)
5. Wen, X.M., Tay, A.A.O., Nee, A.Y.C.: Micro-computer-based optimization of the surface grinding process. J. Mater. Process. Technol. **29**(1–3), 75–90 (1992)
6. Slowik, A., Slowik, J.: Multi-objective optimization of surface grinding process with the use of evolutionary algorithm with remembered Pareto set. Int. J. Adv. Manuf. Technol. **37**(7), 657–669 (2008)
7. Saravanan, R., Asokan, P., Sachidanandam, M.: A multi-objective genetic algorithm (GA) approach for optimization of surface grinding operations. Int. J. Mach. Tools Manuf. **42**(12), 1327–1334 (2002)
8. Baskar, N., Saravanan, R., Asokan, P., Prabhaharan, G.: Ants colony algorithm approach for multi-objective optimisation of surface grinding operations. Int. J. Adv. Manuf. Technol. **23**(5), 311–317 (2004)
9. Aravind, M., Periyasamy, S.: Optimization of surface grinding process parameters by Taguchi method and response surface methodology. Int. J. Eng. Res. Technol. **3**(5), 1721–1727 (2014)
10. Güven, O.: Application of the Taguchi method for parameter optimization of the surface grinding process. Materialpruefung/Materials Test. **57**, 43–48 (2015)
11. Patil, P.J., Patil, C.R.: Analysis of process parameters in surface grinding using single objective Taguchi and multi-objective grey relational grade. Perspect. Sci. **8**, 367–369 (2016)
12. Krishna, A.G.: Retracted: optimization of surface grinding operations using a differential evolution approach. J. Mater. Process. Technol. **183**(2–3), 202–209 (2007)
13. Lee, K.M., Hsu, M.R., Chou, J.H., Guo, C.Y.: Improved differential evolution approach for optimization of surface grinding process. Expert Syst. Appl. **38**(5), 5680–5686 (2011)
14. Krishna, A.G., Rao, K.M.: Multi-objective optimisation of surface grinding operations using scatter search approach. Int. J. Adv. Manuf. Technol. **29**(5), 475–480 (2006)
15. Pawar, P.J., Rao, R.V., Davim, J.P.: Multiobjective optimization of grinding process parameters using particle swarm optimization algorithm. Mater. Manuf. Process. **25**(6), 424–431 (2010)
16. Lin, X., Li, H.: Enhanced Pareto particle swarm approach for multi-objective optimization of surface grinding process. In: Proceedings of the Second International Symposium on Intelligent Information Technology Application, vol. 2, pp. 618–623 (2008)
17. Zhang, G., Liu, M., Li, J., Ming, W.Y., Shao, X.Y., Huang, Y.: Multi-objective optimization for surface grinding process using a hybrid particle swarm optimization algorithm. Int. J. Adv. Manuf. Technol. **71**(9–12), 1861–1872 (2014)
18. Dašić, P.: Comparative analysis of different regression models of the surface roughness in finishing turning of hardened steel with mixed ceramic cutting tools. J. Res. Dev. Mech. Ind. **5**(2), 101–180 (2013)

19. Tonkonogyi, V.M., Rybak, O.V.: Plasma coatings grinding parameters selection for multi-objective optimization of technological process. Mod. Technol. Mech. Eng. **13**, 60–68 (2018)
20. Tonkonogyi, V., Yakimov, A., Bovnegra, L.: Increase of performance of grinding by plate circles. In: NT-2018: New Technologies, Development and Application. Lecture Notes in Networks and Systems, vol. 42, pp. 121–127. Springer, Cham (2019)
21. Lebedev, V., Tonkonogyi, V., Yakimov, A., Bovnegra, L., Klymenko, N.: Provision of the quality of manufacturing gear wheels in energy engineering. In: DSMIE 2018: Advances in Design, Simulation and Manufacturing. Lecture Notes in Mechanical Engineering, pp. 89–96. Springer, Cham (2019)
22. Bhattacharya, M.: Evolutionary landscape and management of population diversity. In: Combinations of Intelligent Methods and Applications. Smart Innovation, Systems and Technologies, vol. 46, pp. 1–18. Springer, Cham (2016)

A Comparative Analysis of the Machined Surfaces Quality of an Aluminum Alloy According to the Cutting Speed and Cutting Depth Variations

Mihail Aurel Ţîţu[1,2](✉) and Alina Bianca Pop[3]

[1] Lucian Blaga University of Sibiu, 10, Victoriei Street, Sibiu, Romania
mihail.titu@ulbsibiu.ro

[2] The Academy of Romanian Scientists, 54, Splaiul Independenţei, Sector 5, Bucharest, Romania

[3] SC TECHNOCAD SA, 72, Vasile Alecsandri Street, Baia Mare, Romania
bianca.bontiu@gmail.com

Abstract. The surface quality obtained by milling operations is one of the most studied aspects in the engineering. The main objective of this scientific paper involves to carry out an experimental research on the end-milling process of an aluminum alloy used in the aerospace industry. In this paper a comparative analysis was carried out on the machined surface quality, using different variations of the cutting speed and cutting depth. The cutting parameters values varies and the experimental program was thought in terms of properly modeled functions. In conclusion, a series of graphs were made based on the experimental data obtained by the surface roughness measurements.

Keywords: Surface quality · Cutting regime · Cutting process · Experimental research · Data processing · Objective function

1 Introduction

Aluminum alloys generally have good machinability condition but this depends on alloying elements, thermal treatments and manufacturing processes. In the research of the authors Tammineni and Yedula, the influence of cutting speed (500–1500 rpm), feed rate (50–70 mm/rot) and cutting depth (0.5–1.5 mm) on the milled surface quality of the Al1050 [1], is studied. Thamban addresses the issue of the cutting forces following the end-milling process of the 6061-T6 aluminum alloy, using for comparison a coated diamond tool and a non-coated carbide tool [2], under a dry cutting condition. The tool wear, tool durability and surface quality during the in Al 2024 machining were studied by Kök. In the turning operation, tools were used with and without coating, and it turned out that the tool durability decreases with the cutting speed increasing in both situations, with the indication that the coated tool has a longer durability [3]. The process parameters influence exerted on the cutting forces, the surface roughness and the tool edge were addressed and analyzed in Gökkaya's research. The material chosen for the study was Al2014 aluminum alloy machined by turning. The author has found

I. Karabegović (Ed.): NT 2019, LNNS 76, pp. 212–218, 2020.
https://doi.org/10.1007/978-3-030-18072-0_24

that the tool edge deposits affect the surface roughness which has low values on high cutting speeds and low feed rates, and the cutting forces are low in high cutting speeds [4]. Other aspects about the chemical composition of aluminum alloys and not only have been addressed in [5–8]. In this paper, the chosen aluminum alloy to conduct the experiments is Al 7136, which is used in the aeronautical industry. This work is a continuation of the previous research [9–14].

2 Research Development

The research method - The experiment was used to conduct the research [15, 16]. The main objective of this research is to make a comparative analysis on the Al7136 surface quality machined by end-milling process using different values of the cutting speed and cutting depth.

The cutting regime - The cutting process parameters and their values are presented in Table 1.

Table 1. Cutting process parameters and their values

v [m/min]	a_p [mm]	f_z [mm/tooth]
570	2	0.04
	2.5	0.06
610	3	0.08
	3.5	0.11
	4	0.14

The workpiece material - The experiments were carried out using the 7136 aluminum alloy used in the aerospace industry, and the workpiece had the dimensions of 500 × 10 × 24.5 [mm], as it is shown in Fig. 1.

Fig. 1. Machining preparation of 7136 aluminum alloy [9]

Fig. 2. SECO tool type R217.69-1616.0-09-2AN used to conduct the experiments [9]

The cutting tool - The cutting tool used to perform the experiments was a standard end-milling cutter of aluminum, SECO R217.69-1616.0-09-2AN with 2 teeth, 16 mm diameter, 100% tool engagement and one set of 2 cutting inserts coded XOEX090308FR-E05, H15, as shown in Fig. 2.

CNC Machine - The CNC machine used to carry out the experiments was the HAAS VF-YT2. The workpieces were each fixed by clamps of three jaw vices on the CNC machine tool table to obtain a maximum rigidity (Fig. 3).

Fig. 3. The workpiece fixing on the CNC machine [9]

Response - After the machining operations, the surface roughness measurements were made on a distance of 5 [mm] using the Mitutoyo SURFTEST SJ-210 rugosimeter. The measured values of the roughnesses are shown in Table 2. Thus, it was possible to compare the surface quality results by combining the process parameters in situations where the cutting speed is 570 [m/min] and 610 [m/min].

Table 2. The R_a values [μm] when cutting speed is 570 [m/min] and 610 [m/min]

Nr. op.	a_p [mm]	f_z [mm/tooth]	R_a [μm] when v = 570 [m/min]	R_a [μm] when v = 610 [m/min]
1	2	0.04	0.442	0.533
2	2.5	0.04	0.473	0.471
3	3	0.04	0.880	0.658
4	3.5	0.04	1.372	0.534
5	4	0.04	1.120	0.506
6	2	0.06	0.465	0.691
7	2.5	0.06	0.509	0.756
8	3	0.06	0.478	0.586
9	3.5	0.06	0.493	0.516

(*continued*)

Table 2. *(continued)*

Nr. op.	a_p [mm]	f_z [mm/tooth]	R_a [µm] when v = 570 [m/min]	R_a [µm] when v = 610 [m/min]
10	4	0.06	0.533	0.524
11	2	0.08	0.470	0.500
12	2.5	0.08	0.488	0.260
13	3	0.08	0.493	0.598
14	3.5	0.08	0.644	0.470
15	4	0.08	0.606	0.458
16	2	0.11	0.429	0.679
17	2.5	0.11	0.427	0.631
18	3	0.11	0.412	0.543
19	3.5	0.11	0.396	0.581
20	4	0.11	0.568	0.511
21	2	0.14	0.369	0.722
22	2.5	0.14	0.393	0.980
23	3	0.14	0.357	0.847
24	3.5	0.14	0.407	0.498
25	4	0.14	0.622	0.490

Using the obtained data it was possible to make comparative graphs between the evolution of the surface roughness on the variation of the cutting speed and the cutting depth in different situations.

3 Comparison Between the Evolution of the Machined Surfaces Quality of 7136 Aluminum Alloy According to the Cutting Speed and Cutting Depth Variations

This paragraph shows the comparative evolution of Al7136's milling surface when the cutting speed is 570 and 610 [m/min], the cutting depth varies, and the feed per tooth retains a constant value.

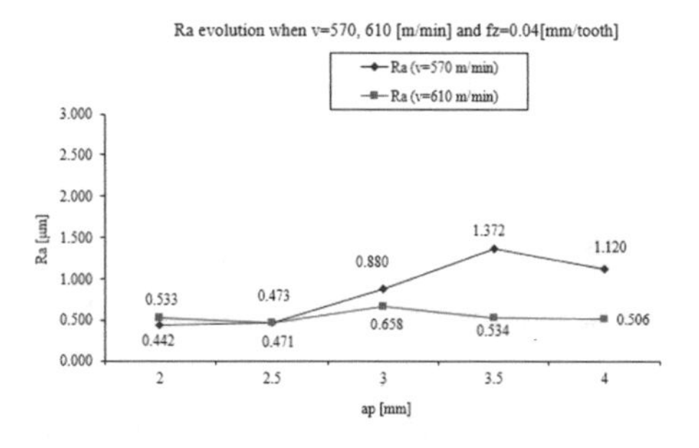

Fig. 4. The R_a evolution under the influence of a_p variation when $f_z = 0.04$ [mm/tooth]

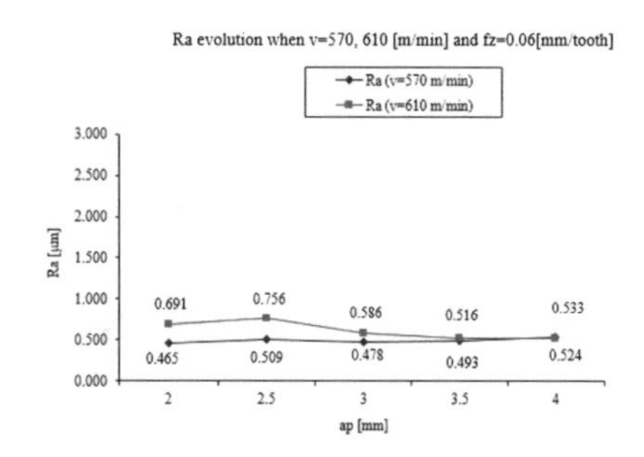

Fig. 5. The R_a evolution under the influence of a_p variation when $f_z = 0.06$ [mm/tooth]

Figure 4 shows the R_a evolution under the influence of variation of v and a_p when $f_z = 0.04$ [mm/tooth]. As it can be seen in this figure, R_a shows higher values on 570 than on 610 [m/min], but in both situations the R_a values increase with the increase of the cutting depth if the feed per tooth is 0.04 [mm]. In Fig. 5, where the feed per tooth is 0.06 [mm/tooth], the R_a values are lower at a higher cutting speed of 610 [m/min]. In this case, the R_a values in both situations decrease with the increase of the cutting depth.

In Fig. 6, the R_a values show insignificant oscillations with close measurement intervals, having slight increases with the increase of the cutting depth.

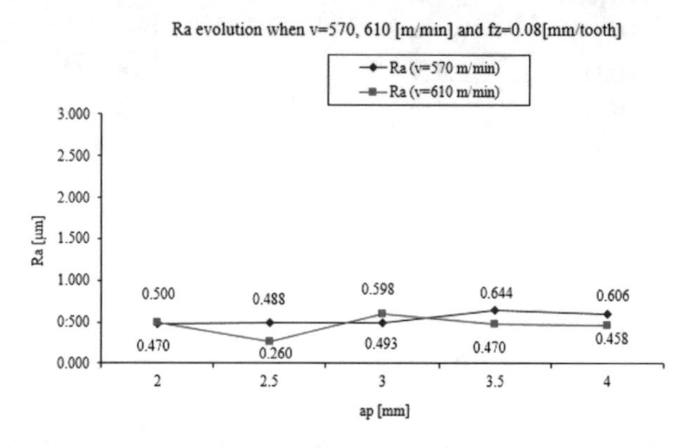

Fig. 6. The R_a evolution under the influence of a_p variation when $f_z = 0.08$ [mm/tooth]

Figure 7 shows the surface roughness R_a evolution when f_z is 0.11 [mm/tooth]. In this case the R_a values are higher at 610 [m/min].

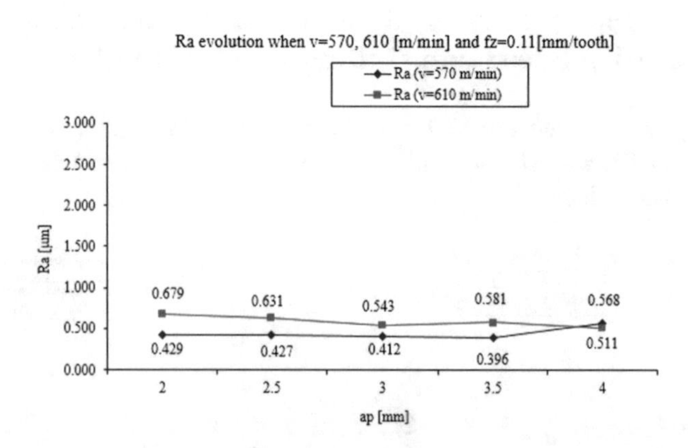

Fig. 7. The R_a evolution under the influence of a_p variation when $f_z = 0.11$ [mm/tooth]

Figure 8 shows the situation when f_z is 0.04 [mm/tooth] and the cutting depth varies. This case is similar to the situation where f_z is 0.06 [mm/tooth].

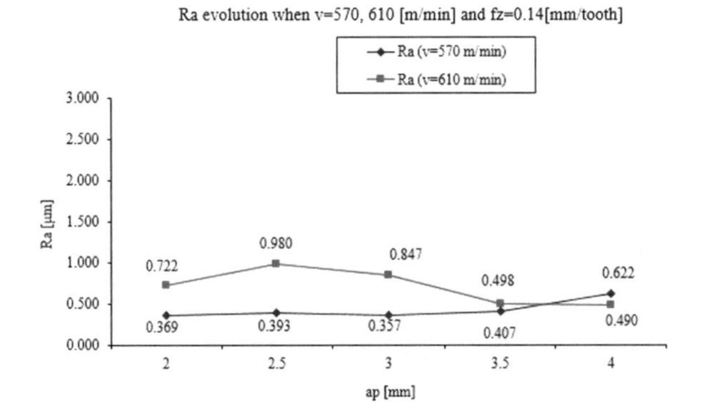

Fig. 8. The R_a evolution under the influence of a_p variation when $f_z = 0.14$ [mm/tooth]

By performing a general analysis of the achieved 5 graphs, it can be seen that the cutting speed along with the cutting depth has some influence on the surface roughness in the sense that at low values of the cutting depth, regardless of the cutting speed, with a low feed per tooth, the R_a values are low, while with the increase of the feed per tooth, the situation changes and the R_a values increase.

4 Conclusions

Based on the analysis of R_a surface quality, obtained by 7136 aluminum alloy end-milling on the variation of the cutting speed and cutting depth, it is found that the highest values of surface roughness measured longitudinally in the direction of the tools feed motion are:

- When the cutting speed is 570 [m/min], the cutting depth of 3.5 [mm] and the feed per tooth 0.04 [mm/tooth], the roughness has a value of 1.372 [µm];
- When the cutting speed is 610 [m/min], the cutting depth of 4 [mm] and the feed per tooth 0.04 [mm/tooth], the roughness has a value of 1.120 [µm].

In conclusion, it can be noticed that with the increase of the cutting depth, there is a slightly downward oscillating tendency of the roughness values with the increase of the feed per tooth.

The increase of the obtained roughness values is caused by the vibrations occured during the end-milling process. The vibrations emergence leads to the rough surfaces obtaining. These vibrations are the effect of the resonance phenomenon given by the contact between the tool and the workpiece and the materials used.

Causes, mode of production and vibration effect could be studied in further research.

References

1. Tammineni, L., Yedula, H.P.R.: Investigation of influence of milling parameters on surface roughness and flatness. Int. J. Adv. Eng. Technol. **6**, 2416 (2014)
2. Thamban, I., Abraham, B.C., Kurian, S.: Machining characteristics analysis of 6061-T6 aluminium alloy with diamond coated and uncoated tungsten carbide tool. Int. J. Latest Res. Sci. Technol. **2**, 553–557 (2013)
3. Kök, M.A.: Study on the machinability of AL2O3 particle reinforced Aluminium alloy composite. Pract. Metallogr. **46**(11), 580–597 (2009)
4. Gökkaya, H.: The effects of machining parameters on cutting forces, surface roughness, Built-Up Edge (BUE) and Built-Up Layer (BUL) during machining AA2014 (T4) alloy. Strojniški vestnik J. Mech. Eng. **56**, 584–593 (2010)
5. Ţîţu, M., Oprean, C.: Management of Intangible Assets in the Context of Knowledge Based Economy, p. 280. Editura LAP Lambert, Germany (2015). ISBN-13 978-3-659-79332-5, ISBN-10 3659793329
6. Dobrotă, D., Dobriţa, F., Petrescu, V., Ţîţu, M.: The analysis of the homogeneity of chemical composition in castings made of Bronze with Tin. Rev. Chim. **67**(4), 679–682 (2016). ISSN 0034-7752, Aprilie 2016, Accession Number: WOS: 000376549200018
7. Dobrotă, D., Ţîţu, M., Dobriţa, F., Petrescu, V.: The analysis of homogeneity of the chemical composition in castings made of aluminum alloy. Rev. Chim. **67**(3), 520–523 (2016). ISSN 0034-7752, Martie 2016, Accession Number: WOS: 000375364800028
8. Bloch, K., Ţîţu, M., Sandus, A.V.: Investigation of the structure and magnetic properties of bulk amorphous FeCoYB alloys. Rev. Chim. **68**(9), 2162–2165 (2017). ISSN 0034-7752, Septembrie 2017, Accession Number: WOS: 000416748800049
9. Bonţiu Pop, A.B.: Surface quality analysis of machined aluminum alloys using end mill tool, Doctoral Thesis, Baia Mare (2015)
10. Pop, A.B., Ţîţu, M.A.: Experimental research regarding the study of surface quality of aluminum alloys processed through milling. MATEC Web Conf. **121**, 05005 (2017)
11. Pop, A.B., Ţîţu, A.M.: Optimization of the surface roughness equation obtained by Al7136 end-milling. MATEC Web Conf. **137**, 03011 (2017)
12. Pop, A.B., Ţîţu, M.A.: The experimental research strategy of the endmilled aluminum alloys. MATEC Web Conf. **112**, 01010 (2017)
13. Pop, A.B., Ţîţu, M.A.: Regarding to the variance analysis of regression equation of the surface roughness obtained by end milling process of 7136 aluminium alloy. IOP Conf. Ser. Mater. Sci. Eng. **161**(1), 012015 (2016)
14. Ţîţu, M.A., Pop, A.B.: Using regression analysis method to model and optimize the quality of chip-removing processed metal surfaces. MATEC Web Conf. **112**, 01009 (2017)
15. Ţîţu, M.A., Pop, A.B.: Contribution on Taguchi's method application on the surface roughness analysis in end milling process on 7136 aluminium alloy. IOP Conf. Ser. Mater. Sci. Eng. **161**(1), 012014 (2016)
16. Ţîţu, M., Oprean, C., Boroiu, Al.: Cercetarea experimentală aplicată în creşterea calităţii produselor şi serviciilor, Colecţia Prelucrarea Datelor Experimentale, 684 pagini. Editura AGIR, Bucureşti (2011). ISBN 978-973-720-362-5

Quality Control Methods for Low Voltage Automotive Wiring

Lucian Gal, Doina Mortoiu, Bogdan Tanasoiu(✉), Aurelia Tanasoiu, and Valentin Muller

Aurel Vlaicu University, Arad, Romania
bogdan.tanasoiu@gmail.com

Abstract. Current industrialization policy created a favourable context to the development of the automotive industry, due to the discovery of the facilities that aluminum materials offer.

Modification of the technology for production of aluminum wiring instead of copper wiring implies altering the quality control methods. The paper present the quality control methods used after the pulling, twisting and extrusion processes for the wiring. After their presentation, a series of optimization steps for the processes are presented.

Keywords: Quality control · Testing methods · Quality optimization

1 Quality Control. Standard Laboratory Conditions

The main aim of quality control is to ensure that the standard laboratory conditions are met. In regards to the employed verification methods, each laboratory is equipped with a thermos-hygrometer as well as a barometer, both of which display continuously the temperature, humidity and atmospheric pressure registered in the room. Depending on the standard requirements, each laboratory is equipped with vents, humidifiers, climate control units, etc.

According to conformity criteria, the values inferred from the standard recommendations are the following:

- Temperature: 23 °C
- Humidity: 50%
- Pressure: 1 Bar (1000 mbar)

2 Quality Control Methods Used in Drawing

The quality methods used in drawing phase include both visual checks as well as dimension control.

The visual tests and checks used for aluminum in the drawing phase is composed of three procedures:

I. Karabegović (Ed.): NT 2019, LNNS 76, pp. 219–228, 2020.
https://doi.org/10.1007/978-3-030-18072-0_25

- *Aluminum spooling*. Here the aluminum wires have to be discrete, and attached at the same level.
- *Aluminum color.* The color of the material needs to be shiny, clear, clean, and without any trace of lubricants.
- *General aspect of aluminum.* The aluminum cable needs to have a cylindrical shape and free of striations.

Operationally to perform the diameter control, a sample of aluminum is collected and checked for any sign of deteriorations. Next, the sample is placed between the two faces of the anvil and the spindle of the micrometer (Fig. 1) to obtain the value registered in the cable. In this particular case, the Quality Department uses the laser micrometer LSDN 650/Y22 – for the purpose of measuring wires and cables with diameters ranging between 29 μm and 650 μm. The LSDN is able to offer highly accurate measurements of diameters and shape without making a direct contat with the wire or cable. To use the LSDN the following steps must be undertaken:

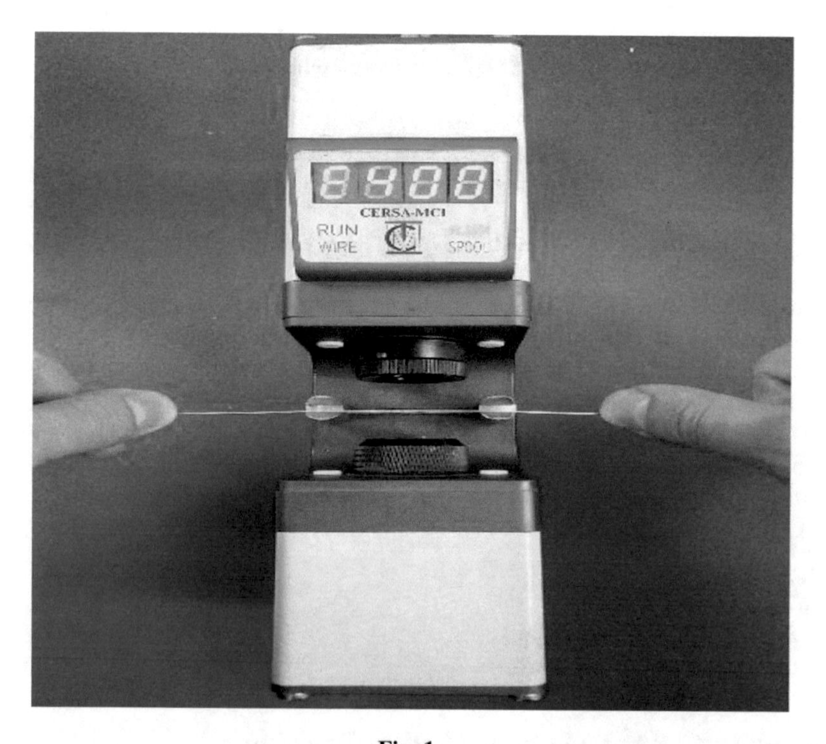

Fig. 1.

1. Starting the micrometer
2. Selecting a sample of an approximate length of 80 mm and placing it in the LDSN support. The sample must be well stretched and the surface should not be directly touched.

3. The passive sample has to be positioned in the support such the laser fascicle is focused on the sample and at the same time it has to remain perpendicular to the capturing plane.

After the correctly completion of all of the previously presented steps, the results are displayed in the LDSN display.

3 Rod (Wire Elongation)

To test the elongation of a wire it is required to use a sample of an approximate length of 250 mm. Similar to the diameter testing, the sample is checked for deterioration. The traction machine is fixed such that the distance between the vice is 200 mm and the speed is that of 100 mm/min. The actual test consists on the wire breakage obtained after the sample is firmly placed between the two vices and the machine is set to start. The obtained elongation result should be equal or greater than 20%.

In this particular case, the EM-401624 (Fig. 2) elongation machine is used.

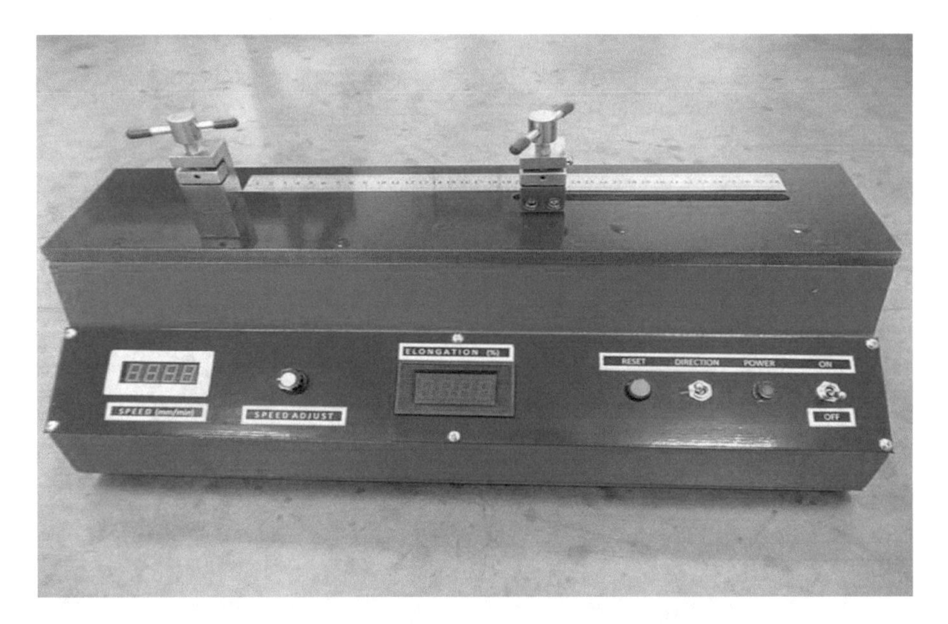

Fig. 2.

The steps involved in this process are the following:

(a) placing the sample in the elongation machine;
(b) identifying the initial length of the aluminum sample by using a ruler (i.e. the distance between the fixed and mobile clamps when it is on the on setting.
(c) the process of elongation is set to start;

(d) measuring the final length of the sample – the distance between the fixed and mobile clamp at the moment of the breakage;
(e) the two lengths are compared (the one indicated by the machine and the calculated elongation).

4 Linear Electrical Resistance

To check the electrical linear resistance the procedure dictates that the following steps must be done:

- a sample of approx. 1.7 m is cut from the wire packet and is checked for any deteriorations
- the sample is put in an acid bath for one minute
- afterwards the conductor is cleaned in distilled water
- the conductor is let to air dry at lab temperature
- following this procedure, the sample is placed between the vices of the electrical resistance measurement device (Fig. 3) in such a way that it placed on all four contact points
- once the value has stabilized, the obtained electrical resistance is compared to the one found in the technical report.

Fig. 3.

5 Control Methods – Aluminium Twisting

Similar to the tests employed in the drawing phase, aluminum twisting employs both a dimensional and visual quality control check.

The visual control is performed by following these steps:

- *Spooling*. The conductor wires need to be discreet and attached at the same level.
- *General aspect and shape of aluminum.* The aluminum conductor needs to be circular, cylindric and to be free of any protruding wires. The twisting phase must be identical.
- *Die diameter*. The die diameter needs to correspond with the technical specifications.
- *The distance between the separation devices.* The distance between the separation devices need to correspond to specifications.
- *Pressure.* The pressure must correspond to the technical specifications.
- *Twisting*. The twisting must correspond to the technical specifications.

6 Dimensional Check

The dimensional check is performed for the following elements:

- The number of wires. Each aluminum wire is carefully counted and compared to the technical specification.
- Linear electrical resistance. For this test, a sample of approx. 1.70 m is cut, and it is checked for any deteriorations and that the twisting has been done correctly. Following the sample is introduced in an acid bath for 1 min; after which the conductor is cleaned in distilled water and left to air dry at lab temperature. Following this procedure, the sample is placed between the vices of the electrical resistance measurement device (Fig. 3) in such a way that it placed on all four contact points. Once the value has stabilized, the obtained electrical resistance is compared to the one found in the technical specifications sheet.
- Welding test. For the welding test 3 samples of approx. 120 mm is cut and the ends are welded with an ultrasonic welding machine. For this particular test, first the peeling test must be conducted (Fig. 4). The maximum force registered at the welding breakage is noted and compared with the value from the technical specifications sheet.

Fig. 4.

7 Test and Verification Methods of Extruded Wires

The control methods for extruded wires used in the auto-motive industry, are based on visual inspections which are then compared to benchmark samples, as well as tests consisting of measurements, monitorization and functionality checks.

The main quality control tests/control methods used in the extrusion phase are the following:

- Isolation diameter; minimum layer of isolation, core shape and isolation between the wires; concentricity (K factor)
- Spark errors, knots and necking
- Visual check
- Number of wires
- Cutting retraction
- Linear electrical resistance
- Memory effect

When conducting the exterior diameter isolation test the following steps are performed:

(a) the isolation is stripped from a sample so that there isolation deformities.
(b) with a blade a tubular sample (±15 mm) is cut vertically from the axis.
(c) The sample is placed vertically on the microscope support and two perpendicular measurements are conducted. The considered diameter dimension will be the average of the two measurements.
(d) The measurements have to fall within the admitted tolerance, according to the technical specifications of the tested wire.

To establish the dimension of the interior (minimum) wall of the insulation, the same sample as the diameter isolation test can be used. In this case, the isolation is carefully removed from the sample so that no deformations are caused. After which with a blade a tubular sample (±15 mm) is cut vertically from the axis. The sample is placed vertically on the microscope support and is rotated such that 6 successive measurements are done until the point with the thinnest isolation is found. The maximum isolation is determined in a similar way. Moreover, in the case of determining the minimum thickness of the wall on a twisted wire, this procedure needs to take place on a flat surface (Fig. 5).

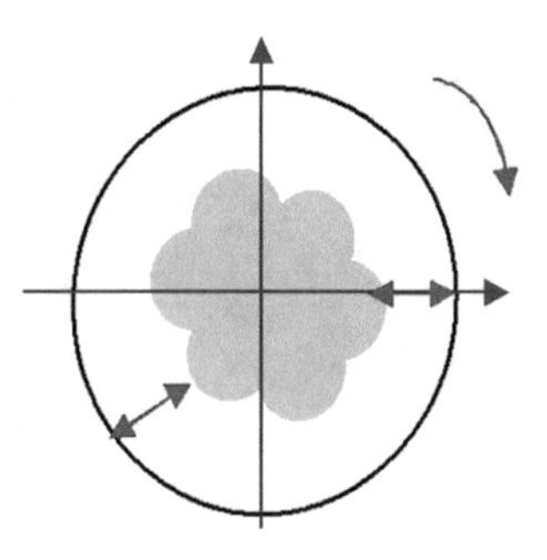

Fig. 5.

The shape of the core and isolation between the wires can be tested by using the same sample as the one used for the previously described tests. The shape of the aluminum cable core has to be gathered and perfectly gathered so that no wires are protruding, especially at the cables with a concentrically composition. Between the conductor wires there mustn't be any visible isolation material and/or any dies (Fig. 6).

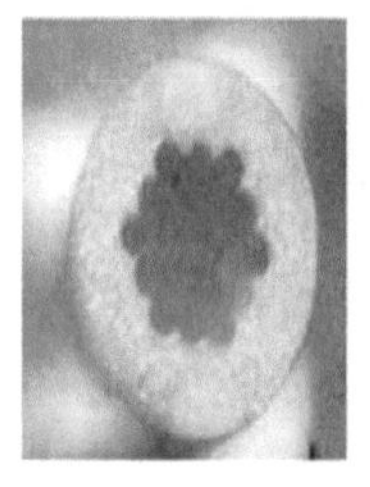

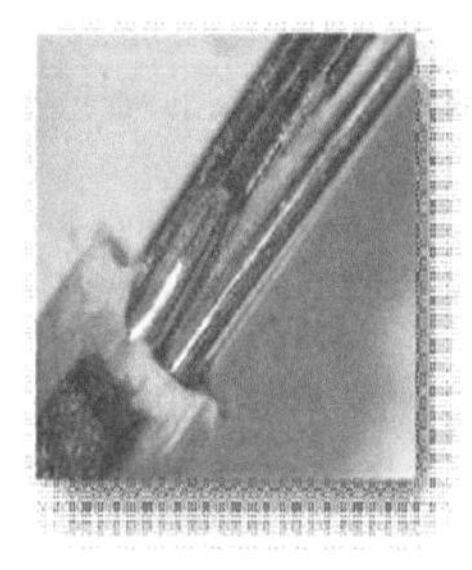
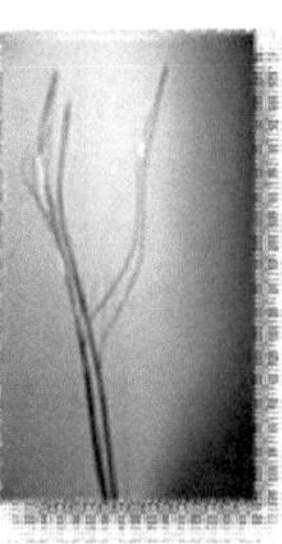

Fig. 6. Shape of the core and insulation between wires (a. basic shape; b. insulation between wires; c. insulation in the middle of wires)

Concentricity value (K factor), recorder in the spool report, must respect the value prescribed for the tested wire. In order to measure the K factor, the values of the minimum and maximum walls from above must be considered (Fig. 7).

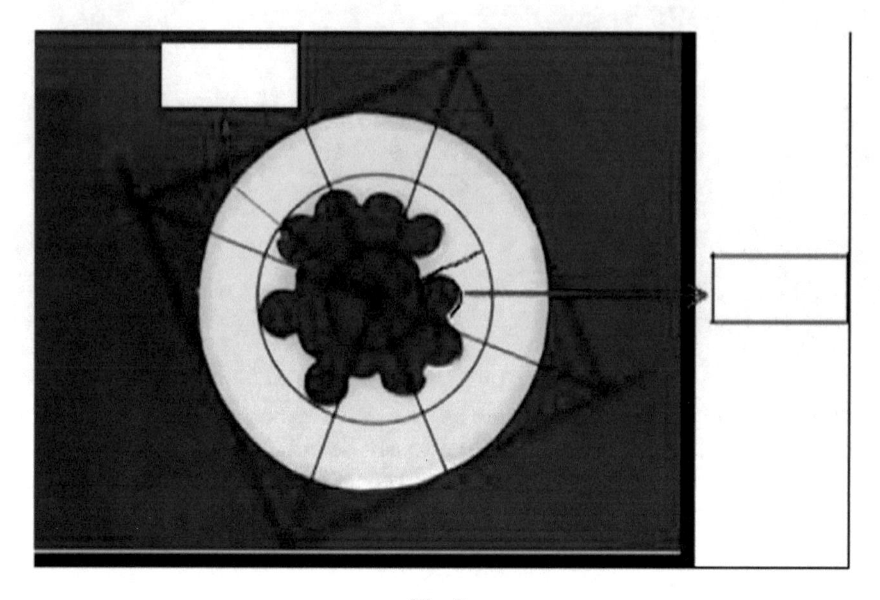

Fig. 7.

In case of a Spark defect (upon appearance of extrusion errors) the statistical reports are checked and validated for each spool. The same process is followed in case of apparition of knots, necking areas.

Visual inspection checks:

- *main color of the insulation*. The color must be well defined on the entire surface of the cable, in accordance with the sample and/or the standard color table. If the cable is produced in skin mode, the process must be checked on a larger length of wire.
- *color, depth and with of secondary band*. The color of the secondary band must be well defined, not show any transparency, and not create any confusion with another color. The width of the band or bands must respect the standard and can be compared to the sample and technical specification of the cable.

The depth of the secondary band must be at most 35%. In order to determine the width of the secondary band, a sample of approximately 100 mm is prepared, the aluminum is then stripped from it and a longitudinal cut is made such that the strip of the secondary band is not cut. The sample is placed under the microscope and the distance is measured.

The memory effect is checked by removing 4 or 5 spires from the spool, holding the sample by one end, and the sample must fall down in a straight line, not a spiral.

8 Quality Optimization

In regards to quality, improvements were made in regards to the labeling and packing of the raw and final products.

If before the labeling was done by simply attaching the labels to the outside of the spools, presently they are introduced in a plastic envelope to protect them from moisture and weather effects in the storage areas and during their transit to the customer, such that the inscribed data remain clear and readable in order to identify the product in case its traceability is necessary. Furthermore, it has been set that final products be protected by cardboard packaging and foil covering, not just by foil covering, and the raw materials just by foil wrapping. An important factor in determining product quality is determining its real resistance. By introducing the aluminum conductor in clorhidric acid solution, before being measured for resistance, the oxide layer is removed and such measurement accuracy is increased (Figs. 8 and 9).

Fig. 8.

The finite product on the spool is protected against humidity at both ends by sealing with thermally contractable lids and the validity term is at maximum 6 months from the date of fabrication.

Fig. 9.

9 Conclusions

The described tests prove the effectiveness of the new aluminum conductor compared to the automotive client requests. Furthermore the tests emphasize the manufacturers' intended production quality, bolstered and checked by the relevant statistical data.

References

1. Cerințe tehnice – Torsadare, E.D.M., Coficab E.E. Arad
2. Condiții standard de laborator, Q.D.M., Coficab E.E. Arad
3. Control Plan - Masterbatch for Incoming Inspection, Q.D.M., Coficab E.E. Arad
4. EM – 4016 Wire Elongation Tester Utilization and Verification, Q.D.M., Coficab E.E. Arad
5. General Presentation of Coficab Eastern Europe, H.R.D.M, Coficab E.E. Arad
6. Identificarea defectelor de echipament din raportul de producție, P.D.M., Coficab E.E. Arad
7. Instrucțiune de alimentare a plasticului cu colorant, P.I.M., Coficab E.E. Arad

Study of the Effect of Transducer Thickness and Direction on the Coercive Force Magnitude

Aleksandr I. Burya[1,2], Ye. A. Yeriomina[2], V. I. Volokh[2], and Predrag Dašić[3,4(✉)]

[1] Ukrainian Technological Academy (UTA), 80000 Kiev, Ukraine
[2] Dniprovsk State Technical University, Kamianske 51918, Ukraine
[3] High Technical Mechanical School of Professional Studies, 37240 Trstenik, Serbia
dasicp58@gmail.com
[4] Faculty of Strategic and Operational Management (FSOM), Str. Jurija Gagarina bb, 11070 Novi Beograd, Serbia

Abstract. The purpose of this work is to determine the influence of the product thickness of metal structure samples, as well as the location of the magnetizing device (transducer) on the magnitude of the coercive force. The experiment was realized as a complete plan of the experiment with repetition at the zero point, and the mathematical model was chosen in the form of a square model of surface response.

Keywords: Transducer · Non-destructive testing · Mathematical modeling · Response surface methodology (RSM)

1 Introduction

Building metal-constructions operate dozens of years and they are exposed to different factors during this time and, as a result, strength properties of its particular elements reduce that lead to micro-damage accumulation and deformation bringing to construction failure and initiation of cracks. Therefore, every year the risk of destruction of metal constructions made of structural iron increases and the problem of the operating condition of the facility is becoming increasingly important in the work of technical experts.

To date, there are physical methods of non-destructive testing, namely: radiographic, ultrasonic, hardness measurement, thermal, electric, magnetic powder, vortex-current, capillary, but none of them allows controlling the structure and evaluating structural and phase changes in the metal. The most effective way to control the structure is the magnetic coercimetric method of non-destructive testing, which is based on the fact that the magnetic properties of steel structures of sheet products operating under cyclic loading are formed under the influence of tensile, compressive, bending loads. However, as in each of the methods of non-destructive testing, there are some drawbacks. In particular, in the case of a magnetically coercimetric non-destructive

I. Karabegović (Ed.): NT 2019, LNNS 76, pp. 229–237, 2020.
https://doi.org/10.1007/978-3-030-18072-0_26

method of a steel sample control with different thicknesses a gradual decrease in the coercive force indices with thickness increasing is observed. This is due to the insufficient resolution of structuroscopes.

We are aware of the anisotropy of magnetic properties: in crystals of all metals, physical and mechanical properties more or less depend on the direction. At the same time, the magnetic and mechanical properties of electrical steel and a number of other sheet materials largely depend on both the crystallographic texture and other factors (chemical composition, manufacturing conditions, grain size, composition and amount of harmful impurities, the condition and surface of the finished sheets, the presence of internal stresses, sheet thickness, etc.). Thus, the type of prevailing texture and the degree of its evidence determine the nature of the steel properties anisotropy [1]. Scientific studies [2–4], etc., cover the investigations on the effect of mechanical stresses, internal defects, and product thickness on the coercive force (Hc), but they do not contain the studies on the effect of the angle setting of the transducer and the thickness of the controlled product on the value of Hc.

2 Materials and Methods

2.1 Materials

Low-carbon steel of Steel3bs grade was used in the work as the most widespread in the manufacture of metal structures in industry. Magnetic structuroscope KRM-TS-K2M (Fig. 1) was used for control the strained and deformed state of the metal structures and determination of the product thickness effect on the coercive force.

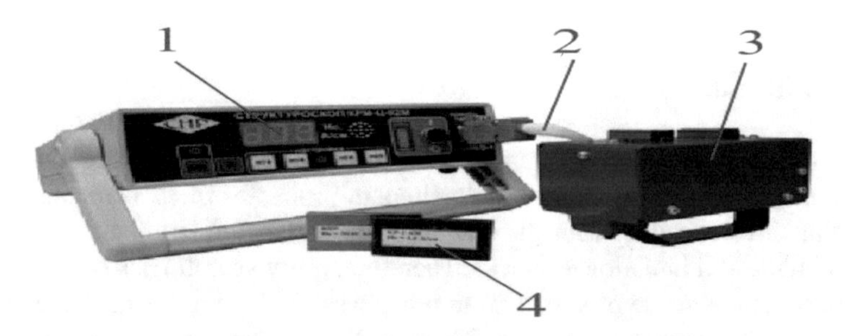

Fig. 1. Appearance of the magnetic structuroscope CRM-TsK-2M: 1 - measurement block; 2 - connecting cable; 3 - transducer; 4 - standard specimen

Method considering the effect of value Hc on the thickness of the controlled product is proposed [5]. However, in practice, it does not give actual results. The disadvantage of the magnetic coercimetric method of non-destructive testing is its sensitivity to local changes of wall thickness.

The authors used an “Experimental sample for structuroscope calibrating” to eliminate the error in determining the coercive force on the objects of control with different thicknesses (Fig. 2).

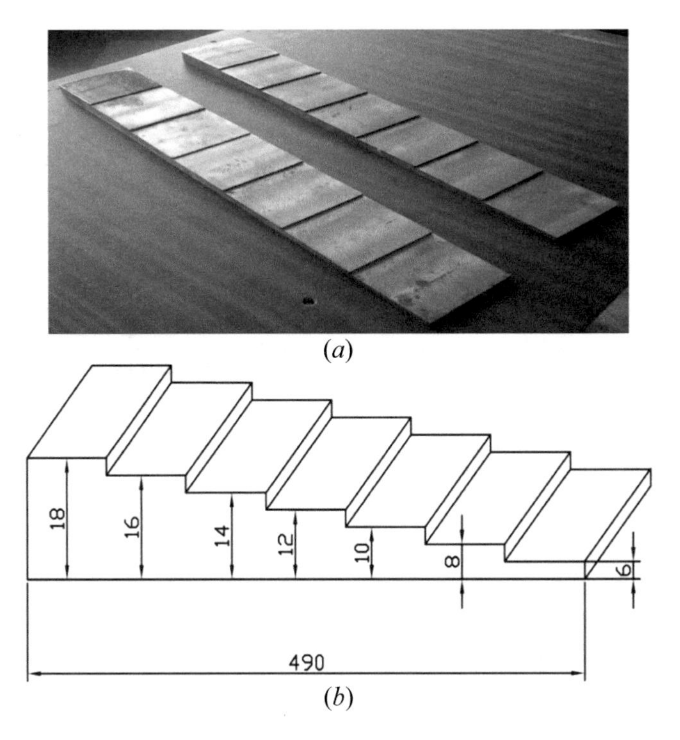

Fig. 2. Appearance (a) and design (b) of the experimental sample with variable thickness from 6 to 18 mm made of steel St. 3

2.2 Methods

The measurements were made in different directions with respect to the experimental sample: -lengthwise, -across, -at an angle of 45 degrees. The measurement results show that the location of the transducer relative to the direction is of decisive importance. In parallel position of the transducer axis relative to the experimental sample Hc increases, while in perpendicular - it decreases slightly. When the instrument axis is positioned at an angle of 45°, a relationship, the character of which can be considered intermediate between the dependencies described above, is observed.

The mathematical description of the process under consideration was proposed to be sought as a regression equation that determines the dependence of the process under study on the thickness and setting angle of the transducer and is represented as a second-order polynomial:

$$y_i = b_0 + \sum_{i=1}^{n} b_i x_i + \sum_{i=j}^{n} b_{i,j} x_i x_j + \sum_{i=1}^{n} b_{ii}^2 x_i^2. \quad (1)$$

Based on the experimental data obtained, the mean value of the response function $\bar{y}_j$ was calculated:

$$\bar{y}_j = \frac{1}{k}\sum_{i=1}^{k} y_{ji}, j = 1, 2, \ldots, N \quad (2)$$

Proceeding from the problem formulation, in order to save the research time, the orthogonal central compositional planning for the second-order experiment of type 2^2 [7]. The measurements were carried out at two levels of each of the parameters (Table 1).

Table 1. Initial data for experiment planning

Parameter	Symbol	Variation step (h)	Levels of variation		
			+1	0	−1
Length	h, mm	6	18	12	6
Angle	θ, °	45	90	45	0

According to the accepted experiment plan, a total of $N = N_я + 2n + 1 = 9$ experiments were carried out, where $N_я$ is the number of experiments in the core of the plan, and n is the number of factors. Each experiment was repeated twice ($k = 2$) in a randomized order to eliminate systemic errors.

Experiment and mathematical modeling was realized out on the basis of design of experiment (DoE) and response surface methodology (RSM) which are described in the papers [9–17].

3 Results and Discussion

The observed data of the experiments are summarized in Table 2 as an optimization parameter $k = 2$.

Table 2. Optimization parameter

No	Coded variables		Full scale		Mean value	Estimated value
	x_1	x_2	h [mm]	θ [°]	$\bar{y}_j$	y_j^p
1	1	1	18	90	1,5	1,4
2	1	−1	18	0	1,9	1,9
3	−1	1	6	90	3,0	2,9
4	−1	−1	6	0	4,3	4,3
5	−1	0	6	45	3,6	3,6
6	1	0	18	45	1,6	1,6
7	0	1	12	90	1,8	1,9
8	0	−1	12	0	2,9	2,8
9	0	0	12	45	2,3	2,3

The variance of the optimization parameter was determined on the results of the experiments in the center of the plan:

$$S_b^2 = \frac{\sum_{j=1}^{n_0} \sum_{i=1}^{k} (y_{ji} - \overline{y_j})^2}{n_0 - 1} \tag{3}$$

The error of the experiment was calculated by the formula:

$$S_{bi} = \sqrt{S_b^2}. \tag{4}$$

Based on the factor experiment, the regression coefficients were calculated, in accordance with the formulas:

$$b_i = \frac{1}{6}\sum_{i=1}^{N} x_i \overline{y_i},$$

$$b_{ij} = \frac{1}{4}\sum_{i=1}^{N} x_{ij} \overline{y_i},$$

$$b_{ii} = \frac{1}{2}\sum_{i=1}^{N} \left(x_i^{'}\right)^2 \overline{y_i},$$

$$b_0 = \frac{1}{N}\sum_{i=1}^{N} x_0 \overline{y_i} - \frac{2}{3}(b_{11} + b_{22}) \tag{5}$$

After calculating all the coefficients, the equation takes the form:

$$y = 2.5 - 1.0 \cdot x_1 - 0.5 \cdot x_2 + 0.3 \cdot x_1^2 - 0.03 \cdot x_2^2 - 0.23 \cdot x_1 \cdot x_2 \tag{6}$$

The statistical significance of the coefficients of the regression equation b_0, b_1, b_2, $b_{1,2}$, b_{11}, b_{22} was evaluated based on the calculation of confidence intervals, taking into account the variance characterizing the error in determining the coefficients of the equation. The confidence interval itself was calculated by the Student's criterion, chosen according to the accepted degrees of freedom (f_1, f_2) and significance level (0.95). For the orthogonal central compositional planning of the experiment, the error coefficients are determined:

$$t = \frac{|b_i|}{S_{bi}}. \tag{7}$$

Confidence interval:

$$\Delta b_i = t_{кр} \cdot S_{bi}. \tag{8}$$

The critical value $t_{\kappa p}$ [18] was chosen for the number of degrees of freedom $N(n-1) = 9$ and the accepted significance level 0.95. It is generally assumed that the regression coefficient is significant if the following condition is satisfied:

$$t_{\kappa p} < t. \tag{9}$$

The coefficients of Eq. (5), except b_2, are statistically significant, therefore, the equation describing the process under study will take the form:

$$y = 2.5 - 1.0 \cdot x_1 - 0.5 \cdot x_2 + 0.3 \cdot x_1^2 - 0.23 \cdot x_1 \cdot x_2 \tag{10}$$

The derived equation was checked for adequacy. For this purpose, we estimated the deviations of the values of the optimization parameter y_j^p calculated from Eq. (10) based on experimental $\overline{y_j}$ for each of the tests, which made it possible to determine the variance of adequacy for an equal number of parallel experiments:

$$S_{a\partial}^2 = \frac{1}{N-B} \sum_{j=1}^{k} (\overline{y_j} - y_j^p)^2 \tag{11}$$

where: B is the number of significant coefficients of the equation. The number of degrees of freedom $f_{a\partial} = 9$ is also concerned with it.

The calculated values of the optimization parameter are presented in Table 2. To determine the adequacy of the mathematical description (11), after calculating the regression coefficients, the degree of correspondence between the obtained model and the theoretical form of the relationship between the input and output parameters under consideration was checked. For this purpose, the Fisher criterion (F_p), was used, which is the ratio of the variance of adequacy $S_{a\partial}^2$ to the variance of the experiment reproducibility S_b^2 (see Table 3) and is calculated by the formula:

$$F_p = \frac{S_{a\partial}^2}{S_b^2} \tag{12}$$

When calculating the Fisher criterion, we had to meet the condition, $s_{a\partial}^2 > s_y^2$, otherwise it is necessary to rearrange the variances, which was done.

Table 3. Calculated values for assessing the adequacy of the equations by the Fisher criterion

S_b^2	S_{bi}	Regression coefficients						
		b_0	b_1	b_2	$b_{1,2}$	b_{11}	b_{22}	$S_{a\partial}^2$
0,0214	0,03	2,5	−1,0	−0,5	0,23	0,3	0,03	0,0097

As at the significance level of 0,95 and degrees of freedom f_1 = 9 and f_2 = 8 for the equation under consideration F_p = 2,19 which is less than the tabulated F_{tabl} = 2,65 [18], then it is adequate.

The derived mathematical model, for clarity, is presented in the form of a response surface. In our case, it illustrates a surface in three-dimensional space and a contour (Fig. 3).

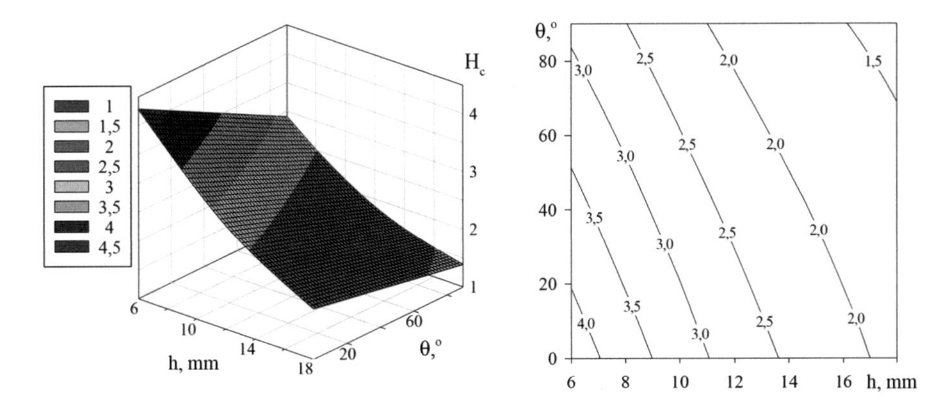

Fig. 3. Dependence of the coercive force value (Hc) on the setting angle of the transducer (θ°) and the thickness of the product (h)

As can be seen from Fig. 3, the maximum of the optimization parameter is at the angle of θ = 19° and thickness of h = 7 mm. Mathematical processing of control results allows to calculate the optimal setting angles of transducers to the monitored surface. The data obtained are summarized in Table 4.

Table 4. Dependence of the coercive force value (Hc) on the thickness of controlled product and setting angle of the transducer.

Product thickness h [mm]	Setting angle of transducer			
	20°	40°	60°	80°
6	3,9	3,7	3,4	3,1
8	3,5	3,3	2,9	2,65
10	3,1	2,8	2,5	2,32
12	2,6	2,4	2,25	2,0
14	2,3	2,2	2,0	1,8
16	2,0	1,85	1,75	1,6
18	1,85	1,75	1,65	1,4

4 Conclusion

A as a result of using the mathematical modeling of the experiment when installing the transducer on the product, it was found that the greatest sensitivity of the coercive force Hc to the thickness is the depth of the maximum magnetization of 7.0 mm when installing the transducer at an angle of 19°. With the thickness increase of the controlled product, the magnetic permeability decreases and, correspondingly, the coercive force Hc decreases. The computed values of the coercive force Hc (Table 4) at the setting angles of the transducer show the optimum values of the coercive force from thickness, which must be taken into account when controlling different thicknesses of products in the form of correction factors.

References

1. Klyuev, V.V., Muzhitskii, V.F., Gorkunov, E.S., Shcherbinin, V.E.: Nondestructive Control: A Handbook. Magnetic Methods of Control, vol. 6. Mashinostroenie, Moscow (2006)
2. Volokh, V.I., Sotnikov, A.L., Vlasenko, N.N.: Control of the stress-strain state of a metal on the coercive force. Mod. Lab. **3**(3), 13–17 (2010)
3. Gubsky, S.A., Sukhomlin, V.I., Volokh, V.I.: Stress state control of steels according to coercive force. In: Mechanical Engineering Col. of Research Papers, 13, pp. 6–10. Ukrainian Engineering and Pedagogical Academy, Kharkiv (2014)
4. Grigorov, O.V., Gubsky, S.A., Kovalenko, D.M., Strizhak, V.V.: Forecasting the residual life of the metal structures of cranes with a thickness of elements over 12 mm, in Lifting transport equipment (Dnepropetrovsk) (2007)
5. Methodical instructions for conducting magnetic control of the stress-strain state of lifting structures and determining the remaining resource. MB 0.00-7.01-05.-X, 77 pp. (2005)
6. Grigorov, O.V., Gubsky, S.A., Okun, A.A., Popov, V.A., Horlo, N.F.: Experimental sample for calibration of structuroscope, Ukraine Patent № 77319 (2013). (in Ukrainian)
7. Nalimov, V.V., Chernov, N.L.: Statistical Methods for Planning Extreme Experiments. Phys. - Math., Moscow, Russia (1965)
8. Blokhin, V.G., Gludkin, O.P., Gurov, A.I., Khanin, M.L.: Modern experiment: Preparation, conduction, analysis of results, Radio and Communication, Moscow, Russia (1997)
9. Anderson, M.J., Whitcomb, P.J.: RSM Simplified: Optimizing Processes Using Response Surface Methods for Design of Experiments. Productivity Press, New York (2005)
10. Anderson-Cook, C.M., Borror, C.M., Montgomery, D.C.: Response surface design evaluation and comparison. J. Stat. Plann. Inference **139**(2), 629–641 (2009)
11. Box, G.E.P., Wilson, K.B.: On the experimental attainment of optimum conditions. J. R. Stat. Soc. Ser. B (Methodol.) **13**(1), 1–45 (1951)
12. Dašić, P.: Comparative analysis of different regression models of the surface roughness in finishing turning of hardened steel with mixed ceramic cutting tools. J. Res. Dev. Mech. Ind. (JRaDMI) **5**(2), 101–180 (2013)
13. Dašić, P.: Research of processed surface roughness for turning hardened steel by means of ceramic cutting tools. In: Proceedings of the 2nd World Tribology Congress (WTC 2001), Vienna, Austria, 3–7 September 2001, M-51-29-055. Österreichische Tribologische Gesellschaft – The Austrian Tribology Society (ÖTG), Vienna (2001)

14. Dašić, P.: The choice of regression equation in fields metalworking. In: Proceedings of the 3rd International Conference "Research and Development in Mechanical Industry" (RaDMI 2003), Herceg Novi, Serbia and Montenegro, pp. 147–158, 19–23 September 2003. (Trstenik: High Technical Mechanical School, Kruševac: Institute IMK "14. October" and Podgorica: Institute of Faculty of Mechanical Engineering)
15. Flaig, J.J.: A new classification of variables in design of experiments. Qual. Technol. Quant. Manag. **3**(1), 103–110 (2006)
16. Montgomery, D.C.: Design and Analysis of Experiments, 8th edn. Wiley, Hoboken (2012)
17. Myers, R.H., Montgomery, D.C., Vining, G.G., Borror, C.M., Kowalski, S.M.: Response surface methodology: a retrospective and literature review. J. Qual. Technol. **36**(1), 53–77 (2004)
18. Bolshev, L.N., Smirnov, N.V.: Tables of mathematical statistics, Science, Moscow, Russia (1965)

A Comparative Analysis of the Machined Surfaces Quality of an Aluminum Alloy According to the Cutting Speed and Feed per Tooth Variations

Alina Bianca Pop[1] and Mihail Aurel Țîțu[2,3](✉)

[1] SC TECHNOCAD SA, 72, Vasile Alecsandri Street, Baia Mare, Romania
[2] Lucian Blaga University of Sibiu, 10, Victoriei Street, Sibiu, Romania
mihail.titu@ulbsibiu.ro
[3] The Academy of Romanian Scientists, 54, Splaiul Independenței, Sector 5, Bucharest, Romania

Abstract. The milling operations are a topical and especially future-oriented methods. The purpose of this scientific paper is to carry out an experimental research referring to the end-milling process of the 7136 aluminum alloy. This paper is a continuation of the own research activity carried out so far, within a company with the field of activity in the aerospace industry. The main objective is a comparative analysis which was carried out on the surface quality - machined on various cutting regimes. The cutting process parameters used in the end-milling process are: the cutting speed, the feed per tooth and the cutting depth. The obtained results of the surface roughness's were measured longitudinally on the tool feed motion using a professional surface tester.

Keywords: Surface quality · Cutting speed · Feed per tooth · End-milling · Experimental research

1 Introduction

The aluminum alloys machinability has been studied and tested most often by turning, milling and drilling operations. The use of high cutting speeds in order to increase the productivity is a real challenge because increasing the cutting speed implies an acceleration of cutting tool wear. Kuttolamadom, tracked the Al6061 alloy's machinability by studying the effect of feed rate on surface roughness [1]. Liou tracked the chip morphology in AL6061 turning using the simulation based on the finite element analysis method [2]. Relevant research in this regard includes the research studies of the Ghan and Ambekar. Here is a study with a brief presentation of the surface roughness analysis obtained by milling [3]. In this paper, there are mentioned the researches of the authors Lou M.S., Chen J.C. and Li C.M., who developed a technique for predicting the surface roughness obtained by end-milling operation of steel using a mathematical model. Then, they studied the effect of cutting speed, feed rate and cutting depth on the roughness obtained by end-milling operation of aluminum sample.

I. Karabegović (Ed.): NT 2019, LNNS 76, pp. 238–244, 2020.
https://doi.org/10.1007/978-3-030-18072-0_27

According to the author's work, Moaz used the finite element modeling to predict the effect of feed rate on surface roughness and cutting forces during the end-milling of the titanium alloy. He carried out a series of tests in which the variable parameter was the feed rate, while the axial cutting depth and the cutting speed remained constant, the process being carried out under the conditions of a dry cutting. The obtained results indicated the possibility of predicting the surface roughness with a 97% accuracy by measuring the cutting force based on the finite element analysis instead of the direct roughness measurement. Also, Mustafa in his research tested the effect of the cutting regime parameters and the length of the aluminum sample on the geometrical tolerances and surface roughness in the turning process. Other aspects of the chemical composition of aluminum alloys and others have been addressed in [4–6]. The 7136 aluminum alloy is the main object of this present study, used in the aerospace industry. This alloy has been also studied in [7–12] and this paper is actually a continuation of the previous research.

2 Research Method

The research method used in the study is the experiment [13, 14]. This paper is a purely technical one made in SC Universal Alloy Corporation Europe SRL, a renowned aerospace company. The purpose of the paper consist on a comparative analysis carried out on the quality of the machined surface of 7136 aluminum alloy, end-milled by various cutting regimes.

Cutting regime - The machining parameters used in the cutting process are as follows: cutting speed (v), cutting depth (a_p) and feed per tooth (f_z). The values taken by the cutting parameters are: the cutting speed: 570 and 610 [m/min]; the cutting depth: 2, 2.5, 3, 3.5 and 4 [mm]; the feed per tooth 0.04, 0.06, 0.08, 0.11 and 0.14 [mm/tooth].

Workpiece material - The research was conducted on the 7136 aluminum alloy, used in the aerospace industry, which is a material with superior properties to other alloys, such as: hardness, high resistance to low weight, low thermal expansion, better mechanical and thermal properties than other alloys, chemical resistance, resistance to corrosive atmospheres, is non-magnetic, good electrical and thermal conductor. For the experiments, the workpiece had dimensions of 500 × 101 × 24.5 [mm].

Cutting tool - The experiments were carried out using a conventional end-milling standard tool on aluminum machining, SECO R217.69-1616.0-09-2AN with 2 teeth, 16 mm diameter, with 100% tool engagement and one set of 2 cutting inserts coded XOEX090308FR-E05, H15.

CNC machine - The CNC machine used to carry out the experiments was the HAAS VF-YT2. The cutting fluid used during the experiment was Blasocut BC 35 Kombi SW mineral oil. The cooling fluid pressure at the machine pump was 8 bar, three flow direction hoses of Ø6 were used.

Response - After the Al7136 milling, the surface roughness measurements were made on 5 mm distance using Mitutoyo SURFTEST SJ-210 with 0.002 [μm] resolution. The areas delimitation of the roughness measurements are graphically represented in Fig. 1.

Fig. 1. Graphical representation of the areas where the R_a measurements were performed

3 Designing the Experimental Program

The values of the surface roughness measurements R_a [μm] are shown in Table 1.

Table 1. R_a values [μm] when cutting speed is 570 [m/min] and 610 [m/min]

Op. no.	a_p [mm]	f_z [mm/tooth]	R_a [μm] v = 570 [m/min]	R_a [μm] v = 610 [m/min]
1	2	0.04	0.442	0.533
2	2.5	0.04	0.473	0.471
3	3	0.04	0.880	0.658
4	3.5	0.04	1.372	0.534
5	4	0.04	1.120	0.506
6	2	0.06	0.465	0.691
7	2.5	0.06	0.509	0.756
8	3	0.06	0.478	0.586
9	3.5	0.06	0.493	0.516
10	4	0.06	0.533	0.524
11	2	0.08	0.470	0.500
12	2.5	0.08	0.488	0.260
13	3	0.08	0.493	0.598
14	3.5	0.08	0.644	0.470
15	4	0.08	0.606	0.458
16	2	0.11	0.429	0.679
17	2.5	0.11	0.427	0.631
18	3	0.11	0.412	0.543
19	3.5	0.11	0.396	0.581
20	4	0.11	0.568	0.511
21	2	0.14	0.369	0.722
22	2.5	0.14	0.393	0.980
23	3	0.14	0.357	0.847
24	3.5	0.14	0.407	0.498
25	4	0.14	0.622	0.490

The obtained measurements were used to determine the comparison of the surface quality evolutions in different situations related to the cutting regimes resulting from the combination of the process parameters in situations where the cutting speed is 570 [m/min] and 610 [m/min]. With the help of the obtained results it was possible to perform comparative graphs between the evolution of the surface roughness on the variation of the cutting speed and feed per tooth in various situations.

4 Comparisons of the Evolution of the Machined Surfaces Quality of an Aluminum Alloy According to the Cutting Speed and Feed per Tooth Variations

Figure 2 shows the comparison of the surface roughness R_a in the situation when the cutting speed is 570 and 610 [m/min], the cutting depth remains constant at 2 [mm] and the feed per tooth varies. As can be seen in these cutting conditions, the R_a values are lower at the cutting speed of 570 [m/min], but they show a slight decrease with the decrease of the feed per tooth in both situations.

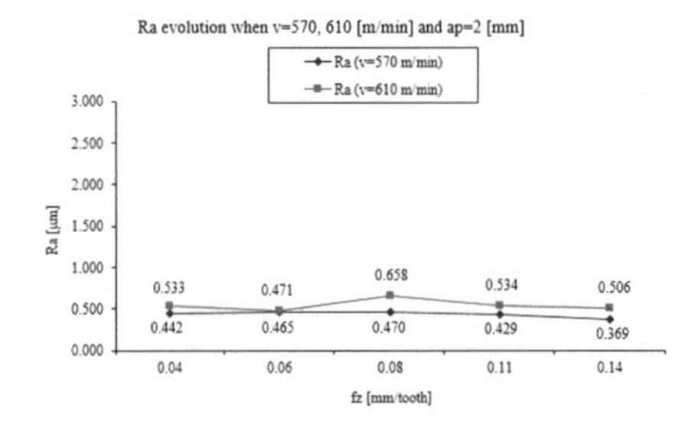

Fig. 2. Evolution of R_a under the influence of f_z variation when a_p is 2 [mm]

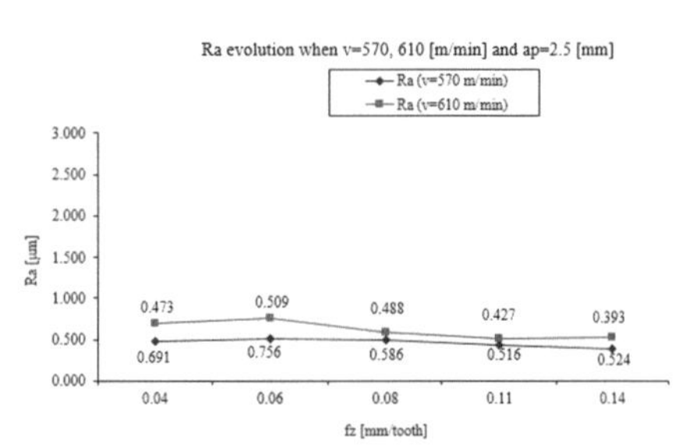

Fig. 3. Evolution of R_a under the influence of f_z variation when a_p is 2.5 [mm]

In the Fig. 3 graph, the situation is similar to the previous one. As it can be seen in the case of a_p with 3 [mm] and f_z with 0.06 [mm/tooth] the R_a shows a sudden decrease of the recorded values from 0.880 [µm] to 0.478 [µm] of 570 [m/min], retaining this decrease, instead at 610 [m/min] the R_a exhibits an initial decrease from 0.500 [µm] to 0.260 [µm] followed by a slight increase to 0.598 [µm] than decreases again (Fig. 4). In the 5th figure the case where the cutting depth remains constant at 3.5 [mm] is shown.

The recorded roughness values of 570 [m/min] on the feed per tooth variation between 0.04–0.14 [mm/tooth] show decreases once decrease fz. But, at the cutting speed of 610 [m/min], the R_a recorded values show insignificant variations. By comparing the two speeds, the measured R_a values are lower on 610 [m/min] (Fig. 5).

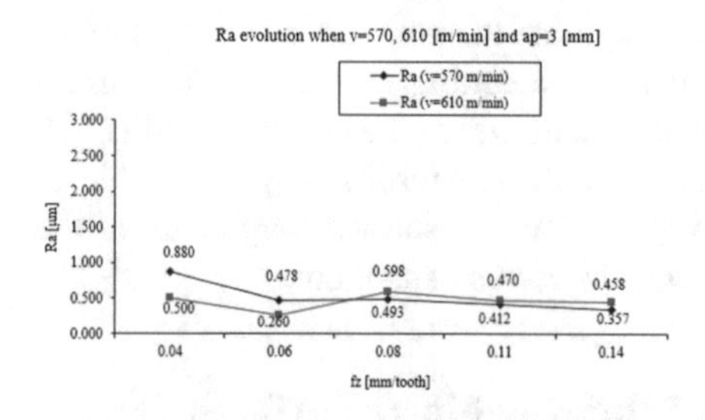

Fig. 4. Evolution of R_a under the influence of f_z variation when a_p is 3 [mm]

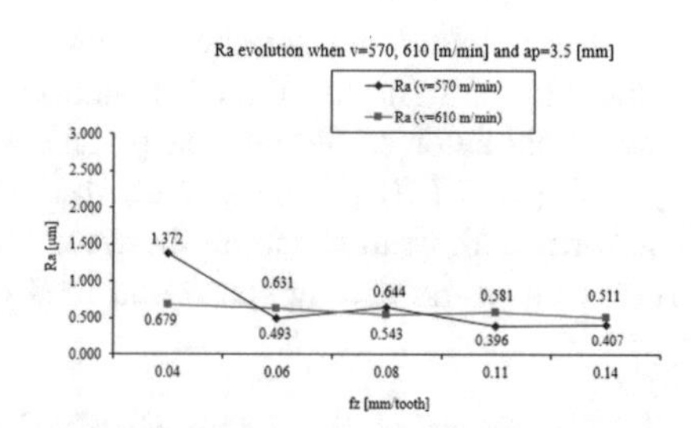

Fig. 5. Evolution of R_a under the influence of f_z variation when a_p is 3.5 [mm]

The Fig. 6 shows the evolution of R_a under the influence of the feed per tooth at the cutting speeds of 570 and 610 [m/min] and 4 [mm] of the cutting depth.

In this case, the R_a values are lower on 570 [m/min]. Therefore, the resulting graphs comparison shows that the highest roughness value is 1.372 [µm] - recorded on 0.04 [mm/tooth] feed per tooth, 3.5 [mm] cutting depth and a cutting speed of 570 [m/min]. In the case of the 610 [m/min] cutting speed, the highest value of R_a is 0.980 [µm], recorded on a_p with 4 [mm] and f_z with 0.06 [mm/tooth]. The lowest R_a values are recorded on 0.357 [µm] when the cutting speed is 570 [m/min] the cutting depth is 3 [mm] and the feed per tooth 0.14 [mm/tooth] and 0.260 v = 610 [m/min], a_p = 3 [mm], f_z = 0.06 [mm/tooth]. From the point of view of the influence of the cutting speed in the analyzed situations, at the center of the cutting interval values, R_a oscillates, therefore, the cutting speed affects roughness only at high depth and low feed where R_a has the highest values. However, with the increase of the feed per tooth at a high cutting depth, the R_a roughness decreases, irrespective of the cutting speed.

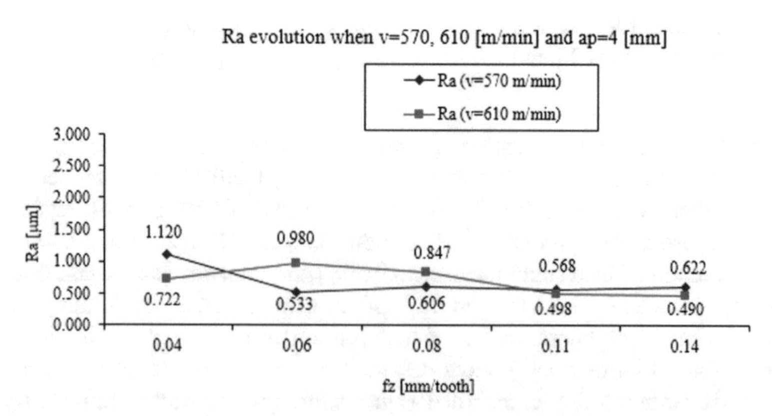

Fig. 6. Evolution of R_a under the influence of f_z variation when a_p is 4 [mm]

5 Conclusions

Based on the surface quality analysis, obtained by end-milling of the 7136 aluminum alloy at the variation of the cutting depth and the feed per tooth when the cutting speed is 570 and 610 [m/min] the highest values of R_a are recorded according to the data presented in Figs. 7 and 8.

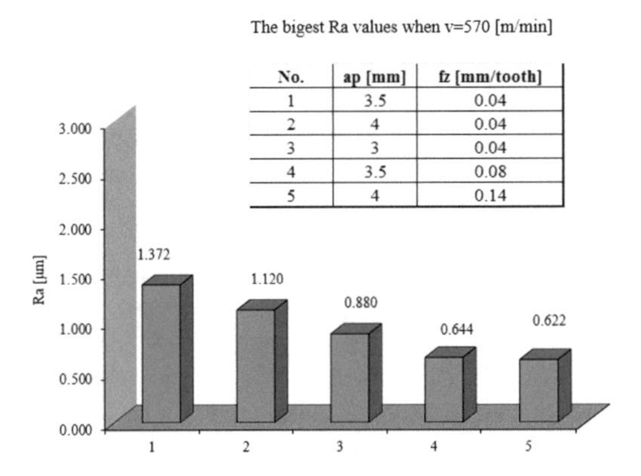

Fig. 7. The R_a highest values when v = 570 [m/min], a_p and f_z varies

Figure 7 shows that the smallest roughness values occurred by Al7136 end-milling when the cutting depth is 3.5 [mm] and the feed per tooth is 0.08 [mm/tooth].

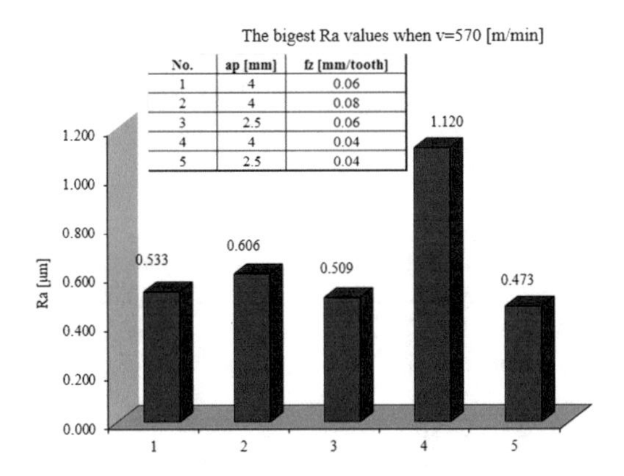

Fig. 8. The R_a highest values when v = 610 [m/min], a_p and f_z varies

Figure 8 shows that the minimum surface roughness value, measured longitudinally in the direction of the cutting tool feed motion is 0.473 [µm] and is recorded on 0.04 [mm/tooth] and 610 [m/min].

By the general analysis of the graphs presented in this research, it can be concluded that the feed per tooth has a direct influence on the surface quality, resulting an increasing tendency of the measured roughness values with the increase of the feed per tooth and the decrease of the cutting depth.

References

1. Kuttolamadom, M.A., Hamzehlouia, S., Mears, L.: Effect of machining feed on surface roughness in cutting 6061 aluminum. SAE. Int. J. Mater. Manuf. **3**, 108–119 (2010)
2. Liou, B.: Modeling and analysis of machining systems. Chip morphology with finite element simulations in high speed turning of 7075-T6 aluminum alloy (2005)
3. Ghan, H.R., Ambekar, S.D.: Review on optimization of cutting parameter for surface roughness, material removal rate and machining time of Aluminium LM-26 alloy. Int. J. Res. Aeronaut. Mech. Eng. **2**(2), 23–28 (2014)
4. Dobrotă, D., Dobrița, F., Petrescu, V., Țîțu, M.: The analysis of the homogeneity of chemical composition in castings made of Bronze with Tin. Rev. Chim. **67**(4), 679–682 (2016). ISSN 0034-7752, Aprilie 2016, Accession Number: WOS: 000376549200018
5. Dobrotă, D., Țîțu, M., Dobrița, F., Petrescu, V.: The analysis of homogeneity of the chemical composition in castings made of aluminum alloy. Rev. Chim. **67**(3), 520–523 (2016). ISSN 0034-7752, Martie 2016, Accession Number: WOS: 000375364800028
6. Bloch, K., Țîțu, M., Sandu, A.V.: Investigation of the structure and magnetic properties of bulk amorphous FeCoYB Alloys. Rev. Chim. **68**(9), 2162–2165 (2017). ISSN 0034-7752, Septembrie 2017, Accession Number: WOS: 000416748800049
7. Țîțu, M.A., Pop, A.B.: Using regression analysis method to model and optimize the quality of chip-removing processed metal surfaces. MATEC Web Conf. **112**, 01009 (2017)
8. Țîțu, M.A., Pop, A.B.: Contribution on Taguchi's method application on the surface roughness analysis in end milling process on 7136 aluminium alloy. IOP Conf. Ser. Mater. Sci. Eng. **161**(1), 012014 (2016)
9. Pop, A.B., Țîțu, M.A.: Experimental research regarding the study of surface quality of aluminum alloys processed through milling. MATEC Web Conf. **121**, 05005 (2017)
10. Pop, A.B., Țîțu, A.M.: Optimization of the surface roughness equation obtained by Al7136 end-milling. MATEC Web Conf. **137**, 03011 (2017)
11. Pop, A.B., Țîțu, M.A.: The experimental research strategy of the endmilled aluminum alloys. MATEC Web Conf. **112**, 01010 (2017)
12. Pop, A.B., Țîțu, M.A.: Regarding to the variance analysis of regression equation of the surface roughness obtained by end milling process of 7136 aluminium alloy. IOP Conf. Ser. Mater. Sci. Eng. **161**(1), 012015 (2016)
13. Țîțu, M., Oprean, C., Boroiu, Al.: Cercetarea experimentală aplicată în creșterea calității produselor și serviciilor, Colecția Prelucrarea Datelor Experimentale, 684 pagini. Editura AGIR, București (2011). ISBN 978-973-720-362-5
14. Țîțu, M., Oprean, C.: Management of Intangible Assets in the Context of Knowledge Based Economy, 280 p. Editura LAP Lambert, Saarbrücken (2015). ISBN-13 978-3-659-79332-5, ISBN-10 3659793329

Reliability of Steel Hall in Zenica Loaded by Snow Load

Rašid Hadžović[1,3(✉)]
and Vahid Redžić[2]

[1] "DzemalBijedic" University of Mostar, 88000 Mostar,
Bosnia and Herzegovina
rasid.hadzovic@unmo.ba
[2] Polytechnic Faculty, University of Zenica, 72000 Zenica,
Bosnia and Herzegovina
[3] Faculty of Civil Engineering, University "DžemalBijedić",
USRC "Mithat Hujdur Hujka", 88 104 Mostar, Bosnia and Herzegovina

Abstract. Snow load is the dominant load for steel structures of hall in Zenica, Bosnia and Herzegovina. Existing structures are calculated in accordance with the JUS regulations that were valid in the former Yugoslavia. The snow load was based on the values proposed in the DIN regulations of 1936 and used until the development of new regulations according to Eurocode 1 in accordance with the EU recommendations. New values of snow load are given by the map of snow loads in digital form. The paper analyzes the snow load values in Zenica according to JUS and EC 1 and certain reliability indices for critical points of the structure.

Keywords: Steel hall · Snowload · JUS · Eurocode 1 · Reliability index

1 Introduction

The paper aim is to check the reliability of the steel hall in Zenica exposed to snow load defined in Yugoslavian standards (JUS) regulations and in Eurocode 1 (EC 1) standards. The reason of this analysis is a recently made map of Bosnia and Herzegovina with snow load values according to EC 1. It is significant to determine reliability of existing construction due to the increase of snow load values in Zenica that is defined by new standards. Due to extreme snowfall in December 1999 and in February 2012 significant number of objects suffered smaller or serious damage. The damage appeared on different kinds of girders and materials but mostly on steel structures because snow is the dominant load on those structures. The previous assumption was that analyzed objects were "absolutely safe", if the analyzed load values were multiplied by the safety coefficient, if the stress levels were under the allowed values and if deflection values were within allowed limits. In this paper the probabilistic analysis of existing steel hall will be done with determining reliability index according to JUS regulations from 1961 and EC 1.

I. Karabegović (Ed.): NT 2019, LNNS 76, pp. 245–251, 2020.
https://doi.org/10.1007/978-3-030-18072-0_28

2 Structural Analysis of Existing Steel Hall

2.1 Calculation of Reliability Index for the Existing Steel Hall in Zenica

The structural analysis [5] of existing steel hall is done according to JUS regulations from 1961 [1] and EC 1 standards [3] for the model of load-bearing structure shown in Fig. 1. Input data for steel structure is presented in Table 1:

Table 1. Input data for existing steel hall.

Metrical location and elevation	Zenica, elevation 316 m
Cover type	Lightweight roof panels
Selected section	2 UPE 240
Girder span	L = 16,5 m
Girder incline	$\alpha = 8^{\circ}$
Beam span	$\lambda = 4,0$ m

In the continuation of this paper the reliability indexes are determined according to characteristic snow load for location Zenica on the previously analyzed construction. The results are compared to the recommended values in EC 1.

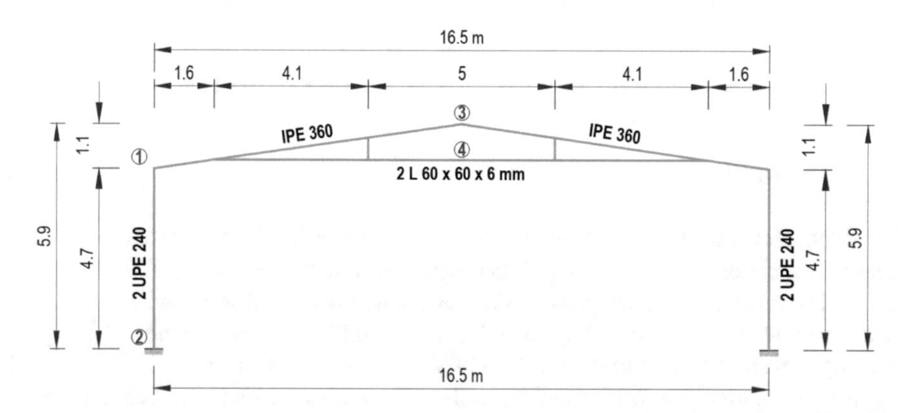

Fig. 1. Model of the load-bearing structure with points relevant for calculation of reliability index.

2.2 The Structural Analysis of Existing Steel Hall in Zenica

Characteristic snow load on the ground for measure location Zenica is 1,53 kN/m^2 [3] and the determined snow load is 0,75 kN/m^2 according to JUS regulations from 1961 [1].

Load analysis for steel hall [2]:

-roof panels	0,10 kN/m^2
- weight of the girders (included by software)	
- dead load on the main girder: 0,10 kN/m^2 * λ = 0,10 kN/m^2 * 4,00 m	0,40 kN/m
- live load (determined snow load according to JUS regulations): S_{norm} = 0,75 kN/m^2 * 4,0 m * 0,8	2,4 kN/m
- live load (snow load according to EC-1) S = 1,53 kN/m^2 * 4,0 m * 0,8	4,896 kN/m

2.3 Results of Structural Analysis

The results of structural analysis will be presented in the following Figures. Results refer to bending moments for: dead load, snow load according to JUS regulations (determined snow load) and for characteristic snow load according to EC 1 (Figs. 2, 3, 4, 5, 6 and 7).

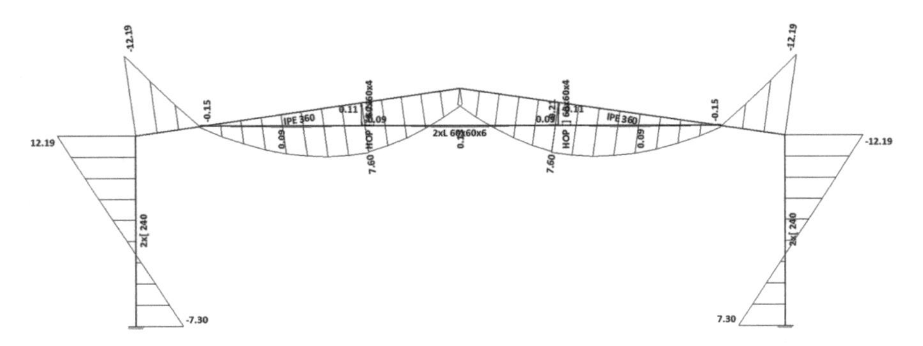

Fig. 2. Bending moment diagram for dead load

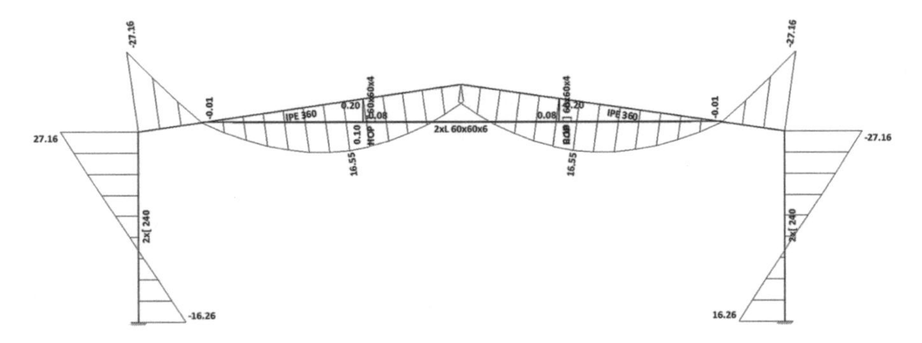

Fig. 3. Bending moment diagram for determined snow load (JUS regulations)

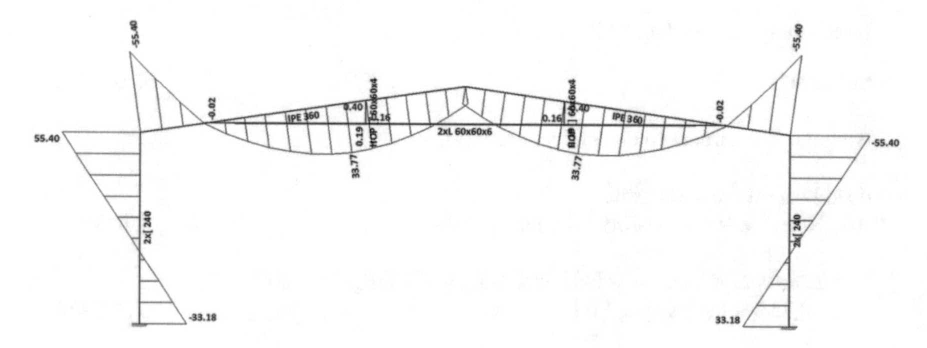

Fig. 4. Bending moment diagram for determined snow load according to EC 1

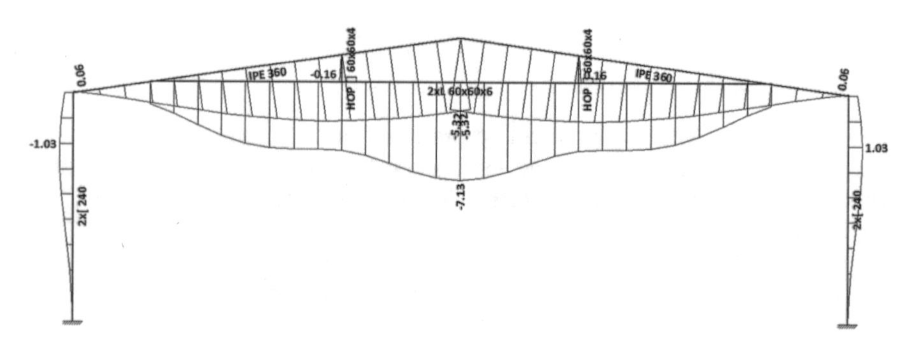

Fig. 5. Deflection diagram for dead load

These results are relevant for determining the reliability of existing steel hall for the ultimate limit state (ULS). Deflections are relevant for the serviceability limit state (SLS) and they are shown for load cases listed above. Software package Radimpex Tower 6.0 is used for structural analysis.

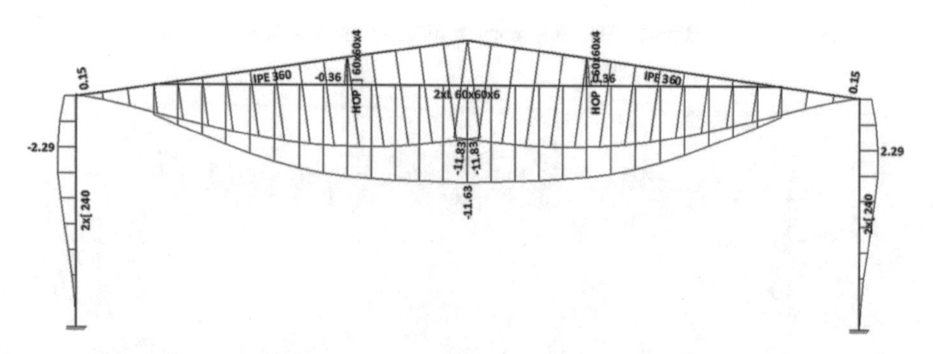

Fig. 6. Deflection diagram for determined snow load according to JUS regulations

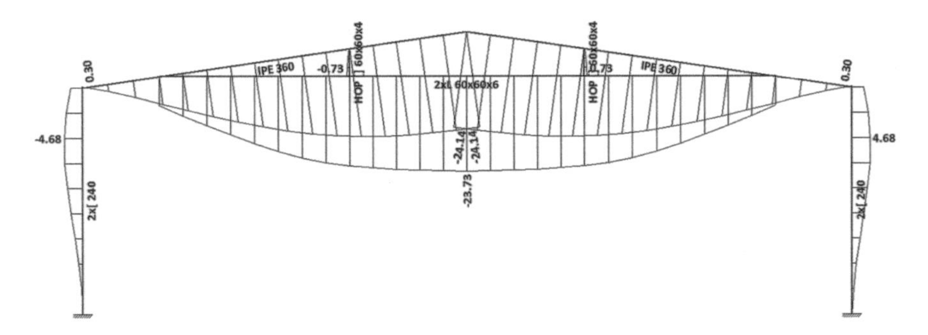

Fig. 7. Deflection diagram for snow load according to EC1

2.4 Probabilistic Analysis of Reliability Index for the Existing Steel Hall in Zenica

Probabilistic analysis considers determining of reliability index according to equations of ultimate limit state (ULS) and serviceability limit state (SLS). Every equation needs to contain load values S and values of construction resistance R in form of Z = R – S.

Base variables X - (R) AND Y- (S) are given in Tables 2 and 3. They are considered random variables described by the average value, standard deviation and distribution. The analysis is done with method First Order Reliability Method – FORM using software package STRUREL (part Comrel). The aim of this analysis is to get values of reliability index higher than recommended in 2.1 and 2.2. Recommended values of reliability index according to EC 1 for return period of 50 years are [1]:

$$\beta_{\text{norm}} = 3,8 \text{ for ULS (ultimate limit state)} \tag{2.1}$$

$$\beta_{\text{norm}} = 1,5 \text{ for SLS (serviceability limit state)} \tag{2.2}$$

Table 2. Base variables of structure resistance [1]

Base variables of resistance [X]				
Variable	Average value	V	Distribution	Description of base variable
X_1	30,9 kN/cm^2	0,1	Weibull	Yield strength
X_2	84,6 cm^2	0,05	Normal	Cross section area
X_3	300 cm^3	0,05	Normal	Moment of resistance
X_4	l/300 = 5,5 cm	0,1	Normal	Limited deflection of girder

Table 3. Base variables of actions on structure

Base variables of actions [Y]				
Variable	Average value	V	Distribution	Description of base variable
Y_1	0,40 kN/m	0,1	Normal	Dead load
Y_2	S_p, S_1, …, S_5	0,05	Gumbell	Snow load

In Tables 4 and 5 are shown values of determined reliability indexes for both load cases [6]. In Table 4 it can be seen that the determined values of reliability index are higher than values 2.1 for ULS and 2.2 for SLS. That means that the construction is safe. In Table 5 it can be seen that the values β are also higher than values 2.1 for ULS and 2.2 for SLS and that means that the construction is safe for this load case as well. For all stated it can be concluded that the existing steel hall is overdesigned which can be clearly concluded from determined reliability indexes that are significantly higher than recommended values according to EC 1 for both limit states. It is determined from structural analysis that this existing steel hall in Zenica can bear characteristic snow load of 4,50 kN/m^2 [6] which is nearly three times higher than recommended snow load according to EC 1. For this snow load values reliability index come to $\beta = 3{,}84$ which is approximately equal to the value 2.1 that is recommended in EC 1.

Table 4. Calculated values of reliability index β for determined load [6]

Points on construction		Value of reliability index β	Probability of failure P_f
Ultimate limit state - ULS	1	11,72	$4{,}62 \times 10^{-32}$
Ultimate limit state - ULS	2	13,18	$5{,}22 \times 10^{-40}$
Serviceability limit state - SLS	3	7,38	$8{,}21 \times 10^{-14}$
Serviceability limit state - SLS	4	5,39	$3{,}45 \times 10^{-8}$

Table 5. Calculated values of reliability index for snow load according to EC 1 [6]

Points on construction		Value of reliability index β	Probability of failure P_f
Ultimate limit state - ULS	1	8,70	$1{,}63 \times 10^{-18}$
Ultimate limit state - ULS	2	12,25	$7{,}99 \times 10^{-35}$
Serviceability limit state - SLS	3	4,51	$3{,}20 \times 10^{-6}$
Serviceability limit state - SLS	4	1,88	$1{,}76 \times 10^{-2}$

Reliability indexes β are inversion function of probability of failure P_f Value of $P_f = 4{,}62 \times 10^{-32}$ for ULS in Point 1, mean 4,62 object will collapse from 10^{32} constructed objects. This is also proof that the steel hall is safe, but also designed to be uneconomical. This hall could have been designed and built with smaller cross-sections of the hall elements.

3 Conclusion

Based on the results of this example the following can be concluded:

- snow load is the dominant load for steel structures,
- determined snow load according to JUS regulations 1961 is lower than defined load according to EC 1

- according to valid regulations snow load does not depend on structure class and type of construction,
- determined by analysis values of reliability index β for snow load according to EC 1 are lower than values of β for snow load according to JUS regulation from 1961 but both values are higher than optimal values recommended in EC 1,
- this existing steel hall in Zenica is safe in aspect of structural capacity and serviceability but it does not satisfy economical condition because it is significantly overdesigned which is concluded from determined reliability indexes for both limit states,

Safety, economy and function are three main conditions that every construction needs to meet. To avoid collapsing structures in the future equalizing of reliability index has to be done and therefore safety level will be equalized on the territory of Bosnia and Herzegovina. Under these circumstances every structure in every part of the country is equally safe independently of the elevation, snow load and snow zone. Through this example it can be seen that in structural analysis economy must be taken in consideration but at the same time girder must satisfy all checks within ultimate limit state and serviceability limit state.

References

1. Hadžović, R., Peroš, B.: Pouzdanost konstrukcija dominantno opterećenih snijegom u Bosnii Hercegovini, GrađevinskifakultetUniverziteta "Džemal Bijedić" u Mostaru, Mostar (2016)
2. Evropski standard, Eurocode 1, Bas EN 1991-1-3:2018: "DejstvanakonstrukcijeDio 1-3 Opterećenjesnijegom", Sarajevo (2018)
3. http://eurokodovi.ba/snijeg/
4. Androić, B., Dujmović, D., Džeba, I.: "Metalnekonstrukcije 1", Sveučilište u Zagrebu, "A. G. Matoš" d.d. Samobor, Zagreb (1994)
5. Androić, B., Dujmović, D., Džeba, I.: "Metalnekonstrukcije 3", Sveučilište u Zagrebu, Zagreb (1998)
6. STRUREL - A Structural Reliability Analysis Program System, RCP Consult (1996)
7. Peroš, B., Hadžović, R.: Analizapouzdanostičeličnihkonstrukcijazaslučajekstremnogopterećenjasnijega u Kantonu Sarajevo, stručni rad, Bihać, RIM (2003)

Database Systems for CAD

Senad Rahimic[1,2](✉) and Mersida Manjgo[1]

[1] University "DžemalBijedić" of Mostar, 88000 Mostar, Bosnia and Herzegovina
senad.rahimic@unmo.ba
[2] Faculty of Mechanical Engineering, University "DžemalBijedić" of Mostar, Univerzitetskikampusbb., 88104 Mostar, Bosnia and Herzegovina

Abstract. The paper presents the methodology for creating a 3D CAD database for standard steel profiles that are available on the market for the purpose of their application for constructions. In this paper we developed a methodology that enables us to create the base of all standard steel profiles in the CAD system, using a file in the "Library Feature Part" format. After designing the construction in the 3D sketch, the "Weldements" command selects the appropriate profile from the database and creates the 3D construction. The "Filet Bead" option is used for the welded join, then the surfaces are selected to be joined and that define the type and thickness of the welded join.

Keywords: Construction · Database · System · CAD · Structural analysis

1 Introduction

Conventional database systems manage data independently of the application, but impose limitations that are inadmissible in CAD environments. On the other hand, object orientation emerged as a programming language paradigm comprising desirable features for CAD, such as rich data modeling and a uniform framework for engineering and system objects [1, 2]. CAD software is also incorporating the ability to gain useful engineering knowledge from the CAD interface. Many modern CAD packages include the ability to attach material properties to the model. The most of the tendencies in an engineer education due to CAD technologies and overall computer use came out from previous analysis. Modern engineer must have certain multidisciplinary knowledge to succeed on the market (Chan 2002).

2 Database for CAD

The traditional design process accounts for 75% of total production costs [1]. Traditional design departmentalized as the engineers would develop the product design and any working and detail drawings associated within the conceptual design. This difference between static relational and dynamic object schema is addressed by Joseph et al. [1]: "Objects in a database are created using 'templates'; these templates are relations in a relational system and classes in a object system. … It is not possible to

I. Karabegović (Ed.): NT 2019, LNNS 76, pp. 252–255, 2020.
https://doi.org/10.1007/978-3-030-18072-0_29

know the correct structure of all templates when the application is first launched". Heiler et al. [2] argue that most object oriented approaches support only IS-A relationships, a kind of generalization specialization relationship, where the object belonging to a specialized class inherits the properties of the general object, and adds its own special features. Other relationships have been proposed, such as component-of, instance-of, version-of, configuration. Even these relationships don't cover all applications. The application semantics vary, and there is a need for user-defined relationships. There is also a need for flexible inheritance mechanisms, in a way which allows the user to specify what properties, operations and constraints should be inherited, and how multiple inheritance conflicts can be resolved. The importance of addressing database problems in such endeavours is well known. However, before problems can be solved they must be clearly defined, and a clear definition of the CAD database problem has yet to be provided. In fact, the CAD database literature bemoans the lack of even a comprehensive functional specification for CAD databases. In this paper, results of research into the problems of CAD databases for steel profiles. More recently, the trend of integrating computer-aided design (CAD) with computer-aided engineering (CAE) has led to a requirement for model simplification. An effective database and database management system is the key to the success of an integrated approach to software engineering applications in general, and Computer-Aided Design (CAD) for structural applications in particular

2.1 Create a Database of Steel Profiles

In this paper, used steel profiles that are defined by the standard according to EN 10219 have been created, a database has been created for all profiles using the dimensions of the specified standard. Figure 1 shows an example of creating a 2D sketch for a straight steel profile.

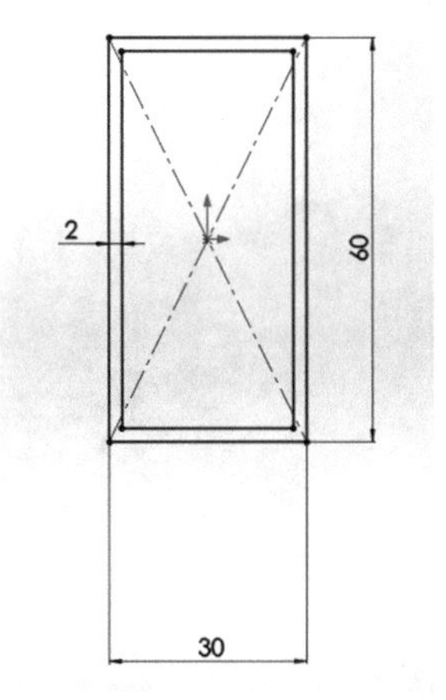

Fig. 1. Creating an example of a steel profile

Figure 2 shows the database access methodology in the CAD system and the possibility of choosing the appropriate steel profile required for a constructive solution. Intelligent CAD systems are systems, which operating autonomously or semi-autonomously in uncertain environments with minimum supervision and interaction with a human operator.

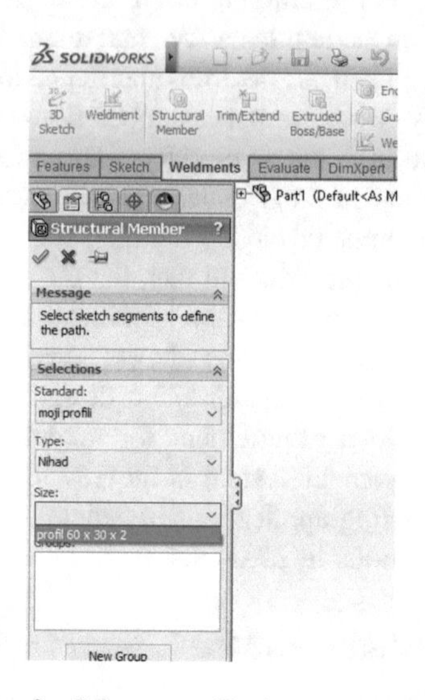

Fig. 2. Select a profile from a database

For a defined 3D sketch, it is necessary to select steel profiles and the type of connection and the CAD system creates a 3D structure with all the features we need for the budget, shown in Fig. 3.

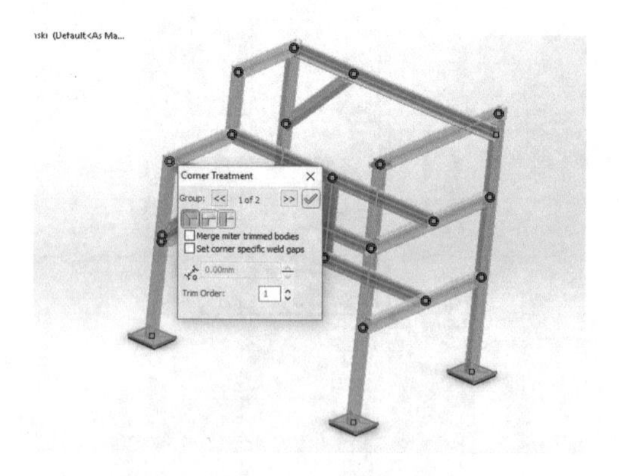

Fig. 3. Created 3D construction

One of the main problems found in the passage from CAD to CAE is the lack of intersecting application space between these two categories of applications [4]. Stress concentration and large displacements are usual problems in the components of the structure of the construction, and that finite element method (FEM) can be a tool to minimize its effects. The goal of this paper is to get results of stresses and displacements of construction structure by using FEM for static simulation, shown on Fig. 4.

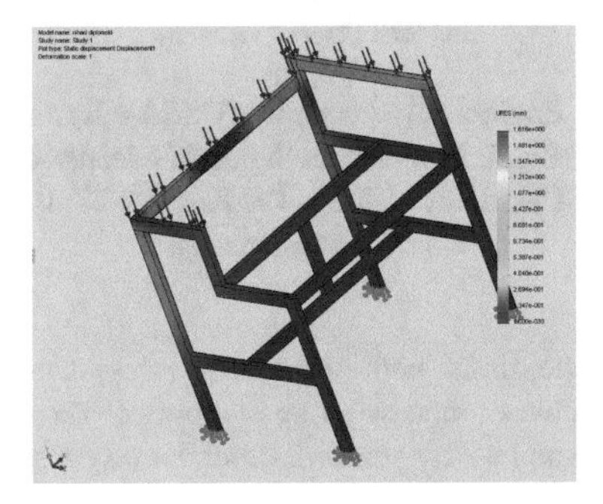

Fig. 4. Static structural analysis

3 Conclusion

This paper has presented the methodology for creating a 3D CAD database for standard steel profiles that are available on the market for the purpose of their application for constructions. In this paper we developed a methodology that enables us to create the base of all standard steel profiles in the CAD system, using a file in the "Library Feature Part" format. After designing the construction in the 3D sketch, the "Weldements" command selects the appropriate profile from the database and creates the 3D construction.

References

1. Chiang, T.A., Trappey, A.J.C., Ku, C.C.: Using a knowledge-based intelligent system to support dynamic design reasoning for a collaborative design community. Int. J. Adv. Manuf. Technol. **31**, 421–433 (2006)
2. Joseph, J.V., Thatte, S., Thompson, C., Wells, D.: Object-oriented databases: design and implementation. Proc. IEEE **79**(1), 42–64 (1991)
3. Heiler, S., Dayal, U., Orenstein, J., Randke-Sproull, S.: An object-oriented approach to data management: why design databases need it. In: 24th ACM/IEEE Design Automation Conference, Miami Beach, FL, 28 June–1 July 1987, pp. 335–340 (1987)
4. Beall, M.W., Walsh, J., Shephard, M.S.: Accessing CAD geometry for mesh generation. In: Proceedings of International Meshing Roundtable, Sandia National Lab. (2003)

Principles of Systemization as a Basis for Designing of Combination Technological Systems

Sergiy Kovalevskyy[1], Raul Turmanidze[1,2(✉)], and Olena Kovalevska[1]

[1] Donbas State Engineering Academy (DSEA), Kramatorsk, Ukraine
[2] Department of Production Technologies of Mechanical Engineering, Georgian Technical University (GTU), Str. Kostava 77, 0175 Tbilisi, Georgia
inform@gtu.ge

Abstract. The article deals with the issues of systematization of factors determining the creation of metalworking equipment. The analysis of the basic laws of the existence and development of production systems is carried out. The main requirements for reconstructed production are given. As a result of completing the production program according to priorities, conditions are created for the maximum value of the rate of formation of the final indicator of production efficiency. The incidence matrix, which is constructed for the technological equipment of machine-building shops of machine-building enterprises, is proposed. The matrix takes into account the parametric and structural components of the main and auxiliary equipment of the site of machining and assembly sites of machines. An updated concept of reconfigured production of an appropriate incidence matrix is formulated. The structure of the assembly plant for the conditions of the reconstructed production is shown.

Keywords: Reconfigured production · Mobile machines · Production system

1 Introduction

The problem of increasing the flexibility of mechanical assembly production, while improving the productivity and quality of production, not only remains relevant, but also has a tendency to the need for constant growth of investments in production funds, the period of renovation of which is steadily decreasing due to the need to ensure the competitiveness of manufactured products.

The promising strategy of loading the production capacity of a modern enterprise is based on the concept of flexible production, capable of quickly adjusting to the production of small-scale production with productivity approaching the indicators of multi-series and mass. But the formation of a range of products to load existing capacity of the company is determined not only by the needs of the market, but also the capabilities of the enterprise. Traditional ways of updating technological equipment include the passing of their life cycle - from foundation to utilization. In this case, the parameters of equipment flexibility remain unchanged and can only be partially changed with regular upgrades.

I. Karabegović (Ed.): NT 2019, LNNS 76, pp. 256–264, 2020.
https://doi.org/10.1007/978-3-030-18072-0_30

Modern machine-building is characterized by constant updating and improvement of metal-cutting machine tools and their manufacturing technologies for the processing of modern products. But the technological capabilities of machine tools do not always correspond to modern ideas about the organization of machine-building production and advanced technologies that require the concentration of operations with simultaneous increase in its productivity. Such conditions, for example, are due to repeated relocation of parts, when the task of ensuring the increased accuracy of positioning of the object of processing and technological, including form-forming movements of the executive body of the machine, which provides an instrumental function [1, 5, 10].

It becomes necessary to ensure the flexibility of production not only to quickly adjust the equipment and equipment available at the enterprise due to multioperation machine tools (machining centers). Such equipment has high indicators of flexibility and technological capabilities, which are achieved through automation and material consistency, which accompanies its complexity and characteristics of accuracy, vibration resistance, dimensions of the working area, etc.

Great prospects for the construction of reconstructed productions are being opened in connection with the use of metalworking machines on the basis of mechanisms with parallel kinematics [4, 9]. Their designs differ rigidity, low metal capacity in comparison with traditional machines of similar working space; the ability to carry large loads through the elements of the farmer; the precision of movement of executive mechanisms and their positioning; the possibility of scaling constructions in a dimensional series; simplicity of the elements of the machine and their final assembly.

The purpose of the work is to analyze and study the possibility of forming reconfigured production in machine-building workshops on the basis of mobile machine tools - works.

2 Testing Methodology

The basis of systematization of factors that determine the creation of metalworking equipment, taking into account the principles of mobility and a high level of automation, can be laid the conditions of system-structural transformation of technology objects. It is necessary to take into account the basic laws of the existence and development of production systems [12].

1. The law of expedient unity of integrity determines the full subordination of the component composition, structure, function of the technological system of the purpose of production. From this law follow special laws, laws and principles: provision of target or optimal parameters of the target function of technological design; target management; expedient functioning without failing the system, as well as the development of the technological system. The fulfillment of this law, the laws and principles proceeding from it, allows us to eliminate the disproportions in the technological system, ensures compliance of the component composition of the system of the entire set (hierarchy) of the goals and the achievement of the most desirable production states.

2. The law of expedient unity of integrity determines the full subordination of the component composition, structure, function of the technological system of the purpose of production. From this law follow special laws, laws and principles: provision of target or optimal parameters of the target function of technological design; target management; expedient functioning without failing the system, as well as the development of the

technological system. The fulfillment of this law, passing from him the laws and principles allows to eliminate disproportions in the technological system, ensures the proportionality and conformity of the component composition of the system of the whole population (hierarchy) goals and the achievement of the most pre-respectful conditions of the production system. The manifestation of the law of expedient unity of integrity is characterized by different levels of correspondence between objects of the structure of the system. It is expressed by the identity of the production structure and structure of the component composition, which serves as a unique multiplier of the efficiency of the production of machine-building products. However, in the technological cycle of production, various tasks related to a single production process must be solved, but they usually provide not one but a few local purposes (bulk manufacturing, machining, assembly, logistics, etc.), which raises additional problems of matching the goals, the separation from them of the main, decomposition of goals for the task, ensuring the compliance of the structure of the set of goals and objectives. The solution to this problem is based on the use of private laws and laws that form the principles of composition and proportionality. The discrepancy between the parts of the whole, its components or structure for the purposes of providing the technological process causes disproportions that reduce the efficiency of production.

The achievement of the required matching is performed under the structural $Z = \{C_1; C_2 \ldots C_n\}$ and the parametric $P = \{p_1; p_2 \ldots p_m\}$ optimization. The degree of correspondence of the components of the structure to the parameters of the objectives of the production system can be represented by the binary relation Q in the direct multiplication of $Z \times P$. The property Q in turn is determined by the incidence matrix

$$
\begin{array}{c c}
 & \begin{array}{cccccc} p_1 & p_2 & p_3 & \cdots & P_m \end{array} \\
Q = & \left[\begin{array}{cccccc} a_{11} & a_{12} & a_{13} & \cdots & a_{1m} \\ a_{21} & a_{22} & a_{23} & \cdots & a_{2m} \\ a_{31} & a_{32} & a_{33} & \cdots & a_{3m} \\ \vdots & \vdots & \vdots & \cdots & \vdots \\ a_{n1} & a_{n2} & a_{n3} & \cdots & a_{nm} \end{array}\right] \begin{array}{c} C_1 \\ C_2 \\ C_3 \\ \vdots \\ C_n \end{array}
\end{array}
$$

where $a_{ij} = 1$, if P_i corresponds to C_j; $a_{ij} = 0$ otherwise.

In the general case, structural-parametric optimization involves identifying the structural constituents of constructions and their parameters on a variety of options. Structural variants and their parameters should be considered together with the criterion of optimality.

The matrix of inactivity can be used to identify the best combination of parameters and structure of technological machines. Such an approach would allow predicting the development of their structures.

Since the matrix of incidence reflects the system of boolean functions, such a matrix can be represented as a graph, where the columns of the matrix are the vertices of the graph, and the rows are its edges. In this case, the graph displays the system of relations of parametric and structural properties of the object.

In order to find the best solution for optimizing the decision of the problem of choosing the combination of the structure, layout and parameters of the metal-working equipment for a lot of nomenclature production, taking into account the many-connected processes that accompany the manufacture of machine-building products from the billets of individual parts to the testing of manufactured knots and machines, it is suggested to search the route of the by-pass vertices of the graph constructed on the incidence matrix. This route is a kind of code, which allows to determine the actual situation, described by the incidence matrix of the layout, structure and parameters of technological machines and the conditions of their application.

As an example, the incident matrix of the form $n \times m = 8 \times 11$ is proposed, which is constructed for the technological equipment of machine-building shops of machine-building enterprises, which takes into account the parametric and structural components of the main and auxiliary equipment by the site of machining and assembly sites (Fig. 1).

```
0, 0, 0, 1, 1, 0, 0, 0, 0, 0, 0
0, 1, 1, 0, 0, 1, 0, 1, 0, 0, 0
0, 0, 0, 0, 0, 0, 0, 0, 0, 0, 0
0, 0, 0, 1, 0, 1, 0, 1, 0, 0, 0
1, 1, 0, 1, 0, 0, 1, 0, 0, 1, 0
1, 1, 0, 0, 0, 0, 0, 1, 1  0, 1
0, 0, 1, 0, 0, 0, 0, 0, 1, 0, 0
0, 1, 0, 1, 0, 0, 1, 0, 0, 1, 1
```

Fig. 1. Matrix of the incidence of the species $n \times m = 8 \times 11$

Depth-first search (DFS) is described by the following algorithm.

1. Select from the unpaced vertices the vertex with the smallest number. If there are no passed vertices, finish the job is the end of the algorithm.
2. Pass the selected vertex and mark it in the array of marks as passed (implemented by a cycle by column).
3. View a list of the incidence of the just passed vertex and to each non-trailing vertex from the list apply recursively points 2 and 3 of the given algorithm (implemented by a loop on a row, embedded in the first loop by column, and another loop on a column, enclosed in a loop by row).
4. Go to step 1.

Full listing of the entire program, which contains an already arranged array, the function of the roundabout graph in depth and the main function with the call of this function. To display the entire roundabout route starting from the first vertex, you must enter the value of the variable from (starting point), one less than the number of the first vertex (in this example it is "−1").

```
#include <iostream>
using namespace std;
const int n = 8;
const int m = 11;
int iArr[n][m] = { 0, 0, 0, 1, 1, 0, 0, 0, 0, 0, 0
                   0, 1, 1, 0, 0, 1, 0, 1, 0, 0, 0
                   0, 0, 0, 0, 0, 0, 0, 0, 0, 0, 0
                   0, 0, 0, 1, 0, 1, 0, 1, 0, 0, 0
                   1, 1, 0, 1, 0, 0, 1, 0, 0, 1, 0
                   1, 1, 0, 0, 0, 0, 0, 1, 1  0, 1
                   0, 0, 1, 0, 0, 0, 0, 0, 1, 0, 0
                   0, 1, 0, 1, 0, 0, 1, 0, 0, 1, 1

bool used[n];
```

```
int j = 0;
int r = 0;
int i = 0;
int k = 0;

void dfs(int t) {

  used[t] = true;

  int p;

  for (i = k; i < n; i++)
  {
          j = r;
          if ((iArr[i][j] != 0) && (!used[i]))
          {
                  used[i] = true;
                  p = i;

                  cout << i << " ";

                  for (j = 0; j < m; j++)
                  {
                          i == p;
                          if (iArr[i][j] != 0)
                          {
                                  r = j;

                                  for (k = 0; k < n; k++)
                                  {
                                          j == r;

                                          if ((iArr[k][j] != 0) &&
(!used[k]))

                                          {

                                                  dfs(i);

                                          }
                                  }
                          }
                  }
          }
  }
}

int main()
{
  for (int i = 0; i < n; i++)
  {
          used[i] = false;
          for (int j = 0; j < m; j++)
                  cout << " " << iArr[i][j];
          cout << endl;
  }
  int from;
  cout << "From >> ";
  cin >> from;
```

```
    cout << "Order: " << endl;

    dfs(from);
    cout << endl;
    return 0;
}
```

As a result of bypassing the graph shown in this example, the following sequence of vertices is displayed: 0, 7, 4, 3, 1, 6, 2, 5.

The above incidence matrix allows us to formulate the principle of conformity of the structure of the production system to its goals as follows: the structure and composition of the production system must ensure the process of achieving the goal in the best way.

Using the revealed structural-parametric sequence of the variant corresponding to the incidence matrix, an updated concept of reconfigured production is formulated.

If we take - the length of the assembly of a mobile machine to work on normalized elements, then the length of assembly equipment of the production site will be equal to (1):

$$T = \sum_{k} t_i + \frac{1}{r} \sum_{r} t_j, \tag{1}$$

where k - the number of successive collected mobile robots; t_i - time spent on the assembly of the i-th machine; r - the number of mobile robots being assembled in parallel; t_j - time spent on assembling a work machine.

Since productivity P is a function of the complexity of manufactured products S and technological capabilities TV of the equipment (2):

$$P = \frac{TV}{S}, \tag{2}$$

In the reconfigured production, the value P will be determined by the speed of "re-assembling" mobile machines, but for a long period of operation of such a production system, the value P will always be the largest possible. As a result of completing the production program according to priorities, conditions are created for the maximum value of the rate of formation of the final indicator of production efficiency. Formalized description of the function is extremely complicated, therefore its research and application should be based on the mathematical model of simulation modeling with subsequent correction of the model based on the results of observations of the practical activity of the reconstructed area (shop).

3 Results and Discussions

Features of machines on the basis of mechanisms with parallel kinematics can predict the evolution of mechanical assembly production in the direction of further increasing the flexibility of production by reconfiguration of machining sites by equipping them with machine tools-robots from normalized elements with a wide-range dimensional series.

On the basis of the analysis, there were identified additional requirements for reconfigurable production:

- high level of automation of main and auxiliary equipment;
- rational use of the working space in the processing of both large and small-sized parts;
- use of the principle “in production areas - only actively used equipment”;
- the possibility of constant updating of morally and physically obsolete equipment in the conditions of extremely limited resources.

The structure of the assembly plant for conditions of reconfigured production should include:

- production platform for the installation of mobile machines in accordance with the technological process with the installation of mobile machines on vibropores;
- the composition of the normalized elements of mobile machines with dense storage with high quality taking into account the acceptance, storage and delivery of the necessary kits to a collection of mobile machines (the best way to handle this function is an automated warehouse);
- site of a collection of mobile robot machines equipped with an automated control system;
- Distributed Management System for Automated Equipment.

4 Conclusion

The paper proposes an approach in which the formation of the composition and volume of the main and auxiliary equipment in the mechanical assembly industry is determined not by the traditions and possibilities of its renewal, but by the range of products, which at the moment is relevant and most in demand. This approach will allow adapting production possibilities to the release of various products.

Replacing the prospect of equipping plants with modern metal-cutting machine tools, machining centers, flexible automated modules, flexible production sites and workshops for the continuous assembly of sites and workshops by mobile robot machines, you can reduce the costs of fixed assets by a few steps. The peculiarity of such development of production capacity of the enterprise is:

- formation of orders structure for products of high demand (selection criterion is the benefit of its production);
- development of technological processes using a lot of possible variants of configuration of mobile robot machines;
- choice of optimal variants of the arrangement of mobile intelligent robot machines;
- design of optimal configuration of each machine with mechanisms of parallel structure and auxiliary equipment;

- development of tasks for the assembly of mobile robot machines from normalized elements and control programs for the manufacture of products with the use of reconfigurable metalworking equipment - mobile robot machines;
- assembly and installation of mobile robot machines in the machining area at designated positions;
- creation of reference models of the working space of each mobile machine tool for the implementation of ideal control programs for managing mechanisms with parallel kinematics on the basis of continuous diagnostics of machine tools;
- production of a given program of production of products in mechanical assembly production (assembly operations are included in the list of works for the provision of production activities of the site, workshop, etc.);
- dismantling (if necessary) previously reconfigurable equipment, storing normalized elements, assembling previously disassembled equipment parts in accordance with a new production task.

Compliance with the given algorithm will allow the company to expand the specialization dictated by previously purchased equipment and unchanged within the period of its payback. In practice, such specialization is transformed into adaptation to provide the conditions for the greatest benefit.

The stated approach will not only reduce the production area by size. Thus, large enterprises can receive the greatest benefit (in absolute terms), simultaneously optimizing the structure of auxiliary industries, in particular, as in [2, 3].

References

1. Clavel, R.: DELTA, a fast robot with parallel geometry. In: 18th International Symposium on Industrial Robot, Lausanne, pp. 91–100 (1988)
2. Jennings, N., Paratin, P., Jonson, M.: Using intelligent agents to manage business processes. In: Proceedings of the First International Conference (The Practical Application of Intelligent Agents and MultiAgent Technology), London, UK, pp. 345–376 (1996)
3. Kristensen, S., Horstmann, S., Klandt, J., Lohner, F., Stopp, A.: Human-friendly interaction for learning and cooperation. In: Proceedings of the 2001 IEEE International Conference on Robotics and Automation, Seoul, Korea, pp. 2590–2595. IEEE (2001)
4. Merlet, J.-P.: Parallel Robots, p. 394. Springer, New York (2006)
5. Bushuyev, V.V., Kholshev, I.G.: Mechanisms of a parallel structure in engineering, No. 1, pp. 3–8. STIN (2001)
6. Koltsov, V.A.: Multifunctional Equipment Based on Parallel Kinematics, p. 131. OmSTU, Omsk (2006)
7. Gavrilov, V.A., Koltsov, A.G., Shamutdinov, A.Kh.: Classification of mechanisms for technological machines with parallel kinematics, No. 9, pp. 28–31. STIN (2005)
8. Kovalevsky, S.V., Kovalevsky, O.S., Korzhov, E.O., Kosheva, A.O.: Diagnostics of Technological Systems and Products of Mechanical Engineering (Using Neural Network Approach): Monograph, p. 186. DSEA, Kramatorsk (2016)
9. Kirichenko, A.M.: Conducting to the zone of processing rigidity and compliance of equipment with mechanisms of parallel structure, No. 59, pp. 205–210. Kyiv Polytechnic Institute, Kyiv, (2010)

10. Kuznetsov, Yu.M.: World trends and perspectives of machine tool development in Ukraine, Problems of physical and mathematical and technical education and science of Ukraine in the context of European integration, pp. 45–55. NPU Drahomanova (2007)
11. Kuznetsov, Yu.M., Kryzhanivsky, V.A., Sklyarov, R.A.: Current state, forecasting and prospects of the development of machines with parallel kinematics. In: Processes of Mechanical Processing in Mechanical Engineering, pp. 320–333. ZHDTU (2005)
12. Selivanov, S.G., Guzairov, M.B.: System engineering of innovative production preparation in mechanical engineering. Mech. Eng., 568 (2012)
13. Strutinsky, V.B.: Theoretical analysis of the stiffness of a six-coordinate mechanism of a parallel structure. Bulletin of the National Technical University of Ukraine, Kyiv Polytechnic Institute, No. 57, pp. 198–207 (2009)
14. Strutinsky, S.V.: Determination of the basic structural parameters of spherical support units of the spatial mechanism by the Monte Carlo method. Mach. Sci. **5**, 37–43 (2007)
15. Strutinsky, S.V.: Mathematical modeling of spatial cross-angle micro-displacements of spherical hinge using recursive bonds. Mach. Sci. **1**, 37–43 (2009)
16. Shamutdinov, A.Kh.: Investigation of classification of multiple-drive mechanisms of parallel kinematics, No. 2, pp. 85–90. Omsk Scientific Herald (2011)

The Influence of Shock Disinfection on Durability of Internal Water Supply

Mario Krzyk(✉) and Darko Drev

Faculty of Civil and Geodetic Engineering, University of Ljubljana,
Jamovaulica 2, 1000 Ljubljana, Slovenia
mario.krzyk@fgg.uni-lj.si

Abstract. Water supply is a potentially exposed to bacterial contamination when it is newly constructed or re-introduced. It must pass tests for chlorine concentration and coliform absence before being put into use. Shock chlorination is usually performed preventively. This is a process of disinfecting internal water supply or plumbing systems by circulating a concentrated chlorine solution throughout the system. Shock disinfection is intended to destroy pathogenic microorganisms. During its implementation very aggressive conditions are present. Under these conditions various metals dissolve, which can cause serious damage to the internal water supply network. Before shock disinfection is carried out it is necessary to assess how it should be applied so that disinfection will be successful without damaging the plumbing installation. Carrying out shock disinfection should take into account all the microorganisms to be destroyed and the materials used that are more or less susceptible to corrosion.

Keywords: Shock disinfection · Internal water supply · Corrosion

1 Introduction

Before putting into use new or reconstructed facilities connected to the public water supply system, shock disinfection of the internal water supply network is necessary. This regulation is based on Articles 4 and 9 of the Rules on Drinking Water, Official Gazette of the RS, No. 19/04, 35/04, 26/06, 92/06, 25/09, 74/15, and 51/17 [1]. To assist water system operators in utilizing the shock disinfection procedure, the National Institute of Public Health (Nacinalniinstitut za javnozdravje - NIJZ) prepared the "Instructions for the Water Supply System Disinfection". The instructions lay down the procedure for disinfection using disinfectants and neutralising agents, as provided in Table 1 [2].

Before using a disinfectant it is necessary to carry out intensive flushing of the part of the network that involves water discharge from the network. The time of flushing should be at least 15 min. The flushing should be applied as shock treatment. If this involves a minor intervention in the in-house network, based on the operator's recommendations to the facility owners, then intensive flushing of part of the network is sufficient.

This instruction does not take into consideration the fact that some microorganisms are resistant to shock chlorination and cannot be eliminated with chlorine. Some

I. Karabegović (Ed.): NT 2019, LNNS 76, pp. 265–271, 2020.
https://doi.org/10.1007/978-3-030-18072-0_31

Table 1. Recommended disinfectants and neutralising agents [2]

Disinfectants	Maximum concentrations	Neutralising agents
Cl_2 - chlorine in gaseous state	50 mg Cl/l	Sulphur dioxide (SO_2)
		Sodium thiosulphate ($Na_2S_2O_3$)
NaClO - sodium hypochlorite	50 mg Cl/l	Sulphur dioxide (SO_2)
		Sodium thiosulphate ($Na_2S_2O_3$)
		Sulphur dioxide (SO_2)
		Sodium thiosulphate ($Na_2S_2O_3$)
$Ca(ClO)_2$ - calcium hypochlorite	50 mg Cl/l	Sulphur dioxide (SO_2)
		Sodium thiosulphate ($Na_2S_2O_3$)

bacteria "hide" into other organisms or among impurities, which makes them harder to destroy. Besides bacteria, disinfection destroys other problematic microorganisms as well (viruses, parasites, fungi, etc.). Figure 1 shows various chlorine concentrations and contact times required to inactivate various types of bacteria.

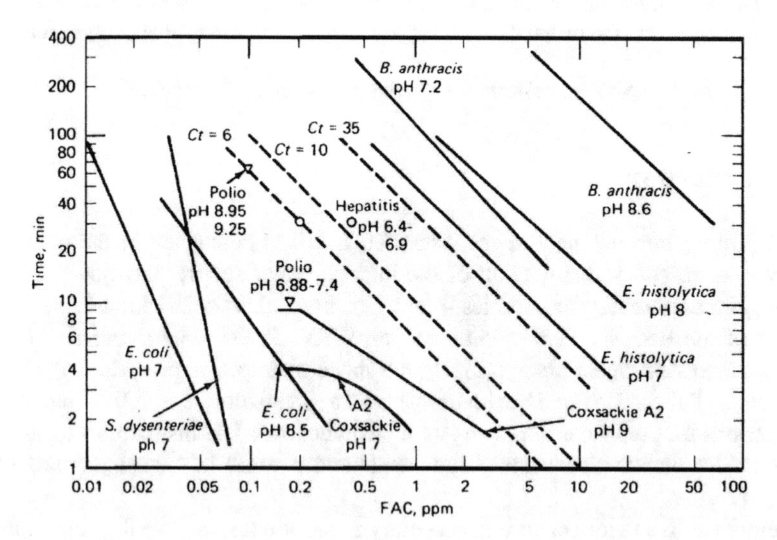

Fig. 1. Disinfection (2-log) of microorganisms by free available chlorine [3]

The effectiveness of chlorine for disinfection depends on the form of chlorine. When chlorine gas is dissolved in water, it reacts with the water to form HOCl and OCl^-. The percentage of hypochlorous acid (HOCl) and hypochlorite ion (OCl^-) is determined by the pH of the water. The HOCl molecule has the ability to penetrate the bacteria, acting as an oxidizing agent. This is possible because HOCl molecules are neutral, while bacteria carry a negative charge. OCl^- is also an oxidising agent, but it is negatively charged and repelled by bacteria. At lower pH there is relatively more HOCl, as shown in Fig. 2. Disinfection is more efficient under these conditions. In slightly alkaline waters a higher disinfectant concentration is necessary to achieve the same disinfection effect. By lowering pH values into slightly acidic conditions the dissolution of metals increases, what causes corrosion of the pipes. Corrosion is the deterioration of a material, especially metal, as a result of chemical or electrochemical reactions on the metal/solution interface [4]. The electrochemical process is the most common in nature and consists of oxidation-reduction irreversible reactions which results in the formation of a corrosion cell [5]. In the environment of low pH zinc, which is widely used as an anti-corrosion agent in steel piping, is readily dissolved. Accordingly, the decision about which disinfection method to use must take into account the materials that the internal water network is built of, so as not to damage it. The selection of materials and disinfection procedures must be considered at the design stage of the facility. State-of-the art methods of building and implementing water installations present new traps in ensuring safe drinking-water supply. The health profession alone cannot cope with the problem of Legionella in the water network. As a result, an engineering approach as well as collaboration and support of technical experts are necessary. Risk management related to the presence of Legionella in drinking water requires ongoing adjustments of the measures taken, the use of new strategies, and trial-and-error learning.

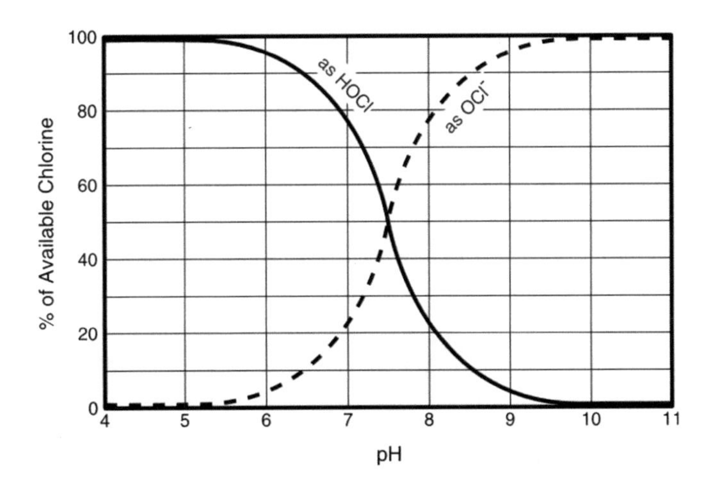

Fig. 2. Chlorine concentration (HOCl and OCl^-) as a function of the pH of the water [6]

2 Materials and Methods

The study used concrete data from the case where we were involved as court experts and evaluators. In case where damages and destruction of internal water supply networks occurred it was necessary to identify the reasons for the situation and find suitable solutions. Water quality measurements were done using validated methods in accredited laboratories. Below we will show the case of internal water supply where corrosion occurred due to improperly carried out disinfection, and we shall analyse the reasons for this.

2.1 The Case of a Destroyed Internal Water Supply Installation in a Hospital

The upgrading of the facilities of an important medical institution involved the installation of galvanised steel pipes. When reviewing the available design documentation it was found that the contractor installed the water installations in line with the design documentation. The water at the inflow into the hospital is hard (>2.5 mmol $CaCO_3$/L) so there is a real risk of scale buildup in the part of the water supply network with hot water. Polyphosphates were used to prevent scale formation in the part of piping carrying hot water. Prior to putting the hospital into service, a shock disinfection procedure according to NIJZ's instructions was undertaken NIJZ (2015). It became apparent that the shock disinfection, following the instructions by NIJZ, was not successful. The bacteria of the Legionella genus could not be destroyed. After repeated shock chlorination, the zinc layer in the water piping was destroyed, resulting in corrosion in the piping. When the zinc coating dissolved, the iron started to dissolve as well. Because appropriate electrostatic conditions are created, the corrosion processes on metals promote microbial contamination of water supply systems. This is because the layer of iron (Fe) oxides has a positive zeta potential, which is beneficial for the growth of bacteria on the piping (Fig. 3). The uneven and contaminated inside surface of the piping became a bacterial breeding ground, protected by biofilm during the various shock disinfection treatments. In the next step biocorrosion could appear. The biocorrosion occurs due to the fixation of bacteria, release of metabolites and formation of biofilms that induce or accelerate the corrosion process [7].

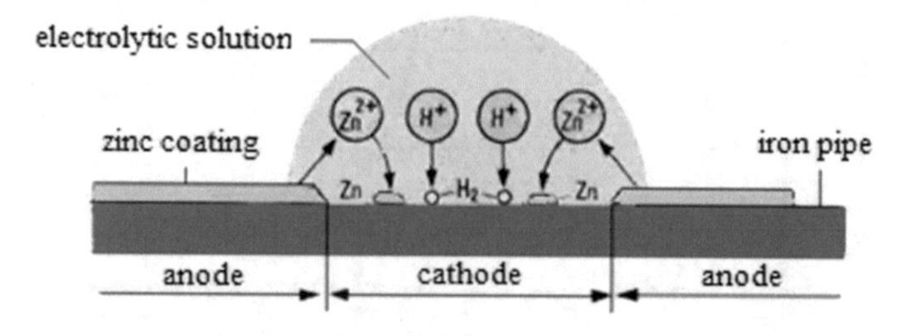

Fig. 3. Corrosion of galvanised steel pipes due to disinfection

Another major problem was that the new facility was connected with old facilities and old water installations. In these parts of the water installation there was no adequate flow in the water system, which provided the ideal hiding place for various microorganisms. Adding polyphosphates under such conditions meant providing nutrients for the bacteria and promoting their growth. The water piping situation is illustrated in Fig. 4. After a few years the entire internal water network in the hospital had to be replaced.

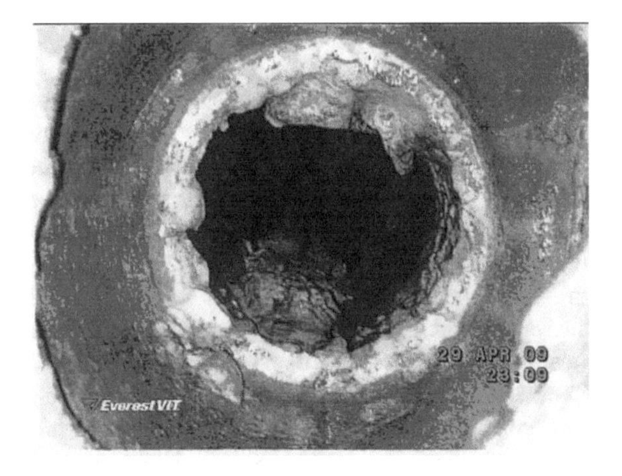

Fig. 4. Corrosion product inside the water installations

Investigations into the development of bacteria on iron, PVC and some other surfaces showed that, due to their electrostatic characteristics, bacteria develop significantly better on iron (Fig. 5). Very few different colony types were isolated from each material with the largest number (nine) recovered from cast iron [8]. The corrosion and biofilm formation additionally contributes to better conditions for development of bacteria and their resistance against disinfectants.

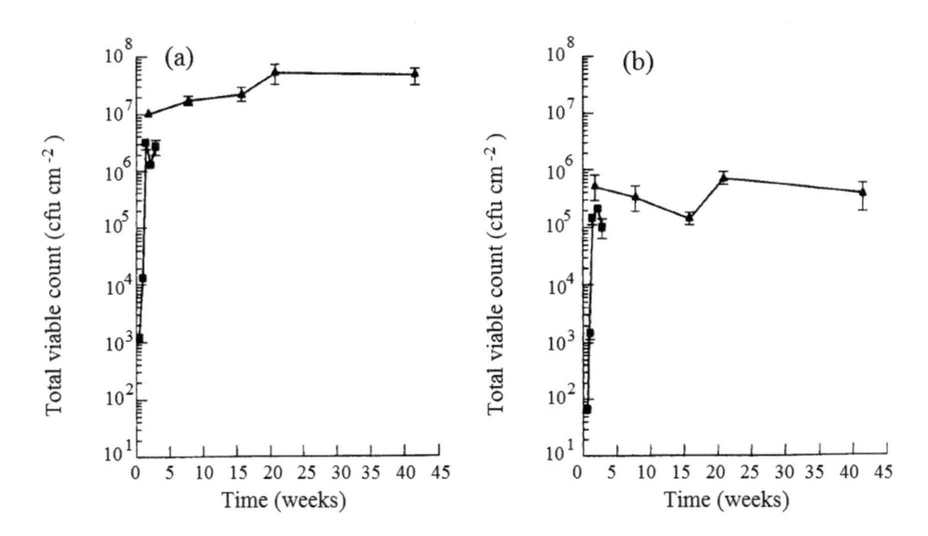

Fig. 5. Bacteria biofilms forming on (a) iron and (b) PVC surfaces [8]

3 Conclusion

Shock disinfection is often required in case of potential risks or in case of internal water supply systems already contaminated with microorganisms. In practice, disinfection is done by heating water or by adding disinfectants in somewhat larger concentrations. Thermal disinfection is a measure to prevent the growth of Legionella in hot water due to their sensitivity to heat. The water at a temperature of 70 °C can destroy Legionella in 10 min and the water at 60 °C in 25 min [9]. The thermal shock efficiency must be controlled through microbial sampling of the water at the outlet. However, such an approach is not always efficient. Some configurations of internal water supply systems do not allow for heating of the entire system. Therefore in practice shock chlorination as a disinfection treatment is often used. The efficiency of chlorination depends on the form of chlorine in the water and the pH values of the water in the water supply network. Chlorine is more effective at low pH values. However, such an environment causes corrosion of galvanised steel piping. The corrosion damages that occur induce the growth of microorganisms, some of which can be dangerous, such as Legionella. In such places large layers of microorganisms in the form of biofilm are developed in a relatively short time. The presence of biofilm and calcium scaling inside the water piping allow for Legionella to resist thermal and chemical disinfection of water and thus undermine the efforts made. In construction and later adaptations and refurbishments of piping, blind, i.e. redundant, ends of piping must be excluded. Managing the problem of disinfection of an internal water supply network requires the collaboration between the facility operator, health and technical professionals, and motivated water users. Collaboration with sanitary epidemiological service is key for reaching a successful solution to the problem. Addressing the problem of the presence of Legionella in the drinking water of the hospital water network provided insight into the scale of the problem and a possible solution.

References

1. Rules on Drinking Water, Official Gazette of the RS, No. 19/04, 35/04, 26/06, 92/06, 25/09, 74/15, and 51/17, in Slovene: Pravilnik o pitnivodi,Uradni list RS, št. 19/04, 35/04, 26/06, 92/06, 25/09, 74/15 in 51/17
2. Instructions for the Water Supply System Disinfection, in Slovene: Navodilo za izvedbodez-infekcijevodovodnegaomrežja, NIJZ – Center za zdravstvenoekologijo, 19 November 2015
3. White, G.C.: Handbook of Chlorination and Alternative Disinfectants. Wiley, New York (1999)
4. Shi, X., Xie, N., Gong, J.: Recent progress in the research on microbially influenced corrosion: a bird's eye view through the engineering lens. Recent Pat. Corros. Sci. 118–131 (2011)
5. Ghali, E.: Corrosion Resistance of Aluminum and Magnesium Alloys: Understanding, Performance, and Testing. Wiley, Canada (2010)
6. http://www.hydroinstruments.com/files/Basic%20Chemistry%20of%20Chlorination.pdf, 15 January 2019

7. Moura, M.C., Pontual, E.V., Paiva, P.M.G., Coelho, L.C.B.B.: An outline to corrosive bacteria. In: Méndez-Vilas, A. (ed.) Microbial Pathogens and Strategies for Combating them: Science, Technology and Education (2013)
8. Kerr, C.J., Osborn, K.S., Robson, G.D., Handley, P.S.: The relationship between pipe material and biofilm formation in a laboratory model system. J. Appl. Microbiol. Symp. Suppl. **85**, 29S–39S (1999). https://doi.org/10.1111/j.1365-2672.1998.tb05280.x
9. Stout, J.E., Best, M.G., Yu, V.L.: Susceptibility of members of the family Legionellaceae to thermal stress: implications for heat eradication methods in water distribution systems. Appl. Environ. Microbiol. **52**, 396–399 (1986)

The Way to a New Definition of the Kilogram

Daut Denjo(✉) and Senada Pobrić

Faculty of Mechanical Engineering, "Džemal Bijedić" University of Mostar, Univerzitetski kampus bb, 88000 Mostar, Bosnia and Herzegovina
daut.denjo@unmo.ba

Abstract. This paper presents a way to a new definition of a mass unit, kilogram, through two very complex experiments which are performed in the leading world metrology institutes. These experiments are Avogadro project and Watt's scales, which are aimed to connecting kilograms with natural constant, Planck's constant. Also, the paper mentioned natural constant, published by CODATA.

Keywords: Metrology · Mass unit kilogram

1 Introduction

At the 26th General Conference for Measures and Weights (CGPM) in november 2018, it was agreed that there are no more artifacts for defining units, such as a prototype of kilogram (for mass scale) or a special isotope mixture of water (for a temperature scale). The most stable that physics can offer are natural constants. They appear in all basic physical equations and so determine the "rules of nature." For over a hundred years, great minds such as James Clerk Maxwell, Ludwig Boltzmann and Max Planck came up with such an idea. And now this idea is useful. Since a metrology institutes around the world as accurately as possible measured the values of the selected natural constants in extremely complex experiments, their values are now determined once and for all in November 2018. On May 20th, 2019, for the World Metrology Day, "the new SI" will come into effect. Even after successful new definitions of basic units of SI system, their realization and transfer remain the primary task. The methods for measuring the constants, according to the currently valid SI, can be used directly to realize the corresponding units according to the new definitions. The current work in this regard is of great importance for the future.

Further development of these techniques, as well as the development of completely new methods, in the future will allow the industry more accurate implementation of units over the extended measuring areas.

2 The New Basis for All Measures

Natural constants in the future will define units in the International System of Units. Seconds and meters are already ahead of other units. Kilograms, ampers, and others in the class of physical base units are trying to make a connection with natural constants.

I. Karabegović (Ed.): NT 2019, LNNS 76, pp. 272–277, 2020.
https://doi.org/10.1007/978-3-030-18072-0_32

This is it what is happening in the world of metrology. The international system of units is faced with a fundamental redefinition: natural constants will serve as a reference quantities in the future for all seven basic units and for all derived units. Sensitive objects such as original kilogram or completely impractical formulations, such as intensity of electric electricity are outdated. Experimental preparations for these new definitions are in full swing in the world, especially in Physikalisch-Technische Bundesanstalt (PTB). On the General Conference for Measures and Weights in Versaj ion November 2018, the amendments were adopted and officially enter into force on May 20, 2019, for the World Metrology Day. Great benefits will have science and high technology. The consumer will not feel any changes in their every day measurements.

The planned new definitions in the SI system connect each basic unit with a defined constant and determine the number values for these seven selected constants. A unique illustration one-to-one exists for a second and a mole. For other basic units, need more than a fixed constant, for example, for example, for the meter, besides the speed of light, the reference frequency of cesium is also required. It is important in the context of new definitions, that the definition of seven constants automatically defines all derived

Table 1. Natural constants with numerical values

Physical quantities	Basic unit	Natural constant
Time	*Second*	$\Delta\nu(^{133}Cs)_{hfs}$ *(Crossing the hyperfine structure)*
	The frequency $\Delta\nu(^{133}\mathrm{Cs})_{\mathrm{hfs}}$ crosses the hyperfine structure of the baseline state of the cesium atom is exactly equal to 9 192 631 770 Hz.	
Length	*Meter*	*c (speed of light)*
	The speed of light in vacuum c is exactly equal to 299 792 458 m s^{-1}	
Mass	*Kilogram*	*h (Planck's constant)*
	Planck's constant h is exactly equal to 6,626 070 15 $\cdot$ 10^{-34} J s. (J s = kg m^2 s^{-1})	
Intensity of electric electricity	*Amper*	*e (Elemental charge)*
	Elemental charge e is exactly equal to 1,602 176 634 $\cdot$ 10^{-19} C. (C = A s)	
Temperature	*Kelvin*	k_B *(Boltzmann's constant)*
	Boltzmann's constant k_{B} is exactly equal to 1,380 649 $\cdot$ 10^{23} J K^{-1}. (J K^{-1} = kg m^2 s^{-2} K^{-1})	
Amount of substance	*Mol*	N_A *(Avogadro's constant)*
	Avogadro's constant N_{A} is exactly equal to 6,02214076$\cdot 10^{23}$ mol^{-1}	
Intensity of light	*Kandela*	K_{CD} *(The photometric equivalent of the radiation)*
	The photometric equivalent of the radiation K_{CD} of the monochromatic radiation, the frequency 540 $\cdot$ 10^{12} Hz is exactly equal to 683 lm W^{-1}	

units. Thus Coulomb (as Amper multiply Second) is a direct multiple of the "elementary charge" constant. There is also an expert group "CODATA - group for fundamental constants" in the United States, which task is to evaluate and accept the values of natural constants determined in physical laboratories around the world. Table 1 gives seven natural constants in the new SI, as well as numerical values published by CODATA 2017.

3 Kilogram and Mol

After nearly 130 years, in the autumn of 2018, the international original kilogram (will be withdrawn). As a meter and a second, the kilogram will be defined over an unchanging natural constant. Avogadro experiment and Watt's scale are the basis of the new definition of kilogram. They represent two independent ways of implementing a new definition of kilogram. In addition, the Avogadro experiment determines how many atoms are located in an almost perfect silicon ball. Scientists at Physikalisch-Technische Bundesanstalt (PTB) diligently count atoms in the 28-silicon (Si) ball. The more precisely they do it, the more accurately they can finally say how much weight silicon atom. So, the jump to the level of natural constants has succeeded. The researchers can determine the Avogadro′s constant and the Planck′s constant and thus have a "recipe" for a new, more stable etalon of mass.

3.1 Avogadro Project

In the beginning there was a desire to using silicon ball of high purity to determine Avogadro's number as accurately as possible, respectively the number of particles in one mole. The international network of research partners from around the world started the "Avogadro project". After that it would be just a small step, that even the kilogram is defined by the exact number of silicon atoms. However, the international community decided for the second way, so the focus of research with years has shifted to determining Planck's constant h.

The best balls consist of 99.999% silicon-28. The material of high purity is the result of a long and costly process of centrifugation and cleaning performed by Russian partners on the project. The Das Institut für Kristallzüchtung in Berlin made a "single crystal" weighing 5 kg, in which the atoms are arranged in a uniform grid - without irregularity, from polycrystalline raw materials with a lot of knowledge and experience. From a single monocrystal, in PTB later produced two silicon balls. The balls have a diameter of about 93.7 mm. The deviation from the perfect shape of the ball is only about 10 nm (1 nm = 10^{-9} m). This almost perfect surface is achieved by a long grinding and polishing process. The volume was determined using the Fizeau interferometer, which at the same time measures thousands of diameters. The volume of the ball, number "AVO28-S5c" is 430.891289 cm^3. Also, the topography of the spherical surface can be precisely recorded in this way. On the ball, a thin layer of oxide, water, and a contaminated carbon layer form naturally. Together, they can be thick to 3 nm. These must be taken into account when measuring the volume of the ball and the mass of the ball. PTB has established an ultra-high vacuum system with combined X-ray fluorescence and photoelectron spectroscopy for determining the chemical composition as well as the density of the mass on the surface of the little balls. The PTB's silicon balls have a mass of 1 kg, which can be measured very accurately, for example, the mass of the ball bearing the symbol "Si28kg01b" is 1.000 012 601 kg and is determined using a vacuum comparator mass "Sartorius CCL1007" with a measurement uncertainty of 6 μg.

3.2 Where Is Planck in Ball

To Planck's constant in the balls Si you come across the following relations:
mass of Si ball = mass of atom Si × number of atoms
mass of Si ball = mass of atom Si × volume of Si ball/volume of atom
mass of atom Si = molar mass of ball material/Avogadro's constant N_A
Avogadro's constant N_A = molar mass's constant/atomic's units of mass m_a
atomic's units of mass = mass of electron/electron relative mass
mass of electron = 2Rydberg's constant (R_∞) × h/speed of light c × constant of fine structure α^2

3.3 Electric Kilogram (Watt's Scale)

It is harder to connect the Planck's constant and the kilogram. Planck's constant h has a unit of measure of kgm/s, where the meter and the second already defined by the definition of the second and determined by the speed of light. If a numeric value h is determined precisely in this unit of measurement, the kilogram will be indirectly defined. A direct way to realize the kilogram, provide Avogadro's project and Watt's scale. The Watt's scale is tons of heavy electromechanical precision equipment, with which two different experiments are performed. On one side, the mass m (for example 1 kg) is held in balance, by balancing the gravitational force $m{\cdot}g$ acting on it (with gravitational acceleration g), with electromagnetic force. This force is generated by passing electrical through the wire in the magnetic field. On the other hand, if in this magnetic field the wire moves at constant speed v, and the induced electrical voltage is measured in it.

The basic equation is obtained from both experiments: $U \cdot I = m \cdot g \cdot v$ (with the Watt unit, therefore the "Watt's scale"), so that the mass m can be precisely determined by precise measurement U, I, g and v. Gravitational acceleration g can be measured extremely accurately with commercial devices. By measuring U and I using Josephson's effect (h/2e and quantum Hall's effect h/e^2), Planck's constant returns to the game so that a direct connection between it and the mass unit of kilogram is made.

$$h = \frac{4mgv}{v_m v_g} \tag{1}$$

where:
v_m and v_g are the frequencies of microwave radiation, which are measured at Josephson's voltage in the first or second experiment (Fig. 1).

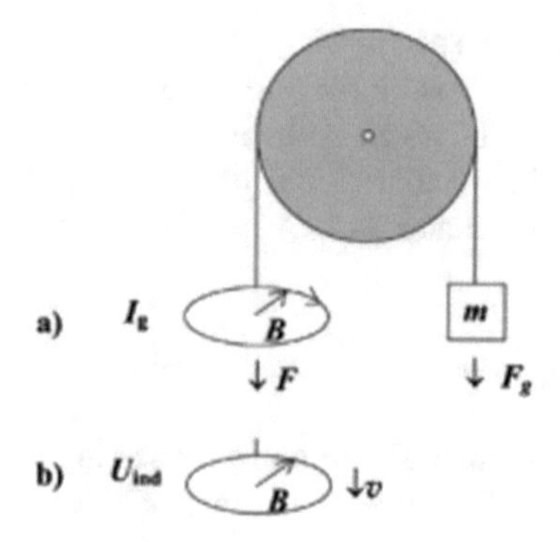

Fig. 1. Principle of Watt's scale (a) static mode; (b) mobile mode

B - radial magnetic field, I_g - coil electric, F_g - force mass weight m, F - electromagnetic force, U_{ind} - induced voltage in coil, v - the velocity of movement of the coil

4 Conclusion

While team of scientists in the United States, as well as scientists in Canada and other countries want to determine the Planck's constant using Watt's scale, German scientists have gone the way to Avogadro's constant, counting atoms in ^{28}Si ball. However, counting each atom individually is not possible - it is not necessary. In the almost perfect crystal, such as monocrystalline 28 Silicon, from which Si balls were formed in PTB, the atoms are regularly arranged in the crystal grid. X-ray makes the grid structure visible and determines the volume of atoms. In the next step, scientists have determined the diameter of the ball. For this purpose, the diameter of the ball is determined by ball interferometer on its surface over a million times. Thus, you can accurately determine the average diameter of the ball up to three atomic's diameters and calculate the volume. The ratio of two volumes gives the number of atoms in the ball. The scientists in PTB can do so well enough that only hundreds of millions of atoms are calculated with a mistake of one to two atoms. Through all these measurements and calculations, researchers know how many moles of silicon exist in their ball and how many atoms are found in mole. This last corresponds to Avogadro's constants. These measurements are equally important with measuring Avogadro's constant. In PTB research works are intensified on measurement the volume of the ball, grid parameters, ratio of isotopes, chemical purity as well as quantitative chemical characterization of the surface. The aim is to suppress the relative uncertainty of the measured value for h apropos N_A below 1.5×10^{-8}.

Determining the value of the Planck's constant is obtained by the so-called Rydberg's constant, very accurate values for the mass of electrons. The mass of the electron is again allowed to be compared with the atomic masses. When Planck's constant is determined, and the atomic masses with a kilogram unit are very well known. Thus, after determining the value for the Planck's constant and the number of atoms in the Avogadro project and also by the Watt's scale, which is called the Kibble scale since

2016, a mass unit of kilogram will be produced. With a variety of methods for the realization of kilogram, by Kibble scale in a total of 13 countries and a completely different Avogadro project, quite a small incompliance appeared. There are differences of up to 70 μg. But optimistic is that errors can be experimentally better limited and reduced. Still, in the future, control measurements will always at the end give that which is not provided during the definition of a kilogram with Watt's scale and the Avogadro project. In the Avogadro project, PTB is now the world leader. They can, almost all by themselves, different required measurements perform. Therefore, PTB has better options for identification where errors can still occur.

References

1. Becker, P., Borys, M., Gläser, M., Güttler, B., Schiel, D.: PTB MITTEILINGEN, Fachjournal der Physikalisch-Technischen Bundesanstal - Das System der Einheiten, 122.Jahrgang, Heft 1, März 2012
2. PTB-Infoblatt - Das neue Internationale Einheitensystem (SI), www.ptb.de, Stand, March 2017
3. Simon, J.: Wo steckt Planck in der Kugel? PTB, Maßtaäbe, Maße für alle, Heft 14, November 2018
4. FrIschmuth, I.: Der Blick in die Kristallkugel, PTB, Maßtaäbe, Maße für alle, Heft 14, November 2018
5. Simon, J.: Naturkonstanten als Hauptdarsteller (2018). www.ptb.de

Real Time Control of Above-Knee Prosthesis with Powered Knee and Ankle Joints

Zlata Jelačić[1](✉) and Remzo Dedić[2]

[1] Faculty of Mechanical Engineering, University of Sarajevo, Vilsonovo šetalište 9, 71000 Sarajevo, Bosnia and Herzegovina
jelacic@mef.unsa.ba

[2] University of Mostar, 88000 Mostar, Bosnia and Herzegovina

Abstract. Passive prostheses, which are mostly used by amputees, enable performance of various activities such as walking on levelled and inclined ground, and even running, riding a bicycle and as of lately swimming. However, performing high power demanding tasks, such as stair ascent, presents a problem, because the lack of muscles makes it impossible to produce the required forces. To perform high demanding power activities, prosthesis must be powered, primarily in its main joints – knee and ankle. In this paper, we present the real time control of newly developed above-knee prosthetic SmartLeg prototype with powered knee and ankle joints. Specialized control unit is developed in order to achieve required kinematics and dynamics to enable it to perform high power demanding activities in more natural manner, especially stair ascent.

Keywords: Powered above-knee prosthesis · Control unit · Real time control · Prototype testing

1 Introduction

Today, most amputees use passive prostheses. Passive means that these devices are not powered in any way and cannot give any additional output of energy to its user. They usually work on using elastic or some other potential energy which is stored during one sequence of gait, and then released in another. This is far more efficient than using peg leg prosthesis, but still not suitable enough to naturally perform common gait activities. Various types of passive prostheses enable performing of different activities such as walking on levelled and inclined ground, and even running, riding a bicycle and as of lately swimming. However, performing high power demanding tasks, such as stair ascent, presents a problem for passive prostheses.

In the process of ascending stairs, during loading response and midstance phase, body is being lifted from one stair to another and only one leg touches the stair and lifts the body, while another is in swing phase (in the air). In this phase the entire load is on one leg, and the body is in such position that moments in the knee and ankle joints are very high and exceed body weight by several times. In case of above knee amputation, the lack of muscles makes it impossible to produce these forces, and passive prostheses which only use stored energy are unable to perform such activities. This means, that in

I. Karabegović (Ed.): NT 2019, LNNS 76, pp. 278–284, 2020.
https://doi.org/10.1007/978-3-030-18072-0_33

order for the prosthesis to be able to perform high demanding power activities, it must be powered, primarily in its main joints – knee and ankle.

2 SmartLeg Prosthetic Device

Our prosthesis is designed in a way to mimic main leg movements in the sagittal plane. Its key features are knee and ankle joints which are movable in sagittal plane and are externally powered. Movements of knee and ankle enable obtaining required kinematics of the prosthesis, and powering system enables its dynamic features [1]. Powered knee enables overcoming large forces that occur during loading response. Having only powered knee in the prosthesis design is not enough, since joint in the ankle is also very important in natural stair ascent. Powered ankle joint enables dorsal and plantar flexion movement of the entire foot, which provides better stabilization of the knee and the entire prosthesis and also provides power needed in push-off phase.

2.1 Hydraulic Power System of the Prosthesis

Our hydraulic power system concept is designed to enable characteristic movements of the prosthesis in sagittal plane while walking and ascending stairs. Since it is far more difficult to achieve, we are concentrated on testing the prosthesis on stair ascend. The base components of our hydraulic system are two hydraulic actuators, one for powering the knee and other for powering the ankle joint, hydraulic power pack unit and accompanying hydraulic installation (Fig. 1). We chose off-the-shelf hydraulic power pack unit because it consists of all needed hydraulic installation components (electrical motor, hydraulic pump, reservoir, appropriate valves, connections etc.) integrated in one whole unit [2].

Fig. 1. SmartLeg prototype in standalone setting

2.2 Prosthesis Control System

In this version, for the first time, six sensors, which are a combination of a gyroscope and an accelerometer, are integrated along with the MPU 60/50 development board. Three sensors are placed on the prosthesis and three on the healthy leg of a subject without amputation or on the healthy leg of an amputee [3]. The sensors serve to map the angles of the leg and more precisely the angles of the knee, ankle and the foot. To facilitate the placement of sensors on the prosthesis and on the leg of a healthy subject, plastic cages were made for each sensor individually through 3D prototyping. The same is done for the electronics (Fig. 2).

Fig. 2. Control sensors and control unit for the real-time prosthetic control

3 Preparation of the Experiment

In order to determine the movement quality of the prosthetic leg prototype with two-way hydraulic drive and integrated hydraulic knee and ankle actuators, experimental testing was carried out. To measure the joint trajectory and tracking error, trakSTAR™ measurement system was used. Hydraulic above-knee prosthesis was tested autonomously in laboratory environment for the purpose of examining its functionality. The results were compared the movement of the subjects without amputation.

3.1 trakSTAR™ Measurement Equipment

The trakSTAR™ electromagnetic 6DoF tracking solutions from Ascension Technology Corporation (an NDI company) provides a cost-effective, high accuracy position and orientation tracking technology for integration into the most innovative and realistic medical training and surgical rehearsal simulators and military flight and gunnery simulation systems (Fig. 3).

Miniaturized, lightweight sensors embedded into medical instruments, helmets or Head Mounted Displays (HMD) track the user's movements in all six degrees of freedom (6DoF), with no line of sight requirements. Every movement within the tracking volume, no matter how subtle or precise, is tracked in real time to exceptional accuracy. This accuracy, combined with low latency and fast update rates, allows sensor position and orientation data to integrate seamlessly with the simulator display to produce interactive training that's remarkably true to life.

A configurable design, small hardware footprint, and ease of integration further complement this 6DoF electromagnetic tracking technology solution known for its high value and high performance [4].

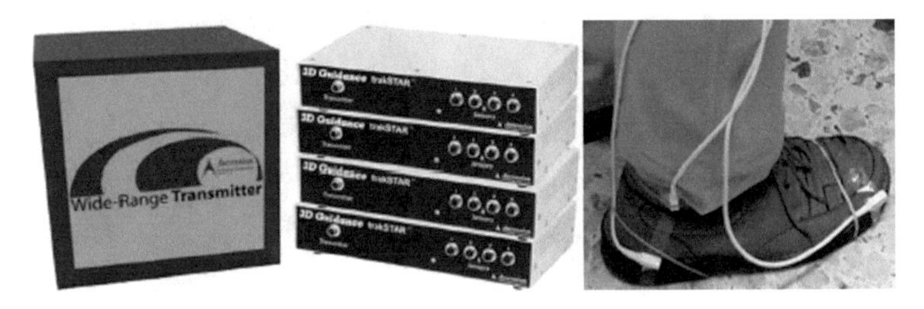

Fig. 3. trakSTAR™ measurement equipment: central unit and sensors [4]

3.2 Prosthetic Leg Installation

For the autonomous testing, the prosthetic prototype was mounted on a specially designed stand. Hydraulic aggregates and power supply are mounted as well. A prosthesis is placed on the base and then connected to a hydraulic aggregates and power system. The prosthetic foot is fixated to the base with a screw in the toe area (Fig. 4).

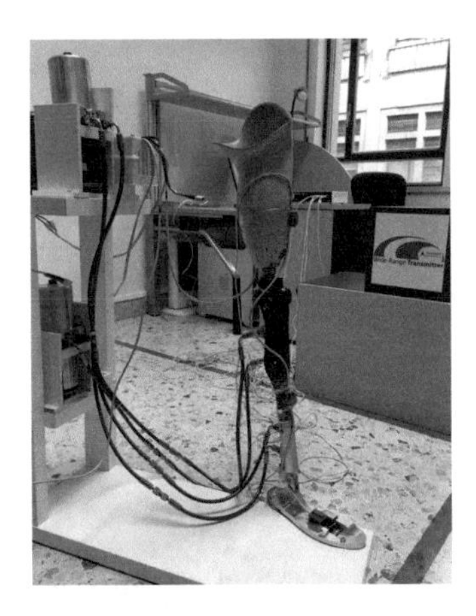

Fig. 4. Mounting of control unit sensors on the prosthetic prototype

3.3 Installation of the trakSTAR™ Measurement System

After the prosthetic prototype was mounted and connected to the hydraulic power and control system, the trakSTAR™ sensors were placed on the characteristic points of the prosthetic leg. trakSTAR™ measurement equipment has four sensors in total and they can be placed either on the healthy leg or the prosthetic prototype at the same time (Fig. 5): sensor 1 is mounted on the hip, sensor 2 is mounted on the knee joint, sensor 3 is mounted on the ankle joint and sensor 4 is mounted on the heel.

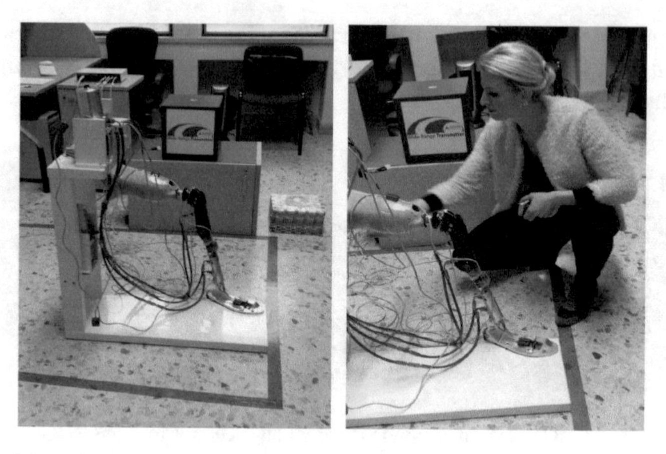

Fig. 5. Mounting of trakSTAR™ measurement sensors on the prosthetic prototype

In order for the prosthetic prototype to mimick the movements of the healthy leg, our control unit and sensors were used for the real time control based on the master-slave concept. On one hand, a set of sensors was placed on the healthy leg, which simulated various climbing stages related to stair ascent. On the other hand, a set of sensors was placed on the prosthetic prototype which performed the same movements as the healthy leg in real time (Fig. 6).

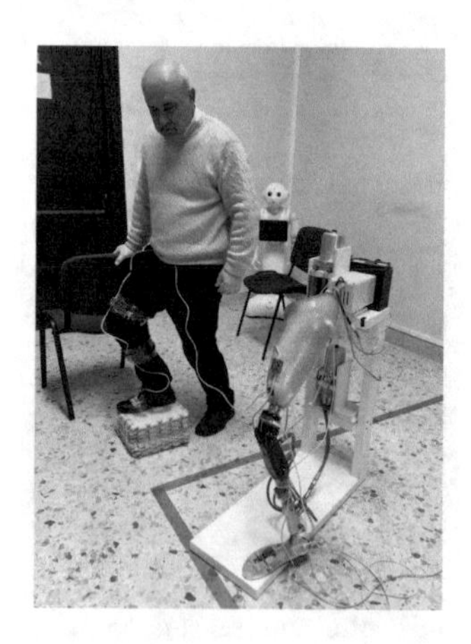

Fig. 6. Control unit with BionicLeg sensors for the control on real time

4 Results and Discussion

After the first contact of the foot with the stairs, the part of the climbing period in which the foot accepts the weight follows, with a strong concentric contraction of the hip and knee muscles (extensors) in order to straighten the leg and lift the body to the second step. This is the phase of dual support where the muscles of gastrocnemius and soleus are expressed. After a double suspension, the phase of one-sided resting followed by the leg muscles of the right-legs of the examinees acting strongly to prevent the pelvic lowering of the unpaired side and to pull the hull laterally over the retractable right leg. At a later stage of suspension, when the body weight is completely on the retractable leg and the knee is stretched, the quadriceps acts isometrically so that the centre of gravity of the body passes in front of the right retractable foot. It is the phase of climbing to the second step, or the step with the second leg to the next step, with the foot bending and the ankle joint is maximally active and shows the greatest deviation from the phase of equilibrium. This is the phase that will be analysed in this paper. The division is primarily made to test the behaviour of the foot and ankle joint at different stages of climbing the stairs.

Below are the results of measurements with trakSTAR equipment on subjects without amputation and the SmartLeg prototype with separate power system in both knee and ankle joints, using the control unit with BionicLeg sensors for the control in real time (Fig. 7).

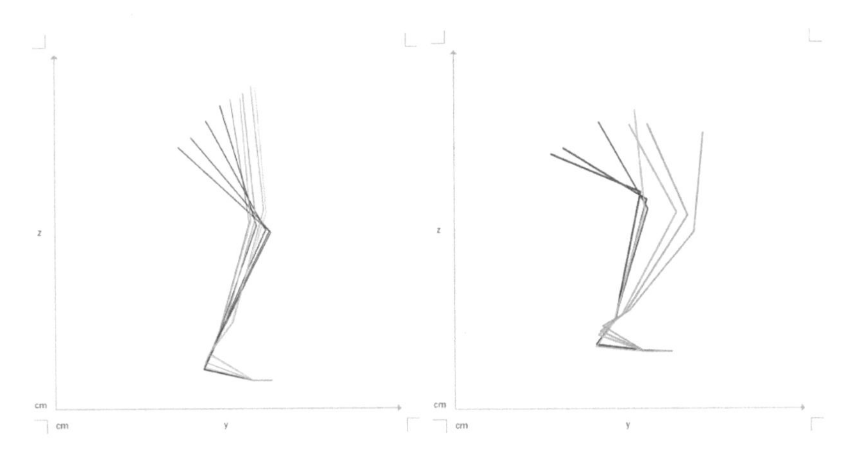

Fig. 7. Comparison of the joint and leg position between a healthy subject (left) and a SmartLeg prosthetic prototype (right) in real time

From the above results it can be clearly seen that the tested prototype which has a separate power system in knee and ankle joint, as well as a newly developed control system, allow fine regulation of movement in real time. We can even see that the prosthetic foot moves although it has only a passive metatarsophalangeal joint. The reason why the thigh of the prosthetic leg lags behind is that there is no counter weight present on the prosthetic leg in autonomous testing setting.

5 Conclusion

With rapid development of different novel prosthetic solutions, laboratory testing of the prosthesis becomes a necessary requirement to verify new concepts prior to their application on human subjects. Our new prosthesis could make the gait of above knee amputees generally more natural, and in the near future with the latest and more precise technology, it might be able to fully mimic the natural human gait.

Our current developments offer an improvement in above-knee prosthetic design but further optimisation is still required. One way to optimise our current prosthetic design is through the integration of power system into the prosthetic's housing. This means that smaller components need to be implemented into a design, but with special attention to their positioning in order not to affect the moments of inertia of the prosthesis.

References

1. Jelačić, Z.: Contact force problem in the rehabilitation robot control design. In: Badnjevic, A. (ed.) CMBEBIH 2017: Proceedings of the International Conference on Medical and Biological Engineering 2017, pp. 193–204. Springer, Singapore (2017)
2. Rupar, M., Jelačić, Z., Dedić, R., Vučina, A.: Power and control system of knee and ankle powered above knee prosthesis. In: New Technologies, Development and Applications. Springer International Publishing AG (2019). ISBN 978-3-319-90893-9
3. Jelačić, Z., Dedić, R., Isić, S., Husnić, Ž.: Matlab simulation of robust control for active above-knee prosthetic device. In: Advanced Technologies, Systems, and Applications III. Lecture Notes in Networks and Systems. Springer International Publishing AG (2019). ISBN 978-3-030-02576-2
4. Ascension Technology Corporation: 3D Guidance trakSTAR WIDE-RANGE, 14 October 2017. http://www.ascension-tech.com/

Computer Science, Information and Communication Technologies, e-Business

METU Smart Campus Project (iEAST)

Şeyda Ertekin[1(✉)], Ozan Keysan[1], Murat Göl[1], Hande Bayazıt[1], Tuna Yıldız[1], Andrea Marr[2], Mehdi Ganji[2], Saeed Teimourzadeh[3], Osman Bülent Tör[3], and Sıla Özkavaf[3]

[1] Middle East Technical University, Ankara, Turkey
sertekin@metu.edu.tr
[2] Willdan Energy Solutions, Anaheim, USA
[3] EPRA Elektrik Enerji, Ankara, Turkey

Abstract. With the rise of urbanization, cities around the world have embraced applications and benefits of leveraging advanced technologies to deliver a range of services while promoting efficient, environmentally friendly, and sustainable eco-systems. By harnessing technology to improve the quality of life of citizens, these advanced technological tools have become critical in transforming urbanized cities across the globe into smart cities. Universities in particular have served as an ideal platform to showcase smart applications to promote smart campuses. This paper presents METU Smart Campus Project which addresses necessary analysis and recommendations for the implementation of a smart and sustainable campus at METU. Scope of the project includes development of a 10 year smart campus roadmap and a plan for implementation of near-term smart and sustainable campus activities. The project will assist METU in planning and implementing the smart intelligence, Energy, Aquatic (Water), Security, and Transportation Campus (iEAST).

Keywords: Smart city · Smart campus · Energy efficiency · Transportation · Smart buildings · Water management

1 Introduction

Smart city projects are meant to leverage the power of Information and Communication Technology (ICT) to provide pertinent information to citizens, businesses and tourists, improve the efficiency of government services, and interconnect new or legacy infrastructure systems such as energy, water, and transportation [1]. These projects are long-term investments that require collaboration across the public and private sector. With the rise of urbanization, cities around the world have embraced applications and benefits of leveraging advanced technologies to deliver a range of services while promoting efficient, environmentally friendly, and sustainable eco-systems [2]. The use of ICT globally has not only become central to how cities function, but also how they are managed. By harnessing technology to improve the quality of life of citizens, these advanced technological tools have become critical in transforming urbanized cities across the globe into smart cities. Universities in particular have served as an ideal platform to showcase smart applications to promote smart campuses.

I. Karabegović (Ed.): NT 2019, LNNS 76, pp. 287–297, 2020.
https://doi.org/10.1007/978-3-030-18072-0_34

The U.S. Trade and Development Agency (USTDA) is providing a grant to the Middle East Technical University (METU) for technical assistance (TA) in support of a Smart Campus Project. The TA will explore the benefits of applying advanced technology solutions to enhance the areas of smart energy, smart transportation, smart water management, smart buildings and ICT, management and control systems. In addition to developing detailed sector strategies in these key areas, the TA would provide technical, operational and financial guidance to serve as a roadmap for implementing the METU Smart Campus Project. The project will assist METU in planning and implementing the smart intelligence, Energy, Aquatic (Water), Security, and Transportation Campus (iEAST).

The aim of the METU Smart Campus Project is to develop a comprehensive smart and sustainable city in a relatively small scale. Implementing smart city projects at the METU in Ankara can provide an early demonstration of the cost effectiveness and social and environmental benefits of smart cities projects, while operating in an environment where most services and infrastructure is owned and operated by one entity, avoiding many of the political and institutional challenges of smart city projects. These projects will serve as a demonstration to the public and private sector how smart city technologies and projects can be deployed throughout Turkey. Ultimately, the goal of the smart campus project is to enable METU to determine how to best achieve integration of campus infrastructure and services through an integrated, centrally managed and operated project. As illustrated in Fig. 1, the project would include four sector components – smart energy, smart transportation, smart water management and smart buildings – as well as one overall ICT management and control component, which would be implemented in the near term (2 to 5 years).

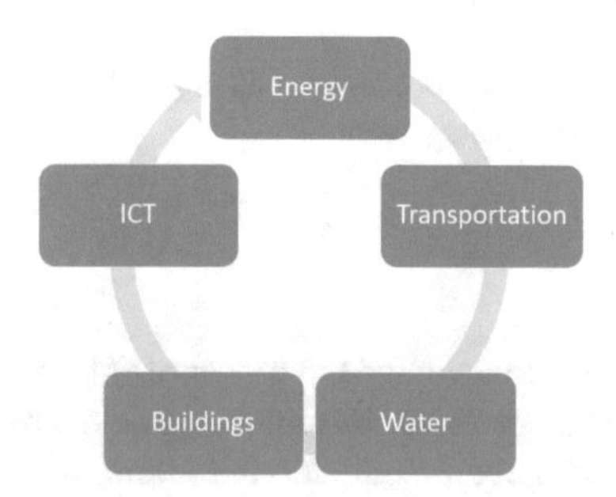

Fig. 1. Sector components to be addressed in the project.

The study is organised as follows. First, summary of METU key figures is made in Sect. 2. Then, technical approach of the Smart Campus Project is presented in Sect. 3. Key observations and initial recommendations in terms of key sectors are addressed in Sect. 4. Note that, since the project is ongoing and planned to be completed by the end of 2019, the recommendations in this study are broad and based on initial observations. Conclusions drawn from the study are summarized in Sect. 5.

2 METU at a Glance

METU is a public technical university located in Ankara, Turkey [3]. The university has an emphasis on research and education in engineering and natural sciences, and offers undergraduate, graduate and doctoral programs. The campus hosts METU Techno-park, which is the largest techno-park in Turkey with 320 R&D companies and 5,500 employees. METU has demonstrated its commitment to a smart campus through a range of projects, including advanced wastewater, energy efficiency, renewable energy, and transportation projects. The administration is committed to planning and budgeting for an integrated smart campus project that will maximize efficiency and cost savings while providing for state of the art and sustainable solutions.

The main campus of METU spans an area of 11,100 acres (4,500 ha), comprising, in addition to academic and auxiliary facilities, a forest area of 7,500 acres (3,000 ha), and the natural lake Eymir. METU is the home of more than 9,500 undergraduate students, graduate students, 1,250 fulltime teaching faculty members, and 2,600 academic staff with 42 undergraduate degree 99 Masters and 62 Ph.D. programs. The METU population includes 26,500 students, 7,000 residents and 5,500 employees. As shown Fig. 2, the campus hosts 14 dormitories, 650 housing spaces for academic personnel, 320 R&D companies as part of the METU Techno-park on the campus, restaurants, cafes, supermarkets, banks, and health-care centres. METU campus includes large forests that provide fresh air not only to campus but also to Ankara city as well. However, emissions from the central heating plant and cars in particular during rush-hour conditions contribute negative environmental impact s on the campus.

Fig. 2. METU campus – academic area view.

METU is a highly selective university – most of its departments accept the top 0.1% of the nearly 1.5 million applicants. METU had the greatest share in national research funding by the Scientific and Technological Research Council for Turkey (TÜBİTAK) [4] in the last five years, and it is the leading university in Turkey in terms of the number of European Union Framework Programme projects and participation. The campus houses 24 interdisciplinary research including Solar Energy Research (GUNAM), Wind Energy Technologies Research (RUZGEM), Biomaterials and Tissue Engineering Research (BIOMATEN), the Marine Ecosystems and Climate Studies

Center (DEKOSIM), and Image Analysis (OGAM). The METU campuses have the characteristics of a small city with the same opportunities for using information technology to enhance workability and sustainability. As a result, METU can serve as a test bed and pilot site for smart city applications across Turkey and in other countries.

3 Project Technical Approach

In the project, after conducting a preliminary review of the existing campus infrastructure, how the campus infrastructure can be upgraded through the deployment of advanced technologies to enable the efficient management of energy, water and transportation sector services within the campus, will be determined (Fig. 3). The Near-Term Plan for the upgrade of the existing infrastructure will lead to more efficient regulation of energy and water usage, improved transportation, and enhancements to the campus's safety and security features through the integration of centralized ICT-providing METU operational staff and their management with enhance visibility, data analytics, and controllability. The project will provide METU with an iEAST vision that captures the key decision makers' priorities for a smart and sustainable campus, and contains a technology and information gap analysis for each of the iEAST sectors. METU iEAST Roadmap will provide direction to help METU and other campuses achieve the iEAST vision, followed by a plan to implement the necessary activities to be performed in within the short-term period (10 years).

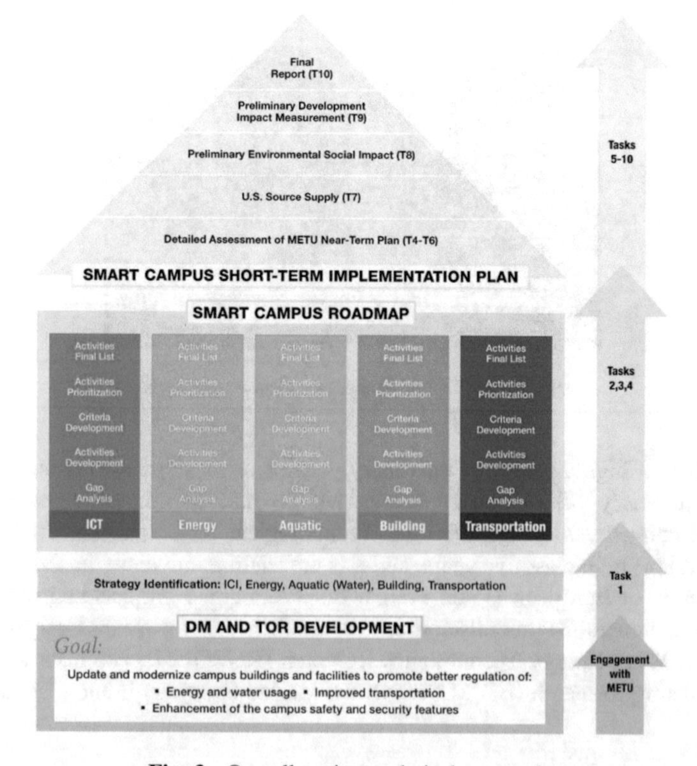

Fig. 3. Overall project technical approach.

4 Initial Observations and Recommendations

4.1 Transportation

The three campus entrances to ODTÜ; (1) İnönü Boulevard, (2) Bilkent Boulevard, and (3) Yüzüncüyıl District are shown in Fig. 4. During morning and evening rush hours, 8:00–9:00 and 16:30–17:30, respectively, commuters use these three gates to enter/exit the campus, which results in a high volume of traffic. In order to mitigate the congestion, alumni are allowed to enter after rush hour. Cars with stickers get into the campus through the same gate as visitors, and alumni with ID leave through these gates, which also increases the congestion. Visits to the Techno-park zone are allowed only from the Bilkent entrance.

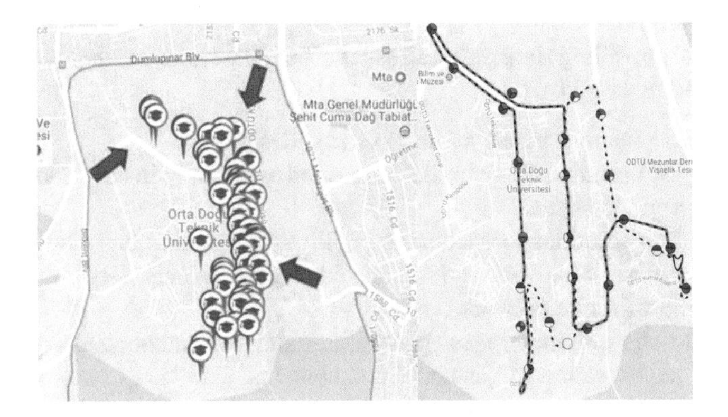

Fig. 4. Three main entrances of the main campus (left) and METU ring route (right).

METU campus' bus fleet operate on a ring road that passes through the campus, as shown in Fig. 4. Buses are old and use liquid hydrocarbon fuels, which impact air quality. A web application communicates the current status of the ring buses.

While it is common for students to hitch-hike from the campus to the city this practice creates congestion when drivers stop to communicate with the students about

Fig. 5. Hitch-hiking on the campus (left) and example of a crowded parking lot (right)

their route (Fig. 5). If the driver's route aligns with the students' route, the driver takes the students. If not, they continue driving without taking the students. This process introduces extra traffic jams and safety concerns, particularly during evening rush hours. This topic will also be addressed in the scope of recommending smart solutions for mitigating traffic "jams" on the campus.

One of the main issues on the METU campus is the crowded parking lots (Fig. 5). In order to address high demand in the parking lots, METU has issued stickers authorizing different levels of parking privileges (for example, stickers assigned to academicians enable parking in any lot, but those for students are restricted, based on sticker type). There is no bike sharing system in the campus. Universities across the world have committed to bike-share systems as part of their sustainable transportation programs, in which students and university personnel can pick up/drop off bikes at different locations on the campus. This is an important topic to be addressed in the scope of the project.

The most prevailing implementations to be addressed in the scope of smart transportation are as follows:

- Smart urban planning transportation systems design
- Technologies that can reduce traffic congestion, especially in car parks and near the university entrance gates
- Transportation solutions to encourage public transportation and ride sharing
- Electric shuttle-bus network and automated electric vehicle fleet
- Bike sharing or rental network
- Security, at the entrance gates, pedestrians and cars authorized entrance-permit detection, identification of personnel and students, detection of security violations
- Mapping the campus road network and transportation system into a multi-layered GIS that is designed to be capable of mapping other aspects of utilities, services and infrastructure at METU

4.2 Energy

The campus is served by a medium-voltage (MV) distribution network loop with radial connections to the loads, shown in Fig. 6. This design provides the campus with adequate power reliability. Most of the buildings are supported by diesel generators, which serve the critical loads with back-up power. The MV layout needs improvement considering demand increase in the campus as well as reliability in case of n-1 contingency [5]. The Main Campus heating load is provided by the central heating plant. Natural gas is utilized to generate hot water and steam for heating purposes. Annual natural gas consumption is approximately 11.000.000 m^3 [6]. There is no central electrical power source on campus and all electricity is provided from the electric grid. A micro-grid with local power generation is among the most promising concepts for smart campuses worldwide and should be a prime consideration for METU. Bilkent University, located next to METU, for example, has a combined heat and power (CHP) system, which is utilized for both heat and electric power supply.

Fig. 6. Medium voltage layout of the campus (left) and 50 kW PV roof-top implementation at Ayaslı Research Center (right)

There is a 50 kW PV panel implementation on the roof of Ayaslı Research Center at the Electrical and Electronics Engineering (EEE) Department (Fig. 6). An R&D project, which is supported by EMRA of Turkey (EPDK), is ongoing in collaboration with Başkent DisCo and METU. It proposes to deploy a micro-grid at the Ayaslı Research Center. In addition, 50 kW PV panels will be installed on the A-Building roof in the EEE department under the support of GAMA Energy Co. in Turkey. These projects will enable a significant infrastructure for micro-grid implementations in METU.

Investigations on the current electrical system of METU and comparing its status with smart-campuses worldwide (e.g., IIT campus as one of the pioneers in smart campus subject [7]) implies that deployment of more renewable-based generation at METU is the first step for moving towards the smart campus. Here, solar generation is of great interest which can be installed both on the buildings and parking lots. Investigations on the current heating system of METU reveals that METU campus has a great potential for installing combined heat and power systems due to availability of heating center. Installing CHP system in METU campus paves its ways towards smart campus notion. CHP systems are comprised, essentially, of three components: a power generator, a heat recovery system, and a thermally activated machine(s). A CHP system may also include electrical equipment driven by the on-site power generator. In some cases, a combination of thermally activated and electrical equipment may make sense or may be the only feasible option due to a limited availability of waste heat or its incompatible quality. Due to availability of heating center, establishment of CHP generation deemed to be feasible considering the following steps:

- The most important costs for realizing CHP generation at METU campus includes turbine and electrical generator cost; substation and associated equipment cost.
- Similar to any other generation, the adequacy of METU distribution grid for hosting the CHP generation should be studied.

- Deployment of CHP unit might increase the fuel cost as more steam would be required. This cost should be compared with the price of energy provided by the local DSCO with the aim of determining the size and rating of CHP unit.
- The power generator of a CHP system can be sized and operated to meet either the thermal load, known as "thermal-load-following" or electrical load, referred to as "electrical-load-following." In the former case, depending on the thermal-to-electrical load ratio, the power supply may be more or less than the demand, leading to the sale or purchase of power from utility, respectively.
- For dispatching the CHP unit, the dependency of generated power to the heat should be taken into account.

Currently, there is no smart LED street light system available in METU campus. The majority of street lights at METU campus are conventional Mercury-vapor or Sodium-vapor lamps. The first step for realizing smart street lighting system is to replace available lamps with LED ones which imposes associated cost. The other main ingredient is communication network which enables command\info sharing between the street lights and the control system. The deployment of energy storage system can be performed with different objectives, say cutting campus energy cost or electrifying a critical load during outages. However due to subsidized price of electrical energy for METU campus, electrification of critical load, say hospital or public safety station, can be considered as the main objective for storage deployment. Establishing energy storage system would be more reasonable after proliferation of distributed energy resources, say rooftop PVs and CHP unit. The size of deployed storage system is the main factor for its investment cost. Currently no demand side management and demand response system is deployed in METU. For METU campus, the loads can be categorized as (1) Educational loads representing the load of educational areas and offices; (2) Residential loads representing the electrical consumption of dormitories. As a demand response program, curtailing building loads can be considered as the heating and cooling system is not electrical. Cost of such an approach includes load curtailment cost. In addition, it might has negative impact on social welfare.

Currently SCADA system is available for heat and water system (as depicted below); however, no SCADA system is available for electrical network. The first step for moving towards realizing micro-grid in METU campus is to construct a supervisory system with the objective of data gathering and issuing control commands. To do so, smart meters should be installed within the campus. The main objective for locating smart meters is to provide maximum coverage while deploying minimum meters. By the smart meters in place, the required data are measured and sent to the campus control system. METU has planned to establish as situation room which provides monitoring and control function for electrical, heat and water systems. From electrical system point of view, the situation room should be equipped with a master control which applies a hierarchical control via SCADA to ensure reliable and economic operation of campus electrical system. It also coordinates the operation of switches, on-site generation, storage, and individual building controllers. Intelligent switching and advanced coordination technologies of the master controller through communication systems facilitates rapid fault assessments and isolations.

4.3 Smart Water Strategy

Water necessary for heating/cooling and a portion of the daily water usage are supplied from Eymir Lake (100 L/s). Thirty percent of the daily water usage is supplied from wells in the campus. The quality of the water in these significant resources should be monitored by smart systems. The most prevailing implementations to be addressed in the scope of smart water management strategy are as follows:

- SCADA applications
- Energy efficient pumping and distribution
- Leak detection systems
- Monitoring of water quality and advanced wastewater treatment
- Gray water storage and reuse in the buildings
- Mapping the campus water and wastewater network into a multi-layered GIS that is designed to be capable of mapping other aspects of utilities, services and infrastructure at METU

4.4 Smart Buildings Strategy

There is no available building level data on campus. During the site visit there were multiple discussions about the value of sub-metering each building, if only perhaps for the period of this study. Obtaining information regarding the electricity distribution system and estimated campus loads is critical to adequately sizing renewable generation assets on campus. The most prevailing implementations to be addressed in the scope of smart building strategy are as follows:

- Smart heating, ventilation and air conditioning (HVAC)
- Energy efficiency improvements, including lighting and real-time energy consumption monitoring and building automation systems
- Building security
- Mapping the campus buildings into a multi-layered GIS that is designed to be capable of mapping other aspects of utilities, services and infrastructure at METU

4.5 ICT Strategy

METU was the first entity in Turkey to be connected to the internet. METU controls 53,000 user accounts which are served by on-campus 160 servers. All of these servers are stand-alone servers which have been designed for specific purposes. To date, the capacity of these 160 servers has not been shared. More than 75 K emails are being sent or received daily. Currently the METU network serves 2,737,656 online visitors with 3,500 TB download and 1,500 TB upload. The main network feed's band with is 3 Gbps and campus which is equipped with an additional 1 Gbps redundant feed. As shown in Fig. 7, the whole campus is connected using both single (Yellow), and multiple mode fiber (Red) optics.

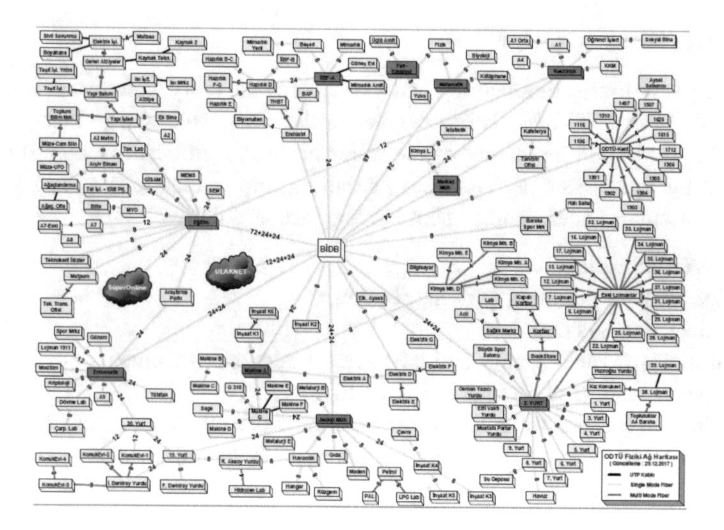

Fig. 7. Campus wired network infrastructure

The most prevailing implementations to be addressed in the scope of smart ICT strategy are as follows:

- Communications network
- Data collection, aggregation, processing, storage and analytics
- A platform for data and analytics which can be utilized for development of additional services and applications by METU stakeholders;
- Public Wi-Fi
- GIS mapping tools and analytics
- Improved decision support systems information channels between different layers of management and stakeholders of the campus (e.g., academics, students, employees etc.), such as:
  – Real-time incident reporting tools
  – Efficient information broadcasting, including in case of emergency situations
  – Dashboards for senior management and department/division management
- Centralized management and operations
- Cross utilization of key systems such as SCADA, GIS, and other capabilities to allow the common use of these systems to deliver multiple smart campus capabilities

5 Conclusions

METU Smart Campus Project will create a smart METU campus under these categories: Smart Energy, Smart Transportation, Smart Water Management, Smart Buildings, and ICT Management and Control. Advanced and state-of-art ICT systems will be required to utilize newly developed cloud data centers, big data analytics and

Internet of Things (IOT) to achieve integration of campus infrastructure and services, maximize efficiency, lower the campus' carbon footprint and improve citizen collaboration and well-being.

Acknowledgement. This study is supported by U.S. Trade and Development Agency (USTDA) under the Grant Agreement between USTDA and METU on September 22, 2017. The study is supported in part by TUBITAK 1509-Uluslararası Sanayi Ar-Ge Projeleri Destekleme Programı (Project no/name: 9180003/Multi-layer aggregator solutions to facilitate optimum demand response and grid flexibility (SMART-MLA) under ERA-NET Smart Grids Plus 2017).

References

1. Dinh, D.V., et al.: ICT enabling technologies for smart cities. In: 20th International Conference on Advanced Communication Technology (ICACT). IEEE (2018)
2. Du, S., Meng, F., Gao, B.: Research on the application system of smart campus in the context of smart city. In: 8th International Conference on Information Technology in Medicine and Education (ITME). IEEE (2016)
3. METU (2018). www.metu.edu.tr
4. TUBITAK (2018). www.tubitak.gov.tr
5. Mendes, A., Boland, N., Guiney, P., Riveros, C.: (N-1) contingency planning in radial distribution networks using genetic algorithms. In: IEEE/PES Transmission & Distribution Conference and Exposition: Latin America (T&D-LA). IEEE (2010)
6. METU Water and Heating Directorate (2018). http://isim.metu.edu.tr
7. Shahidehpour, M.: Micro-grids for enhancing the economics, reliability, and resilience of smart cities – an IIT experience. In: Smart Grid Conference (SGC). IEEE (2014)

Solving Agile Software Development Problems with Swarm Intelligence Algorithms

Lucija Brezočnik(✉), Iztok Fister Jr., and Vili Podgorelec

Faculty of Electrical Engineering and Computer Science,
University of Maribor, Koroška cesta 46, 2000 Maribor, Slovenia
lucija.brezocnik@um.si

Abstract. This paper outlines a short overview of swarm intelligence algorithms that are used within the software engineering area. Swarm intelligence algorithms have been used in many software engineering tasks, e.g., grammatical inference or mutation testing. However, their presence in the agile software development field is still awakening. As there are some promising results of solving different problems of agile software development with swarm intelligence, this paper discusses such problems and the proposed solutions within the last decade. Based on the results we propose a systematic classification of swarm intelligence algorithms according to problems within agile software development, i.e., next release problem, risk, software design, software cost estimation, and software effort estimation. Afterwards, we present papers that fall in the scope of the proposed classification, and provide highlights of each paper for researchers, conducting research in this and associated fields. In this manner, we provide some conclusions for each of the classified problem groups, and, in the end, we review the guidelines for the future.

Keywords: Agile software development · Swarm intelligence · Optimization · Search-based software engineering

1 Introduction

Swarm intelligence or, simply, SI algorithms, are a sub-branch of Computational Intelligence. Loosely speaking, swarm intelligence algorithms are methods that are inspired mostly by nature. In other words, they concern the collective, emerging behavior of multiple, interacting agents who follow some simple rules [7]. These agents might be considered as unintelligent. Interestingly, when working together, the whole system of multiple agents may show some self-organization behavior (collective intelligence). The history of swarm intelligence goes back to 1989 when Beni [4] coined this term. Expansion of these algorithms began after the 1990 s. On the one hand, researchers showed that they are very useful when solving continuous optimization problems, while, on the other, they also behave well in discrete optimization problems. Interestingly, there are also a lot of practical applications that are based on SI algorithms in the real world. During the past decades, many swarm intelligence algorithms were also applied in the domain of Software Engineering, along with evolutionary algorithms [12]. Mostly, researchers were concentrated on solving

I. Karabegović (Ed.): NT 2019, LNNS 76, pp. 298–309, 2020.
https://doi.org/10.1007/978-3-030-18072-0_35

problems such as the development of mutation testing [13], grammatical inference [21], and test effort estimation [27]. In contrast, too little attention was devoted to the swarm intelligence algorithms in the domain of agile software development. According to our literature research, we saw that in recent years more applications had been proposed in the literature. For that reason, the primary missions of this paper are:

- to present a short overview of this vital research field,
- to review this research field and classify problems that are solved by swarm intelligence algorithms,
- to study why swarm intelligence algorithms are useful for solving problems within the agile software development research field, and
- to determine guidelines for the future of this research field.

The structure of this paper is as follows: Sect. 2 acquaints the reader with the fundamentals of swarm intelligence, while Sect. 3 presents agile software development methods. Section 4 discusses problems in agile software development that were tackled by swarm intelligence algorithms. Section 5 is devoted to the future of this field, while the paper is concluded with a summary of SI methods in the agile software development field.

2 Core Fundamentals of Swarm Intelligence

Let us imagine bees when searching for nectar, or ants when building anthills, or even fireflies when mating during the summer nights. At first sight, we can say that they are pure individuals that would like to survive in their natural habitat. However, this observation is not correct. Although these individuals are considered as unintelligent, they cooperate in each aspect. These characteristics can be conceived as swarm intelligence. Swarm intelligence involves the collective, emerging behavior of multiple, interacting agents who follow some simple rules. While each agent may be considered as unintelligent, the whole system of multiple agents may show some self-organizing behavior and, thus, can behave as a kind of collective intelligence [7].

Nowadays, many algorithms have been developed by drawing inspiration from swarm intelligence systems. Roughly speaking, there are probably more than 100 SI algorithms, due to the popularity of SI research. However, some researchers have recently warned that some algorithms might have roots in existing algorithms [26]. Among the most well-established SI algorithms are: Particle Swarm Optimization (PSO), Ant Colony Optimization (ACO), Artificial Bee Colony Optimization(ABC), Firefly Algorithm (FA), Cuckoo Search (CS) and Bees Algorithms (BA).

Even though there is a bunch of SI algorithms, all of them follow the SI framework presented in Fig. 1. All SI algorithms are population-based. Therefore, the first step in the algorithm is a random generation of the initial population and evaluation of this population. Later, in the main loop, the individuals in the search space are moved towards the best individuals, while the best-evaluated individuals are selected for the next generation.

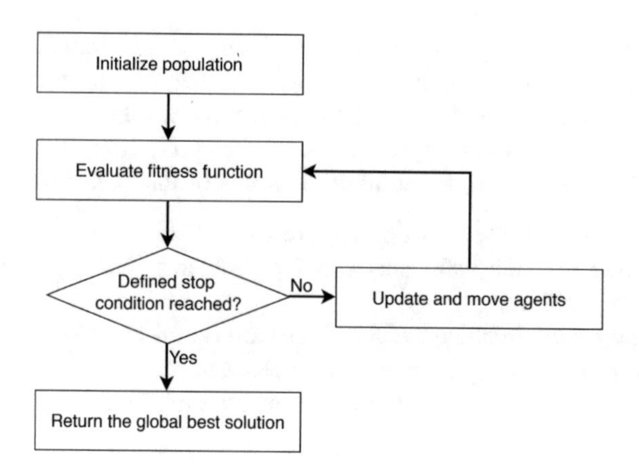

Fig. 1. Swarm intelligence framework [7].

3 Agile Methods

Nowadays, we can hardly find non-agile software companies, i.e., companies that do not utilize agile practices in their product, the project development, or both. The main reason for that is because they want to accelerate product delivery, enhance the ability to manage changing priorities, and increase productivity [29]. Thus, we can see that the biggest problem of the traditional approaches is their incapability to respond to the constant flow of changes quickly even though this is the most important thing for the customers. Consequently, many agile methodologies were proposed and introduced in companies all around the world. The most frequently used are still Scrum and some custom Scrum hybrids (62%), followed by Scrumban (8%), and Kanban (5%) [29].

The question that arises here is why do we need to include swarm intelligence techniques into agile software development? The answer is quite simple. Firstly, agile software development is more than just code writing and testing. Many optimization problems occur already at the beginning of a project, i.e., planning. What are the functionalities that must be developed in the first iteration? How will we evaluate the effort of the tasks? How long will it take us to develop a specific task? These are just a few questions that project managers are dealing with daily. However, the advantage of those questions is that we can describe and present them as optimization problems. After the problem is described mathematically, we can tackle it with many SI algorithms to find the optimal solution for the given problem. Although optimal solutions are mostly hard to find because of the multiple conflicting objectives such as lack of data and benchmarks, the research on this field is awakening. In Sect. 4, we present some solutions where researchers introduced SI algorithms to specific agile software design problem. Search methodology is defined in Subsect. 4.1 and the obtained results in Subsect. 4.2.

4 Methods and Applications

4.1 Search Methodology

The methodology for searching the relevant literature for this paper is the following. Firstly, we defined the search term. We combined the most used SI algorithms (PSO, ACO, ABC, CS, BA, and CaSO) with the software development problem (NRP, R, SD, SCE, and SEE). For example, while searching for papers that applied the PSO algorithm to the next release problem, we used the following search term: ((“PSO”OR “particle swarm optimization” OR “particle swarm optimisation”) AND (“Next Release Problem”)).The search was limited to the four major databases, i.e., Springer Link, IEEE Explore, Google Scholar, and Science Direct. Next step combined pre-screening of the results, where we eliminated redundant and inappropriate results. This approach resulted in 21 selected papers that are presented in detail in Subsect. 4.2.

4.2 Results

Agile software development problems tackled with the SI algorithms can be divided into the following five groups:

- **Next Release Problem (NRP)** is present in software development companies all the time. In this phase, features should be selected that must be developed in the next release. The selection process is very hard, since multiple constraints must be taken into account, such as cost, time, dependent requirements, client satisfaction, and reliability. The goal is to find the optimal solution for the given restrictions;
- **Risk (R)** To satisfy requirements of quality software, risk factors must be well defined and prioritized to avoid any overpayment of costs or money. Thus, the authors try to apply techniques for risk factor prioritization;
- **Software Design (SD)**, where software designers try to find good designs of software in the early stages of the software development process. The authors tackle this problem with different interactive and non-interactive approaches;
- **Software Cost Estimation (SCE)** is a process in which the required time and cost are predicted. When dealing with this kind of problem, the authors, in most cases, try to tune the parameters of the Constructive Cost Model (COCOMO);
- **Software Effort Estimation (SEE)**, as the name implies, is a process in which the amount of effort required to develop a product increment is estimated. Such estimations are, in literature, done in many cases with Case-Based Reasoning (CBR) and fuzzy logic for simulation of the uncertainty factors.

Based on our literature research, only six different SI algorithms were applied to solve the mentioned problems, i.e., PSO, ACO, ABC, CS, BA, and Cat Swarm Optimization (CaSO). An overview of the findings, in which eleven journal articles, seven book chapters, and three papers from conference proceedings have been included, is presented in Tables 2, 3 and 4, while Fig. 2 presents the distribution of SI algorithms across different software development tasks, and Fig. 3 presents the dynamics of the occurrence of specific algorithms in the literature over time[1]. As we can see from Tables 2, 3 and 4, each paper is described with the base SI algorithm, the algorithms and/or methods used to compare the obtained results of the proposed SI algorithm with, and some remarks on the proposed SI algorithm for a given software development problem.

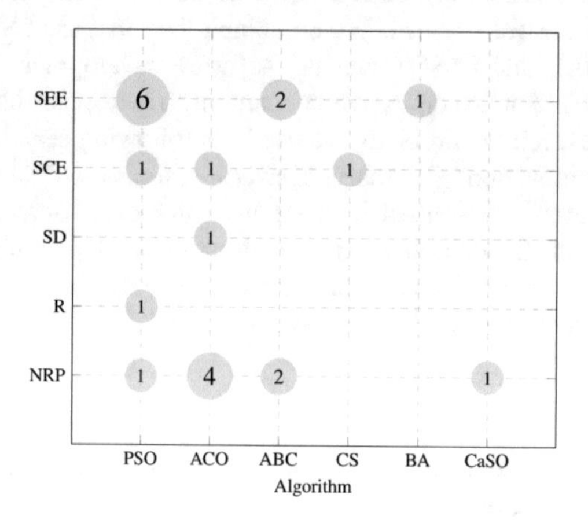

Fig. 2. Bubble plot for the number of papers regarding two variables: Algorithm (x-axis) and software development problem (y-axis).

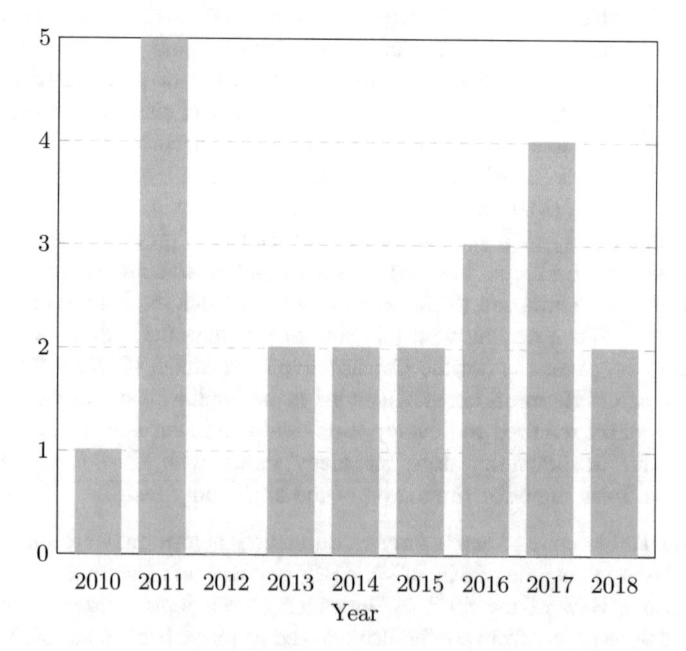

Fig. 3. Analyzed papers that appeared in a particular year.

[1] Note that only the last nine years were considered in this study.

The papers addressing the NRP are listed in Table 2. The proposed SI algorithms, aimed at optimizing the selection of features to be developed in the next release, are focused primarily on finding the optimal solution while fulfilling all given constraints (time, interaction, cost, and budget thresholds, effort boundaries, constraints regarding requirements, Scrum task allocation). Within this category, ACO and ABC are used predominantly, while being primarily compared to genetic algorithms, simulated annealing, hill climbing, GRASP, and NSGA-II, as well as to manual optimization.

Table 1 lists the papers which address the SCE problems. When compared to NRP, the number of research papers within this category is much smaller. ACO, PSO, and CS algorithms have been used here to best estimate the required time and cost of software development projects. Interestingly, all the proposed SI algorithms within this category have been, besides to the COCOMO model and genetic programming, compared to other SI algorithms, which suggests that there is a lack of existing cost estimation methods available.

The papers listed in Table 3 address the SEE problems in software development. The PSO algorithm is used predominantly for problems within this category, followed by the ABC and BA algorithms. Similar to the SCE problems, also here, the fuzzy logic is applied in some cases to handle the uncertainty in effort estimation. Besides providing the absolute effort estimation, the SI algorithms are also used to reduce the difference between actual and predicted effort when using some other prediction methods or techniques. Within this category, the proposed SI algorithms are, in general, compared with the highest number of other existing effort estimation approaches, including the COCOMO model, analogy-based estimation, genetic algorithms, artificial neural networks, case-based reasoning, multiple regression, regression towards the mean, stepwise regression, as well as to a number of other SI algorithms, mostly ABC (and PSO, if the proposed algorithm is not based on it).

Finally, Table 4 lists the papers which address some other software development problems (SD, R). The research on these topics is scarce, as we have been able to find only two such papers. Like other software development problems, however, also here the two most commonly used SI algorithms have been used – PSO and ACO. They have both been compared to methods from the decision analysis area, which combine objective and subjective measures to find a solution that best utilizes the given goals.

To summarize the most important findings, we see that not many SI algorithms were applied to solve software development problems. Out of those algorithms which were, PSO still predominates (39%), followed closely by ACO (33%) (See Fig. 2). However, regardless of the used algorithm, it seems that, for uncertainties in software development problems (especially in SEE and SCE) authors simulate mostly with the use of fuzzy logic.

Table 1. Overview of the SI algorithms used for the SCE problems.

Ref	Base algorithm	Compared to	Remarks
[28]	ACO	PSO, GP, RMSE	Proposed algorithm combined with TSP for SCE. Results were evaluated with three datasets in terms of Root Mean Square Error (RMSE)
[16]	CS	COCOMO, KNN, CUCKOO-KNN	Proposed fuzzy inference system combined with Cuckoo optimization algorithm. Results were evaluated with tera-PROMISE datasets in terms of improved accuracy and cost estimation
[24]	PSO	COCOMO	Proposed model for COCOMO parameters' tuning using multi-objective PSO with objectives (mean absolute relative error and prediction). Results were evaluated with the COCOMO dataset

* KNN–k-Nearest Neighbors; TSP–Traveling Salesman Problem

Table 2. Overview of the SI algorithms used for the NRP problems.

Ref	Base algorithm	Compared to	Remarks
[23]	ABC	MOTLBO	Proposed algorithm with objectives (minimum cost, maximum client satisfaction, minimum time consumption and maximum reliability) and constraints (time threshold, interaction and cost threshold). Results were evaluated in terms of hyper-volume indicator, spread indicator and number of non-dominated solutions
[8]	ABC	ACO, NSGA_II, GRASP	Proposed algorithm with objectives (cost and satisfaction) and constraints (types of interaction). Results were evaluated with two real life datasets with the in terms of hyper-volume indicator, spread indicator and number of non-dominated solutions
[11]	ACO	NIACS	Proposed single-objective formulation for the interactive version of the NRP with the budget constraints and incorporated user preferences. Results were evaluated with three real life datasets in terms of budget constraints and user preferences
[10]	ACO	GRASP, NSGA-II	Proposed multi-objective ACS for requirements' selection. Results were evaluated with two real life datasets in terms of hyper-volume indicator, spread and spacing indicators, and number of non-dominated solutions. Highlighted were problems with crossover and mutation operations in NSGA-II in NRP
[9]	ACO	GA, SA	Proposed method for the NRP problem with dependent requirements. Results were evaluated with the 72 synthetic datasets in terms of quality and execution time

(*continued*)

Table 2. (*continued*)

Ref	Base algorithm	Compared to	Remarks
[14]	ACO	GA, SA, FHC, ACO	Proposed hybrid ACO method with incorporated local search to improve solution quality. Results were evaluated with five synthetic datasets in terms of solution quality and execution time
[15]	CaSO	Synthetic dataset	Proposed multi-objective collaborative scheduling model for NRP. Results were evaluated with dataset in terms of product development time and costs
[5]	PSO	Manual allocation	Proposed method for Scrum task allocation problem with constrains. Results were evaluated with the real life internal project

* FHC–First Found Hill Climbing; SA–Simulated Annealing; CaSO–Cat Swarm Optimization

Table 3. Overview of the SI algorithms used for the SEE problems.

Ref	Base algorithm	Compared to	Remarks
[19]	ABC	ABC, COCOMO II	Proposed algorithm with the teaching-learning mechanism applied to the ABC algorithm. Results were evaluated with a NASA software project dataset in terms of SEE. Highlighted faster convergence than ABC
[18]	ABC/PSO	ABC, PSO, regression	Proposed method based on velocity and story point factors, where parameters are optimized using PSO. Results were evaluated on dataset in terms of accuracy of predicted results
[3]	BA	CBR, GA, RTM, other papers	Proposed method to adjust the retrieved project efforts and find the optimal number of analogies by using BA. Results were evaluated on six datasets in terms of different performance measures. Search capability of the BA applied to overcome the local tuning problem of effort adjustment
[17]	PSO	CART, SWR, MLR, ANN, ABE	Proposed Analogy-Based Estimation (ABE) algorithm combined with PSO. Results were evaluated with the three real life datasets in terms of accuracy of the SEE
[30]	PSO	CBR methods	Proposed optimized weights of CBR methods with PSO. Results were evaluated with the two datasets in terms of three quality metrics, i.e., mean magnitude of relative error, median magnitude of relative error and Pred(0.25)
[2]	PSO	UCP, TPA	PSO applied to UCP and TPA to reduce the difference between actual and predicted effort. Results were evaluated with the cases from two papers and compared regarding UCP or TPA

(*continued*)

Table 3. (*continued*)

Ref	Base algorithm	Compared to	Remarks
[22]	PSO	SEE models	Proposed algorithm that applies fuzzy logic to obtain uncertainty in EE and PSO for parameters' tuning. Results were evaluated with ten NASA software projects on the basis of the VAF, MARE, and VARE
[20]	PSO	COCOMO	Proposed algorithm for COCOMO parameters' tuning using multi-objective PSO. Results were evaluated with Magnitude of relative error and prediction level

* ANN–ArtificialNeuralNetwork; CBR–Case-BasedReasoning; MLR–Multipleregression; RTM–Regression Towards the Mean; SWR–Stepwise Regression; UCP–Use Case Points; TPA–TestPointAnalysis; VAF–VarianceaccountedFor; MARE–MeanAbsoluteRelativeError; VARE–Variance Absolute Relative Error

Table 4. Overview of the SI algorithms used for some other (SD, R) software development problems.

Ref	Base algorithm	Compared to	Remarks
[25]	ACO	IEA	SD problem. Proposed multi-objective ACO search steered jointly by an adaptive model that combines subjective and objective measures. Results were evaluated by the experts
[1]	PSO	AHP	R problem. Proposed method for optimization of the project duration by using a current optimal risk factor with PSO. Results were evaluated with ten agile software development projects

* AHP–Analytic Hierarchy Process

5 Future Paths

Although the research in the software development area was more focused on the GA algorithms in the past, some problems arise regarding the GA fundamental phases. According to the authors in [10], the most obvious problems using GA are crossover and mutation operations, especially in the NRP when considering restrictions. Therefore, researchers try to find some other ways to solve software development problems. Recently, research using SI algorithms for software development problems has increased. Nevertheless, there are still some challenges that must be addressed.

One of the most significant problems is test data. The optimal way is to use data that were obtained from a real-life scenario, but we often do not have such access. Therefore, online repositories should be prepared with multiple projects. For each project, parameters should be defined, such as the number of iterations, requirements, dependencies between requirements, planned vs. actual software development process (can also be in the form of a Burndown chart). Furthermore, projects could be classified by difficulty, e.g., projects with multiple dependencies are harder to solve than those with fewer. If real-life data could not be obtained, a dataset generator for the systematic generation of instances should be provided, as was also highlighted by the authors in [8].

Benchmarks for each project could be defined if we refer to repositories. With such benchmarks, we could facilitate the work of researchers who propose some novel algorithm and want to check given solutions briefly.

As far as the algorithms themselves are concerned, other SI algorithms should be applied to the already mentioned problems. After that, a study can be conducted with an emphasis on which of the SI algorithms performed the best for the specific software development problem. Moreover, it would also be sensible to check various hybridization of the SI algorithms, as was pointed out by Kuhat and Thi MyHanh [19]. With hybridization, we can take advantage of the powerful features of more than one individual algorithm and find potentially better solutions.

A subfield worth exploring is also class distribution skews and underrepresented data in software defect prediction [6].

6 Conclusion

Agile software development is present in many software development companies worldwide. Software companies make use of various agile methods to improve the productivity of teams and write better code with fewer bugs. Recently, some researchers even improved these methods by combining them with artificial intelligence methods. In this paper, we made a short overview of swarm intelligence algorithms that are applied within the agile software development field. The systematic outline shows that swarm intelligence methods are beneficial for solving agile software development tasks. In line with this, we can expect more solutions that are based on swarm intelligence algorithms in the future.

Acknowledgements. The authors acknowledge the financial support from the Slovenian Research Agency (Research Core Funding No. P2-0057).

References

1. Agrawal, R., Singh, D., Sharma. A.: Prioritizing and optimizing risk factors in agile software development. In: 2016 Ninth International Conference on Contemporary Computing (IC3), pp. 1–7 (2016)
2. Aloka, S., Singh, P., Rakshit, G., Srivastava, P.R.: Test Effort Estimation-Particle Swarm Optimization Based Approach, pp. 463–474. Springer, Heidelberg (2011)
3. Azzeh, M.: Adjusted Case-Based Software Effort Estimation Using Bees Optimization Algorithm, pp. 315–324. Springer, Heidelberg (2011)
4. Beniand, G., Wang, J.: Swarm Intelligence in Cellular Robotic Systems, pp. 703–712. Springer, Heidelberg (1993)
5. Brezočnik, L., Fister, I., Podgorelec, V.: Scrum task allocation based on particle swarm optimization. In: Korošec, P., Melab, N., Talbi, E.-G. (eds.) Bioinspired Optimization Methods and Their Applications, pp. 38–49. Springer International Publishing, Cham (2018)

6. Brezočnik, L., Podgorelec, V.: Applying weighted particle swarm optimization to imbalanced data in software defect prediction. In: Karabegović, I. (ed.) New Technologies, Development and Application, pp. 289–296. Springer International Publishing, Cham (2019)
7. Brezočnik, L., Fister, I., Podgorelec, V.: Swarm intelligence algorithms for feature selection: a review. Appl. Sci. **8**(9) (2018)
8. Chaves-González, J.M., Pérez-Toledano, M.A., Navasa, A.: Software requirement optimization using a multiobjective swarm intelligence evolutionary algorithm. Knowl.-Based Syst. **83**, 105–115 (2015)
9. de Souza, J.T., Maia, C.L.B., do Nascimento Ferreira, T., de do Carmo, R.A.F., de Brasil, M. M.A.: An AntColony Optimization Approach to the Software Release Planning with Dependent Requirements, pp. 142–157. Springer, Heidelberg (2011)
10. delSagrado, J., del Águila, I.M., Orellana, F.J.: Multi-objective ant colony optimization for requirements selection. Empirical Softw. Eng. **20**(3), 577–610 (2015)
11. do Nascimento Ferreira, T., Arajo, A.A., Neto, A.D.B., de Souza, J.T.: J.T.: Incorporating user preferences in ant colony optimization for the next release problem. Appl. Soft Comput. **49**, 1283–1296 (2016)
12. Harman, M.: The current state and future of search based software engineering. In: 2007 Future of Software Engineering, pp. 342–357. IEEE Computer Society (2007)
13. Jia, Y., Harman, M.: An analysis and survey of the development of mutation testing. IEEE Trans. Softw. Eng. **37**(5), 649–678 (2011)
14. Jiang, H., Zhang, J., Xuan, J., Ren, Z., Hu, Y.: A hybrid ACO algorithm for the next release problem. In: The 2nd International Conference on Software Engineering and Data Mining, pp. 166–171. IEEE (2010)
15. Jiang, J.-J., Yang, X., Yin, M.: Cooperative control model of geographically distributed multi-team agile development based on MO-CSO. In: Proceedings of the 2nd International Conference on E-Education, E-Business and E-Technology, ICEBT 2018, pp. 121–125, New York, NY, USA. ACM (2018)
16. Kaushik, A., Verma, S., Singh, H.J., Chhabra, G.: Software cost optimization integrating fuzzy system and COA-Cuckoo optimization algorithm. Int. J. Syst. Assur. Eng. Manag. **8**(2), 1461–1471 (2017)
17. KhatibiBardsiri, V., Jawawi, D.N.A., Hashim, S.Z.M., Khatibi, E.: A PSO-based modelto increase the accuracy of software development effort estimation. Softw. Qual. J. **21**(3), 501–526 (2013)
18. Khuat, T., Le. M.: A Novel Hybrid ABC-PSO algorithm for effort estimation of software projects using agile methodologies. J. Intell. Syst. 1–18 (2017)
19. Khuat, T., My Hanh, L.: Applying teaching-learning to artificial bee colony for parameter optimization of software effort estimation model. J. Eng Sci. Technol **12**(5), 1178–1190 (2017)
20. Manga, I., Blamah, N.: A particle swarm optimization-based framework for agile software effort estimation. Int. J. Eng. Sci. (IJES) **3**, 30–36 (2014)
21. Mernik, M., Hrnčič, D., Bryant, B.R., Sprague, A.P., Gray, J., Liu, Q., Javed, F.: Grammar inference algorithms and applications in software engineering. In: 2009 XXII International Symposium on Information, Communication and Automation Technologies. ICAT 2009, pp. 1–7. IEEE (2009)
22. Prasad Reddy, P.V.G.D., Hari, C.V.M.K.: Fuzzy Based PSO for Software Effort Estimation, pp. 227–232. Springer, Heidelberg (2011)
23. Ranjith, N., Marimuthu, A.: A multi objective teacher-learning-artificial bee colony (MOTLABC) optimization for software requirements selection. Indian J. Sci.Technol. **6** (2016)

24. Rao, G.S., Krishna, C.V.P., Rao, K.R.: Multi Objective Particle Swarm Optimization for Software Cost Estimation, pp. 125–132. Springer International Publishing (2014)
25. Simons, C.L., Smith, J., White, P.: Interactive ant colony optimization (iACO) for early lifecycle software design. Swarm Intell. **8**(2), 139–157 (2014)
26. Sörensen, K.: Metaheuristics–the metaphor exposed. Int. Trans. Oper. Res. **22**(1), 3–18 (2013). https://doi.org/10.1111/itor.12001
27. Srivastava, P.R., Varshney, A., Nama, P., Yang, X.-S.: Software test effort estimation: a model based on cuckoo search. Int. J. Bio-Inspired Comput. **4**(5), 278–285 (2012)
28. Venkataiah, V., Mohanty, R., Pahariya, J.S., Nagaratna, M.: Application of Ant Colony Optimization Techniques to Predict Software Cost Estimation, pp. 315–325. Springer, Singapore (2017)
29. VersionOne. VersionOne 12th Annual State of Agile Report (2018)
30. Wu, D., Li, J., Liang, Y.: Linear combination of multiple case-based reasoning with optimized weight for software effort estimation. J. Supercomput. **64**(3), 898–918 (2013)

Optimization Based on Simulation of Ants Colony

Mihailo Jovanović[1(✉)] and Ermin Husak[2]

[1] Faculty of Management Herceg Novi, Zemunska 143, Meljine, 85348 Herceg Novi, Montenegro
mihajovanovic30@gmail.com
[2] Technical Faculty, University of Bihać, 77000 Bihać, Bosnia and Herzegovina

Abstract. Natural processes optimize life on earth for thousands of years, so people are inspired by many problem-solving techniques in nature. Meta-heuristics inspired by natural processes and systems have become a very active field of research in recent years. One of the most popular methods is Ant Colony Optimization (ACO). In this paper is considered the application of Ant Colony Optimization in the case of the Traveling Salesman Problem (TSP). Different cases, with a different number of ants (population size) with a different number of iteration using software simulation, are considered. It is shown that Roulette Wheel Selection has some impact on the speed of the result. On the other hand, with more ants in each iteration, we get more constructed solutions, which increases the probability of finding a better solution.

Keywords: Ant Colony Optimization · ACO · Software · Traveling Salesman Problem – TSP

1 Introduction

Heuristics or heuristic solutions are called algorithms that ensure that in polynomial time we find enough good solutions, without the guarantee that it will be optimal. The heuristic approach to the problem is the empirical search or optimization method that most often solves the problem, but without evidence for it that scientists would accept [1, 2]. These methods are used when classical methods are overdue or when problems can not be solved, and sometimes they are the only way to solve them, as they often start from a more realistic model of a problem than an exact science. Heuristic algorithms are algorithms that do not give the best solution, but at best provide a good enough solution within a reasonable time period [3]. We use heuristic methods in solving poorly structured problems for which there is no precise algorithm (job shop problem, different scheduling problems), in solving problems for which precise algorithms can be set, but would not be efficient for large dimensions of problems (traveling salesman problem, graph coloring problem, knapsack problem) and etc.

Natural processes have been optimizing life on earth for thousands of years, so people have found inspiration for many of the problem-solving techniques in nature. Some of the most famous ones were found in animals that live in flocks or swarms, such as ants, bees, birds. The best known examples are genetic algorithms, ant colony

I. Karabegović (Ed.): NT 2019, LNNS 76, pp. 310–316, 2020.
https://doi.org/10.1007/978-3-030-18072-0_36

optimization methods, abbreviated ACO. ACO belongs to the metaheuristics of collective intelligence or flock (Swarm Intelligence), which are becoming more and more relevant because they give competing results. Beside from ACO methods, popular algorithms are inspired by particle swarm behavior, artificial bee colony, glowworm swarm optimization, the artificial immune system, and others [4].

2 Ant Colony Optimization (ACO)

Swarm intelligence is the ability of a group to find a greater number of better solutions to a particular problem than would be individually succeeded by individuals in that group.

Swarm intelligence is a system consisting of a population of agents interacting with one another at the local level and with the environment. Agents follow simple instructions and rules, although there is no centralized supervision of agents at the local level. "Random" interactions between them lead to the emergence of "intelligent" behavior globally, which is unknown to individual agents.

Optimization by simulating an ant colony appears for the first time in the work of Marco Dorigo [5], during the nineties of the twentieth century. The inspiration for this kind of optimization was found in an experiment in 1989, which was carried out by Goss with Argentinian ants [6]. Setting two roads between the ants and food sources, it was noticed that the ants quickly "recognized" a shorter path to food and eventually began to use it alone.

Ant Colony Optimization (ACO) is inspired by the behavior of an ant in search of food. As they move towards the source of food, these ants leave behind a trace of pheromones to mark the path to food that the other ants will follow. If sense a stronger trace of pheromones, they prefer to go after him, but for the weaker trail. The strength of the pheromone trace depends on two factors. One is the evaporation of the pheromone. So, the more time has elapsed since placing pheromones, this is a weaker trail. This fits well with the logic of finding food, since that food that is more remote from the anthill is thus less likely to be traced in relation to some nearby location. Another factor of the difference between the traces of the pheromone is the number of ants that have set it. The more ants have previously decided on a particular path, the greater the likelihood that the road is good and therefore the other ants follow it.

If there is an obstacle on the road that divides the existing path into two uneven routes (Fig. 1), the ants initially do not know which side is shorter. Therefore, it is an equal probability that ant ants will follow one of two paths.

As one side is shorter, on that side the trace of the pheromone will soon become stronger, as it does not get on the shorter side to evaporate as much as it evaporates on the longer side. As they feel that the trace of the pheromone on one side is stronger, more ants decide for it, and soon all the ants decide on a shorter path.

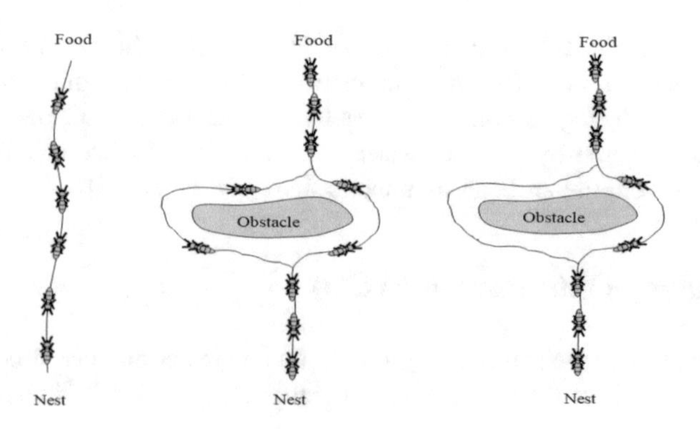

Fig. 1. Optimal path

3 Traveling Salesman Problem (TSP)

One of the most commonly used examples for explaining the application of the ACO algorithm is the so-called Traveling Salesman Problem (TSP). The data is given by a set of location-cities and their mutual distance, and the aim is to find the shortest route by which each city will be visited once and only once (we need to find Hamilton's cycle with the minimum weight of fully connected graph). With the realization of this algorithm, each node represents a city, each branch represents the distance between cities, while the pheromone represents a variable that "skillful ants" can read and change. Since we apply this algorithm to the problem of a traveling salesman (TSP), we will take a graph whose nodes represent the cities that we need to visit. This graph is called a construction graph or solution. With TSP it is possible to switch from each city to each other (fully connected graph), the number of nodes is equal to the number of cities, while the city's length between the two nodes is proportional to the distance between the two cities.

The value of the variable that represents the pheromone will change during the operation of the algorithm, and given that we have taken the problem of a traveling salesman, their values will be inverse to the lengths of the branch (Fig. 2).

ACO is an iterative algorithm wherein each iteration a number of artificial ants appear. The ants build the solution by moving from a randomly chosen city (the node of the graph) and during each subsequent iteration they move along the branch of the graph. The ant goes to the next node that has not been previously visited, and in the meantime remembers the path to which it was moving. An ant will form a solution when has finished visiting each node of a given graph. The higher the value of the variable that represents the pheromone and heuristic information, the greater the probability that the ant will go this way. When, at the end of the iteration, all ants form their own solution, the values of the pheromones related to branch graphs are updated, in order to indicate the ants in the next iteration for a better solution than the previous one. The values of the pheromones of each branch are reduced for a certain percentage during each iteration, which would result in evaporation. Then, each branch receives an additional amount of pheromone, proportional to the quality of the solution to which it belongs. The algorithm repeats itself until the stopping condition is met [7–11].

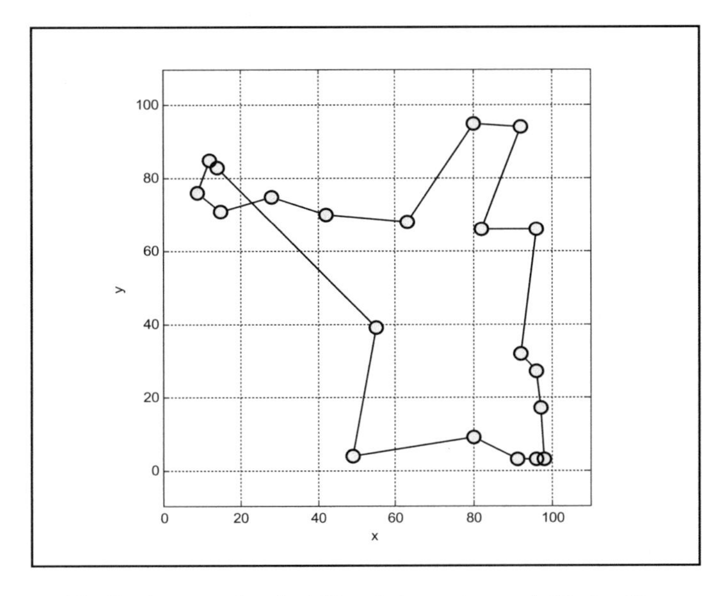

Fig. 2. An example of a TSP solution using an ACO algorithm

4 Results

The ACO algorithm is written in MATLAB software. Using the above software, analyses were carried out with a number of iterations (MaxIt) and a number of ants (nAnt). During the iteration, for the given number of ants, the appearance of the graph changes, showing always the best solution that ants-agents have achieved up to that point (until this iteration). After completing the given number of iterations, the application gives a construction graph as well as a diagram of the best solutions through iterations (Figs. 3, 4, 5, 6). Best Cost means the shortest path length.

Figure 3 shows that Best Cost reaches 385.8789 already in 59 iterations. The number of iterations is set to 80, while the number of ants is 10.

Figure 4 shows the travel path and Best Cost diagram for the same number of ants as in the first case (10), and for the given number of iterations 280. The algorithm gave a shorter solution than the previous one, which at the same time represents the best solution for this problem with Best Cost = 362.038.

In the cases shown in Figs. 5 and 6, an equal number of ants is set (nAnt = 100). Each hundred of them makes their construction graph in each iteration. In the case of Fig. 5, we quickly arrive at the optimal solution Best Cost = 362.038, already in the twelfth iteration. This suggests that the random selection of movements (Roulette Wheel Selection) still has a certain influence on time needed to get optimal result.

On the other hand, with more ants in each iteration we get more solutions built, which increases the probability of finding a better solution.

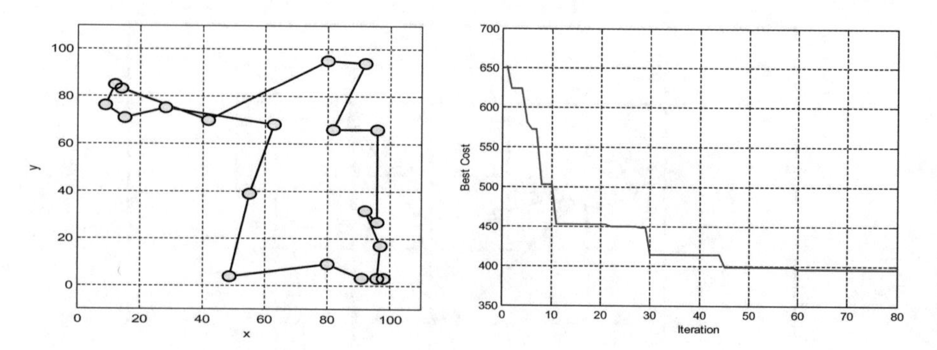

Fig. 3. Salesman movement path and Best Cost diagram for 80 iterations and 10 ants

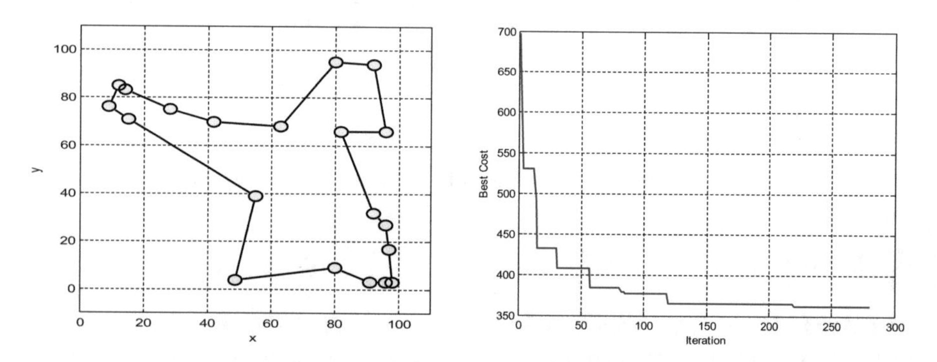

Fig. 4. Salesman movement path and Best Cost diagram for 280 iterations and 10 ants

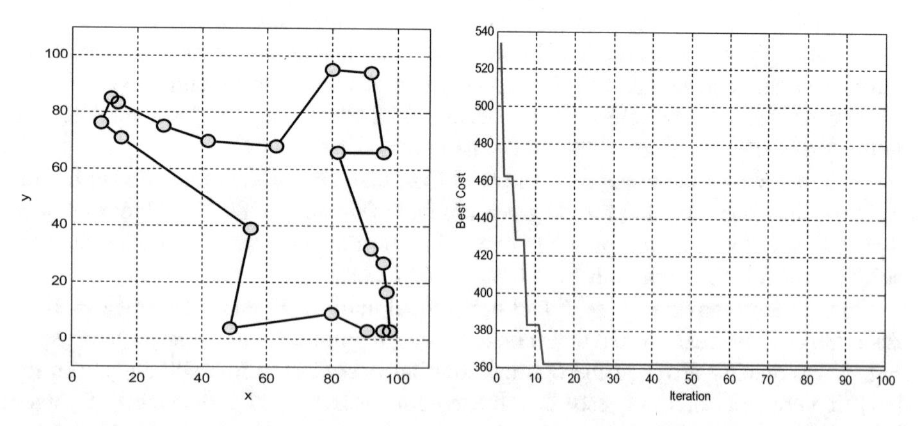

Fig. 5. Salesman movement path and Best Cost diagram for 100 iterations and 100 ants

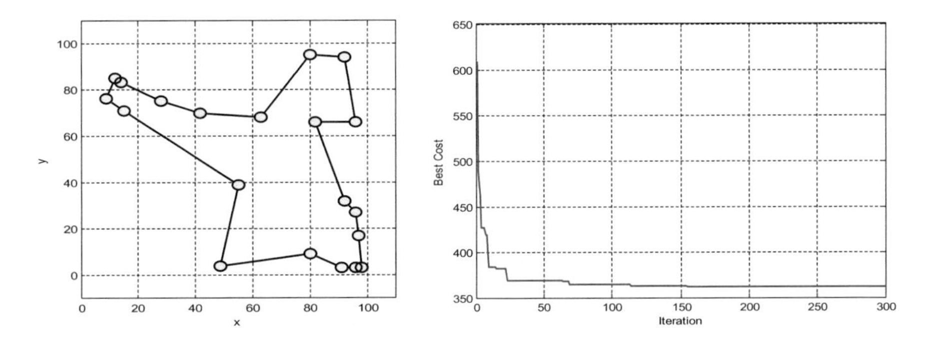

Fig. 6. Salesman movement path and Best Cost diagram for 300 iterations and 100 ants

5 Conclusion

We have explained the workflow of the ACO algorithm through one of the most commonly used examples for the application of this algorithm, like Travelling Salesman Problem. We have presented the application of the ACO algorithm implemented in MATLAB for solving and analyzing the Traveling Salesman Problem (TSP), as well as the results of the simulation. Based on the results of the simulation, it can be concluded that in principle with the increase in the number of iterations a better solution is likely to be obtained. Likewise, with multiple ants in each iteration, more built solutions are obtained, which increases the probability of finding a better solution. On the other hand, the simulation results showed that the random selection of motion (Roulette Wheel Selection) has a certain impact on the speed of finding optimal results.

The results of the ACO application on TSP show that in properly distributed cities, as with circularly located cities, the algorithm converts very quickly and an optimal solution arrives. Repeatedly ran algorithms provide identical solutions. The reason for this behavior of the algorithm is that in circularly distributed cities there are no locally optimal solutions, and therefore optimization ends very quickly.

References

1. Hotomski, P.: Sistemi veštačke inteligencije. Tehnički fakultet “Mihajlo Pupin”, Zrenjanin (2006)
2. Amaldi, E., Capone, A., Malucelli, F.: Optimization models with power control and algorithm (2003)
3. Schauer, C., Hu, B.: Heuristic optimization techniques. A lecture held on University of Technology Vienna WS (2011)
4. Carić, T.: Optimizacija prometnih procesa. Sveučilište u Zagrebu (2014)
5. Dorigo, M., Maniezzo, V., Colorni, A.: Ant System: Optimization by a colony of cooperating agents. IEEE Trans. Syst. Man Cybern.-Part B **26**(1), 29–41 (1996)
6. Goss, S., Aron, S., Deneubourg, J.-L., Pasteels, J.M.: Self-organized shortcuts in the Argentine ant. Naturwissenschaften **76**, 579–581 (1989)

7. Dorigo, M., Caro, D.G., Gambardella, L.M.: Ant algorithms for discrete optimization. Artif. Life **5**(2), 137–172 (1999)
8. Fidanova, S.: ACO algorithm for MKP using various heuristic information. In: Dimov, I., Lirkov, I., Margenov, S., Zlatev, Z. (eds.) Numerical Methods and Applications. LNCS, vol. 2542, pp. 434–330. Springer-Verlag, Berlin (2003)
9. Dorigo, M., et al.: Ant System: An Autocatalytic Optimizing Process. Politecnico di Milano (1991)
10. Stützle, T, Hoos, H.H.: Improving the Ant System. A detailed report on MAX –MIN Ant System, Technical report, AIDA-96-12, TU Darmstadt
11. Dorigo, M., Gambardella, L.M.: Ant Colonies for the Traveling Salesman Problem. BioSystems **43**, 73–81 (1997)

Performance Analysis of the HDLC Protocol-NRM Mode

Zlatan Jukic[1,2(✉)]

[1] Faculty of Electronics and Computer Engineering, HTL Rankweil, Negrellistraße 50, 6830 Rankweil, Austria
zlatan.jukic@htl-rankweil.at
[2] TU Wien, Vienna Uniersity of Technology, Wien, Austria

Abstract. HDLC (High-level Data Link Control) protocol is the most fundamental error-and flow control procedure used in data communications for a single physical channel. Moreover HDLC is bit-oriented synchronous, data-framing, switched and non-switched data link layer (layer 2) protocol. Apart from the classical HDLC standard of ISO, the generic recovery mechanisms of this protocol constitute the basis for a whole family of data link protocols designed for wired and wireless systems. The following work presents a series of detailed analytic studies on performance modelling and evaluation of an HDLC-controlled link wherein various retransmission options and traffic scenarios are considered. The goal of the analysis is the mean link throughput and mean frame flow time computation with respect to a set of link and protocol parameters, as frame length, window size, duration of time-outs for a given bit-error probability, channel load and distance between communicating stations.

Keywords: Mean link throughput · Mean frame flow time · Bit-error probability · Time-outs

1 Introduction

The HDLC protocol is defined by ISO (TC97/SC6). The current standard ISO 13239, is used for point-to-point and point-to-multipoint logical configurations over full-duplex and half-duplex data links. The fact triggered more analytic and simulative studies on HDLC itself, as well as on the generic retransmission techniques investigated ad hoc. HDLC incorporates the basic (fundamental) error and flow mechanisms (flow control procedure and error detection and correction procedure) used in data communications for a single physical channel. Uses a header with control information and a trailing cyclic redundancy check character (which is usually 16 or 32 bits in length).

The protocol uses the services of a physical layer, and provides reliable communications path between the transmitter and receiver (i.e. with acknowledged data transfer). HDLC is connection-oriented protocol that supports synchronous and asynchronous communication, code-transparent data communication between stations and between host and remote stations. Moreover, the protocol was introduced to

I. Karabegović (Ed.): NT 2019, LNNS 76, pp. 317–331, 2020.
https://doi.org/10.1007/978-3-030-18072-0_37

packet-switched and circuit-switched networking in the recommendations of X.25 and ISDN (Integrated Services Digital Network) architectures [4].

2 Related Work

The chosen method was developed by the researches W. Bux and H.L. Truong (Institute of Switching and Data Technics, University of Stuttgart) together with K. Kuemmerle (IBM Zurich Research Laboratory). It was published in 1979 [2] and extended gradually in [3, 5–8], whereby the most complete treatment is provided in Truong's report [2].

Starting with the paper, a queueing model for HDLC protocol, in 1979 [2] the authors completely described the HDLC behavior with a M/G/1 model in all three operating modes, ABM, NRM and ARM (all the results are comprehensively compared in [6]. Zorzi and Rao found upper and lower bounds on the throughput for go-back-n operating in a Markovian error channel [9].

A new scheme of Benelli [1] assumes aggregation of multiple frames in a single block. Similar techniques using the selective repeat in intra-block were developed by Nakajima and Serizawa, who considered improvements for asymmetrical channels [10], and Hayashida et al., whose amendments should minimize delays in a slotted time system [11]. Cumulative acknowledgments could find valuable applicability for point-to-multipoint satellite communications as investigated in [12].

3 HDLC Operation Modes

The HDLC protocol can operate in three basic modes: NRM (Normal Response Mode) ARM (Asynchronous Response Mode) and ABM (Asynchronous Balanced Mode).

3.1 Normal Response Mode (NRM)

In this mode, the secondary station is not allowed to send any frame until it gets a permission to do so from the primary station. The permission is in form of a command (supervisory frame) with the P-bit set to one. The secondary station send a response when and only when, it is instructed to do so by the primary station. The connection may be point-to-point or multipoint. NRM is an unbalanced configuration.

3.2 Asynchronous Balanced Mode (ABM)

The Asynchronous Balance Mode is designed exclusively for full duplex point-to-point connections. Each station performs the role of both primary and secondary functions. ABM is used in balanced configuration for computer to computer communications and for connections between a computer and a packed switched data network.

3.3 Asynchronous Response Mode (ARM)

In this mode, the secondary station is allowed to send responses at any time. The secondary station initiates a transmission without receiving permission from the primary station. ARM is used in unbalanced configurations for point-to-point connections and makes more efficient use for full duplex connections.

3.4 Three Types of HDLC Stations

Primary Station (P-Station). Has full rights to control the operation of the link. Frames sent by primary station are known as commands. Frames sent by primary station are known as commands, Fig. 1.

Secondary Station (S-Station). Operates under control of the primary one issuing frames called responses. The primary station keeps a separate logical link connection with each secondary station in the system, Fig. 1.

Combined Station (C-Station). It has the features of the primary and the secondary at the same time sending commands as well as responses, Fig. 1.

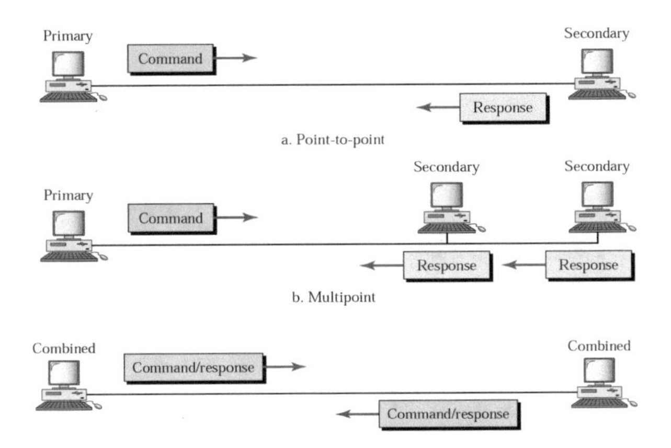

Fig. 1. Point-to-point connection, each station works as primary and secondary (combined station)

3.5 HDLC Frame Structure (Types of HDLC Frames)

HDLC frames are distinguished:

1. **Information frames** or (I-frames) Only this HDLC frame contains user information. I-frames transport user data from the network layer. In addition they can also include flow and error control information piggybacked on data.

2. **Supervisory frames** (S-frames) The S-frames contain no information field are used for flow and error control whenever piggybacking is impossible or inappropriate, such as when a station does not have data to send. S-frames do not have information fields.
   (a) **Receive-Ready (RR)** RR-frames are returned by a station in order to announce that it is capable of receiving a next I-frame or to acknowledge one or more I-frames.
   (b) **Receive-Not-Ready (RNR)** An RNR-frame acknowledges positively an I-frame, as RR does, but also asks the communication partner to suspend any further transmission.
   (c) **Reject (REJ)** An REJ-frame with N(R) = n indicates a negative acknowledgment for the nth I-frame sent. In this case, not only this one but also all other, yet not acknowledged, I-frames with the N(S) greater than n must be retransmitted.
3. **Selective Reject (SREJ)** A return of an SREJ-frame with N(R) = n requires retransmission only of the nth I-frame and no other as with REJ.
4. **Unnumbered frames** U-frames are used for various miscellaneous purposes.

3.6 Format of HDLC Frames (I-Frame, S-Frame)

See Figs. 2 and 3.

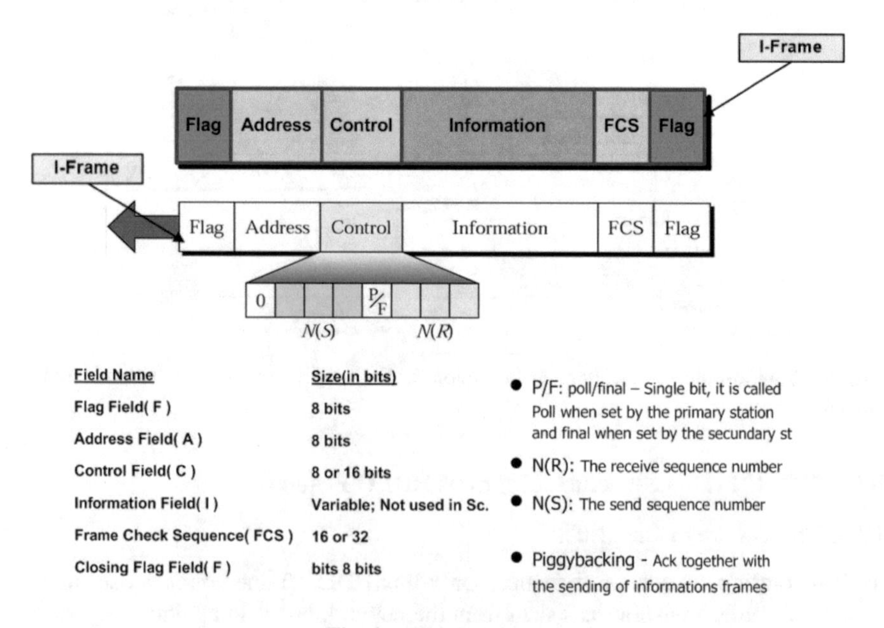

Fig. 2. HDLC, I-frame

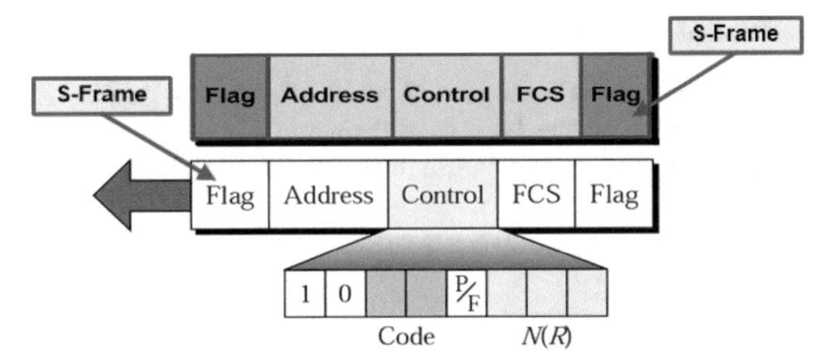

00 – RR (Receave Ready)
01 – RNR (Receave Not Ready)
10 – REJ (Reject) – reject
11 – SREJ – selective reject
N(R) – request sequence number

Fig. 3. HDLC, S-frame

3.7 Sequence Numbers

See Fig. 4.

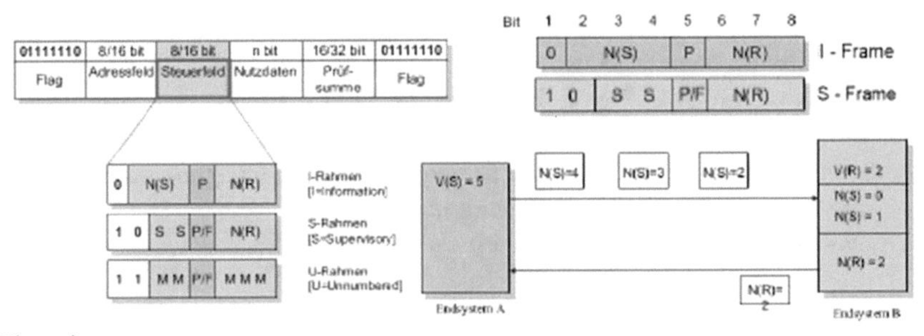

Where is:
N(S) send sequence number
N(R) receive sequence number
P/FP/F Bit, Polling/Final-Bit
S supervisory function bits
M modifier function bits
V(S) send state variable
V(R) receive state variable

Fig. 4. HDLC, sequence numbers

3.8 The Idea of the Position Number

The position number W is a random variable, where W = 1, 2, …, w, …, (M − 1), indicating the position of one I-frame in the cycle of (M − 1) at the beginning of its virtual service time. The mean of virtual service time is given by:

$$E[T_v(w)] = E[T_0(w)] + p_F E[T_1(w)] + \frac{p_F^2}{1 - p_F} E[T_2(w)]$$

It means that an I-frame does not have a position number until its service time starts, although it could be already transmitted one or more times. As a consequence, a conditional virtual service time $T_v(w)$ must be analysed that involves this position number in its realization. It is related to the mean virtual service time by: $E[T_v]$

$$E[T_v] = \sum_{w=1}^{M-1} \propto (w) E[T_v(w)]$$

3.9 Determination of the Probability $\propto(w)$

Determination of the position number W for $t_{ack} > (M - 2)\, t_I$, whereby M = 4 and $T_v^{(w)}$ is the virtual service time of an I-frame with position number w: Fig. 5 shows the position number for error-free transmission. In Fig. 6 I-frame I, 2, 0 with position number 3 is destroyed, transmission with errors. Since the frame-errors are independent with probability p_F, the above defined process can be described by a discrete Markov state diagram presented in Fig. 7. With probability $(1 - p_F)$ the states are perpetually passed through, from 1 up to (M − 1). If an error takes place with probability p_F in an state, the transition is made invariably to state 2. This state diagram can be solved for $\propto (w)$ as shown in [5] with the following result:

Derivation of the equation for $\propto (w)$:

$$\propto (1) = (1 - p_F) \propto (M - 1)$$

$$\propto (2) = (1 - p_F) \propto (1) + p_F$$

$$\propto (3) = (1 - p_F) \propto (2)$$

$$\propto (4) = (1 - p_F) \propto (3) = (1 - p_F)^2 \propto (2)$$

$$\propto (w) = (1 - p_F)^{w-2} \propto (2)$$

$$\propto (M - 1) = (1 - p_F)^{M-3} \propto (2)$$

$$\propto (1) = (1 - p_F)^{M-2} \propto (2)$$

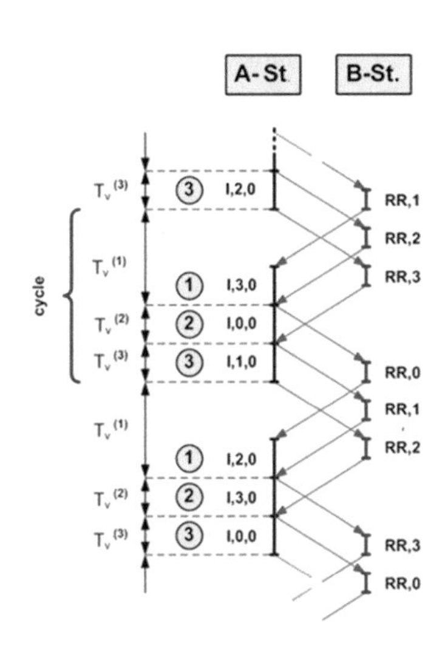

Fig. 5. HDLC-ABM mode, one-way transmission, Case (a) $t_{ack} > (MOD\text{-}2)t$, MOD = 4 error-free transmission

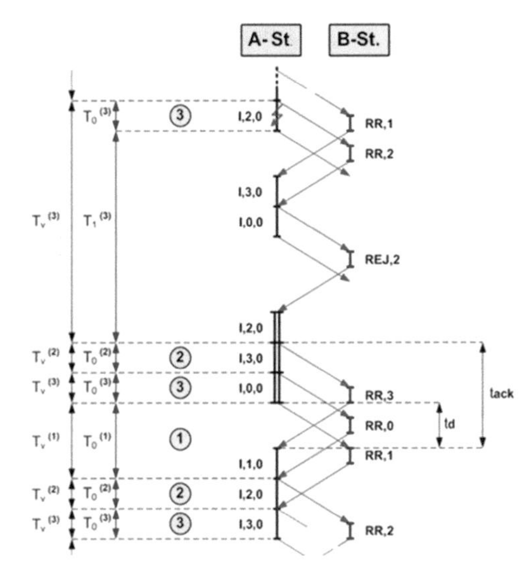

Fig. 6. HDLC-ABM mode, one-way transmission, Case (a) $t_{ack} > (MOD\text{-}2)t$, MOD = 4, transmission with errors

$$\propto(2) = (1 - p_F)^{M-1} \propto(2) + p_F$$

$$\propto(2) = \frac{p_F}{1 - (1 - p_F)^{M-1}}$$

$$\propto(1) = \frac{p_F(1 - p_F)^{M-2}}{1 - (1 - p_F)^{M-1}}$$

$$\propto(w) = \frac{p_F(1 - p_F)^{w-2}}{1 - (1 - p_F)^{M-1}}$$

$$\propto(w) = \begin{cases} \frac{p_F(1-p_F)^{M-2}}{1-(1-p_F)^{M-1}} & \text{for} \quad w = 1 \\ \frac{p_F(1-p_F)^{w-2}}{1-(1-p_F)^{M-1}} & \text{for} \quad w = 2, 3, \ldots, M-1 \end{cases}$$

The probability $\propto(w)$ that a considered I-frame gets the position number W = w, and with substitution z = MOD − 1, the probability $\propto(w)$ is:

$$\propto(w) = \begin{cases} \frac{p_F(1-p_F)^{z-1}}{1-(1-p_F)^{z}} & \text{for} \quad w = 1 \\ \frac{p_F(1-p_F)^{w-2}}{1-(1-p_F)^{z}} & \text{for} \quad w = 2, 3, \ldots, z \end{cases}$$

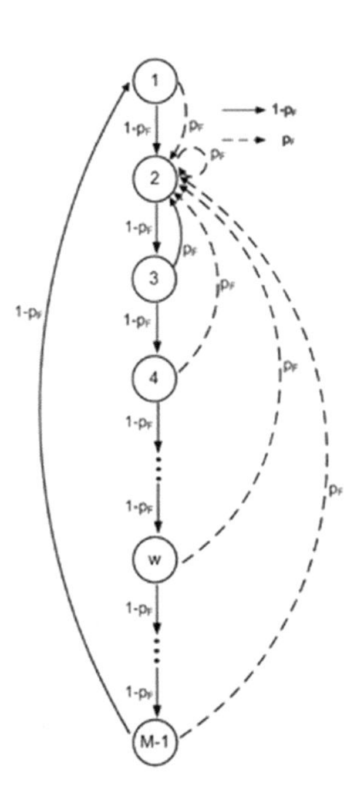

Fig. 7. State transition diagram for determination of position number w

$\propto$ (w) is the probability that a considered I-frame gets the position number w. We define now n as an auxiliary variable for this case distinction:

$$n = \left\lceil \frac{t_{ack}}{t_I} \right\rceil$$

We defined now 4 following traffic scenarios or 4 traffic cases:
Case (a): $n > MOD - 1$
Case (b): $n = MOD - 1$
Case (c): $MOD - 1 > n > (MOD - 1)/2$
Case (d): $n \leq (MOD - 1)/2$

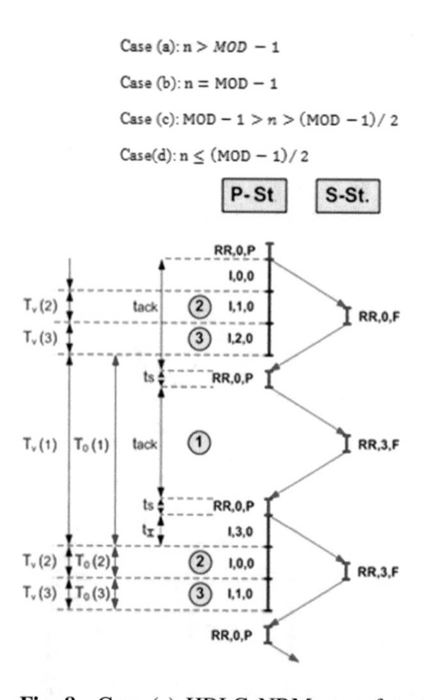

Fig. 8. Case (a) HDLC NRM error-free transmission one-way transmission from P to S, MOD = 4, n = 4

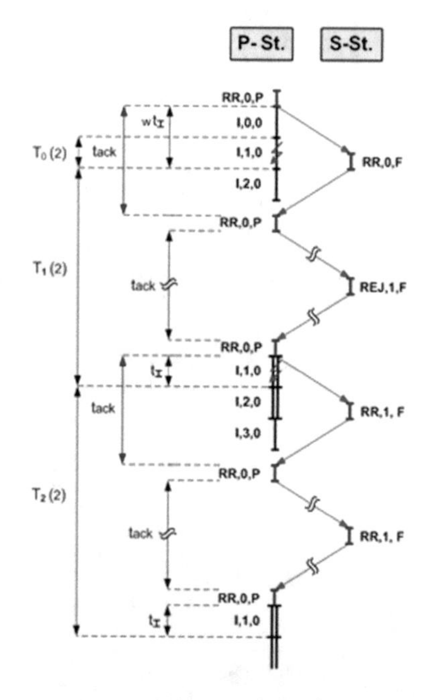

Fig. 9. Case (a) Determination of $T_1(w)$ and $T_2(w)$ HDLC NRM transmission with error, n = 4 one-way transmission from P to S, MOD = 4

Case (a), Figs. 8 and 9:
Determination (Calculation) of *time components* $T_0(w)$, $T_1(w)$, $T_2(w)$:

$$n > MOD - 1$$

$$T_0(1) = 2t_{ack} + 2t_s + t_I - (MOD - 1)t_I$$

$$E[T_0(w)] = \begin{cases} 2t_{ack} + 2t_s - (M-2)t_I & \text{for} \quad w = 1 \\ t_I & \text{for} \quad w = 2, 3, \ldots, M-1 \end{cases}$$

$$E[T_1(w)] = t_{ack} - wt_I + t_s + t_{ack} + t_s + t_I = 2t_{ack} + 2t_s - (w-1)t_I$$
$$\text{for} \quad w = 1, \ldots, M-1$$

$$E[T_2(w)] = t_{ack} - t_I + t_s + t_s + t_{ack} + t_I = 2t_{ack} + 2t_s$$
$$\text{for } w = 1, \ldots, M-1$$

Case (b), Figs. 10 and 11:
Determination (Calculation) of *time components* $T_0(w)$, $T_1(w)$, $T_2(w)$:

$$n = MOD - 1$$

$$E[T_0(w)] = \begin{cases} t_{ack} + 2t_s + t_I & \text{for} \quad w = 1 \\ t_I & \text{for} \quad w = 2, 3, \ldots, M-1 \end{cases}$$

$$E[T_1(w)] = (M-1)t_I - wt_I + t_s + t_{ack} + t_s + t_I = t_{ack} + 2t_s + (M-w)t_I$$
$$\text{for } w = 1, \ldots, M-1$$

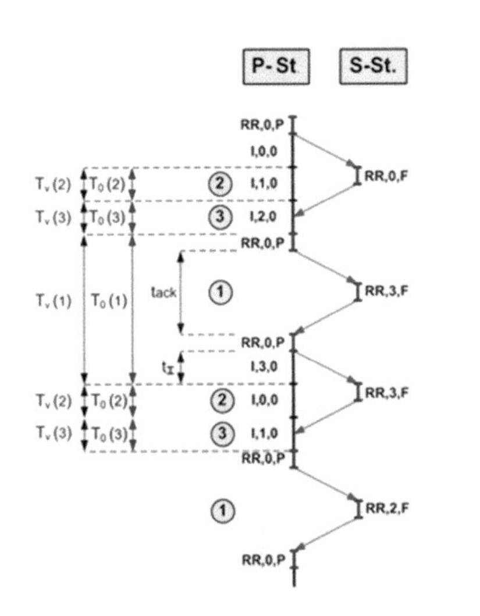

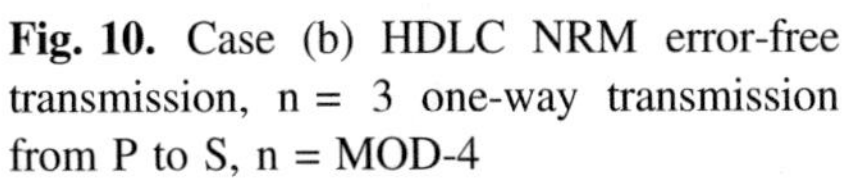

Fig. 10. Case (b) HDLC NRM error-free transmission, n = 3 one-way transmission from P to S, n = MOD-4

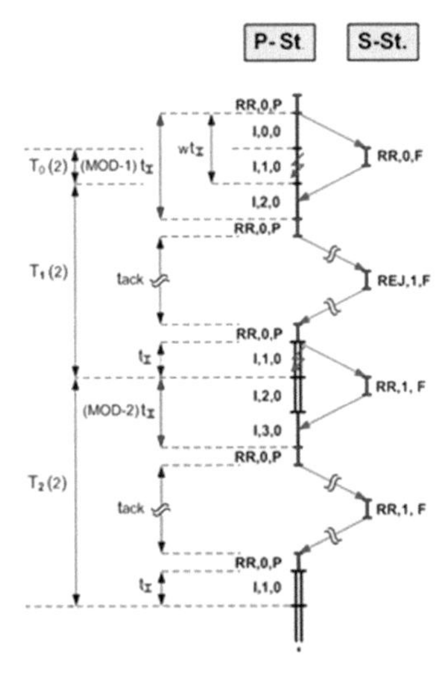

Fig. 11. Case (b) Determination of $T_1(w)$ and $T_2(w)$ HDLC NRM transmission with error, n = 3 one-way transmission P to S, n = MOD-1

$$E[T_2(w)] = (M-2)t_I + t_s + t_{ack} + t_s + t_I = t_{ack} + 2t_s + (M-1)t_I$$
$$\text{for } w = 1, \ldots, M-1$$

Case (c), Figs. 12 and 13: MOD − 1 > n > (MOD − 1)/2
Determination (Calculation) of *time components* $T_0(w)$, $T_1(w)$, $T_2(w)$:

$$E[T_1(1)] = t_{ack} - (M-1-n)t_I + t_s + t_I = t_{ack} + t_s - (M-2-n))t_I$$
$$\text{for } w = 1, \ldots, M-1$$

Virtual transmission time of I-frame with $w = n+1$ is delayed by transmission a RR-frame with a P-Bit: $E[T_0(n+1)] = t_I + t_s$

$$E[T_0(w)] = \begin{cases} t_{ack} + t_s - (M-2-n)t_I & \text{for} \quad w = 1 \\ t_I + t_s & \text{for} \quad w = n+1 \\ t_I & \text{for else} \end{cases}$$

for $w = 1, 2, \ldots, n$ is following equation

$$E[T_1(w)] = t_{ack} - (n-w) + t_s + t_{ack} + t_s + t_I = t_{ack} + 2t_s + (n-w+1))t_I$$

for $w = 1, \ldots, n$

$$E[T_2(w)] = (n-1)t_I + t_s + t_{ack} + t_s + t_I = t_{ack} + 2t_s + n\ t_I$$

for $w = n+1, \ldots, M-1$

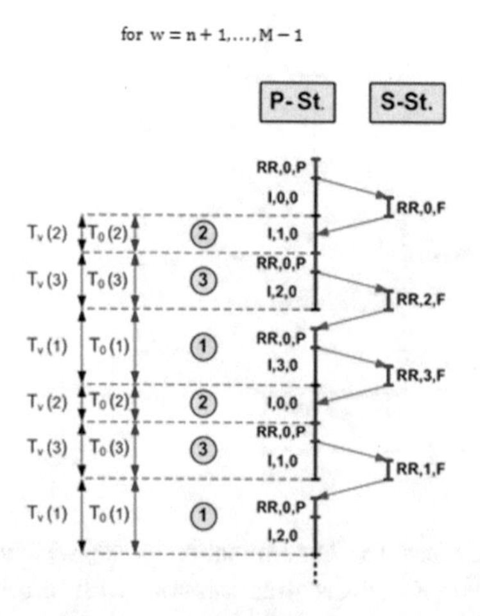

Fig. 12. Case (c) HDLC NRM error-free transmission, one-way transmission from P to S, M = 4, n = 2, MOD-1 > n > (MOD-1)/2

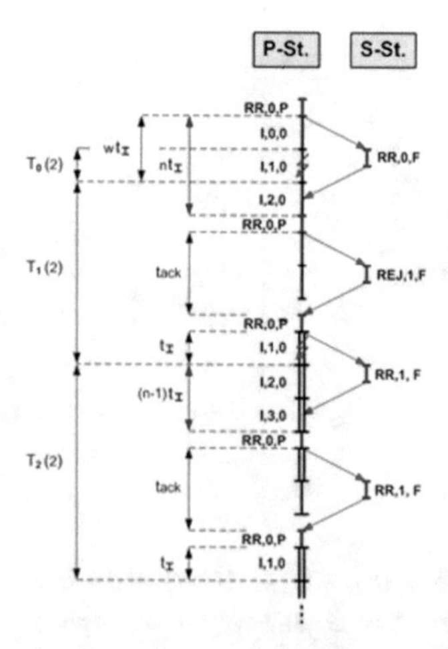

Fig. 13. Case (c) Determination of $T_1(w)$ and $T_2(w)$ HDLC NRM transmission with error, n = 3, w = 2 one-way transmission from P to S, MOD = 6

If $w > n$, than applies Fig. 14

$$E[T_1(1)] = nt_I + t_s + t_{ack} - (wt_I + t_s) + t_s + nt_I + t_s + t_I$$
$$= t_{ack} + 2t_s + (2n - w + 1))t_I$$

for $w = n + 1, \ldots, M - 1$

$$E[T_2(w)] = (n - 1)t_I + t_s + t_{ack} + t_s + t_I = t_{ack} + 2t_s + nt_I$$
$$\text{for} \quad w = n + 1, \ldots, M - 1$$

Case (d), Figs. 15 and 16:

$$n \leq (MOD - 1)/2$$

Determination (Calculation) of *time components* $T_0(w), T_1(w), T_2(w)$:

$$E[T_0(w)] = \begin{cases} t_I + t_s & \text{for} \quad w = 1 \\ t_I & \text{for} \quad w = 2, 3, \ldots, n \end{cases}$$

$$E[T_1(1)] = (n - w)t_I + t_s + nt_I + t_s + t_I = (2n - w + 1))t_I + 2t_s$$
$$\text{for } w = n + 1, \ldots, M$$

$$E[T_2(w)] = 2nt_I + t_s$$

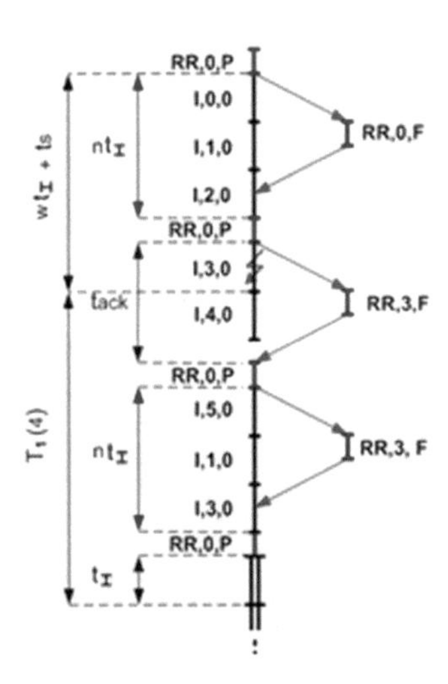

Fig. 14. Case (c) Determination of $T_1(w)$, M = 6, n = 3 HDLC NRM transmission with error, w = 4 one-way transmission from P to S

4 Two Cases for the Direction from S to P Case (a), Acknowledgement Time Is Long

Figure 17 case (e) and Fig. 18 case (f)

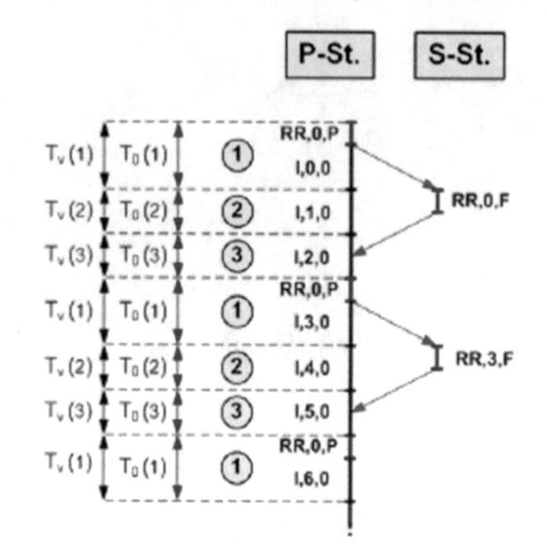

Fig. 15. Case (d) HDLC NRM error-free transmission, one-way transmission from P to S, MOD = 8, n = 3

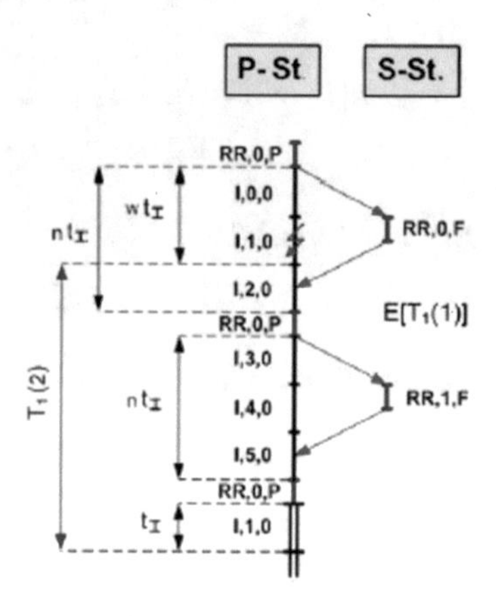

Fig. 16. Case (d) Determination of $T_1(w)$ HDLC NRM transmission with error, n = 3, w = 2 one-way transmission from P to S, MOD = 8

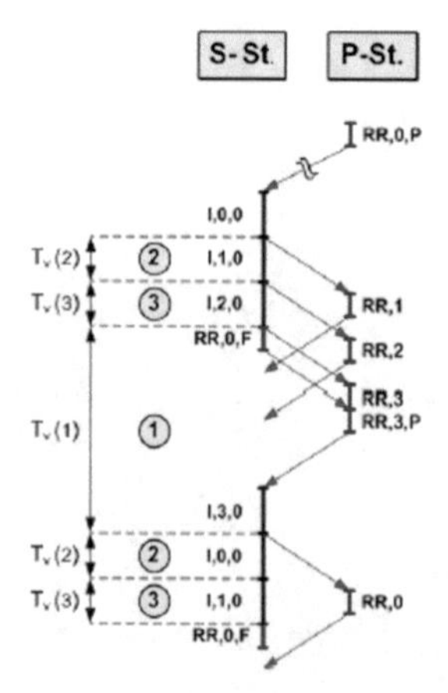

Fig. 17. Case (e) HDLC NRM error-free transmission, one-way transmission from P to S, M = 4

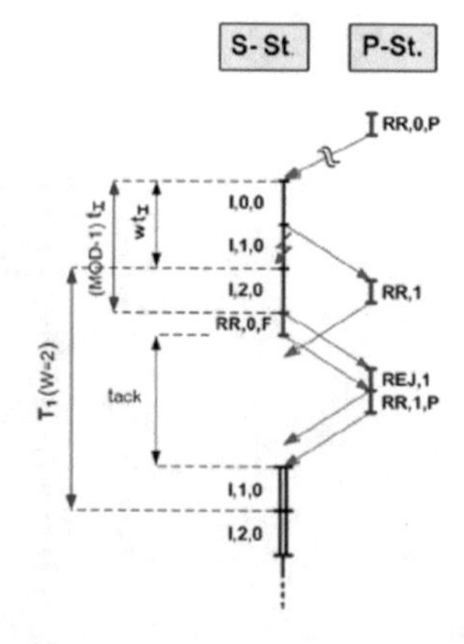

Fig. 18. Case (f) Determination of $T_1(w)$ HDLC NRM transmission with error, M = 4, one-way transmission from S to P

Determination of *time components* $T_0(w)$, $T_1(w)$, $T_2(w) t_{ack} > (M-2)t_I$

$$E[T_0(w)] = \begin{cases} t_{ack} + t_I + t_s & \text{for} \quad w = 1 \\ t_I & \text{for} \quad w = 2, 3, \ldots, M-1 \end{cases}$$

$$\begin{aligned} E[T_1(w)] &= (M-1)t_I - wt_I + t_{ack} + t_s + t_I \\ &= (M-w)t_I + t_{ack} + t_s \quad \text{for } w = 1, \cdots, M-1 \end{aligned}$$

$$E[T_2(w)] = (M-1)t_I + t_{ack} + t_s$$

5 Results of Throughput and Flow Time Evaluation

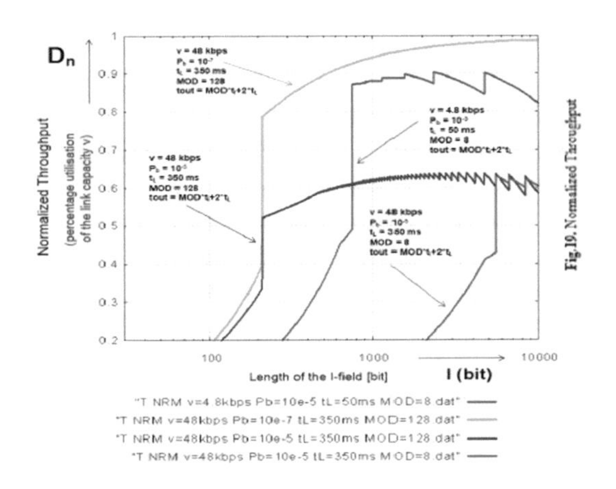

Fig. 19. Normalized throughout

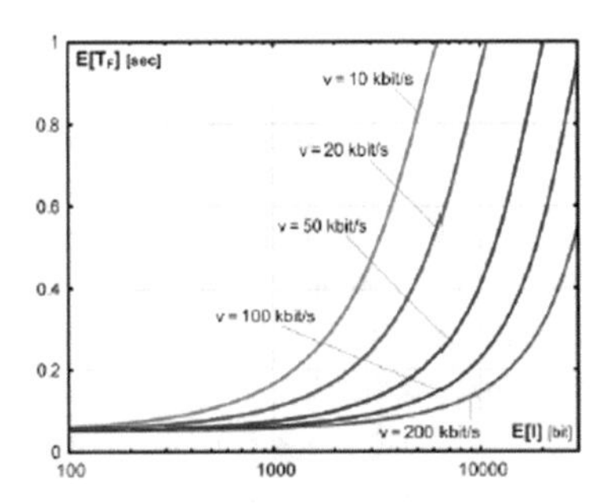

Fig. 20. Mean flow time t_p = ms, v = 50 kbit/s

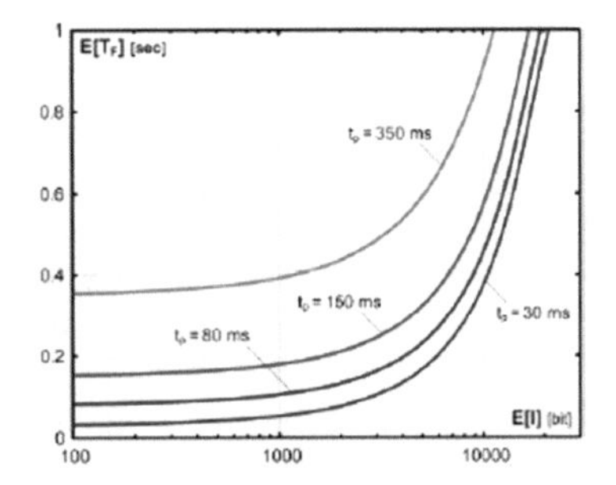

Fig. 21. Mean flow time $p_E = 10^{-5}$, v = 50 kbit/s

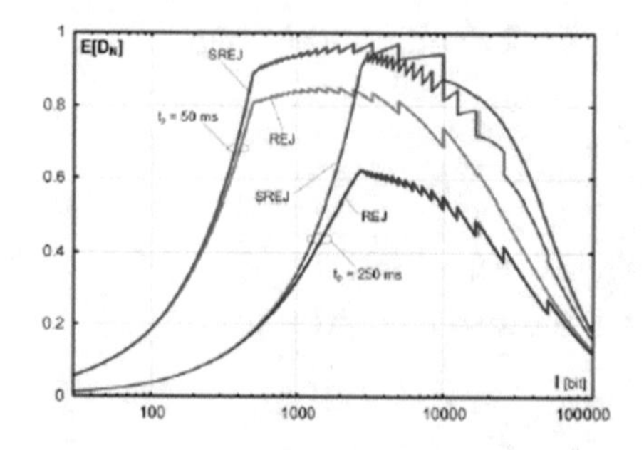

Fig. 22. Mean normalized throughput $p_E = 10^{-4}$, v = 100 kbit/s, M = 20

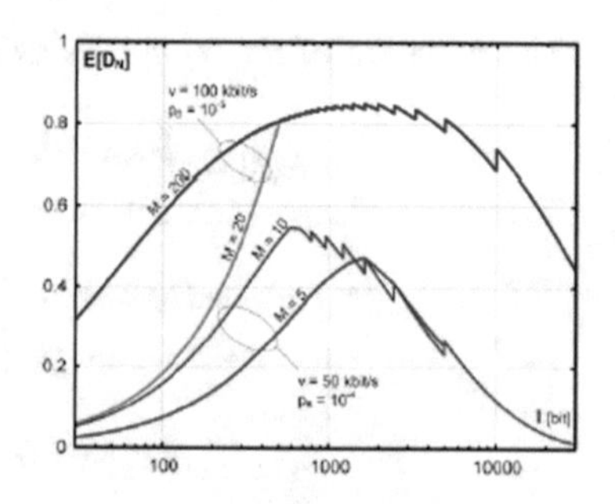

Fig. 23. Mean normalized throughput in function of paylead length t_p = 50 ms in case of the same bit rate v and bit error probability p_E

6 Conclusions

In Fig. 20, it is shown how the mean flow time decreases while increasing the bit rate speeding up the physical transmissions of I-frames and S-frames. Figure 21 shows that the propagation delay is a constant fraction of the flow time irrespective of the other parameters. For very short I-frames this delay constitutes a major fraction of this performance measure. As the I-frames become larger and are more likely to be destroyed, it also contributes to flow time due to more frequent recovery actions.

NRM, Figs. 19, 22, 23 has a lower throughput for short block lengths due to the relatively large amount of control bits. Within the I-frame the additional bits (48 bits for MOD = 8) are used for control purposes. This is also observed for long blocks due to the increase of frame error probability. In long blocks the additional bits can be neglected, but the frame error probability p_F increases causing frequent repetitions. The throughput is lower due to additional control bits in the area to the left of the maximum. The direction from P to S station loses additional capacity due to the transmission of the S-frames. The direction from P to S station in this area shows a lower throughput than the opposite direction.

The throughput is determined by the frame error probability in the area to the right of the maximum. The corrupted I-frame is repeated by means of a REJ-frame in the transmission direction from S to P station. The P station can send the REJ-frame only when it has recognized a sequence error. Due to that an improvement in the throughput is observed considering this transmission direction. Higher transmission rate and shorter blocks result in a frequent interruption of I-frames (after (M-1) I-frames). The throughput is very low considering 48 kBit/s and short block lengths.

The effect of the Modulo-rule is more visible in the direction from P to S station than in the opposite direction. No difference between those two directions is observed when long blocks and small bit error probability is used. The increase of the Modulo-value from 8 to 128 can improve the throughput. The throughput for the direction from S to P station has a strong decrease when block lengths are larger than 5000 bits. The larger the Modulo-value the longer the S station needs to reach the Modulo limit. The error situation is much longer at Modulo 128 than at Modulo 8.

References

1. Benelli, G.: A selective ARQ protocol with a finite-length buffer. IEEE Trans. Commun. **41**(7), 1102–1111 (1993)
2. Bux, W., Truong, H.L.: A queueing model for HDLC-controlled data link. In: Proceedings of Symposium Flow Control in Computer Networks, Varsailles, pp. 287–306 (1979)
3. Bux, W., Truong, H.L.: High Level data link control - traffic considerations. In: Proceedings of the 9th International Teletraffic Congress, vol. 2, Session 44, Torremolinos, session 17 (1979)
4. Stallings, W.: Networking Standards, A Guide to OSI, ISDN, LAN, and MAN Standards. Addison-Wesley, Boston (1993)
5. Bux, W., Kuemmerle, K., Truong, H.L.: Balanced HDLC procedures. IEEE Trans. Commun. **28**(11), 1889–1898 (1980)
6. Bux, W., Kuemmerle, K., Truong, H.L.: Data link-control performance: comparing HDLC operation modes. Comput. Netw. **6**, 37–51 (1982)
7. Raith, T., Truong, H.L.: Analysis of the normal response mode with full-duplex transmission - a new calculation method and simulation. In: Proceedings of the 10th International Teletraffic Congress, Session 3.4, Paper 2, Montreal, session 3.4, paper 2 (1983)
8. Goeldner, E., Truong, H.L.: A Simulation study of HDLC-ABM with selective and non-selective reject. In: Proceedings of the 10th International Teletra
9. Zorzi, M., Rao, R.R.: On the use of renewal theory in the analysis of ARQ protocols. IEEE Trans. Commun. **44**(9), 1077–1081 (1996)
10. Nakajima, N., Serizawa, M.: Retransmission control scheme in asymmetrical channel National Convention Record, B-468, Japan (1996)
11. Maeda, A., Sugimachi, N., Hayashida, H., Fujii, S.: Traffic characteristics of Go-Back N-ARQ scheme with selective repeat in intra-block. Electron. Commun. Jpn, Part 1 **86**(3), 9–16 (2003)
12. de Lima, H.M., Duarte, O.C.: Point-to-multipoint SR ARQ scheme with acumulative acknowledgement for satellite communications. Comput. Netw. ISDN Syst. **30**, 1311–1325 (1998)

Optimisation of Project Duration and Costs Using Over-Working of Executors and Equipment When There Are Limitations for Their Maximum Units

Omer Kurtanović[1(✉)] and Fatih Destović[2]

[1] Faculty of Economics, University of Bihac, 77000 Bihać, Bosnia and Herzegovina
adsami@bih.net.ba

[2] Faculty of Educational Sciences, University of Sarajevo, 71000 Sarajevo, Bosnia and Herzegovina

Abstract. This paper presents the engagement of resources of the type of workers and equipment on the project with their adopted normal working hours and the maximum allowed overtime work. At the same time, the two criteria, the duration and costs of the project are minimised, by introducing the overtime work of these resources and by importing their available quantities. It is noted that overtime work costs are generally higher than normal work hours and the required reduction in the duration of the project should be achieved with minimum additional costs. The problem has the final number of Pareto-optimal solutions. An illustrative example of a hypothetical project is solved using an interactive "analytic - project management software" approach.

Keywords: Project Management · Project duration · Project cost · Regular work · Overtime · Available resources · Multicriteria optimization · Pareto-optimal solutions · Software

1 Introduction

It is widely known that Project Management (PM) is an extremely important scientific discipline. Many authors considered different aspects of planning and managing complex projects, e.g. [2–7]. Since there is no project without requirements for certain resources, the use of labour, equipment, mechanisation, materials, money and others was particularly exposed. It is understood that the use of PM software, such as MS Project and Primavera Planner, support leveling of "Work type" resources to reduce resource requirements to the available limits. It is considered that it is possible to provide the necessary "Material type resources" in a timely manner and enough money for the project plan based on the available amount of "Work type" ones. If there are restrictions on other types of resources, it is necessary to apply the appropriate interactive process "Analyst-PM software" for planning: (a) materials, (b) the project liquidity, or (c) material and liquidity (e.g. [8]). The dependence of the duration of the activity on the engagement of the "Work type" resource was discussed in [10], and the

I. Karabegović (Ed.): NT 2019, LNNS 76, pp. 332–341, 2020.
https://doi.org/10.1007/978-3-030-18072-0_38

dependence of the duration of the project on the availability of this type of resource was presented in [11] (parametric analysis) and [9, 14] (vector parametric analysis). Resource costs were discussed in more detail in [12, 13] from the standpoint: type (for regular working hours, overtime, fixed costs, etc.), variability during project realisation, and allocation in time. To consider multiple criteria in the project, as in this paper, general rules of multi-criteria optimisation (e.g. [1]) should be applied. Selected multi-criteria problems in the Project Management process were detailed in [5].

Below is a shortening of the duration of the activity and the project by introducing the overtime work of the "Work type" resource, requiring a minimal increase in the total cost of the project. Moreover, a modification of the general method PERT/COST is performed.

2 Types of Resources on the Project

Two types of resource types and their characteristics are conveniently interpreted on the project with the number of p activities identified A_j in the nodes of the network diagram, (Precedence Diagraming Method – PDM), $j \in J = \{1,\ldots,p\}$ [9, 14].

Non-consumable or permanently present, i.e. renewable m resources R_i of "*Work* type", (as workers with necessary qualifications, working machines with certain characteristics, etc.), $i \in I = \{1,\ldots,m\}$. They determine the time of the activities and of the project.

Consumable or non-renewable q resources M_k of "*Material* type" are various types of materials, parts, and the like, used by "Work type" resources to perform certain activities, $k \in K = \{1,\ldots,q\}$. It is commonly assumed that these resources can be obtained in the necessary quantities and with the dynamics defined by the resources "*Work*".

It's convenient to consider the following terms: T_{p}, t, v, $T = \{1,\ldots,T_{\mathrm{p}}\}$-duration of the project, absolute time units (days) $t \in T$ for analysis of project elements, their number and their set; W_{ij}, $W_{ij}^{(1)}$, $W_{ij}^{(2)}$, I_j-the total amount of work (hours) of resources R_i to perform A_j, normal time of regular work, overtime work and an indexes set of resources, $W_{ij} = W_{ij}^{(1)} + W_{ij}^{(2)}$, $i \in I_j \subseteq I$, $j \in J$; $w_{ij}(t)$, $w_{ij}^{(1)}(t)$, $w_{ij}^{(2)}(t)$, $n_{ij}(t)$, $a_{ij}(t)$, $a_{ij}^{(1)}(t)$, $a_{ij}^{(2)}(t)$-total amount of work (intensity from the time point of view), regular work, overtime work, planned number of units (intensity from the point of view of resources' units), regular hours and overtime hours for R_i on A_j in $t \in T$; $a_{ij} = a_{ij}^{(1)} + a_{ij}^{(2)}$; t_{ij}, t_j-duration (days) of engaging R_i on A_j and duration of A_j. In order to simplify the problem, let's assume constant values for all elements given as functions of time $t: w_{ij} = w_{ij}(t), w_{ij}^{(1)} = w_{ij}^{(1)}(t), w_{ij}^{(2)} = w_{ij}^{(2)}(t)$, $n_{ij} = n_{ij}(t)$, $a_{ij} = a_{ij}(t)$, $a_{ij}^{(1)} = a_{ij}^{(1)}(t)$, $a_{ij}^{(2)} = a_{ij}^{(2)}(t)$, $i \in I_j, j \in J, t \in T$.

The total time W_i to complete the activity A_j is equal to the sum of the work for regular working hours $W_j^{(1)}$ and the overtime work $W_j^{(2)}$ of all its resources R_i, expressed by (1). The formulas (2) and (3) are valid. If A_j uses only one category of R_i, then followed $t_j = t_{ij}$. Otherwise, for several categories of resources, t_j determines a longer t_{ij} using (4).

$$W_j = W_j^{(1)} + W_j^{(2)},\ W_j^{(1)} = \sum \{ W_{ij}^{(1)} | i \in I_j \}, W_j^{(2)} = \sum \{ W_{ij}^{(2)} | i \in I_j \}, j \in J \quad (1)$$

$$w_{ij} = (a_{ij}^{(1)} + a_{ij}^{(2)}) n_{ij} \Rightarrow \left\{ \begin{array}{ll} n_{ij} = w_{ij}/(a_{ij}^{(1)} + a_{ij}^{(2)}), & \text{given } w_{ij}, a_{ij}^{(1)}, a_{ij}^{(2)} \\ a_{ij}^{(1)} + a_{ij}^{(2)} = w_{ij}/n_{ij}, & \text{given } w_{ij}, n_{ij} \end{array} \right\}, i \in I_j, j \in J \quad (2)$$

$$t_{ij} = W_{ij}/w_{ij} = W_{ij}/(a_{ij} n_{ij}), i \in I_j, j \in J \quad (3)$$

$$t_j = \max_{i \in I_j} \{ t_{ij} \} = \max_{i \in I_j} \{ W_{ij}/w_{ij} \} = \max_{i \in I_j} \{ W_{ij}/[(a_{ij}^{(1)} + a_{ij}^{(2)}) n_{ij}] \}, j \in J \quad (4)$$

3 Minimisation of the Project Duration and Costs

There are several types of costs on the project, but it is appropriate to explain the problem of optimising the project duration and costs by considering only two types of costs for the "Work type" resources. It should be considered that the same category of resources has unique costs of engagement in each of its activities. It may be assumed that this does not impair the generality of the problem.

Let's mark it with J_i a set of indexes i of those activities A_j using the observed resource R_i of "Work type", $J_i \subseteq J$, $i \in I$. Let's assume that the time units t on the project are expressed in days. Moreover, daily costs per resource unit in activity follow formulas (5): $c_{ij}^{(1)}$, $c_{ij}^{(2)}$, c_{ij}-costs of R_i on A_j for regular working hours $a_{ij}^{(1)}$, overtime work $a_{ij}^{(2)}$ and in total amount. The number of resource units n_{ij} determines the analogue daily values (6): $C_{ij}^{(1)}$, $C_{ij}^{(2)}$, C_{ij}-costs for R_i on A_j with $a_{ij}^{(1)}$, with $a_{ij}^{(2)}$ and in total. Finally, daily costs can be determined for R_i regarding all A_j all those who demand that resource (7): $c_i^{(1)}$, $c_i^{(2)}$, c_{ij}-costs with $a_{ij}^{(1)}$, with $a_{ij}^{(2)}$ and in total. By summarizing the individual types of daily costs of all resources, the corresponding types of daily project costs are generated. Further summing from the point of view of time units, the total cost of the project is calculated by type of cost (regular working time, overtime) and consolidated for both types of costs. However, as the project makes its activities more relevant, it is more appropriate to observe the corresponding daily costs (8) from the point of view of activities A_j based on all of its resources R_i : $\underline{c}_j^{(1)}$, $\underline{c}_j^{(2)}$, $\underline{c}_j$-the cost of engaging resources with regular working hours, for overtime and for total time.

$$c_{ij}^{(1)} = c_i^{(1)} a_{ij}^{(1)}, c_{ij}^{(2)} = c_i^{(2)} a_{ij}^{(2)}, c_{ij} = c_{ij}^{(1)} + c_{ij}^{(2)}, i \in I, j \in J_i \quad (5)$$

$$C_{ij}^{(1)} = c_i^{(1)} a_{ij}^{(1)} n_{ij},\ C_{ij}^{(2)} = c_i^{(2)} a_{ij}^{(2)} n_{ij}, C_{ij} = C_{ij}^{(1)} + C_{ij}^{(2)}, i \in I, j \in J_i \quad (6)$$

$$c_i^{(1)} = \sum \{ C_{ij}^{(1)} | j \in J_i \}, c_i^{(2)} = \sum \{ C_{ij}^{(2)} | j \in J_i \}, c_i = c_i^{(1)} + c_i^{(2)},\ i \in I \qquad (7)$$

$$\underline{c}_j^{(1)} = \sum \{ C_{ij}^{(1)} \,|\, i \in I_j \}, \underline{c}_j^{(2)} = \sum \{ C_{ij}^{(2)} \,|\, i \in I_j \}, \underline{c}_i = \underline{c}_i^{(1)} + \underline{c}_i^{(2)},\ j \in J \qquad (8)$$

Further, the total costs of the activities A_j are calculated as the product of the daily costs $\underline{c}_j$ and duration t_j, $j \in J$. The costs of the project, Cp are the sum of the costs of all activities given as (10).

The general mathematical model of minimising the duration and cost of a project with limitations on the resources of "type Work" can be given by formulas (9)–(12).

$$\min_{GD} Tp = \sum \{ t_j | j \in CP \} \qquad (9)$$

$$\min Cp = \sum\nolimits_{j \in J} \underline{c}_j t_j \qquad (10)$$

subject to:

$$\sum\nolimits_{j \in J_i} n_{ij}(t)\, h_{ij}(t) \leq n_i(t), i \in I,\ t \in T \qquad (11)$$

$$h_{ij}(t) = \begin{Bmatrix} 1, & \text{if } A_i \text{ uses } R_i \text{ in term } t \\ 0, & \text{othervise} \end{Bmatrix}, j \in J_i,\ i \in I,\ t \in T \qquad (12)$$

With (9) the duration of the project, Tp is minimised by observing each Critical Path (CP) as the sum of duration t_j of the corresponding continuous series of critical activities (CA)A_j from the beginning to the completion of the project. In the minimisation request Tp it was noted that one should consider the Gant Diagram of the project (GD) that does not depend on the Network Diagram form and is necessary for interpreting the constraints (11) (12). Another criterion (10) for minimising the cost of the project, Cp was explained before. The inequalities (11) refer to each resource R_i and each time unit t. Now $n_{ij}(t)$ expresses the number of units for R_i at A_j in the time unit t. While the binary value $h_{ij}(t)$ has the purpose of identifying whether R_i is used for A_j in t. For each t, the sum of the required R_i units on the left side can not be greater than the number of units of that resource, $n_i(t)$ on the right. The characteristic of $h_{ij}(t)$ is explained by (12).

The modern PM software solves the single-criterion model (9) (11) (12) and defines the minimum duration of the project (9 with the available resources of "Work type", $n_i(t)$ in the Leveling process (11) (12). Then the costs of the project (10) are calculated. The solution of the model (9)–(12) with two criteria can be done using the ε-*Constraint* method [1, 5]. One of the criteria is optimised and the other transfers to the limits with the required upper limit (1st Problem: min Cp, 2nd Problem: min Tp). The basic method PERT/COST without constraints (11) (12) is simply implemented as 1st Problem, by minimising Cp (10) and Tp (9) is set as limitation in formula (13). The required upper limit T_0 for the duration of the project is iteratively reduced, starting from Tp with leveled resources when they do not use overtime.

$$Tp = \sum \{ t_j | j \in CP \} \leq T_0 \tag{13}$$

Thus, a single-criterion model (10)–(13) is to be solved. This is done indirectly: by introducing overtime work of a resource it is necessary to shorten the activities to provide the smallest growth of Cp and achieve $Tp \leq T_0$ from (13). The method of solving is heuristic type and relatively complex, and is carried out in the interactive process of "Analyst – PM software". The analyst enters the necessary data and requests that the software determines the project plan by leveling resources that respects the limits (11). Namely, the finite iteration number $k = 1, 2, \ldots, n$ is defined, which will give a finite set of n^* Pareto-optimal solutions or Efficient Solutions $[Tp^{(k)}, Cp^{(k)}]$, $n^* \leq n$. The solution is Pareto-optimal only if the growth of one criterion decreases the other, and vice versa. If the observed solution changes both criteria in the same direction comparing to any other Pareto solution, then the observed solution is not Pareto-optimal and is to be rejected. The solution to the application is the Efficient Solution chosen by the decision-maker as the most acceptable in the given business conditions.

$$a_{ij}^{(2,k)} > 0 \Rightarrow \left\{ t_j^{(k)} < t_j^{(k-1)} \wedge Tp^{(k)} < Tp^{(k-1)} \wedge Cp^{(k)} > Cp^{(k-1)} \right\}, i \in I_j, j \in J, k = 1, 2, \ldots \tag{14}$$

4 The Illustrative Example

A hypothetical project consists of $p = 10$ tasks A to J type of 'Fixed Work' using $m = 3$ categories of workers, Resource R_1 to R_3 (Table 1). Based on data for workers, Res. Work W_{ij}[hrs] and Units n_{ij}[Peak/day], the initial (normal) duration of the activity was calculated. Duration $t_j^{(0)}$[days] with regular working hours $a_{ij}^{(1)} = 8$[hrs/days] when there is no overtime work ($a_{ij}^{(2,0)} = 0$). MS Project software is used, Project Start 1.7.2019 adopted and Standard Calendar used: 5 working days a week, the start of a working day at 08:00, 1 h break from 12:00 to 13:00 and ending at 17:00. Without limitations on resources, the minimum duration of the project $Tp_{min} = 18$ [days] is achieved. There are two critical paths (CP), A-D-G-J and B-E-H-I, if the calculation of multiple paths CP is performed. Leveling resources, according to the basic rule without the individual assignments on a task or(and) the splits in a remaining work, with available number of workers Max Units $[n_i^{(0)}] =$ [6, 5, 4] expressed in [Units/day] determines the conditioned minimum project duration $Tp^{(0)} = Tp* = 26$ [days]. Remain the same CP.

Unit resource costs for $a_i^{(1)}$, 'Resource Cost Standard Rate' or 'Rsc. Std. Rate' i.e. 'Std. Rate', $[c_i^{(1)}] = [3{,}00;2{,}50;2{,}00]$ expressed in [Money Units/hrs] e.g.[m.u./hr] determine the minimum total cost of the project $Cp_{\min} = Cp^{(0)} = 4.216{,}00$ [m.u.],

Table 1. Initial data

Activity		Predecessors	Duration [days]	Work [hrs]	Resource R_1		Resource R_2		Resource R_3	
j	Name				Work	Units	Work	Units	Work	Units
			$t_j^{(0)}$	W_j	$W_{1j}^{(0)}$	n_{1j}	W_{2j}	n_{2j}	W_{3j}	n_{3j}
1	*A* (-)	–	4	160	160	5	–	–	–	–
2	*B* (-)	–	5	160	–	–	160	4	–	–
3	*C* (*A*)	*A*	3	120	48	2	72	3	–	–
4	*D* (*A*)	*A*	5	160	–	–	–	–	160	4
5	*E* (*B*)	*B*	1	32	32	4	–	–	–	–
6	*F* (*C*)	*C*	3	240	144	6	72	3	24	1
7	*G* (*D*, *E*)	*D*, *E*	6	192	–	–	192	4	–	–
8	*H* (*E*)	*E*	6	384	192	4	192	4	–	–
9	*I* (*H*)	*H*	3	112	80	5	–	–	32	2
10	*J* (*G*, *H*)	*G*, *H*	3	48	–	–	–	–	48	2
Max Units (Std. Rate; Ovt. Rate)					6 (3,00; 3,90)		5 (2,50; 3,25)		4 (2,00; 2,60)	

regardless of whether $Tp_{\min}$ or $Tp^{(0)}$ is observed. Unit resource costs $c_i^{(2)}$ for $a_i^{(2)}$, 'Ovt. Rate' are higher by 30% of $c_i^{(1)}$ and are equal to $[c_i^{(2)}] = [3{,}90;3{,}25;2{,}60]$ [m.u./hr].

According to Note 1 to Note 5, the solution process is carried out in the iterations $k = 0,1,\ldots,7$ (Table 2). For $k = 0$ the initial data is used. On critical activities $(CA)A_j$ from the previous iteration $(k-1)$, in the current iteration $(k > 0)$ for R_i are introduced 'Ovt. Work' $W_{ij}^{(2,k)}$ under condition $a_i^{(2)} \leq 3$ [hr/day]. The software calculates $t_j^{(k)}$ and corresponding costs, and with a leveling of resources determined $Tp^{(k)}$. To the values defined above, is now being introduced and the ordinal numbers (k) of iterations, e.g. $t_j^{(k)}, \Delta t_j^{(k)}, W_{ij}^{(1,k)}, W_{ij}^{(2,k)}, \Delta Tp^{(k)}, Tp^{(k)}$, etc. Also, instead of index j for A_j the names of activities are used, e.g. $t_2^{(k)}$ and $W_{i,2}^{(1,k)}$ in activity B with $j = 2$ is replaced with $t_B^{(k)}$ and $W_{i,B}^{(1,k)}$. The [days] and [d] marks as well [hrs] and [hr] or [h] are equally used.

Note 1. Dependencies between values at R_i and A_j. The workers' overtime in the current iteration (k) reduces the regular time of these workers and their activities from the previous iteration $(k-1)$ according to the formulas given above (1) to (4), so the duration of the activities and the project should be minimised with a minimum increase of the costs are achieved by the conditions of formula (14) and obtain an Efficient solution. Actually, the introduced $a_{ij}^{(2,k)}$ result in $W_{ij}^{(2,k)} = a_{ij}^{(2,k)} \cdot n_{ij} \cdot t_j^{(k)}$ and must $\Delta t_j^{(k)} < 0$ to form $t_j^{(k)} = t_j^{(k-1)} + \Delta t_j^{(k)} < t_j^{(k-1)}$, $W_{ij}^{(1,k)} = W_{ij} - W_{ij}^{(2,k)} < W_{ij}^{(1,k-1)}$, $Tp^{(k)} = Tp^{(k-1)} + \Delta Tp^{(k)} < Tp^{(k-1)}$ with $\Delta Tp^{(k)} < 0$ and $Cp^{(k)} = Cp^{(k-1)} + \Delta Cp^{(k)} > Cp^{(k-1)}$ with $\Delta Cp^{(k)} > 0$. It is not permissible for the project to shrink less than the activity, or reduce activity more than necessary. The basic requirement to achieve is $|\Delta Tp^{(k)}| \geq |\Delta t_j^{(k)}|$.

Table 2. Iteration of the introduction of overtime work

Iter. k	Activ	$t_j^{(k-1)}$ [d]	$\Delta t_j^{(k)}$[d]	$t_j^{(k)}$[d]	Ovt. Work $W_{ij}^{(2,k)}$[h]				Project				Resources
					R_1	R_2	R_3	Σ	$\Delta Tp^{(k)}$	$Tp^{(k)}$	$\Delta Cp^{(k)}$	$Cp^{(k)}$	$\sum c_{ij}^{(2,k)}$
0	All	$t_j^{(0)}$	-	$t^{(0)}$	-	-	-	-	-	26,00	-	4.216,00	-
1	J	3	−0,75	2,25	-	-	12	12	−0,75	25,25	7,20	4.223,20	31,20
2	B	5	−1,00	4,00	-	32	-	32	−1,00	24,25	24,00	4.247,20	104,00
3	C	3	−0,75	2,25	12	18	-	30	−0,75	23,50	24,30	4.271,60	105,30
4	A	4	−0,25	3,75	10	-	-	10	-0,25	23,25	15,00	4.286,50	39,00
	B	4		3,75	-	8	-	8					26,00
5	F	3	−0,50	2,50	24	12	4	40	+0,50	23,75	33,00	4.319,50	143,00
6		2,50	0,50	3,00	−24	−24	−24	−40	−3,50	20,25	−33,00	4.286,50	−143,00
7	G	6	−1,00	5	-	32	-	32	−1,00	19,25	24,00	4.310,50	104,00
				Σ	22	90	12	124	Iter. 6,7: Max Units[5,7,4] from 15.7.2019				

Values without changes are transferred to the next iteration, e.g. $t_j^{(k)} = t_j^{(k-1)}$ if $\Delta t_j^{(k)} = 0$.

Note 2. Shortening one CA (iterations k = 1,2,3,7). There is a *CP* with A_j that is *CA* or it is common for multiple *CP*. (i) If the selected *CA* uses one resource, firstly observe R_3 with $c_3^{(2)} = 2{,}60$ [m.u./h], then R_2 with $c_2^{(2)} = 3{,}25$ and R_1 with $c_1^{(2)} = 3{,}90$. (ii) When two resources are used, use the order $\{R_2, R_3\}$, $\{R_1, R_3\}$, $\{R_1, R_2\}$. (ii) Finally, if it is possible shorted the *CA* with all three resources $\{R_1, R_2, R_3\}$. If more R_i are observed, the formulas (3) and (4) must be respected with introducing $a_{sj}^{(2)}$ to each R_s so that its $t_{sj}^{(k)}$ is not without any need less than $t_{ij}^{(k)}$ of those R_i which determine $t_j^{(k)}$, $s \in I_j$.

Note3. Reduction of more CA (iteration k = 4). They are on the different *CP*. Each can use one resource, but some of them or all can use multiple resources. For each resource should be introduced the minimum $W_{ij}^{(2)}$ in order to, for the same time, $\Delta t_j^{(k)}$ shrink $t_j^{(k-1)}$ in the all observed *CA* and shrink $Tp^{(k-1)}$ for $|\Delta Tp^{(k)}| \geq |\Delta t_j^{(k)}|$.

Note4. Multiple shortening of the same CA (iteration k = 4). The same A_j can be shortened in several iterations. Here is B with R_2 shortened in k = 2 and k = 4. Therefore k = 4 is implemented with $W_{2,B}^{(2,2)} + W_{2,B}^{(2,4)} = 32 + 8 = 40$ [h], but additional costs are calculated only for $W_{2,B}^{(2,4)}$.

Iteration k = 5. Only F using all three resources can be shortened. The displayed values of $W_{i,F}^{(2,5)}$ determine $\Delta t_F^{(5)} = -0{,}50$ [d], $\Delta Tp^{(5)} = 0{,}50$ [d] and $\Delta Cp^{(5)} = 33{,}00$ [m. u.]. As the growth occurs of $Tp^{(5)}$ and $Cp^{(5)}$ in relation to k = 4, was not obtained Efficient solution with k = 5 and such is to be discarded. The process of resolution ends.

Iteration $k = 6$. Some additional workers may be needed. A careful analysis of the 'Resource Graph' and 'Resource Usage' report states that the software determines $Tp^{(5)}$ growth in leveling due to insufficient resources available $[n_i^{(5)}] = [6, 5, 4]$ for the period $t \geq 15.7.2019$. Make it possible to provide an additional 2 R_2. Set $[n_i^{(6)}] = [6, 7, 4]$ and cancel the data for F from $k = 5$ with $W_{i,F}^{(2,6)} = -W_{i,F}^{(2,5)}$. With that F is no longer CA, while G that was not CA becomes CA. Due to $Tp^{(6)} < Tp^{(4)}$ and $Cp^{(6)} = Cp^{(4)}$ it was concluded that $k = 4$ has Non-Efficient solution, and is rejected.

Iteration $k = 7$. G is shortened. It is necessary to partially level the resources for $t \geq 15.7.2019$. From the standpoint of the resources available, all A_j become CA, there are no those that can be shortened and the problem resolving ends. $Tp^{(7)} < Tp^{(6)}$ and $Cp^{(7)} > Cp^{(6)}$ indicates that an Efficient solution was obtained.

Note 5. Unlike the basic PERT/COST method when $\Delta Tp^{(k)} = \Delta t_j^{(k)}$, the exposed modification with limits $n_i^{(k)}$ for R_i can have $\Delta Tp^{(k)} \leq \Delta t_j^{(k)}$, or $|\Delta Tp^{(k)}| \geq |\Delta t_j^{(k)}|$ with $\Delta Tp^{(k)} < 0, \Delta t_j^{(k)} < 0$. This is the result of leveling the resource with $n_i^{(k)}$. It is, of course, desirable to realise $\Delta Tp^{(k)} < \Delta t_j^{(k)}$ with aim to reduce as much as possible $Tp^{(k)}$ in relation to the shortening of $t_j^{(k)}$. This case does not occur directly but indirectly for $k = 5$ with $\Delta t_F^{(5)} = 0{,}50$ [d] achieved $\Delta Tp^{(5)} = -3{,}50$ [d].

The starting available workers $[n_i^{(k)}] = [6, 5, 4]$ have five Efficient solutions $[Tp^{(k)}, Cp^{(k)}]$ if the process is carried out with $k = 0,..,4$. By introducing additional workers $[n_i^{(k)}] = [5, 7, 4]$ two solutions of this type are determined with $k = 6,7$. As the solution with $k = 4$ is no longer an Efficient solution, the problem has only six Efficient solutions with $k = 0,1,2,3,6,7$. (Table 2, Fig. 2). The project plan for the appropriate solution determines the daily needs for the workers on the project. It is characteristic that $n_2^{(7)} = 7$ for$k = 7$ is used only for $t \geq 19.7.2019$. The timings in non-integer numbers $^{(7)}$[d] determine the completion of the project on 26.7.2019. at 10:00. At the same time, there are A_j having Start or (and) Finish and are not at the beginning or end of working hours (8:00, 17:00) (Fig. 1).

	Task Name (Predcss)	Duration	Work (Total)	Regular Work	Owt. Work	Start	Finish	Cost (Total)	Reg. Cost	Ovt. Cost	Total Slack	Iteration
0	- Proj-Iter7	19,25 days	1.608 hrs	1.484 hrs	124 hrs	1.7.19 08:00	26.7.19 10:00	4.310,50	3.901,00	409,50	0 days	
1	A (-)	3,75 days	160 hrs	150 hrs	10 hrs	1.7.19 08:00	4.7.19 15:00	489,00	450,00	39,00	0 days	4: R1
2	B (-)	3,75 days	160 hrs	120 hrs	40 hrs	1.7.19 08:00	4.7.19 15:00	430,00	300,00	130,00	0 days	2: R2
3	C (A)	2,25 days	120 hrs	90 hrs	30 hrs	4.7.19 15:00	8.7.19 17:00	348,30	243,00	105,30	0 days	3,4: R1,2
4	D (A)	5 days	160 hrs	160 hrs	0 hrs	4.7.19 15:00	11.7.19 15:00	320,00	320,00	0,00	0 days	
5	E (B)	1 day	32 hrs	32 hrs	0 hrs	4.7.19 15:00	5.7.19 15:00	96,00	96,00	0,00	0 days	
6	F (C)	3 days	240 hrs	240 hrs	0 hrs	19.7.19 08:00	23.7.19 17:00	660,00	660,00	0,00	0 days	5,6: R1-3
7	G (D,E)	5 days	192 hrs	160 hrs	32 hrs	17.7.19 08:00	23.7.19 17:00	504,00	400,00	104,00	0 days	7: R2
8	H (E)	6 days	384 hrs	384 hrs	0 hrs	9.7.19 08:00	16.7.19 17:00	1.056,00	1.056,00	0,00	0 days	
9	I (H)	2 days	112 hrs	112 hrs	0 hrs	17.7.19 08:00	18.7.19 17:00	304,00	304,00	0,00	0 days	
10	J (G,H)	2,25 days	48 hrs	36 hrs	12 hrs	24.7.19 08:00	26.7.19 10:00	103,20	72,00	31,20	0 days	1: R3

Jul '19
24 1 8 15 22 2

Fig. 1. Iteration k = 7 Project Plan with Max Units [4, 5, 7] starting July 15, 2019

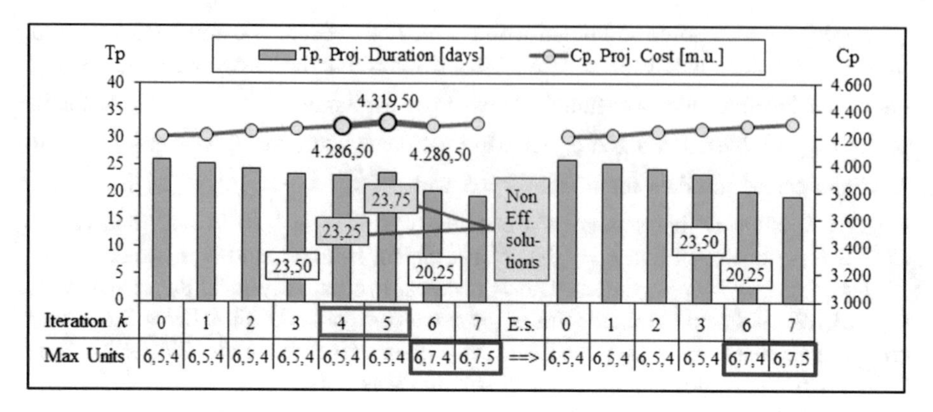

Fig. 2. The solutions of problem, Iteration k = 4,5 has Non-efficient solutions

5 Conclusion

There is a problem of planning project with two criteria, duration and costs of the project, when there are limitations on the "Work type" resources and Overtime Work is introduced by them. A specific procedure for finding Pareto-optimal solutions has been developed as an interactive process of "Analyst - PM software". The analyst defines Resource Overtime Work by respecting the appropriate rules and requires the software to minimise the duration of the project by leveling the appropriate resources. The incurred costs will be calculated.

References

1. Matthias, E.: Multicriteria Optimization-Lecture Notes in Economics and Mathematical Systems, vol. 491. Springer, Heidelberg (2000)
2. Goordpasture, C.J., Quantitative Methods in Project Management, J. Ross Publishing, Inc. (2004)
3. Lester, A.: Project Management, Planning and Control. Elsevier, Amsterdam (2007)
4. Hendrickson, C.: Project Management for Construction, Department of Civil and Environmental Engineering, Carnegie Mellon University, Pittsburgh (2008). (Chapter 4. Labor, Material and Equipment Utilization). http://pmbook.ce.cmu.edu/
5. Kurtanović, O.: Višekriterijumska optimizacija u procesu upravljanja projektom (Multicriteria optimization in project management process). Ph.D. thesis, Faculty of Natural Sciences and Mathematics, University of Sarajevo, Bosnia and Herzegovina (2012). (in Bosnian)
6. Meredith, J.R., Mantel, Jr., S.J., Shafer, S.M., Sutton, M.M.: Project Management in Practice, 5th edn. Wiley, Hoboken (2013)
7. Pašagić, H., Kurtanović, O.: Kvantitativne metode u ekonomiji (Quantitative methods in economics). I izdanje, Bihać (2017). in Bosnian
8. Božilović, Z., Nikolić, N.: Planning of material resources and liquidity on project with available work resources. In: Annals of the University of Oradea, Fascicle of Management and Technological Engineering, Issue #2, pp. 33–38, August 2016. http://www.imtuoradea.ro/auo.fmte/

9. Kurtanović, O., Dacić, L.: Support parametric vector analysis of available resources for minimisation of project duration – four varieties of conditions. In: Karabegović, I. (ed.) New Technologies, Development and Application, NT 2018. Lecture Notes in Networks and Systems, vol. 42, pp. 590–596. Springer, Cham (2019). https://link.springer.com/chapter/10.1007/978-3-319-90893-9_69
10. Kurtanović, O., Nikolić, N.: Determination of activity duration in dependence of 'Work' type resources engagement and project time minimization by applying MS Project software. In: Proceedings, ICDQM 2013, 4th International Conference Life Cycle Engineering and Management, 27–28 June 2013, Belgrade, Serbia, pp. 77–87 (2013)
11. Kurtanović, O., Nikolić, N.: Parametarska analiza raspoloživih resursa za problem minimizacije trajanja projekta (Parametric analysis of the available resources to the problem of minimizing the duration of the project), Zbornikradova, ICDQM 2016, 19. Međunarodnakonferencija Upravljanjekvalitetomipouzdanošću, 29–30 June 2016, Prijevor, Serbia, pp. 401—407 (2016). (in Bosnian)
12. Božilović, Z., Nikolić, N.: Zapažanja 1. O promenljivim troškovima na projektu - Vrste troškova (Observations 1. About variable costs on the project - Types of costs). In: Zbornik radova, ICDQM 2017, 20th International Conference – Dependability and Quality Management, 29–30 June 2017, Prijevor, Serbia, pp. 273–278 (2017). (in Serbian)
13. Kurtanović, O., Nikolić, N.: Zapažanja 2. O promenljivim troškovima na projektu - Pravila alokacije isplata troškova (Observations 2. About variable costs on the project-The rules of allocation of payment of the cost in time). In: Zbornik radova, ICDQM 2017, 20. ICDQM 2017, 20th International Conference – Dependability and Quality Management, 29–30 June 2017, Prijevor, Srerbia, pp. 279–285 (2017). (in Serbian)
14. Kurtanović, O., Dacić, L.: Support parametric vector analysis of available resources for minimisation of project duration – four varieties of conditions. In: 4th International Conference NEWTECHNOLOGIESNT-2018, Development and Application, 14–16 June 2018, Sarajevo, Bosnia and Herzegovina, pp. 590–596 (2018)

Mobile Application mPodaci (mData)

Suad Sućeska(✉)

71.000 Sarajevo, Bosnia and Herzegovina

Abstract. Mobile application mPodaci (mData) has purpose to get up-to-date data of clients by smartphone. It is written for mobile operating system Android, based on programming language JAVA. It is able to send the data using two ways: email and SMS. It also means that data could be sent using two different kinds of connections: Internet connection, used by email, and phone connection, used for sending SMS messages. The application supports Android from version 15 to the latest.

Keywords: Mobile application · Mobile user · Android · Email · SMS message

1 Introduction

Today, contact data of companies are changing very quickly. To have an up-to-date database is not easy. Although firms are obliged to deliver any change of contact data to the institutions in which are registered, they do not do that, and the data which can be obtained from these institutions does not correspond to those in use. There are several ways for update this data. Some of them require direct access to the database with contact data, which is not in agreement with the security policies of some firms and institutions. Mobile application mPodaci (mData) is also one of the ways to get up-to-date contact data from firms. This application sends entered contact data to a specific email or SMS number.

2 About Application

By reviewing company contact data in the database, it has been noted that a large number of company contact data have a mobile phone number. Some of these companies only have this contact information. Mobile application mPodaci is made to enable firms sending up-to-date contact data. It can be downloaded from the following Web address:

http://aplikacije.suads.com.ba/mPodaciE/mPodaci.apk.

The application is made in JAVA for the Android with minimum version 15. It's download, of course, has to be made with a smartphone with an active Internet connection (Wi-Fi, …). After downloading, the application needs to be installed on the smartphone, which is done by clicking the button Install during the procedure. On all prompts during the download and installation procedures should be answered positively.

I. Karabegović (Ed.): NT 2019, LNNS 76, pp. 342–346, 2020.
https://doi.org/10.1007/978-3-030-18072-0_39

2.1 Description of Work of the Application

The main activity shown in Fig. 1 is obtained after starting of the mobile application mPodaci. This activity contains an action bar, which consists of the name of the application and activity, and the menu (three vertical points). Below that are the texts of the data entry notification and 11 fields for entering the contact data of the company: ID, Company name, Company address, Company postal code, Company city, Company telephone number, Company fax number, Company email, Company contact person, Company mobile phone, Company Web address. Below these fields, at the bottom of the activity, are the buttons: Send email, Send SMS and Close [1].

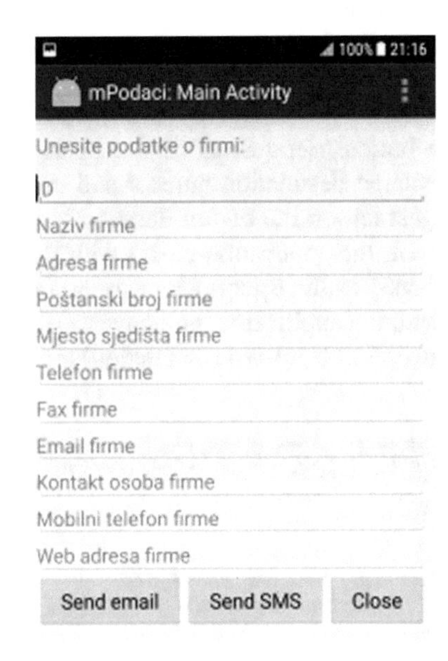

Fig. 1. Main activity of mobile application mPodaci [5, 7]

The click on the menu (three points) in the action bar gives three items: Uputstvo (Instructions), Info, and Close. This menu is shown in Fig. 2 [4].

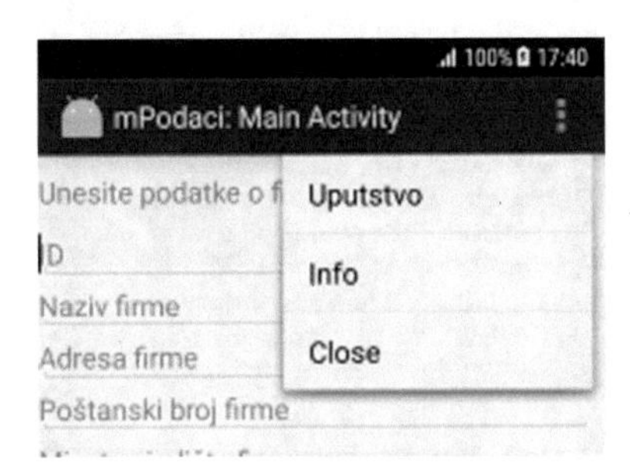

Fig. 2. Menu of the application mPodaci [5]

To send up-to-date contact data of a company, first fill in the fields: ID, Company name, Company address, Company postal code, Company city, Company telephone number, Company fax number, Company email, Company contact person, Company mobile phone, Company Web address. Fields without data are left blank. When click on the field, the virtual keyboard appears. [2] If it blocks some fields, it is necessary to touch the form, then drag the finger up or down to display the required field or button [3]. After filling out the fields with company contact information, they can be sent by email or SMS. To send data via email, you need to have an established Internet connection (Wi-Fi, …). After established connection, tap the button 'Send email'. From offerings of email clients, which appear afterwards, tap one configured. This opens the email client with a ready-made message that contains destination email and contact data in the appropriate form. Click on the button Send sends the message to the specified email address. Data received via email will serve to update a companies contact database. If the data is to be sent by SMS, after entering the data in the appropriate fields, tap the button 'Send SMS'. After opening an application to send SMS messages with the entered destination number and the corresponding formatted data in the message text, just tap on the button Send.

After the message is sent, the application closes with the tap on the button Close next to the button 'Send SMS' or the option Close in the menu (Fig. 2).

Before starting work with the application, you have to read the instruction which is obtained by clicking on the option Uputstvo (Instruction) located in the menu shown in

Fig. 3. The dialogue Uputstvo (Instruction) of mobile application mPodaci [7]

Fig. 2 [6]. You can also scroll the contents of the instruction in vertical direction. The instruction is closed by tap on the button Close (Fig. 3).

General application information can also be viewed from this menu. They are obtained by clicking on the option Info in the menu. The dialogue 'Info' could be closed by click on the button Close.

2.2 Usage

Mobile application mPodaci (mData) is aimed for making update of company contact data in companies database. Application installation file is made in the form of APK (Application Package Kit), which is common for the installation files of Android mobile applications. The file is not offered through Google's Google Play Store, which is usually used for download of Android mobile applications. The installation file of this application, mPodaci.apk, can be set to any Web address specified by a company that requests contact information. This company puts a notice on its Web site with a link to the address of the application's installation file. Firms, whose contact data is not up-to-date, have to access this address by smartphone, download the installation file, and install the application. It can be used to send company's current contact data after any changes. The email address and SMS number to which the data is sent are coded in the application (Fig. 4).

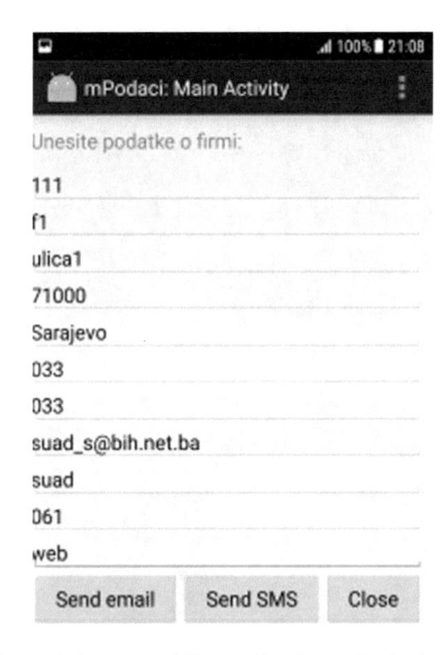

Fig. 4. The main activity of mobile application mPodaci with filled in fields

3 Conclusion

Mobile application mPodaci (mData) enables sending company contact data using smartphone via email and SMS. These methods use two different connections: Internet connection (email) and telephone connection (SMS). Data is sent to a specific email address and phone number. The data obtained is used to update the contact information of the company in the database. The application is written for the Android operating system minimal version 15.

References

1. Fain, Y.: JAVA 24-Hour Trainer, 2nd edn. Wiley Publishing Inc., Indianapolis (2015). (Kindle Edition. Wrox)
2. McWherter, J., Gowell, S.: Professional Mobile Application Development. Wiley, Indianapolis (2015). Kindle Edition. Wrox
3. Advanced Java Programming, LINK group 1998 (2017)
4. Java Web Programming, LINK group 1998 (2017)
5. Introduction to Android Application Development, LINK group 1998 (2017)
6. Data Driven Android Application Development, LINK group 1998 (2017)
7. Advanced Android Application Development, LINK group 1998 (2017)

Efficiency of the Bosnian-Herzegovinian Economy

Vjekoslav Domljan[1(✉)] and Ivana Domljan[2]

[1] SSST Sarajevo, Hrasnička cesta 3a, 71000 Sarajevo, Bosnia and Herzegovina
vjekoslav.domljan@ssst.edu.ba
[2] University of Mostar, 88000 Mostar, Bosnia and Herzegovina

Abstract. The aim of this paper is to analyse the issue of efficiency its components for Bosnia and Herzegovina (BIH) and a group of its comparator countries.

We estimate a global production/efficiency frontier and a frontier for small Central and South East European countries (comparator countries) from 1950 to 2014 and from 1996 to 2014 respectively.

The KWH ("Kumbhakar-Wang-Horncastle") decomposition of efficiency using Stochastic Frontier Analysis (SFA) is applied to the respective countries to evaluate the effects of changes in efficiency. Comparing the efficiency of comparator countries and BIH, and recognizing country heterogeneity, we examine whether there is the evidence of efficiency convergence, i.e. whether BIH moves toward the respective efficiency frontier.

According to our research, BIH is lagging behind some comparator countries (Switzerland, Austria, Slovenia and Macedonia) after the global crises of 2007.

Keywords: Cobb-Douglas parametric frontier model · Efficiency · Bosnia and Herzegovina

1 Introduction

Why is typical for a European country to be a high income country but not for BIH and ten other countries? Conceptually, development accounting can be thought of as quantifying the relationship:

$$\text{Income} = \text{F(Factors, Efficiency)} \tag{1}$$

The starting point of growth accounting is an aggregate production function[1], typically given as the Cobb-Douglas parametric frontier model with constant returns to scale:

$$Y = AK^{\alpha}H^{1-\alpha} \tag{2}$$

where Y is total real output, K is the stock of physical capital, H represents human capital stock, and A represents shifts of the production frontier and denotes efficiency with which factors are used [1, 2].

[1] The terms productivity and efficiency are used interchangeably, but they do not have always the same meaning.

I. Karabegović (Ed.): NT 2019, LNNS 76, pp. 347–354, 2020.
https://doi.org/10.1007/978-3-030-18072-0_40

By assuming H = Lh, the 'quality adjusted' human capital stock, where h represents the average human capital and L represents the number of employees, the production function can be rewritten as:

$$Y = AK^{\alpha}(Lh)^{1-\alpha} \quad (3)$$

In per-worker terms the aggregate production function can be rewritten as

$$y = Ak^{\alpha}h^{1-\alpha} \quad (4)$$

where y is total real output per worker and k is the capital-labour ratio, i.e. k = K/L.

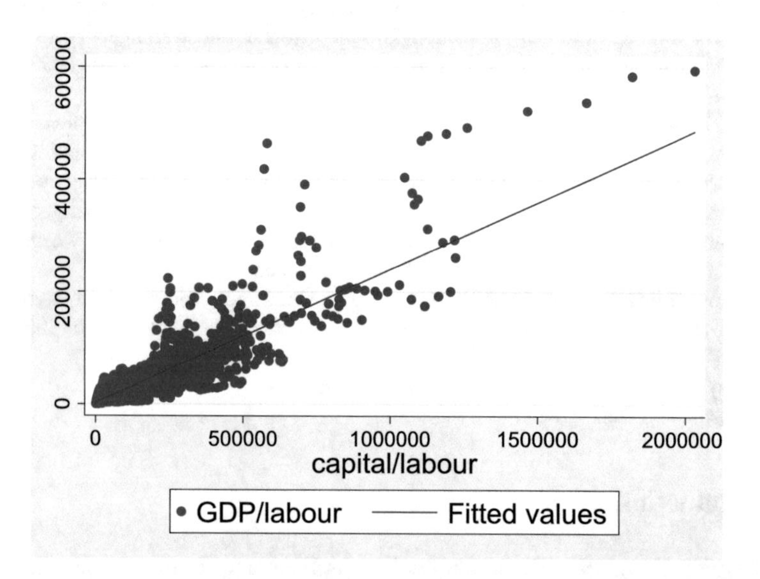

Fig. 1. Productivity and the capital intensity; world, 1950–2014 (in 2011US$)

As Fig. 1 demonstrates, the large capital/labour ratio (capital intensity) of an economy, the large GDP/labour ratio (the productivity) of an economy. Over time, the economy tends to gain a higher capital intensity as it seeks to reach a higher productivity from investing into capital. Actually, differences in productivity levels could be primarily explained by differences in per capital input [4].

The aim of this paper is to explain why BIH extracts less output than most European countries from its factors of production.

2 Conceptual Framework and Methodology

Since its introduction by Aigner et al. [5] stochastic frontier estimation has been extensively used to estimate technical efficiency in applied research. Among panel data models, which are the focus of this paper, the inefficiency specification used by Battese and Coelli [6] and Green [7] are most frequently used in empirical studies [8].

In a standard panel data model, the focus is mostly on controlling country effects (heterogeneity due to unobserved time-invariant factors). However, Kumbhakar et al. [8] have introduced a new model which fills several gaps in the standard panel data models by decomposing the time-invariant country effect and a persistent technical inefficiency. The presence of such effects can be justified, for example, by making an argument that there are unobserved time-invariants inputs that are not inefficiency. The model is specified as [8, 9]:

$$y_{it} = \alpha_o + f(x_{it}; б) + \mu_i + v_{it} - \eta_i - u_{it} \quad (5)$$

where μ_i are random firm effects that capture unobserved time-invariant inputs.

The model has four components two of which (η_i and u_{it}) are inefficiency and the other two are country effects and noise (μ_i and v_{it}). These components appeared in other models in various combinations but not all at the same time in one model. Estimation of the model can be done in a multi-step procedure, for which purpose the model in (5) is rewritten as

$$y_{it} = \alpha_0^* + f(x_{it}; б) + \alpha_i + \varepsilon_{it} \quad (6)$$

where a $\alpha_0^* = \alpha_0 - E(\eta_i) - E(u_{it})$; $\alpha_i = \mu_i - \eta_i + E(\eta_i)$; and $\varepsilon_{it} = \upsilon_{it} - u_{it} + E(u_{it})$.

With this specification α_i and ε_{it} have zero mean and constant variance.

In a nutshell, it is possible to examine whether inefficiency is persistent over time or it is time-varying. The following questions related to the time-invariant individual effects is whether the individual effects represent (persistent) inefficiency, or whether the effects are independent of the inefficiency and capture (persistent) unobserved heterogeneity. Related to this is the question whether the individual effects are fixed parameters or are realisations of a random variable. Comparing the efficiency of countries, and recognising country heterogeneity, it is possible to examine whether there is evidence of efficiency convergence, i.e. whether a country moves toward the global/group frontier or its relative inefficiencies remain unchanged. It is possible to find out the rate of efficiency change, whether the rate of frontier shift is significantly over time [10]. If the persistent inefficiency component is large for a country, then it is expected to function with a relatively high level of inefficiency over time, unless some changes in development policy take place. Thus, the high value of u_i is of more concern form a long-term point of view because of its persistent nature. The advantage of the present specification is that it enables to test the presence of the persistent nature of technical inefficiency without imposing any parametric form of time-dependence [10].

3 Model and Results

In this paper we apply the stochastic frontier framework, developed in [8] to model output levels for BIH and other countries to examine how big is a component of the BIH's inefficiency, and whether it has been persistent over time.

We measure the gap between BIH's efficiency on the one hand and the global production frontier and comparator countries production frontier on the other hand.

We use version 9.0 of the Penn World Tables [PWT 9.0] [3] that provides data on relative levels of income, output, input and productivity for 182 countries over the period 1954–2014.

We measure output from PWT 9.0 as real GDP per worker in international dollars (i.e. in PPP-this variable is called *rgdpe - expenditure-side real GDP at chained PPPs (in mil. 2011US$).*

As for human capital stock we use as proxy the number of employees – this variable is called *emp - number of persons engaged (in millions).* We do not attempt to augment our measure of human capital to reflect its quality-typically accomplished by taking into account the average years of schooling and the return on education - due to lack of data on BIH in existing cross-country datasets [2, 11].

As far as the stock of physical capital is concerned we use the variable called *rkna-capital stock at constant 2011 national prices (in mil. 2011US$).*

Table 1 summarises descriptive statistics of the data used in the efficiency estimation. Due to missing values, we have to drop some countries from analysis. In the end, the panel data is unbalanced: the 178 countries contribute totally 8,244 observations.

Table 1. Descriptive statistics of study variables

No.	Variable	Obs	Mean	Std. Dev	Min	Max
(A) World (1950–2014)						
1	empl	8,244	14.2	56.5	.00118	798.4
2	rgdpna	8,244	289351	1062296	64.9	17100000
3	rkna	8,244	1025229	3518452	317.7	67600000
(B) Comparator countries (1996–2014)						
4	empl	267	2.48857	1.664416	.18977	5.17549
5	rgdpna	267	131254.9	129475.2	5692	5.17549
6	rkna	267	566227.4	590532.6	16619	1701097

We used the Cobb–Douglas parametric frontier model $Y = AK^{\alpha}L^{1-\alpha}$ but we did not linearize our data immediately by taking logarithms because we would then have got two different estimates of α. The coefficient of ln K would give us one estimate and the coefficient of ln L, which would be an estimate of $(1-\alpha)$ would enable us to calculate another. Instead, we divide both sides by L and rewrite the function as

$$Y/L = A\ (K/L)^{\alpha}\upsilon \tag{7}$$

including a disturbance term, υ.

The key steps in our stochastic frontier analysis (SFA) analysis are: (i) choosing a functional form for A and (ii) accurately measuring output and factors. Efficiency is treated as a residual.

In this form the function can be seen to relate output per worker to capital per worker, and we now linearize by taking logarithms [12]:

$$\mathrm{Ln}\ (Y/L) = \ln A + \alpha \ln(K/L) + \log \upsilon \tag{8}$$

Based on the data presented in Table 1 and the model 7, we developed three models, two for analysing BIH in the framework of the global economy and one for analysing BIH within the framework of comparators countries.

Fitting this to the data in Table 1, we developed global model 1, the results of which are shown in Table 2.

Table 2. Global model 1

VARIABLES	Model 1
Ln k	0.764***
	(0.00590)
Constant	1.358***
	(0.0712)
Observations	8,244
No. of countries	178

Standard errors in parentheses
*** $p < 0.01$, ** $p < 0.05$, * $p < 0.1$

As the Cobb-Douglas function is a special case of the more general formulation

$$Y = AK^{\alpha}L^{1-\alpha} \tag{9}$$

With no link between the output elasticities [12], we developed the global model 2, the results of which are shown in Table 3.

Table 3. Global model 2

VARIABLES	Model 2
Ln K	0.739***
	(0.00619)
Ln L	0.347***
	(0.0109)
Constant	1.575***
	(0.0731)
Observations	8,244
No. of countries	178

Standard errors in parentheses
*** $p < 0.01$, ** $p < 0.05$, * $p < 0.1$

For both models we carried out Housman test that shows that is more convenient to use random-effects panel data model than fixed-effects panel data model.

Using the global model 2 we estimated the efficiency of the BIH economy (see Fig. 2).

Figure 2 shows that, on average, efficiency of the BIH economy is higher than global efficiency. The BIH average efficiency (OTE_klh) estimated for the period 1990–2014 (the first inputs to PWT 9.0 for BIH are related to the year of 1990) to be 69.7% while the global average efficiency is estimated to be 65.1%. Persistent efficiency (TE_P_klh) of the BIH economy is 83.1%, higher than 77.6%, i.e. than global average of persistent efficiency.

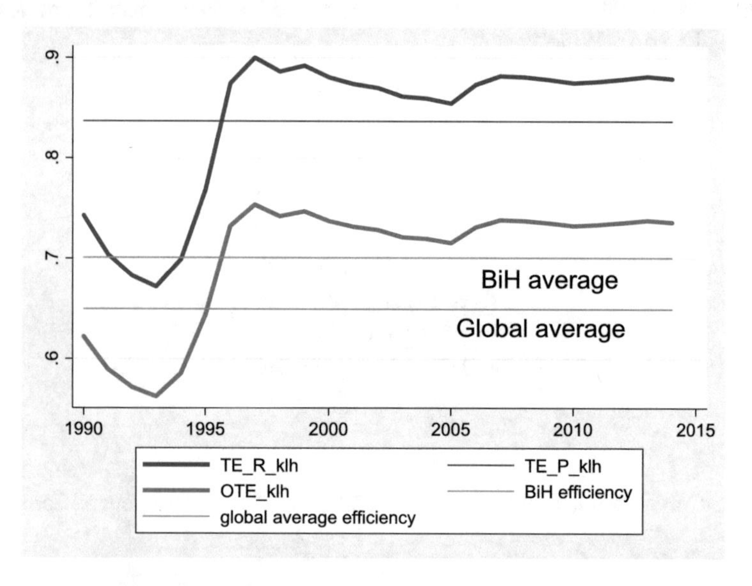

Fig. 2. BIH efficiency in the global framework, 1990–2014

Table 4. Regional model 1

VARIABLES	Model 3
Ln k	0.710***
	(0.0276)
Constant	2.161***
	(0.333)
Observations	267
No. of countries	14

Standard errors in parentheses
*** $p < 0.01$, ** $p < 0.05$, * $p < 0.1$

Using the regional model 1 (see Table 4) we estimated the efficiency of the BIH economy (Fig. 3).

It is important to underline the limits of development accounting. It is based on a Cobb- Douglas parametric frontier model presuming that returns on scale are constant, that efficiency differences are factor neutral etc.

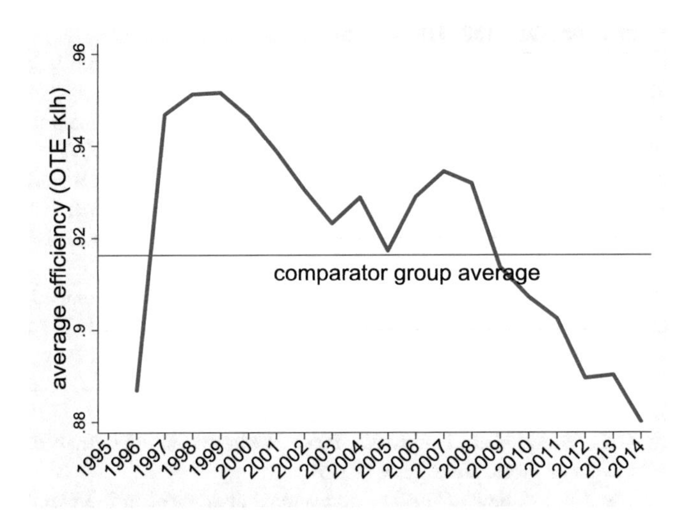

Fig. 3. BIH efficiency in the regional framework, 1996–2014

In addition, in the above estimates some important factors are not taken into consideration. For instance, numerous research studies adjust the quality of human capital by tanking into estimation differences inhuman capital originating from differences in the quantity of education. However, this data is not available for BIH.

4 Conclusions

We have examined - comparing the productivity of countries, in our case in the terms of GDP per worker - whether there is evidence of economic growth convergence, that is, whether BIH moves toward the comparator countries production frontier.

The KLH model enabled us to take into account some heterogeneity by introducing individual (unobservable) effect, say that is time-invariant and country specific, and not interacted with other variables.

According to our research BIH is lagging behind some comparator countries (Switzerland, Austria, Slovenia and Macedonia) after the global crises of 2007. Actually, BIH is a trap of structural transformation, not been able to increase capital accumulation.

References

1. Caselli, F.: Accounting for cross-country in come differences. NBER Working Paper No. 10828 (2004)
2. Abdelkader, K., et al.: Bosnia and Herzegovina: selected issues, IMF Country Report No. 10/347 (2010)
3. Feenstra, R.C., Inklaar, R., Timmer, M.P.: The next generation of the Penn World Table. Am. Econ. Rev. **105**(10), 3150–3182 (2015)
4. Jorgenson, D.W., Vu, K.: Information technology and the world economy. Scand. J. Econ. **107**(4), 631–650 (2005)
5. Aigner, D., Lovel Knox, C.A., Schmidth, P.: Formulation and estimation of stochastic frontier production function models. J. Econometrics **21**(1), 21–37 (1977)
6. Battese, G.E., Coelli, T.J.: A model for technical inefficiency effects in a stochastic frontier production function for panel data. Empirical Economics **20**(2), 325–332 (1995)
7. Green, W.: Fixed and random effects in stochastic frontier models. J. Prod. Anal. **23**(1), 7–32 (2005)
8. Kumbhakar, S.C., Lien, G., Hardaker, J.B.: Technical efficiency in competing panel data models: a study of Norwegian grain farming. J. Prod. Anal. **41**(2), 321–337 (2014)
9. Colombi, R., Kumbhakar, S.C., Martini, G., Vittadini, G.: Closed-skew normality in stochastic frontiers with individual effects and long/short-run efficiency. J. Prod. Anal. **42**(2), 123–136 (2014)
10. Kumbhakar, C.S., Wang, H.-J., Horncastle, P.A.: Stochastic Frontier Analysis Using Stata. Cambridge University Press, New York (2015)
11. Barro, R.J., Lee, J.W.: A new data set of educational attainment in the world, 1950–2010. J. Dev. Econ. **104**(2013), 184–198 (2013)
12. Dougherty, C.: Introduction to Econometrics. Oxford University Press, New York (1992)

Challenges of Applying Blockchain Technology

Savo Stupar(✉), Mirha Bičo Ćar, and Elvir Šahić

School of Economics and Business, University of Sarajevo,
Trg oslobođenja 1, 71000 Sarajevo, Bosnia and Herzegovina
{savo.stupar,mirha.car}@efsa.unsa.ba

Abstract. Due to the fact that modern computers have "perfect memory" and can process a huge number of transactions through a computer network very quickly, the idea of a decentralized currency system is not a new thing and it has long been a dream of advocates of the concept of anonymous digital money. Satoshi Nakamoto has come up with a genius idea to create a system that is a digital abstraction book of the balance and make that system public for all practical transactions related to accounts and transactions, which, when simplified, are reduced to mere addition and subtraction operations. He used the digital media feature to make perfect own copies, so he distributed this copy of the balance book to every computer in the network (decentralized system) and obtained a unique (structurally unchangeable) and global (public) book of balance. Thus, a revolutionary blockchain technology was created, which initially enabled and supported the functioning of the first digital crypto currency - bitcoin. Blockchain technology is an amalgam of several different and equally revolutionary technological achievements in the field of cryptography and computer networks. The aim of this paper is, using Bitcoin as example, to briefly explain the functioning of this technology and its suitability for use in many other areas of human activity, especially those that have a problem of distrust and the possibility of fraud.

Keywords: Blockchain technology · Digital cripto currency · Bitcoin · Mining · Hash

1 Introduction

Any commodity exchange, whether it is a trade of goods between companies or people, is based on transactions. A transaction can be defined as the transfer of ownership of goods (things, money, securities) from one entity to another, whereby one entity (initiator the transaction) loses the right to own the goods, and the one to whom the goods are given, acquires the right of ownership over it. The different names for the transaction are the purchase or sale, payment or repayment, provision or use, debt or claim, increase or decrease, receipt or giving, etc. All today's services are based on this concept. However, the key problem that every transaction carries is distrust. A solution to this problem is through a history found in mediation, that is, in a third entity, called a mediator. The best example of this are banks. Lately, globalization of the market enabled banks to expand their business beyond the borders of the state and some banks become larger and more powerful than some countries. The technology that allows the

I. Karabegović (Ed.): NT 2019, LNNS 76, pp. 355–364, 2020.
https://doi.org/10.1007/978-3-030-18072-0_41

liberalization of the transactional business model by eliminating intermediaries, increasing the security and speed of transactions, reducing costs, eliminating the possibility of misuses and frauds, etc., is blockchain technology. Because it is introducing revolutionary changes in the current approach to transaction implementation, blockchain technology is considered a technology of the future.

2 The Emergence of Bitcoin

We can not talk about blockchain technology without talking about Bitcoin. When Satoshi Nakamoto (whose identity is not yet known: some claim it's a person, some that's a team of people), published the document "Bitcoin: A Peer to Peer Electronic Cash System" [1] already in 2008, which he described "A true version of electronic money through a network of equal users" called Bitcoin, blockchain technology is released to the public. Not long after the document was released, Bitcoin was offered "open source"[1] to the public in 2009. Blockchain had answer for the question of digital distrust because it records important information in the public space without allowing information to be changed or deleted. Blockchain is transparent, timed and decentralized.

"Blockchain to Bitcoin, is what Internet is to e-mail. A large system that can be upgraded by applications. The currency is only one," [2] said Sally Davies, FT Technology Reporter.

2.1 What Is Bitcoin?

For Bitcoin various descriptive terms are used, such as: virtual and digital money; virtual, digital, electronic, synthetic and crypto currencies. Unfortunately, there is currently no single definition.

US FinCEN [3] and the European Central Bank [4] have classified Bitcoin as a virtual currency. The National Bank of China has classified Bitcoin as something that "originally is not a currency, but is an investment object [5]". The German court characterized Bitcoin as a unit of measure [6]. The Finnish government [7] classified Bitcoin as a commodity. Bitcoin.org, a Bitcoin wiki portal, defined it as [8]: "BitCoin is a payment method based on the concept of a digital crypto currency that functions without any central authority or third party as a creditor." As can be seen from the above, the definitions differ significantly. However, what is an indisputable fact is that Bitcoin is a new technology or more precisely a protocol, and that any other function that Bitcoin performs just derives from its technical characteristics. Generally speaking, the protocol is an arranged process that needs to be followed in a particular situation.

This crypto-currency is not based on a gold background, there is no country of origin, and no state or central bank of any state stands behind it, nor does the central bank of a state union.

[1] Open source software with license for modification, reconstruction and enhancement.

2.2 How Does Bitcoin Work?

This is a question that is very often asked, and it even more often causes confusion. The creator of Bitcoin, Satoshi Nakamoto, once said, "I'm sorry to have to be a wet blanket. But describing Bitcoin's general audience is damn hard. There is nothing I could compare it to" [9].

The most important thing is that a new user can start using Bitcoin without fully understanding the technical details in the same way that someone can use the program without being a programmer. It is also necessary that the user installs the Bitcoin Electronic Wallet (E_Wallet) application on a computer or smartphone, which will then generate his first Bitcoin address. Transfer Bitcoin from one address to another, or bank-based "from one account to another", is in practice similar to sending and receiving e-mails. As when sending an e-mail, the user who sends it needs to know the address of the user to whom the message is sent, so the user sending the Bitcoins must know the address of the user to whom they are sending it to.

Transactions in the network are linked to the user through a private and public key system. The public key, is publicly available and sent together with a network transaction, and the private key is known only to the user. The public key at the same time represents the address of the wallet (or the number of the safes), and the private key allows the user (who knows the private key) to access the wallet resources. Bitcoin's wallet also offers other benefits, such as the automatic generation of QR code[2] [10] that a smartphone can scan with its camera, which will automatically launch a payment request. For example, You want to pay lunch at a restaurant that receives BTCs. You come to the cashier where your waiter gives you an invoice, but this invoice is in the form of a QR code. You take out your smartphone on which you have the Bitcoin wallet installed (as well as the appropriate funds at your address) and bring your smartphone camera to "screen" the QR code on the screen (or you get a request through bluetooth) and the process is automatically initiated: the wallet application asks you "Do you want to make a transaction in the specified value to the restaurant address (address)?" By clicking the "YES" button, the transaction was done and you paid your bill. This is nothing unusual, and is already used in restaurants and cafes all over the world, and recently in region of SEE, neighboring Slovenia, Serbia and Croatia [11, 12].

3 Blockchain Concept

Even today there are people who believe that Bitcoin is the same as Blockchain technology, although these are two different concepts. Those who already spotted differences in 2014 and noticed that blockchain can be used for other things than crypts, have begun to invest in research and search for various implementations of this

[2] The QR Code (Quick Response Code) is a type of two-dimensional barcode originally designed for the automotive industry in Japan. Barcode is an optical tag that the machine can read and contains information about the product to which it is attached [10].

technology. On its basis, blockchain is a digital, open and decentralized “book” that permanently records all transactions between the two sides since the beginning of the use of the application without the need for mediation or authentication by a third party. This process is very efficient and it is anticipated that the costs of transactions will be drastically reduced in the future.

Blockchain is a technology that, besides logging Bitcoin’s transaction, records transactions of all other crypts, as well as records of transactions of any property, apartment, home, land, car, information on the origin of real estate, cars, medical data, landlord, cadastre, records of diplomas, etc. All transactions in any area applicable to essential transaction elements (such as in the example of a cryptoid account number from which you are being paid or a unique payroll identification number, the amount to be paid, the account number to which you are paid or the unique the payee identification number, etc.) are recorded digitally and stored in a single file whose copy is distributed to a large number of computers (servers or nodes connected to the network) and is publicly available to all members of the network. One transaction is recorded in one record and packed in a block transaction. Each subsequent transaction is added and packed in the same block.

The key idea of the blockchain technology is that this unique central register replaces copies of blockchain files on a large number of powerful computers (servers), which will be located in different locations, which are called nodes. The owner of these powerful servers can be anyone, that is, anyone who signs up for such a job and provides accessibility and the ability to update the record (blockchain file) to all participants in the transaction at any time of the day. Due to the fact that this job requires significant investment in equipment, and the payment of electricity costs, computer administration and maintenance, it is natural that owners of these powerful servers (miners) will not apply for such a job just like that. In order for someone to deal with these jobs, there must be an appropriate reward for entering and carrying out such a job. The main task of the miner is to confirm the transactions of the participants in the transaction and to ensure their validity.

3.1 Hashing and Mining [13, 15, 16]

Miner can also be an individual, but nowdays, mining pools will be engaged in mining business entities that have a larger number of computers or so-called mining pools, which organize the work of small individual (home) miners [13]. Mining Pool acts as one user on the network, but distributes internal work to all its members, which then shares the earned bitcoins in proportion to the power of their computers. In order for the BTC transaction between the users to be confirmed, it is necessary to pack it into the block - the basic element of the blockchain. The Hash[3] value of the block is calculated

[3] Hash function is any function that can convert digital input data of arbitrary size into a set of digital data of a fixed size (hash value). A quality hash function is that one which has small differences in input data result in very large differences in output data. Hashing is the process of getting hash values [14, p. 10].

so that the algorithm-based miner selects a certain number of transactions received from the network and then hashed them with the standard SHA256 algorithm[4].

Two very important questions to answer are: 1. How to ensure that all copies of a blockchain file, which are dispersed on nodes, are identical at all times? and 2. How are these transactions executed at all? Each of these servers keeps on the hard disk the most up to date copy of the blockchain file with all transactions since the beginning of the application's use. Every time a new transaction occurs, it is currently updated on each of these servers (nodes). Due to the fact that the possible malicious action of changing or deleting a transaction should be done not only on one (central) computer but on all copies of the blockchain file, the procedure of decoding the transaction records, or copying the record file to a large number of computers, it is practically impossible to execute such malicious activity [15, p. 15; 65], [16, p. 2].

The initiation (intention to execute) of a transaction is done by one of the participants by sending a message with the basic elements of the transaction to all the nodes (miners) in the network via its computer. In this way, an initiated transaction, but since it has not yet been validated (not valid), can not be recorded in the register of all transactions (blockchain file). The servers of all miners on the Internet receive information about the intention to execute this transaction. They first check whether the transaction initiator has enough funds in his account, so that the program "traverses" a complete blockchain chain, that is, "reads" all the transactions (one by one) he ever had to calculate his account balance, and verified that the amount of the initiated transaction (which is transferred to another account) is greater than the calculated balance of the account. In this way, it is checked that the transaction initiator has sufficient funds in the account to be able to execute the transaction. It is easy to conclude that there is no record of a transaction in the transaction register with a recorded balance of a user's account, but this situation is recalculated every time on the basis of all the transactions he has ever had. In this way, with the help of blockchain technology, another (higher) level of security and transaction control is achieved. If someone was maliciously trying to change the available balance on the user's account, he would have to change all transactions on that account ever done, which is practically impossible [16].

Thus, the first step in performing a transaction is to initiate a transaction, which controls whether the transaction initiator has enough funds in the account to be able to execute the transaction. The second step in this process is the bidding that the first major will succeed in installing the block of transactions to which the transaction belongs to the blockchain file (transaction register) and to permanently add it as the next block in the transaction chain. In order for the new transaction to be definitely recorded (entered) in the register of all transactions, and adding a new link (block transaction) in the chain, it is necessary for the majors (randomly generating) to begin searching for the appropriate number. This number can not be calculated by any formula (algorithm). It is the number that only "passes" in the chain link, linking the

[4] The feature of the sha256 hash (Secure Hash Standard) function is that regardless of the size of the input, one letter or whole book, always as a result, it gives a 256-bit number. This 256-bit number originally found in the binary zeros and ones format is shown in its hexadecimal version, which is 64 hex characters (24 = 16; 64 * 4 = 256). The characteristic of a good hashing function, is its collision resistance (collision is getting the same results for different input values).

last verified transaction with the new transaction. It can only be obtained by accidental guess. This can be practically done only with the help of a computer program, random number generation by random function, and checking whether this number matches (passes).

All yet unverified transactions performed over a specific time period, which usually lasts 10 min, are collected and placed in one block. The information stored in that block is used by miners by applying the mathematical formula to them, converting that information into something much shorter, seemingly random series of numbers and letters, which we call hash. This hash is stored together with a block at the end of the chain of blocks. Hashes have some specific characteristics. It is very easy to produce hash from data such as bitcoin block, but it's almost impossible to find out which data is looking only for hash. It is very easy to produce hash from a large amount of data, but each hash is unique. If only one letter or number is changed in the input string (message), hash is completely changed. However, miners use not only a transaction block to generate hash, but also use some other data. One of these data is the hash of the last block stored in a block of blocks. Because the hash of each block is generated using a hash block before it, that block becomes a digital seal or signature. He confirms that this block, and every block behind it, is valid, because if someone tried to change it, everyone in the network would know it and would not allow it [16, p. 2].

If someone attempts to falsify a transaction by changing a block that is already stored in a block of blocks, it would change the hash of that block. If somebody went to check the authenticity of the block by running the hash function on it, it would find that this hash does not respond to that block, which is already stored next to it in the block of blocks. Therefore, it would be known that this block is counterfeit. Because the hash of each previous block is used to create the next block in the chain, changing one block would cause the next block to change. This means that changing any block would cause a chain reaction that would extend all the way to the end of the chain.

It is precisely the invention of the appropriate digital signature of the new block proof of work in the Proof of Work blockchain system. The miners compete who will digitally sign a new block and add it to the chain because it carries valuable prizes: 1. the fixed value of the new Bitcoin prescribed by the bitcoin protocol and 2. the variable value of the bitcoin that the network users decided to hook up on their transactions to would motivate miners to verify their transaction [16, p. 2] [17, p. 3]. The SHA256 algorithm is again used to verify the block, and it is performed over the following data – block index + *hash* od previous block + data (BTC transactions) + *timestamp* (of new block of candidates) + *nonce*.

The problem is that with the help of the computer it is very easy to create a hash from the input data. Because of this, the Bitcoin network had to make things heavier, otherwise everyone would create thousands of hashes every second, and all bitcoins would be "spoiled" for time card. The Bitcoin protocol deliberately aggravates the situation by means of something called Proof of Work [16, p. 4]. Bitcoin protocol will not accept any kind of hash. "The Hash of the new block must be executed according to the current specification of the bitcoin algorithm, and this parameter is called difficulty (in free translation - the weight of the calculation). If the difficulty dictates that the hash of the new block must initially have four zeros, the hardware accounts for a new hash until the first time it comes to a compatible value. Of all the values that are hashing, one is allowed to change, which is nonce. The value ranges from zero, in each of the hash recalculation cycles it increases by one, which gives a different value to the final hash. If a miner manages to find the correct hash for a new block before a network fellow sends him his own correct block, he adds it to his local blockchain and sends it back to the network. If a new block arrives from the network before, the calculations are interrupted, the list of transactions received from those included in the new block is cleared, a new list of transactions for verification is created, and the calculations for the next block begins" [13].

It is clear why miners compete who will first calculate a block that continues blockchain, because it is precisely this competition that guarantees the safety and independence of the system to an acceptable level. The Difficulty algorithm is otherwise variable, and is calculated based on the verification speed of the previous 2,016 blocks. An algorithm that adjusts the difficulty for the goal to verify the new block roughly every 10 min, [18, p. 178] thus maintaining the computational complexity and intensity of the complete process. Maintaining the complexity and intensity of the process of verification of each new block of transactions is extremely important because it greatly complicates the possibility of fraud within the network.

When the miners definitely verify the block, they are then added to the block chain, which contain the previous transactions. Therefore, the miners' job is to verify and record transactions in the General ledger. This creates a chain of all transactions. That's why blockchain can be defined as a list of all transactions that have occurred since the introduction of a specific crypt, sorted by the time the transaction was created [18, p. 177].

One block of transactions in the blockchain consists of the following structural elements - block index (block number in the complete chain), hash values of the previous block (digital signature of the previous block), timestamp (data when the block is generated), bitcoin transaction data, hash the values of the current block and nonce, a very important number indicating how intense the miner needs to calculate to obtain a valid or corresponding hash for the current block [18, p. 169].

Since the strengths of the blockchain technology have proven to work with cryptoworks, its massive application in many other areas begins.

3.2 Application of Blockchain Technology

The application of technology, which was created primarily to support the logging of bitcoin cryptools transactions, has no limits, but is the most effective in those areas of human activity involving different types of transaction records (land, housing, car, insurance, contracts, diploma, other crypts, etc.) where transparency is needed, it eliminates the possibility of fraud, malversation, bribery and corruption and is evicted or bypassed by an intermediary who usually owns the said records (bank, state, lawyers, courts, educational institutions, insurance companies, sellers, etc.). We will provide some other application [customized according to 19, 20 and 21]:

1. Smart contracts are one of the possible variants of improving the concept of blockchain, in which instead of the transaction, the program code is entered into the blockchain. The reason why the program code is entered into blockchain is one feature of blockchain which prevents the information stored in it from being changed (in this case, the program code) and do not depend on the trust between the parties entered into the contract [19];
2. Application in the banking service for interbank transactions, which would eliminate the current broker - SWIFT service and be more acceptable to all banks which are using this service, because it would be more efficient and it would have significantly lower transaction costs [20, p. 13].
3. Apply land register records (cadastre and gruntu) where one entered transactions become publicly available and verifiable for all citizens and stakeholders [20, p. 14].
4. Apply in the electoral voting record in the elections, where each voter would have the opportunity to check if his voice really counted on the number of votes won by the candidate or party for which he voted [19];
5. Application in healthcare, for patient medical records, where it is possible and necessary to use chains with protected access to data [21];
6. Application in education, diploma and certification records, which would significantly reduce the possibility of misuse and falsification of diplomas [20, p. 16];
7. Application to supply chain records [19];
8. The management of the records of groceries of organic origin;
9. Control of personal data, used for marketing purposes;
10. Sharing personal computer resources;
11. Records of payments within computer games [19].

4 Conclusion

In this final part of the paper, we need to answer the question: "What is the revolutionary character of blockchain technology compared to other data processing technologies? By applying blockchain technology in any area, **it is forbidden for any time to delete or modify any transaction that is entered in the blockchain file of the specific application.** If the blockchain file of each application (the registry of all transactions in that application), which has been created since the beginning of that application, is stored and kept in one place, or more exactly one computer (centralized), regardless of all types and levels of protection against unauthorized access, there is a real possibility of modifying or deleting transactions from that file, or misuse by computer experts (hackers). Another, no less significant possibility of misuse is the possibility for those who have entrusted users (participants in transactions) to keep records of their account (e.g. the bank) or their property (the state) or their contract (lawyer or notary), that is, the owner the computer on which the blockchain file is located and which is managed by the user's records, completely nonetheless whether intentionally or out of negligence changes or deletes some transactions. The two abovementioned misuse are disabled if the file with the records of all transactions ever created is stored on a large number of computers (nodes) in a decentralized way because someone who would want to modify a transaction in that file would have to change that transaction in each of the copies located on a large number of computers. If it succeeds (which is unlikely), there is another, even more secure level of protection: "By changing only one character in the data about transaction, the hash (digital signature of the transaction block) would be changed and would not correspond to the original hash entered in the next block of transactions in a blockchain file, and therefore could not be entered in place of the original block" [18, p. 169].

The basic characteristics of each transaction record, which uses blockchain technology are: transparency, publicity, the unchangeability of one recorded transactions, and thus security and trust in the accuracy of the records, elimination of misuse, elimination of any intermediary (third party), and consequently reduction of management costs records.

Because of these characteristics, taking into account the opinions of many information technology experts, one might conclude that the emergence of blockchain technology today is the same as the emergence of the Internet in the 1990s. This technology provides unprecedented, almost utopian possibilities of change for the better in all areas of human activity. But whether it will become the ruling technology of the future, it does not depend on those who have created it, nor from those who are developing it further, nor from those who know its possibilities and who are supporting its further development.

References

1. Nakamoto, S.: Bitcoin: a peer-to-peer electronic cash system, [bitcion.org] (2008). https://bitcoin.org/bitcoin.pdf. Accessed 13 Jan 2019
2. Marr, B.: A very brief history of blockchain technology everyone should read. Forbes Magazine [Forbes.com] (2018). https://www.forbes.com/sites/bernardmarr/2018/02/16/a-very-brief-history-of-blockchain-technology-everyone-should-read/#682f67b97bc4. Accessed 20 Jan 2019
3. Financial Crimes Enforcement Network: Statement of Jennifer Shasky Calvery, Director, Financial Crimes Enforcement Network, US Department of the Treasury, 19 November 2013. www.fincen.gov/news/testimony/statement-jennifer-shasky-calvery-director-financial-crimes-enforcement-network. Accessed 22 Jan 2019
4. European Central Bank: Virtual Currency Schemes, Frankfurt (2012). https://www.ecb.europa.eu/pub/pdf/other/virtualcurrencyschemes201210en.pdf. Accessed 26 Jan 2019
5. CNBC: Bitcoin can be an asset but not a currency, says China central bank adviser CNBC, 06 July 2017. https://www.cnbc.com/2017/07/06/bitcoin-can-be-an-asset-but-not-a-currency-says-china-central-bank-adviser.html. Accessed 26 Jan 2019
6. Clinch, M.: Bitcoin recognized by Germany as 'private money', CNBC, 19 August 2013. https://www.cnbc.com/id/100971898. Accessed 22 Jan 2019
7. Pohjanpalo, K.: Bitcoin Judged Commodity in Finland After Failing Money Test, Bloomberg, 20 January 2014. https://www.ecb.europa.eu/pub/pdf/other/virtualcurrencyschemes201210en.pdf. Accessed 27 Jan 2019

8. Bitcoin.org: What is Bitcoin? (2014). https://bitcoin.org/en/faq#general. Accessed 17 Jan 2019
9. Satoshi Nakamoto Institute: Bitcoin Talk Re: Slashdot Submission for 1.0 (2010). https://satoshi.nakamotoinstitute.org/posts/bitcointalk/167/. Accessed 08 Jan 2019
10. Chang, J.H.: An introduction to using QR codes in scholarly journals. Sci. Editing **1**(2) (2014). https://www.escienceediting.org/journal/view.php?number=24. Accessed 29 Jan 2019
11. Aljazeera Balkans: Stvarna kupovina virtuelnim novcem, 12 September 2014. http://balkans.aljazeera.net/vijesti/stvarna-kupovina-virtuelnim-novcem. Accessed 26 Jan 2019
12. Bitcoin Forum: Tko prihvaća bitcoin kao sredstvo plaćanja, 03 April 2015. https://bitcointalk.org/index.php?topic=1011490.0. Accessed 29 Jan 2019
13. Arunović, D.: Što je u stvari blockshain i kako radi? BUG (2018). https://www.bug.hr/tehnologije/sto-je-u-stvari-blockchain-i-kako-radi-3011. Accessed 24 Jan 2019
14. Swathi, E., Vivek, G., Rani, G.S.: Role of hash function in cryptography. Int. J. Adv. Eng. Res. Sci. (IJAERS) (2016). https://www.researchgate.net/publication/312242372_Role_of_Hash_Function_in_Cryptography. Accessed 24 Jan 2019
15. Karame, G.O., Androulaki, E.: Bitcoin and Blockchain Security. Artech House, Boston, London (2016)
16. Gupta, S., Sadoghi, M.: Blockchain transaction processing. In: Sakr, S., Zomaya, A. (eds.) Encyclopedia of Big Data Technologies. Springer, Cham (2018)
17. Kiayias, A., Koutsoupias, E., Kyropoulou, M., Tselekounis, Y.: Blockchain Mining Games, ERC project CODAMODA (2016). https://arxiv.org/pdf/1607.02420.pdf. Accessed 24 Jan 2019
18. Antonopoulos, A.M.: Mastering Bitcoin. O'Reilly Media, Sebastopol (2010)
19. Minović, M.: Blockchain tehnologija: mogućnosti upotrebe izvan kripto valuta. Conference paper, Infotech 2017, At Aranđelovac, Srbija (2017). https://www.researchgate.net/publication/318722738_BLOCKCHAIN_TEHNOLOGIJA_MOGUCNOSTI_UPOTREBE_IZVAN_KRIPTO_VALUTA. Accessed 02 Feb 2019
20. Crosby, M., Nachiappan, Pattanayak, P. Verma, S., Kalyanaraman, V.: Block chain technology beyond bitcoin, Sutardja Center for Entrepreneurship & Technology Technical Report (2015). https://scet.berkeley.edu/wp-content/uploads/BlockchainPaper.pdf. Accessed 29 Jan 2019
21. Ekblaw, A., Azaria, A., Halamka, J.D., Lippman, A.: A case study for blockchain in healthcare: "MedRec" prototype for electronic health records and medical research data. White Paper (2016). https://pdfs.semanticscholar.org/56e6/5b469cad2f3ebd560b3a10e7346-780f4ab0a.pdf. Accessed 02 Feb 2019

Intelligent Transport Systems, Logistics, Traffic Control

Intelligent Mobility

Sadko Mandžuka[(✉)]

Faculty of Traffic and Transport Sciences, University of Zagreb,
Vukelićeva 4, 10000 Zagreb, Croatia
sadko.mandzuka@fpz.hr

Abstract. Modern traffic problems can no longer be solved solely by the physical construction of new roads or reconstructions of existing roads. In this respect, significant scientific and research efforts have been made over the past twenty years, to address the problem of transport using new information and communication technology resources and novel knowledge on how to run such complex systems and processes. This new area of classical traffic engineering, called Intelligent Transport Systems (ITS), demonstrates a new approach and application of advanced management and technical-technological solutions to achieve greater safety, efficiency and reliability of transport, while reducing environmental and social impacts (reduction pollutant emissions, noise and the like). Modern "upgrade" of ITS (especially in an urban environment) is Intelligent Mobility, which is focused on the quality of the mobility itself and viewed from the perspective of the end-user. Intelligent Mobility is one of main piers of Smart City concept. The basic features of these new concepts are presented in the paper.

Keywords: Intelligent Mobility · Intelligent Transport Systems · Smart City · Information-communication technologies · Sustainability

1 Introduction

The apparent constraints of a classical approach to the development of the transport system, in accordance with the principles of scientifically-based creation of a policy of management and sustainable development, have led to the demand for new harmonized solutions in road and other transport branches. The essence of ITS is systematic management and information-communication solutions built into traffic infrastructure, vehicles, control centres etc., [1, 2].

The successful development and implementation of complex systems such as ITS cannot be based on a classical development cycle that assumes that input requirements are well defined, and that technology will not significantly change during the development cycle. As proof of this claim, numerous complex system projects can be used as well as initial ITS projects that are realized in various countries. Users cannot express clearly the requirements or realize the possible implications of new technical/technology solutions. The dynamic development of information and communication technologies constantly expands the possible solutions. As a special problem in implementation of intelligent transport systems, it is possible to use some of the existing resources built within the existing road transport infrastructure. This approach offers a series of possible

I. Karabegović (Ed.): NT 2019, LNNS 76, pp. 367–376, 2020.
https://doi.org/10.1007/978-3-030-18072-0_42

savings, but also potential dangers for non-functional future solutions. In this respect, it is necessary to carefully consider the possibilities of existing systems in the foreseeable life cycle of the future intelligent transport system.

Intelligent Mobility, which is focused on the quality of the mobility itself and viewed from the perspective of the end-user. Intelligent mobility (or Smart mobility) is usually most important pier of smart city concept. The basic objective of the Smart City concept is to achieve the viability of the city's life, based on balance between social and economic development. Successful management of the environment is a precondition for such a conceived approach. For example, many of the negative impacts of transport on the environment, on society or on the economy are not "internal" costs incurred directly by whoever travels (through the payment of fuel, motorway fees, car taxes, tickets for public transport etc.), but external costs, which fall on the whole community. According to some estimates, these external costs would represent about a third of the total costs of the transport system, [3].

The concept of Smart City is presented in Sect. 2 In Sect. 3 an overview of the Intelligent Transport Systems is given. The basic features of Intelligent mobility concept are described in Sect. 4. Paper is finished with the conclusion remarks in Sect. 5.

2 Smart City

There are a lot of definitions of Smart City. Most commonly these definitions are set depending on the viewpoint of the author. However, today the most commonly used term is the Smart City In this sense, the following definitions are best known:

Smart city as a high-tech intensive and advanced city that connects people, information and city elements using new technologies in order to create a sustainable, greener city, competitive and innovative commerce, and an increased life quality, [4].

Smart cities will take advantage of communications and sensor capabilities sewn into the cities' infrastructures to optimize electrical, transportation, and other logistical operations supporting daily life, thereby improving the quality of life for everyone, [5].

A smart city is based on intelligent exchanges of information that flow between its many different subsystems. This flow of information is analysed and translated into citizen and commercial services. The city will act on this information flow to make its wider ecosystem more resource efficient and sustainable. The information exchange is based on a smart governance operating framework designed to make cities sustainable, [6].

A city connecting the physical infrastructure, the IT infrastructure, the social infrastructure, and the business infrastructure to leverage the collective intelligence of the city, [7].

A smart city infuses information into its physical infrastructure to improve conveniences, facilitate mobility, add efficiencies, conserve energy, improve the quality of air and water, identify problems and fix them quickly, recover rapidly from disasters, collect data to make better decisions, deploy resources effectively, and share data to enable collaboration across entities and domains, [8].

Smart Cities initiatives try to improve urban performance by using data, information and information technologies (IT) to provide more efficient services to citizens, to monitor and optimize existing infrastructure, to increase collaboration among different economic actors, and to encourage innovative business models in both the private and public sectors, [9].

Also, other terms are also used: Intelligent City, Digital City, etc. In Smart City concept there are more components: Smart Infrastructure (SI), smart mobility (SM), smart living (SL), smart parking (SPk), smart governance (SG), smart environment (SE), smart homes (SH), smart technology (ST), smart economy (SEc), and finally smart people (SP), see Fig. 1.

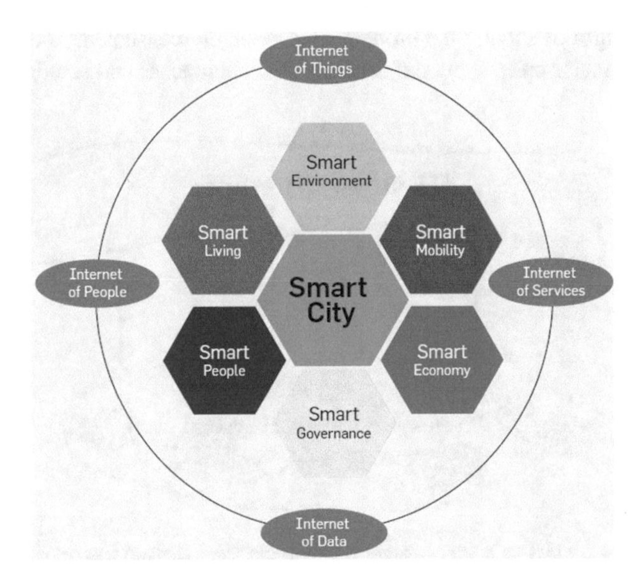

Fig. 1. Smart city model [10]

To achieve goals of Smart City, there are mature technologies today. First, this is the Internet of Things (IoT) and technology related to the Big Data Concept. The IoT is the network of interconnected devices including computers, smartphones, sensors, smart buildings, connected vehicles, various actuators and wearable devices. It can be divided in four main components: the "things", the local area network, the Internet, and the cloud.

The large amount of data is generated by thousands of these sensors and devices. In addition, a significant source of information is also the various social networks in which data from different groups and individuals are exchanged. Traditional data-processing technologies are not efficient, and this disadvantage is solved by Big data-based methodologies.

3 Intelligent Transport Systems

3.1 Definition

ITS can be defined as a holistic, control and information-communication (cybernetic) upgrade of the classical traffic and transport system that achieves significantly improvement in performance of traffic flow and transport operations, more efficient transport of passengers and goods, improving traffic safety, comfort and protection of passengers, lower environmental pollution, etc., Fig. 2, [2]. ITS has the meaning of a new critical concept that changes the approach and the trend of transport science and transport technology of people and goods to address the growing problems of traffic congestion, environmental pollution, transport efficiency, safety and protection of people and goods in traffic. ITS changes the dominant paradigm of solving the problem that is largely consumed. The growing traffic problems in all major cities show the need for new approaches and solutions.

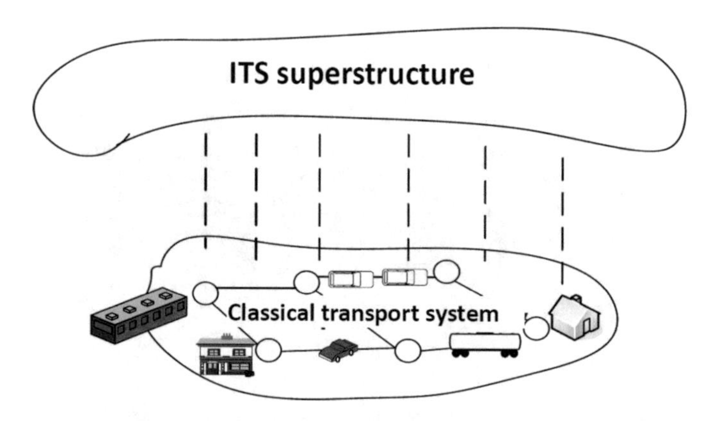

Fig. 2. ITS as a superstructure of the classical transport system [2]

During the development and implementation of new ITS projects, there are basically three approaches to estimate the effects and benefits of ITS: comparative review of the experience of others with the various extrapolation of ITS results; carrying out a pilot project and estimation the probable benefits in the concrete context and using a simulation model.

3.2 Basic ITS Domains and Services

ISO (International Standardization Organization) has set the initial standardization of ITS services focused on road traffic (ISO TR 14813-1). It was last corrected in 2015. Defined domains and their corresponding services are:

I. Traveller information
   (Pre-trip information; On-trip information; Route guidance and navigation; Route guidance and navigation on-trip; Trip planning support; Travel services information)

II. Traffic management and operations
(Traffic management and control; Transport related incident management; Demand management; Transport infrastructure maintenance management; Policing/enforcing traffic regulations)
III. Vehicle services
(Transport-related vision enhancement; Automated vehicle operation; Collision avoidance; Safety readiness; Pre-crash restraint deployment)
IV. Freight transport
(Administrative functions: Commercial vehicle preclearance, Commercial vehicle administrative processes, Automated roadside safety inspection, Commercial vehicle on-board safety monitoring; Commercial functions: Freight transport fleet management, Intermodal information management, Management and control of intermodal centers, Management of dangerous freight)
V. Public transport
(Public transport management; Demand responsive and shared transport)
VI. Emergency
(Transport-related emergency notification and personal security; After theft vehicle recovery; Emergency vehicle management; Emergency vehicle pre-emption; Emergency vehicle data; Hazardous materials and incident notification)
VII. Transport-related electronic payment;
(Transport-related electronic financial transactions; Integration of transport-related electronic payment services)
VIII. Road transport related personal safety;
(Public travel security; Safety enhancements for vulnerable road users; Safety enhancements for disabled road users; Safety provisions for pedestrians using intelligent junctions and links)
IX. Weather and environmental conditions monitoring;
(Environmental conditions monitoring)
X. Disaster response management and coordination;
(Disaster data management; Disaster response management; Coordination with emergency agencies)
XI. National security.
(Monitoring and control of suspicious vehicles; Utility or pipeline monitoring)

The European standardization of ITS covers the following Areas of ITS:

I. Electronic Fee Collection
II. Emergency Notification and Response – Roadside and In-Vehicle Notification
III. Traffic Management – Urban, Inter-Urban, Parking, Tunnels and Bridges, Maintenance and Simulation, together with the Management of Incidents, Road Vehicle Based Pollution and the Demand for Road Use
IV. Public Transport Management – Schedules, Fares, On-Demand Services, Fleet and Driver Management
V. In-Vehicle Systems – includes some Cooperative Systems
VI. Traveller Assistance – Pre-Journey and On-Trip Planning, Travel Information
VII. Support for Law Enforcement

VIII. Freight and Fleet Management
IX. Provide Support for Cooperative Systems – specific services not included elsewhere, e.g. bus lane use, freight vehicle parking…
X. Multi-modal interfaces – links to other modes when required, e.g. travel information, multi-modal crossing management

National ITS architectures can include domains and their corresponding services that are not explicitly mentioned in existing taxonomies.

3.3 ITS Architecture

ITS architecture represents fundamental organization of intelligent transport system that contains the components, their relationships and connections to environment, and their design and development principles, observing the entire life cycle of the system, [2]. Figure 3 shows the basic aspects of realization ITS project using FRAME architecture principles.

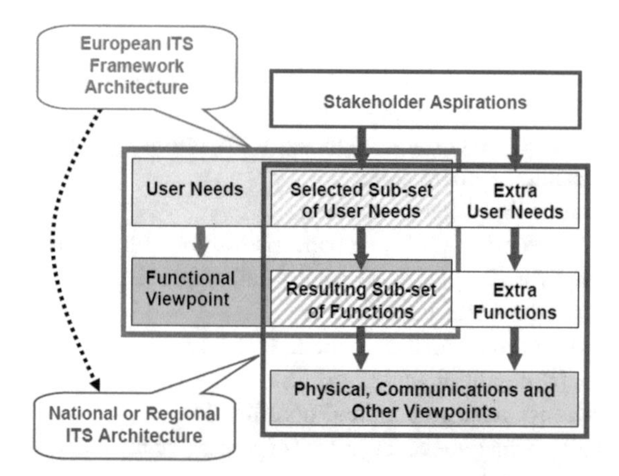

Fig. 3. The basic point of view of ITS architecture, [11]

The first step in the development of ITS architecture is clear and unambiguous define the needs/requirements of users (interest groups). Thereafter, a research of the functional aspect of functions (more and lower levels) is required to meet the requirements and to realize the interfaces with the outside world over terminators or actors. Planned data flows can be presented separately or as part of a functional (logical) architecture. Physical ITS architecture defines and describes ways in which parts of a functional architecture can be connected to form physical entities. The basic feature of a physical entity is that they can provide one or more services provided to users and that they can be physically realized. The process of creation involves physical and/or virtual (information) entities such as roads, telematic devices, software, etc. Physical systems, subsystems and modules perform communication via wired and wireless media with defined data protocols. The communication aspect can be viewed separately from physical architecture, only as communicational architecture. ITS architecture is a primary requirement and element of ITS planning and compliance development many ITS application. Architecture specifies how different components are interacted with so that concrete transport and traffic problems are addressed in a context. Architecture gives the General Framework for planning, designing and implementation integrated systems in a given spatial-temporal coverage.

4 Intelligent Mobility

Smart mobility aims at making transport systems more intelligent, more flexible and adept using ICT: from managing complex transportation systems in a cooperative way to supports decision making on how to travel and to organize the planned activities in smarter, greener and environmentally sustainable ways, [12].

The main areas of Intelligent mobility can be divided on:

(a) Big data for Smart mobility–for effective analysis and utilization of data from urban domains, [13]
(b) Driving Safety and Automated Driving–provide technology for safe and secure mobility, allowing cars to interact with other vehicles and the infrastructure around them,
(c) Innovation in Public transport – use ICT to cost reduction and transport effectiveness, service quality and traveller's satisfaction, etc.
(d) Advanced City logistics – for efficiency transport of goods in urban environment, including smart routing.
(e) Smart roads and infrastructure – for better traffic management optimizes energy consumption, reduce traffic congestion and improve traffic flow.
(f) Sharing and Urban Mobility – shared transport systems include car sharing, bicycle sharing, carpools, and vanpools. Besides that, multimodal systems can use different and optimally combined transport modes within the trip chain in a seamless way to approach greater sustainability in urban transport.
(g) Electric Mobility – provides a key to the sustainable redesign of mobility that is climate and environment friendly, efficient and allows to save resources.
(h) Green Mobility – minimizes the environmental impact caused by the transportation sector without impacting the growth momentum.
(i) Transport cybersecurity – reducing the consequences of possible malfunctions caused by cyber-attacks in transport systems
(j) Smart Payment–implemented to overcome the limitation of the conventional payment methods by revamping the payment method via parking meter and other technologies, [14]

For example, search for a free parking space is one of the leading causes of congestion in city centres [15]. Circulation on an urban traffic network does not increase the total travel time of this vehicle alone, but also of all other vehicles that must slow down or stop due to the congestion caused. This overall negative impact of this phenomenon, such as fuel consumption, emissions, stress and driver, and significantly increased the production of noise and compromising safety. ITS has enabled the provision of dynamic information and the creation of such systems that can display real-time information, fault-proof, capable of providing information on each street and off-street parking space, identifying vehicles, allowing billing with various means of payment, providing the collected and processed information to all stakeholders, worrying about the cyber-security and appropriate authorities, Fig. 4. Information can be displayed at anytime and anywhere (desktop or laptops, handhelds, tablets, smart phones, vehicle devices, variable content traffic signs, various displays). Evident and measurable parking-related problems, which primarily relate to congestion, environmental pollution and safety justify relatively high investment in intelligent parking systems. Rational management of parking spaces and a quality set strategy and parking policy can significantly affect the quality of life in local communities and customer satisfaction. These systems as a subset of intelligent mobility applications become components of the vision of smart cities. Intelligent parking systems will soon become an indispensable part of the new infrastructure projects of larger city centres as a recognizable part of the Internet of things concept.

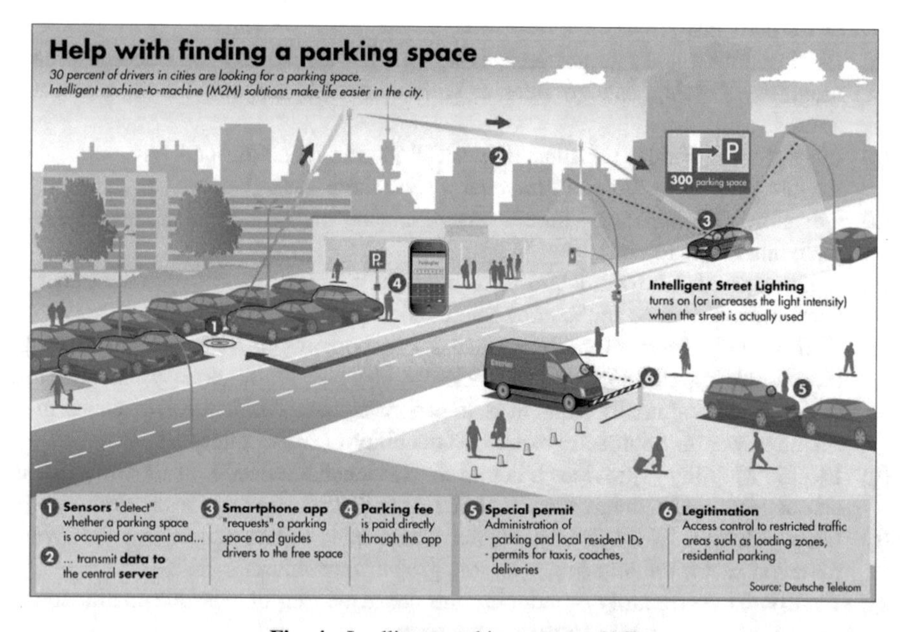

Fig. 4. Intelligent parking system, [16]

It is important to notice the difference between ITS and Intelligent Mobility. ITS focuses on applications and infrastructure related services. The area of Intelligent Mobility is focused on the quality of the mobility itself and viewed from the perspective of the end-user. It is in fact a kind of superstructure of ITS. In that sense, the Smart mobility has a wider and citizens-oriented aspect while still being strongly transport technology driven and can be defined as:

$$SM \supset \{SuM \cup ITS \cup AV \cup TC\,2.0 \cup SMG\} \quad (1)$$

where:

SuM – sustainable mobility,
ITS – Intelligent Transport System,
AV – autonomous vehicles and related infrastructure,
TC 2.0 – traffic control 2.0, including the control of traditional multimodal traffic,
SMG – smart mobility governance, [17].

Significant role in the Intelligent mobility concept will have cooperative traffic systems and various forms of automated and autonomous vehicles. Cooperative traffic systems can be defined in narrow sense as a problem of communication of a moving entity (vehicle) with road infrastructure and/or other moving entities. In this narrow sense the following communications were recognised: V2V – vehicle to vehicle, V2I – vehicle to infrastructure, V2P – vehicle to pedestrian, I2P – infrastructure to pedestrian, [18, 19].

5 Conclusion

Intelligent mobility as an upgrade to ITS approach is a promising paradigm that solves the needs for end-user (passengers and drivers) in a personalized way. It is human-oriented approach which takes many factors into account. The journey is optimized both from the perspective of the end-user but with respect for other social and environmental requirements. This is not just technology but also the demands of real and sustainable development of cities and larger regions. With proper integration Intelligent mobility and other Smart Cities components (Smart Infrastructure, Smart Life, Smart Governance, Smart Economy etc.), results will have synergetic effects.

Acknowledgment. The research presented in this paper is supported with University of Zagreb Program funds Support for scientific and artistic research (2018) through the project: "Impact of the application of cooperative systems and autonomous vehicles on traffic and society".

References

1. Mandžuka, S., Žura, M., Horvat, B., Bićanić, D., Mitsakis, E.: Directives of the European Union on Intelligent Transport Systems and their impact on the Republic of Croatia. Promet-Traffic Transp. **25**(3), 273–283 (2013)
2. Bošnjak, I.: Intelligent Transport Systems 1, Faculty of Traffic and Transport Sciences, Zagreb (2006). (in Croatian)
3. Staricco, L.: Smart mobility: opportunità e condizioni. J. Land Use Mob. Environ. **6**(3), 342–354 (2013)
4. Bakıcı, T., Almirall, E., Wareham, J.: A smart city initiative: the case of Barcelona. J. Knowl. Econ. **2**(1), 1–14 (2012)
5. Chen, T.M.: Smart grids, smart cities need better networks. IEEE Netw. **24**(2), 2–3 (2010)
6. Anavitarte, L., Tratz-Ryan, B.: Market Insight: 'Smart Cities' in Emerging Markets. Gartner, Stamford (2010)

7. Harrison, C., Eckman, B., Hamilton, R., Hartswick, P., Kalagnanam, J., Paraszczak, J., Williams, P.: Foundations for smarter cities. IBM J. Res. Dev. **54**(4), 1–16 (2010)
8. Nam, T., Pardo, T.A.: Conceptualizing smart city with dimensions of technology, people, and institutions. In: Proceedings 12th Conference on Digital Government Research, College Park, MD, 12–15 June 2011
9. Marsal-Llacuna, M.L., Colomer-Llinàs, J., Meléndez-Frigola, J.: Lessons in urban monitoring taken from sustainable and livable cities to better address the Smart Cities initiative. Technol. Forecast. Soc. Chang. (2014)
10. Khatoun, R., Zeadally, S.: Smart cities: concepts, architectures, research opportunites. Commun. ACM **59**(8), 46–57 (2016)
11. Planning a Modern Transport System, A Guide to Intelligent Transport System Architecture. https://frame-online.eu
12. Smart mobility. http://www.umbertopernice.com/expertises/smart-mobility
13. Nuaimi, E.A.: Applications of big data to smart cities. J. Internet Serv. Appl. **6**(1) (2015)
14. Faria, R.: Smart mobility: a survey. In: International Conference on Internet of Things for the Global Community (IoTGC). IEEE (2017)
15. Shoup, D.C.: Free parking or free markets. Access Mag. **38**, 28 (2011)
16. Smart parking system. https://erticonetwork.com/pisa-and-deutsche-telekom-launch-6-month-smart-city-pilot-project-to-optimize-city-parking
17. Šemanjski, I., Mandžuka, S., Gautama, S.: Smart mobility. In: International Symposium ELMAR. IEEE (2018)
18. Mandžuka, S., Ivanjko, E., Vujić, M., Škorput, P., Gregurić, M.: The Use of Cooperative ITS in Urban Traffic Management, Intelligent Transport Systems: Technologies and Applications. Wiley, New York (2015)
19. Mandžuka, S.: Cooperative systems in traffic technology and transport. In: International Conference New Technologies, Development and Applications, pp. 299–308. Springer (2018)

Performance Evaluation of Two Computational Approaches for Vehicle Collision Simulation

Clio G. Vossou(✉), Dimitris V. Koulocheris(✉), and Kiriakos P. Kapetis(✉)

Vehicles Laboratory, School of Mechanical Engineering, National Technical University of Athens, Zografou Campus, IroonPolytexneiou 9, 157 80 Athens, Greece
klvossou@mail.ntua.gr, dbkoulva@central.ntua.gr, kyrkapetis@gmail.com

Abstract. A vehicle collision can be divided in three distinct time phases, the pre-collision, the collision and the post-collision phase. Usually during a traffic accident reconstruction the collision and post-collision phases are investigated in order for the accident reconstructionist to draw conclusions concerning the causes and the events that lead to the vehicle collision. The investigation of both phases is usually a repetitive procedure which terminates when the investigation results match the physical evidence drawn from the accident scene. The objective of the analysis of the collision phase is the determination of the velocities of both vehicles prior and post collision. For the computational simulation of the collision phase two main approaches exist in the literature, the energy-based approach, developed by McHenry, and the momentum based one, developed by Brach, both in the late 1970s. The objective of the analysis of the post collision phase is the reconstruction of the trajectories of both vehicles from the point of collision to the point of rest. For the computational simulation of the vehicle trajectories different approaches exist, such as their approximation using geometric curves and the application of the equations of motion for each vehicle after collision. In the present paper two algorithms for vehicle collision reconstruction have been set up in Matlab. Each one utilizes a different approach for both the collision and the post collision phase. In more details the momentum-based approach has been coupled with geometric approximation of the trajectories while the energy-based approach has been coupled with the equations of motion for the post-collision phase. Both algorithms incorporate a suitable optimization method in order to provide optimized results in terms of collision geometry, collision physics and post-collisional trajectories of the vehicles. In order to evaluate the performance of both algorithms, the vehicle collisions described in details in the RICSAC database have been used. The results of each algorithm are compared with each other as well as with the measured quantities existing in RICSAC database.

Keywords: Vehicle collision · Genetic algorithms · RICSAC database · Planar Impact Mechanics · Crush energy

I. Karabegović (Ed.): NT 2019, LNNS 76, pp. 377–385, 2020.
https://doi.org/10.1007/978-3-030-18072-0_43

1 Introduction

The collision of two vehicles is a common type of traffic accident, which can lead to injury or even casualties for the passengers. The reconstruction of traffic accidents is crucial to safety improvements in vehicles either through electronic systems which provide better steering or through structural alterations which enhance the crashworthiness of the vehicle. Traffic accident reconstruction involves the qualitative and quantitative estimation of the way such an accident occurred utilizing engineering, scientific and mathematical laws as well as data and physical evidence collected from the accident scene, the involved vehicles and the passengers or witnesses. Often a traffic accident reconstruction is performed using computational simulations.

The event of a vehicle collision can be considered in three distinct time phases, the pre-collision, the instantaneous collision and the post-collision phase. In the literature, different approaches exist for the computational simulation of each phase. Two broad approaches are usually utilized for the simulation of the collision phase, the first one is based on the conservation of linear and/or angular momentum and the second one is based on the conservation of energy. The momentum-based models are these defined by Brach [1] and Ishikawa [2] while the model considering the conservation of energy is this defined by McHenry [3]. As far as the computational simulation of the post–collision phase is concerned geometrical approximation [4], use of vehicle models [5] and use of the motion equations for each vehicle [6] have been proposed in the literature.

In the present paper two computational algorithms for traffic accident reconstruction have been implemented in the programming environment of Matlab®. The first one utilizes the momentum-based approach for the simulation of the collision phase along with the geometrical approximation for the post–collision phase. The deterministic optimization method of SQP, implemented in Matlab® through fmincon function, is incorporated in the algorithm in order to provide results closer to the evidence provided in the accident scene. The second algorithm utilizes the energy-based approach for the simulation of the collision phase and the use of motion equations for the post–collision phase. In this algorithm the stochastic optimization method of the Genetic Algorithm existing in the optimization toolbox of Matlab has been utilized in order to provide results optimized with respect to the physical evidence. Both algorithms have been tested in the vehicle collisions composing the RICSAC database. The results provided by the algorithms have been compared to each other and to experimentally measured quantities as provided by the RICSAC database [7].

2 Traffic Accident Reconstruction Algorithms

In this section the traffic accident reconstruction algorithms set up in Matlab® are described in detail. Each algorithm has one module for the simulation of the collision phase and one for the simulation of the post–collision phase. Both algorithms incorporate an optimization method, coupled with a different module of the algorithm.

2.1 Algorithm A

Algorithm A combines the Planar Impact Mechanics (PIM) model as described by Brach et al. for the simulation of the collision phase to the geometrical approximation of the trajectory of the vehicles, as described by Struble et al., for the simulation of post–collision phase. It is an algorithm that reconstructs the accident inversely in time.

The PIM model is based on Newton's second law and the principle of impulse – momentum and it consists of six algebraic equations [8–10]. Briefly, these equations correlate (a) six initial velocity components (three for each vehicle – V_{ix}, V_{iy}, Ω_i), (b) six final velocity components (three for each vehicle – V_{fx}, V_{fy}, Ω_f), (c) two vehicle inertial properties (I, m) and (d) five parameters describing the collision geometry (d_a, d_b, d_c, d_d, Γ). The first three equations, Eqs. 1–3, are provided considering the conservation of linear and angular momentum.

$$m_1 \cdot \left(V_{1fx} - V_{1ix}\right) + m_2 \cdot \left(V_{2fx} - V_{2ix}\right) = 0 \tag{1}$$

$$m_1 \cdot \left(V_{1fy} - V_{1iy}\right) + m_2 \cdot \left(V_{2fy} - V_{2iy}\right) = 0 \tag{2}$$

$$I_1 \cdot (\Omega_{1f} - \Omega_{1i}) + I_2 \cdot (\Omega_{2f} - \Omega_{2i}) + m_1 \cdot (d_b + d_d) \cdot \left(V_{1fy} - V_{1iy}\right) + m_2 \cdot (d_a + d_c) \cdot \left(V_{2fx} - V_{2ix}\right) = 0 \tag{3}$$

The latter equations are provided considering the three impact coefficients which simulate the physics of the collision. These coefficients are the coefficient of restitution, e (Eq. 4), the equivalent friction of coefficient, μ (Eq. 5) and the moment coefficient of restitution e_m (Eq. 6).

$$\begin{aligned} &\left(V_{1fy} - d_d \cdot \Omega_{1f} - V_{2fy} - d_b \cdot \Omega_{2f}\right) \cdot sin\Gamma + \left(V_{1fx} - d_c \cdot \Omega_{1f} - V_{2fx} - d_a \cdot \Omega_{2f}\right) \cdot cos\Gamma = e \cdot \\ &\left[\left(\left(V_{1iy} - d_d \cdot \Omega_{1i} - V_{2iy} - d_b \cdot \Omega_{2i}\right) \cdot sin\Gamma\right) + \left(V_{1ix} - d_c \cdot \Omega_{1i} - V_{2ix} - d_a \cdot \Omega_{2i}\right) \cdot cos\Gamma\right] \end{aligned} \tag{4}$$

$$m_1 \cdot \left(V_{1fy} - V_{1iy}\right) \cdot (cos\Gamma - \mu \cdot sin\Gamma) + m_2 \cdot \left(V_{2fx} - V_{2ix}\right) \cdot (sin\Gamma + \mu \cdot cos\Gamma) = 0 \tag{5}$$

$$\begin{aligned} (\Omega_{1f} - \Omega_{2f}) \cdot (1 - e_m) = -e_m \cdot &\left[\left((\Omega_{1f} - \Omega_{1i}) - m_1 \cdot d_c \cdot \frac{\left(V_{1fx} - V_{1ix}\right)}{I_1} + m_1 \cdot d_d \cdot \frac{\left(V_{1fy} - V_{1iy}\right)}{I_1}\right.\right. \\ &\left.\left. -\left(\Omega_{2f} - \Omega_{2i}\right) - m_2 \cdot d_a \cdot \frac{\left(V_{2fx} - V_{2ix}\right)}{I_2} + m_2 \cdot d_b \cdot \frac{\left(V_{2fy} - V_{2iy}\right)}{I_2}\right)\right] \end{aligned} \tag{6}$$

The geometrical approximation method is based on physical evidence present on the accident scene, such as tire marks and debris, which allow the reconstructionist to assume points on the trajectory of the vehicles moving from the collision point to the point of rest. These points are considered as key points and a geometrical curve, in this case a cubic spline, is interpolated through them [4]. Both the points of collision and rest are considered as key points. Following the interpolation of the trajectories of the vehicles with two splines, their velocities on each key point are calculated using the conservation of kinetic energy. Since the velocities of the vehicles at their points of rest are known and equal to zero, Eq. 7 is applied successively for each trajectory segment.

$$V_i = \sqrt{V_{i+1}^2 + 2 \cdot g \cdot (DragF_i \cdot \Delta S_i + Z_{i+1} - Z_i)} \tag{7}$$

In Eq. 7 g is the acceleration of gravity, ΔS is the length of the trajectory segment and Z the height of the road, with respect to the point of collision. Furthermore, the coefficient *DragFi* represents the energy loss through each trajectory segment and it is calculated as $DragF_i = \mu \cdot \sqrt{(LF_i \cdot \cos(\theta_{crab_i}))^2 + sin^2(\theta_{crab_i})}$, where μ is the friction coefficient of the road surface and θ_{crab_i} is the slip angle. The slip angle is calculated as $\theta_{crab_i} = \frac{1}{2}(\psi_i + \psi_{i+1}) - \theta_i$ where, ψ is the direction of the vehicle in the Cartesian reference system and θ is the angle of the trajectory segment.

The inputs of Algorithm A are the known coordinates of the points of rest of the both vehicles. The geometrical approximation of the trajectory of each vehicle up to the point of collision is performed and the velocity of each vehicle on all key points is calculated with Eq. 7. The trajectories of the vehicles as well as the calculated velocities refers to the center of mass of the vehicle. For the geometrical approximation 20 key points have been used. Then, using the velocities of the vehicle at the point of collision as input, the PIM model is implemented providing the initial velocities of each vehicle as output. The SQP optimization method, in the form of the fmincon function of Matlab® is coupled with the PIM model having as design variables the impact coefficients and the initial velocities of the vehicles. The cost function of the SQP

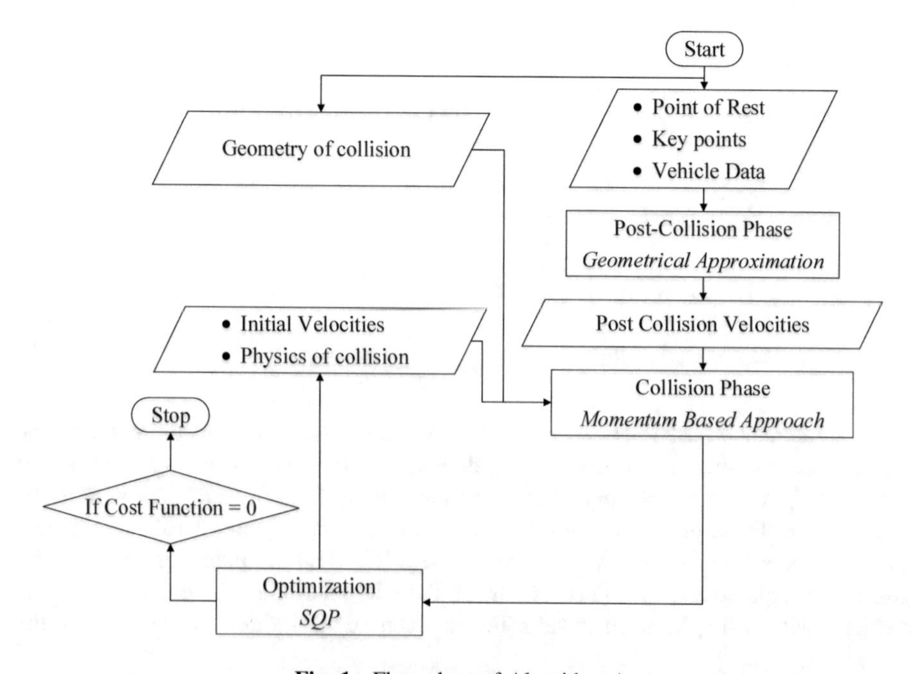

Fig. 1. Flow chart of Algorithm A

optimization method is a least square type of function matching the initial velocities of the vehicles to experimental estimates. In Fig. 1 the flow chart of Algorithm A is presented.

2.2 Algorithm B

Algorithm B combines the energy-based approach described by McHenry et al. for the simulation of the collision phase and the use of the equations of motion for each vehicle for the simulation of the post–collision phase.

According to the energy-based approach, the residual deformation of the vehicles due to collision is used in order to calculate the change in velocity, ΔV_i, sustained by each vehicle during a collision [11]. The necessary information for the estimation of the crush energy loss during collision, consists on the measurements of the deformation of both vehicles and the experimentally determined crush stiffness coefficients for each vehicle [12].

According to Campbell, for impact speeds above approximately 9 m/s, a near linear relationship between the impact speed and crush depth can be obtained. Moreover, given that in a barrier impact all the kinetic energy of the vehicle at impact is converted into residual crush, there is linear relation between force F and crush deformation C, suitable to describe the force per unit width (Eq. 8) where A and B are coefficients of unit width properties obtained from barrier crash test data [11, 13].

$$F = A + B \cdot C \tag{8}$$

The work produced from the residual deformation of the vehicle can be determined by integrating Eq. 8 with respect to the crush deformation and the damage width (L).

$$E = \int_0^L \int_0^C (A + B \cdot C) dC dL \tag{9}$$

In order to numerically integrate Eq. 9 the crush area is divided, usually, into 6 crush zones, defined by a series of crush measurements C_i and Eq. 10 is produced.

$$E_C = \frac{L}{5}\left[\frac{A}{2} \cdot (C_1 + 2 \cdot C_2 + 2 \cdot C_3 + 2 \cdot C_4 + 2 \cdot C_5 + C_6) + \frac{B}{6}(C_1^2 + 2 \cdot C_2^2 + 2 \cdot C_3^2 + + 2 \cdot C_4^2 + 2 \cdot C_5^2 \right.$$
$$\left. + + C_6^2 + C_1 \cdot C_2 + C_2 \cdot C_3 + C_3 \cdot C_4 + C_4 \cdot C_5 + C_5 \cdot C_6) + 5 \cdot G\right] \tag{10}$$

In Eq. 10, the coefficients A, B and G are the coefficients of unit width properties obtained from barrier crash test data for each vehicle.

Knowing the absorbed energy through residual deformation for each vehicle, the linear velocity change ΔV_i of each vehicle is determined using Eq. 11a and Eq. 11b [11].

$$\Delta V_1 = \sqrt{\frac{2 \cdot \gamma_1 \cdot (E_1 + E_2)}{M_1 \cdot \left(1 + \frac{\gamma_1 \cdot M_1}{\gamma_2 \cdot M_2}\right)}} \, and \, \Delta V_2 = \sqrt{\frac{2 \cdot \gamma_2 \cdot (E_1 + E_2)}{M_2 \cdot \left(1 + \frac{\gamma_2 \cdot M_2}{\gamma_1 \cdot M_1}\right)}} \tag{11a, 11b}$$

In Eq. 11a and Eq. 11b, M_i is the mass of vehicle i, $\gamma_i = \frac{k_i^2}{k_i^2 + h_i^2}$, where k_i is the radius of gyration and h_i are the moment arms of the resultant collision force on each vehicle. For the value of h_i, the line of action of the impulse during collision, which coincides with the Principal Direction of Force (PDOF), needs to be defined [11, 13]. PDOF is, usually, visually estimated from the residual damage of each vehicle.

For the post–collision phase the equations of motion for each vehicle and Newton's second law are used in order to estimate the velocity of each vehicle. In more details, the two forces acting on each wheel are calculated using Eq. 12 (longitudinal force) and Eq. 13 (circumferential force).

$$F_L^i = \mu m_t^i g\cos\left(a^i\right) RR \tag{12}$$

$$F_C^i = \mu m_t^i g\sin\left(a^i\right) \tag{13}$$

In Eqs. 12 and 13, μ is the tire to ground frictional coefficient, g is the acceleration of gravity, m the weight supported by each wheel, RR is the rolling resistance corresponding to the percentage of blocked wheel and a_i is the angle defining the direction of the force in the coordinate system of the vehicle. The rolling resistance stands for the percentage of the longitudinal frictional force between the tire and the ground and has a value in the interval of [0–1]. The resultant force of each wheel is the geometrical sum of its longitudinal and circumferential force [6]. The index i, with discrete values in [1, 4] corresponds to the wheel number. The resultant velocity of each wheel is calculated and using Newton's second law and the deceleration of the vehicle is calculated. For small time steps, Δt, Newton's second law is solved in the inertial coordinate system and the deceleration of the vehicle is calculated. Using numerical integration, the velocity of the vehicle and consequently the coordinates of the center of mass of each vehicle, are evaluated using Eqs. 14 and 15, respectively.

$$U_{i+1} = U_i + a_i \cdot \Delta t \tag{14}$$

$$X_{i+1} = X_i + U_i \cdot \Delta t \tag{15}$$

This procedure is repeated until the velocity becomes nearly zero ($<10^{-2}$), leading to the coordinates of the rest point of the vehicle.

Algorithm B starts from the collision phase, implementing the energy-based approach calculating the velocities of each vehicle after collision moving to the point of rest of the vehicles, calculating the coordinates of the point of rest of each vehicle. The Genetic Algorithm optimization method is implemented through the optimization toolbox of Matlab® and has as design variables the initial velocities of the vehicles and the coordinates of the point of rest of each vehicle. In Fig. 2 the flow chart of Algorithms B is presented.

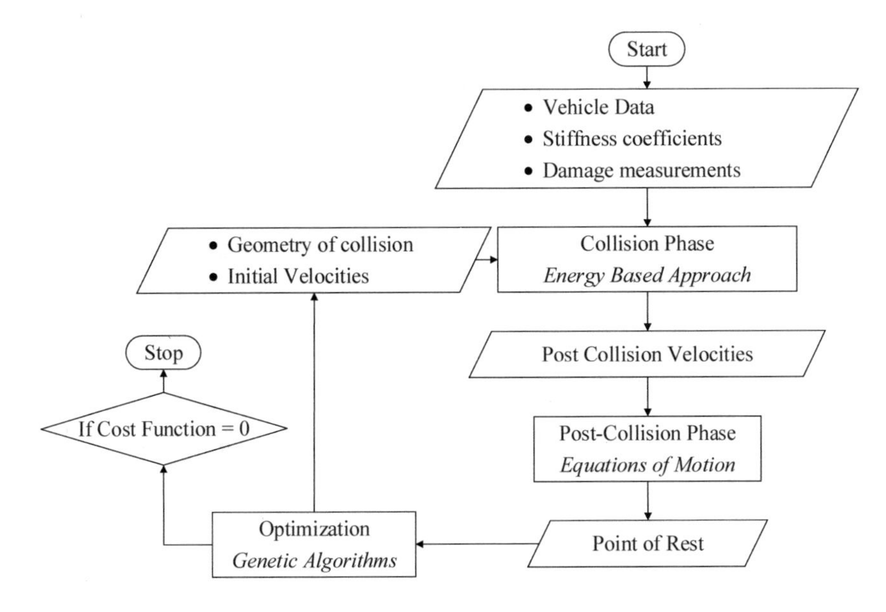

Fig. 2. Flow chart of Algorithm B

2.3 RICSAC Database

In the 1970s, a research project, named the "Research Input for Computer Simulation of Automobile Collisions" (RICSAC), was set up and funded by the National Highway Traffic Safety Administration (NHTSA). This project resulted in a test matrix of 12 full – scale vehicle collisions. In order to monitor each collision in detail cameras, accelerometers, potentiometers and gyroscopes were attached to each vehicle [14]. Moreover, the final test reports include objective information on the impact speeds, vehicle weights, vehicle dimensions, weight distributions, spin-out trajectories and positions of rest [7]. RICSAC database contains vehicle collisions engaging six different vehicles included in four categories of vehicle sizes. The tests can be classified into four impact configurations (IC1 – IC4) according to the relative orientation of the vehicles at the time of collision.

3 Results

The initial velocities calculated from both algorithms are presented in Table 1. RICSAC test No. 2 is omitted since no experimental values exist in the RICSAC database due to loss of data.

In Table 1 is obvious that for the same RICSAC test Algorithms A and B provide different results for the initial velocities of both vehicles. The mean % of absolute difference between the results of the two algorithms is 16% for Vehicle 1 and 34% for Vehicle 2.

Table 1. Initial velocities for both vehicles with both algorithms

RICSAC test no.	Algorithm A		Algorithm B	
	V1 (m/s)	V2 (m/s)	V1 (m/s)	V2 (m/s)
1	8.66	8.71	6.78	5.74
3	10.82	1.09	8.13	1.46
4	17.34	1.11	15.66	0.13
5	16.42	0.79	17.02	0.26
6	8.93	9.30	10.83	9.28
7	12.92	12.01	13.33	15.25
8	4.91	7.29	5.7	7.37
9	9.62	11.22	8.48	9.44
10	17.74	15.40	12.89	11.85
11	8.14	10.52	8.05	6.95
12	14.26	14.88	8.9	8.18

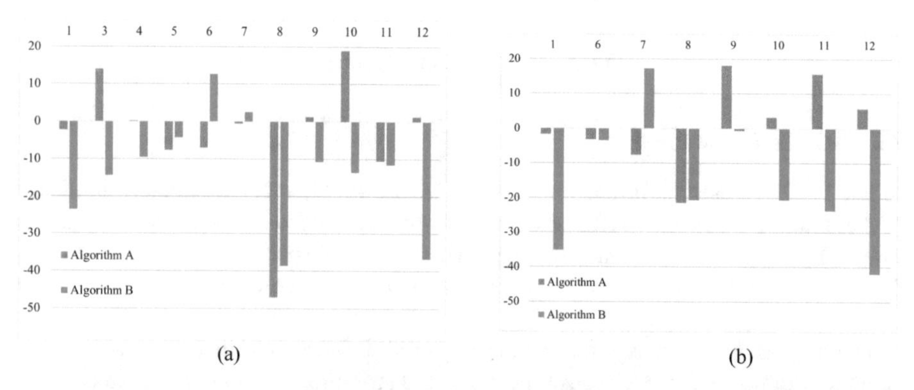

Fig. 3. % difference between measured and calculated values of initial velocities for (a) Vehicle 1 & (b) Vehicle 2.

In Fig. 3 the % of error of the initial velocities calculated by each algorithm compared to these measured and documented in the RICSAC database is presented.

In Fig. 3 is obvious that both algorithms underestimate the initial velocities in most of the tests. In Fig. 3b the results of tests 3 to 5 are omitted since in this IC Vehicle 2 was stopped prior to collision. The maximum error for Vehicle 1 is 47% for Algorithm A in RICSAC test 8 while for Vehicle 2 is 41.90% in RICSAC test 12.

4 Conclusion

In the present paper two traffic accident reconstruction algorithms implementing the main approaches for the simulation of the collision and the post–collision phase, existing in the literature, have been implemented and tested in the RICSAC tests database. Both algorithms converge into feasible solutions in reasonable amount of computational time. However, algorithm A seems to approximate the experimentally measured velocities better with mean absolute error of 10% for both Vehicle 1 and 2. For Algorithm B the same errors are 16% and 19%, respectively.

References

1. Brach, R.M., Brach, R.M.: A review of impact models for vehicle collision (No. 870048). SAE Technical Paper (1987)
2. Ishikawa, H.: Impact model for accident reconstruction-normal and tangential restitution coefficients (No. 930654). SAE Technical Paper (1993)
3. McHenry, R.: CRASH3 users guide and technical manual. NHTSA, DOT Report HS, 805, 732 (1981)
4. Struble, D.E.: Automotive Accident Reconstruction: Practices and Principles. CRC Press, Boca Raton (2013)
5. James, M.E., Ross, H.E.: HVOSM User's Manual (No. TTI-2-10-69-140-9 Intrm Rpt.). Texas Transportation Institute, Texas A & M University (1974)
6. Reed, W.S.: Automobile Accident Reconstruction by Dynamic Simulation. The Center (1983)
7. Jones, I.S., Baum, A.S.: Research Input for Computer Simulation of Automobile Collisions, vol. IV: Staged Collision Reconstructions. DOT HS-805, 40 (1977)
8. Brach, R.: An impact moment coefficient for vehicle collision analysis. SAE Transactions, pp. 30–37 (1977)
9. Brach, R.M.: Impact analysis of two-vehicle collisions. No. 830468. SAE Technical Paper (1983)
10. Vossou, C.G., Koulocheris, D.V.: A computational model for the reconstruction of vehicle collisions. Mob. Veh. Mech. **44**(3), 27–42 (2018)
11. McHenry, B.G.: The algorithms of CRASH. In: Southeast Coast Collision Conference, pp. 1–34 (2001)
12. Brach, R.M., Welsh, K.J., Brach, R.M.: Residual crush energy partitioning, normal and tangential energy losses (No. 2007-01-0737). SAE Technical Paper (2007)
13. Neades, J.G.J.: Developments in road vehicle crush analysis for forensic collision investigation (2011)
14. McHenry, B., McHenry, R.: RICSAC-97 A Re-evaluation of the reference set of full scale crash tests (No. 970961). SAE Technical Paper (1997)

Development of TSCLab: A Tool for Evaluation of the Effectiveness of Adaptive Traffic Control Systems

Daniel Pavleski[1], Daniela Koltovska Nechoska[2], and Edouard Ivanjko[3(✉)]

[1] Traffic Department of the City of Skopje, Skopje 1000, Republic of Macedonia
[2] Faculty of Technical Sciences, St. Kliment Ohridski University, 7000 Bitola, Republic of Macedonia
[3] Department of Intelligent Transportation Systems, Faculty of Transport and Traffic Sciences, University of Zagreb, Vukelićeva Street 4, 10000 Zagreb, Republic of Croatia
edouard.ivanjko@fpz.hr

Abstract. Adaptive Traffic Control Systems (ATCS) have been widely implemented for urban traffic control due to their capability to alleviate congestion. ATCS adjust the signal programs of signalized intersections in real time according to the measured fluctuations of traffic flow. This results in an improvement of the efficiency of traffic operations of urban networks. The process of evaluating the effectiveness of complex ATCS is challenging and presents an open problem. The most important issue is to identify whether the ATCS fulfills the goals and needs it was envisioned to achieve. For this, different measures of effectiveness with in-depth insights into the traffic situations of the controlled signalized intersection are required. In this paper, development of TSCLab (Traffic Signal Control Laboratory), a MATLAB based tool for evaluation of ATCS is presented. TSCLab can gather and visualize relevant data, which describe the performance of ATCS (green time duration, maximum green time utilization ratio, percent of arrived vehicles on green, etc.) in real time, in a VISSIM based microscopic simulation environment using different traffic scenarios. It can also process the gathered data to evaluate the effectiveness of the analyzed ATCS after simulation. To proof the capabilities of TSCLab, the effectiveness of the UTOPIA/SPOT ATCS using an isolated signalized urban intersection as the use case has been evaluated.

Keywords: Intelligent Transport Systems · Adaptive Traffic Control System · Isolated signalized urban intersection · Evaluation of effectiveness

1 Introduction

Today's modern large cities have an Urban Traffic Control (UTC) center. In such a UTC center all data about traffic flows of the managed urban transport network are gathered and signal programs for intersections are changed. In the past, only fixed signal programs were used. Their drawback is that they cannot take into account

I. Karabegović (Ed.): NT 2019, LNNS 76, pp. 386–394, 2020.
https://doi.org/10.1007/978-3-030-18072-0_44

changes in traffic demand that occur during the day. This results with congestion and non-optimal use of the controlled transport infrastructure. Today adaptive or learning based traffic control systems from the domain of Intelligent Transport Systems (ITS) are used for adjusting signal programs that can cope with changes in traffic demand [1, 2]. Such systems can improve the Level of Service (LoS) for all traffic entities or just specific ones (public transportation (PT), emergency vehicles, transit traffic). Thereby, learning based traffic control systems can, in the long term, learn traffic control policies capable of solving every day, seasonal and incident related changes in traffic demand [3].

Prior implementation of an ATCS in an urban environment, it has to be evaluated using realistic traffic scenarios. This is important in order to assess the possible improvement of LoS and to analyze the cost-benefit ratio before a costly upgrade of the transport infrastructure [1, 4, 5]. Mostly used approach for this is software-in-the-loop where a microscopic traffic simulator in combination with an ATCS is applied [1, 2, 5]. The advantage of such an approach is that various Measures of Effectiveness (MoE) can be gathered an analyzed prior a real-life implementation ensuring a good functioning of the ATCS [3]. The evaluation of ATCS is still an open area. Many researchers address this problem in order to implement an appropriate framework, ensure realistic traffic scenarios from different world regions, enable in-depth behavior analysis of the managed urban transport network and define appropriate MoEs [1–3, 5]. This paper also tackles the problem of evaluating ATCS using the software-in-the-loop approach. It presents a continuation of previous work of the authors in which the microscopic traffic simulator VISSIM was connected with the UTOPIA/SPOT ATCS [6] and used to compare basic MoEs of UTOPIA/SPOT to those of a fixed time control strategy [7]. In order to improve the evaluation framework used in [6, 7], the TSCLab tool has been developed. It is a MATLAB based tool for evaluation of ATCS that enables in-depth analysis of a variety of MoEs and traffic scenarios for isolated signalized urban intersections [8]. Thereby, the post-simulation in-depth analysis of chosen MoEs can be done in TSCLab alleviating the evaluation process, unlike current approaches that require manual processing of gathered simulation data.

This paper is organized as follows. The second section describes the most important measures for analyzing the performance of ATCS. Basic description of the evaluated UTOPIA/SPOT ATCS is presented in the third section. The fourth section describes the architecture of the developed TSCLab application. The continuing fifth section describes the simulation framework applied to evaluate UTOPIA/SPOT. Details about the use case, obtained results and short discussion about them are given in the sixth section. Conclusion and future work description ends the paper.

2 ATCS Performance Analysis

In the absence of information for the performance of adaptive traffic signal control, the quality of signal operations cannot be determined, and the functionality of the control strategies cannot be validated [1, 9]. Therefore, the monitoring of these so-called "live" systems i.e. the monitoring of their real-time "responses" to certain traffic state and a specific objective function is crucial. The available literature describes a variety of

MoEs that can be used for ATCS performance analysis [10–12]. According to [10], each MoE is denoted as a candidate since it is not necessary to calculate or compare all of the measures to validate the functionality of a system. The research study [11] documented an extensive portfolio of performance measures for evaluating traffic signal systems with emphasis on performance measures obtained from high-resolution data and from external travel time measurements. The report [12] presents the next step toward implementing and using performance measures to manage adaptive traffic signal systems.

Most of the mentioned research have been used as a starting point and base for selection of MoEs for ATCS performance analysis. Selected measures that have been used for ATCS performance evaluation in this paper are presented in Table 1. Each of the selected MoE has a different level of difficulty in implementation or interpretation and more details regarding ATCS evaluation are described in continuation.

Table 1. Selected measures for ATCS performance evaluation

Tasks	Measures for performance evaluation
Signal timing	Cycle length and Green time duration
	Maximum green time utilization ratio
Throughput	Arrived vehicles per cycle
	Served vehicles per green signal
Capacity	Green/Red occupancy ratio
	Queue length
	Delay
	Stops
Progression quality	Percent of arrived vehicles on green signal
	Platoon ratio and Arrival type

3 UTOPIA/SPOT

UTOPIA has a two-level distributed architecture. The upper level consists of a central subsystem responsible for medium and long term forecasting, and control over the whole managed area. At this level, the traffic light reference plans and also the criteria needed for the adaptive coordination are calculated dynamically. In addition, a continuous diagnostic activity is carried out for the whole network. The lower level consists of a network of multi-function units working as local controllers. These are interconnected, and each is responsible for the management of one intersection. Local controllers determine in real time the sequence and optimum length of traffic light phases using the coordination criteria established by the upper level, traffic measurements detected locally and information received from the controllers of adjacent intersections.

In Torino, UTOPIA/SPOT has resulted in reduced travel times of 2–7% for PT vehicles and 10% for cars [1]. Field tests of UTOPIA/SPOT in Skopje have shown 20% travel time reduction in peak hours [4]. But detailed insight about how UTOPIA achieves this reduction in Skopje is missing and this presents the main motivation for developing TSCLab [8]. Especially, since TSCLab can be used to evaluate different parts of the transport network currently not managed by UTOPIA and be possible augmented to include MoEs difficultly measured in real-life situations including connection to different ATCS.

4 Architecture of TSCLab and Evaluation Framework

The impact of the signal operations of ATCS is often underestimated or overlooked. The consequence is that is not possible to know, with certainty, the efficiency of the applied control strategies. In order to gather and visualize relevant data, which describe the performance of ATCS in real time, a new tool was developed using a VISSIM based microscopic simulation environment. Its name is TSCLab (Traffic Signal Control Laboratory) and is a MATLAB based tool developed for ATCS evaluation [8]. The developed simulation framework is given in Fig. 1.

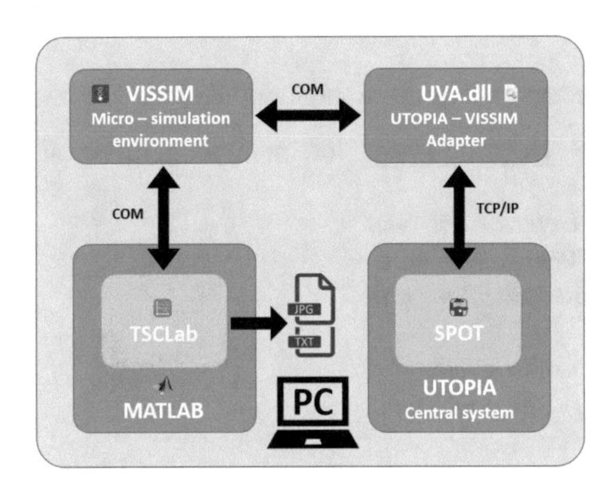

Fig. 1. Simulation framework used for the evaluation [8]

The main graphical user interface of the TSCLab tool is shown in Fig. 2 and consists of three main components. First *Measurement assignment* in which objects defined in the VISSIM model such as Nodes, Data collection Measurements, and Queue Counters, which will be accessed for the reading of data generated by VISSIM, can be specified. In this part values for Detector Distance to Stop line can be specified also. Second *Traffic signal control assignment* in which objects defined in VISSIM model such as Traffic Light Controller, Signal Groups, Detector ports, and Detectors, which will be accessed for the reading of data generated by VISSIM, can be specified. In this part values for Max Green Time can be specified. Third *Simulation* in which parameters for simulation in a VISSIM model such as Start/End of Time Period and Resolution can be specified. Additionally, the main graphical user interface contains 5 menus: (i) Micro-simulation environment VISSIM; (ii) Urban traffic control system UTOPIA/SPOT; (iii) Monitoring of performances; (iv) Results; and (v) Help. More details on the interface can be found in [8].

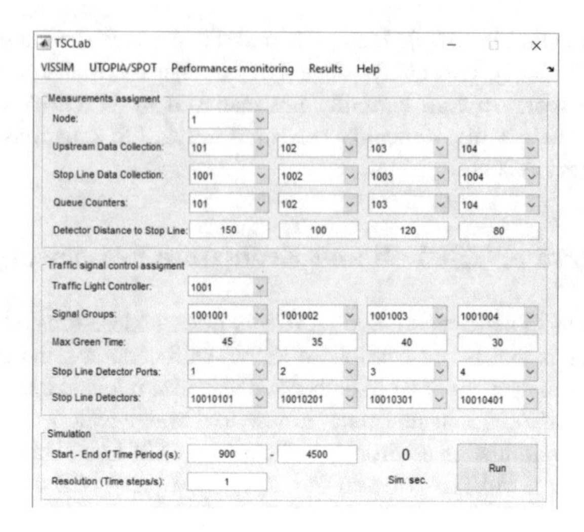

Fig. 2. Main graphical user interface of TSCLab [8]

The Performance monitoring menu monitors the following elements:

- <u>*Signal Groups & Detector events*</u> for monitoring of signal times and detector occupancy;
  - *Signal Times/Detector occupancy*;
  - *Signal Times/Vehicle arrivals*;
  - *Max green time utilization ratio*;
- <u>*Throughput*</u>;
  - *Vehicle arrivals per cycle*;
  - *Served vehicles per green time*;
- <u>*Capacity*</u>;
  - *Green/Red occupancy ratio*;
  - *Vehicle Queue link profile*;
- <u>*Progression*</u>;
  - *Percent of vehicle arrivals on green time*;
  - *Vehicle Platoon ratio & Arrival type*.

The monitoring of signal time changes and vehicle arrivals is important for an in-depth analysis of the traffic situations of the controlled signalized intersection and to adjust the signal timing of each phase to optimize an objective function. The corresponding diagram is presented in Fig. 3. It can be seen that the bottom red right line describes the end of the red-yellow signal of the previous cycle, green line presents the end of the green signal, yellow line describes the end of the yellow signal, and upper red line describes the end of a particular signal cycle. The blue circles present vehicle arrivals during the cycle time period detected at a certain distance from the stop line.

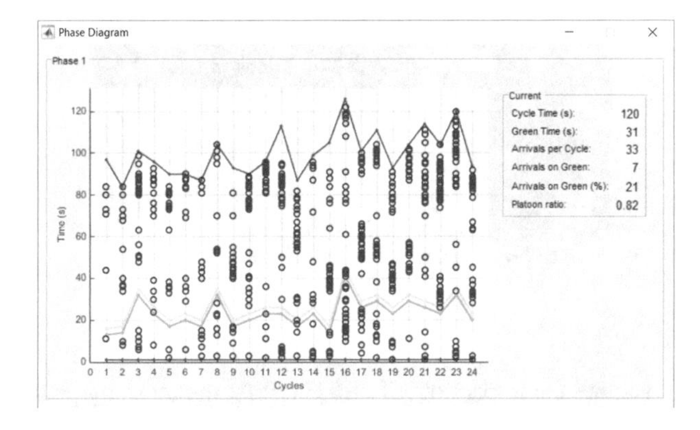

Fig. 3. Diagram of signal time changes and vehicle arrivals monitoring [8]

The Results menu consists of: (i) Signal timing; (ii) Throughput; (iii) Capacity and (iv) Progression. An example of output data related with signal timing is given in Fig. 4. In it, the variability of green time duration and maximum green time utilization ratio per signal group provided for every phase during one adaptive control cycle can be observed.

Signal Timing Results

	1	2	3	4	5	Min	Max	Average
Cycle length (s)	97	84	101	96	90	84	135	105.6364
Green time duration - SG 1 (s)	12	13	31	22	16	12	40	24.6667
Green time duration - SG 2 (s)	8	8	8	8	11	7	20	11.4545
Green time duration - SG 3 (s)	41	26	26	26	26	25	41	26.9697
Green time duration - SG 4 (s)	9	10	10	13	10	7	22	15.6061
Max green time utilization ratio - SG 1	0.2222	0.2407	0.5741	0.4074	0.2963	0.2222	0.7407	0.4568
Max green time utilization ratio - SG 2	0.2667	0.2667	0.2667	0.2667	0.3667	0.2333	0.6667	0.3818
Max green time utilization ratio - SG 3	0.9111	0.5778	0.5778	0.5778	0.5778	0.5556	0.9111	0.5993
Max green time utilization ratio - SG 4	0.3000	0.3333	0.3333	0.4333	0.3333	0.2333	0.7333	0.5202

Fig. 4. Output data related to signal times [8]

5 Use Case and Evaluation Results

To proof the capabilities of the developed TSCLab tool it has been applied to evaluate the effectiveness of the UTOPIA/SPOT ATCS using an isolated signalized urban intersection as the use case. Intersection denoted as I2, which is a part of the urban network located in the wider central area of Skopje, Macedonia, was selected for testing (Fig. 5). Considering the fact that all signalized intersections in Fig. 5 are controlled by the ATCS UTOPIA, UTOPIA was connected to VISSIM trough the UTOPIA VISSIM Adapter (UVA). With this connection, UTOPIA manages the traffic signals for the simulated road network and VISSIM provides the needed traffic data from the sensors. Both, traffic signal commands and sensor measurements are refreshed every second [12]. The UTOPIA system is used as a "black box" control unit connected to the traffic network simulated in VISSIM using UVA.

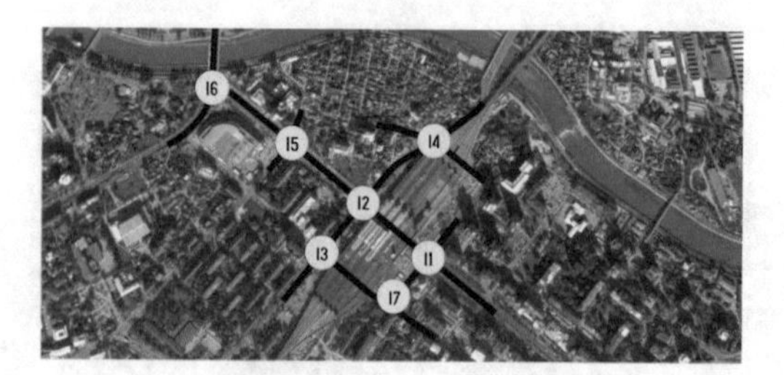

Fig. 5. Study area [6]

In order to evaluate the impact of ATCS UTOPIA, the morning peak hour (7:15 to 8:15) in a typical working day was chosen for analysis. Traffic flow data were collected using manual traffic counting and automatic data collection from inductive loop detectors. Data from inductive loop detectors were obtained from the traffic management and control center of Skopje. The calibration procedure described in [13] was applied for the calibration of the VISSIM model. The VISSIM Wiedemann 74 model was used and the following parameters were changed in the default urban (motorized) driver behavior: (i) minimum look ahead distance was increased to 30 m; and (ii) average standstill distance was decreased to 1.5 m for all link types. In order to more accurately reflect the driver behavior, a new desired speed distribution ranging from 40 to 90 km/h was defined for the free flow speeds on the basis of collected data and used in simulations. It is characteristic that only 15% of drivers are moving with speed under the mandatory speed limit. The saturation flow was selected as the parameter for validation of the model. According to [13], the modeled saturation flows values should be within 10% of observed values. Because the saturation flows appear to be modeled incorrectly uniformly across the network, the parameters of the global "driver behavior" model: average standstill distance, additive part of safety distance and multiplicative part of safety distance were adjusted to comply with the calibration and validation procedure.

Table 2. Results from TSCLab for measures related to signal timing

Peak hour period	Approach	Cycle length (s)	Green time duration (s)	Max green time utilization ration
07:15 - 08:15	1	108	25.32	0.47
	2		13.02	0.40
	3		27.43	**0.60**
	4		16.25	**0.54**

Table 2 shows the average values of variability of the green time duration and maximum green time utilization ratio per intersection approach provided in a signal cycle length. From the presented results, one can conclude that for the morning peak hour, max green time utilization ration is highest on third and fourth approach; on the other hand, arrived vehicles of all approaches are not served during the cycle time duration. Also, total number of arrived vehicles of all approaches are not served during the cycle time.

6 Conclusion and Future Work

The aim of this research paper is to provide a performance evaluation of ATCS applied in today's UTC centers. For this, a simulation framework was developed to enable a software in the loop simulation of the adaptive traffic control UTOPIA using the microscopic simulator VISSIM. It contains a new MATLAB based tool named TSCLab: Traffic Signal Control Laboratory. The developed tool can gather and visualize relevant data, which describe the performance of ATCS (like green time duration, maximum green time utilization ratio, percent of arrived vehicles on green, etc.) in real time, in VISSIM based microscopic simulation environment using different traffic scenarios. Using these data an in-depth analysis of the effectiveness of the evaluated ATCS is possible for isolated intersection based traffic scenarios.

To proof the capabilities of the developed TSCLab, it was applied to evaluate the UTOPIA/SPOT ATCS on a road network in Skopje using the micro-simulation environment VISSIM. An isolated signalized urban intersection was the use case. The gathered data enabled an in-depth analysis of the traffic situation on the chosen insolated intersection. Consequently, TSCLab provides the operators in the traffic management and control center a possibility to test the control strategy and if it is necessary, to change the algorithm parameters prior to being implemented on the field.

Future work on this topic will be related to the augmentation of TSCLab to enable evaluation using a simulation model containing more controlled signalized intersections in an urban network and comparison of different control strategies. Additionally, adding interfaces to other commercially available or open source traffic simulators and ATCS will be considered also.

Acknowledgement. The authors would like to thank the companies PTV Group and SWARCO MIZAR S.r.l., the Traffic management and control center of the City of Skopje, and the Faculty of Transport and Traffic Sciences, University of Zagreb for supporting the work published in this paper.

References

1. Wahlstedt, J.: Evaluation of the two self-optimising traffic signal systems Utopia/Spot and ImFlow, and comparison with existing signal control in Stockholm, Sweden. In: Proceedings of the 16th International IEEE Annual Conference on Intelligent Transportation Systems (ITSC 2013), pp. 1541–1546 (2013)
2. El-Tantawy, S., Abdulhai, B., Abdelgawad, H.: Multiagent reinforcement learning for integrated network of adaptive traffic signal controllers (MARLIN-ATSC): methodology and large-scale application on Downtown Toronto. IEEE Trans. ITS **14**(3), 1140–1150 (2013)

3. Michailidis, I.T., Manolis, D., Michailidis, P., Diakaki, C., Kosmatopoulos, E.B.: Autonomous self-regulating intersections in large-scale urban traffic networks: a Chania City case study. In: 2018 5th International Conference on Control, Decision and Information Technologies (CoDIT), pp. 853–858 (2018)
4. SWARCO MIZAR S.p.A.: ATM Skopje Preliminary Before/After Study, Turin, Italy (2014)
5. Dakic, I., Mladenović, M., Stevanović, A., Zlatkovic, M.: Upgrade evaluation of traffic signal assets: high-resolution performance measurement framework. Promet Traffic Transp. **30**(3), 323–332 (2018)
6. Pavleski, D., Nechoska Koltovska, D., Ivanjko, E.: Evaluation of adaptive traffic control system UTOPIA using microscopic simulation. In: Proceedings of 59th International Symposium ELMAR-2017, pp. 17–20 (2017)
7. Pavleski, D., Nechoska Koltovska, D., Ivanjko, E.: Evaluation of adaptive and fixed time traffic signal strategies: case study of Skopje. In: Book of abstracts of Second International Conference Transport for Today's Society (2018)
8. Pavleski, D.: Adaptive traffic signal control performances evaluation in micro-simulation environment. Master thesis, Faculty of Technical Sciences, Bitola, Macedonia (2018). (in Macedonian)
9. Samadi, S., Rad, A.P., Kazemi, F.M., Jafarian, H.: Performance evaluation of intelligent adaptive traffic control systems: a case study. J. Transp. Technol. **2**(3), 248–259 (2012)
10. Gettman, D., et al.: Measures of effectiveness and validation guidance for adaptive signal control technologies, US Department of Transportation, Federal Highway Administration (2013)
11. Day, C., et al.: Performance Measures for Traffic Signal Systems: An Outcome-Oriented Approach. Purdue University, West Lafayette (2014)
12. Day, C., Bullock, D.M., Li, H., Lavrenz, S.M., Smith, W.B., Sturdevant, J.R.: Integrating Traffic Signal Performance Measures into Agency Business Processes. Purdue University, West Lafayette (2015)
13. Smith, J., Blewitt, R. (eds.): Mayor of London: Traffic modelling guidelines, TfL Traffic manager and network performance best practices, Transport for London (2010)

Advanced Applications for Urban Motorway Traffic Control

Martin Gregurić[1(✉)], Sadko Mandžuka[1], and Krešimir Vidović[2]

[1] Faculty of Traffic and Transport Sciences, University of Zagreb, Vukelićeva 4, 10000 Zagreb, Croatia
martin.greguric@fpz.hr

[2] Ericsson Nikola Tesla, Krapinska 44, 10000 Zagreb, Croatia

Abstract. Nowadays urban motorways cannot fulfill their originally projected purpose as urban bypass due to congestions. Congestions are caused by recurrent traffic from the urban area, which tries to bypass controlled traffic intersection in the same urban area, and non-recurrent transit traffic. The problem cannot be solved by constructional build-up since they became surrounded by the urban and traffic infrastructure. This requires new approaches to make the traffic flows on them more efficient and safer. Modern information-communication technologies and advanced traffic control algorithms are introduced as an only valid approach for mentioned problems. This study proposes the latest approach of coordination between controlling on-ramp flows with ramp metering (RM) and Dynamic Route Guidance Information Systems (DRGIS), which reroute vehicles from congested parts of the motorway. In the next decades, road transport will undergo a deep transformation with the advent of autonomous vehicles, which are about to drastically change the way we commute. This paper also provides a quick overview of the Intelligent Speed Adaptation (ISA) implementation in the context of autonomous vehicles and connected driving.

Keywords: Traffic control · Ramp metering · Route guidance · Autonomous vehicles

1 Introduction

Urban motorway can be considered as the former bypass of the larger urban area, which became heavily integrated with the traffic network of the urban area. Therefore, urban motorway has a direct connection with the arterial roads of urban traffic network through numerous and mutually close on- and off-ramps. Intense traffic demand generated at the urban motor way on-ramps can significantly reduce speed and increase the density of the mainstream flow near those on-ramps. That can lead to downstream congestions near overburdened on-ramps, what consequently induce queues on them. If those on-ramp queues are large enough, they can spill over into the urban traffic arterials ("spillback effect") and induce heavy congestions on urban traffic network. It is possible to conclude that on- and off-ramps are places where it is possible to make a significant impact on motorway mainstream traffic flow and at on-ramp queues [1, 2].

I. Karabegović (Ed.): NT 2019, LNNS 76, pp. 395–400, 2020.
https://doi.org/10.1007/978-3-030-18072-0_45

Ramp metering (RM) provides restriction for the vehicle which has the intention to access mainstream flow from on-ramp by using special traffic lights [2, 3]. A most important part of the RM system along with the detectors, appropriate traffic lights, and other signalization is the RM algorithm which determines the "access rate reduction" for the respective on-ramp flow [4]. The "access rate reduction" is computed according to the traffic parameters of the mainstream flow and number of vehicles which are queueing at the particular on-ramp. It is important to emphasise that RM provide control only over the vehicles which are already part of the queue at the particular on-ramp. In the case when the mainstream density near on-ramp is close to the maximal capacity of this section, there is no more space to adopt additional vehicles from a nearby on-ramp. In that case, on-ramp queue can grow proportional to the on-ramp traffic demand, what makes RM obsolete, and consequentially induce "spillback effect". It is possible to conclude that RM cannot provide control over the on-ramp traffic demand. In order to alleviate that problem, RM must be set in coordination with other motorway control methods such is variable speed limit control (VSLC) or route guidance system. VSLC in coordination with the RM can reduce motorway upstream mean speed if the congestion near particular on-ramp occurs. That type of coordination slows down incoming mainstream vehicles to the place of congestion and gives extra space to accommodate vehicle from congested on-ramps into the mainstream. It is possible to conclude that this coordination does not efficiently solve the problem if the traffic demand at congested on-ramp is too high. In that case, it is necessary to reroute vehicles which have the intention to use congested on-ramp. This can be done with the application of the Dynamic Route Guidance Information Systems (DRGIS). The problem of coordination of ramp metering and DRGIS is presented in Sect. 2. In Sect. 3 quick overview of the impact of autonomous vehicles on traffic control is given. Paper is concluded with the conclusion remarks in Sect. 4.

2 Coordination of Ramp Metering and DRGIS

DRGIS is used to inform drivers about current or expected travel times and queue lengths so that they may reconsider their choice for a certain route [5]. Mentioned system transfer information about current/predicted travel times for critical parts of urban motorway system to the drivers by special DRGIS messaging signs, or it can be integrated as an advisory service within the existing GPS routing system. In Fig. 1 it is possible to see a conceptual example of a system which enables coordination between DRGIS and RM.

Special DRGIS messaging signs can be located near entry points to the on-ramps from the urban traffic network as it is the case in node N3 at Fig. 1. At Fig. 1 on-ramp OR1 is denoted as congested, DRGIS will advise drivers about long travel times at this on-ramp what will motivate drivers to use L3 link for accessing motorway by using on-ramp OR2 instead OR1. Furthermore, in Fig. 1 nodes N1 and N2, are places where DRGIS signs are placed on the motorway itself so the drivers can obtain information about eventual downstream congestions, which in our case occurs at motorway section H3. DRGIS system will advise drivers to leave motorway by using one of the available downstream off-ramps if congestion in the H3 motorway section is strong enough. In

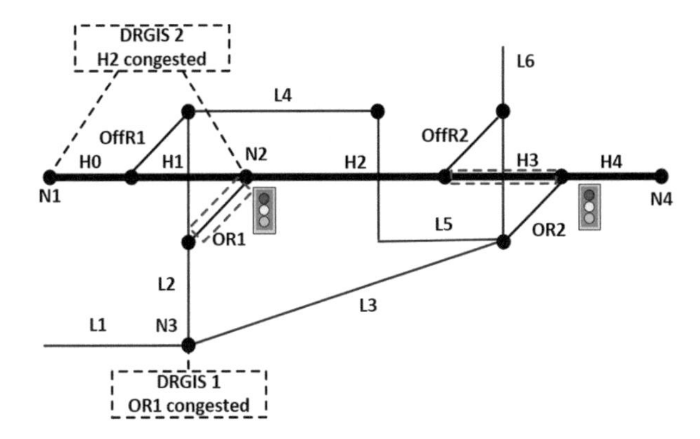

Fig. 1. A conceptual example of a system which enables coordination between DRGIS and RM

the case presented in Fig. 1 those alternative routes begin with the off-ramps OffR1 and OffR2. Drivers at motorway section H0 can choose the off-ramp OffR1, in order to avoid congestions at on-ramp OR1 and motorway section H3. In continuation of their new route drivers can access motorway through the uncongested on-ramp OR2 by using links L4 and L5. In the other hand drivers in motorway section H1, can choose a route to the off-ramp OffR2 and leave motorway in order to avoid congested motorway section H3.

It is possible to conclude that DRGIS can directly impact on traffic demand at the urban traffic system by informing the drivers about travel times on its crucial segments. Those information's will suggests drivers not to choose the route which contains segments of urban motorway with lower travel times or to avoid using congested on-ramps. Reduced traffic demand on congested urban motorway section or at congested on-ramp in coordination with the adequate ramp metering control strategies can prevent "spillback effect" and increase overall throughput of the urban motorways.

In [5] it is proposed that DRGIS in coordination with the RM compute optimized travel times(as a control signal) which are optimized in combination with and at the same time as the RM control signals. Model predictive control (MPC) is used as the optimization engine. This approach is proposed because instantaneous travel times are not reliable due to the stochastic traffic dynamics. In the other hand, predicted travel time, as another common output of DRGIS, is not suitable because resulting rerouting by the drivers may result in a traffic distribution that is not always optimal with respect to the total network performance [5]. In [6], a multi-class and multi-objective combined RM and routing control strategy is proposed, in order to reduce the total travel time and the total emissions in the motorway system in a balanced way. This study is based on the total travel time and total emissions prediction. Since this approach is based upon multi-class control strategy it is possible to specify different control policies for the different classes of vehicles, i.e. cars, trucks, and other types of vehicles that can be of interest for the considered application case [6]. This is especially important in the case when it is necessary to adequately reroute large platoons of trucks due to their negative impacts at on-ramp queues and their increased emissions while they are queuing.

Technically, for these purposes different technologies can be used for cooperative systems, such as for example public mobile networks and the like [7, 8]. In that sense, the application of the future 5G mobile network is of great importance. The source of information for these needs is for example the urban traffic management center, including traffic incident management center [9].

3 Impact of Autonomous Vehicles on Traffic Control

Penetration of upcoming autonomous vehicle on the urban traffic network will have a huge impact on the traffic flows and control over them, and it will happen gradually. This is the reason why is very important to estimate the impact of autonomous vehicles on the one of the most basic traffic parameter such is traffic speed on urban motorway [10, 11]. Nowadays, speed on urban motorways is controlled by VSLC. Those VSLC applications are oriented on the infrastructural traffic signs for posting appropriate speed limits such as Variable Message Signs (VMS). Speed limits are computed in response to the prevailing traffic conditions which is measured by the static traffic detectors. It is possible to conclude that the effectiveness of VSLC is related to the driver compliance rate to the posted speed limit value on VMS. [12] it is shown that the VSLC system becomes inefficient if drivers do not adjust the speed of the vehicle according to the posted speed limit value. Those problems can be efficiently alleviated by higher penetration rate of the autonomous vehicle.

An autonomous vehicle with their numerous sensors and connectivity possibilities with other vehicles and infrastructure can be considered as moving traffic sensors (FCD – Floating Car Data). Data from those vehicles can be used to produce a more comprehensive and accurate picture of the traffic flows based on which, it will be possible to compute desired speed profiles for much smaller time frames compared to the current VSLC. Furthermore, it is possible to use autonomous vehicles as the one form of the information routers which will pass information about current speed profiles to the vehicles which are too far from the broadcasting range of the roadside unit. All mentioned technical advantages of autonomous vehicles are used for the implementation of the Intelligent Speed Adaptation (ISA). An autonomous vehicle with ISA will automatically reduce its current speed or just inform the driver about necessary action according to the acquired speed limit from ISA roadside unit. Even with a smaller penetration rate of autonomous vehicles with ISA, speed homogenization effect of classic VSLC that enable reductions of fluctuations in traffic parameters, i.e. speed differences of vehicles between lanes and within the same lane, can be improved [13]. The architecture block scheme of a system that presents communication between autonomous vehicles and traffic management system of ISA is presented in Fig. 2.

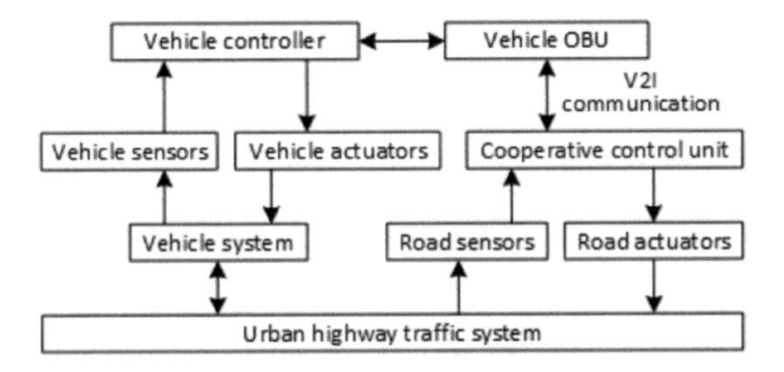

Fig. 2. Block scheme of an architecture for communication between a vehicle and the traffic control system of ISA [7]

4 Conclusion

Nowadays, urban motorways cannot fulfill their originally projected Level of Service (LoS) due to congestions caused by local urban and transit traffic. Urban motorways are fully integrated with the urban infrastructure and urban traffic network so there is no more space for constructional expansion of existing motorway capacities. The only solution for dissolving congestions is the application of advanced traffic control strategies and information systems for urban mobility. Latest approaches propose coordination between traffic control system such as RM and traffic information system such as a route guidance system. On the other hand, ISA can provide a feasible solution for the problem of a higher percentage of vehicles that do not comply with the posted speed limit in the upcoming era of autonomous vehicles. Abilities of ISA regarding the more accurate computing of the speed profiles for each vehicle based on the more comprehensive traffic picture of overall motorway can be extended, if it is placed in cooperative environment with the motorway infrastructure.

Acknowledgment. The research presented in this paper is supported with University of Zagreb Program funds Support for scientific and artistic research (2018) through the project: "Impact of the application of cooperative systems and autonomous vehicles on traffic and society".

References

1. Mandžuka, S., Žura, M., Horvat, B., Bićanić, D., Mitsakis, E.: Directives of the European Union on Intelligent Transport Systems and their impact on the Republic of Croatia. Promet-Traffic Transp. **25**(3), 273–283 (2013)
2. Gregurić, M., Ivanjko, E., Mandžuka, S.: The use of cooperative approach in ramp metering. Promet-Traffic Transp. **28**(1), 11–22 (2016)
3. Mandžuka, S., Horvat, B., Škorput, P.: Development of ITS in Republic of Croatia. In: Proceedings of 19th ITS World Congress, ERTICO, Wien (2012)
4. Gregurić, M., Ivanjko, E., Mandžuka, S.: Cooperative ramp metering simulation. In: 2014 37th International Convention on Information and Communication Technology, Electronics and Microelectronics (MIPRO), Opatija, pp. 970–975 (2014)

5. Karimi, A., Hegyi, A., De Schutter, B., Hellendoorn, H., Middelham, F.: Integration of dynamic route guidance and freeway ramp metering using model predictive control. In: Proceedings of the 2004 American Control Conference, Boston, Massachusetts, pp. 5533–5538, June–July 2004
6. Pasquale, C., Sacone, S., Siri, S., De Schutter, B.: A multi-class ramp metering and routing control scheme to reduce congestion and traffic emissions in freeway networks. In: Proceedings of the 14th IFAC (CTS 2016), Istanbul, Turkey, pp. 329–334, May 2016
7. Mandzuka, S., Kljaić, Z., Skorput, P.: The use of mobile communication in traffic incident management process. J. Green Eng. **1**(4), 35–42 (2011)
8. Vidović, K., Mandžuka, S., Brčić, D.: Estimation of urban mobility using public mobile network. In: ELMAR, 2017 International Symposium, pp. 21–24. IEEE (2017)
9. Škorput, P., Mandžuka, S., Jelušić, N.: Real-time detection of road traffic incidents. PROMET-Traffic Transp. **22**(4), 273–283 (2010)
10. Talebpour, A., Mahmassani, H.: Influence of connected and autonomous vehicles on traffic flow stability and throughput. Transp. Res. Part C Emerg. Technol. **71**, 143–163 (2016)
11. Makridis, M., Konstantinos, M., Ciuffo, B., Raposoa, M.A., Toledob, T., Thiela, C.: Connected and automated vehicles on a freeway scenario. In: Effect on Traffic Congestion and Network Capacity, Proceedings of 7th Transport Research Arena TRA 2018, 16–19 April 2018, Vienna, Austria (2018)
12. Gregurić, M., Mandžuka, S.: The use of cooperative approach in intelligent speed adaptation. In: Proceedings of 26th Telecommunications Forum (TELFOR 2018), Belgrade, Serbia, 20–21 September 2018
13. Wang, M., Daamen, W., Hoogendoorn, S.P.: Connected variable speed limits control and vehicle acceleration control to resolve moving jams. In: Transportation Research Board 94th Annual Meeting (2015)

Applying Unmanned Aerial Vehicles (UAV) in Traffic Investigation Process

Pero Škorput(✉), Sadko Mandžuka, Martin Gregurić, and Maja Tonec Vrančić

Faculty of Traffic and Transport Sciences, Vukelićeva 4, 10000 Zagreb, Croatia
pero.skorput@fpz.hr

Abstract. Traffic accident investigations are technical and technological processes which essentially contain two opposed demands. The first demand refers to the need of the shortest clearance time in order to prevent secondary incidents, to bring back traffic to its full operative capacity and to completely normalize the traffic flow. The second demand refers to investigation as an act which purpose is to determine the elements of criminal offense of causing a traffic accident or a traffic offense. The investigation today, in time context, is a depended process that requires relatively longer time period to gather evidence, make necessary measurements, describe traces etc. In this paper will be shown the possibilities of digitalization and three-dimensional (3D) modelling of traffic accident sight. Innovative technologies will be described, such as making three-dimensional (3D) models of the actual place where traffic accident happened, using Unmanned Aerial Vehicle (UAV) and photogrammetric procedures.

Keywords: Digitalization · UAV · Investigation · Traffic accident · 3D model

1 Introduction

Official bodies that participate in the process of traffic accident investigation, such as police, judiciary or criminalistics, traffic and other professions, deal on daily bases with demands like quickest traffic normalization with also securing enough time for complete investigation process. The basic need or request is referred to the shortest time for clearing traffic accident location in order to prevent secondary incidents, and to bring back road to its full traffic flow profile. This action would bring back traffic flow to its completely normalized state in short period of time.

Equally important request refers to investigation as an investigative action with the purpose to determine the elements of criminal offense of causing a traffic accident or a traffic offense. Investigation is time depended process that requires relatively longer time period to gather evidence, make necessary measurements, describe traces etc. Requests mentioned are essentially opposed because competent authorities need to perform the investigation in shortest possible time period to bring traffic to its full profile, and at the same time insure enough time to perform a quality investigation.

In this science paper an application of innovative support technologies for traffic accident investigation is suggested. With those technologies it is possible to perform quality and comprehensive investigation in significant shorter time period. Innovative

I. Karabegović (Ed.): NT 2019, LNNS 76, pp. 401–405, 2020.
https://doi.org/10.1007/978-3-030-18072-0_46

technologies include the application of UAVs, that are already in use in similar areas of work, and also geo-referencial documenting of full area of event in a form of three-dimensional dot cloud. Afterward, from this cloud, it's possible to measure traces, position and distance between vehicles and objects themselves and between them and the traffic road geometry, etc.

2 Literature Review

In this chapter, several articles related to the application of UAV will be presented. Mohammed et al. [1] in their paper explain various UAV applications in smart cities, including monitoring of traffic flow for measuring and detecting floods and natural disasters by using wireless sensors. They also explain challenges and issues of UAV usage such as safety, privacy and ethical uses.

The paper "Application of unmanned aerial vehicles in logistic processes" [2] describe the organization of logistic processes that nowadays require synergistic optimization effects of physical processes and application of innovative technologies.

Floreano and Wood [3] review the science, technology and the future of small UAVs. They show the advantages of small UAVs and the future possibilities for their implementation in various fields. They explain the important socio-economic impacts of small UAVs such as fixed-wing UAVs with a long flight time which could provide bird's-eye-view images and a communication network for rescuers on the ground, rotorcrafts with hovering capabilities which could inspect structures for cracks and leaks.

The use of Unmanned Aerial Vehicles for reducing negative impacts of forest fire was explored in paper "The use of Unmanned Aerial Vehicles for forest fire monitoring" [4]. Through research it was shown how UAVs can contribute to reducing probability of errors made by tactics on the ground and in air, reaction time, accuracy in decision making and load of people and equipment in peak days. Detail described proposed system architecture including module for communication, data receiving module, video play module, fire detection module and GIS display module. Such systems can contribute to reducing the probability of errors, shortening reaction time, increasing accuracy in decision making, and shortening load of people and techniques in peak days.

The complexity of traffic incident management demands decomposition of process to four phases of management [5]. Traffic accident investigation is the first activity in clearance phase and its duration significantly effects on all phases that follow. The duration of investigation depends on a type of traffic accident. Investigation is a process where competent authorities take action in order to determine and clarify important factors of procedure. During the investigation it is important to exactly describe and sketch the sight of wider and narrow area of accident, and that photos taken precisely as possible show accident scene and position of all tracks.

Different gadgets for gathering and documenting relevant data are used during the investigation, such as chalk or paint for marking tracks, photocamera for recording track, measuring trolley for measure the distance between tracks, total station, etc. Technology of traffic accident investigation experiences improvement with application of new technologies. The methodology of investigation can be significantly improved with digitalization of traffic accident investigation process.

3 Potentials of Digitalization and 3D Traffic Accident Location Modelling

For the purposes of this science paper, simulation of traffic accident was staged on polygon. From the height of 30 m and 50 m sets of geo-referencial photographs were made, and were used to make 3D test model of the traffic accident location. UAV DJI Phantom 4 was used for the purpose of the experiment. It was assisted with appropriate mobile application and in circular trajectory around the accident location it took about 30 angled photographs at the height of 30 m. In order to achieve more visibility and surface of 3D model, a second set of straight-lined flights combination was given on an altitude of 50 m, where UAV also took angled photographs of experimental accident location. The flight preparation time was about 10 min in total. UAV Phantom 4 weights around 1,5 kg, which means it's light enough for one person to make flight preparation and control to it. After flight plan it took less than 10 min to complete the UAVs flight, and during the flight total of 61 geo-referencial photographs were made. From these photographs it is possible to generate and to visualize computer 3D dot cloud. Based on UAVs known positions, during the shooting it calculates the position of every dot in 3D dot cloud. This given cloud of dots is necessary to fill with texture in the following computer actions. On generated 3D and orthophoto computer models of traffic accident sight it's possible to make measurements, as shown in Fig. 1.

Fig. 1. Measuring on computer 3D model and orthophoto model of accident location

For the purposes of this paper, classic field measurements were made on the location of simulated traffic accident, along with the measurements on 3D and orthophoto computer models. Measurements are shown in Table 1.

Table 1. Comparation of manual measurement method and computer 3D model measurement

Measurement	3D model [m]	Real meas. [m]	Error [m]	Error [%]
Vehicle A width	1,72	1,68	−0,04	−1,02%
Vehicle A length	4,13	4,02	−0,11	−2,74%
Vehicle B width	1,95	1,84	−0,11	−5,98%
Vehicle B length	4,6	4,65	0,05	1,08%
Object A length	0,82	0,8	−0,02	−2,50%
Object B length	0,82	0,8	−0,02	−2,50%
Wheel span of vehicle A	2,55	2,61	0,06	2,30%
Pavement 1 width	5,62	5,5	−0,12	−2,18%
Pavement 2 width	5,59	5,5	−0,09	−1,64%
Pavement 3 width	5,59	5,5	−0,09	−1,64%
Lane width	2,49	2,5	0,01	0,40%
Stopping distance	21,64	21,6	−0,04	−0,19%

Usage of UAVs and 3D tool for the reconstruction of traffic accident saves time and costs while serving satisfying results. Generated results are permanently available in the shape of a file, with possibility of conducting measurements at any time. Real events are saved as 3D objects with a centimetre precise detailed data.

Also, it takes less training to operate with the data, and they are permanently preserved in order to reconstruct accident scene of court expert demands.

4 Discussion and Conclusions

In this paper, as one of the first digitalization elements of traffic accident investigation process, building a system based on UAV and 3D modelling programme tools is recommended. Precondition for development of this system is definition of methodology base for the application of UAVs, photogrammetric techniques and modern tools for making three-dimensional computer models of traffic accidents. Just on such methodological base, it's possible to develop a system prototype for traffic accident investigation and also making of appropriate programme interface towards other tools of reconstruction and accident dynamics expertise. On system prototype, measurable indicators (or benefits) can be determined, regarding enhancement of quality and quantity of gathered evidences and reducing the investigation time spend. By digitalization of process and application of innovative support technologies for traffic accidents, time needed to reconstruct a traffic accident is significantly reduced. Since application of UAVs significantly reduces the need for physical attention of person in charge, also is significant a reduction of injury risks for employees of police, court and other services during the investigation process. By applying innovative technologies, whole area of traffic accident location, including the smallest details such as potsherd, breaking trails and similar, can be documented in time interval lower than 10 min. This significantly reduces the time needed for reopening the road, gather detail information and for practically instant documentation of accident parameters. In a way presented, the number of mistakes during the investigation is reduced and brought down to nearly zero. With high availability level of suggested technological solution, is insured the sustainability of digitalization process impacts because of extremely low implementation prices and simplicity of usage. Realization of system prototype opens a possibility of future research and improvement of traffic accidents investigation processes. Research in the field UAV application in other phases of traffic incident situations can be based on the realization and testing suggested prototype. UAVs can also be used in a similar way for phases of road incident verifications after the incident takes place, as well as in other processes of police domain. The results of digitalization of investigation process in suggested segment give positive impacts on the whole community, but the most valuable impact is the potential of reducing accident investigation time, enhancement of investigation quality and increase of safety for individuals conducting the investigation.

Acknowledgment. The research presented in this paper is supported with NATIONAL ROAD TRAFFIC SAFETY PROGRAM OF THE REPUBLIC OF CROATIA 2011–2020, through the project: "Application of innovative support technologies for traffic accidents investigations". This paper is based on presentation Digitalization of Traffic Accident Investigation Processes [6].

References

1. Mohammed, F., Idries, A., Mohamed, N., Al-Jaroodi, J., Jawhar, I.: UAVs for smart cities: opportunities and challenges. In: International Conference on Unmanned Aircraft Systems (ICUAS), pp. 267–273 (2014)
2. Škrinjar, J.P., Škorput, P., Furdić, M.: Application of unmanned aerial vehicles in logistic processes. In: International Conference "New Technologies, Development and Applications", pp. 359–366. Springer, Cham (2018)
3. Floreano, D., Wood, R.J.: Science, technology and the future of small autonomous drones. Nature **521**(7553), 460–466 (2015)
4. Škorput, P., Mandžuka, S., Vojvodić, H.: The use of unmanned aerial vehicles for forest fire monitoring. In: Proceedings of 58th International Symposium ELMAR-2016, pp. 93–96 (2016)
5. Škorput, P., Mandžuka, S., Jelušić, N.: Real-time detection of road traffic incidents. PROMET-Traffic Transp. **22**(4), 273–283 (2010)
6. Škorput, S., Mandžuka, S., Gregurić, M.: Digitalization of traffic accident investigation processes. In: Proceedings of 11th International Conference DKU-2018 (2018)

IoT Concept in Cooperative Traffic Management

Miroslav Vujic[1(✉)], Sadko Mandzuka[1], and Luka Dedic[2]

[1] Faculty of Traffic and Transport Sciences, University of Zagreb, Vukeliceva 4, 10000 Zagreb, Croatia
miroslav.vujic@fpz.hr

[2] HR-10000, 10000 Treskavicka, Zagreb, Croatia

Abstract. In order to improve the quality of traffic in urban areas it is necessary to establish communication between infrastructure and vehicles. Basic concept is defined through cooperative approach where traffic information is exchanged between vehicles and vehicles and vehicles and infrastructure. Because of large amount of data, new communication technologies are crucial for data saving and analysis. Internet of Things is a plausible concept for data storage, but the intention of this research is to define possible wireless communication technologies that can be used in urban traffic management. Many ITS solutions demand different performance of wireless technologies (speed of connection establishment, bandwidth, etc.), so certain scenarios will be described in this research.

Keywords: Internet of Things · Intelligent transport systems · Traffic environment · Traffic control

1 Introduction

Urban traffic network has stochastic nature because of it the system can provide vast amount of data for better traffic control. It is necessary to enable the basis for data collection and analysis. New concept in data collection and storage is Internet of Things (IoT) concept which has the main goal to create an environment that allows the real time communication between traffic entities through Internet network [1]. IoT concept was developed by Radio Frequency Identification development community because of the rapid growth of mobile devices, analysis of data, etc. [2]. Various traffic scenarios have different requirements regarding data transmission. According to [3], ZigBee, Bluetooth and Wi-Fi are designed for short-range wireless communication with low power solutions and for usage in cooperative systems (more specifically, vehicle-to-vehicle and vehicle-to-infrastructure communication). Communication Access for Land Mobile (CALM) is defined as architecture which comprises a set of standards which was created within ISO Technical Committee 204 Working Group and the main scope is to provide a communication architecture for high speed ITS communication using one or more protocols which can enable transparent connectivity over any possible wireless communication media [4]. Also, various papers describe different scenarios regarding the usage of wireless communication between vehicles and infrastructure [5, 6].

I. Karabegović (Ed.): NT 2019, LNNS 76, pp. 406–410, 2020.
https://doi.org/10.1007/978-3-030-18072-0_47

2 IoT Concept

IoT concept can be defined as a global infrastructure which can enable advanced services and establishment of communication based on existing information and communication technologies (ICT). With the expansion of ICT and physical objects (mobile devices, in-vehicle devices, etc.), it was necessary to develop a new concept where exchange of data is possible between all participants (people, infrastructure and things). Vision of IoT considers presence of objects (things) that enables communication through wired and wireless connections to reach certain goals [7]. The fundamental concept of IoT is shown on Fig. 1.

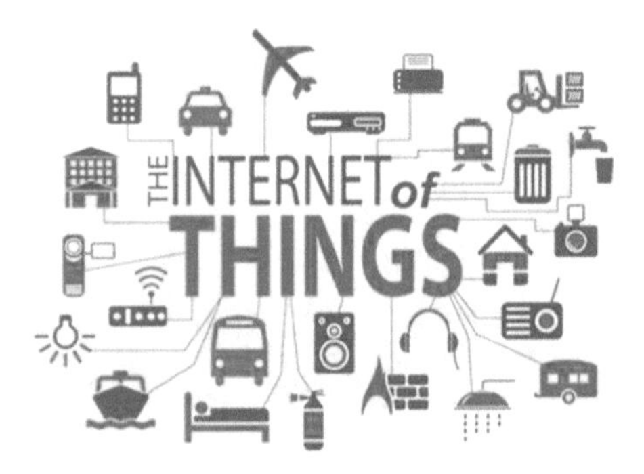

Fig. 1. Fundamental concept of IoT [7]

According to [7] main characteristics of IoT are defined:

- **interconnectivity** – any device can be connected to other device or infrastructure,
- **heterogeneity** – devices and infrastructure can be based on different hardware and software platforms and through different networks,
- **dynamic Changes** – the state of devices can change in time (connected/disconnected, location, speed, etc.),
- **wide-scale** – there is no limitations of number of devices connected or interconnected.

Also, IoT enables communication with objects in human environment by extending their interaction with different information (current location, speed, estimated travel times, etc.). According to [2], two basic modes of communication within IoT concept can be defined. First is communication between persons and objects where users generate communication with available devices to gather specific information, and second is communication between objects, where information is exchanged between objects mostly without human intervention.

In the traffic network, objects are physical things (vehicles, pedestrians, roadside objects, etc.) and they have the ability to interact between themselves in order to exchange traffic data mostly for the improvement of traffic network quality [2, 8].

2.1 IoT Reference Model

IoT reference model includes the concepts and definitions how the model itself can be built. It consists of all sub-models that define general IoT architecture. Communication between sub-models is shown on Fig. 2.

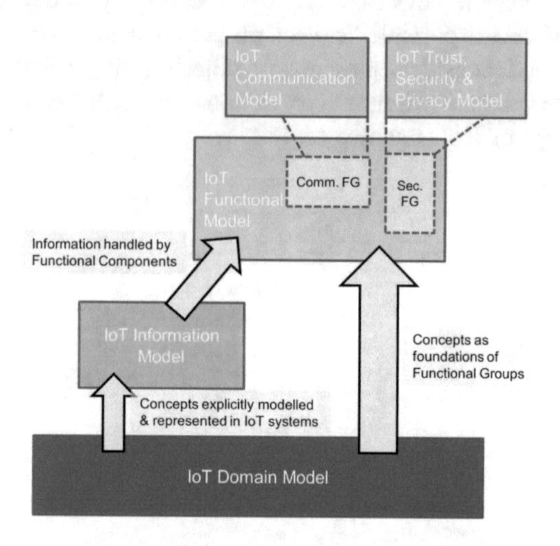

Fig. 2. Interaction of sub-models in the IoT reference model [9]

As depicted in Fig. 2, IoT domain model describes all concepts that are relevant in the IoT. All other models are based on Domain model and it is mandatory for every scenario that can be established via IoT. Arrows in Fig. 2 show that concepts of "lower" model are the basis for "higher" model in hierarchy.

2.2 Application of IoT in Traffic

As said before, traffic network generates vast amount of data such as number of vehicles, speed, traffic flow, OD matrices, etc. Those data can be stored and analyzed for prediction of traffic situation during certain period. According to saved data, traffic management can be improved. Also, there are other intelligent transport system (ITS) services and solutions that can be integrated in an ideal traffic network management (traveler information, incident management, etc.). All these integrated ITS services can improve traffic network performance which can be recognized through reduction in travel times, increase of vehicle speeds, reduction of vehicle emissions, noise, etc. [10, 11]. Also, the extension of integrated system can be recognized with multimodal travel guide, where users can access traffic information for different modes of transport in one unique service [13]. Main features of cooperative traffic management are presented in the next chapter.

3 IoT Application in Cooperative Traffic Management

Cooperative approach includes establishment of communication between three main traffic entities: vehicles, infrastructure and driver, but overall there are two main communication channels. First is communication between vehicles (vehicle-to-vehicle communication – V2V), and second is between vehicles and infrastructure (vehicle-to-infrastructure communication –V2I) [12, 14]. As said before various ITS services can be integrated in one complete system of traffic management and control, but the need of unique data collection platform is needed. IoT is such a platform which can integrate and disseminate traffic data to all relevant entities.

In this research cooperative model for future application of urban traffic management as part of EU FP7 STREP project Intelligent Cooperative Sensing for Improved Traffic Efficiency – ICSI is considered. System is based on Wireless Sensor Networks (WSNs) and vehicular networks. This implies that it should be able to collect and process a large amount of sensed data in scalable and reliable way. The ICSI system is composed by various components (Fig. 3): control centres, road-side units, on-board units and personal devices. IoT concept can be used in this proposed system because of the possibility to collect big amount of data and to forward it to whoever needs that information.

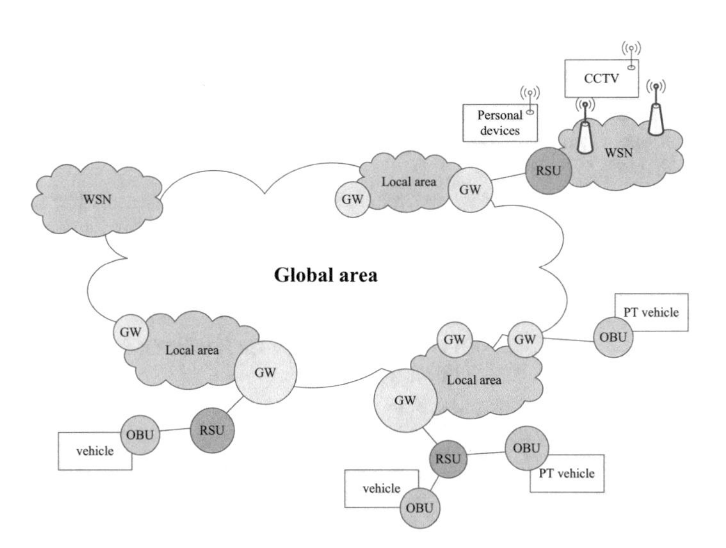

Fig. 3. ICSI system architecture [14]

Long-term evaluation and data analysis is done in control and management centre with collected all statistical data from the traffic network. But, the focus is to enable secure system with protection of user information and vulnerability of the system [15].

4 Conclusion

Wireless communication is essential for upgrading urban traffic management control. Establishment of real time communication between vehicles and infrastructure is necessary so that fixed traffic signal control can be upgraded with adaptive traffic control based on traffic data gathered directly from traffic network. Different communication technologies can be used for different ITS services regarding performances. In this paper the analysis of wireless communication technologies regarding technical demands was presented with some possible applications in already developed ITS solutions and services. In further research, proposed wireless communication technologies will be tested and according to real world experience, for different ITS solutions, optimal technology will be proposed.

References

1. Perakovic, D., Husnjak, S, Cvitic, I.: IoT infrastructure as a basis for new information services in the its environment. In: 22nd Telecommunications Forum Telfor (TELFOR). IEEE (2014)
2. Patel, K.K., Patel, S.M.: Internet of Things IOT: definition, characteristics, architecture, enabling technologies, application & future challenges. Int. J. Eng. Sci. Comput. **6**(5), 6122–6135 (2016)
3. Kirusnapillai, S., et al.: Deploying wireless sensor devices in intelligent transportation system applications. In: Intelligent Transportation Systems. InTech (2012)
4. Thierry, E., Nebehaj, V., Søråsen, R.: CVIS: CALM proof of concept preliminary results. In: 9th International Conference on Intelligent Transport Systems Telecommunications, (ITST). IEEE (2009)
5. ISO, TC. 204. Intelligent transport systems-Communications Access for Land Mobiles (CALM)-Architecture, ISO 21217. International Organization for Standardization (2013)
6. Budi, A., von Arnim, A.: TRACKSS approach to improving road safety through sensors collaboration on vehicle and in infrastructure. In: IEEE 68th Vehicular Technology Conference, 2008. VTC 2008-Fall. IEEE (2008)
7. Hua, C., Lin, Y.: A roadside ITS data bus prototype for intelligent highways. IEEE Trans. Intell. Transp. Syst. **9**(2), 344–348 (2008)
8. Vujic, M., Skorput, P., Celic, J.: Wireless communication in cooperative urban traffic management. Pomorstvo **29**(2), 150–155 (2015)
9. Bertin, E., Crespi, N., Magedanz, T.: Evolution of Telecommunication Services. Springer, Heidelberg (2013)
10. D1.5: Internet of Things – Architecture, Final Architectural Reference Model for IoT v3.0
11. Mandzuka, S., Skorput, P., Vujic, M.: Architecture of cooperative systems in traffic and transportation. In: 23rd Telecommunications Forum Telfor (TELFOR). IEEE (2015)
12. Vujic, M., Mandzuka, S., Greguric, M.: Pilot implementation of public transport priority in the city of Zagreb. Promet-Traffic Transp. **27**(3), 257–265 (2015)
13. Škorput, P., Mandžuka, B., Vujić, M: The development of cooperative multimodal travel guide. In: 22nd Telecommunications Forum Telfor (TELFOR). IEEE (2014)
14. D1.1.1: Use Cases Definition and Analysis, FP7 ICSI - Intelligent Cooperative Sensing for Improved Traffic Efficiency (Grant Agreement Number: 317671) (2014)
15. Škorput, P., Vojvodić, H., Mandžuka, S.: Cyber security in cooperative intelligent transportation system. ELMAR, 2017 International Symposium (2017)

Application of New Technologies to Improve the Visual Field of Heavy Duty Vehicles' Drivers

Nadica Stojanovic[1](✉), Ivan Grujic[1], Jasna Glisovic[1], Oday I. Abdullah[2,3], and Sasa Vasiljevic[4]

[1] Faculty of Engineering, University of Kragujevac, Sestre Janjic 6, 34000 Kragujevac, Serbia
nadica.stojanovic@kg.ac.rs
[2] Department of Energy Engineering, College of Engineering, University of Baghdad, Baghdad, Iraq
[3] Hamburg University of Technology, Hamburg, Germany
[4] High Technical School of Professional Studies, 34000 Kragujevac, Serbia

Abstract. The World Health Organization has declared pedestrians, cyclists and motorcycle, moped drivers and passengers, as vulnerable categories of participants in traffic. Some of the reasons are that they often can be found in a blind spot of the cars, vans and heavy duty vehicles. Therefore, the drivers of such vehicles can't notice them. Determination of the visual field of heavy duty vehicles' drivers, is performed in the paper by applying Catia V5 software, module Ramsis (in case if only rear-view mirrors are used). Based on this approach, the reasons for using cameras and sensors instead of a rear-view mirror are reflected in the reduction of blind spots and reduction of the aerodynamic drag coefficient, the numerical simulation was obtained by using Ansys/Workbench 14.5 software.

Keywords: Blind spot · Visual field · Ramsis · Rear-view mirrors · Cameras and sensors · Ansys

1 Introduction

It can be considered the heavy duty vehicles are main key to transport the goods. Heavy duty vehicle (category N3) is the vehicle for which the highest allowed mass is higher than 12t. Category N motor vehicles are projected and constructed for cargo transportation.

Motor vehicles must have two rear-view mirrors. Mirrors that are outside the vehicle must have area (surface) at least 150 cm^2 if the surface mirror is round, and 300 cm^2 for mirrors with flat surface. Mirror surface radius can't be less than 80 cm.

Heavy duty vehicles that are first time registered in the Republic of Serbia after 01. July 2011, and don't have camera for recording the space behind the vehicle, must too have implemented device for sound signalization on rear side of the vehicle that will activate automatically with reverse gear ratio [1].

I. Karabegović (Ed.): NT 2019, LNNS 76, pp. 411–421, 2020.
https://doi.org/10.1007/978-3-030-18072-0_48

In order to reduce the risk of accidents involving the blind spots, the European Union (EU) implemented Directive 2003/97/EC, which substantially increased the field of view available from the mirrors of new trucks sold in the EU from January 2007 [2]. Statistics reveals that the visual problems are the main reasons for the high number of road accidents.

It can be seen the fields of vision in the Fig. 1 are [3]:

Fig. 1. Fields of vision [4]

- Class II - Main rear-view device on the driver's side and the passenger's side.
- Class IV Wide-angle view device on the driver's side and the passenger's side.
- Class V close-proximity view device.
- Class VI front-view device.

Way and Reed [5] presented the method for physically measuring the driver's view field in rear view mirrors. A portable coordinate measurement apparatus (FARO Arm) is used to measure the mirror locations, contours, and curvature. Measurements of the driver's field of view from the cab of a combination vehicle show restrictions in critical areas around the vehicle, and are consistent with the results from the crash data. Direct fields of view using the different mirrors in one truck, a tractor-semitrailer with a conventional cab (hood out in front) and an integral sleeper [6], shown in Fig. 2.

The straight blue lines in Fig. 2 show the boundaries of obstructions by the left and right planar mirrors, the A-pillars and rear of the cab, and the hood of the vehicle. The light blue curves outline the view provided by left and right door-mounted convex mirrors and the light orange curves show the view that using fender-mounted convex mirrors. Solid lines illustrate the field at the ground level, while the dotted lines were located with 1.2 m above ground level. The driver has direct view forward, above the level of the hood and inside the A pillars. To the left and right, areas of the driver's field of view are obstructed by the door, the A-pillars, and the right and left planar mirrors. Particularly to the right, the diagram shows that without a hood-mounted right convex mirror, the driver is unable to view the lane immediately to the right of the cab. The diagram includes drawing shows that a typical passenger vehicle can fit fairly comfortably in the area that the driver can only view using a fender-mounted mirror. The door structure and hood almost entirely occlude the driver's view of this area. Only a small part of the left rear of the vehicle would be visible in the right door-mounted

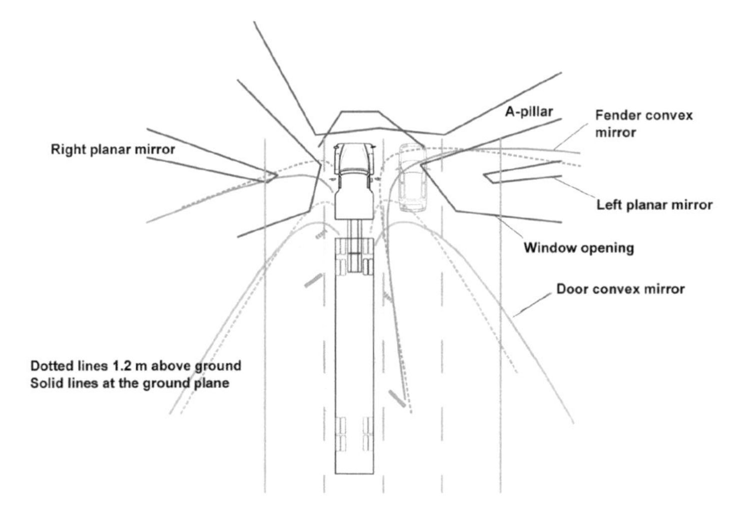

Fig. 2. Mirror and direct fields of view in one truck [6]

convex mirror, and a small piece of the hood of the vehicle would be visible to the driver through the window opening. To the left, the driver has direct vision through the window opening to the lane immediately left of the cab, but has to rely on mirrors for the lane to the left behind the cab. Thus, even with a full set of mirrors properly adjusted, there is a significant blind zone to the right front of the truck cab. One can observe that the quality of the image in the mirrors is not addressed here, but that could also affect the driver's ability to identify other vehicles or road users around the vehicle.

The project that implemented by Summers kill and Marshall was provided the objective comparisons of blind spots size between 19 vehicle models which are the top selling in the UK. This was allowed the key design features for the cabs to be analysed to determine what causes the size and location of blind spots. The identified largest blinds spots were associated with direct vision from the cabs, and the project has shown considerable variability in the size of direct vision blind spots between vehicles which have been shown to be associated with the height of the cab above the ground [7].

Visibility through the windows of most vehicles is restricted due to the required structure of the vehicle, the most manufacturers and users incorporate a series of mirrors to enhance the driver visibility and to reduce the blind spots. Paper of Ball and others represents the procedures for calculating and modelling the driver visibility for commercial vehicles. The primary techniques presented require access to the vehicle; also the paper presented the techniques to analysis the visibility through photogrammetry and 3-D computer models. Both for the vehicle and any mirrors were incorporated onto the vehicle [8].

The main purpose of this paper is to represent the blind spots on one heavy duty vehicle, if mirrors are used. Furthermore, it was developed a numerical analysis to determine accurately the drag coefficient for the mirrors, and for vehicles that have cameras instead of mirrors. Finally, it was presented the full details and justifications about the reasons to use the sensors and camera usage on one heavy duty vehicle.

2 Visual Field Determination

It was build 3D model of the truck to analyze the visual field as shown in Fig. 3. More the one view of the truck was illustrated in Fig. 3 (isometric and front view).

Fig. 3. 3D model of the heavy duty vehicle

Based on the size of truck, it could be said, that the truck drivers are not endangered category of traffic participants. The problem that occurs on truck is incorrectly adjusting of the mirror as shown in Fig. 4. Therefore, it can be happened that the truck driver can't see other traffic participants, for example two-wheelies. If two-wheelies are found in area between truck trailer and shaded surface, the truck driver couldn't see him in rear-view mirror. In case of passing from one track to another or during the overtaking the slower vehicle, the truck would see that in other track are no vehicles and it will start with the overtaking. In such situation, the traffic accident can happen between heavy duty vehicle and two-wheelie.

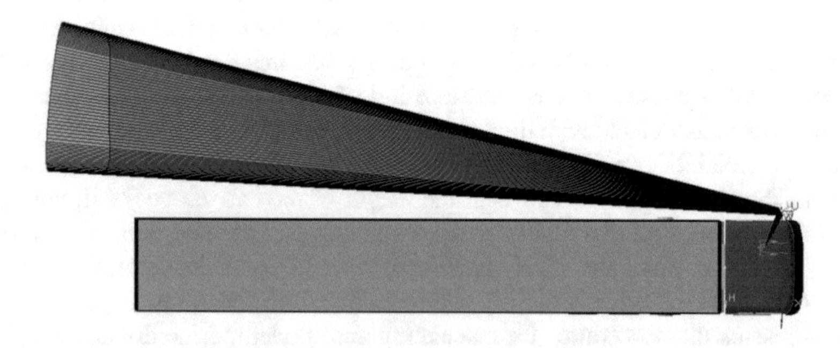

Fig. 4. Visual field in the case of incorrectly mirror adjustment

On the accident risk scale, it was ranked that the motorcyclists are exposed to the highest level of risk. Together with pedestrians and motorcyclists represent the most vulnerable category of traffic participants [9].

One more critical area at trucks is the area in the front of the truck. Ina case, when the truck is stopping to let the pedestrian to cross the street. If the truck doesn't have a front mirror because of its height and shape, the truck driver can't see whether the pedestrian crossed or not, and especially if he misses the child to cross the road. So, it's important for such one vehicle to have the front mirror, in order to cover the area in the front of the truck, which is shown in Fig. 5. In order to cover the shown area, it is necessary to be the size mirror equal to D150 mm, which has spherical shape with 500 mm radius. Of course the correctly adjustment is necessary to cover the area. The position of the mirror depends on the driver, more accurate from the height of the look.

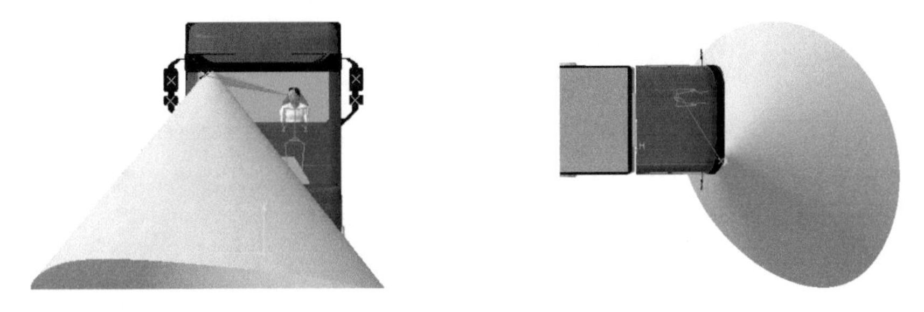

Fig. 5. Fields of vision – Class VI

The areas at the left and the right sides of the driver are shown in Figs. 5, 6, 7 and 8. The bigger mirror at the left side of the driver is spherical, and it covers the area as shown in Fig. 6. While, the small mirror is spherical shape, that covers the area which illustrated in Fig. 7. It‘s very important to cover as much as possible the area of the truck. In other words, to be more accurate that the driver have insight as much as possible at area around the truck in order to avoid the accident.

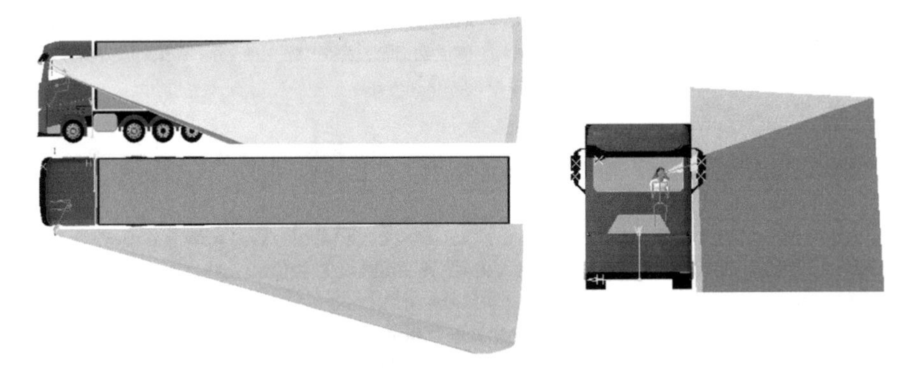

Fig. 6. Visual field by looking at bigger mirror on the driver side

As it must be covered, as large as possible area around the truck on the left side, it must be on the right side as shown in Fig. 8. The problem that can occur is the aggressive drive of two-wheelies, which usually end up on the both sides, the left and the right sides of the truck. The most often accidents are happening when the truck is stopping in the traffic light and he wants to turn right. Also at his right side when the two-wheelie is stopped, with moving of truck and two-wheelies, truck can hit the

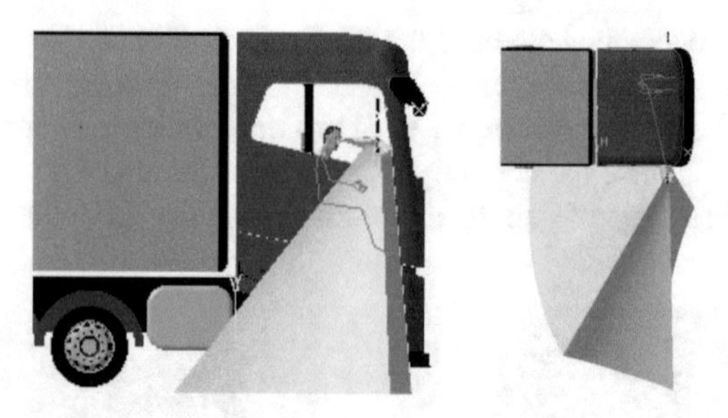

Fig. 7. Visual field by looking at smaller mirror on the passenger side

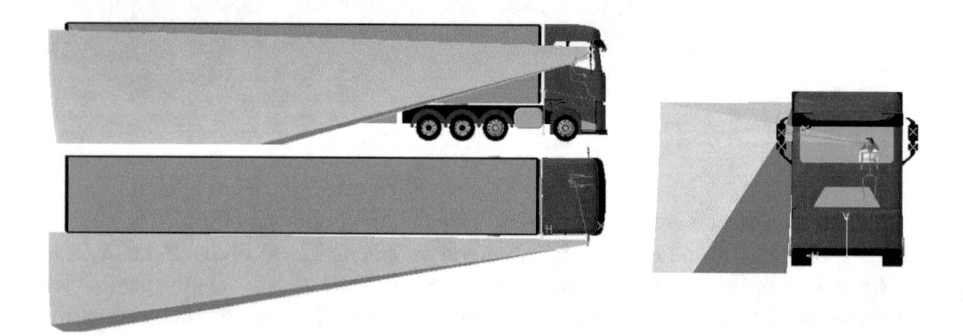

Fig. 8. Visual field by looking at bigger mirror on the passenger side

two-wheelie. Therefore, a mirror will be required to cover the surface as shown in Fig. 7. The event sequence could be as follows: the driver of the two-wheelie falls and in the worst case scenario the truck is overtaking it.

3 Determination of Aerodynamicity of Rear-View Mirrors

Owing to the cube shape, trucks have the big drag coefficient. The new development of vehicles tends to make the drag coefficient of vehicles is minimum as much as possible, because the drag coefficient have a significant influence on the fuel consumption. The drag coefficient value only for semi-truck is 0.608, and for truck with trailer is 0.704. Which means that the drag coefficient of trailer increased by 15.8% [10].

Figure 9 shows that the mirror illustrated in Fig. 10, will increasing the drag coefficient with 0.2%. Furthermore, it can be observed the value of drag coefficient of the mirror as shown in Fig. 10, and for MirrorCam is shown in Fig. 11.

The drag coefficient has been determined only for rear-view mirror, and for the MirrorCam, it will not be observed for the full truck. The value of the drag coefficient or the rear-view mirror is found to be 1.52, while for MirrorCam is found to be 0.43, which is 71.7% less than the classic rear-view mirrors. Figure 12 represents the pressure distribution on the rear-view mirror and MirrorCam surface, on which direct affect the air currents. The maximum pressures are approximately equal; therefore the

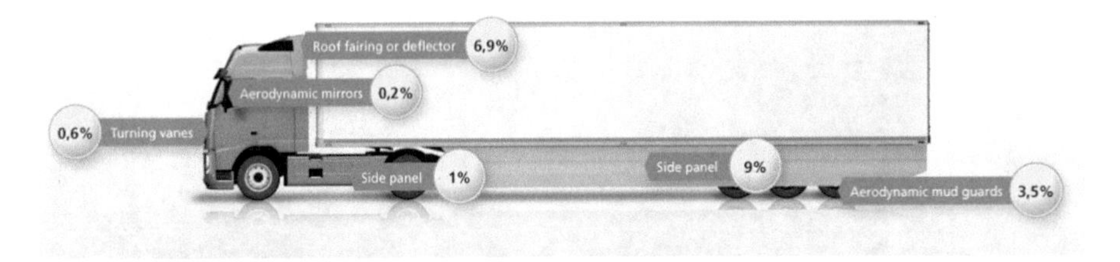

Fig. 9. Discover how aerodynamics contribute to saving fuel [11]

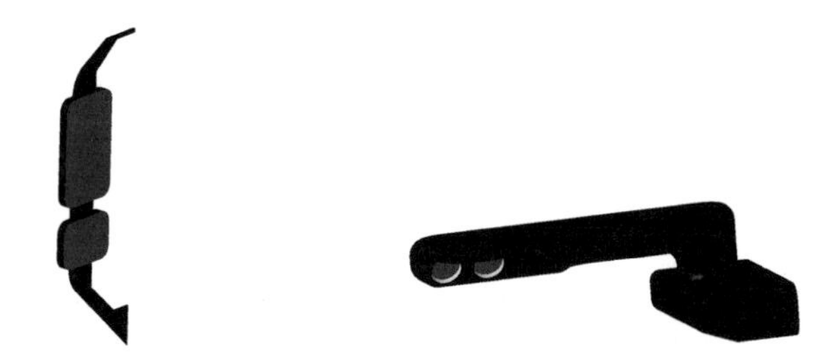

Fig. 10. Rear-view mirror

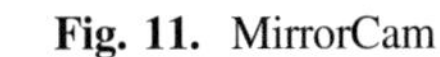

Fig. 11. MirrorCam

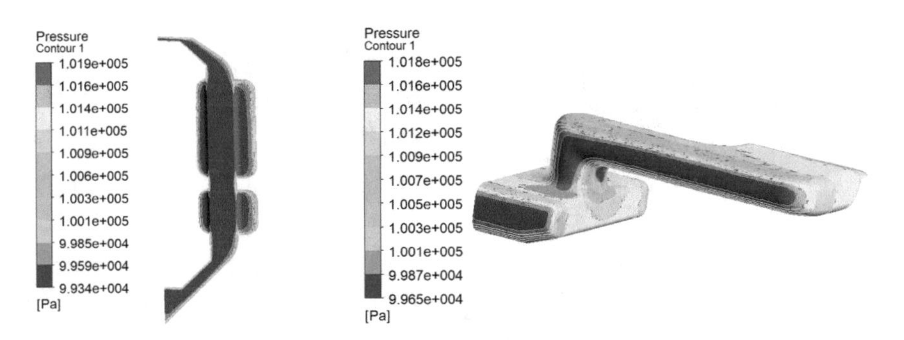

Fig. 12. Pressure distribution on the rear-view mirror and MirrorCam surface

surface of the rear-view mirror is big and flat, in order to obtain a uniform pressure distribution and the lowest level of the pressure to resist the effect of pressure that subjected on the mirror.

Air turbulence that occurs on the rear side of the rear-view mirror and MirrorCam are exhibited in Fig. 13. The value for the Turbulence kinetic energy is higher for the rear-view mirror, this is happened because of the drag coefficient is higher than the others.

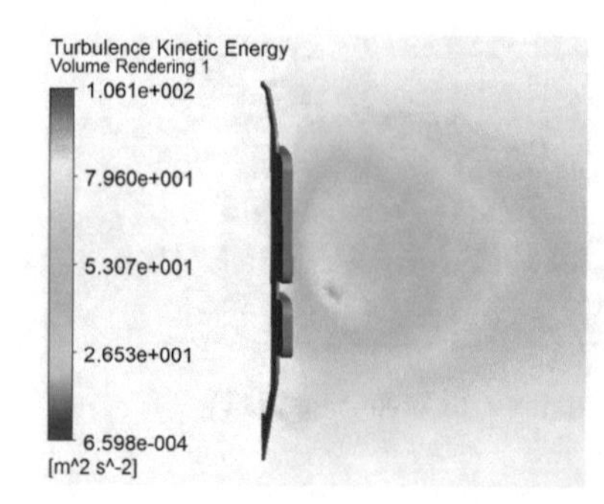

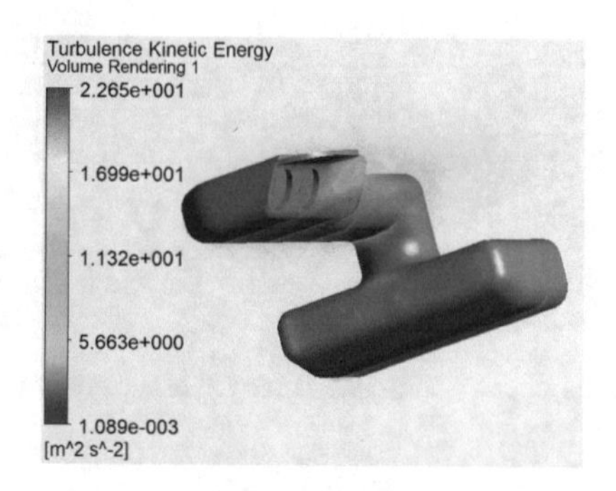

Fig. 13. Turbulence kinetic energy

4 The Application of the New Technologies at Heavy Duty Vehicles

MirrorEye consists of two camera units, a cable harness and two 12.3-inch monitors [12] as shown in Fig. 14.

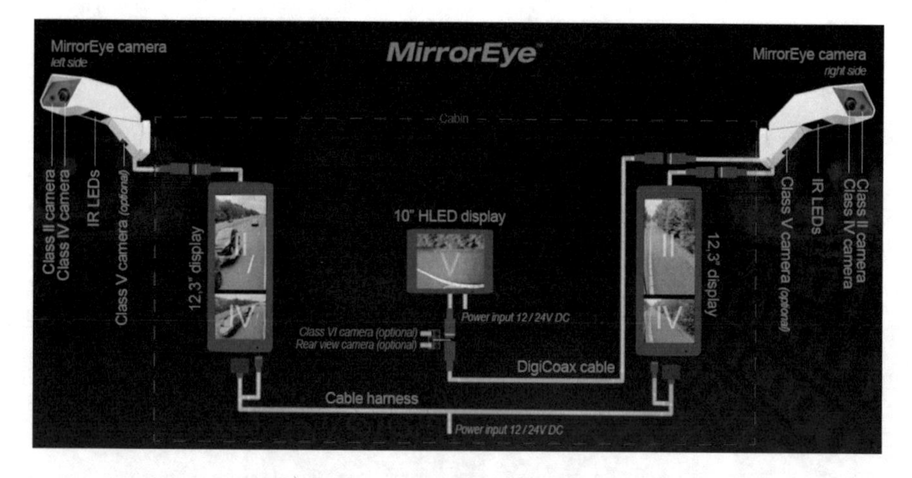

Fig. 14. The components of MirrorEye [12]

Camera Units - For MirrorEye, Orlaco uses the two-component camera set for classes II and IV. The cameras are in the camera unit that is mounted on the body of the truck using a vehicle-specific interface. Class V cameras can be integrated into the camera units as an option.

Cable Harness - The camera images are sent to the monitors via the cable harness.

Monitors - The real-time HD images from the cameras are displayed by using a split screen on two 12.3-inch monitors as shown in Fig. 14. Just as with conventional mirrors, the view of class II is shown at the top of the monitor and the view of class IV is at the bottom. The images from the class V camera are displayed on an optional, by using the separate 10-inch monitor.

Brigade's complementary range of safety devices helps to prevent collisions by assisting the driver whilst protecting workers, pedestrians and cyclists. The similar vehicle that is analysed in this paper, where it was applied new technologies that shown in Fig. 15, where every items of the sensors and cameras in heavy duty vehicle are marked, and the description of these items can be summarized as follows:

- 1–6 → Camera Monitor Systems;
- 7–9 → Reversing and Warning Alarms;
- 10–12 → Ultrasonic Obstacle Detection and
- 13 → Mobile Digital Recording.

The safety that is necessary to achieve by this system is separated in three levels. Every sensor/camera provides a certain level of safety, it can be seen the details of these levels in Table 1.

Fig. 15. Places for sensors and cameras implementation on the heavy duty vehicle [12]

Table 1. Level safety each of the sensors/cameras for typical vehicle type [13]

Safety level	1	2	3	4	5	6	7	8	9	10	11	12	13
Level 1	x	x	x				x						
Level 2				x				x	x	x			
Level 3					x	x					x	x	x

Where the levels are:

Level 1 - Minimum recommended safety
Level 2 - Increased recommended safety
Level 3 - Ultimate recommended safety

Backeye®360 is an intelligent camera monitor system designed to assist low-speed manoeuvring by providing the driver with a complete surround view of the vehicle in real time.

Brigade offers a choice of 360° technologies, both of which work with four ultra-wide-angle cameras that each one covers one full side of the vehicle with a viewing angle of over 180°. High-mounted on the front, rear and sides, the calibrated cameras capture all of the surrounding area including the blind spots of the vehicle. The four live images are simultaneously sent to an electronic control unit (ECU) where they are instantly processed, combined, blended and stitched. The distortion from the wide-angle camera lens is also corrected before delivering a clear, single, smooth, real-time image onto the driver's monitor, as shown in Fig. 16.

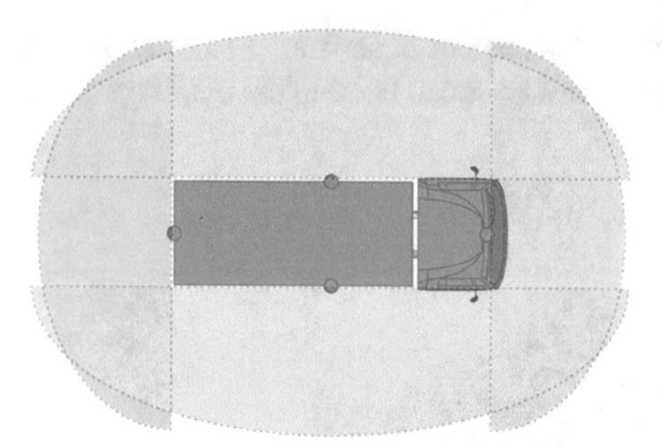

Fig. 16. Backeye®360 [14]

5 Conclusions and Remarks

It can be considered the adjustment of rear-view mirrors correctly is main key to improve the safety factor and to avoid the potential accident, where the correct adjustment of rear-view mirrors will provide a good insight around the heavy duty vehicle. Also, this adjustment of mirrors will help the driver to obtain the information about situation around vehicle. However, the rear-view mirrors create will increase the resistance during vehicle movement, which is not the case with MirrorCam. So, in this case, it's beater to use MirrorCam. One more reason to use the sensors and cameras is to reduce the effort of the driver in focusing when using rear-view mirrors, which can be tired. While by using the cameras, it will be much easier and less effort needed it for the driver. So in the case of some the unreasonable actions the system will react and would not allow the occurrence of a traffic accident.

Furthermore, this research paper can be considered the foundation to analyze the view field, for a larger population of people. In this work, a comparison was presented between two different cases of driver; the first case when use the mirror and the second case when use the MirrorCam. Finally, the most important recommendation for future work is should to analyze accurately the effect of applying MirrorCam on the number of accidents, in order to find if by applying of MirrorCam the number of accidents will be reduced. This research is preliminary research that will be followed by further research that will study deeply all the effective factors to obtain the optimal visual field of heavy duty vehicles' drivers.

Acknowledgments. This paper was realized within the researching project "The research of vehicle safety as part of a cybernetic system: Driver-Vehicle-Environment" ref. no. TR35041, funded by Ministry of Education, Science and Technological Development of the Republic of Serbia.

References

1. The Official Gazette of the Republic of Serbia, No. 40/12, 102/12, 19/13, 41/13, 102/14, 41/15, 78/15, 111/15, 14/16, 108/16, 7/17 (ispravka) i 63/17 Rulebook on the Classification of Motor Vehicles and Trailers, and Technical Conditions for Vehicles in Road Traffic
2. Knight, I.: A study of the implementation of Directive 2007/38/EC on the retrofitting of blind spot mirrors to HGVs, PUBLISHED PROJECT REPORT PPR588 (2011)
3. Regulation No. 46 Uniform provisions concerning the approval of devices for indirect vision and of motor vehicles with regard to the installation of these devices
4. Orlaco. https://www.orlaco.com/files/vision%20systems/MirrorEye/truck/fov-truck-orlaco.png. Accessed 12 Dec 2018
5. Way, M.L., Reed, M.P.: A method for measuring the field of view in vehicle mirrors. In: SAE International, 2003-01-0297 (2003)
6. Blower, D.: Truck mirrors, fields of view, and serious truck crashes, Report No. UMTRI-2007-25 (2007)
7. Summerskill, S., Marshall, R.: Understanding direct and indirect driver vision from heavy goods vehicles, Summary report (2016)
8. Ball, J.K., Danaher, D.A., Ziernicki, R.M.: A method for determining and presenting driver visibility in commercial vehicles. In: SAE International, 2007-01-4232 (2007)
9. Jevtić, V.: Speed as an indicator of motorcyclists' safety. Doctoral dissertation, University of Belgrade, Faculty of Transport and Traffic Engineering (2015)
10. Bayındırlı, C., Akansu, Y.E., Salman, M.S.: The determination of aerodynamic drag coefficient of truck and trailer model by wind tunnel tests. Int. J. Automot. Eng. Technol. **5**(2), 53–60 (2016)
11. WUWT – Watts Up With That. https://wattsupwiththat.files.wordpress.com/2009/11/semi-truck-savings.png. Accessed 17 Dec 2018
12. Orlaco. https://www.orlaco.com/mirroreye-for-trucks. Accessed 02 Dec 2018
13. Brigade Electronics Group Plc. https://brigade-electronics.com/vehicles/articulated-truck-safety-products/. Accessed 17 Dec 2018
14. Brigade Electronics Group Plc. https://brigade-electronics.com/products/backeye360/. Accessed 17 Dec 2018

The Impact of the Seatback Angle on the Appearance of the Driver's Discomfort

Slavica Mačužić(✉) and Jovanka Lukić

Faculty of Engineering, University of Kragujevac,
Sestre Janjic 6, 34000 Kragujevac, Serbia
slavicamacuzic89@gmail.com,
lukicj@kg.ac.rs

Abstract. Car model, color, design, motor power, driving performance, is just some of the factors that influence the purchase of a new car. However, one of the most important task of vehicle manufactures is driver comfort. In the initial phase of designing a car, a posture of vehicle driver must be carefully considered, because a long-term driving can affect driver performance, especially in an inadequate posture. In this paper, a virtual environment of middle-class car was created. Different angles of the seat back are considered, with the aim of determining the driver body discomfort. Anthropometrics characteristics of different male and female populations were used. The analysis was carried out for two conditions: during driving and resting. Software package Ramsis was used to perform posture analysis of "mannequin" in driving and resting condition. The obtained values of fatigue and discomfort, as body parts discomfort, were different for all kinds of subjects.

Keywords: Anthropometry · Driver · Modeling · Discomfort · Ramsis

1 Introduction

The time in which we live today and the traffic situations we are in require a different use of the car. The car has become a basic tool for human life, whether it is used for personal or business purposes. Special attention is paid to the comfort of driver and passengers. The driver suffers the most loads during the drive. For this reason, it is necessary to have a comfortable seat so that the driver's fatigue is as low as possible. The car seat plays an important role in improving the comfort of the car. The discomfort is mostly associated with biomechanical factors involving muscular and skeletal systems [1]. Good seat is essential for the prevention of back pain, as well as other illnesses that arise from a poor standing position. Driver position during driving is a complex study in relation to the traditional seating position. For the design of a comfortable car seat, it is necessary to consider the aspects that affect the driver's comfort. Author De Looze [2] described that car seats should be designed according to the contours of the human body. He concluded that the dimensions of the seat, hardness or softness of the material have an impact on the human body during driving. Author Bridger [3] concluded that at the same time ergonomic and anthropometric criteria are applied in order to satisfy the driver's comfort. Seat comfort and discomfort are defined as independent subjects linked by various factors. Discomfort is associated with

I. Karabegović (Ed.): NT 2019, LNNS 76, pp. 422–428, 2020.
https://doi.org/10.1007/978-3-030-18072-0_49

biomechanics and fatigue factors, and comfort with a sense of well-being and aesthetics. Comfort can be presented as a static and dynamic. Static comfort refers to the correct position of the seat-back angle. Two factors that affect the comfort of the static seat comfort are the distribution of the seat-back pressure and the seat-back angle. In the case of dynamic comfort, the transfer function affects the driver's seat. Seat transfer function is related to the vibrations that appear on the driver's seat. Vibrations produced on the car seat create driver discomfort and reduce the perception of observation [4, 5]. Excessive exposure to vibrations can negatively affect the human body. Comfort impacts differently on different populations. Many research studies the impact of the backrest on driver comfort [6, 7]. In this paper, digital human models are used to determine discomfort and fatigue in different populations. A comparative analysis of discomfort and fatigue for different angles of sitting and for different populations was performed. The Catia V5 R18 and Ramsis software packages were used in this analysis.

2 Methods

Digital human models help in defining vehicle design and the position of various components in the vehicle [8]. That's why they have been used for testing in the automotive industry for many years. The final appearance of the vehicle from an ergonomic point requires information which populations will use it, and therefore it is important to know the anthropometric characteristics of the driver. This paper describes the impact of the seatback of the seat on the occurrence of discomfort. The analysis was performed for three different seating angles of 90°, 100° and 105° to eleven different population populations. The first task was to create a car interior using the Catia V5 R18 software package. The second task was to put a model in the vehicle and set boundary conditions (Fig. 1). For this purpose, Ramsis software package and geometric model of the model were used.

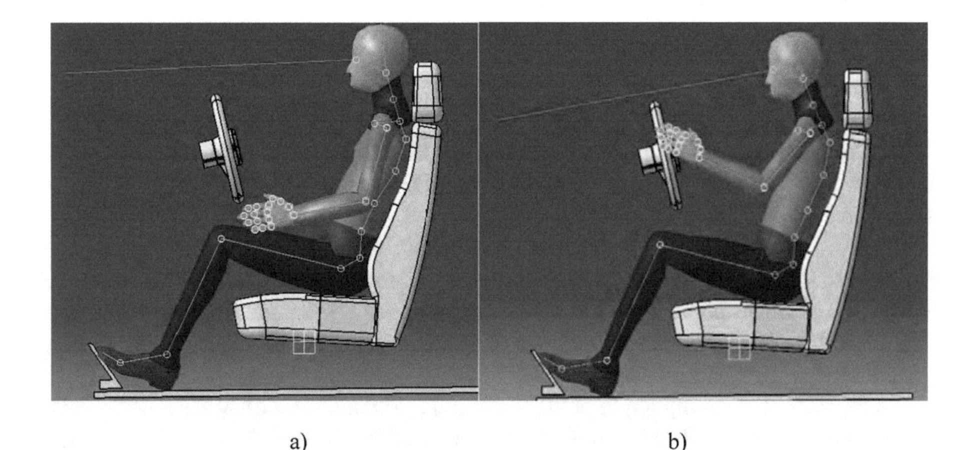

a) b)

Fig. 1. Modeled environment in the vehicle; (a) resting condition, (b) driving condition

Anthropometric data for eleven different populations of 1% female and 99% male populations are shown in Table 1. Table 2 shows anthropometric data for eleven different populations of 5% female and 95% male populations. The populations belonged to the age group of 18–70 years. The analysis was performed for two different states, resting and driving conditions. On the following diagrams, with 1 was marked driving condition, while with 2 marked resting condition.

Table 1. The anthropometric data of 1% female and 99% male populations [9]

Population	Height (mm)		Sitting height (mm)		Foot length (mm)		Shoulder breadth (mm)	
	1%	99%	1%	99%	1%	99%	1%	99%
West Africa	1402	1833	720	906	206	286	341	462
North India	1412	1805	750	940	199	278	305	429
Eastern Europe	1502	1885	814	980	217	293	354	506
North Europe	1541	1952	823	1020	217	288	353	514
Australia	1521	1933	810	1000	212	286	331	506
South East Europe	1485	1865	790	970	212	293	358	499
Central Europe	1518	1845	903	1010	212	298	357	509
South East Africa	1442	1815	750	930	202	288	348	479
Middle East	1496	1838	780	967	214	288	348	486
South India	1351	1755	723	897	194	273	318	449
North Asia	1469	1874	780	986	199	285	341	479

Ramsis software uses different human models and thus provides accurate analysis and comfort assessment. Fatigue and discomfort are defined by numerical values based on experiments. Comfort is displayed with a value of 2.5, while the occurrence of discomfort is defined with a value above 5.5.

3 Results

Using the Ramsis software package, an analysis of the interaction between driver and seat was performed. An example of the obtained fatigue and discomfort values is shown in Fig. 2.

The difference between the red and the yellow column indicates how much the sitting position is comfortable or uncomfortable. Comparative results of discomfort of 1% and 5% female population, for a seat angles of 90°, 100° and 105° are shown in Fig. 3a and b. It can be clearly seen from Fig. 3a that the smallest feeling of discomfort gave the angle of seating of 90°.

In the case of a seating angle of 90° (Fig. 3a) the smallest value of discomfort have population from South East Europe, Middle East and South India. The maximum value of the discomfort for these populations was 5. The seating angle of 100° gives the

Table 2. The anthropometric data of 5% female and 95% male populations [9]

Population	Height (mm)		Sitting height (mm)		Foot length (mm)		Shoulder breadth (mm)	
	5%	95%	5%	95%	5%	95%	5%	95%
West Africa	1440	1785	741	881	212	278	355	450
North India	1450	1765	771	919	205	270	315	415
Eastern Europe	1540	1845	831	959	225	285	371	489
North Europe	1585	1910	846	999	227	280	367	498
Australia	1565	1885	831	979	220	280	345	489
South East Europe	1525	1825	811	949	220	285	372	485
Central Europe	1560	1805	826	989	220	288	376	495
South East Africa	1480	1775	771	909	210	280	362	465
Middle East	1529	1800	801	944	222	280	360	469
South India	1395	1715	746	874	200	265	330	435
North Asia	1504	1820	801	961	207	275	355	465

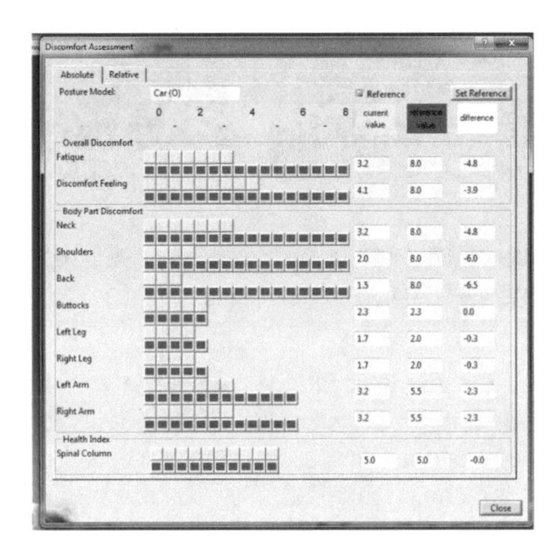

Fig. 2. Example of fatigue and discomfort values for one population

smallest discomfort for North Asia and Eastern Europe populations (max value 5.6). Third case, the seating angle of 105° gives the smallest values for Eastern Europe, North Europe and Australia populations.

From Fig. 3b is interesting to note that the 5% female population have the smallest value of discomfort for the seating angle of 105°. In this case, the least discomfort are felt by the populations from West Africa, Australia and South East Africa. Maximal values of discomfort does not exceed 4.8 for all populations.

Comparative results of fatigue of 1% and 5% female population, for a seat angles of 90°, 100° and 105° are shown in Fig. 4a and b.

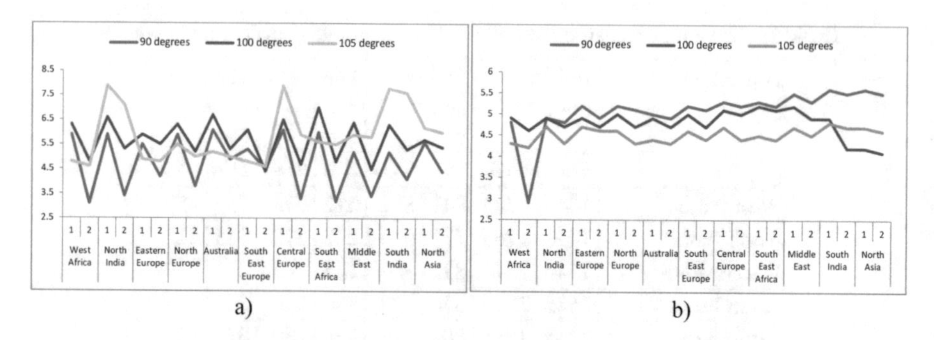

Fig. 3. Comparative results of discomfort for a seating angles of 90°, 100° and 105°: (a) 1% female population, (b) 5% female population

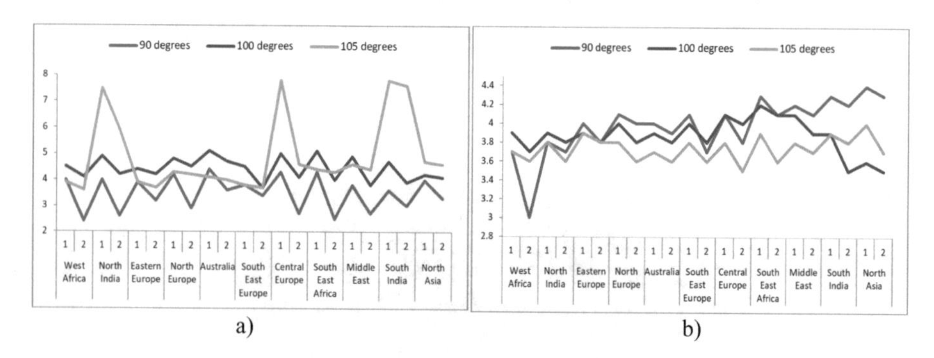

Fig. 4. Comparative results of fatigue for a seating angles of 90°, 100° and 105°: (a) 1% female population, (b) 5% female population

Figure 4a shows that the slightest fatigue causes the seatback angle of 90° for the case of 1% female population, on the example of the populations South India, Middle East, South East Europe, Eastern Europe etc. whose values go to 3.8.

In the case of 5% female population, Fig. 4b, the best choice of angle of the seatback is an angle of 105°. This angle causes least fatigue in the following populations: Australia, West Africa, Middle East etc. (maximum values was 3.7)

Comparative results of discomfort of 95% and 99% male population, for a seat angles of 90°, 100° and 105° are shown in Fig. 5a and b.

In the case of a population of 95% less discomfort provides an angle of 90° (Fig. 5a), whose values go to 4.2, while in the case of 99% of populations, better results give an seatback angle of 105° whose maximum value does not exceed 3.9 for all populations (Fig. 5b).

Figure 6a and b shows comparative results of fatigue of 95% and 99% male populations.

From Fig. 6a it can be seen that the best seating angle, from the aspect of fatigue, is a 100° angle. The smallest fatigue has populations from North Europe, Australia, South East Europe, Central Europe etc., whose fatigue values do not exceed 2.5. In the case of 99% male populations, Fig. 6b, less fatigue value gave an angle of 105°, compared

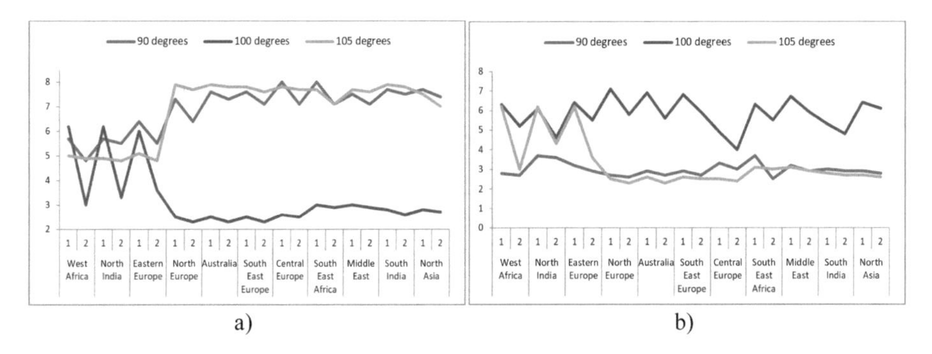

Fig. 5. Comparative results of discomfort for a seating angles of 90°, 100° and 105°: (a) 95% male population, (b) 99% male population

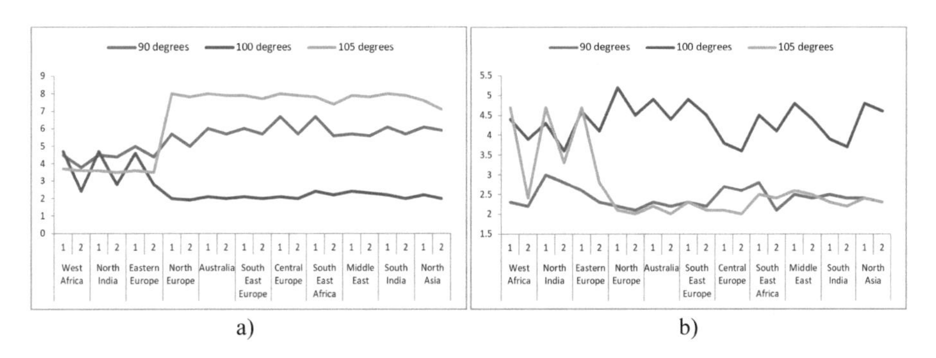

Fig. 6. Comparative results of fatigue for a seating angles of 90°, 100° and 105°: (a) 95% male population, (b) 99% male population

with 90°. For this example of analysis, low fatigue levels have populations from North Europe, Australia, South East Europe and Central Europe whose fatigue values do not exceed 2.2.

4 Conclusion

The aim of this paper was to perform a comparative analysis of fatigue and discomfort during driving for different angles of sitting. Digital human models were used. Ramsis software package was used for analysis. An analysis was carried out on eleven different populations of 1% and 5% female population and 95% and 99% male population.

In the case of an analysis of the discomfort of 1% of the female population, it was concluded that the 90° angle provides relatively lower maximum discomfort values of 5, while in the case of 5% female populations better seating angle was 105° with maximal discomfort values go to 4.8. Comparing these results with male results, in both cases, for 95 and 99% male population, the best angles were 90° and 105°, but with slightly lower discomfort values of 4.2 and 3.9, respectively.

The analysis of the fatigue of female populations, for 1% and 5%, showed that the best seating angles were 90° and 105°, respectively, with the highest fatigue values of 3.8 and 3.7, respectively. In the case of fatigue analysis of male populations, for 95% and 99%, it was concluded that 95% of the population corresponds to a seating angle of 100°, with a maximum fatigue value of 2.5, and a 99% population corresponds to an seating angle of 105°, with the highest fatigue value of 2.2.

Using this software, car makers have the opportunity, before the first prototype, to graphically display how the model interacts with its environment inside the car. The benefits of such research during the design of the car can do a lot to save money and predict where manufacturers can sell their products.

Acknowledgements. This research supported by the Ministry of Education, Science and Technological Development of Republic of Serbia through Grant TR 35041.

References

1. Helander, Z.L.M., Drury, C.: Identifying factors of comfort and discomfort. J. Hum. Factors Ergon. Soc. **38**, 377–389 (1996)
2. De Looze, M.P., Kuijt-Evers, L.F.M., Van Dieën, J.: Sitting comfort and discomfort and the relationships with objective measures. Ergonomics **46**(10), 985–997 (2003)
3. Bridger, R.S.: Introduction to Ergonomics. Taylor & Francis Group, Boca Raton (2009)
4. Vink, P.: Comfort and Design, Principles and Good Practice. CRC Press, Boca Raton (2004)
5. Helander, M.G., Zhang, L.: Field studies of comfort and discomfort in sitting. Ergonomics **40**, 895–915 (1997)
6. Giacomin, J., Quattrocolo, S.: An analysis of human comfort when entering and exiting the rear seat of an automobile. Appl. Ergon. **28**(5/6), 397–406 (1997)
7. Buckle, P., Fernandes, A.: Mattress evaluation—assessment of contact pressure, comfort and discomfort. Appl. Ergon. **29**(1), 35–39 (1998)
8. Seidl, A.: Das Menschmodell RAMSIS: Analyse, Synthese und Simulation dreidimensionaler Körperhaltungen des Menschen, PhD-thesis Technische (1994)
9. http://dined.io.tudelft.nl/en. Accessed 10 Dec 2018

Achieving Optimal Stiffness of Planar Multilink Mechanisms Structure by Members Shape Modification

Denijal Sprečić[(✉)], Džemal Kovačević, Jasmin Halilović, and Edis Nasić

University of Tuzla, 75000 Tuzla, Bosnia and Herzegovina
denijal.sprecic@untz.ba

Abstract. FE stress and strain simulations (in the remaining text FEA), for shape optimisation, of individual parts or assemblies are widely used. FEA should simulate either real load conditions (if possible), either load conditions applied during lab.-testing (which is much more often situation in automotive industry). FEAs are shortening shape optimisation duration, but if one tries to achieve absolutely credible simulation of physical load condition, FEA can be relatively long lasting, too [1]. In the present paper is shown an example of shape optimization for stiffness improvement for multilink automotive Bonnet Hinges. There are shown objective reasons why it is very difficult, longlasting and resource consuming to perform accurate FEA for specific assembly and given Load Case. Here one initial design (physically available) was chosen. Simplified/fast FEA for initial design have been performed. Comparing initial design FEA result, experimental measurement result and goal value, reference value for FEA have been obtained. Hinges design have been varied, fast FEA for each variation have been performed, until reference value is reached. Procedure defined has resulted in a successfully optimized design.

Keywords: Planar multilink mechanisms · FE simulation · Stiffness optimization

1 Introduction

This paper describes an example of fast performed multilink Bonnet Hinges (Fig. 1) shape optimization for stiffness improvement. Shape optimization is performed with goal to improve the stiffness of Car Bonnet at high driving velocities. At high velocities (in this case, over 200 km/h), air flow on the Bonnet results with forces puling Bonnet upwards and/or pushing it downward [2, 3], deforming Bonnet Hinges/Bonnet, therefore Bonnet is oscillating in the vertical direction. If Bonnet Hinges are not stiff enough, these oscillations can cause small damages of Car body, and become noticeable to the Car driver, causing sense of discomfort, as well. Car Body (BIW-body in white) has to fulfill number of requirements, such as smooth going and stable property, requirements of body structure and sealing for acoustic and NVH (Noise, Vibration and Harshness), materials requirements of safety for crash performance (front crash, rear crash, side crash and roll, others), performance and technical requirements of body covers for geometry and dent resistance, performance and technical requirements of fatigue and

I. Karabegović (Ed.): NT 2019, LNNS 76, pp. 429–435, 2020.
https://doi.org/10.1007/978-3-030-18072-0_50

antirust for some typical parts, protection requirements of BIW for roll, relational requirements of body for satisfy difference of profile, torsion stiffness and bending stiffness requirements of BIW for improving performance etc. [4]. Car Body with examined Bonnet Hinges assembled, have been tested in the Wind-tunnel, and test that have shown unsatisfactory stiffness, therefore shape optimisation utilizing FEA, have been performed.

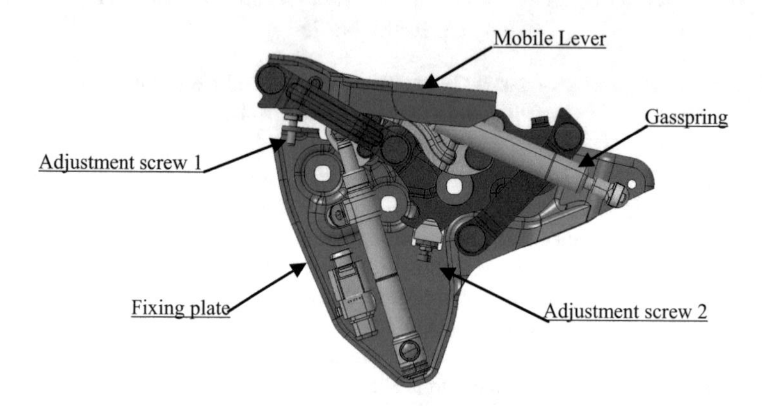

Fig. 1. Examined multilink Bonnet Hinge - Initial design

Finite element analysis is a highly reliable engineering approach and valid tool for product development. It reduces time spent on design and validation processes. Thus, the cost of design decreases. It has several advantages such as analyzing complex geometry, anisotropic and nonlinear material, etc. The accuracy of results of finite element analysis depends on using real mechanical values of material, using proper boundary conditions, and meshing smoothness [5]. Performing FEA for multilink mechanisms, compared to FEA for single parts, or even two-link mechanisms, can have several aggravating issues. One of such assemblies is automotive Bonnet Hinge. These issues arise primary from function and design of mechanisms pivot points. Pivot point function is to insure relative rotation between links. To insure this function there has to be either fabricated clearance [6], or some sort of bushings assembled in the pivot points. Bonnet Hinge Levers are commonly metal (in examined case steel). Bonnet Hinge pivot points are made using combination of Rivets and Bushings (Fig. 2), made from various composites (steel plate/Teflon, brass mesh/Teflon, etc.). These materials have different elasto-plastic properties, which should be considered when performing FEA [7]. Additionally, during assembling Bushing is deformed (calibrated), where deformation varies, because assembling is performance adjusted.

During assembling Bonnet Hinges are connected on the Bonnet and then on the Car body. Afterwards, Gas Springs are connected to Hinges Mobile lever and Fixing plate, then Bonnet stiffness is adjusted by adjusting Bonnet Hinges Adjustment Screws. Gas Springs function is to optimize hand-force during opening and closing of the Bonnet, but their force influencing Bonnet stiffness too. Forces acting on Hinges from Gas Springs and Adjustment Screws pretension Hinges (remove possible clearances, and deform Hinges), which influences Hinges and Bonnet behaviour during basic function, and behaviour when Bonnet is subjected to external loads as well.

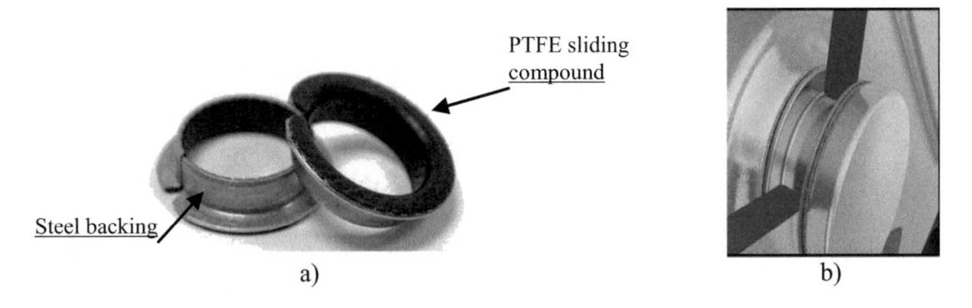

Fig. 2. Bushing assembled on Lever: (a) not assembled, (b) assembled

For these reasons precise FEA for multilink mechanisms, such are multilink Bonnet Hinges, is complex and long-lasting to prepare and to perform. One should make number of experiments and measurements, to obtain exact data (forces and deflections) how Gas Springs and Adjustment Screws are impacting Hinges, overall. This FEA, perhaps, should be performed by non-linear solver, to describe Bushings properties exactly. Bushings properties should be determinates by number of experiments as well, for each material/dimensions combination.

To save time examined Hinges have been optimized utilizing FEA in linear solver Elfini, combined with simple comparative analysis. This simplification was based on several bases:

- deformation of levers is elastic (proportional).
- test force is always the same, which should result with similar deformation of PTFE layer in bushings (nonlinear material), and similar cancellation of clearances, therefore approximately same impact on each design, and same error for each FEA.
- Gas Spring and Adjustment screws are also similar for all design, which lead to same conclusion as above.
- Therefore if simulated deflection (FEA) decreases for certain percentage, physical assembly stiffness should improve proportionally.

2 Experimental Work

In automotive industry different tests are performed, ranging from static deformation tests, experimental modal analysis to operational testing on laboratory test benches and the road [8]. Regarding test in the Wind-tunnel is long-lasting to prepare, unavailable on all locations, and expensive; for preliminary stiffness tests simplified method is utilized, where (Fig. 3): X Z

- Pair of Bonnet Hinges are connected to Car body (or specialized Test-rack) and Bonnet (or specially adopted frame),
- Assembly is pretensioned using Adjustment Screws,
- Appropriate test forces (dragging and compressing) are applied on the Bonnet, and displacement of rear-end of the bonnet is measured.

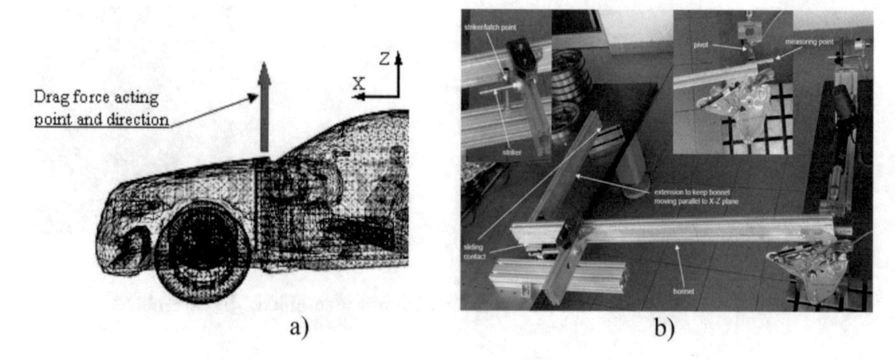

Fig. 3. Z-direction deformation measurement: (a) on the Car body, (b) on the Test-rack

This testing method is much more suitable for FEA, and therefore used. To minimize FEA duration, following procedure was adopted:

- Drag load in the Z-direction (Z-axis in automobile global coordinate system), have been chosen as more significant.
- Displacement of the Initial Bonnet Hinge design (design with the Pedestrian Safety System, which had an insufficient stiffness) was taken as a referent value.
- Short Connection Lever design was detected as most influencing Bonnet Hinge stiffness.
- Short Connection Lever shape variation was performed, where for each variation FEA was performed.
- Bonnet Hinge displacement value readings, for each variation, were compared with the referent value (R).
- Version with smallest reading, with reasonable workability, was chosen as an optimal version.

FEA setup is shown on Fig. 4:

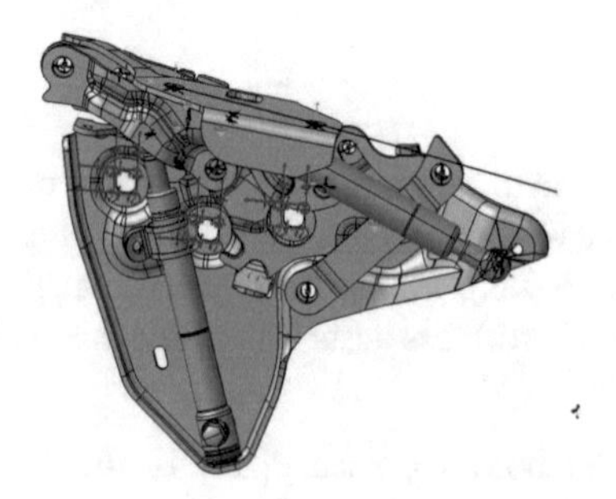

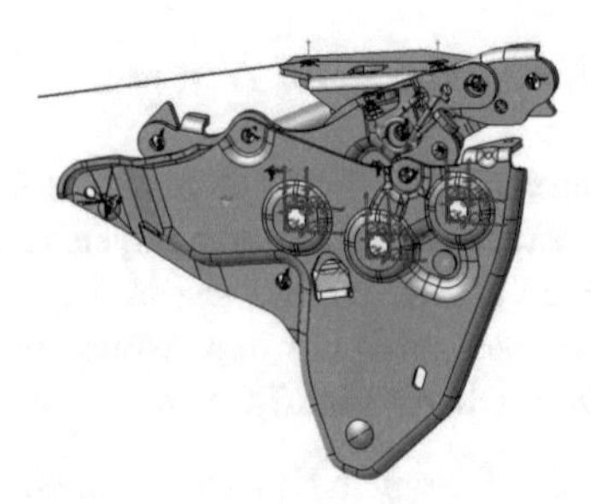

Fig. 4. FEA setup

3 Results and Discussion

Stiffness of Bonnet Hinges examined, has been tested on Test-rack too, and result shown insufficient stiffness; Measured displacement value, for initial design was:

$$M_I = b_I \cdot G = 1{,}638 \cdot G, \tag{1}$$

where G is goal value, defined by costumer standards

FEA for Bonnet Hinges Initial design, have been performed. Displacement component in the Z-axis direction reading, from FEA is marked as “Z_I”, and reference value (R - which is defined as goal FEA reading for Z-axis direction displacement reading) is obtained from relation:

$$Z_I = b_I \cdot R \Rightarrow R = Z_I / b_I \tag{2}$$

Design with FEA displacement in the Z-axis direction reading equal to “R”-value should have satisfactory stiffness.

After Bonnet Hinge displacement reference value was calculated, Bonnet Hinge Short Connection Lever shape was varied and FEA has been performed, for each version. When various versions of Short Connection Lever were designed, two criteria have been taken in account:

- Space available (in all possible Bonnet positions).
- Workability (Short Connection Lever should be suitable for stamping production process).

In the Table 1 Bonnet Hinges FEA displacement readings (“Z_i”) are displayed, for designs with different Short Connection Lever (defined by the value of correlation factor: $b_i = Z_i/R$). From Table 1 is visible that versions 14 and 15 resulting with Bonnet Hinge best stiffness. FEA results for these two designs are given on Fig. 5. Version 14 have given the best result, but for further proceeding and Prototype production version 15 have been chosen. Because of small production quantity, concept is chosen to primarily Laser-cut, and then deform material, for which version 14 was not suitable. It was expected that part stiffness at physical testing will be bit better than one predicted by FEA (3,6% above Goal value), taking in account Gas spring, and Adjustment screws influence, which is smaller if Bonnet Hinge design is stiffer. Figure 6 shows version 15 workability parameters (Fig. 6).

Table 1. Hinge displacement with different Short Connection Lever shapes

Version (i)	1	2	3	4	7	8	11	12	13	14	15	16
b_i	1,079	1,141	1,118	1,220	1,146	1,158	1,181	1,177	1,187	0,970	1,036	1,077

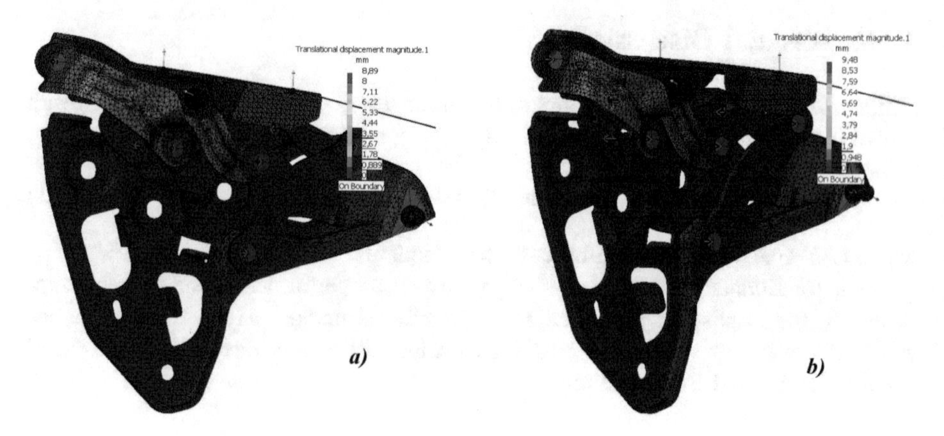

Fig. 5. FEA results – displacement: (a) Version 14, (b) Version 15

Research on automotive part optimization should be preceded by the following preliminary processes: identifying structural characteristics such as tensile strength and yield strength through strength tests and checking the test results against finite-element analysis (FEA) results to validate the parts [9, 10]. After Prototypes were produced, with Short Connection Lever version 15, and tested with drag force in the Z-direction), measured displacement was $\mathbf{M_{Proto} = 0{,}938} \cdot G$ (where G is goal value), which have been satisfactory. Prototype stiffness in the Wind-tunnel test has been satisfactory, as well.

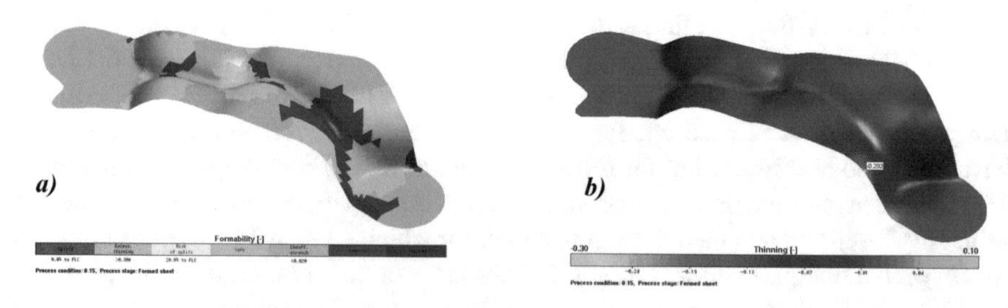

Fig. 6. Optimized Short Connection Lever: (a) formability, (b) thinning

4 Conclusion

As explained, performing FEA for multilink mechanisms, compared to FEA for single parts, or even two-link mechanisms, can have several aggravating issues. This issues are even more magnified for multilink automotive Bonnet Hinges, caused by a Gas spring forces and Adjustment screws forces impact. In the examined example, FEA has been set with significant simplifications, which have multiply shortened simulation duration. Time saving effect is bigger because FEA has been performed 16 times. Simplifications have been carefully chosen with goal to have impute errors approximately proportional to result. This way it was possible to compare FEA results with

Initial Bonnet Hinge design displacement, and set a reference value which should be achieved in FEA for one modified design, to consider it sufficiently optimized. For the design chosen as optimal, by FEA displacement 3,6% larger then goal value was predicted, where displacement measured afterwards on physical test was 6,2% smaller then goal value. Measured displacement difference from predicted value has been satisfactory small, and it satisfied requirement.

References

1. Kovačević, D., Sprečić, D., Halilović, J., Nasić, E.: Automotive Bonnet Hinges shape optimization using linear FE simulation & comparative method. J. Trends Develop. Mach. Assoc. Technol. **21**(1), 109–112 (2018). ISSN 2303-4009
2. Balzer, L.A.: Atmospheric turbulence encountered by high-speed ground transport vehicles. J. Mech. Eng. Sci. ImechE **19**(5), 227–235 (1977)
3. Aljure, D.E., Lehmkuhl, O., Rodríguez, I., Oliva, A.: Flow and turbulent structures around simplified car models. Int. J. Comput. Fluids **96**, 122–135 (2014)
4. Ma, M., Lu, H.: Design, evaluation methods and parameters of automotive lightweight. In: SAE-China, FISITA (eds.) Proceedings of the FISITA 2012 World Automotive Congress. LNEE, vol. 196, pp. 965–975. Springer, Heidelberg (2013). https://doi.org/10.1007/978-3-642-33738-3_2
5. Yılmaz, T.G., Tüfekçi, M., Karpat, F.: A study of lightweight door hinges of commercial vehicles using aluminum instead of steel for sustainable transportation. Sustainability - Open Access Journal **9**, 1661–1669 (2017)
6. Chen, X., Jiang, S., Deng, Y., Wang, Q.: Nonlinear dynamics and analysis of a planar multilink complex mechanism with clearance. J. Shock Vibr., 1–17 (2018). Article ID 6172676
7. Ge, C., Dong, Y., Maimaitituersun, W.: Microscale simulation on mechanical properties of Al/PTFE composite based on real microstructures. Open Access Journal – Materials **9**, 590–605 (2016)
8. Helsen, J., Cremers, L., Mas, P., Sas, P.: Global static and dynamic car body stiffness based on a single experimental modal analysis test. In: International Conference on Noise and Vibration Engineering, pp. 2505–2521 (2010). ISBN: 978-90-73802-87-2
9. Yang, Y., Jeon, E.S.: Tailgate hinge stiffness design using topology optimization. Int. J. Appl. Eng. Res. **11**, 9776–9781 (2016)
10. Dogan, S., Guven, S., Karpet, F., Yilmaz, T.G., Dogan, O.: Experimental verification and finite element analysis of automotive door hinge. In: ASME 2014 International Mechanical Engineering Congress and Exposition, Volume 11: Systems, Design and Complexity, Montreal, Canada, pp. 39295–39299 (2014)

New Technologies in the Field Energy: Renewable Energy, Power Quality, Advanced Electrical Power Systems

Developments in Solar Powered Micro Gas Turbines and Waste Heat Recovery Organic Rankine Cycles

Jafar Alzaili, Martin White, and Abdulnaser Sayma(✉)

Centre for Compressor Technology,
City University of London, London ECIV OHB, UK
a.sayma@city.ac.uk

Abstract. This main objective of this paper is to present recent developments and future challenges in two distributed power generation technologies that have the potential to play an important role in the future low carbon power generation. The first is parabolic solar dish systems powering a micro gas turbine by focusing solar energy to a focal area to heat the air in a Brayton cycle. The use of micro gas turbines can lead efficient, reliable and cost-effective technology. The second technology is small scale organic Rankine cycles (ORCs) that can be used to generate electricity from low grade heat, either generated as waste from industry processes and thermal plants, or from concentrated solar power. Although large scale ORCs have been successfully commercialised, there is still research and development required to achieve wide commercialisation at small scale, particularly regarding expanders.

Keywords: Micro-Gas Turbines · Organic Rankine Cycles · Distributed power generation · Turbo expanders · Turbomachinery · Concentrated solar power

1 Introduction

Distributed power generation refers to variety of technologies that generate electricity at or near the point of consumption. It may serve single domestic or commercial premises or multiples in a micro grid arrangement. In addition to significant reduction in cost due to reduced grid infrastructure requirements and the associated transmission losses, it allows for significant efficiency gains through multiple generation such as combined heat and power or combined cooling, heat and power. It also improves the security of energy supply and energy system resilience. Environmental benefits can be maximized through the utilization of renewable energy resources. This paper addresses technological developments in distributed concentrated solar power systems based on two prime movers. In addition to the above, distributed CSP multiple generation systems offer carbon free solutions to power requirements for remote and off-grid locations which is of particular importance to developing countries helping to improve the quality of life for poor communities.

I. Karabegović (Ed.): NT 2019, LNNS 76, pp. 439–452, 2020.
https://doi.org/10.1007/978-3-030-18072-0_51

Concentrated Solar Power (CSP) systems use mirrors arranged to focus the direct sunlight onto a receiver where the temperature of a working fluid rises. The working fluid then transfers its energy to a prime mover. Parabolic dish and solar towers have the potential to increase the temperature of the working fluid in excess of 800 °C which is needed for certain types of prime movers such as gas turbines [1], while parabolic trough systems concentrate solar power onto a liner receiver raising temperatures to about 380 °C. Solar towers are not suitable for low power range of tens of kilo Watts [2], hence parabolic solar dish concentrators are the most suitable in small scale applications. Internal combustion reciprocating (IC) engines are not a suitable contender for power generation from CSP because of inherent features in their basic engine cycle and design principles. Stirling engines have been proposed as a possible choice. These are piston type reciprocating engines that operate with an external heat source, and thus they are suited for CSP. Stirling engines use pressurised hydrogen as the working fluid at high pressure. Mainly because of this feature, they suffer from a number of technical problems affecting life and reliability, such as issues with cylinder seals, Hydrogen leakage, hot spots in the heater and difficulties with part-load control. Such problems lead to system complexity and increased weight and cost.

Gas turbines are used extensively in power generation as well as in aero-applications. Micro-gas turbines (MGTs) are a term loosely used for gas turbines producing power from several hundred Watts to about 1 MW. Nowadays, they are commercially available in the power range from 30 kW up to 1 MW. Small MGTs producing less than 25k. We are not commercially available yet. Compared with Sterling engines currently dominate small scale CSP power generation research and development projects, MGTs offer the potential for higher reliability, longer engine life, lower noise and vibrations and, reduced maintenance costs. Hence the main focus in this paper on high temperature CSP systems is on MGTs.

The first attempt to use an MGT in a CSP prime mover seems to be the work done by United States Department of Energy during the 1980's [3]. However, the work has not been continued. Six and Elkins [4], among others, attempted the concept by adopting turbochargers technology. More recently, purpose designed MGT engine was adopted to examine the performance of the CSP systems based on MGT [5]. It is worth noting that some projects for Stirling engine based CSP failed mainly because the cost of the prime mover hardware was too high [6]. The EU funded project OMSoP [8], completed recently, aimed at demonstrating the solar dish technology operating an MGT.

For CSP systems, the MGT presents relatively small part of the overall capital cost compared to the solar dish (concentrator). Thus, optimising the MGT for high efficiency and the related increase in production cost could be outweighed by the reduction in the dish size and thus overall cost of the system. The variable solar insolation requires the MGT to operate efficiently at a much wider range than conventional technology. This has a significant impact on the rotor-dynamic design of the shaft-bearing arrangement and mechanical design. Control strategy of the MGT needs to consider that thermal input to the system cannot be used as a control parameter. Integration with thermal energy storage allows for such systems to provide electricity around the clock without backup equipment. The main challenge with thermal storage is to provide high temperature output from thermal storage systems usually in excess of 800 °C.

The second CSP option considered in this paper is using lower temperature receivers that are typically based on parabolic trough concentrator. These are cheaper than solar dish systems and have lower working fluid temperatures that are not typically adequate for gas turbine or steam cycles and hence Organic Rankine Cycles are an obvious choice as a prime mover.

Over recent years, the organic Rankine cycle (ORC) has been established as a suitable technology for the conversion of low-temperature heat between 80 and 400 °C into electricity [8]. An ORC follows the same principle as steam power plants in which a heat source is used to evaporate a pressurised working fluid, which is then expanded to generate shaft power and hence electricity. However, in an ORC a refrigerant is used instead of steam, which means it is possible to more effectively utilise low-temperature heat sources. This makes ORC technology a viable option for power generation from low-temperature solar energy, in addition to waste-heat sources, geothermal and biomass. Compared to steam cycles, ORCs have a relatively simple construction and require little maintenance, which makes the technology an ideal candidate for power generation in remote areas. However, despite the successful commercialisation of ORC technology for power outputs above a few-hundred kW [9], the widespread implementation of the technology at the small-scale has been restricted due to high specific-investment costs. Lemmens [10] reported module specific-investment costs of around 7,000 €2014/kW for systems with power outputs below 100 kW, which is representative of values reported elsewhere [8]. Another challenge facing small-scale ORC systems is the identification of a suitable expander technology that can achieve a high efficiency. Moreover, in the same study Lemmens [6] reported that for the system considered the pump, expander and generator contributed to 69% of the overall purchased equipment cost. Therefore, the identification of an expander-generator assembly that is efficient and cheap to manufacture in large volumes could facilitate a significant step towards small scale, economically viable ORC systems.

This paper will outline recent developments in the CSP technologies utilising micro gas turbines and organic Rankine cycles focusing on specific elements of the system that were considered critical to achieving the performance versus cost requirements.

2 Solar Dish Micro Gas Turbine Systems

2.1 The Micro Gas Turbine (MGT)

MGTs in the power range below 100 kW typically use recuperated Brayton cycle. In this cycle, air pressure is increased in the compressor, heat is added at constant pressure before expanding in a turbine producing shaft power that derives the compressor and an electricity generator. Compared to larger gas turbines, they have low-pressure ratios and hence efficiency is usually enhanced by using the hot exhaust gas existing the turbine to preheat the compressed air existing the compressor, in a heat exchanger termed a recuperator, and hence reducing thermal energy input. Figure 1 shows the main thermodynamic transformations characterising a recuperated Brayton Cycle. The ideal reference processes are characterised by an isentropic compression $(1 - 2')$ an isobaric heat recovery $(2' - 3')$ and heating $(3' - 4')$ and finally an isentropic expansion

(4′ – 5′). The thermodynamic transformations are, in practice, irreversible and non-adiabatic leading to deviation from the ideal behaviour as shown in the real behaviour processes shown on the Temperature-Entropy (T-S) diagram (Fig. 1) of the cycle from 1 to 5. Moreover, the recuperator is not able to fully recover all the possible extractable exhaust heat and is usually characterised by an effectiveness around 85–90%.

In a solar powered MGT, air is heated in a solar receiver by absorbing the energy from the concentrated solar irradiation. A solar powered micro gas turbine system has been recently demonstrated with the project, optimised microturbine solar power – OMSOP, funded by the European Commission 7th framework programme. A schematic of the system is shown in Fig. 2 which incorporates a recuperated cycle micro gas turbine rated at 6 kWe. The main micro gas turbine parameters are given in Table 1. The subsections below describe the main design features of a solar powered MGT system components.

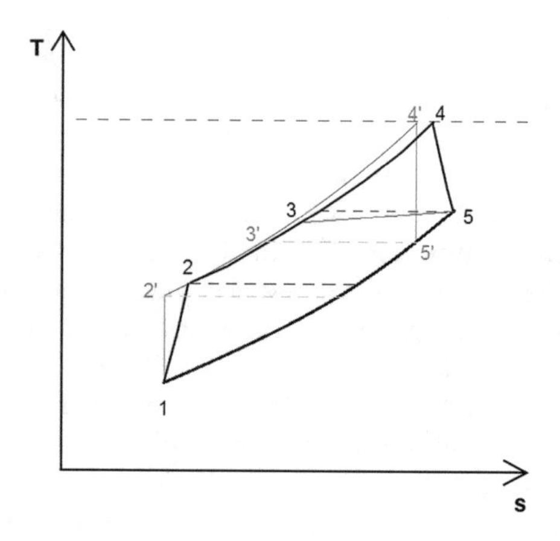

Fig. 1. T-S Thermodynamic diagram representing a recuperated Brayton-Joule cycle.

Table 1. OMSoP plant's main specifications.

Parameter	Value
Turbine inlet temperature	800 °C
Compressor efficiency	74%
Turbine efficiency	80%
Recuperator effectiveness	85%
Dish area	42 m^2
Rotational speed	130 krpm
Thermal efficiency	22.3%

2.2 System Components Design Challenges

Turbomachinery: CSP-based MGTs are required to operate in wide range of rotational speeds and loading due to the variation in the solar irradiaiton. To keep the overall efficiency of the system close to its highest possible level for the wide operating range, it is necessary that the turbomachinery components (turbine and the compressor) operate at high efficiency for a wider range of conditions, which is not typical in conventional designs. Typically, the turbine and compressor are designed in such a way that they have their maximum efficiency within a narrow range of operation dictated by the need to have the highest possible peak efficiency. For CSP applications however, it is more desirable to design them in such a way that their off-design performance is high. This may require the added complication of variable geometry or sacrificing some of the peak efficiency or both. Most of the current micro gas turbine designs use one centrifugal compressor and one radial turbine arrangement. An alternative approach is to use two-stage compressors and two-stage turbines in order to reduce the rotational speed and improve the dynamic behaviour of the micro gas turbine allowing, for example, to use ceramic components for the turbine [11]. Lower stage loading also benefits component's efficiencies although this may be compromised by the ducting losses. Turbine efficiency can be improved using end wall features such as groves that disrupt secondary flow patterns and reduce wall drag [12]. These could also be tailored to reduce the over-the-tip leakage losses. Figure 3 shows an example of end wall riblets installed at the hub of a radial turbine rotor. Experimentally validated computational results show the increase in turbine efficiency for various riblet arrangements in Fig. 4.

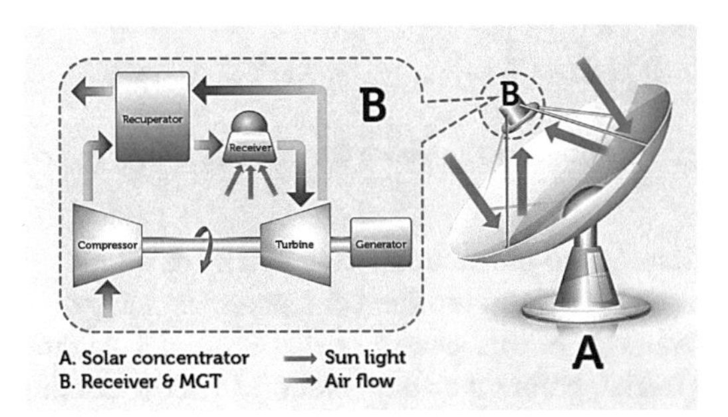

Fig. 2. OMSoP plant scheme (www.omspo.eu).

Mechanical Arrangement: A challenge is associated with the variable solar insolation requiring the MGT to operate efficiently at a much wider range than conventional technology. The wide operating range also has a significant impact on the rotor-dynamic design of the shaft-bearing arrangement and mass distribution. At least three different mechanical arrangements can be considered for the MGT components (compressor, bearings, turbine and the electricity generator). These are cantilevered design, generator-in-middle design and coupled shaft design.

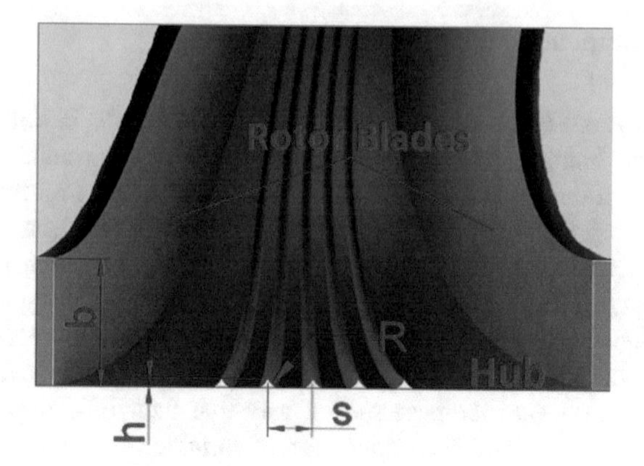

Fig. 3. Ribletgeometry at turbine hub

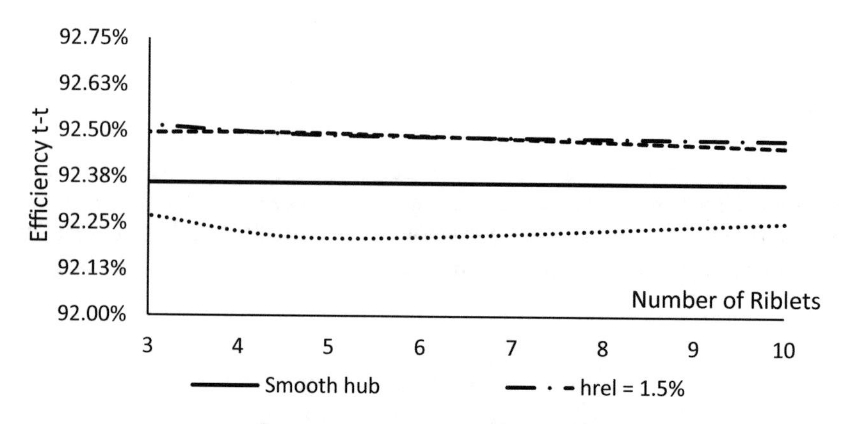

Fig. 4. Turbine performance for different riblets geometry

Cantilever arrangement is the most common in small-scale micro gas turbines. In this arrangement, the assembly of the turbine-compressor is hanged from one end of the rotor. The main advantage of this design is that no major cooling is needed for the generator as the air intake, before the compressor, can act as cooling system. Although this could affect the efficiency, it can reduce the size and cost of the accessories. However, for the power range of solar dish micro gas turbines, it is not practical to have a stable cantilever design for a wide range of operating rotational speeds.

A coupled shaft arrangement where two separate shafts are coupled, one containing the high speed generator and the other shaft including the compressor and turbine impellers is possible and has more stable rotodynamic. This however has the disadvantage of higher cost and potential reliability issues of the coupling.

A generator-in-middle arrangement is compact, leads to rotodynamic stability, but has the challenge of thermal management due to the close proximity of the high-speed generator and bearing to the turbine [13].

Solar Receiver: A solar receiver is a device that can capture the solar energy coming from the dish and transfers it to the working fluid. For the MGT solar dish application the air can be directly heated inside the receiver without having to resort to another heat transfer fluid. Different layouts can be conceptualized, one of the most promising arrangement is the Air tube cavity receiver. As shown by Fig. 5, an air tube cavity receiver is mainly composed from a cylindrical cavity surrounded by an insulator. At the bottom of the cavity an optical splitter can be present to readdress incident sun rays and at the top of the cavity, a quartz glass can be present to minimise heat transfer between the receiver and the ambient, reducing losses.

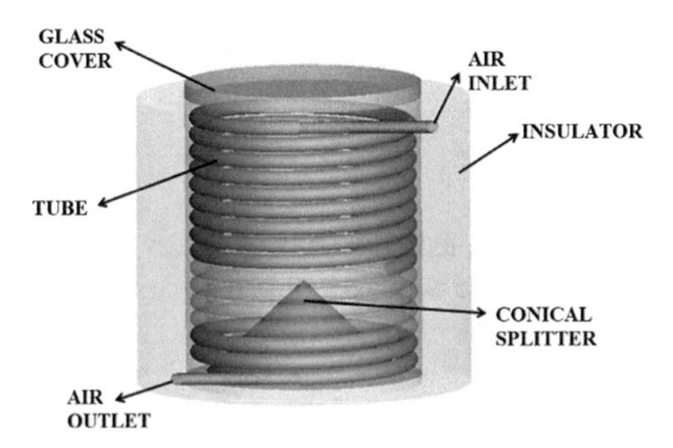

Fig. 5. Air tube cavity receiver components.

Another concept that can be used by this application is the one adopted by the OMSoP project, called volumetric receiver. In this arrangement, the inlet air is heated impinging on the absorber surface. This arrangement is pipe-less and then able to minimize the thermal stresses that characterise other arrangements (Fig. 6).

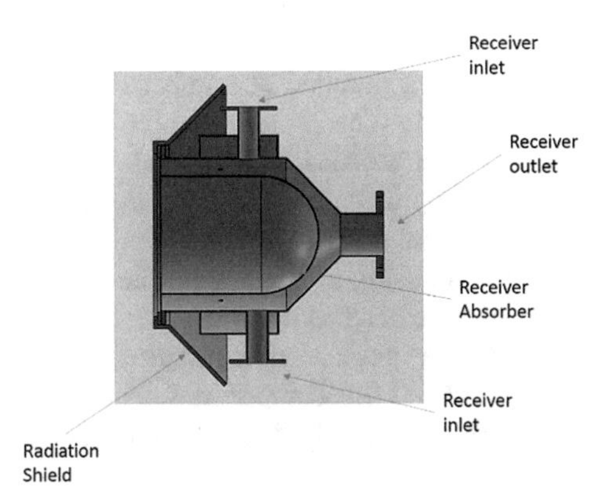

Fig. 6. Volumetric cavity receiver schematic

Control Strategies: The control strategy of the MGT needs to be adapted to the fact that the thermal input to the system cannot be used as a control parameter. In a conventional MGT the fuel (or heat input for the external-fired MGTs) are used to control the MGT, namely the rotational speed, Turbine Inlet Temperature (TIT) and Turbine Exit Temperature (TET). For a CSP based MGTs this option is not possible as the incoming solar power to the receiver cannot be practically controlled. For example, controlling the dish position in order to adjust the amount of thermal input power would not be an option due to the much slower dynamic response of the dish movement mechanism than that of the MGT dynamic response that is orders of magnitude apart. A feasible option to control the MGT in such system is by adjusting the power taken (or given) to the HSG. This would result in controlling the TIT, TET and rotational speed.

3 Organic Rankine Cycles

To realise the widespread implementation of small-scale ORC technology it is necessary to reduce system costs. One way to do this is to develop ORC components, such as the turbine, which can operate efficiently with different working fluids. Consequently, the focus on advancement in this paper will be on some recent advancements in this field. ORC turbine design is complicated since organic fluids do not obey the ideal gas law, thus requiring complex equations of state. Furthermore, at operation close to their critical point, organic fluids exhibit non-classical fluid dynamic behaviour where classical fluid dynamics could be inverted leading to expansion shocks and compression fans [13, 14]. In addition to this, low values for the speed of sound can also result in supersonic flows at the rotor inlet and possibly at the rotor outlet. For rotor inlet Mach numbers only slightly exceeding unity, a conventional converging stator blade can be implemented with the required supersonic conditions being achieved in the rotor-stator interspace [15]. However, at higher Mach numbers, supersonic stators are required. These are constructed with a subsonic converging section that accelerates the flow to the choked conditions followed by a diverging section that expands the flow isentropically to the desired Mach number. A number of papers have addressed the design of these stators, for example Wheeler and Ong [16] who develop a stator design model using the Method of Characteristics, and Pasquale et al. [17] who construct the stator using Bezier splines and optimise the geometry using a genetic algorithm.

Alongside the development of design methods, there have also been studies on supersonic flows in ORC stators using computational fluid dynamics (CFD) [18–21]. These simulations couple either research flow solvers, or commercial CFD codes, with suitable equation of state. In the absence of experimental data for validation, these solvers were evaluated qualitatively by comparisons with design models, or experimental data for ideal gas turbines. Recently, Galiana et al. [14] presented experimental results investigating trailing edge losses in ORC turbines in a Ludwieg tube.

Clearly, the added complexities of using non-ideal working fluids means it is not suitable to apply conventional ideal gas turbine performance prediction models, such as similitude theory, directly to ORC turbines. Some researchers have attempted this, for example [22], but the unsuitability of using simplified versions based on the ideal gas law has been demonstrated recently [23]. The similitude concept has been correctly applied to ORC turbines [24], but this is generally for turbine design, rather than off-design performance predictions. Previously, the authors have investigated using similitude theory for the performance prediction of subsonic ORC turbines and developed a modified similitude model [25]. Equation 1 is a simplified form of this model assuming there is no change in the geometry, where Δh_s. is the isentropic enthalpy drop across the turbine, η is the turbine isentropic efficiency, $\dot{m}$ is the mass flow rate, N is the rotational speed, and a^*, ρ^* and μ are the density, speed of sound and viscosity at the stator throat. Recently, using this model it has been shown how the economy-of-scale of small subsonic ORCurbines could be improved through appropriate working fluid selection [26].

$$\left[\frac{\Delta h_s}{a^{*2}}, \eta\right] = f\left(\frac{\dot{m}}{\rho^* a^*}, \frac{\rho^* a^*}{\mu}, \frac{N}{a^*}\right). \tag{1}$$

The aim here is to extend the concept of similitude to investigate working fluid replacement in supersonic ORC turbines. This is done through the design and analysis of two converging-diverging nozzles for R245fa and Toluene using a minimum length nozzle model.

3.1 Design of a Supersonic Nozzle for ORC Turbines

To design a supersonic nozzle a 2D minimum length nozzle design model has been developed. The model is a minimum length Method of Characteristics (MoC) design model, but is coupled to real gas equations of state using REFPROP [27] to account for non-ideal fluid properties. For a differential fluid element undergoing a supersonic expansion, the amount of flow turning, $d\theta$, is given by Eq. 2, where c is the fluid velocity. For a perfect gas with a constant ratio of specific heats, Eq. 2 can be integrated analytically to determine the Prandtl-Meyer function. However, for a non-ideal gas REFPROP can be used to integrate Eq. 2 numerically. This numerical integration is performed once at the start of the design process, and then the MoC design model proceeds in the same manner as for a perfect gas.

$$d\theta = \sqrt{\mathrm{Ma}^2 - 1}\frac{dc}{c} \tag{2}$$

Before considering an ORC turbine stator, the flow within 2D converging-diverging nozzles is studied. This is done not only to validate the design model, but also as a preliminary investigation into scaling effects in supersonic flows of organic fluids. Using the total inlet conditions and Mach numbers defined in Table 1 for R245fa and Toluene, the MoC model was used to design the diverging sections of two converging-diverging nozzles. These two working fluids were selected since they are suitable working fluids for low temperature and high temperature ORC applications respectively. The turbine total inlet conditions were defined as 95% and 101% of the critical pressure and temperature respectively to ensure operation near the critical point where non-ideal gas effects are expected. The mass flow rates and rotor inlet conditions correspond to a rotor diameter of 80 mm.

Table 2. Thermodynamic cycles and turbine designs used for this study.

	T_{01} [K]	P_{01} [kPa]	$\dot{m}$ [kg/s]	Ma_2	P_2 [kPa]	b_2 [mm]
R245fa	431.4	3469	1.76	1.4	1138	2.38
Toluene	597.7	3920	0.45	1.7	667	1.2

The accuracy of the MoC model depends on the number of lines used to construct the expansion fan, and also the number of elements used during the numerical integration of Eq. 2 to obtain the Prandtl-Meyer function. Therefore, a sensitivity study was completed during the nozzle design process using n = 50, 100 and 250 for both of these parameters. It was found that 100 elements were sufficient, with maximum errors for the wall coordinates of less than 0.05% and 0.25% for R245fa and Toluene respectively, relative to the coordinates obtained for 250 elements. The resulting nozzle designs are shown in Fig. 7. The parameters used to construct the converging part of the nozzles are shown in Fig. 8 and Table 2. The throat width, o_{th}, is calculated by Eq. 3 where b_2 is the rotor inlet blade height defined in Table 2, Z_n is the number of turbine stator vanes (Z_n = 16 assumed), and ρ^* and a^* are the static choked throat conditions and are calculated assuming an isentropic expansion from the total inlet conditions to the nozzle throat. Equation 4 gives the design Reynolds number where D_h is the hydraulic diameter based on the throat dimensions (Table 3).

$$o_{th} = \frac{\dot{m}}{\rho^* a^* b_4 Z_n} \tag{3}$$

$$Re_d = \frac{\rho^* a^* D_h}{\mu} \tag{4}$$

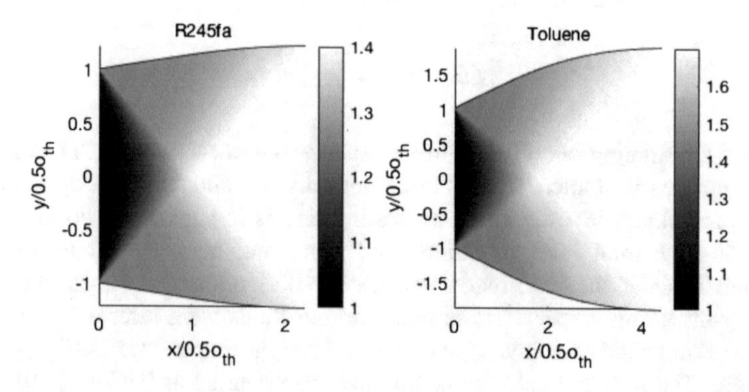

Fig. 7. Minimum length supersonic nozzles for expansion or R25fa to Ma = 1.4 and Toluene to Ma = 1.7 obtained from MoC model

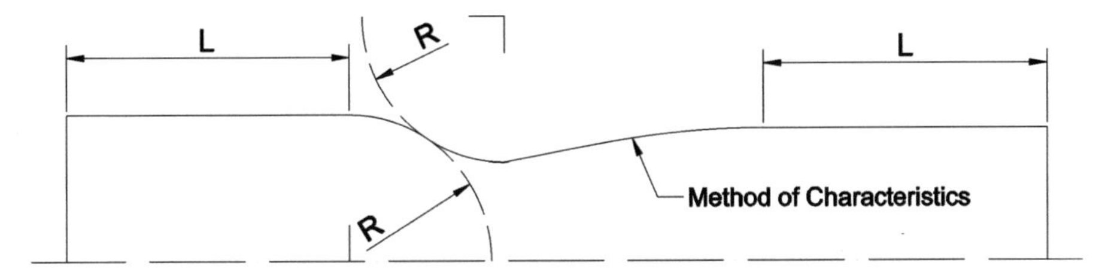

Fig. 8. Geometry of the converging diverging nozzles

Table 3. Design parameters for the converging-diverging nozzles.

	o_{th} [mm]	R [mm]	L [mm]	Re_d
R245fa	2.97	2.23	8.91	2.52×10^6
Toluene	1.89	1.41	5.66	1.09×10^6

3.2 CFD Simulation Setup and MoC Validation

A steady-state CFD simulation was setup in ANSYS CFX to validate the minimum length MoC designs for both R245fa and Toluene. The inlet conditions were set to the turbine total inlet conditions, the outlet was set to a supersonic outlet and the k-ω SST turbulence model with automatic wall function was selected. The meshes used for the R245fa and Toluene nozzles were constructed from 1.6×10^5 elements. For each nozzle design two additional meshes, consisting of 4×10^4 and 3.6×10^5 elements, were also constructed. It was found that the percentage error in the nozzle efficiency predicted by the original mesh compared to the finest mesh was less than 0.1% for both the R245fa and Toluene nozzle, thus confirming the suitability of the original meshes. Working fluid properties were accounted for by generating fluid property tables using REFPROP prior to the simulation, and a sensitivity study considering the resolution of

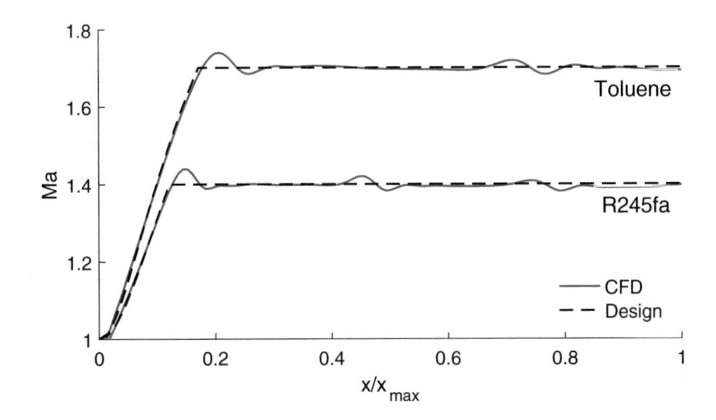

Fig. 9. Validation of the MoC nozzle design model: comparison between the midline Mach number distributions obtained using the MoC model and CFD simulation.

these fluid property tables was also conducted, with fluid property table sizes of 50, 100 and 250. The results show that the percentage errors in the nozzle efficiency predicted for a table size of 100 compared to a table size of 250 were 0.07% and 0.04% for the R245fa and Toluene nozzles respectively. The results from the CFD simulations for the R245fa and Toluene nozzles, both with a mesh size of 1.6×10^5 elements and fluid property table size of 100, are shown in Fig. 9. In this figure the MoC midline Mach number distributions are also compared to the CFD results.

Overall, the MoC model has produced supersonic nozzles that expand the two working fluids to the desired Mach numbers. Weak shocks are found downstream of the diverging supersonic sections and this leads to small bumps in the midline Mach number distribution. This effect is attributed to the nozzle wall boundary layers, which effectively reduce the nozzle flow area. Boundary layer effects are not considered within the MoC model, therefore explaining the small discrepancies observed. The model could be improved by extending the nozzle wall by the boundary layer displacement thickness. However, given that there is only a small discrepancy this modification was deemed unnecessary.

4 Conclusion

This paper highlighted recent advancements in two distributed power generation technologies that can be tailored for generating electricity from concentrated solar power. Firstly, main aspects and some design challenges in micro gas turbines used in parabolic dish concentrated solar power generation were addressed. The main features related to the required wide operating range of the MGT were highlighted, particularly for turbomachinery design and mechanical arrangement. It was shown that turbine efficiency can be improved by the introduction of profiled features at the turbine rotor hub such as riblets. The challenges related to ensuring rotodynamic stability and trade-offs with compactness and reliability of the design were outlined. Specific feature required in the control system were also outlined.

This second part investigated the effect of changes in the inlet conditions and working fluid on the performance of supersonic converging-diverging nozzles used in ORC turbo expanders. Two supersonic nozzles have been developed using a minimum length Method of Characteristics design model and the performance of these nozzles has been verified using CFD simulations completed using ANSYS CFX. Predictions for the nozzle performance were made using a modified similitude model, and also by conserving the amount of flow turning (the Prandtl-Meyer function) within the nozzle. Whilst the modified similitude model accurately predicted the nozzle mass flow rate, it is unsuitable to predict the nozzle outlet conditions. However, it was found that conserving the Prandtl-Meyer function can lead to accurately predicting the nozzle outlet conditions to within 2% provided the percentage change in k is less than 10%.

References

1. Pavlovic, T.M., Radonjic, I., Milosavljevi, D., Pantic, L.: A review of concentrating solar power plants in the world and their potential use in Serbia. Renew. Sustain. Energy Rev. **16**(6), 3891–3902 (2012)
2. Buck, R., Bräuning, T., Denk, T., Pfänder, M., Schwarzbözl, P., Téllez, F.: Solar-hybrid gas turbine-based power tower systems (REFOS). J. Solar Energy Eng. **124**(1), 2–9 (2002)
3. English, R.E.: Technology for Brayton-cycle space power plants using solar and nuclear energy. NASA-TP-2558 (1986)
4. Six, L., Elkins, R.: Solar Brayton engine/alternator set. In: Parabolic Dish Solar Thermal Power Annual Program Review, pp. 23–36 (1981)
5. Dickey, B.: Test results from a concentrated solar microturbine Brayton cycle integration. ASME GT2011-45918 (2011)
6. Sinai, J., Sugarmen, C., Fisher, U.: Adaptation and modification of gas turbines for solar energy applications. ASME GT2005 (68122) (2005)
7. Alzaili, J., Sayma, A.: Challenges in the development of micro gas turbines for concentrated solar power systems. In: 8th International Gas Turbine Conference, Brussels, Belgium (2016)
8. Quoilin, S., Van Den Broek, M., Declaye, S., Dewallef, P., Lemort, V.: Techno-economic survey of organic Rankine cycle (ORC) systems. Renew. Sust. Energy Rev. **22**, 168–186 (2013). https://doi.org/10.1016/j.rser.2013.01.028
9. www.orc-world-map.org. Accessed 23 Apr 2018
10. Lemmens, S.: Cost engineering techniques and their applicability for cost estimation of organic Rankine cycle systems. Energies **9**(7), 485 (2016). https://doi.org/10.3390/en9070485
11. Vick, M.J., Heyes, A., Pullen, K.: Design overview of a 3 kW recuperated ceramic turboshaft engine. ASME GT2009-60297 (2009)
12. Khader, M.A., Sayma, A.I.: Drag reduction within radial turbine rotor passages using riblets. Proc. Inst. Mech. Eng. Part E J. Process Mech. Eng. (2018). https://doi.org/10.1177/0954408918819399
13. Thompson, P.: Fundamental derivative in gas-dynamics. Phys. Fluids **14**(90), 1843 (1971)
14. Galiana, F.J.D., Wheeler, A.P.S., Ong, J.: A study on the trailing-edge losses in organic Rankine cycle turbines. In: ASME Turbo Expo 2015, 15–19 June, Montreal Canada (2015)
15. Moustapha, H., Zelesky, M.F., Baines, N.C., Japiske, D.: Axial and Radial Turbines, Concepts. NREC, Inc. (2003)
16. Wheeler, A.P.S., Ong, J.: The role of dense gas dynamyics on organic Rankine cycle turbine performance. J. Eng. Gas Turb. Power **135**(10), 9 (2013)
17. Pasquale, D., Ghidoni, A., Rebay, S.: Shape optimization of an organic Rankine cycle radial turbine nozzle. J. Eng. Gas Turb. Power **135**(4), 13 (2013)
18. Hoffren, J., Talonpoika, T., Larjola, J., Siikonen, T.: Numerical simulation of real-gas flow in a supersonic turbine nozzle ring. J. Eng. Gas Turb. Power **124**(4), 9 (2002)
19. Harinck, J., Turunen-Saaresti, T., Colonna, P., Rebay, S., van Buijtenen, J.: Computational study of a high-expansion ratio radial organic Rankine cycle turbine stator. J. Eng. Gas Turb. Power **132**(5), 6 (2010)
20. Harinck, J., Pasquale, D., Pecnik, R., Buijtenen, J.V., Colonna, P.: Performance improvement of a radial organic Rankine cycle turbine by means of automated computational fluid dynamic design. Proc. Inst. Mech. Eng. Part A J. Power Energy **227**(6), 637–645 (2013)

21. Wheeler, A.P.S., Ong, J.: A study on the three-dimensional unsteady real-gas flows within a transonic ORC turbine. In: ASME Turbo Expo 2014, 16–20 June, Dusseldorf, Germany (2014)
22. Cameretti, M.C., Ferrara, F., Gimelli, A., Tuccillo, R.: Employing micro-turbine components in integrated Solar-MGT-ORC power plants. In: ASME Turbo Expo 2015, 15–19 June, Montreal, Canada (2015)
23. Wong, C.S., Krumdieck, S.: Scaling of gas turbine from air to refrigerants for organic Rankine cycles using similarity concept. J. Eng. Gas Turb. Power **138**(6), 10 (2015)
24. Astolfi, M., Macchi, E.: Efficiency correlations for axial-flow turbines working with non-conventional fluids. In: 3rd International Seminar on ORC Power Systems, 12–14 October, Brussels, Belgium (2015)
25. White, M., Sayma, A.I.: The application of similitude theory for the performance prediction of radial turbines within small-scale low-temperature organic Rankine cycles. J. Eng. Gas Turb. Power **137**(12), 10 (2015)
26. White, M., Sayma, A.I.: The impact of component performance on the overall cycle performance of small-scale low temperature organic Rankine cycles. In: 9th International Conference on Compressors and their Systems, London, UK (2015)
27. Lemmon, E.W., Huber, M.L., McLinden, M.O.: NIST Standard Reference Database 23: Reference Fluid Thermodynamic and Transport Properties-REFPROP, Version 9.1. National Institute of Standards and Technology, Standard Reference Data Program, Gaithersburg (2013)

Laboratory Research of the Influence of Pulsating Flow of Flue Gases at the Heat Transfer

Nihad Hodzic(✉), Sadjit Metovic, Rejhana Blazevic, and Senid Delic

Faculty of Mechanical Engineering, University of Sarajevo,
Vilsonovo setaliste 9, 71000 Sarajevo, Bosnia and Herzegovina
hodzic@mef.unsa.ba

Abstract. Compared to classical combustion methods, pulse combustion is a qualitative step forward also in terms of a more intensive heat transfer resulting from turbulent pulsating flue gas flow. This phenomenon is to some extent experimentally dealt through extensive research of pulse combustion characteristics of gas fuel in water-cooled burner of simple and modular geometry with aerodynamic valves. During the research, the heat exchanger is considered as cocurrent or countercurrent depending on the flow of working fluids, whereby the data on which the dependence of heat transfer intensity from the pulsating flue gas flow to the water as heat receiver can be established. It has been shown that the heat transfer coefficient in the pulse flow depends on the burner geometry - the distance between the end of the resonant pipe and the heat exchanger inlet, and the thermal load of the burner. For certain burner geometries and thermal loads, the heat transfer coefficient in pulsating flue gas flow is more than 2 times the heat transfer coefficient in developed turbulent flow without pulsations.

Keywords: Burner · Pulse combustion · Pressure pulsations · Heat transfer

1 Introduction

Due to the growing need to increase the efficiency of conversion of primary energy from fuels while reducing the negative impacts of this process on the environment, it is necessary to develop new and to improve existing methods of high-efficiency low-waste combustion. Also, there is a continuing need for the construction of plants or elements of plants in which the heat released from combustion will be transformed as rational as possible. Significant possibilities in this regard are provided by pulse combustion process that, due to its performance, is classified as high-efficiency low-waste energy conversion process. In comparison with conventional combustion methods, pulse combustion means a qualitative step in the following: more efficient combustion, suction and thrust effects due to pulsations, more efficient heat transfer, reduction of nitrogen oxide emissions, more flexible combustion, self-cleaning of heating surfaces on the flue-gas side. However, despite of a century of research, several shortcomings have prevented the widespread use of this technique for commercial purposes. The main reasons for the insufficiently widespread application of the process

I. Karabegović (Ed.): NT 2019, LNNS 76, pp. 453–460, 2020.
https://doi.org/10.1007/978-3-030-18072-0_52

are intense noise, which is almost unavoidably generated during pulse combustion, but also insufficient knowledge of the values and character of the change of parameters with time, [1].

By studying the contribution of flue gas pulses to the intensification of heat transfer a lot of papers have been produced, for example [2]. In this paper it is stated that at a water heating plant with a pulse combustion burner, the total degree of utilization of primary energy from fuel ranged from 84 to 92.5%. It has also been shown that the coefficient of heat transfer depends on the frequency of the phenomenon and that, in the case of a pulsating flow of the heat transmitter (for certain conditions and in the case that the heat receiver is water), it is 4 to 7 times higher than the corresponding coefficient for the forced convection and for the stationary flow.

In this paper, the application of a burner for pulse combustion of gas fuel (LPG, propane-butane) with aerodynamic air-intake valves, with water-cooled combustion chamber zone and underlying resonant tube is given.

2 Experimental Rig and Measuring Equipment

Basically, the pulse combustion burner consists of: combustion chamber, air-intake valves and resonant pipe. During the operation of the burner, an asymmetric flow of media along the burner is provided, thanks in particular to the mechanical or aerodynamic valves mounted on the air suction side. Namely, during the expansion of combustion products, during and after combustion, which is necessarily intermittent, the products generally expand through the resonant pipe further into the system or the environment, Fig. 1.

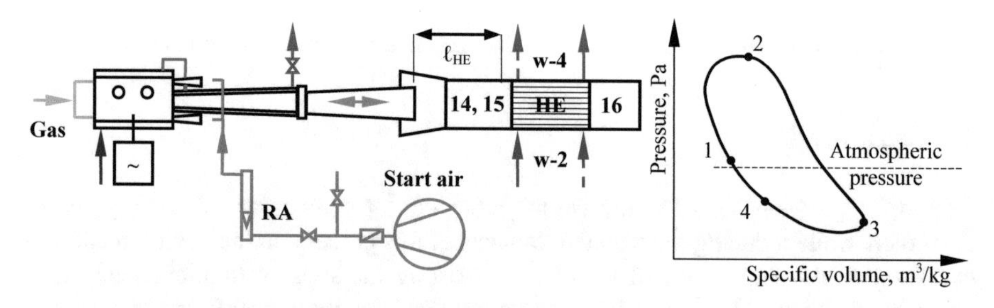

Fig. 1. Schematic representation of the water-cooled pulse combustion burner (left) and the theoretical working cycle in the p-v diagram, HE - heat exchanger, [1, 3].

The laboratory research of the pulse combustion characteristics, including the investigation of the influence of pulsations of flue gas flow on heat transfer, was carried out at the Mechanical Engineering Faculty in Sarajevo. It's a simple, but robust, modular type burner, without moving parts and which, after entering the self pumping mode, works without support from the outside. This means that there is no forced air supply in the combustion chamber nor the forced removal of the resulting flue gases nor the forced ignition of the reactants mixture, Fig. 2, [1].

Fig. 2. Laboratory at the Mechanical Engineering Faculty Sarajevo: water-cooled pulse combustion burner connected to flue gas duct in which a heat exchanger is installed (during the research the heat exchanger was coated with a layer of insulation $\delta = 50$ mm)

The heat exchanger consists of a tube bundle with one flow passage for both working fluids. Flue gases flow through nineteen properly distributed straight copper tubes measuring $\varnothing 22 \times 1 \times 376$ mm. The heat receiver is water that flows around the tubes. With the choice of the inlet or outlet of the water from the heat exchanger, it is possible to establish a co-current or counter-current heat exchanger, [1].

To measure the pressure pulsations of the flue gases at the inlet of the heat exchanger (measuring point 14), the pressure transmitter Kristal CERALINE-S, type SEN-8700, is mounted. To measure the flue gas temperature, thermocouples type K (NiCr-Ni, measuring points 15 and 16) were used, while thermocouples type J (copper-constant, measuring points w-2 and w-4) were used on the water side. To measure the flow of water through the heat exchanger, a flowmeter was used: ams, (Qn1.5 AH-B). During the measurements, the data acquisition DEWE-BOOK®[1] and DEWESoft® software were used. At the exit of the heat exchanger, in addition to temperature measurements, samples of flue gases were also taken out with the aim of analysis and determining the coefficient of excess air, i.e. flow rate of flue gases through the heat exchanger.

From all modular burner geometry variations three geometries that enable automatic operation are selected: label G:22 (small combustion chamber, short air-intake valves, basic resonant tube - MKT), label G:41 (medium combustion chamber, long air-intake valves, basic resonant tube - SDT), label G:43 (medium combustion chamber, long air-intake valves, medium resonant tube - SDS). For each investigated burner geometry, measurements were made for three different thermal loads within the range of possible burner operation - the selected limit thermal loads do not represent the work of certain geometric forms of the burner at the very limits of stopping working. The flow of water through the heat exchanger was kept constant in all test regimes in this part of the study and amounted to 0.4 kg/s, [1, 3].

1 ® Software is licensed.

3 Results of the Research with Analysis

An example of the flue gas pressure pulsations directly at the inlet of the heat exchanger is given in Fig. 3. Presented pressure changes were recorded during the investigation of the burner geometry labeled G:22 and the thermal load of the burner was 170 kW.

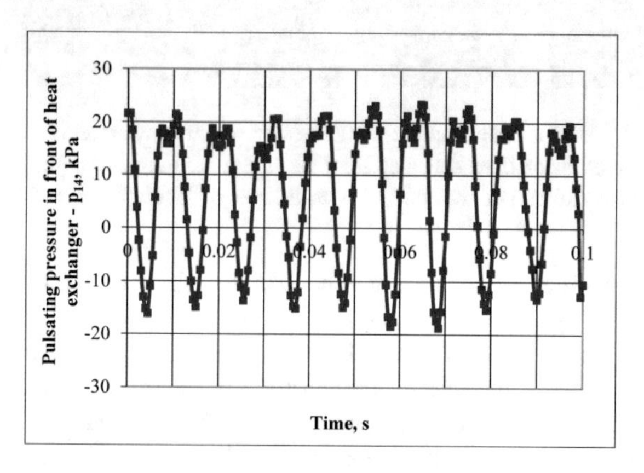

Fig. 3. An example of the flue gas pressure pulsations directly at the inlet of the heat exchanger; burner geometry: label G:22; thermal load of the burner: 170 kW; response: $p_{14uk} \approx 35$ kPa, $p_{14}^{+} \approx 20$ kPa, $p_{14}^{-} \approx 15$ kPa, $f \approx 96$ Hz

From the results of the study, which is illustrated in the previous figure, it can be seen that due to the distance from the inlet into the heat exchanger (measuring point 14) to the end of the burner resonant pipe (ℓ_{HE}), significant deformation or weakening of the pressure pulsations occur, especially in the overpressure zone.

The position of the open end of the resonant pipe, in all test regimes during this part of the research, was always the same relative to the flue gas duct regardless of the burner geometry - the degree of junction sealing was: $\psi_d = 45{,}25$. Thus, for each observed burner geometry, the distance from the end of the resonant pipe to the inlet of the heat exchanger was $\ell_{HE} = 1.7$ m.

The heat transfer coefficient for the heat exchanger, i.e. the heat transfer coefficient on the flue gas side during pulsating flow is calculated by the following expression and for practical reasons it is possible to ignore the heat resistance on water side and through the wall of the clean pipe:

$$k_p = \frac{1}{\frac{1}{\alpha_p} + \frac{\delta_{Cu}}{\lambda_{Cu}} + \frac{1}{\alpha_w}} = \frac{\dot{Q}}{A_{HE}\Delta t_{ln}} = \frac{\dot{m}_w c_w (t_{w-4} - t_{w-2})}{A_{HE}\Delta t_{ln}} \approx \alpha_p, \ \frac{W}{m^2K} \tag{1}$$

Cocurrent Fluid Flow (A): For all three burner geometry forms, the dependence of the total pressure deflection directly in front of the heat exchanger (p_{14uk}) from the thermal load of the burner is shown in Fig. 4. In Fig. 5 the dependence of the heat transfer coefficient in the pulsating flue gas flow (α_p) from the burner thermal load is presented.

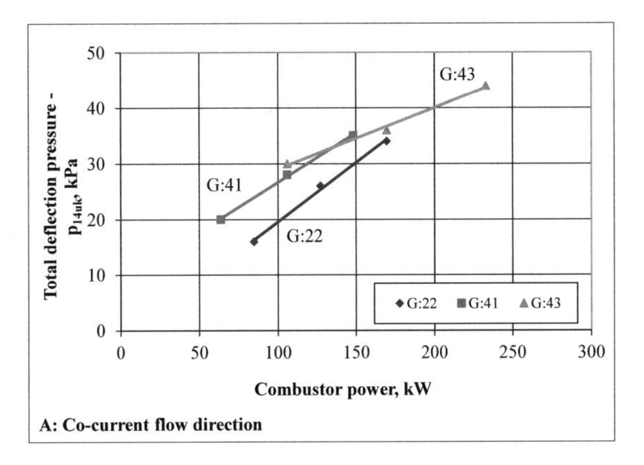

Fig. 4. The dependence of the total pressure deflection directly in front of the heat exchanger (p_{14uk}) from the thermal load of the burner

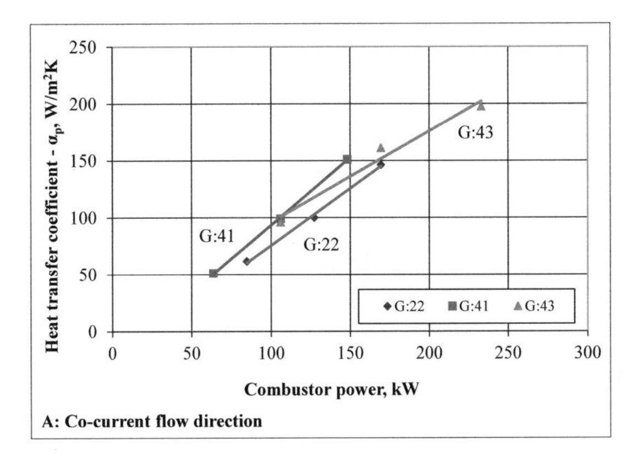

Fig. 5. Dependence of the heat transfer coefficient in the pulsating flue gas flow (α_p) from the thermal load of the burner

Based on the previous results, it can be concluded that there is a correlation between the total flue gas pressure deflection (p_{uk}) and the heat transfer coefficient (α_p). It is noticeable that the heat transfer coefficient significantly depends on the burner geometry and generally increases with the increase of the thermal load of the burner (Q_f). In addition, it is important to note that the frequency of the process (f) depends largely on the geometry of the burner. Thus, the general dependence of the heat transfer coefficient on the side of the pulsating flue gases can be defined as:

$$k_p \approx \alpha_p = \alpha_p\left(\dot{Q}_f, p_{uk}, f, \ell_{HE}, \ldots\right), \frac{W}{m^2K} \tag{2}$$

If we compare the values of the heat transfer coefficient for the same thermal power of the burner and similar burner geometries, e.g. the burners label G:41 and G:43 at thermal power of about 106 kW, it is concluded that, in addition to the total pressure deflection: p_{14uk}(G:41) = 28 kPa < p_{14uk}(G:43) = 30 kPa - Fig. 4, and the frequency of the process (whose ratio in this case is: f(G:41) = 91 Hz > f(G:43) = 88 Hz) has a certain influence on the heat transfer coefficient: α_p(G:41) = 99,0 W/m^{2}K > α_p(G:43) = 96,1 W/m^{2}K - Fig. 5.

Countercurrent Fluid Flow (B): In analogy to previous research, measurements were made for the same geometries and thermal loads and for the same water flow rate through the exchanger. The results of these measurements are presented in Fig. 6.

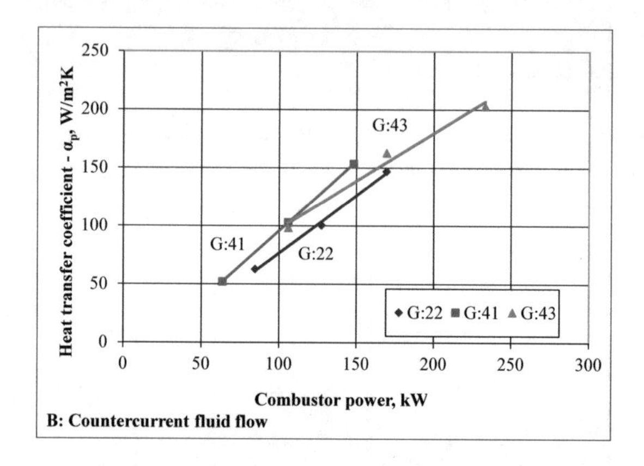

Fig. 6. Heat transfer coefficient of the pulsating flow (α_p) depending on the thermal load and the burner geometry

In comparison to the values of cocurrent fluid flow, here the values of the heat transfer coefficient are somewhat higher - for example, at a thermal load of about 148 kW and for burner geometry labeled G:41: α_{pA} = 151 W/m^{2}K < α_{pB} = 158 W/m^{2}K. In addition, a more favorable logarithmic mean temperature difference of the fluid was recorded, which ultimately contributed to somewhat higher thermal power of the heat exchanger (Q) compared to the thermal power for cocurrent fluid flow - e.g. Q_A = 16,6 kW < Q_B = 16,8 kW, which confirmed the previously known fact that countercurrent heat exchangers are somewhat more efficient under the same other conditions.

Comparison of α and α_p: The heat transfer coefficient in the turbulent flue gas flow through the pipe (d_u) and without pulses (α) is calculated on the basis of the expression:

$$\mathrm{Nu} = \frac{\alpha d_u}{\lambda} = 0,021\varepsilon_\ell \mathrm{Re}^{0,8}\mathrm{Pr}^{0,43} \tag{3}$$

which is valid for turbulent flow in a straight pipe or channel for Reynolds number (Re) in the interval of $1 \cdot 10^4 < \mathrm{Re} < 5 \cdot 10^6$. In this case Re is in the interval of $[0.9 \div 2] \times 10^4$, and the correction coefficient due to the length of the pipe is ε_ℓ = 1,1.

α is calculated for three thermal loads of the burner and both cases of fluid flow and then compared with the corresponding heat transfer coefficient for the pulsating flue gas flow (α_p) - comparison of these results is given in Fig. 7. The results refer to the burner geometry labeled G:43. On the basis of these results, for the cocurrent flow of working fluids, the ratio between the heat transfer coefficient in the pulsating flue gas flow and the forced convection without pulses $R_A = (\alpha_p/\alpha)_A$ is calculated and presented in the same diagram. From the results it can be seen that for geometry G:43 and the observed range of thermal load of the burner (106 ÷ 233 kW), the ratio R_A increases slightly and at the burner thermal load of 233 kW $R_A = 2{,}1$. Trend of changes of the ratio $R_B = (\alpha_p/\alpha)_B$ is similar.

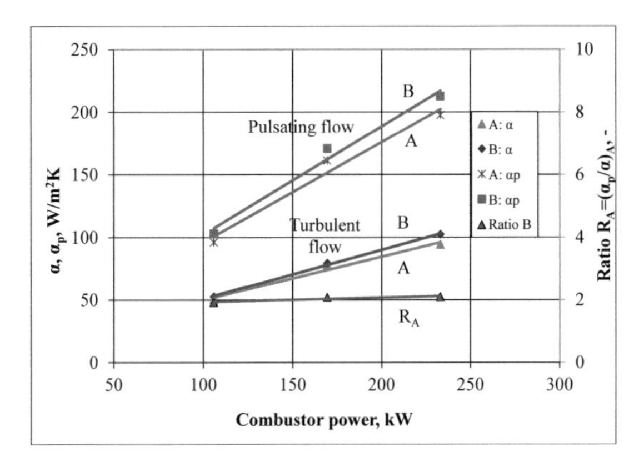

Fig. 7. Heat transfer coefficient for the pulsating flue gas flow (α_p) and in developed turbulent flow without pulsations (α) depending on burner thermal load

4 Conclusion

From the results obtained during the study of the influence of the flue gas pressure pulsations on heat transfer, part of which is presented in this paper, the following conclusions are drawn:

- the heat transfer coefficient in a pulsating flue gas flow depends significantly on the geometry of the burner which actually defines the limits of stopping work of the combustion process, i.e. defines the range of the possible thermal power of the burner,
- the form of the flue gas pressure pulsations changes significantly and deforms with increasing the distance from the free end of the resonant pipe of the burner, especially in the overpressure zone. The overall amplitude of these pulsations also weakens. From this it follows that the heat transfer coefficient in a pulsating flue gas flow is significantly dependent on the distance of the free end of the resonant pipe from the inlet to the heat exchanger,
- with the increase of the thermal load of the burner, the heat transfer coefficient in the pulsating flue gas flow is also increasing,

- there is a direct correlation between the total flue gas pressure deflection and the heat transfer coefficient in a pulsating flow (proportional dependence),
- in addition to the influence of the total pressure deflection on the heat transfer coefficient in the pulsating flue gas flow, a certain influence has the frequency of the combustion process, which is again determined by the geometry, the thermal load of the burner and the intensity of the cooling of the burner,
- it has been shown for certain geometries and thermal loads of the burner, that the heat transfer coefficient in a pulsating flue gas flow is more than 2 times greater than the heat transfer coefficient in developed turbulent flow without pulsations,
- in the case where possible, it is recommended to use an countercurrent flow heat exchanger with the goal of achieving slightly higher thermal power of the heat exchanger or to use smaller and more compact versions of the exchanger in relation to the cocurrent flow fluid flow.

References

1. Hodžić, N.: Laboratory investigation of the possible application of pulsating combustion in large-scale boilers (in Bosnian). M.Sc. Thesis, University of Sarajevo, Faculty of Mechanical Engineering Sarajevo, Sarajevo (2007)
2. Hargrave, G.K., Kilham, J.K., Williams, A.: Operating characteristics and convective heat transfer of a natural-gas fired pulsating combustor. J. Inst. Energy **59**(439), 63–69 (1986)
3. Smajević, I., Hodžić, N., Hanjalić, K.: Aerovalved gas pulse combustor for enhancement of heat transfer in large scale solid-fuel boilers. In: 10th International conference Turbulence, Heat and Mass Transfer 8, © 2015 Begell House, Inc., Sarajevo, 20–25 September 2015 (2015)
4. Belles, F.E.: Pulse combustion and its application. Am. Gas Assoc. Monthly **62**(11), 20–22 (1980)
5. Hanjalić, K., Smajević, I.: Further experience in using detonation waves for cleaning boilers heating surfaces. Int. J. Energy Res. **17**, 583–595(1993)
6. Hodžić, N., Metović, S., Smajević, I.: Pulsating combustion burner on the boiler model: an experiment supported by numerical investigation. In: 10th International Research/Expert Conference "Trends in the Development of Machinery and Associated Technology" TMT 2006, Lloret de Mar, Spain, 11–15 September 2006 (2006)
7. Smajević, I., Hanjalić, K., Hodžić, N.: Possibility for application of the pulse combustors for cleaning flue gas side boiler heating surfaces – experience in using and new development. VGB Techniche Vereinigung der Grosskraftwerks-betreiber E.V., Merkblatt für Poster-vorträge, Düseldorf, 10–12 Oktober 2000, A5 (2000)
8. Smajević, I., Hodžić, N.: Pulse combustor as a potential tool for improving the energy efficiency of the power plants. In: The 5th International Conference on Environmental and Material Flow Management "EMFM 2015" Zenica, B&H, 05–07 November 2015 (2015)
9. Putnam, A.A., Belles, F.E., Kentfield, J.A.C.: Pulse combustion. Prog. Energy Combust. Sci. **12**(1), 43–79 (1986)
10. Mullen, J.J.: The pulse combustion furnace-how it is changing an industry. ASHRAE **26**(7), 28–33 (1984)

Development, Significance and Possibilities of Application Cofiring of Coal with Biomass in Thermal Power Plant in BOSNIA and Herzegovina

Izet Smajevic[1] and Anes Kazagic[2](✉)

[1] University of Sarajevo, 71000 Sarajevo, Bosnia and Herzegovina
[2] JP Electric Power Industry B&H d.d., Sektor za strateški razvoj, Vilsonovo šetalište 15, 71000 Sarajevo, Bosnia and Herzegovina
a.kazagic@epbih.ba

Abstract. Above 30% of power production in Europe is still coming from coal. More than 600 coal or lignite fired power plants produce electricity all over the Europe, with more capacity located in Germany, Poland, Bulgaria, Greece, the Czech Republic and South East Europe. Various ways of reducing CO_2 from coal-based power generation are currently in certain phases of research, development and demonstration. Many of them involve biomass co-firing. Co-firing coal with biomass can be carried out directly (in the same combustion chamber), indirectly (after pre-treatment), in parallel (separate combustion), and completely (full conversion to biomass). Opportunities for retrofitting are co-firing of biomass and complete retrofits (full conversion). This paper summarize recent activities carried out in Bosnia and Herzegovina, namely JP Elektroprivreda BiH and Faculty of Mechanical Engineering of University in Sarajevo, to develop and implement solutions of retrofitting large thermal power plants in Bosnia and Herzegovina with biomass. Target is achieving the power production in thermal power plants in Bosnia and Herzegovina to be sustainable in long-term view.

Keywords: Cofiring · Biomass · Coal · Thermal power plants · Sustainability

1 Introduction

Coal is still an important energy source in many countries as it is cheap and available. More than 600 coal or lignite fired power plants produce electricity all over the Europe, with more capacity located in Germany, Poland, Bulgaria, Greece, the Czech Republic and South East Europe. Various ways of reducing CO_2 from coal-based power generation are currently in certain phases of research, development and demonstration. Many of them involve biomass co-firing. Transition of traditional coal-fired power stations into multi-fuel power plants is ongoing. Multi-fuel operation of coal fired power stations, running co-firing different kind of biomass with coal, is nowadays mainly done to provide fuel mix diversity in order to reduce CO_2 emissions, improve security of supply and reduce operational costs by fuel cost optimization. In the last decade, significant progress was made in the utilization of biomass in coal-fired power

I. Karabegović (Ed.): NT 2019, LNNS 76, pp. 461–467, 2020.
https://doi.org/10.1007/978-3-030-18072-0_53

plants. Over 250 units worldwide have either tested or demonstrated co-firing of biomass or are currently co-firing on a commercial basis, as reported by KEMA [1]. Coal is often replaced with up to 30% of biomass by weight in pulverised coal based power plants. Most of these projects refer to co-firing biomass with high-rank coal (both bituminous and anthracite), while availability of projects on biomass co-firing with low-rank sub-bituminous coal and lignite is more scarce, like the project involving Greek lignite reported by Kakaras [2]. Furthermore, progress is made in application of different types of municipal solid waste as a fuel in coal-based power plants (solid recovered fuel - SRF or refuse derived fuel - RDF, including their gasification), or even co-firing with sewage sludge, see [3]. Examples of biomass co-firing can be found in other industries as well, like in the example of biomass co-firing in the cement industry, reported by Mikulcic et al. [4].

Nevertheless, research, development and demonstration projects and technologies which include combination different types of biomass co-fired with coal have to be investigated, to achieve sustainable solution for solid-based power plants. Over the last decade many research studies were conducted in order to investigate the biomass co-firing phenomenon, but only few of them included a multi-fuel concept. As an example, Wang et al. evaluated the combustion behaviour and ash properties of a number of renewable fuels, like rice husk, straw, coffee husk and RDF derived from municipal waste [5]. The work used a drop tube furnace to evaluate the combustion behaviour and ash properties of biomass, waste derived fuels, pine and coal. Kupka et al. investigated the ash deposit formation during the process of co-firing coal with sewage sludge, saw-dust and refuse derived fuels in a drop tube furnace, to optimize biomass co-firing blends [6]. Williams et al. investigated the emission of pollutants from solid biomass fuel combustion [7]. A combination of fuels can give rise to positive or negative synergy effects, of which the interactions between S, Cl, K, Al and Si are the best known, which may give rise to or prevent deposits on tubes [8], or may have an influence on the formation of dioxins [9]. Biomass can further be used for reburning in order to reduce NOx emissions [10].

Although different types of energy crops have been recognized as significant energy resource in future low carbon society, cofiring energy crops with coal or other types of biomass have not been widely, if at all, investigated so far.

This work presents experimental investigation into co-firing Bosnian low-rank coal with woody biomass and harbacus energy crops Miscanthus, by a multi-fuel pulverised combustion concept. Emissions and ash-related problems were already investigated in case of Bosnian low rank coal, also in co-firing Bosnian low rank coal with waste woody biomass (Kazagic et al.) [8, 11], where some specific benefits and synergy effects of biomass co-firing with that coal type have been observed. The same coal type, used in co-firing with biomass and natural gas, was subject of research into reburning [10].

The experimental study presented here is aimed to investigate possibilities for the co-firing coal and biomass with energy crops Miscanthus at higher co-firing ratio in operation of large coal-based Boilers.

2 Experimental

The paper presents a research study on ash-related problems and emissions during cofiring low-rank Bosnian brown coal and lignite with different kinds of biomass: in this case woody sawdust and harbacus energy crops Miscanthus. Lab-scale drop tube PC furnace was used for the lab tests, varying fuel portions in a fuel mix at high co-firing ratio; with up to 30%w of the woody sawdust and up to 10%w of Miscanthus in fuel blends tested. The tests should optimize the process temperature, air distribution (including OFA), fuel portions and fuel distributions (including reburning), as function of emissions of SO_2 and NO_x as well as combustion process efficiency, estimated through the ash deposits behaviours, CO emissions and unburnt.

In parallel, trial run was carried out at Kakanj power station, operated by EPBiH power utility, to check operation of biomass co-firing in real conditions.

2.1 Fuel Sampling

<u>*Coal Sampling and Transport.*</u> Coal samples have been prepared in Tuzla Thermal power plant (namely coals TET5 and TET7) and in Kakanj Thermal power plant (namely coals TEK6 and TEK8), operated by EPBiH power utility. Approximately 30 kg of each coal are prepared for the lab-tests. The coal samples were taken in pulverized form, capturing the coal dust from the channel behind the mill, to have realistic test fuel size distribution which fully corresponding to real situation in large boiler.

<u>*Biomass Sampling and Transport.*</u> The forest residues have been taken from the forestry from the region of Middle Bosnia and East Bosnia, from potential supplier of the forest residues. The wood chips have been made from these forest residues, and after that they have been chipped and then pulverized in the laboratory mill.

The miscanthus has been taken from the trial field of miscanthus in Butmir (Bosnia) in November 2015. The chips from miscanthus have been made from these samples in February 2016 and they have been pulverized in the laboratory mill.

2.2 Lab-Scale Test Demonstration

The 20 kW lab-scale furnace, the electrically heated entrained PF flow furnace depicted in Fig. 1, designed and installed at Faculty of Mechanical Engineering of Sarajevo University, was used for the laboratory experimental tests. Various types of coal and sawdust were used. The appropriate particle sizes of the mixtures were obtained using a laboratory hammer mill. The pulverized fuel particles were fed into the furnace by means of a volumetric-type feeder, equipped with a speed controller, allowing mass flow in the range of 0.25–5.25 kg/h. The air for combustion from the blower was split into primary air, secondary air, tertiary air and over-fire air (OFA). The first three air portions were introduced into the furnace over a burner placed at the top, enabling downward flow of the air-fuel particles mixture. The final air portion, or OFA, was used to investigate the air stage combustion, by simulating the OFA system used in large boilers. It was consequently introduced directly into the reaction tube (see Fig. 1). The excess air ratio was adjusted by controlling the air flow in each particular air line at constant fuel flow rate. For more details, see [8–11].

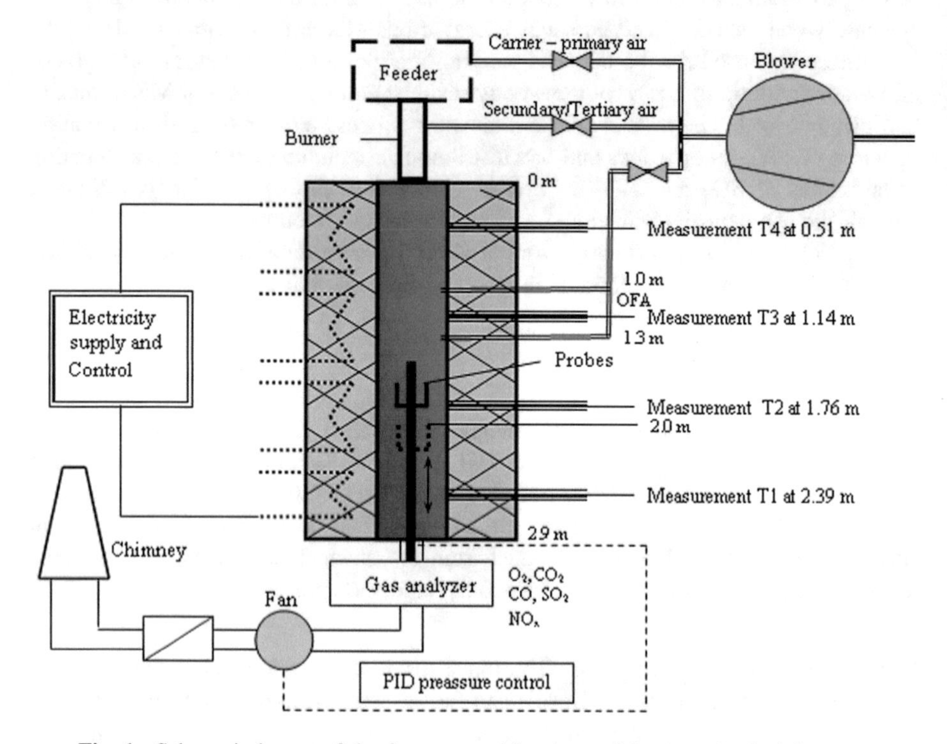

Fig. 1. Schematic layout of the furnace used for the co-firing tests in the laboratory

2.3 Trial Run in Large Boiler

Based on the laboratory research findings, a trial co-firing was completed at the Kakanj power station unit 5 (110 MWe), which is equipped with PF boiler with slag tap furnace, hammer mills and low NOx swirl burners, Fig. 2, and fuelled with brown coal from the nearby coal mines of Middle Bosnia. The trial run was adapted to the brown coal - spruce sawdust mixtures with the sawdust weight ratio of 0%, 5% and 7% (fuel blends U100, U95B5 and U93B7, respectively). Spruce sawdust was supplied from the nearby sources by a contracted supplier. The blend of coal and sawdust is prepared at the coal depot, where the fuels are mixed by coal excavators before being transported to the boiler. The fuel mixture was introduced into the furnace over the hammer mills and the coal burners. The load was varied between 70 and 100% of maximal load, operating the unit at a gross electric power output of 75, 90 and 105 MW.

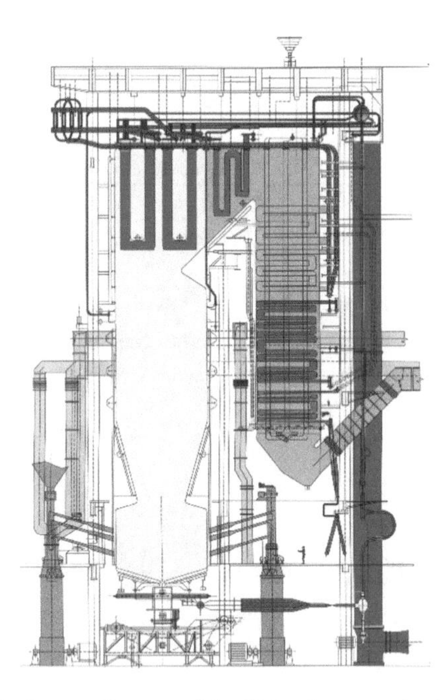

Fig. 2. Boiler with slag tap furnace of the Kakanj power plant Unit 5 (110 MWe)

3 Results and Discussion

3.1 Slagging

Multi-criteria evaluation of slagging/deposition propensity was applied, estimating the shape, state and structure of the ash deposit samples as well as identifying the alkali redistribution. The results for selected 12 co-firing fuel combinations tested impose reasonable expectation that the coal/biomass/Miscanthus co-firing blends could be successfully run under certain conditions not producing any serious ash-related problems. Evaluation of slagging for selected 12 co-firing regimes, conducted in laboratory during 2015, is given in Table 1.

With regard to biomass co-firing on Large Boiler of Kakaknj TPP Unit 5, the conclusion is that there is no deterioration in the form of deposits in the co-firing trials compared to the trial of "reference coal". Indeed, deposits from the co-firing trials are more brittle. Specifically, the form of ash deposits in this zone of the boiler (the furnace outlet) is normal and acceptable; they are more crisp and easy-to-remove. They do not represent a threat to the stability of operation or the efficiency of the boiler. This correlates with the findings of the laboratory tests and the suggestion that deposits will not be a problem if 7–10%w sawdust is blended with coal used in large power stations.

Table 1. Evaluation of slagging

	No.	Temperature, °C	1250	1450
		λ_1/λ	0.95/1.20	0.95/1.20
		Co-firing regime	Evaluation of slagging	
TET	1	TET5:WB = 93:7%w	Low	
	2	TET5:WB = 85:15%w	Low to Medium	
	3	TET5:M = 93:7%w	Low	
	4	TET7:WB = 85:15%w	Low to Medium	
	5	TET7:WB:M = 80:13:7%w	Low to Medium	
	6	TET7:WB:M = 75:15:10%w	Low to Medium	
TEK	7	TEK6:WB = 85:15%w		Very Strong*/Medium**
	8	TEK6:WB = 75:25%w	Low to Medium	
	9	TEK6:M = 93:7%w	Low	
	10	TEK8:WB = 75:25%w	Low to Medium	
	11	TEK8:WB:M = 85:8:7%w	Low to Medium	
	12	TEK8:WB:M = 75:15:10%w	Low to Medium	

*If considered to be used in Boiler with dry bottom furnace, **If considered to be used in Boiler with slag-tap furnace

3.2 Emissions

Both for the brown coal and lignite co-firing with woody sawdust and Miscanthus, SO_2 emissions were slightly higher when higher content of woody biomass was used, confirmed both in laboratory and on Large Boiler. Oppositely, higher Miscanthus percentage in the fuel mix slightly decreases SO_2 emissions, as suggested by laboratory tests. Both for the lignite and brown coal based fuel combinations, NO_x emissions generally decreased with an increase biomass co-firing rate. For the lignite/biomass/Miscanthus co-firing, in the staged air combustion scenario (overall excess air ratio was 1.10 while excess air in the primary combustion zone was kept at 0.95) and combustion temperature of 1250 °C, NO_x emissions were between 310 and 375 mg/m_n^3 at 6% O_2 dry, while for the brown coal/biomass/Miscanthus fuel blends (having higher content of Nitrogen), also with OFA switched on, NO_x emissions were between 640 and 760 mg/m_n^3 at 6% O_2 dry for the temperature range of (1250 to 1450) °C. At this, CO emissions increased a bit with an increase co-firing rate, which suits well to the trend of NO_x emissions.

4 Conclusion

The results for selected 12 co-firing fuel combinations tested in laboratory impose reasonable expectation that the coal/biomass/Miscanthus co-firing blends could be successfully run under certain conditions not producing any serious ash-related problems. With regard to biomass co-firing on Large Boiler of Kakaknj TPP Unit 5, the conclusion is that there is no deterioration in the form of ash deposits in the co-firing trials compared to the trial of "reference coal". Both for the brown coal and lignite co-firing with woody sawdust and Miscanthus, SO_2 emissions were slightly higher when higher content of woody biomass was used, confirmed both in laboratory and on Large Boiler. Oppositely, higher Miscanthus percentage in the fuel mix slightly decreases SO_2 emissions, as suggested by laboratory tests. Both for the lignite and brown coal based fuel combinations, NO_x emissions generally decreased with an increase biomass co-firing rate.

The experimental study results suggest that running low rank Bosnian coal types with woody sawdust and Miscanthus shows promise at higher co-firing ratios for pulverized combustion in large Boilers of thermal power plants Tuzla and Kakanj.

References

1. KEMA: Technical status of biomass co-firing. IEA Bioenergy Task 32, Netherlands (2009)
2. Kakaras, E.: Low emission co-combustion of different waste wood species and lignite derived products in industrial power plants. In: XXXII Krafwerkstechnisches colloquium: Nutzung schwieriger brennstoffe in kraftwerken, Dresden, pp. 37–46 (2000)
3. Wischnewski, R., Werther, J., Heidenhof, N.: Synergy effects of the co-combustion of biomass and sewage sludge with coal in the CFB combustor of Stadtwerke Duisburg AG. VGB Power Tech. **86**(12), 63–70 (2006)
4. Mikulcic, H., von Berg, E., Vujanovic, M., Duic, N.: Numerical study of co-firing pulverized coal and biomass inside a cement calciner. Waste Manag. Res. **32**(7), 661–669 (2014)
5. Wang, G., Silva, R.B., Azevedo, J.L.T., Martins-Dias, S., Costa, M.: Evaluation of the combustion behaviour and ash characteristics of biomass waste derived fuels, pine and coal in a drop tube furnace. Fuel **117**, 809–824 (2014)
6. Kupka, T., Mancini, M., Irmer, M., Weber, R.: Investigation of ash deposit formation during co-firing of coal with sewage sludge, saw-dust and refuse derived fuel. Fuel **87**(12), 2824–2837 (2008)
7. Williams, A., Jones, J.M., Ma, L., Pourkashanian, M.: Pollutants from the combustion of solid biomass fuels. Prog. Energy Combust. Sci. **38**(2), 113–137 (2012)
8. Kazagic, A., Smajevic, I.: Synergy effects of co-firing of woody biomass with Bosnian coal. Energy **34**(5), 699–707 (2009)
9. Leckner, B.: Co-combustion: a summary of technology. Thermal Sci. **11**(4), 5–40 (2007)
10. Hodzic, N., Kazagic, A., Smajevic, I.: Influence of multiple air staging and reburning on NOx emissions during co-firing of low rank brown coal with woody biomass and natural gas. Appl. Energy **168**, 38–47 (2016)
11. Kazagic, A., Smajevic, I.: Experimental investigation of ash behaviour and emissions during combustion of Bosnian coal and biomass. Energy **32**(10), 2006–2016 (2007)

Nozzle Optimization of Dual Thrust Rocket Motors

Mohammed Alazeezi[1] and Predrag Elek[2(✉)]

[1] Emirates Advanced Research & Technology Holding LLC, Abu Dhabi, UAE
[2] Faculty of Mechanical Engineering, University of Belgrade, Kraljice Marije 16, 11120 Belgrade, Serbia
pelek@mas.bg.ac.rs

Abstract. Optimizing the nozzle of a solid propellant rocket motor plays an essential rule in the overall performance of the motor. In this paper, the investigation of an optimization model of dual thrust propellant rocket motors will be presented. Due to having two phases of thrust in this type of rocket motors, determination of the rocket nozzle expansion ratio is a non-trivial problem. The idea is to use a simple, fixed length and expansion ratio, convergent-divergent nozzle, which provides the highest total impulse of the motor. Usual assumptions for an ideal rocket motor have been used. The optimization model was developed in MATLAB and calculations has been performed using previously obtained interior ballistic and other relevant data for a dual thrust solid propellant rocket motor.

Keywords: Optimization · Convergent-divergent nozzle · Rocket propulsion · Dual thrust

1 Introduction

A typical solid propellant rocket motor (SPRM) is a device where chemical energy is transformed to kinetic energy in its combustion chamber which is used also as propellant reservoir. The combustion chamber has cylindrical shape (the most common shape). The SPRM main components are: propellant grain, case (cylindrical section), nozzle, igniter, forward closure, insulation and attachment skirt. This device has many design approaches, especially in casing configuration when it's used for specific purpose. The solid propellants used mainly are either double-base or composite type of propellants. The required mass of propellant grain is determined by total projectile impulse, type of propellant, nozzle efficiency and aerodynamic characteristics. Motor casing is made of steel, aluminum or titanium alloy. Forward closure of the casing can be made from the same material with cylindrical section. Nozzle shape and motor total mass effect SPRM efficiency directly. Burn time or burn duration is defined as the action time minus the ignition time and the time required for thrust tail-off and this effect propulsive performance [1].

Dual thrust rocket motors (DTRM) are also solid propellant rocket motors which are designed to have two levels of thrust: a high level thrust (boosting) and low level thrust (sustaining) thrust [2]. This device has various designs the simplest of them have

I. Karabegović (Ed.): NT 2019, LNNS 76, pp. 468–477, 2020.
https://doi.org/10.1007/978-3-030-18072-0_54

two propellants differ in geometry and composition, but they are placed in the same chamber. Theoretically, high boost-to-sustain thrust ratios can be attained by adjusting the chamber pressure: high pressure at boosting and then low pressure at sustaining. This operation conditions may lead to instabilities in the boosting phase. There is a solution for this conflicting problem which is to separate the two grains with an orifice like nozzle.

Nozzle is a device designed to control flow rate, speed, direction, mass and the pressure of the stream passing through them. For the propellant after getting heated it first converges at the throat of the nozzle and then expands in the divergent part. The convergent divergent nozzle (also known as de Laval nozzle) used to convert thermal energy generated in the combustion chamber to kinetic energy. The nozzle converts low velocity, high pressure and temperature into high velocity, low pressure and temperature. The general range of exhaust velocity is 2…4.5 km/s. For de Laval nozzle the inlet Mach number is less than 1, convergent section accelerate flow to sonic velocity at the throat and then it accelerated to supersonic velocity at the diverging section. At the converging part the density of air (compressible fluid) is difficult to change with the drop-in pressure. In the diverging part, when Mach number becomes larger than one, the density decreases quite faster and that is how velocity increase with the increase in cross-sectional area.

2 Optimization Model

In dual thrust rocket motors, there is a significant change of pressure between the two phases which requires a careful attention when designing the nozzle. In a single phase thrust profile rocket motors, especially for neutral burning, it is straightforward when it comes to the nozzle design since the chamber pressure is not changing as in the case when we have dual phase (Fig. 1).

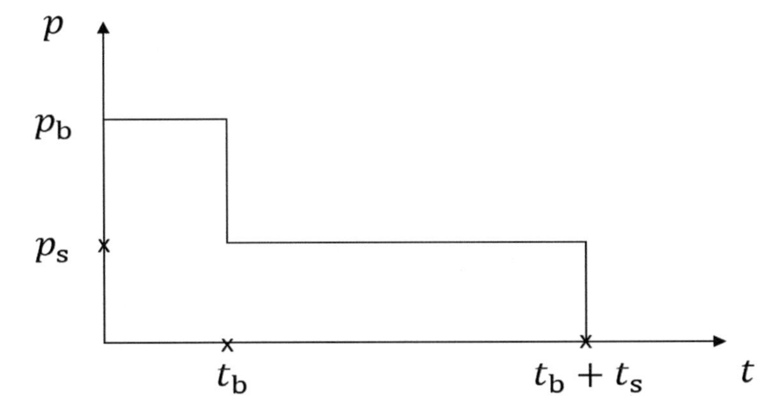

Fig. 1. An illustration a dual phase rocket motor pressure profile

The optimum design of a nozzle for dual thrust motors could be achieved via two methods. The first method is to have two parts of nozzle that one of the parts eject after completing the first phase of the grain's burning, and the second part remains for the rest of the motor's operation. The second method is two have only one piece of convergent-divergent nozzle which could satisfy the requirements of the two phases. This paper will focus on the second method that an optimization model will be developed for the nozzle design of a dual thrust rocket motors.

The boost phase has a higher value of chamber pressure than the sustaining phase; therefore, it requires a long divergent part of the nozzle to expand the exhaust gas from that high pressure to lower one. The exit area for the sustain phase will be lower than the boost phase, so an optimum exit area has to be found for the design, which would make the exit pressure equal to the ambient pressure. Figure 2 gives a good illustration of how there should be an optimum exit area between the two-phases exit areas.

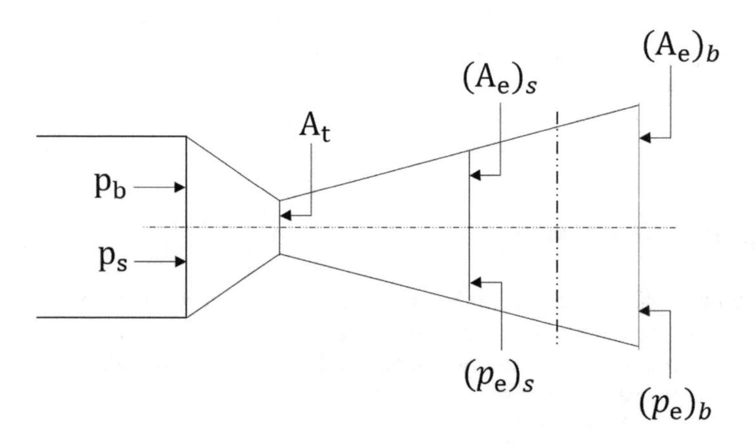

Fig. 2. Half section of the nozzle of rocket motor demonstrates the exit areas of the two phases

The goal from optimizing the nozzle is to achieve the maximum total impulse I_t that a dual thrust rocket motor could deliver. The total impulse is defined as the average thrust times the total time of firing, so it could be calculated by integrating the area and thrust curve from the thrust versus time profile. It's known from theory that the total impulse is proportional to the total energy released by all the propellant in a propulsion system.

$$I_t = \int_0^{t_b + t_s} F\,dt \tag{1}$$

where, t_b is the burning time of the boosting phase and t_s is the burning time of the sustaining phase and usually t_b is less than t_s. Assuming the thrust F is constant and neglecting the start and stop transients, I_t can be presented in the following form for the boosting and sustaining phases:

$$I_t = F_b t_b + F_s t_s \tag{2}$$

The thrust could be determined as a function of chamber pressure p, throat area A_t and thrust coefficient c_F, Eq. (3).

$$F = c_F p A_t \tag{3}$$

The right side of the previous equation could be substituted for thrust in Eq. (2), which makes I_t in the following form:

$$I_t = [(c_F)_b p_b t_b + (c_F)_s p_s t_s] A_t \tag{4}$$

The total impulse equation could be simplified furthermore by considering that the burning time and chamber pressure would be given parameters prior the design. Therefore, we can define the parameter π dividing the sustain pressure by the boost pressure Eq. (5):

$$\pi = \frac{p_s}{p_b} \tag{5}$$

The second parameter τ is the ratio of the boost phase burning time by the sustain phase burning time, Eq. (6):

$$\tau = \frac{t_b}{t_s} \tag{6}$$

The final expression of I_t is shown in Eq. (7), where p_b, t_s and A_t could be treated as given constants hence the parameters inside the brackets $[(\tau(c_F)_b + \pi(c_F)_s)]$ are what need to be maximized in order to maximize I_t:

$$I_t = p_b t_s (\tau(c_F)_b + \pi(c_F)_s) A_t \tag{7}$$

Another function is formed that includes only what needs to be maximized inside the total impulse equation, and it could be seen that it depends on the given input parameters and thrust coefficient.

$$f(p_e) = \tau(c_F)_b(k, p_b, p_a, (p_e)_b) + \pi(c_F)_s(k, p_s, p_a, (p_e)_s) \tag{8}$$

The thrust coefficient c_F is an important parameter in the nozzle design considerations. It was mentioned in Eq. (3) where it could be found in terms of chamber pressure, throat area and thrust, and it is a dimensionless multiplication factor. Most rocket motors could deliver a thrust coefficient in the range 1.5…1.7. Thrust coefficient could be determined using Eq. (9) as well. The thrust coefficient is a function of adiabatic constant of combustion products k, expansion ratio ϵ, and the pressure ratio across the nozzle $\frac{p_e}{p_o}$. The optimum thrust coefficient is the peak value when the pressure at the nozzle's exit p_e is equal to the atmospheric pressure p_a [3].

$$c_F = \Gamma(k)\sqrt{\frac{2k}{k-1}\left[1-\left(\frac{p_e}{p_o}\right)^{\frac{k-1}{k}}\right]} + \left(\frac{p_e}{p_o}-\frac{p_a}{p_o}\right)\frac{A_e}{A_t} \quad (9)$$

The adiabatic gas constant k has a high influence on the thrust coefficient and on the expansion ratio as well, where function Γ is defined by

$$\Gamma = \sqrt{\mathrm{k}}\left(\frac{2}{\mathrm{k}+1}\right)^{\frac{\mathrm{k}+1}{2(\mathrm{k}-1)}} \quad (10)$$

The ratio of exit area A_e to throat area A_t is basically the nozzle area expansion ratio ϵ Eq. (11).

$$\epsilon = \frac{\mathrm{A_e}}{\mathrm{A_t}} \quad (11)$$

The expansion ratio can be also expressed in the following form:

$$\epsilon = \left(\frac{2}{\mathrm{k}+1}\right)^{\frac{1}{\mathrm{k}-1}}\sqrt{\frac{\mathrm{k}-1}{\mathrm{k}+1}}\left(\frac{\mathrm{p_o}}{\mathrm{p_e}}\right)^{\frac{1}{\mathrm{k}}}\frac{1}{\sqrt{1-\left(\frac{\mathrm{p_e}}{\mathrm{p_o}}\right)^{\frac{\mathrm{k}-1}{\mathrm{k}}}}} \quad (12)$$

The expansion ratio is an important parameter for designing the nozzle. For an adapted (matched) nozzle, the exit pressure p_e equals to the pressure $p_a(p_a = 101325\ \mathrm{Pa} \approx 1\mathrm{bar})$, so the expansion ratio could be found. The optimum value of the expansion ratio is the one that gives the highest total impulse of the motor. Since we have two phases of burning, ϵ will be calculated separately for each phase to find the optimum value in between that would give the highest total impulse.

After finding the optimum expansion ratio, the pressure ratio could be found which would lead to finding all the other missing parameters of the optimization cycle.

The optimization model was implemented in MATLAB [4] to solve it numerically, and the outputs will be presented in the results and analysis section. The flow data of the optimization model is illustrated in Fig. 3.

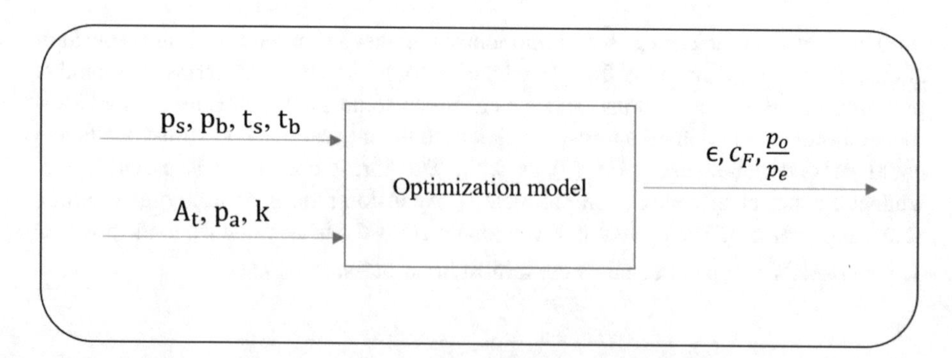

Fig. 3. Flow chart of the optimization model showing the inputs and outputs

3 Results and Analysis

There are several input parameters in the model, which are picked upon the desired mission of the rocket motor [5]. First, a general case has been performed, and then other cases will be performed to see the influence of changing the parameters τ and π. The parameters were chosen and shown in Table 1. The pressure ranges from 20 bar to 130 bar which gives $\pi = 0.15$. In addition, the burning time ratio τ shall be a value between 0 and 1. Several values of k are used to study its influence on the optimization model.

Table 1. Baseline input parameters of the optimization model

Parameter	Value
p_b	130 bar
p_s	20 bar
t_b	1.5 s
t_s	18.5 s
d_t	40 mm
k	1.15, 1.20, 1.25, 1.30

The expansion ratio for the boost and sustain phases is calculated for each value of k using Eq. (12), and the results are shown in Table 2. It could be seen that the expansion ratio is higher when the value of k is lower. The optimum ϵ will fall between ϵ_b and ϵ_s. Additionally, the expansion ratio of boost phase decreases rapidly as k increases; however, the expansion ratio of the sustain phase changes at lower rate.

Table 2. Expansion ratios of the boost and sustain phases for different values of k

	ϵ_b	ϵ_s
k = 1.15	16.22	3.84
k = 1.20	14.36	3.59
k = 1.25	12.83	3.37
k = 1.30	11.56	3.18

After finding the expansion ratio based on the assumption that $p_e = p_a$, the pressure ratio could be calculated again to find the exit pressure for the two phases. Expansion ratio as a function of pressure ratio is shown in Fig. 4. The expansion ratio is obviously proportional to the pressure ratio.

One of the optimization model steps is to find the thrust coefficient, and it was calculated using Eq. (9). Similarly to the expansion ratio, c_F was found for different values of k. It could be seen that the thrust coefficient is proportional to the pressure ratio (Fig. 5).

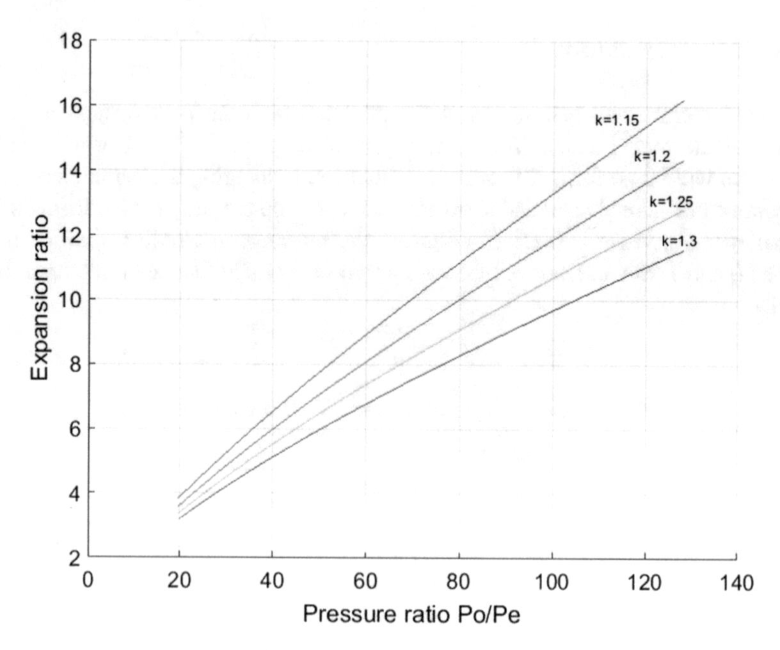

Fig. 4. Expansion ratio of an adapted nozzle as a function of the ratio of chamber pressure to exit pressure, for different values of k

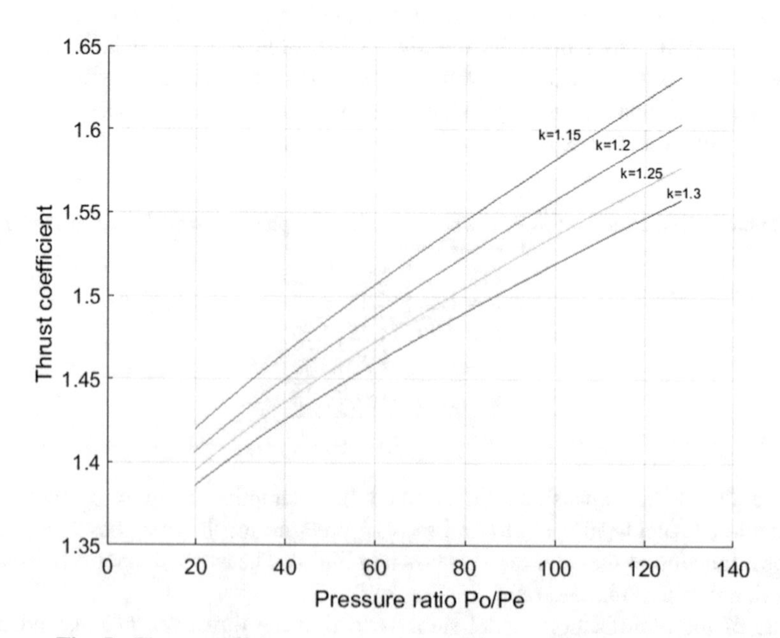

Fig. 5. Thrust coefficient versus pressure ratio for different values of k

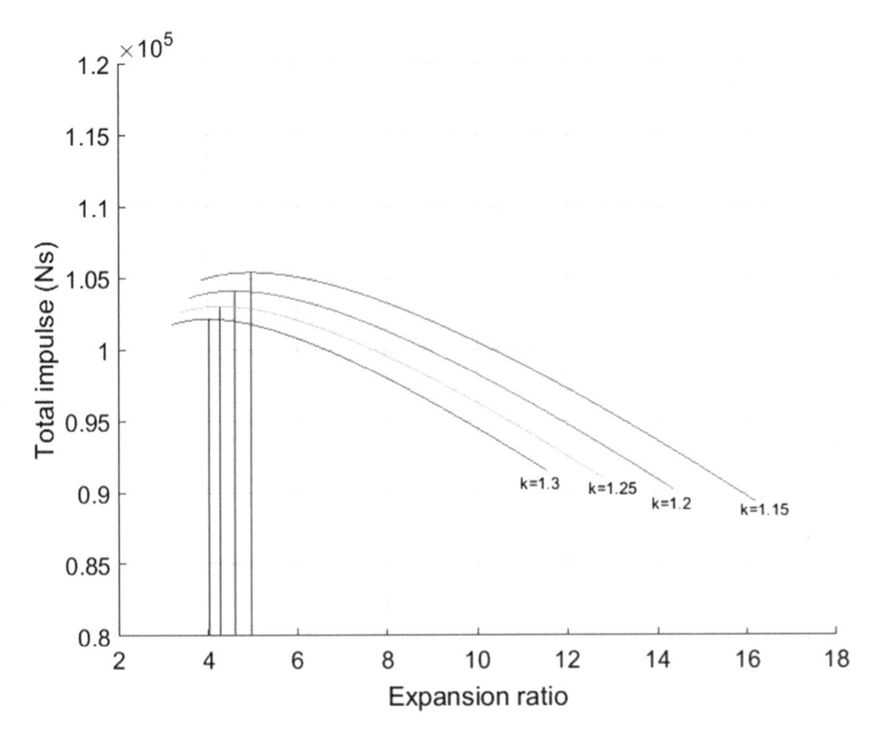

Fig. 6. Optimum expansion ratios that corresponds for the highest total impulse for different values of *k*

An optimum value between ϵ_1 and ϵ_2 will give the highest I_t which grants the best performance of this type of motor. The total impulse versus expansion ratio function was used to determine the optimum value of the expansion ratio ϵ. The optimum expansion ratio for each adiabatic gas constant was found to the corresponding highest value of total impulse which is shown in Fig. 6, and the results are presented in Table 3. The total impulse is the highest when k is the lowest, which results with higher expansion ratio. The optimum ϵ is found to be closer to ϵ_s because of the sustain phase long duration compared to the boost phase.

Table 3. Values of optimum expansion ratio for different values of k that corresponds to the highest total impulse value

	Maximum I_t (Ns)	Optimum ϵ
k = 1.15	105360	4.96
k = 1.20	104070	4.61
k = 1.25	103000	4.27
k = 1.30	102110	4.02

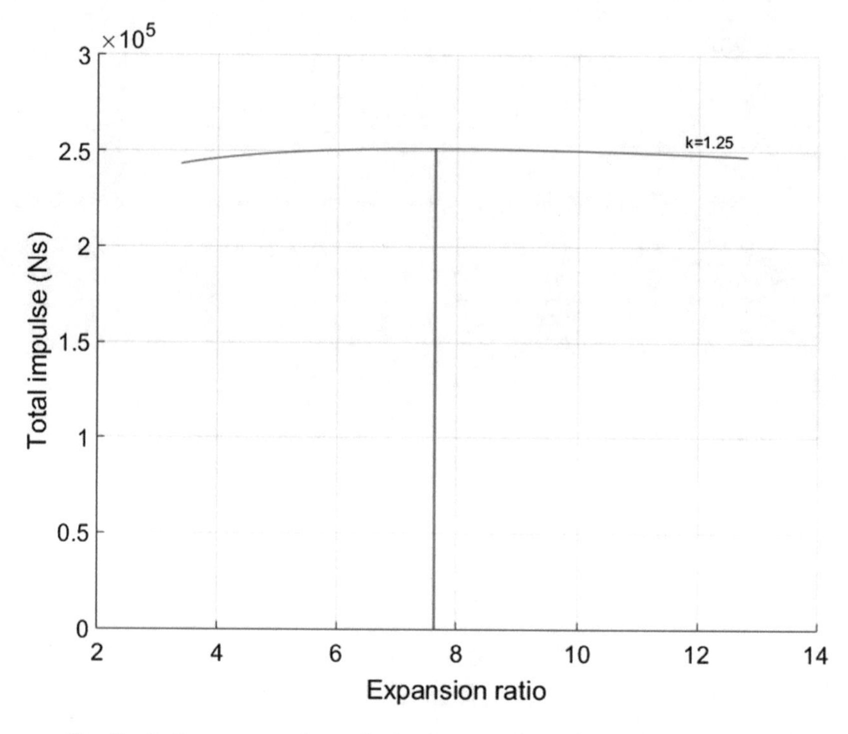

Fig. 7. Optimum expansion ratio for the case: $k = 1.2, t_b = 8s$ and $t_s = 12s$

It was seen that the optimum expansion ratio is closer to the expansion ratio of the sustain phase when the burning time ratio τ is closer to 0. However, when the ratio is increased towards 1, the optimum ϵ falls into the middle of the total impulse curve where it is the highest. This case could be seen in Fig. 7, where $t_b = 8\,s$ and $t_s = 12\,s$ gives $\tau = 0.67$. The value of k was set in the optimization model to be 1.25. The optimum expansion ratio was found to be 7.63 which corresponds to $I_t = 251280\,Ns$. Comparing the previous results to Table 3 showed a high percentage of difference that is expressed in Table 4.

Table 4. Percentage of difference between optimum expansion ratio and total impulse for two burning time ratios

	$\tau = 0.08$	$\tau = 0.67$	% Difference
Maximum I_t (Ns)	103000	251280	83.7
Optimum ϵ	4.27	7.63	56.5

4 Conclusion

The nozzle is one of the main parts in rocket motors, and optimizing it adds a great value to the overall design, especially when there are two phases of burning. In this paper, an optimization model of the nozzle was developed for dual thrust rocket motors. The aid of MATLAB was used for the optimization model to obtain the results numerically. A set of input parameters was chosen to validate the optimization model, and some inputs were altered to check the model's limitations. In a dual thrust rocket motor, it could be summarized that the optimum design of a nozzle could be defined based on the burning time ratio τ, that it is closer to the sustain phase nozzle design when τ is close to 0, and it gets closer to the boost phase when τ increases. Further research could be promising and executed on this topic by performing a flow analysis on the nozzle using computational fluid dynamics (CFD), and it would be challenging to do such a thing for a dual thrust rocket motor.

References

1. Zandbergen, B.T.C.: Typical Solid Propellant Rocket Motors, Delft University of Technology, Faculty of Aerospace Engineering (2) (2013)
2. Sutton, G., Biblarz, O.: Rocket Propulsion Elements. Wiley, Hoboken (2011)
3. Pandey, K.: Analysis of thrust coefficient in a rocket motor. Int. J. Eng. Adv. Technol. **1**(3), 30–33 (2012)
4. MATLAB R2015a: Mathworks (2015)
5. Alazeezi, M., Elek, P.: Analytical and numerical burnback analysis of end burner grain with cylindrical cavity. In: 8th International Scientific Conference on Defensive Technologies – OTEH 2018, Belgrade, Serbia, October 2018

Re-engineering of the Elements of Small Hydro Turbines

Edin Šunje[1(✉)], Amar Leto[1], Adis Bubalo[2], and Safet Isić[1]

[1] Faculty of Mechanical Engineering, University Campus, University "Džemal Bijedić" of Mostar, 88000 Mostar, Bosnia and Herzegovina
edin.sunje@unmo.ba

[2] JP Electric Power Industry BiH d.d. Sarajevo, HE, Neretvi, Jablanica, Bosnia and Herzegovina

Abstract. Rehabilitation and modernization of existing hydro power plants and its elements is necessary to keep it concurrent on more and more demanding energy market. One of the challenges that enterprises are often met is lack of technical documentation of turbine elements. Modern approach to creation of technical documentation for elements of small hydro turbine has been presented. Geometry of penstock valve, guide vanes and draft tube influence on hydro aggregate efficiency. Beside the direct influence on efficiency, in this example the geometry of draft tube is in the same time an input parameter for selection of characteristics and geometry of new hydro turbine that should replace an existing one. The existing dimensional restrains should be respected. Geometry of turbines elements has been scanned using 3D scanner. As the result of scanning process, we obtained a point cloud that represent input data for further CAD software processing and creation of technical documentation as a final goal.

Keywords: Re-engineering · 3D scanning · Technical documentation · Turbine · Guide vanes · Draft tube

1 Introduction

During the maintenance of production system, machines, assemblies one need to replace certain components of machine. Sometimes it is very difficult to find technical documentation of broken component. Technical documentation is the first step to replacement component production. When no technical documentation is available the procedure is conducted reversible, technical documentation is made upon existing component. It requires extraction of information about geometry to develop the part drawing for duplication and manufacture [1], [4], [5], [7]. The reverse engineering method can be manual, or computer aided. The manual approach involves use of measuring instruments. Similar, computer aided method can be contact or non-contact method. The example of contact method is coordinate measuring machine, where the probe on the tip of measuring device must make contact to part geometry. The non-contact type usually employs laser beam or X-ray for scanning or measurement. Non-contact devices include optical, laser, Computed Tomography (CT), Magnetic Resonance Imaging (MRI), etc., usually work very well with freeform surfaces, e.g., biological parts [2]. 3D Scanning technology can

I. Karabegović (Ed.): NT 2019, LNNS 76, pp. 478–485, 2020.
https://doi.org/10.1007/978-3-030-18072-0_55

be used for digitalization of machine parts, but the full benefit can be achieved when it is combined with rapid prototyping technologies. In first case parts are produced indirectly from CAD file through machining process, while in second case the part is produced directly from the CAD file, e.g. 3D printing. In this paper indirect method has been used, and two cases have been analyzed.

2 Re-engineering of Guide Vanes of Small Hydroturbine

Re-engineering of guide vanes of small hydro turbine with Francis runner has been presented in this chapter. During a long-term exploitation, guide vanes have been damaged due to corrosion and cavitation [8]. That damage occurred because of poor material selection. As mentioned earlier, no documentation existed for the guide vanes.

The procedure of re-engineering was included dismounting of turbine assembly, non-destructive testing of guide vanes, determination of chemical composition of material, 3D scanning; conduct a dimensional control to determine dimensional and geometrical tolerances. Technical documentation of guide vanes has been created as a final step of the procedure. Only part of procedure that includes 3D scanning will be presented (Fig. 1).

Fig. 1. Spare guide vane

2.1 3D Scanning

For the scanning process Zeiss Comet L3D 8M scanner has been used. Scanning device is shown on Fig. 2. The device resolution is 0.02 mm with 8 MP camera incorporated. The scanning area of device is 80 × 60 mm (near) and 565 × 425 mm (far). Normally, when 3D scanner is used to capture a physical part into digital form, an individual 3D scans of all sides of part should be captured. After 3D scanning it is necessary to post-process the scan data, as well as aligning and merging the individual scans together. All outlaying scan data points and noises have been eliminated in post-process operation. The process is similar topstitching of individual scans together in order to make a digital 3D model of the object.

Fig. 2. Zeiss Comet L3D scanner.

Using appropriate CAD software solid model has been generated upon polygonal mesh. Based on solid model [3], technical documentation of guide vane has been generated. Model of guide vane is shown on Fig. 3.

As a final step of re-engineering process technical documentation of guide vane has been created. Beside the dimensional and geometrical tolerances drawing includes guide vane profile coordinates to provide profile machining on NC machine (Fig. 4).

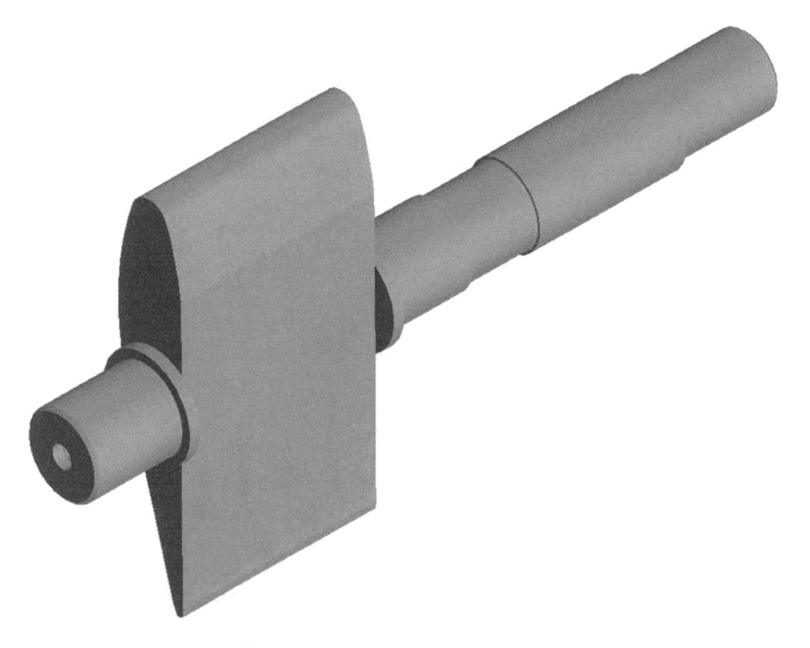

Fig. 3. Solid model of guide vane

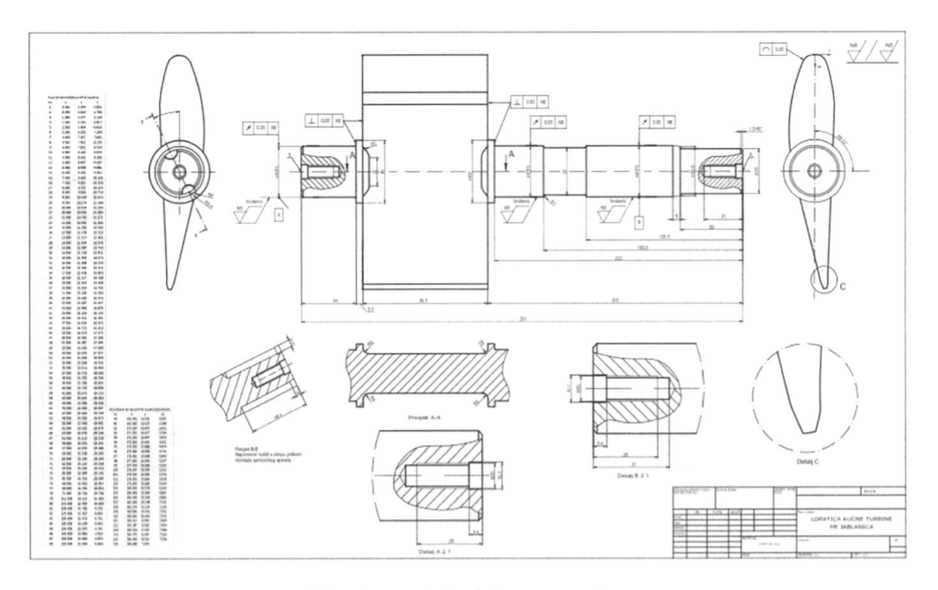

Fig. 4. Technical documentation

3 Re-engineering of Draft Tube of Small Hydroturbine

In this chapter the scanning process of draft tube on small hydro turbine has been presented. Unlike previous example, no turbine dismounting has been allowed. According to restriction scanning process has been conducted at the turbine location.

Fig. 5. Small francis hydroturbine

The goal of this case study was to determine radius and curvature of the elbow marked with an arrow on Fig. 5. The scanning procedure is described in previous chapter. Due to equipment and environmental restriction only, part of elbow has been scanned. Partial scans have been post-processed in Solid Works software where the radius and curvature values of the elbow are determined. Matting spray applied on surfaces before scanning process to prevent reflection on elbow flanges and surfaces (Fig. 6).

Figure 7 shows an individual scan of elbow surface. Upon this surface model its simple to determine radius and curvature of elbow. On the base of elbow dimensions determined by 3D scanning, the final technical documentation of the turbine has been done. Figure 8 shows final turbine disposition drawing.

Fig. 6. Matting spray application

Fig. 7. Elbow surface part

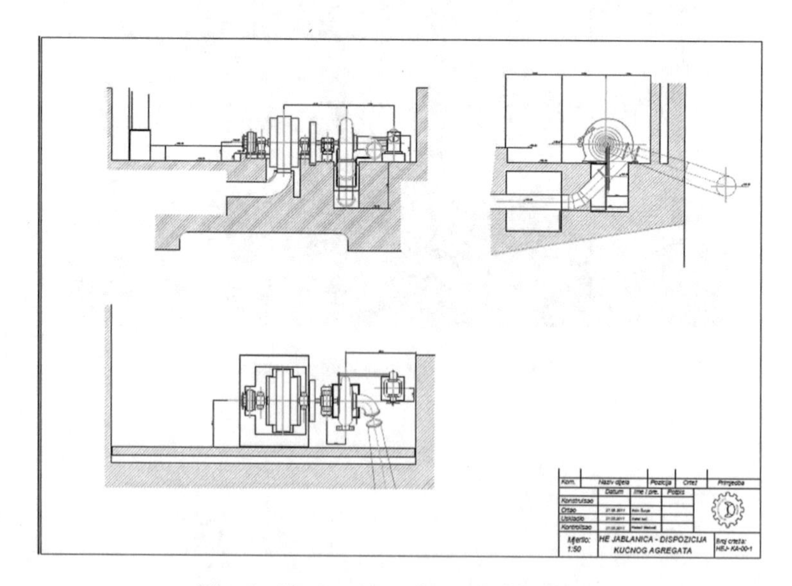

Fig. 8. Final turbine disposition drawing.

4 Conclusion

Using a new technology such as tridimensional scanning and printing it is possible to do re-engineering of some machine, assembly, part. Those techniques have made re-engineering process to become fast, easy to conduct compare to conventional techniques and methods. Full benefit of this techniques can be achieved when it is combined with other production or measuring techniques. It can be concluded:

- 3D scanning can be used in laboratory and on a site,
- Scanners have high accuracy level, but it is necessary to check results with conventional methods.

References

1. Hussain, M.M., Rao, S.C., Parsad, K.E.: Reverse engineering: point cloud generation with coordinate measuring machine for part modeling and error analysis. Asian Res. Publ. Netw. J. Eng. Appl. Sci. **3**(4), 1 (2008)
2. Hubert, A.C., Anosike, N.B., Adamu, A.J.: Computer aided reverse engineering and rapid prototyping of motorcycle rear hub. Int. J. Eng. Tech. **5**(9) (2015). ISSN: 2049-3444 © 2015 – IJET Publications UK
3. Topčić, A., Tufekčić, Dž., Cerjaković, E.: Razvoj proizvoda, Mašinski fakultet Tuzla (2012). ISBN 978-9958-31-074-4
4. Pašić, S., Isić, S., Tiro, D.: Use 3D technology in the rapid re-engineering of construction. In: First International Research Conference NT-2014, pp. 85–92 (2014). ISSN 2303-5668
5. Zivkovic, S., Cerce, L., Kostic, J., Majstorovic, V., Kramar, D.: Reverse engineering of turbine blades kaplan's type for small hydroelectric power station. In: 15th CIRP Conference on Computer Aided Tolerancing - CIRP CAT 2018 (2018)
6. Teran, L.A., Roa, C.V., Muñoz-Cubillos, J., Aponte, R.D., Valdes, J., Larrahondo, F., Rodríguez, S.A., Coronado, J.J.: Failure analysis of a run-of-the-river hydroelectric power plant. Elsevier-Eng. Fail. Anal. **68**, 87–100 (2016)
7. Koirala, R., Chitrakar, S., Panthee, A., Neopane, H.P., Thapa, B.: Implementation of computer aided engineering for francis turbine development in Nepal. Int. J. Manuf. Eng. **2015**, 9 (2015). 509808
8. Technical documentation of for reconstruction of small hydro-aggregates in machine room of HP Jablanica, Bosnia and Herzegovina (Faculty of Mechanical Engineering in Mostar) (2015)

Development of the System for Oil Vapor Drainage from Bearing Housings of Big Hydroaggregates

Safet Isić[1(✉)], Amar Leto[1], Mensud Điddelija[2], and Edin Šunje[1]

[1] Faculty of Mechanical Engineering, University Campus, University "Džemal Bijedić" in Mostar, 88000 Mostar, Bosnia and Herzegovina
safet.isic@unmo.ba

[2] JP Elektroprivreda BiH d.d. Sarajevo, HE na Neretvi, Jablanica, Bosnia and Herzegovina

Abstract. During the operation of large hydro aggregates, oil is heated in the axial and radial bearings and the production of oil vapor appears. The clearance in the gaskets of the bearings increase during the multi-year work and often allows the passage of oil vapor to the surrounding space and condensation on the equipment. Penetration of the oil vapors in the generator part and their condensation on the generator's equipment influences to the unit readiness and requires more frequent cleaning and degreasing. For this reason, it is of vital importance for the hydro aggregates presence of the functional system for oil vapor drying directly from the housing of axial and radial bearings. This paper presents the structural solution of oil vapor drying systems on vertical hydro aggregates with the Kaplan turbine.

Keywords: Oil steam · Oil drainage system · Air flow

1 Introduction

The axial bearing of hydro aggregate is a very important assembly for reliable and functional work of hydro aggregates [1, 2]. It is loaded by a large-scale axial force consisting of the weight of the rotating parts of the hydro aggregate: - a runner with blades, a turbine shaft, a rotor of a generator and a pressure force of the water column above blades of a runner [3].

During the operation of large hydro aggregates, oil is heated in the axial and radial bearings and the production of oil vapor appears in the housing. Because of the large dimension of hydraulic turbine elements, the constructive clearance in the gaskets of the bearings increase during the multi-year work of the hydraulic unit and often allow the passage of oil vapor to the surrounding space and their condensation on the electrical and mechanical equipment. An oiled atmosphere can be a health problem for staff due to inhalation during monitoring and maintenance. The use of oil vapors in the generator part and their condensation on the generator's equipment influences the drive's readiness and requires more frequent cleaning and degreasing. In addition, condensed oil paints on equipment and moving areas cause a security problem for daily

I. Karabegović (Ed.): NT 2019, LNNS 76, pp. 486–493, 2020.
https://doi.org/10.1007/978-3-030-18072-0_56

staff movements as well as overhauling of equipment. For this reason, it is of vital importance for the hydro aggregate to have installed a functional system for oil vapor drainage directly from the casing of axial and radial bearings. The drainage system should cause the stimulated air flow from the housing. Transport of the oil vapor is going through the pipe system from the bearing housing - it condensed on the pipes in the system and collect in the reservoir with the controlled discharge of the condensed oil. Flow in the system should not cause a significant depression in the bearing housings for avoiding suction dirt and moisture in the bearings. This paper presents the structural solution of oil vapor draining system on vertical hydro aggregate no. 1 with the Kaplan turbine of HPP Grabovica, designed to carry an axial load of 1060 tons (Fig. 1).

1. Upper (generator) radial bearing
2. Axial bearing
3. Turbine radial bearing

Fig. 1. A common design of aggregates with Kaplan turbine, with the location of bearings.

2 Data Collection and System Requirements

The oil vapor drainage system should permit the removal of oil vapor from the axial and generator radial bearing housing through the pipe installation, the condensation of the oil vapors during transportation through the pipes and the collection of condensed oil in the oil separator with the valve for discharging. Because of periodic maintenance and dismantling of the unit equipment, the pipe installation of the oil vapor drainage system should consist of segments connected through screw connections. The pipe fitting connections should be adapted to the existing openings on the axial and radial bearing housings and holes on the shaft gasket - if such openings and holes already exist (Table 1.).

Oil vapor extraction needs to be supported by a centrifugal radiator fan designed for oil vapors and oiled air. The fan position and pipe installation location should be

Table 1. Technical data of axial and upper radial (generator) bearing

Item	Value
Bearing type	Segment slide bearing
Number of segments	16/12
The outer diameter of conical support of bearings	Φ 2700 mm/1250 mm
The diameter of conical support of axial bearing	Φ 1400 mm
Total load of the axial bearing	1060 MPa
Total oil volume	5000 l/1650 l
The maximum temperature of the oil film	60 °C
Maximum working oil temperature	70 °C

adapted to the existing construction with no interference in access to the existing equipment. Pipe installation should be made of CrNi steel in order to reduce the possibility of pipe corrosion caused by contaminated oil.

Data required for overall system design are: housing dimensions and position, shaft diameter, positions of holes and openings in the housing, total vertical distance to the air outlet to atmosphere and distribution of the equipment of the aggregate.

3 System Solution Representation

3.1 Pipe System

The general overview of one possible solution of the oil drainage system for vertical hydro aggregate with Kaplan turbine is shown in Fig. 2. This system covers oil vapor drainage only from axial and generator radial bearing, The main reason is that constructive solution of hydro aggregates makes difficult to include turbine radial bearing (or down radial bearing) in pipe system of oil vapor drainage system. Covering this bearing in the drainage system, if possible, completely removes oil vapor from bearing housing and turbine and generator space.

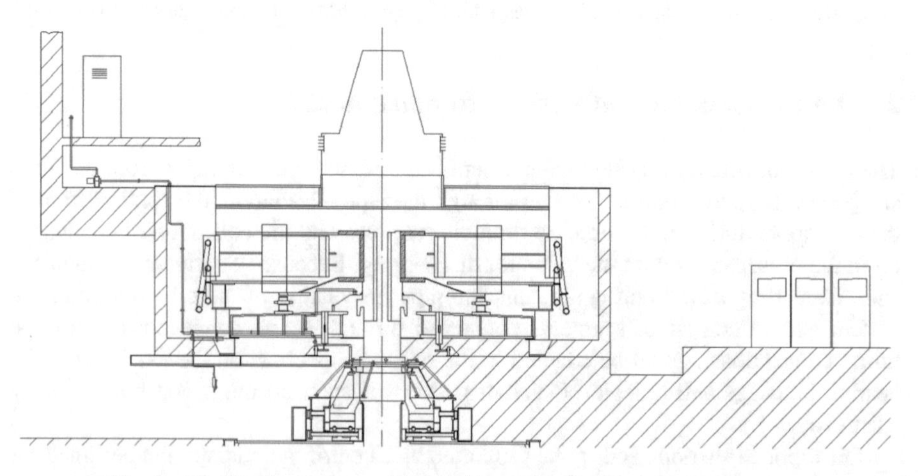

Fig. 2. Distribution and general position of oil drainage system elements

Axial bearing produces the most oil vapor which tends to penetrate through the shaft gasket at the top of the housing. Because of that, pipe system is connected to the housing at maximum diameter by four pipes equally distributed over housing circumference. This kind of distribution of the pipe connections reduces the penetration of the oil vapor through the shaft gasket. To collect any vapor before penetration through the gasket, pipes are put through holes in gasket. All pipes connected to axial bearing housing are joined by segmented circular pipe placed around the shaft (Fig. 3).

From circular pipe starts the main pipe which leads vertically outside of turbine space and connects to the pipe branch from the housing of upper radial bearing. From this connection, main pipe pass by one of upper radial bearing supports to mid space of the generator and its concrete cover construction. In this mid-space, there is also placed generator of the water cooling system, and the position of the pipelines of the draining system is desirable near this system to improve condensation of oil vapor. The main pipe continues vertically to the cover of generator space. System fan is placed near the position of exhausting of the air to the atmosphere. This position enables the maintenance and replacement of the fan without stopping the aggregate. This fan position enables the system to function even if the fan is removed during some maintenance by the natural flow of hot oil vapor.

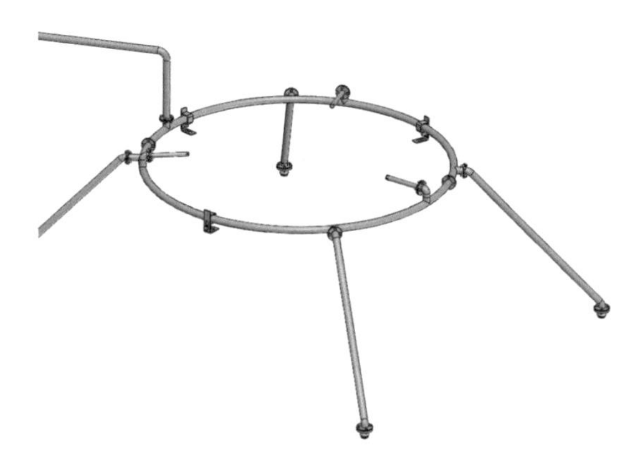

Fig. 3. Collecting ring for connections to the housing of axial bearing.

Figure 4 shows the general elements of the oil drainage system:

- a collecting ring,
- connecting hoses on the axial and radial bearings housing,
- connections to the turbine shaft gasket,
- anchorage of the upper bearing,
- the main oil drain pipe,
- fan for stimulation of the air and oil vapor flow.

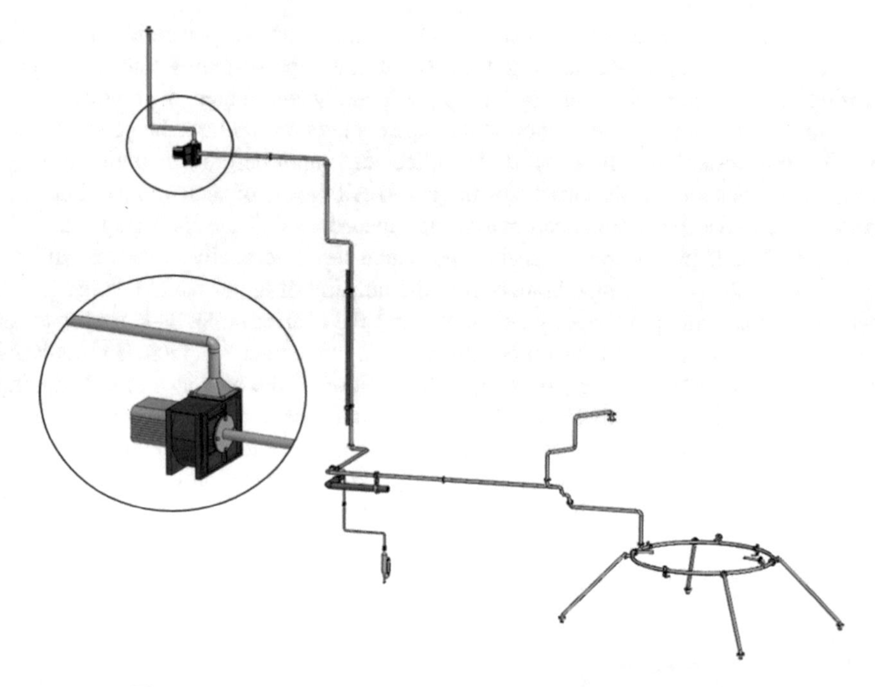

Fig. 4. The complete system for oil drainage on HPP Grabovica.

3.2 Condensed Oil Extraction and Fan Flow Stimulation

The equipment used for extraction of the condensed oil and for flow simulation is shown on Fig. 5.

The tank is made of stainless steel CrNi1910 with a capacity of approx. 1 L. A spherical valve is used for temporary manual discharge of the tank the staff. A condensed oil level indicator is mounted on the tank.

In order to stimulate the flow of oil vapor through the piping system, a centrifugal fan is installed. The fan is powered by a single phase electromotor with output power less than 0.25 kW. The small dimensions of the fan housing enable its placing in intermediate space where the existing turbine installations are located. It is recommended that fan motor is multispeed or frequently driven motor, which enables fine adjusting of the flow to the different working condition of the aggregate.

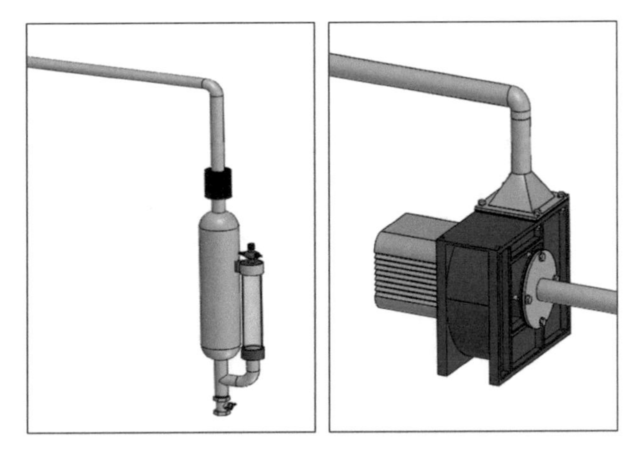

Fig. 5. Stainless steel tank and centrifugal fan

3.3 Practical Implementation of the Vapor Draining System

The presented draining system is implemented on unit 1 of HPP Grabovica with the following characteristics:

Nominal power of the aggregate	58.5 MW,
Nominal power of the generator	65 MVA,
Flow	190 m^3/s,
Rotation speed	150 rpm,
Moment of inertia of the generator	5200 Mpm^2.

The system is used to replace old draining system made of plastic pipes and without stimulated flow by a fan, which was insufficient to remove all oil vapor. Figure 6. shows part of the pipe system of the old drainage system.

Fig. 6. Part of the old oil vapor draining system without stimulated flow

Figures 7, 8, 9 and 10 show elements of the implementation of the new oil drainage system: connections to the bearing housing (axial and radial) and oil tank.

Fig. 7. Pipe installation for vapor drainage from the housing of axial bearing.

Fig. 8. Collecting ring with pipes and connections

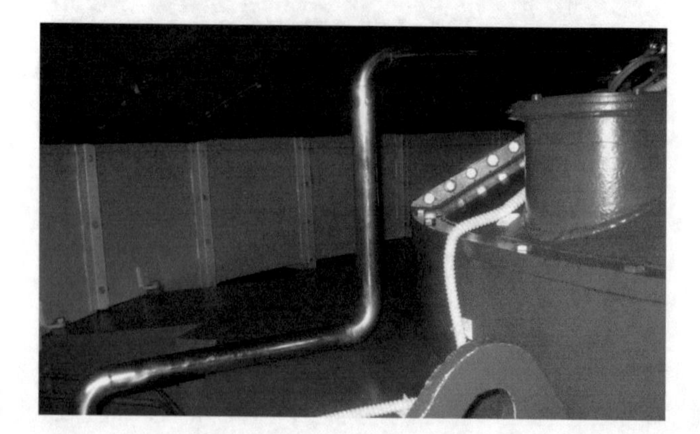

Fig. 9. Pipe installation for vapor drainage from the housing of upper (generator) radial bearing.

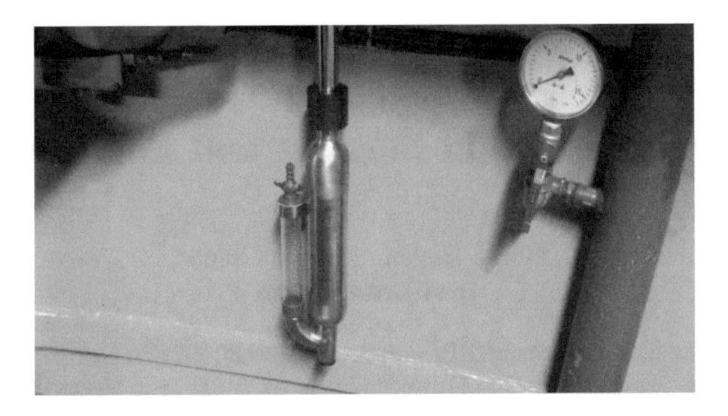

Fig. 10. Oil tank with level indicator and manual valve for oil discharging.

4 Conclusion

During maintenance of hydro aggregates application of new constructive solutions could be used, which will improve the operation of certain mechanical assemblies. One of these new constructive solutions could be a system for draining of oil vapor from the housing of axial and radial bearings of a large hydroaggregates. This system improves health condition for a staff which monitors unit operations or do some maintenance tasks. This system also reduces condensation of oil vapor on the electrical and mechanical equipment, what makes easier access to the equipment during maintenance and monitoring and reduce the possibility of problems in the generator caused by oil vapor.

References

1. Technical documentation of HP Grabovica
2. Technical documentation of HP Grabovica turbines (Litostroj Ljubljana)
3. Isić, S., Đidelija, M.: Maintenance of axial bearing of kaplan turbine, Mach. Des. **9**(4), 151–154 (2017). ISSN 1821-1259
4. Isić, S., Leto, A.: Project documentation of system for drainage of oil vapor. Faculty of Mechanical Engineering in Mostar (2016)

Consideration of Opportunities for the Optimization of Heat Energy Consumption in Industry and Energetics

Stojan Simić(✉), Goran Orašanin, Dušan Golubović, Davor Milić, and Krsto Batinić

Faculty of Mechanical Engineering, University of East Sarajevo, Vuka Karadžića 30, 71123 Istočno Sarajevo, Bosnia and Herzegovina
stojans@modricaoil.com

Abstract. One of the priorities of modern industrial production is the optimization of heat energy consumption. The use of different technical solutions can reduce the heat energy consumption in industry and energetics. The paper considers how the optimization of heat energy consumption is influenced by the following technical solutions: increasing the efficiency of the boiler, returning the condensate to steam boilers and using the evaporator, setting the process parameters of combustion in industrial furnaces, heat insulation of the reservoirs, vessels and installations, application of heat pumps and the use of renewable energy sources and waste materials. Each of the considered technical solutions leads to a reduction in the consumption of heat energy and emissions of waste gases into the atmosphere.

Keywords: Emission · Optimization · Heat energy · Industry · Energetics

1 Introduction

Rationalization of the energy use is a lasting process that comprises its production and consumption as well the protection of human environment. The characteristics of contemporary power plants is high degree effectiveness with increasingly less unit-based energy consumption for the needs of technological process. In the area of industry various technical-technological measures are undertaken aiming at decreasing the energy losses and increasing the energy effectiveness. One of the ways to reduce negative impacts and to positively influence sustainable development is the use of energy effectively. The increase of energy effectiveness leads to a reduction in the consumption of energy for the production of a product, service provided or an activity performed. The consequence is decrease of thermic contamination of the human environment, and in case of obtaining the energy by fossil fuels combustion, also the decrease of the emission of gasses which cause greenhouse effect.

In the paper the measures to optimize heat energy consumption in the industry and energetics are presented.

I. Karabegović (Ed.): NT 2019, LNNS 76, pp. 494–503, 2020.
https://doi.org/10.1007/978-3-030-18072-0_57

2 Overview of the Measures Aiming at the Optimization of Heat Energy Consumption

In the structure of the total energy consumption in the industrial companies, heat energy participates with 40 to 60%. The variety of equipment, its out-of-dateness, great losses in the distribution and use, possibility to use waste heat and so on indicate the need to comprehensively consider and analyze the use of heat energy. The appropriate activities are undertaken in the area of industry and energetics aiming at the decrease of energy losses and the increase of energy effectiveness. In this paper the following energy effectiveness measures will be considered: increasing the efficiency degree of the boiler, returning the condensate to steam boilers and using the evaporator, setting the process parameters of combustion in industrial furnaces, heat insulation of the reservoirs, vessels and installations, application of heat pumps and the use of renewable energy sources.

2.1 Increasing the Efficiency Degree of the Boiler

In the steam-boilers the energy consumption can be reduced by the appropriate measures. The manufacturers of thermo-electric equipment design and make new boilers in accordance with the best available techniques, so that the efficiency degree of these boilers is greater than 95%. In the industry and energetics there is a significant number of boilers which are still in use, and they were put into operation thirty and more years ago. The efficiency degree of these boilers is far less in comparison to the efficiency degree of the newly designed boilers.

In the boilers the energy consumption can be reduced by the appropriate measures. Adjustment of the boiler capacity to the needs of the energy consumers enables saving of heat energy for 1 to 2%. By the adjustment of stehio-metric parameters of the liquid fuel burner combustion in the boilers the decrease of the fuel consumption is achieved for 0,5 to 1%. The increase of flow and adequate distribution of the air in the combustion zone have great influence on the improvement of combustion in the boiler. This is realized by the use of ventilators with differently set hinged lids, as well as by setting the boiler's burner. The improvement of combustion directly influences the increase of the boiler's effectiveness (increase of the steam temperature as well the speed of heat shift in the boiler) [1].

By using the automatic silt cleaning of the boilers the losses at the silt cleaning are decreased and they are up to 2%.

2.2 Returning the Condensate to Steam Boilers and Using the Evaporator

After water-vapor is being used for various purposes in the steam plant, it changes to condensate, which is essentially a high quality warm water. If it does not come to soiled condition during the process, the condensate is ideal to be used for the feed-water of the boiler. In real terms it is impossible to return the entire condensate, a part of the steam can be used in the process inside the plant such as the air moisturizing and steaming. Generally there are also water losses inside the boiler, for example, when silt cleaning is performed.

Returning the condensate represents a relatively great potential for saving the energy inside the boiler-plant. The condensate has the accumulated heat and proportionally 1% less fuel is needed for the 6 °C higher temperature in the feed-reservoir.

In case the condensate under the pressure is being returned to the reservoir, the steam occurs rapidly. That steam should undergo condensation in order that the energy and water are renewed in the reservoir. The classical way to do that is bringing the condensate through the perforated pipe, but more contemporary and effective way for that is the use of fast-condensing deaerator where chemically prepared water, the retuned condensate and expandered steam are mixed.

The system for the use of waste condensate from the production process should be carefully designed, specially taking care of the condensate cleanness. The power plants which supply the process with the heat energy usually have several units for the water-vapor production of various capacities and pressure values. The overheated steam that is produced functions as the power of turbo-generator, while the lower pressure steam is used for the production process needs. In this way the condensate of different quality is obtained. The condensate that is returned from the process, regardless of its current cleanness, can always be soiled with hydrocarbons which could cause huge operating problems by entering into the pipe system of the steam generator. The condensate collected after the condensation in the turbine engine is mainly clean and suitable to be used again.

The practice has shown that these two kinds of condensate (the clean one from the power plant and the potentially contaminated from the process) is not recommendable to be collected into a common vessel. When collecting the condensate it is needed to follow up its quality by periodical laboratory analysis and especially by the processing measuring instruments for the control of the substance of oil in the condensate. In the return flow of the condensate it is desirable to install the instrument for the condensate processing by active coal.

After resolving the condensate collection from the process in a satisfactory way, as well the way of its returning and using in the production process of steam of the needed pressure, balance making and the optimization of evaporator use from the condensate system could commence. In view of this, firstly a reliable and safe condensate returning should be provided, and after that usability of its considerable heat is to be resolved in as effective way as possible. The increase of share of condensate in the feed-water reduces the expenses reflected in:

- decrease of the production and capacities for chemical preparation of water, and
- decrease of heat losses.

From the practical experiences so far the conclusion is that heat losses in the condensate systems are never less than 10%, and often they exceed 30% of the total heat losses.

In the Table 1 there are outlined limits of certain characteristics that the waste condensate should fulfill in order to be used in the steam-boiler of the power plant in a company in Bosnia and Herzegovina.

Table 1. Limits of the waste condensate characteristics [2]

Characteristics	Unit of measure	Limits
Substance of oil and grease	mg/L	0,5
pH Value	–	6,5 ÷ 8,5
Electro-conductivity	μS/cm	20 ÷ 50
Total hardness	°dH	0
Substance of silicon – dioxide (SiO_2)	mg/L	0
Substance of iron	mg/L	0,05

The use of the steam obtained from evaporation of boiling condensate represents relatively great opportunity for decreasing the heat consumption. Percentage of the condensate evaporation depends on the pressures in front of and behind the separator and the degree of the condensate cooling. If the condensate is not cooled the percentage of evaporator is the greatest. In front of the condensate separator there is always higher pressure than the pressure behind the separator. When the separator goes to the open position, the condensate flows from the area with higher pressure to the area of lower pressure. Due to that, a rapid decrease of the condensate pressure occurs. If the condensate in front of the separator is on the temperature above the boiling temperature for the pressure that is behind the separator, the additional evaporation of the condensate will occur. The steam that develops is called the steam evaporator (secondary steam). Upon its development in the condensate conductor, the evaporator flows together with the condensate to the collection reservoir of the condensate, where it flows out through the outlet-pipe to the atmosphere. Flow-out of the evaporator with the condensate is the loss in the heat energy and feed-technical water which is impermissible in the plants.

The evaporator almost always occurs, regardless the type of the separator of condensate. Mainly the least quantity of the evaporator appears when the thermostatic separators of condensate are applied, but there are many cases when the instrument is such that some other type of a separator of condensate must be applied. The separators of condensate represent the necessary part of every system with the steam. Their proper operation is of vital significance for the adequate steam and condensate management because they enable maximal use of the latent heat which influence the regular operation of instruments that use the steam in their operation process, as well as maximal use of the considerable heat.

2.3 Setting the Process Parameters of Combustion in Industrial Furnaces

In the industrial furnaces that operate at high temperatures, the waste heat of smoke gases is from 15 to 55% of the quantity of energy of the fuel used.

Thermo-technical characteristics of the tube furnaces are tested for the case when different number of burners was in the operation. During the testing of thermo-technical characteristics of the tube furnace, the soft paraffin with 46580 kJ/kg of lower heat power was used as fuel. The average consumption of the soft paraffin amounted to 260 kg/h.

In the Table 2 there are outlined the testing results of thermo-technical characteristics of the refinery tube furnace when operating with the different number of burners.

Table 2. Testing results of thermo-technical characteristics of the refinery tube furnace [3]

No.	Magnitude	Unit of measure	Value	
			When 6 burners operate	When 4 burners operate
1.	Temperature of smoke gases at the exit of furnace	°C	530	525
2.	Coefficient of the air excess	–	2,06	1,88
3.	Losses of the heat with smoke gases	%	44,64	40,48
4.	Losses of the heat in the furnace	kJ/h	3,95	3,65
5.	Degree of the furnace use	%	50,36	54,52
6.	Flow of smoke gases from the furnace	m^3/h	4800	4060

The value of degree of the tube furnace use is influenced by the temperature of smoke gases at the exit of furnace that amounts to 530 °C as well the share of oxygen in the smoke gases (11%). Useful heat load of the furnace amounts to 80% of the designed value. The degree of the furnace use is extremely low and amounts to 50,5% and can be maximally increased for 15 to 20% in anticipation of a reduction of the quantity of unnecessary air into the water of smoke gases and decrease of the exiting smoke gases temperature.

Switching over from the operating regime with six burners to the operating regime with four burners and by the regulation of the way of combustion the degree of the tube furnace use is increased for 4,2% and the fuel consumption is decreased for 15 kg/h. Under these conditions the increase of fuel pressure in front of nozzle for 3 bar has occurred. During the furnace operation with three burners there has occurred a complete opening of the burner which regulates fuel feeding to the furnace that resulted in non-possibility to sustain the raw material temperature at the exit of furnace within the needed limits (decrease for 2 to 5 °C).

2.4 Heat Insulation of the Reservoirs, Vessels and Installations

Limited quantities and increasingly more expensive energy led towards that the heat insulation becomes more significant factor in the industry and energetics and the construction of residential and business premises alike. From the aspect of energy effectiveness the basic task of the heat insulation in the industrial and power plants is reduction of the heat losses. Lately the insulation materials were developed and with their use almost the ideal heat insulation is provided, with the heat losses being almost reduced to zero.

When selecting heat insulation in the industry and energetics it is necessary to pay attention to the following distinctive features of the heat insulation materials: heat conductivity, coefficient of the heat flow, heat resistance, density, compressiveness, durability, sensitivity to water and moisture, fire-resistance and that it is adequate sound isolator.

The temperature of the wall of pipe or vessel insignificantly differs from the temperature of a medium in the vessel or the one that flows through the pipeline. Consequently the significant losses of heat into the environment might occur due to uninsulated pipes and vessels. The insulation of boilers, reservoirs, vessels, pipelines and pipe armature is made aiming at reduction of the heat losses in the industry and energetics. The pit and glass wool and polyurethane-foam have the widest application as insulation materials in this area.

Selection of the material is linked with the foreseen place and the way of installment. Apart from the distinctive features mentioned, very often the price of the material is also an important factor for the decision-making. The offer of the heat insulation materials at the market in the recent years is rich. Besides the common and verified classical heat insulation materials, the new i.e. alternative heat insulation materials also come to the market and manufacturers often present them as the environmental materials. These insulation materials are in the phase of exploitation testing and they have not yet reached more significant application in countries in the region.

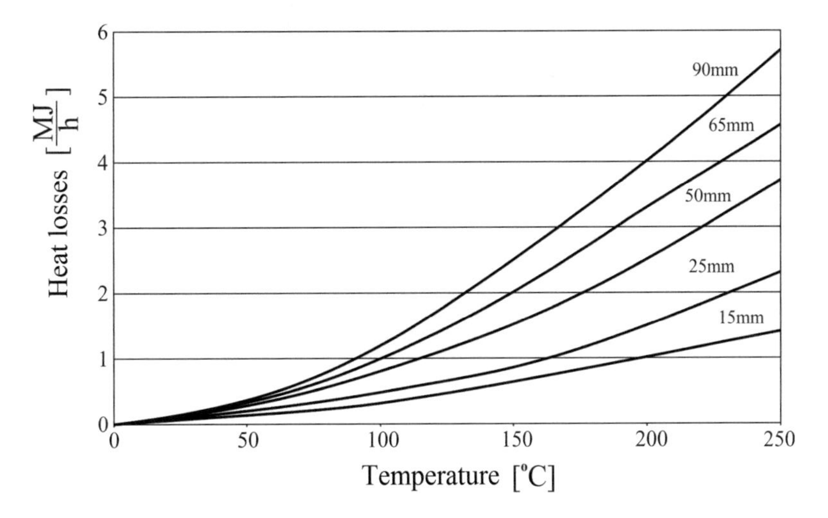

Fig. 1. Estimation of the heat losses through 1 m of the uninsulated pipe

On the Fig. 1 there is outline of the heat loss process through 1 m of the uninsulated pipe depending on the pipe diameter and the temperature of the operational fluid in the pipe.

For the uninsulated pipeline of 100 mm denominating diameter with the flow of overheated water-vapor of 10 bar pressure through the pipeline, the heat losses amount to 1052 W/m. If this pipeline is heat insulated with calcium-silicate (CaSiO4) of 65 mm thickness, the heat losses are decreased and amount to 73 W/m. In the heat uninsulated pipe armature (valves, bolts etc.) the heat losses are equivalent to the heat losses of uninsulated pipeline of the 1,1 to 1,3 m length. When there are uninsulated pipeline peripheries the heat losses are equivalent to the heat losses of uninsulated pipeline of the 0,3 to 0,4 m length.

On the Fig. 2 there are outlined the results of the performed temperature measurements on surface of the pipeline with the external diameter of 60,3 mm and 50 mm thickness of insulation. The overheated water-vapor of 3,6 bar pressure flows through the pipeline. The pipeline was heat insulated five years ago. The results of the performed measurements by the thermo-vision camera show that the temperature on the surface of the heat insulation is lower than the temperature on the pipe surface for approximately 139 °C.

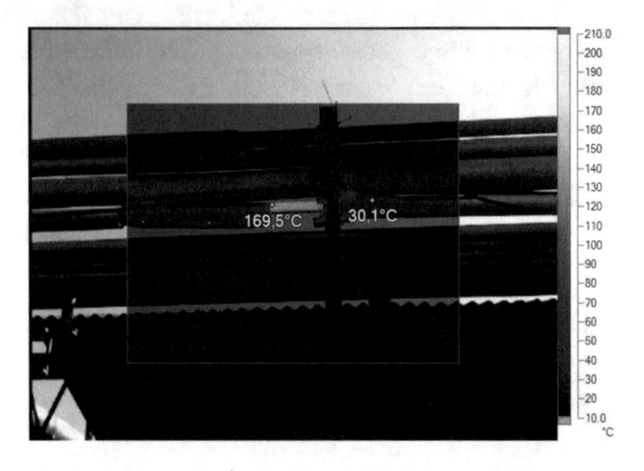

Fig. 2. Outline of the temperature measurement by the thermo-vision camera of the pipeline with and without insulation

The following factors influence the heat insulation effectiveness: compactness (congestion) of the insulation material that is achieved during fitting, presence of water in thermo-insulation, temperature of the environment, atmospheric impacts etc.

2.5 Heat Pumps

The heat pump is used for heating or for the heating and cooling. It can be combined with all the existing heating systems. The advantages of the heat pump's use are the following [4]:

- the costs of heating, cooling and preparation of warm sanitary water are decreased for 75%;
- there are no costs for constructing reservoirs for fuel, chimneys and gas connection;
- quiet and noiseless operation;
- there are no costs for fueling, cleaning of boilers and chimneys;
- independence from the fossil fuels rise in prices (coal, wood, raw oil, gas);
- human environment is not polluted.

For example, the heat pump of water-water system of 550 to 880 kW heat power, depending on the temperature and flow of water on the evaporator whose heat source is the returning cooling water or well water, is used for the heating of construction facilities of 4265 m^2 area that are situated within a factory in the region. Taking into consideration the spent quantity of electric power as well crude water needed for the heat pump operation, during the heating season from October to April 645 MWh of heat energy was produced. The use of the heat pump for heating instead of water-vapor, where mazout is used as the energy source, the consumption of mazout is decreased for approximately 45% that is 38,4 t during the heating season. Besides the economic effect which relates to the decrease of the consumption of mazout there has been also decreased the emission of the greenhouse gases that would occur by combustion of mazout in the steam-boiler.

On the Fig. 3 there is the scheme of the heat pump operation of water-water system.

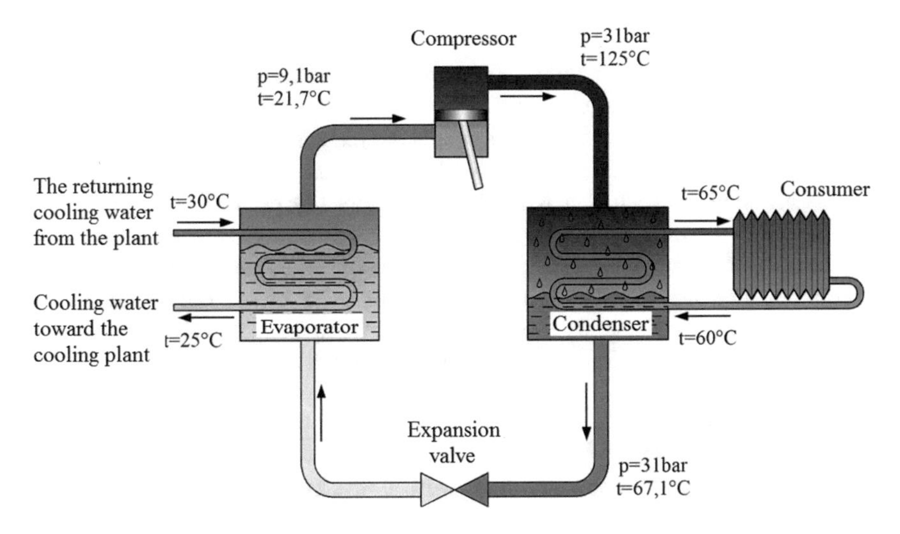

Fig. 3. Scheme of the heat pump operation [4]

2.6 The Use of Renewable Energy Sources and Waste Materials

In the industry and energetics lately there is more and more awareness raising on the need for the use of renewable energy sources, and all in response to the faster increase of the fossil fuel prices and the foresight on their draining in the near future. At the beginning of 2008 the European Union adopted the strategy stipulating that in comparison to 1990 it is needed to provide the decrease of emission of the gases which cause greenhouse effect for at least 20% by the end of 2020, the improvement of energy effectiveness of 20% and participation of the renewable energy sources of 20% in the total energy consumption.

Biomass and biogas are the renewable energy sources which have, and also will have in the following period a significant application as the fuel in the industry and energetics. One of the advantages of the renewable energy sources is that during the exploitation they do not emit the greenhouse gases. Besides the fact that biomass CO2 is neutral, the advantages of its use in the industry and energetics in comparison to the solid fossil fuels are the following:

- preserving the stockpile of the fossil fuels,
- less emission of polluters to the atmosphere (SOx, NOx, CO, particles, etc.),

Biogas is derived from the process of anaerobic digestion of the organic remains from the agriculture and industry. Depending on the application condition it is needed to purify the biogas from carbon-dioxide, hydrogen-sulfide and moisture. By the use of the biogas as a fuel, the consumption of the natural gas can be significantly reduced.

By the use of the waste materials for the heat energy production the two aims are achieved at the same time; the energy is produced and the waste material is taken care of. In order that the waste materials are used for the energy purposes firstly it is necessary to make their preparation. When used as a fuel in the industry and energetics, the waste materials are mixed in the appropriate proportion with the primary fuel. In the recent time there is a greater number of the processing furnaces and boilers designed in the way that the waste materials could be combusted in them. In order that the waste materials could be used as an alternative fuel in the industrial and energetics plants it is needed to obtain the necessary consent by the authorized state institution. If the waste materials are used as the fuel it is needed to additionally perform monitoring of the smoke gases. In the industry of cement, oil industry, thermo-electric power plants and agricultural production the increase of the waste materials application for the heat energy production was registered in the last twenty years or so. The following waste materials are the most frequently used in the industry and energetics for the heat energy production: waste pneumatics, waste oils, oil sludge and silt.

3 The Effects of the Optimization of Heat Energy Consumption to the Environment Conclusion

Until recently one of the rear measures for the air protection was the construction of high chimneys which were providing, thanks to their height, spreading out of contaminating components to large areas and by this their dilution on the needed values. By the optimization of the heat energy consumption the emission of the contaminating components to the air is decreased, and at the same time the economic effect is achieved.

In the industrial plants relatively great energy savings are achieved, especially when the heating devices are well connected to the production process, so that a part of the waste heat from the production process can be usefully used for other purposes. If the temperature level of the waste heat is sufficiently high, the waste heat can be used directly for the needs of heating the premises, preparation of consumable warm water or other production processes. The great fuel savings can be achieved by installing the contemporary low-temperature and condensate boilers which have far greater efficiency degree and better regulation aiming at adjusting to the current heat needs. By decreasing the fuel consumption for the heat energy production there is a direct impact to the decrease of the waste gases quantity, and by this as well to the contaminating components that are emitted to the air.

4 Conclusion

The variety of equipment, its out-of-dateness, great losses in the distribution and use, possibility to use waste heat in the industry indicate the need that the use of heat energy is considered in details.

In the paper there has been considered which technical solutions are most frequently used, aiming at reduction of the heat energy consumption in the industry and energetics. Each of the technical solutions considered influences in varying degrees the reduction of the heat energy consumption which is besides the energy aspect also significant from the environment protection aspect.

References

1. Yang, J.-H., Kim, A.J., Hong, J., Kim, M., Ryu, C., Kim, Y., Park, H.Y., Baek, S.H.: Effects of detailed operating parameters on combustion in two 500-MW coal-fired boilers of an identical design. Fuel **144**, 145–156 (2015)
2. Simić, S., Stanojević, M., Džudželija, Ž.: Considering the possibilities to use waste condensate aiming at rationalization of the energy consumption in refineries. In: SMEITS Beograd, The 27th International Congress on Processing Industry, Processing 2014, Belgrade, 22–24 September 2014 (2014)
3. Elaborate on testing the tube furnace operation, Oil Refinery Modriča (2010)
4. Simić, S., Džudželija, Ž., Ganilović, D.: The use of waste heat by the heat pump in the Oil Refinery Modriča. In: SMEITS Beograd, The 28th International Congress on Processing Industry, Processing 2015, Inđija, 04–05 June 2015, pp. 84–89 (2015)
5. Nyers, J., Nyers, A.: COP of heating-cooling system with heat pump. In: EXPRES 2011, 3rd International Symposium on Exploitation of Renewable Energy Sources, 11–12 March 2011, pp. 17–21 (2011)
6. Wang, D., Sueyoshi, T.: Climate change mitigation targets set by global firms: overview and implications for renewable energy. Renew. Sustain. Energy Rev. **94**, 386–398 (2018)

Reconstruction of Vacuum System on Central Fans on Molins Super 9

Emir Krivić(✉)

Mechanical Engineering, Tobacco Factory Sarajevo, Zagrebačka 25, 71000 Sarajevo, Bosnia and Herzegovina
krivic.emir@bih.net.ba

Abstract. In this experimental work, possible solutions to the problem of damage to the impeller of the centrifugal (radial) fan are provided. Due to the impact of the impeller, a breakdown (damage) of the casing occurred, and then to the bending of the shaft of the fan assembly. In the first part of the paper, as a proposal of the solution, the reconstruction of the axle mounting on the basis of the installation of double bearings of the same type and the structural change of the blades of the rotor was done. In the second part, the process of reconstructing the impeller of the fan with modified production technology was worked out to solve the problem of balancing the mass. The study defines the methodology to be followed through the design documentation and the Instructions on the assembly process, which must be strictly observed. By complying with the prescribed technological procedure, the lifetime of the fan will undoubtedly be prolonged.

Keywords: Centrifugal fan · Impeller (rotor) · Blade wear · Effort · Balancing of rotational masses · Loss of vacuum

1 Introduction

Concept and definition of the fan

The fan is an electrically powered device that is used for airflow for the purpose of achieving physical comfort (primarily for heating, but also for other purposes such as ventilation, exhaust, cooling or any transport of gases). Mechanically, the fan may be any device with a blade or more that rotates and uses to produce air currents. The fans produce high-volume and low-pressure air flows in relation to gas compressors producing high pressures with a relatively low volume. The fan blade will often rotate during air flow exposure, and devices that use it, such as wind turbines, often have a similar fan design. Usually they are used for climate control, cooling systems, personal comfort (e.g. table electric fan), ventilation (e.g. exhaust fan), dust removal (e.g. vacuum cleaner), drying or providing air for the fire [1–12]. Cleaning of ventilation channels and their inspection presents popular applications of modern robotics [1].

Types of fans: Fans can be constructed in many different ways. In the household you can find fans that can be placed on the floor or table, hang on the ceiling, installed in a window, wall, roof or chimney. They can be found in electronic systems such as computers where cooling circuits, or in various appliances, such as dryers or heaters.

I. Karabegović (Ed.): NT 2019, LNNS 76, pp. 504–513, 2020.
https://doi.org/10.1007/978-3-030-18072-0_58

They are used for cooling in air conditioners, and in SUS motors where they are driven by a belt or directly by a motor. Fans create a breeze of fresh air, but do not lower the temperature directly [12]. Three main types of fans: axial, centrifugal (radial) and transverse flow (tangential).

Centrifugal (typical) fan

The centrifugal fan has a movable component (rotor) consisting of a central shaft around which a set of blades or ribs is placed. Centrifugal fans blow out the air at the right angle from the air supply and turn the same to the air outlet. The rotor turns around causing the air to enter the fan near the shaft and move the steering shaft from the shaft to the hole on the fan housing. The basic construction consists of a hub to which the blades are positioned axially, which precisely allows the axial entrance and the radial outlet of the working medium. Centrifugal fans produce more pressure for the given volume of air and are used where higher pressure is needed, such as for example, hair driers, air mattress pump, etc. Usually they are noisier than axial fans [10]. Ventilators are still the primary mode of cooling primarily due to price. In addition to the price there is also the simplicity of the system.

The fan has several advantages compared to air conditioners:

- simple installation, low consumption, easy maintenance, portability and affordable price.

The assembly of the fan is very simple and usually a plain screw-driver and a few minutes of work are enough. One of the biggest advantages of a fan in comparison to the climate is low consumption, but I note that the consumption itself depends on the size and power of the fan. The point is as follows: even with the largest fan, the consumption will be lower than in the weakest climate. Since the fan clearly shows impurities, which is easy to remove, the fan is generally healthier to use than air conditioners. The fan is easily portable. Apart from being able to dislocate from one room to another, it is easily detachable and portable to another destination. Although there are portable air conditioners, it is still a bit harder to operate with them than with fans. Classic air conditioners stay where we install them [11].

Disadvantages of the fan: The fan will never be able to cool the room as well as the climate. The reason is that the climate and fan work differently. The air conditioner takes warm air from the room and turns it into a cold, while the fan only allows for better air circulation in the room [11].

2 System Function

The work defines the cycle of a large fan with central dust separation and the reconstruction of the air transport system-vacuum system as a proposal of the solution. The large fan system has the function of maintaining a suppression in the cycle of cigarette production with the task of sucking and applying tobacco to the cigarette paper tape. Transport air with fine tobacco particles passing through the tape, then through the fan and the silencer is sent to the piping system towards the central bag filter. The original factory solution for the Molins machine is that it has the separate particle separator with free venting of pure air into the atmosphere [5].

Problems and problem analysis: After a certain time of operation of a large fan, the impeller (rotor) (see Fig. 2) of the fan has been damaged. The breakdown of the enclosure occurred due to the blow of the fan rotor. The impeller hit the chassis, as the ball bearing on the side of the impeller collapsed. Because of the consequences of the impact, there was a certain bending of the shaft (see Fig. 1, revision of the drawings MS9-006).

Rolling bearings (ball and roller) due to loads are subject to fatigue, resulting in fine cracks, and the surface is peeling. The weight of the bearing depends on the load, the speed, the frequency and the direction of the load, and at the same time is a function of the speed, and it is determined only by the laboratory. The same bearings, the same load and the life span range may have a ratio of 1:40 due to the inhomogeneity of the material. The dynamic power of wearing C is the maximum load for 500 h of operation and 33.3 min^{-1} and 10^6 revolutions [5, 6]. Mounting conditions are correct. The wearing power of the bearing is inversely proportional to the lifetime of the bearing.

The breakdown of the bearings could have come for the following reasons:

I. Insufficient constructional solution for axle loading in relation to load. The solution is given as a beam with overhang with loads on the same one (one pulley is a drive pulley, and on the other fan impeller).
II. Incorrectly calculated force on the fan rotor, and hence the dynamic bearing load, which influenced the reduction in the lifetime of the bearings.
III. Inbalance of rotational masses that increased over time due to uneven wear of blades.
IV. Incorrect installation of the system from the aspect of the need to rotate the rotary axis of rotational masses.
V. Insufficient constructional design of the blades from the aspect of resistance and the impeller of the fan from the aspect of balancing the masses.
VI. Longer work than foreseen and factory fault in the bearing.

The disruption in the performance of the under pressure resulted in poor quality tobacco application on the tape, and for this reason the proper operation of the large fan is an important segment in the production (Table 1).

Table 1. Distribution of the fan according to the size of the pressure

Pressure size	Fan type	Unit of measure
6–100	Low pressure	mmH_2O (VS) – column of water
100–300	Secondary pressure	
>300	High pressure	
0.1–30 atp	Blowers	atp – technical atmosphere

Based on the theoretical knowledge, I concluded that it is a centrifugal, that is, a radial high-pressure fan [8]. The fan impeller (see Fig. 6, drawing MS 9-010A), besides physical damage due to impact, had unevenly worn blades. The blades are flat with a tilting backward, and the direction of rotation is opposite to the direction of the clock movement. The tip of the blade is much more worn out of the root due to erosion caused by small (large) particles of tobacco (the input surface has been damaged due to turbulent flow of the working medium).

3 Research-Investigation

For this research a differential manometer was used as a measuring device for precise measurements of speeds and pressures at the sites of suction and thrust and own acquired, long-term engineering knowledge. There is no doubt that in the near future the application of such industrial and service robots [2] will increase the reliability of the manufacturing process, reduce the time to create the finished product, and enable adapting and precision in performing tasks that exceed human capabilities. The success of technology involves the advancement of technical solutions in the field of automation of technological processes and the application of intelligent systems in various industrial branches, which also includes our metal processing industry [4].

Intake system: This system is divided into two chambers of different values of sub-assemblies:

- Suction chamber left p = 35 in. H_2O (889 mm H_2O = 9307.5 Pa)
- Suction chamber right p = 30 in. H_2O (762 mm H_2O = 7978 Pa) [6].

The manometers are not calibrated, and the indication of the value of the suppression is shown on the manometers. Effort (supply height) is an increase in energy expressed by the height of the liquid's puff (fluid), that is, the energy that the unit mass of the fluid gets through the working space of the pump [8]. Differential pressure gauge is a device that measures the speed and pressures at the point of suction and thrust, and when the Δp starts to fall, the fan no longer works with the necessary pressure. Since the wheel blades are discharged unevenly, there is a loss and failure of bearings, resulting in a loss of vacuum [7].

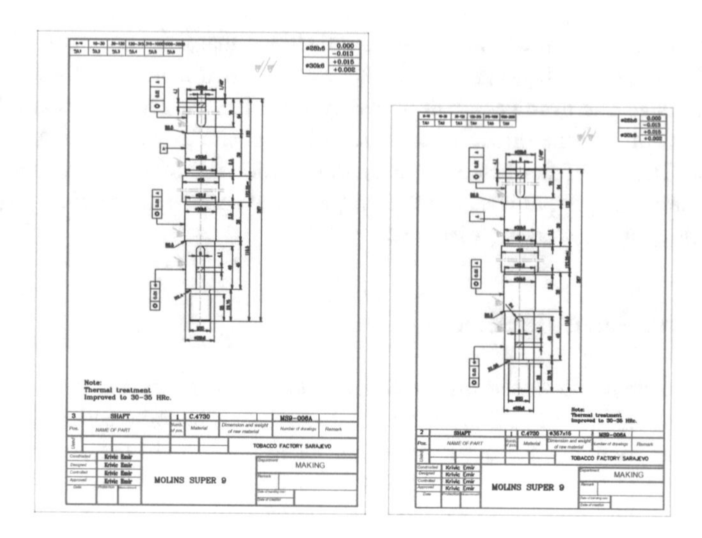

Fig. 1. Worksheet drawings of the fan shaft [5]

Based on the given analysis and the fact that the exact calculation of the force on the fan's impeller's can not be done without Software that can not be provided without investment, and the fact that no conditions have been created for the experimental measurement of the force component, I propose two solutions to this problem:

- Reconstruction of the axle mounting based on the installation of double bearings of the same type and structural shifting of the blades to reduce the resistance, as well as the reconstruction of the impeller with the changed technology of making the solution of the problem of balancing the masses.
- Axis mounting with the installation of radial-axial bearings type B7206C.TPA.P5 with oil lubrication with structural blade change and reconstruction of the impeller to solve the problem of balancing the mass.

Both solutions require the balancing of the rotational masses after the installation of the pulley, as well as the correct installation of the system from the aspect of the rotation of the rotary axis and the direction of the lateral plane of the impeller and the pulley on the shaft, respectively the coaxiality of their axes. Shafts as one of the main structural elements are exposed to different forms of dynamic loads, which with the characteristic of the shaft can cause critical conditions [3].

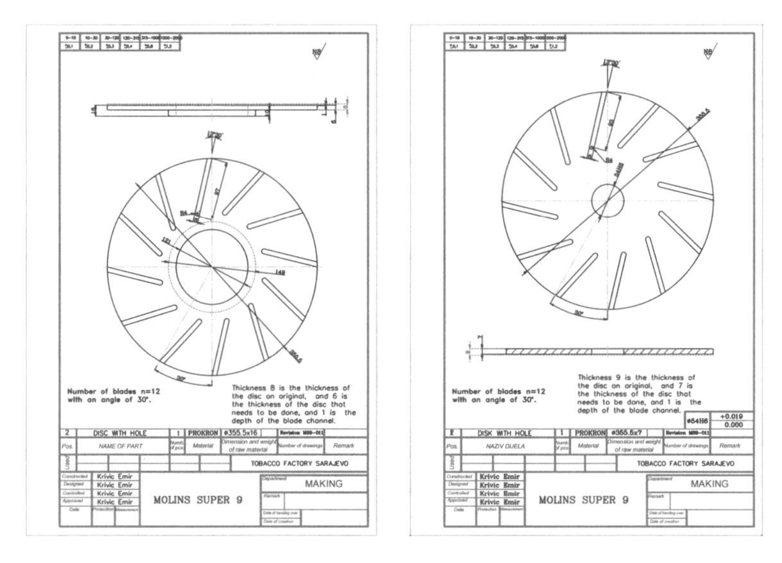

Fig. 2. Worksheet drawings of the disc with a hole and a hub disc [6]

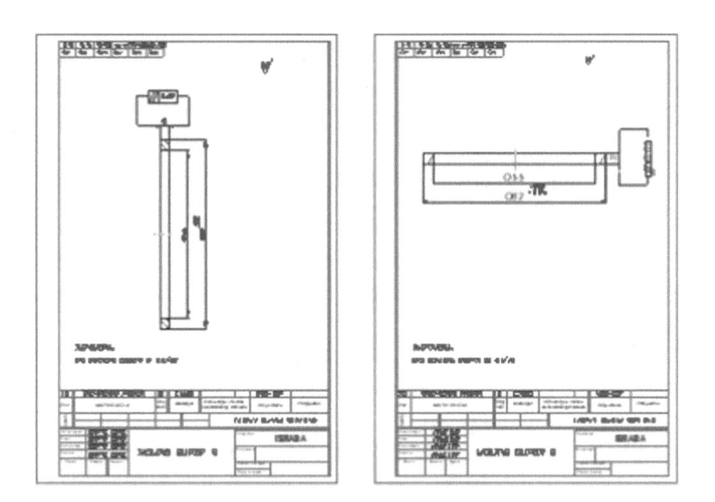

Fig. 3. Worksheet drawings of distance rings [5]

4 Results and Discussion

As the blades of the impeller are unevenly discharged, the mass loss and failure of the bearings result, resulting in a loss of vacuum. Therefore, I propose a measurement in the following way:

The newly installed fan rotor is to measure Δp and monitor its change during operation, as the decrease in Δp will indicate that the wear of the blades has occurred. Worn away blades were caused by mass unbalance at the working speed n = 7000 min^{-1}, resulting in the occurrence of large centrifugal forces, which creates an additional burden on the bearings and creates the damage. It is not the goal that the bearing lasts longer, but that the fan meets the technological parameters prescribed by the documentation, which are the flow of the medium Q = 35.3 m^3/min at pressure p = 145 mm H_2O [5, 6]. After considerable effort reduction Δp it is necessary to stop the fan, remove the rotor and measure Δp to determine the critical point on which the fan does not meet the technological parameters prescribed by the manufacturer.

In order to prevent further wear of the blades, and therefore the breakage of the bearings, I propose the following: The impeller fan and work blades (see Fig. 7) are always made of the same material, because different materials do not wear out. All of the above points to the conclusion that the fan operation should be monitored to define a critical point. The indication of damage to the blades and the appearance of the vibrations of the complete fan, precisely because of the unbalance of the mass of the blades, and the occurrence of the change in effort Δp can indicate a recent premature breakage of the bearings or the failure (malfunctions) of the fan.

Since from the point of view of maintenance of the system I give priority to the first possible solution (see Fig. 5), within this work, a construction solution for the dual bearings, as well as instructions for the assembly process and balancing of the rotary masses, is given. The second solution is basically based on the use of radially-axial bearings and bearing covers with sealing seals and oil-pouring and discharging openings.

Manual (instructions guide) - Instructions before installation and mounting bearings:

1. Mount the bearings by trimming the thickness of the distance ring (see Fig. 3) on the subassembly, so that the assembly is made without gap and axial strain on the bearing rings.
2. Before installing the hollow shaft in the hub (see Fig. 5) in the fan housing, perform a technological (temporary) assembly of the sub-assembly of the axle of the fan with the impeller and the fan pulley.
3. After assembling the subassembly from point 2, carry out balancing of the rotary masses.
4. Before final assembly, check the installation of the fan housing and the horizontal axis of the fan.
5. After finishing, check the balancing of the rotary masses.
6. Tightening the belt should be done according to the manufacturer's instructions, in order not to slip and do not create a large force acting on the bearings. A correctly tensioned belt should have a stroke of 6 mm (see Fig. 4) at half the range between the pulleys.

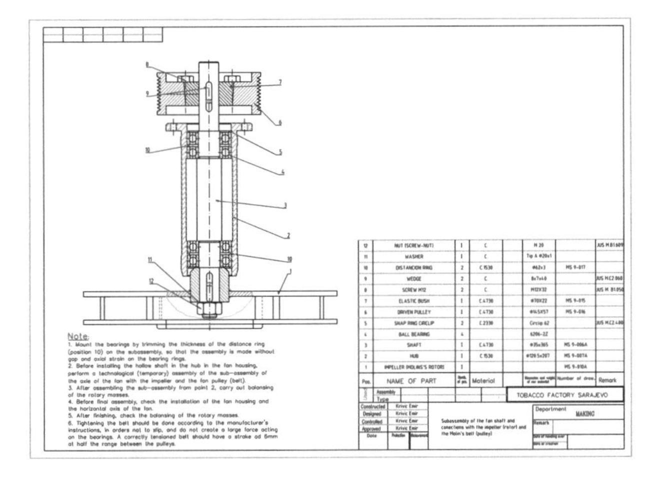

Fig. 4. Subassembly of the fan shaft and connections with the rotor and Molins fan pulley [6]

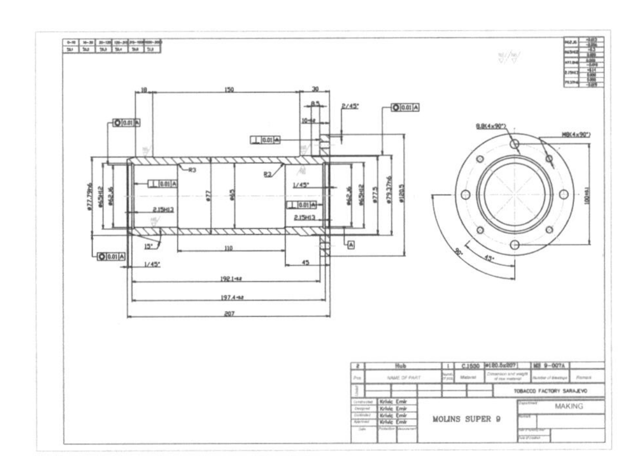

Fig. 5. Worksheet drawings of the hub [5]

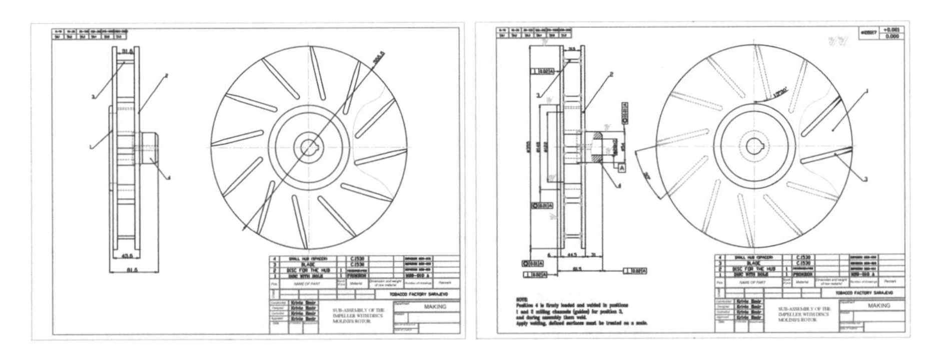

Fig. 6. Sub-assembly drawings of the impeller [6]

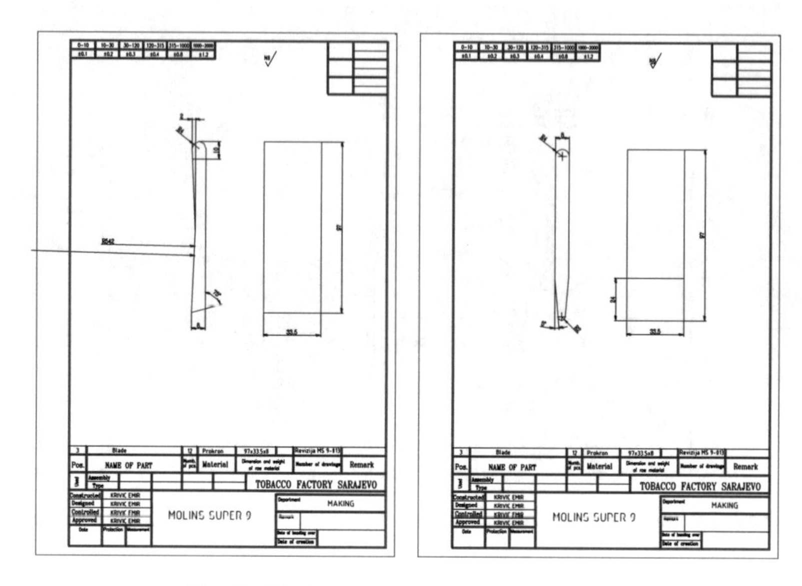

Fig. 7. Worksheet drawings of fan blades [5]

5 Conclusion

Due to the impossibility of an exact calculation of the force on the impeller fan, I did not carry out the calculation of the dynamic load and lifetime of the bearings, but I proceeded from the assumption that the manufacturer of the equipment made a correct calculation, on the basis of which he made the choice of bearings.

This solution will prolong the lifespan of the bearings, and a less wear on the blades will be expected, but the problem of wearing blades and loss of effort will still be present, and my instructions should be followed in the exploitation process.

Both proposed solutions will extend the life of the fan, if comply with the requirements are given on the construction documentation and the installation instructions. Particular attention should be paid to balancing the rotating masses, as well as to the horizontal rotation of the fan in the process of installing the device on the stand.

References

1. Bubanja, M., Markus, M.M., Đukanović, M., Vujović, M.: Robot for cleaning ventilation ducts. University of Montenegro Podgorica, Montenegro
2. Karabegović, I., Mirza, R.: Automation of the welding process by use of industrial robots. University of Bihać, Bosnia and Herzegovina, University of Engineering and Technology, Lahore, Pakistan
3. Husak, E., Haskić, E.: Analysis for torsional vibration of the engine connected with propeller through pair of gears. University of Bihać, Bihać, Bosnia and Herzegovina

4. Doleček, V., Karabegović, I.: Roboti u industriji, Bihać. Društvo za robotiku Bosna i Hercegovina, Tehnički fakultet Bihać (2008)
5. Maintenance book and instructions for monitors parts # 1 "Molins Super 9", England
6. Book of maintenance and repair of spare parts # 2 "Molins Super 9", England M
7. Ristić, B.: Pumps and Fans. Scientific Paper Belgrade, Belgrade (1987)
8. http://www.rgf.rs/predmet/RO/V%20semestar/masine%20i%20uredjaji%20za%20eksploataciji%20nafte%20i%20gasa/Predavanja/11-Pumpe.pdf
9. Academic Thought, Mechanical Manual, Belgrade (2007)
10. https://www.nabava.net/clanci/saveti/prednosti-i-nedostaci-ventilatora-u-usporedbi-s-klimom-306t6
11. https://www.nabava.net/clanci/saveti/prednosti-i-nedostaci-vodenog-hladenja-za-pc-777t6
12. https://en.wikipedia.org/wiki/Ventilator

Parametric Analysis of Silica Fume Effects on the High Strength Concrete Composition

Azra Kurtović(✉) and Naida Ademović

Faculty of Civil Engineering, University of Sarajevo,
Patriotske lige 30, 71000 Sarajevo, Bosnia and Herzegovina
azra.kurtovic1@gmail.com

Abstract. The silica fume is obtained by the reduction of high purity quartz by coal in electrolytic furnaces in the production of silicon and ferrosilicon alloys and consists of very fine spherical particles containing at least 85% of amorphous silicon dioxide. It is used as a mineral additive of type II for production of cement concretes. The paper presents results obtained from an experiment campaign. Influence parameters and their significance in the optimization of the concrete mixture composition prepared with mineral additives and admixtures in terms of the cement amount, water/binder ratio and compaction were analyzed. After obtaining satisfactory results for the usability of the foreseen silica fume, a choice of stone aggregate was conducted with respect to the stability of the concrete mixture and the strength of concrete up to 28 days. In the analysis of the experimental results, it was emphasized that the compaction grade of the concrete mixture is the primer influencer on the formation of the structure of concretes of higher strength.

Keywords: Silica fume · High strength concrete · Optimization · Mineral additives · Admixtures

1 Introduction

Microsilica was first collected in 1947, while the first tests were conducted already several years later, specifically from 1950 to 1952. From the industrial recovery in the early 1970's there has been an increase use of microsilica in concrete followed by the development of adequate standards. More than 20 million m^3 of microsilica concrete is produced annually [1]. Application of silica fume as a mineral additive to concrete has been in the focus of research for over two decades. The publication of the European standard was certainly a very important step, confirming the importance of this material and the necessity of setting standard criteria.

Silica fume represents very fine particles of amorphous silicon dioxide collected as a by-product of the smelting process used to produce silicon metal and ferro-silicon alloys. These industries were firstly located in Scandinavia, North America and Canada where silica fume has been employed as concrete additives for years. Around the beginning of 1980 a new chapter in the history of concrete technology was opened with the introduction of silica fume under the brand name "Microsilica". Silica fume is represented as dusty particles which are finer by a factor of 100 than concrete grains.

I. Karabegović (Ed.): NT 2019, LNNS 76, pp. 514–522, 2020.
https://doi.org/10.1007/978-3-030-18072-0_59

Their high share of amorphous silicic acid and extremely high specific surface gives this puzzolan enormous reactivity. The puzzolanic effect is based on the fact that the highly alkaline porous surface of the fresh binder material makes the silicon and aluminum oxide of the puzzolansto react with the same products as happens after the hydraulic hardening of cement (CSH and CAH phases) [1]. However, these new products have a much higher compressive strength compared to the original material. These physical, chemical-mineralogical and puzzolanic characteristics altogether make the concrete significantly denser and resistant to pressure [2]. Today other names are used for silica fume like condensed silica fume and microsilica.

How does actually microsilica transform concrete? First of all, crystals produced by the hydration reaction grow away from each cement particle. Then finer, stronger crystals grow from both cement particle and "Nucleation Centres" of well dispersed microsilica. This is followed by a pozzolanic reaction; SiO_2 reacts with $Ca(OH)_2$. Once the hydration is complete, the extended crystal structure remains weak and permeable. As a result, a strong, impermeable dense microstructure develops; dense paste structure with strong aggregate bond - improving continually over the years.

In order to improve concrete characteristics concrete additives are used. Inorganic and organic substances are added to the mixture. In [3] and [4] the inorganic additives are divided into two groups:

- Type I. Inert additives e.g. stone powder; according to [5] or pigments [6]
- Type II: puzzolanic or latent hydraulic additives e.g. fly ash [7], Ground granulated blast furnace slag [8] or Silica fume [9].

The inert additives (Type I) first of all improves the grain structure and the workability of the concrete. These fillers are typically used for self-compacting concrete (SCC).

Additives Type II works on a chemical basis, they exert a much greater influence on the material and its characteristics. Their influence is implemented through the hydration process of cement, thus increasing its strength.

2 Experimental Work: Selection of the Petrographic Aggregate Type for Concrete Composition

Bearing in mind that type of rock mass from which the aggregate is produced as well as the physical and physical-mechanical properties of aggregates have a significant effect on the properties of concrete, during the planning process and experimental campaign, the investigation and later on acceptance of the fractionated aggregate was primarily done.

An analysis of the available documentation of the control testing of the potential production of quarries was carried out, with particular reference to the results of the aggregates' mechanical properties, as well as physical and physical-mechanical characteristics of the stone from which the concrete aggregate is crushed.

Based on the studied documentation, a selection of three fractionated aggregates for the preparation of higher strength concrete was performed. All selected aggregates were of natural nature: from riverbed and rock mass, one of sedimentary (limestone) origin and the other of eruptive (diabase) origin. On the sampled aggregate fractions, control standard tests were performed. Geometrical properties (granulometric composition by sieving method and grain shape - shape index) and physical properties of aggregates (determination of density and coefficient of absorption) were checked. In addition, on the rock mass samples, a test of apparent density (volume mass with pores and cavities) and compressive strength of the stone from which the aggregate was produced were conducted. All tested aggregates met the required technical specifications as per the above standards.

Once tests on individual aggregate fractions were carried out, design granulometric curve mixture was conducted. The maximum aggregate size of the aggregate (D_{max}) has a significant effect on achieving the structure of higher strength concrete. Therefore, concrete with D_{max} = 8 mm and D_{max} = 16 mm was produced. Results of the characteristic data for stone, fractionated aggregates and designed granulometric curves 0–16 mm and 0–8 mm are given in Table 1 and Fig. 1.

Table 1. Results of the characteristic stone and fractional aggregate properties

Aggregate type	Apparent density of stone [kg/m^3] [10]	Compressive strength [N/mm^2] [11]	Absorption of unclean aggregates [%] and grain density [kg/m^3] [12]	Participation of the declared factions in the designed mixture 0–16 mm [%]	Participation of the declared factions in the designed mixture 0–8 mm [%]
Eruptive rock	min. 2915 max. 3019 aver. 2945	min. 92.8 max. 150,0 aver. 119,3	0–2 mm [4,1%] [2959 kg/m^3]	0–2 mm [35%]	0–2 mm [50%]
			2–4 mm [1,2%] [2940 kg/m^3]	2–4 mm [15%]	2–4 mm [15%]
			4–8 mm [0,6%] [2912 kg/m^3]	4–8 mm [15%]	4–8 mm [35%]
			8–11 mm [0,8%] [2914 kg/m^3]	8–11 mm [20%]	
			11–16 mm [0,7%] [2923 kg/m^3]	11–16 mm [15%]	
Sediment rock	min. 2719 max. 2783 aver. 2750	min. 2719 max. 2783 aver. 2750	0–4 mm [3,1%] [2761 kg/m^3]	0–4 mm [46%]	0–4 mm [65%]
			4–8 mm [0,4%] [2757 kg/m^3]	4–8 mm [14%]	4–8 mm [35%]
			8–16 mm [0,8%] [2765 kg/m^3]	8–16 mm [40%]	0–4 mm [65%]

(continued)

Table 1. (*continued*)

Aggregate type	Apparent density of stone [kg/m^3] [10]	Compressive strength [N/mm^2] [11]	Absorption of unclean aggregates [%] and grain density [kg/m^3] [12]	Participation of the declared factions in the designed mixture 0–16 mm [%]	Participation of the declared factions in the designed mixture 0–8 mm [%]
River Neretva	No data. Material for the river debris		0–4 mm [2,4%] [2835 kg/m^3]	0–4 mm [50%]	0–4 mm [68%]
			4–8 mm [0,7%] [2740 kg/m^3]	4–8 mm [15%]	4–8 mm [32%]
			8–16 mm [0,2%] [2751 kg/m^3]	8–16 mm [35%]	

All designed granulometric aggregate curves for the production of concrete have a continuous flow and are located in the area “3”, suitable for concrete mixing according to the valid technical specifications [13].

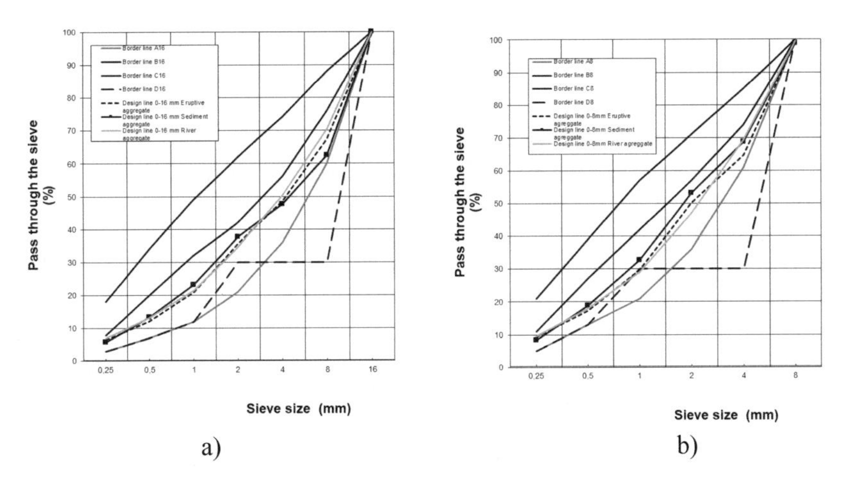

Fig. 1. (a) Designed granulometric curves 0–16 mm; (b) Designed granulometric curves 0–8 mm

3 Choice of Other Components for Concrete Mixing

The appropriate choice of the type and amount of cement has to be done regarding the class, hydraulic heat and chemical resistance which is dependent on the type of concrete being designed. In this case the only condition was the achievement of higher concrete grades, Portland cement type CEM I class 52.5 with normal strength gain

(N) was chosen. Verification of standard properties corresponded to declared properties by the manufacturer. Concrete preparation was carried out with water from the city water supply which, without further testing, can be used for concrete production.

With respect to the chemical additives, taking into account the necessity of reducing the water-cement factor, a super-plasticizer based on high-concentrate poly-carboxylated (PCE) for concretes with extended time of installation and faster strength development was used.

The first innovative high-performance plasticizers based on polycarboxylate ethers were produced in the 1990's in Asia. These additives offer extremely strong liquefying properties and/or reduce the water requirement. The liquefaction lasts considerably longer than previously used plasticizing agents (ligno-, naphtaline and melamine sulphonates) have been able to achieve. Added to which, specially composed PCE polymers virtually eliminate the side effect of retardation, unlike these classic concrete additive agents. As a result of this higher concrete strength can be obtained during the early stage which was previously unattainable. Without reservation, this development can be seen as a quantum leap in concrete additive technology and, as a result, in modern-day concrete technology.

The addition of a PCE-based high-performance plasticizer produced significant improvements in both the fresh and hardened concrete [2].

During the first phase of the experimental work (i.e. in the selection phase of concrete aggregates), the filter silicate fume microsilica conformed to [9], density of 250 kg/m^3 to 350 kg/m^3. At this stage in order to better monitor the effects of a particular type of aggregate, it was very important to know the other components and they had to be of controlled composition, such as factory control of cement production, chemical and mineral additives and the application of water from the city water supply.

In the second phase of the experiment, with the determined components the evaluation of usability and determination of activity index (activity index – measurement of the effect of silica fume on the compressive strength of mortar) of the silica fume from the B.S.I. d.o.o. plant in Jajce were conducted.

Since the quality requirement for the application of the silicate fume [9] is given in respect to the chemical composition, an analysis of declared values with regard to the content of SiO_2, SO_3, Na_2Oeq, Cl^- and free CaO was done. After obtaining a positive opinion, the required physical properties (specific areas and activity index) were checked.

The microsilica is highly reactive in cement-binding systems. In hardened concrete, it contributes to the improvement of impermeability/compactness, strength and resistance against chemical and mechanical activities. Addition of the microsilica to the concrete mixer in both cases was done before the cement. The maximum permissible dosage of the microsilica is in compliance with [4] of 10% by weight of cement. The activity index is determined [9] as the ratio (in percent) of the compressive strength of standard mortar bars, prepared with 90% test cement plus 10% of silica fume per mass of total binder, to the compressive strength of standard mortar bars prepared with 100% test cement, tested at the same age (28 days). Preparation of standard mortar bars and determination of the compressive strength was carried out in accordance with the method described in [14]. The mortar containing silica fume was mixed with an amount of superplasticizer conforming to [15] in order for the mortar to have a consistency equivalent to the reference mortar when tested by the flow table method.

4 Analysis of Test Results of Concrete Mixtures and Hardened Concrete

Concrete mixes were individually tailored for each type of aggregate in a laboratory mixer, strictly respecting the order of component materials dosage into a mixer. The materials were dosed into the mixer in the following sequence: first pouring aggregate by the fractions (large, small), then adding the mineral additive - the silica fume. Dry mixture of aggregates and silica fume was stirred for about 1 min, then cement was added. After that 2/3 of water was poured, while with the last third amount of water chemical additive was added. The prepared concrete mix in the laboratory mixer after adding the chemical supplement was stirred for 3 min.

For each aggregate three compositions with different amounts of cement were made.

The followings elements were taken into account for the final selection of the fractionated aggregate:

- the contribution to the stability of concrete, from the point of view that in the formation of the structure of concrete of higher strength the inclination of concrete to segregation – bleeding will have an effect as a special case of sedimentation.

In this type of segregation of the concrete mixtures, apart from the separation of water which is floating on the surface, it may happen that water is separated beneath the larger grain aggregates, which then prevents the connection of the cementitious stone with the aggregate (transit zone). This is of great importance for concrete with higher strengths especially with the application of mineral additive – silica fume.

- Aggregates packed of the selected granulometric mixture and the homogeneity of the concrete mixture determined by the visual inspection upon cessation of machine operation and by measuring the concreteness of concrete with consistency;
- Scattering of compressive strength of concrete. In all the concretes, the extraordinary sensitivity of the composition to the selected components was noted in respect to the way and duration of the compaction, which was obviously more demanding and more complex in this type of concrete than the normal concrete.
- Strength increase with an accent on the compressive strength after of 28 days.

Testing of rheological properties of fresh concrete (consistency, temperature, air entrained and density) as well as hardened concrete (density and compressive strength) were performed on the prepared test specimens. Properties of hardened concrete were determined on three cubes of 150 mm long edge for concrete aged 7 and 28 days, and additionally on three cylindrical (diameter 150 mm and height 300 mm) samples at the age of 28 days. However, after testing the cylindrical samples, it was found that when the concrete mixture was poured in the mold, was insufficiently compressed by the concrete vibrator and the results were discarded as impracticable (a decrease in strength was 35%–50% for concrete with granulation of 0/16 mm and about 30% for concrete with granulation of 0/8 mm in comparison to the cube samples). These results are not shown in this paper. During the preparation of the cylindrical samples it was noted that the duration of vibration has a direct impact on the compressive strength of concretes of higher strength.

Insufficient compaction of concrete results in drastic differences in the obtained strengths of concrete of higher strength. It is necessary to take care that the samples are made with extreme care. Namely, it is necessary to approach the sample compaction in such a way that it is being poured in the mold in layers, and each layer of concrete has to be compacted until the operator by visual inspection is assured that almost all captured air is completely vanished from the concrete. Of course, in this way the compacting of concrete lasts considerably longer.

After analyzing the obtained experimental results, the river fractional aggregate was selected.

5 Concretes with Silica Fume B.S.I. d.o.o. Jajce

As the concrete for the selection of aggregates was made with the harmonized silica fume of the Microsilica, in the further stage, it was necessary to carry out the preparation of concrete with the same aggregates but with the application of silica fume from B.S.I Jajce d.o.o.

One of the goals was also to confirm the decision which was made, i.e. whether the same visual observations and the choice of the best aggregate would be identical with the silica fume BSI from Jajce. Therefore, concrete mixtures were made with all three aggregates in question. One mixture with 0/16 mm and the other with 0/8 mm granulation was made. Concrete mixtures were done with the same composition but with modification of the silica fume type. The workability of the comparable concrete mixtures was the same, actually the same measure of consistency by flow table was obtained [16].

On the other hand, these tests had to confirm the efficiency, or activity of silica fume from B.S.I d.o.o. Jajce on a concrete mixture. It should be noted that for this testa prerequisite was confirmation of the required activity on cement mortar as defined by the standard [9]. The silica fume was dosed into the concrete mix in an amount of 10% on the cement mass. The sequence of dosing of the component materials in the mixer was completely the same as for the preparation of concrete for the selection of aggregates and the dosing of a matched microsilica. Due to the fact that the amount of silica fume was doubled the required amount of water had to be reduced, (maintenance of water cement factor), resulting in an increase of chemical additive to the concrete.

Results of individual tests of concrete samples are given in Fig. 2. Concrete samples designated with letter E represent eruptive aggregates-Diabase; letter S represent sedimentary aggregates-Limestone; letter R stands for river aggregate. Amount of cement varies and for E1/A, S1/A, R1/A amounts to 480 kg/m^3 of concrete; E2/A, S2/A, R2/A, E2/J, S2/J, and R2/J amounts to 510 kg/m^3 of concrete; E3/A, S3/A, R3/A, E3/J, S3/J, and R3/J amounts to 535 kg/m^3 of concrete. Letter J represents concretes made with silica fume from B.S.I d.o.o. Jajce and A represents the conformity silica.

The obtained test results confirmed the suitability of the silica fume produced in B. S.I. d.o.o. from Jajce for the production of concrete of higher strengths and the advantage of the use of river aggregate in concrete production.

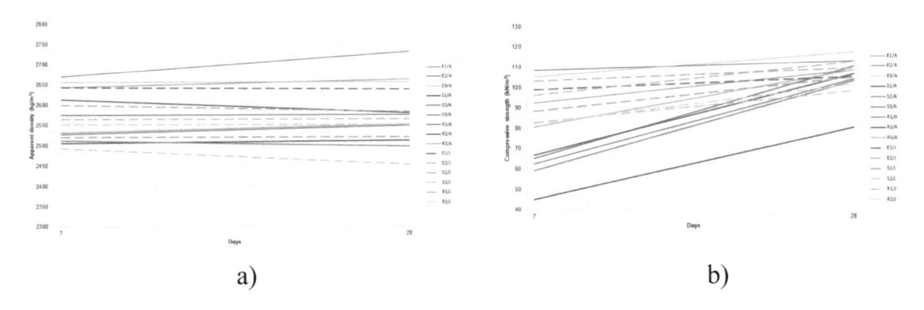

a) b)

Fig. 2. (a) Apparent density of different aggregates and different amount of cement (b) Compressive strength of different aggregates and different amount of cement

6 Conclusion

On the basis of the experimental campaign it can be concluded that increasing the amount of cement in concretes of higher strength above 460 kg/m^3 of concrete is rather insignificant in respect to the compressive strength. Furthermore, the petrographic type of rocks (eruptive, sedimentary and river aggregates) has a very small influence on the final (28 days) value of the compressive strength for this type of concrete (concrete of higher strength). However, early strengths of concrete (up to 7 days) are significantly higher for eruptive aggregates-Diabas in rest to the other petrographic rock types utilized in this experiment. We would like to emphasize the important influence of compaction degree of concrete for the formation of concrete's microstructure.

References

1. Finnet, J.: The supply and use of Elkem Microsilica in sprayed concrete applications for London cross rail project. In: 7th International Symposium on Tunnels and Underground Structures in South-East Europe, Zagreb, Croatia (2017)
2. Dedicated to Innovation 50 years Mc-Bauchemie. Müller GmbH&Co. KG, First Printing 11 2011
3. BAS EN 206-1:2002+ BAS EN 206-1/A1:2006: Concrete – Part 1: specification, performance, production and conformity. Institute for Standardization Bosnia and Herzegovina, Sarajevo, Bosnia and Herzegovina (2006)
4. BAS EN206+A1:2018: Concrete - specification, performance, production and conformity. Institute for Standardization Bosnia and Herzegovina (2018)
5. BAS EN 12620+A1:2009: Aggregates for concrete. Institute for Standardization Bosnia and Herzegovina, Sarajevo, Bosnia and Herzegovina (2009)
6. BAS EN 12878:2015: Pigments for the colouring of building materials based on cement and/or lime - Specifications and methods of test. Institute for Standardization Bosnia and Herzegovina, Sarajevo, Bosnia and Herzegovina (2015)
7. BAS EN 450-1:2013: Fly ash for concrete. Part 1: definition, specifications and conformity criteria. Institute for Standardization Bosnia and Herzegovina, Sarajevo, Bosnia and Herzegovina (2013)

8. BAS EN13263-1+A1:2009: Silica fume for concrete. Part 1: definitions, requirements and conformity criteria. Institute for Standardization Bosnia and Herzegovina, Sarajevo, Bosnia and Herzegovina (2009)
9. BAS EN 15167-1:2009: Ground granulated blast furnace slag for use in concrete, mortar and grout - Part 1: definitions, specifications and conformity criteria. Institute for Standardization Bosnia and Herzegovina, Sarajevo, Bosnia and Herzegovina (2009)
10. BAS EN 1936:2009: Natural stone test methods – determination of real density and apparent density, and of total and open porosity. Institute for Standardization Bosnia and Herzegovina, Sarajevo, Bosnia and Herzegovina (2009)
11. BAS EN 1926:2009: Natural stone test methods – determination of uniaxial compressive strength. Institute for Standardization Bosnia and Herzegovina, Sarajevo, Bosnia and Herzegovina (2009)
12. BAS EN 1097-6:2004: Tests for mechanical and physical properties of aggregates – Part 6: determination of particle density and water absorption ventilated oven. Institute for Standardization Bosnia and Herzegovina, Sarajevo, Bosnia and Herzegovina (2004)
13. HRN 1128:2007: Concrete – guidelines for the implementation of HRNEN 206-1. Institute for Standardization Bosnia and Herzegovina, Sarajevo, Bosnia and Herzegovina (2007)
14. BAS EN 196-1:2006: Methods of testing cement – Part 1: determination of strength. Institute for Standardization Bosnia and Herzegovina, Sarajevo, Bosnia and Herzegovina (2006)
15. BAS EN 934-2:2010: Admixtures for concrete, mortar and grout- Part 2: concrete admixtures – definitions, requirements, conformity, marking and labelling. Institute for Standardization Bosnia and Herzegovina, Sarajevo, Bosnia and Herzegovina (2010)
16. BAS EN 1015-3:2004: Methods of test for mortar for masonry – Part 3: determination of consistence of fresh mortar (by flow table). Institute for Standardization Bosnia and Herzegovina, Sarajevo, Bosnia and Herzegovina (2004)

Estimation of the Minor-Loss at the Junction of a Headrace Tunnel and a Surge Tank with an Orifice

Adis Bubalo[1(✉)], Muris Torlak[2], and Safet Isić[3]

[1] Faculty of Mechanical Engineering, University "Džemal Bijedić" of Mostar, University Campus, 88000 Mostar, Bosnia and Herzegovina
a.bubalo@hotmail.com
[2] University of Sarajevo, 71000 Sarajevo, Bosnia and Herzegovina
[3] University "Džemal Bijedić" in Mostar, 88000 Mostar Bosnia and Herzegovina

Abstract. The main goal of the investigation presented in this paper is to analyse influence of the orifice size on the water level oscillations in a surge tank arising during load rejection and water hammer in a hydropower plant. An approach to achieve it, which is adopted in this study, is to develop an adequate 1D model allowing a fast and reliable simulation of the hydraulic processes in the headrace tunnel and the surge tank. For such a model, estimation of the minor loss at the junction of the surge tank and the headrace tunnel has a crucial role. The calculated water level oscillations show a good agreement with the values measured in an existing surge tank with orifice. The results obtained with the same model for a set of different orifice diameters yield conclusions on the surge tank response in the case of modified orifice design.

Keywords: Hydropower plants · Surge tank · Orifice · Simulation · Minor losses

1 Introduction

In derivation hydropower plants the surge tanks have very important role. They protect hydropower plant components from possible water hammer, they neutralize effects of strong pressure rise and pressure drop in penstocks. A surge tank is an additional storage space or reservoir constructed between the main storage reservoir and the power house [1, 2]. Appropriate shapes and dimensions of a surge tank are decisive for its proper functioning, so that reasonable attention is paid to these during design of a new power plant or in modernization of existing old power plants. One of design variants includes orifice placed at the junction of the surge tank and the headrace tunnel. The orifice size has very significant influence on water surface oscillations in the surge tank. To analyse this phenomena an adequate 1D model is developed [3] that allows a fast and reliable simulation of the hydraulic processes in the headrace tunnel and the surge tank. For such a model, estimation of the minor loss at the junction of the surge tank and the headrace tunnel has a crucial role. The minor loss, which is of

I. Karabegović (Ed.): NT 2019, LNNS 76, pp. 523–529, 2020.
https://doi.org/10.1007/978-3-030-18072-0_60

primary concern in this paper, is triggered by a characteristic shape strongly influenced by the annular orifice at the bottom of the tank.

This paper describes combined use of analytical calculation of minor loss in the junction and numerical solution of two ordinary differential equations to obtain the water level in the surge tank. The analytically calculated minor loss coefficients take into account variable flow direction, and they are implemented in the algorithm published earlier [3]. The method is validated by comparison with the available experimental data for a surge tank of hydropower plant Jablanica in Bosnia-Herzegovina. The calculated water level oscillations show a good agreement with the measured values. This allows application of the method to analysis of water level oscillations in surge tanks of similar shapes at various operating conditions.

2 Method of Calculation

A typical derivation hydropower plant system with the reservoir, the intake tunnel, the surge tank with orifice, and the penstock is shown in Fig. 1. The momentum equation for the water flow in the headrace tunnel of this system can be written in the form [4]:

$$\frac{L}{g}\frac{dQ_{\mathrm{t}}}{dt} = A_{\mathrm{t}}(-z - h_{\mathrm{f}} - h_{\mathrm{j}} - h_{\mathrm{dia}}) \tag{1}$$

where L is the tunnel length, g is the gravitational acceleration, Q_{t} is the water flow rate in the tunnel, t is the time, A_{t} is the cross-sectional area of tunnel, z is the water level in the surge tank above the reservoir level, h_{f} is the head loss caused by friction in tunnel, h_{j} is the head loss in the junction, h_{dia} is the orifice head loss.

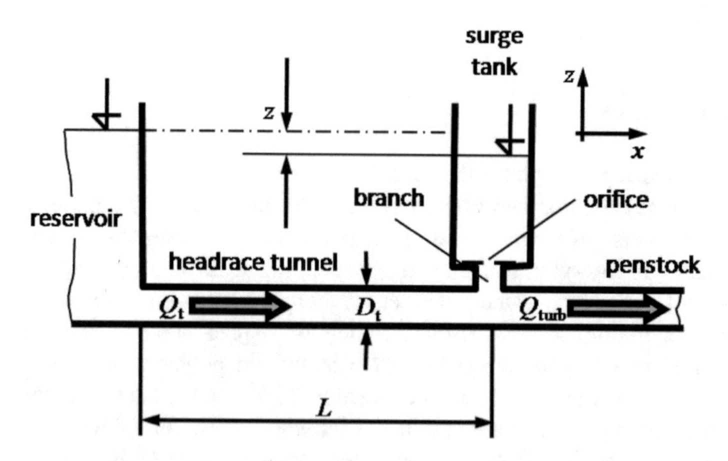

Fig. 1. Schematic view of a typical hydropower plant intake system

Friction loss in the tunnel is calculated using Darcy-Weisbach equation:

$$h_{\mathrm{f}} = \lambda \frac{L v_{\mathrm{t}}^2}{2 g D_{\mathrm{t}}} \qquad (2)$$

where λ is the friction coefficient, v_{t} is the flow velocity in the tunnel (proportional to the flow rate Q_{t}), and D_{t} is the tunnel diameter.

Minor head loss in the junction between the tunnel and the surge tank is given by the formula:

$$h_{\mathrm{j}} = \xi_{\mathrm{j}} \frac{v_{\mathrm{t}}^2}{2g} \qquad (3)$$

where ξ_j is the empirically estimated loss coefficient in the junction (like a T-branch).

The continuity equation for the surge tank reads:

$$\frac{dz}{dt} = \frac{1}{A_{\mathrm{v}}} (Q_{\mathrm{t}} - Q_{\mathrm{turb}}) \qquad (4)$$

where A_{v} is the cross-sectional area of the surge tank, and Q_{turb} is the water flow rate through the penstock toward the turbines.

Equation (4) is coupled with Eq. (1) through the quantities Q_{t} and z. In the algorithm employed here [3], the geometric data (L, A_{t}, A_{v}) are assumed to be known, as well as the flow rate toward the turbines Q_{turb}. First, Eq. (1) is solved for Q_{t}, then Eq. (4) is solved for z, delivering the information on the water level in the surge tank, and the process is repeated several times within each time step until the convergence to support the inter-equation coupling.

The head loss at the orifice is defined by the equation in the same form as Eq. (3):

$$h_{\mathrm{dia}} = \xi_{\mathrm{dia}} \frac{v_{\mathrm{dia}}^2}{2g} \qquad (5)$$

where ξ_{dia} is the minor-loss coefficient of the orifice, and v_{dia} is the velocity of the water approaching the orifice from either side (i.e. from the branch or from the tank). Its value is easily obtained from the continuity law considered in the junction and the resulting water flow rate toward the surge tank shaft ($Q_{\mathrm{t}} - Q_{\mathrm{turb}}$)

The minor-loss coefficients ξ_{j} and ξ_{dia} are usually obtained experimentally or numerically for standard shapes and installation conditions at varied flow rates or flow velocities. Based on so obtained data, various diagrams or formulae are derived for use in engineering practice [5–7].

The individual local head losses can also be substituted by *integral minor loss* h_{ξ} representing the entire connection between the tunnel and the surge shaft (the T-junction including the branch pipe and the orifice, followed by a cross-section expansion, such as appearing in real surge tank geometries), which can be interpreted as a serial connection of the individual contributions:

$$h_{\xi} = h_{\mathrm{j}} + h_{\mathrm{dia}} \tag{6}$$

or written in the same form as the individual minor-loss components, Eqs. (3) and (5):

$$h_{\xi} = \xi \frac{v_{\mathrm{t}}^2}{2g} \tag{7}$$

where ξ is the integral minor-loss coefficient at the junction with orifice. Estimation of ξ in this case requires sophisticated methods, such as experimental measuring or computer simulations, e.g. *computer fluid dynamics* – CFD methods [3, 8]. Such an approach might be preferred due to expected better accuracy. In this paper, however, applicability of the analytical estimation of energy loss in the junction using the individual contributions, Eqs. (3) and (5), whose successful application would save time significantly, is tested.

Various simplified cases of minor-loss coefficient estimation can be found in standard engineering literature, such as for a cross-section expansion, a contraction, or an orifice. Figure 2 shows a combination of these types of minor head loss. For an orifice within a pipe expansion (or contraction, depending on the flow direction), as shown in Fig. 2, minor loss coefficient can be calculated using a formula given by [5, 6], which is here expanded in order to account for the definition of the loss coefficient given by Eq. (5) via the approaching flow velocity v_{dia}:

$$\xi_{\mathrm{dia}} = \frac{A_1^2}{A_0^2}(1 + 0,707\sqrt{1 - \frac{A_0}{A_1}} - \frac{A_0}{A_2})^2 \tag{8}$$

Minor loss coefficient for T-junction ξ_{j} is also taken from literature [5].

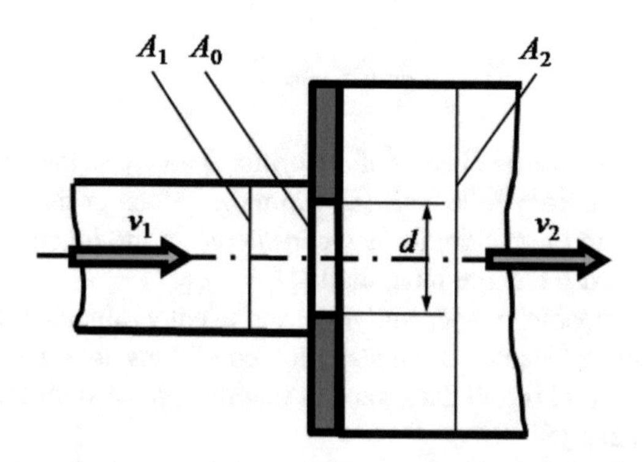

Fig. 2. Orifice in a pipe with cross expansion/contraction

3 Results

The results presented in this section describe the influence of the surge tank orifice diameter on the water level oscillations during the turbine units shut-off. The calculations are done for the case of HPP Jablanica, surge tank no. 2. The headrace tunnel diameter is 6,3 m, the length is 1950 m, the diameter of the branch pipe connecting the headrace tunnel and the shaft is 5 m, and the diameter of the surge shaft is 13 m. The tunnel axis is positioned 217 m a.s.l around the junction. The initial conditions are: water flow rate through the headrace tunnel of 84,8 m^3/s and the water level in the surge tank 267,9 m a.s.l. (which is equivalent to the elevation of nearly 51 m above the tunnel axis). According to the condition of the turbines shut off, the value of Q_{turb} arising in Eq. (4) is adopted to be zero. Note that the surge tank considered here has an 80 m long horizontal side chamber, whose inner diameter is 7 m; it is located between 269 m a.s.l. and 276 m a.s.l.

Table 1. Integral minor loss coefficients at the junction

	d = 3,6 m	d = 3,3 m	d = 3 m	d = 2,5 m
ξ_{dia} (inflow)	7,44	11,34	17,65	39,71
ξ_{dia} (outflow)	229,19	375,25	621,77	1524,16
ξ_j	1,5	1,5	1,5	1,5

The analysis is done for the set of surge tank orifice diameters of d = 3,6 m (the existing diaphragm for which the experimental measurements are available), d = 3,3 m, d = 3 m, and d = 2,5 m. The minor loss coefficient ξ_{dia} is calculated using Eq. (8) for different orifice diameters. The results are shown in Table 1 both for the case of the water flowing into the surge shaft (denoted by “inflow”) and the case of the water flowing from the surge shaft to the tunnel (denoted by “outflow”). The calculation is done with A_1 corresponding to the branch pipe and A_2 corresponding to the surge shaft in the case of “inflow”, and opposite in the case of “outflow”; A_0 is the cross-section area of the orifice opening.

The minor loss coefficients from Table 1 are employed to calculate the water level oscillation in the surge tank by numerical solution of Eqs. (1) and (4) as described in [3]. The results are shown in Fig. 3, where the effect of the orifice diameter variation is clearly observed. The maximum water level becomes smaller with the smaller inner diameter of the orifice, which is the consequence of the stronger damping and the energy loss. Note that the sudden changes in the slope of the curves shown in the diagram, seen at the elevation of about 269 m a.s.l., are caused by the water flow from the surge shaft to the side chamber and vice versa.

Strong damping triggered by small orifice diameter causes slightly longer oscillation period, as expected. However, water rising in the surge shaft is smaller in that case, hence the overflow to the side chamber is also smaller (filling the side chamber and its discharge do not take long time), so that the total oscillation periods – including both

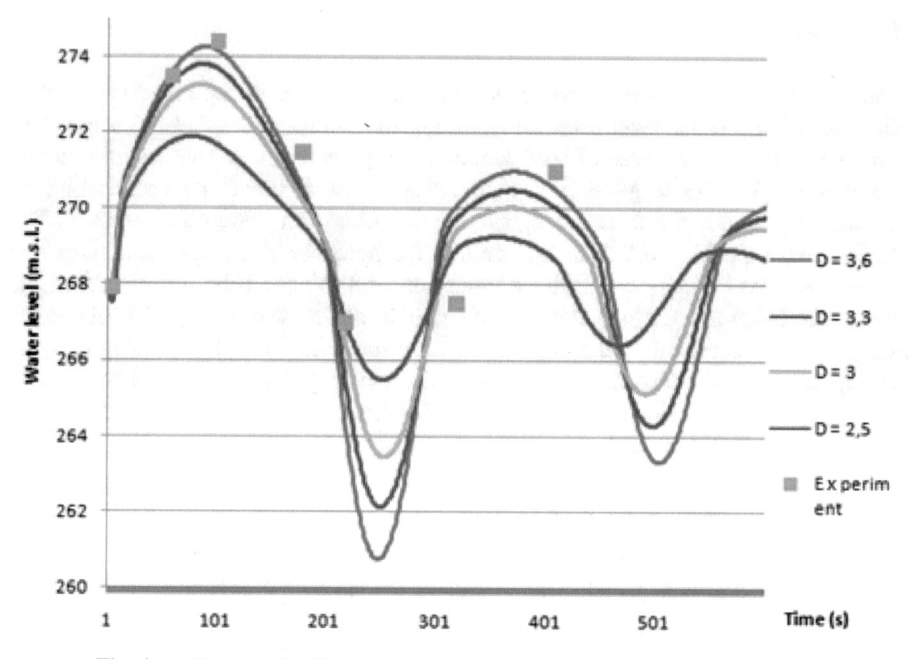

Fig. 3. Influence of orifice diameter on water level oscillation in surge tank

water level rising/sinking, and filling/discharge of the side chamber, become shorter at smaller orifice diameter, especially after the second oscillation cycle.

In addition to the computational results, the measurements obtained in the real surge tank, with the orifice whose inner diameter is 3,6 m, are shown in the same diagram for comparison [9]. The agreement of the results is reasonably good. However, one can see a small time lag in the measured values, particularly after the first oscillation cycle. This indicates that in reality there is a slight additional damping triggered by the energy losses while the water surface moves below the equilibrium level.

Table 2 summarizes the extreme values of the water levels in the first two cycles of oscillation. It can be seen that in the case of the largest orifice diameter the positive amplitude, achieved during the water rising in the shaft from the initial, equilibrium level to the maximum one, is approximately halved in the second cycle, while it is reduced to about one third in the case of the smallest orifice diameter. After the second cycle the decay of amplitudes is not so strong (not shown in the diagram).

Table 2. Influence of the orifice diameter on water level oscillation

Orifice diameter (m)	First maximum water level (m a.s.l.)	First minimum water level (m a.s.l.)	Second maximum water level (m a.s.l.)	Second minimum water level (m a.s.l.)
2,5	271,88	265,5	269,3	266,5
3	273,25	263,48	270,01	265,15
3,3	273,79	262,14	270,5	264,3
3,6	274,25	260,75	271	263,32

Interestingly, in the case of the largest orifice diameter the negative amplitudes, developing during the water sinking in the shaft from the equilibrium level to the minimum one, are stronger than the positive ones. The smallest orifice diameter leads to an opposite result in the first cycle: the positive amplitude is larger than the negative one, while in the second cycle, the positive and the negative amplitude are roughly the same. All this confirms considerably stronger damping of the flow in the shaft caused by the smaller orifice diameter.

4 Conclusions

Analytically estimated values of the minor loss coefficients are combined with numerical solving the equations for transient flow in a headrace tunnel with a surge tank. The results are compared with the measurements conducted in an existing tank with orifice. Agreement of the time-dependent water level for the given orifice size is reasonably good, providing an acceptable model for further examination of design variations of surge tanks of the same type. The calculated water level variations at different orifice diameters are shown. As expected, the results demonstrate clearly stronger energy losses of the flow in the surge shaft, and herewith lower water oscillations amplitudes, at smaller orifice diameters. Thus, applying the model used in this work a proper orifice size can be determined for a desired water level amplitude reduction.

References

1. Wylie, E.B., Streeter, V.L.: Fluid Transients. McGraw-Hill, New York (1978)
2. Pickford, J.: Analysis of Surge. Macmillan and Co Ltd., London (1969)
3. Torlak, M., Šeta, B., Bubalo, A.: Comparative use of FVM and integral approach for computation of water flow in a coiled pipe and a surge tank. In: Erpicum, S., Dewals, B., Archambeau, P., Pirotton, M. (eds.) Proceedings of the 4thIAHR Europe Congress Sustainable Hydraulics in the Era of Global Change, Liege, Belgium (2016)
4. Chaudhry, M.H.: Applied hydraulic transients, 3rd edn. Springer, Heidelberg (2014)
5. Idelchik, I.E.: Handbook of Hydraulic Resistance. Atomic energy commission and National Science foundation, Washington DC (1966)
6. Cengel, Y.A., Cimbala, J.M.: Fluid Mechanics, Fudamentals and Application. McGraw-Hill, New York (2006)
7. Hachem, F., Nicolet, C., Duarte, R., De Cesare, G., Micoulet, G.: Hydraulic design of the diaphragm's orifice at the entrance of the surge shaft of FMHL pumped-storage power plant. In: Proceedings of 2013 IAHR Congress. Tsinghua University press, Bejing (2013)
8. Torlak, M., Bubalo, A., Džaferović, E.: Numerical analysis of water elevation in a hydropower plant surge tank. In: Proceedings of 6th IAHR meeting of the Working Group "Cavitation and Dynamic Problems in Hydraulic Machinery and Systems", Ljubljana, Slovenia, 9–11 September, pp. 339–346 (2015)
9. Vaillant, Y.: Jablanica, additional hydraulic study in view of optimizing the operation of the plant, ANDRITZ HYDRO, internal report (2013)

Renewable Energy Sources and Corelated Environmental Systems

Svetlana Stevović[1(✉)], Dušan Golubović[2], and Slađana Mirjanić[3]

[1] Innovation Center of the Faculty of Mechanical Engineering in Belgrade, Kraljice Marije 16, Belgrade, Serbia
sstevovic@mas.bg.ac.rs

[2] University of East Sarajevo, Lukavica, Bosnia and Herzegovina

[3] University of Banja Luka, Banja Luka, Bosnia and Herzegovina

Abstract. A sustainable development and holistic approach are an imperative for all countries, especially those in developing, with a goal to alleviate global warming effects and minimize cross-border environmental pollution. Development strategy starts bz analyses of their technically feasible energy potential for world emission trading market and for New Energy Policy 2050. This paper presents the research results of current state of renewable energy sources in Serbia, as well as opportunities for incorporating the world and European legislation, new technology, knowledge, and investments in the energy sector of the country. Conclusions and discussion are given in the contecst of possible development for entier region of Balcan countries. Several examples of initial investments in renewable energy sources are reported as smart practice.

Keywords: Renewable energy sources · Environmental policy · Energy law · Developing countries · Greenhouse gas emission · Feed-in tariffs

1 Introduction

Utilization of renewable energy sources (RES) has proven to reduce dependence on imported fossil fuels, introduce new technologies and reinforce local industry capabilities, create new jobs, and significantly decrease greenhouse gas emissions. Therefore, utilization of RES in Serbia has become a critical segment of the country's environmental policy and energy law. A document entitled "The Energy Sector Development Strategy of the Republic of Serbia by 2015" [3] has been prepared to outline basic objectives of the new energy policy in accordance with the Energy Law [4]. The document determines principal directions in the development of energy sectors (i.e., energy production vs. energy consumption) and establishes multiple instruments for the realization of key priorities in the operation, business activity, and development of the entire energy system of Serbia. The selection of the objectives, priorities, and legislative instruments is primarily based on a broad political consent to adjust the whole energy system of the country to its recent economic development, especially its anticipated integration into the European Union. This document also sets a timeframe for the adoption of said instruments, such that the overall changes in the energy sector are carried out according to long-term political, socioeconomic, and environmental

I. Karabegović (Ed.): NT 2019, LNNS 76, pp. 530–540, 2020.
https://doi.org/10.1007/978-3-030-18072-0_61

strategies. Particular attention is paid to fundamentals, legislative framework, and feasibility of RES utilization in Serbia, a developing country that is so close to a highly developed region [5].

2 Technically Feasible RES in Serbia

Technically feasible RES potential of Serbia is significant [6]. It is estimated at 4.3 million toe (Mtoe) annually (toe or *ton of oil equivalent* is the amount of energy released by burning 1,000 kg of crude oil; it is equivalent to 42 gigajoules). This annual potential includes 2.7 Mtoe of biomass energy (63%), 0.6 Mtoe of unused hydropower (14%), 0.2 Mtoe of geothermal energy (5%), 0.2 Mtoe of wind energy (5%), and 0.6 Mtoe of solar energy (14%). Hydropower and biomass energy represent the most important RES as underlined in the strategy of the long-term energy development of Serbia [3].

2.1 Biomass

Serbia has a considerable biomass potential [7]. The territory of Serbia covers 88,360 km^2, of which about 30% is covered with forests, while about 55% of the territory is arable land. This is a very unique setting compared to many European countries. The northern part of Serbia, the province of Vojvodina together with territories along the rivers Sava and Danube, is flat and fertile. This region is the main source of agricultural products and biomass waste, especially waste from crop farming. Wheat and corn production is present in hilly regions as well, from the north to the south of Serbia (Fig. 3). The agricultural biomass waste originates from cereals, mostly wheat, barley and corn, as well as from industrial crops, such as sunflower, soya, and rapeseed. In addition, there are numerous farms for livestock breeding, where liquid manure is considered as biomass waste. The main location for fruit growing is a hilly region on the south, with most common types of fruit being plums, apples, cherries, peaches, and grapes.

Serbia also belongs to countries relatively rich in forests. The forest area lies mainly on the south, but also to the east and west. In 14 out of 145 counties, forests cover more than 45% of the territory. About two thirds of forests are state owned, while the rest are privately owned. About half of all forests are pure deciduous trees, only 5% are pure coniferous trees, while the rest are mixed forests. Main deciduous forest species are beech and oak. Spruce is the main species among conifers.

The annual amount of energy that could be obtained using biomass in Serbia is estimated at 2.7 Mtoe, of which 1.0 Mtoe is wood biomass potential (tree felling and wood biomass remnants during primary and/or industrial treatment), and over 1.7 Mtoe is agricultural biomass (wastes of agriculture and farming crops, including liquid manure). Current studies indicate that there is a great potential for biodiesel production from rapeseed, sunflower, and soya [8].

2.2 Biogas

The available amount of liquid manure from the poultry and livestock industries enables biogas production with an estimated energy content of 42,200 toe [8]. This amount of liquid manure combined with a suitable portion of biomass from agricultural residues allows for installation of biogas power plants with a maximum projected capacity of 80 MW (megawatts). The production of biogas from liquid manure is also important from an ecological perspective. In addition to its primary application for biogas production, anaerobic digestion of manure results in a nutrient-rich digestate which can be used as a fertilizer. Due to pronounced defragmentation of valuable farmland in Serbia, it is recommendable that the manure be collected from multiple farms and treated in a central anaerobic digester.

2.3 Biofuels

There are favorable conditions for biofuel production in Serbia, particularly bioethanol and biodiesel. Bioethanol is primarily produced from molasses (a viscous byproduct of the processing of sugar cane or sugar beets into sugar) and cereal crops. However, the available amount of molasses does not match the current production needs. A total capacity of sugar refineries is about 200,000 metric tons (*metric ton* = 1,000 kg) of molasses annually, while the usable amount for sugar processing is only 50,000 metric tons. The rest of 150,000 metric tons can be used for various purposes, including bioethanol production. The missing quantities of molasses for bioethanol production would need to be imported from foreign markets, which is often subject to fluctuating import quotas, tariffs and trade agreements. On the other hand, Serbia has strong agriculture sectors where cereal crops production outpaces total cereal demand. Consequently, there exists relatively significant potential for bioethanol production from the cereal surplus. It is estimated that 330,000 metric tons of cereal crops are needed for the production of 100,000 metric tons of bioethanol annualy, which is approximately one third of the cereal surplus, or only 2 to 4% of the entire cereal crops production. Other alternative feedstocks for bioethanol production existing in Serbia are reed canary grass, jerusalem artichoke, and potato. It is estimated that about 100,000 hectares (ha = 2.47 acres) of marginal land in Serbia can be used for growing reed canary grass and jerusalem artichoke, which could produce 3 million metric tons of ethanol annually.

Major feedstocks for biodiesel production in Serbia are vegetable oils from oil crops (sunflower, soybean, and rapeseed) and waste edible oil. The total farmland planted with oil crops is estimated at 668,800 ha, with 350,000 ha potentially available for biodiesel production. Table 1 shows the potential biodiesel production of Serbia per hectare (ha).

Table 1. Potential annual biodiesel production from oil crops in Serbia

Oil crop	Average yield (ton/ha)	Seed oil content (%)	Biodiesel Production	
			(kg/ha)	(litre/ha)
Sunflower	1.79	40	716	816
Soybean	2.25	18	405	460
Rapeseed	1.69	36	608	690

Potential biodiesel production in relation to percentage of crop type planted on the available farmland (350,000 ha) in Serbia is presented in Table 2.

Table 2. Potential annual biodiesel production for different crop types in Serbia

Percent of farmland by crop type	Potential biodiesel production (metric ton)
100% Rapeseed	212,800
70% Rapeseed + 30% Sunflower	224,140
50% Rapeseed + 50% Sunflower	231,700
30% Rapeseed + 70% Sunflower	239,260
100% Sunflower	250,600
100% Soybean	141,750

The consumption of edible oils in the Republic of Serbia is about 16 litres per capita. This amount could annually provide about 10,000 metric tons of waste edible oils suitable for biodiesel production.

Recognizing this potential, the Royal Embassy of the Netherlands in Serbia and the Serbian Ministry of Mining and Energy developed a Government-to-Government (G2G) project that aims at developing a biomass action plan for the production and utilization of sustainable biomass, including biofuels in Serbia, thus bringing Serbia closer to the European Union biomass legislation [7]. This project is implemented by the Serbian Ministry of Mining and Energy and SenterNovem, an agency of the Ministry of Economic Affairs of the Netherlands, which resulted in excellent cooperation on bioenergy between Serbia and the Netherlands. In order to foster the presence of Dutch biomass companies in the Western Balkan, there are currently plans to organize a Dutch-Serbian Bioenergy Day with a strong participation from companies from Serbia, the Netherlands, and other countries in the Western Balkan.

2.4 Small Hydropower Plants

According to the Energy Law, the term "small hydropower plant" (SHPP) denotes a hydropower facility of up to 10 MW installed capacity. The total hydropower potential of waterways in Serbia, including the most suitable locations for the installation of SHPPs, is identified in the National Cadastre of SHPPs [9]. It is, however, possible to construct SHPPs in other locations, with the approval of the Ministry of Mining and Energy, based on a feasibility analysis of each particular site. Such approval is generally granted to a new site on the same watercourse when hydrologic, geologic, land use, ecologic, and other conditions substantially differ from those identified in the National Cadastre.

Based on [9], the list of selected counties with the highest SHPP potential in Serbia is done. SSHPPs could ultimately generate about 5% of the total production of electrical energy in Serbia (34,400 GWh/year in 2006), or about 15% of the total hydropower production (10,900 GWh/year).

The results of fifteen feasibility studies recently conducted for the purpose of SHPP construction in Serbia suggest that only about 10% of the most attractive locations from the National Cadastre could be utilized in the existing technological and market conditions. In the future, it will be necessary to revisit the register of probable SHPP sites and prepare a more precise inventory of feasible locations for planning purposes [10].

2.5 Geothermal Energy

Serbia is relatively small in size (about 80,000 km^2), but its geological and tectonic structures are very complex. As a consequence, its geothermal characteristics are unique. On two-thirds of the Serbian territory, values of the heat flow density are greater than those for the continental part of Europe [12].

Geothermal sources have been identified in more than 60 counties. Their groundwater temperature is usually up to 40 °C (104 °F). There are six counties with the water temperature above 60 °C (140 °F): Vranje, Šabac, Kuršumlija, Raška, Medveđa i Apatin. The average flow rate from existing geothermal wells is about 20 litre/s. The flow rate exceeds 50 litre/s in several locations (Bogatić, Kuršumlija, Pribojska Banja, Niška Banja), while in one location it exceeds 100 litre/s (Banja Koviljača) [13].

The total thermal power that could be obtained from all geothermal sources in Serbia is estimated at 220 MW, with an annual thermal energy production of 7,650 TJ (TJ = terajoules = 10^{12} joules), which would replace about 180,000 toe/year. These geothermal sources can potentially be utilized to generate thermal energy for different purposes: electricity production in geothermal power plants, thermal bathing in medical spas (balneotherapy), sanitary water heating, heating of greenhouses, etc. Currently, Serbia uses only about 10% of its actual geothermal potential.

In January 2010, Southern European Exploration (SEE, a subsidiary of Canadian RCC) was granted an exploration permit for additional geothermal sources at Vranjska Banja [14]. This is the first permit of its kind to be issued to a private company in Serbia (SEE has also filed applications for three additional energy permits in the Vojvodina Province of northern Serbia).

2.6 Wind Energy

Global wind energy potential of Serbia has been estimated at 1,300 MW [15]. This assessment was based on the wind measurement data (speed and direction) collected by the Republic Hydrometeorological Service of Serbia. The most promising locations for the installation of wind turbines are: Midžor Planina, Suva Planina, Vršački Breg, Stara Planina, Deli Jovan, Krepoljin, Tupižnica, Juhor, and Jastrebac. High resolution measurements (of wind speed and direction) are needed for a complete assessment of technically feasible wind power potential at the above locations. Some preliminary results of the tall-tower wind measurements (at a 50-m height above ground) collected by the Serbian Energy Efficiency Agency (SEEA) are presented in Table 3.

Table 3. Tall-tower wind measurements in selected locations

Location	Average wind speed at 50-m height (m/s)		Extrapolated wind speed at 80-m height (m/s)	
	6 months	12 months	6 months	12 months
Veliko Gradište	3.6	3.5	3.8	3.7
Negotin	5.2	5.8	5.6	6.1
Titel	4.7	4.7	5.0	5.0

There are plans in the works to carry out comprehensive high-resolution wind measurements at a 50-m height in order to prepare the Wind Atlas of Serbia. This document will indicate general areas where a high wind resource may exist. It will be valuable to wind energy developers and potential wind energy users because it will allow them to choose a general area of estimated high wind resource for more detailed examination. A siting document, such as that written by Hiester and Pennell [16], can ultimately assist a potential user in going from wind resource assessment to particular site selection.

2.7 Solar Energy

The average daily solar radiation intensity over Serbia's territory in winter ranges from 1.1 kWh/m^2/day (kilowatt-hours per square meter per day) in the north to 1.7 kWh/m^2/day in the south. The above range is higher in summer, between 5.9 and 6.6 kWh/m^2/day. The annual average insolation over Serbia ranges from 1,200 kWh/m^2/year in the northwest to 1,550 kWh/m^2/year in the southeast [18]. Since the solar energy utilization rate is directly proportional to solar panel efficiency (maximum 40%), technically feasible solar energy in the Republic of Serbia is about 550 kWh/m^2/year. According to the 2002 Census, Serbia has about 2.5 million households. If every fifth household put up a rooftop solar system with a minimum surface area of 4 m^2, this would provide a total annual solar energy production of 1,750 GWh/year. The majority of this amount would be used for electricity consumption, while a portion could replace fossil fuels utilized for sanitary water heating, resulting in decreased carbon emissions by 2.3 million tons per year.

2.8 Cogeneration of Heat and Power (CHP)

Cogeneration (a.k.a. combined heat and power, CHP) is the use of a heat engine or a power station to simultaneously generate both electricity and useful heat. As a result of the positive experience of the DHPB project, substituting coal with solid biomass is being considered for individual boiler facilities used to heat kindergartens and schools in Belgrade, wherever technically feasible. In a very short period of time and with relatively low initial investment, current consumption of over 100,000 tons of coal for heat generation could be replaced by solid biomass. This can also have a positive impact on local employment opportunities, given that increased consumption of solid biomass must be accompanied by its domestic production, including construction and upgrade of pellet mill facilities.

3 Environmental Policy and Subsidies for RES in Serbia

The overall objective for 2012 of the Republic of Serbia is to enhance its power generation from RES by 7.4% or 735 million kWh compared to 2007 [23]. In this respect, the Ministry of Mining and Energy has prepared a set of changes and amendments to the National Implementation Program of Strategy for the Development of Energy Sector in Serbia [3], adopted a regulation on acquiring the status of privileged power producers in September 2009 [24], and a decree on incentive measures (feed-in tariffs) regarding power generation from renewable sources and by means of combined heat and power (CHP) systems [25].

According to [24], the privileged electric power producers are defined as those that (1) use renewable energy sources or a separated fraction of the communal waste in the electric power generation process; (2) produce electric power in power plants regarded to be small power plants pursuant to the Energy Law; or (3) cogenerate electrical and thermal energy, provided that they met the criteria related to energy efficiency. The privileged power producers are entitled to feed-in tariffs valid at the moment of submission of the request for status acquiring or renewal. Feed-in tariffs, or renewable energy payments, are a policy mechanism designed to encourage the adoption of RES and to help accelerate the move toward grid parity. They typically include three key provisions: (1) guaranteed grid access; (2) long-term contracts for the electricity produced; and (3) purchase prices that are methodologically based on the cost of renewable energy generation. Under a feed-in tariff, an obligation is imposed on regional or national electric grid utilities to buy renewable electricity from all eligible participants. Feed-in tariffs (renewable energy payments) adopted in Serbia per kilowatt hour of electricity, generated from renewable or CHP, are changed by time.

The tariffs were guaranteed and fixed during a 12-year period (from January 2010 to January 2012). The level of a purchase price is set to provide invested capital returned in 12 years, covering all operating costs incurred during the same period. In case of power plants that had been in operation before the application of feed-in tariffs, proposed tariffs are valid for the shortened period of time. Feed-in tariffs for old power plants that have been out of commission for at least five years will be separately defined, in order to encourage their revitalization and re-entering into operation.

4 International Legislative Framework of Serbia Related to RES

In pursuit of its national environmental policy, the Republic of Serbia is fully committed to honoring international obligations towards renewable energy production, reduction of greenhouse gas emissions, and enhancement of its global energy efficiency [10]. Serbia is in the original group of countries to officially sign up to the Copenhagen Accord [26] under which several developed and developing countries (including the United States, Canada, Australia, Brazil, China, India, and South Africa) acknowledged climate change as one of the greatest challenges of our time, outlined intentions and commitments on carbon emissions, and pledged support for technology transfer.

This accord has produced three key outcomes - raising climate change concerns to the highest level of government; political consensus on the long-term, global response to climate change; and negotiations which brought the set of decisions to implement rapid climate action closer to completion.

Serbia is also one of the founders of the first International Renewable Energy Agency [27], officially established on January 26, 2009. As of today, 141 countries and the European Union signed the Statute of the Agency (amongst them are 15 American, 47 African, 33 Asian, 37 European, and 9 Australia/Oceania states). Mandated by these governments worldwide, IRENA will promote the widespread and increased adoption and sustainable use of all forms or renewable energy. Acting as the global voice for renewable energies, it will facilitate access to all relevant renewable energy information, including technical data, economic data, and renewable resource potential data. IRENA will also share experiences on best practices and lessons learned regarding policy frameworks, capacity-building projects, available finance mechanisms, and renewables related energy efficiency measures.

The Republic of Serbia acceded to the Energy Community Treaty (Treaty) in December 2006 [28]. The Treaty establishes the Energy Community between the European Union and the South-East European region consisting of Albania, Bosnia and Herzegovina, Bulgaria, Croatia, Former Yugoslav Republic of Macedonia, Montenegro, Serbia, Romania, and Turkey. It provides for the creation of an integrated market in natural gas and electricity in South-East Europe which will create a stable regulatory and market framework capable of attracting investment in gas networks, power generation and transmission networks, so that all parties have access to the stable and continuous energy supply that is essential for economic development and social stability. The Treaty also enables a regulatory framework to be set up, permitting the efficient operation of energy markets in the region, including RES. Article 20 of the Treaty obligates each contracting party to implement the Directive 2001/77/EC of the European Parliament [29] on the promotion of electricity produced from RES in the internal electricity market and the Directive 2003/30/EC of the European Parliament [30] on the promotion of the use of biofuels or other renewable fuels for transport.

The new Directive 2009/28/EC of the European Parliament [31] on the promotion of the use of energy from RES endorses a mandatory target of a 20% share of energy from RES in overall Community energy consumption by 2020, as well as a mandatory 10% minimum target to be achieved for the share of RES (biofuels, electricity, and hydrogen) in transport petrol and diesel consumption by 2020. This directive aims at facilitating cross-border support of energy from RES without affecting national support schemes. Following the implementation of the directive, Member States may make arrangements for the statistical transfer of a specified amount of RES from one Member State to another and may cooperate with third countries on all types of joint projects regarding the production of electricity from RES. A great impetus for all the contracting parties is that a certain amount of energy from RES produced in the territory of one participating Member State may count towards the national overall target of another participating Member State.

The Treaty (Article 13) also calls on all the parties to recognize the importance of the Kyoto Protocol (Protocol) and to accede to it. The Protocol is an international agreement linked to the United Nations Framework Convention on Climate Change [32]. It sets binding targets for 37 industrialized countries and the European Community for reducing greenhouse gas emissions. The Protocol (Article 2) affirms that each party, in achieving its quantified emission limitation and reduction commitments, in order to promote sustainable development, shall implement and/or further elaborate policies and measures in accordance with its national circumstances.

5 Conclusions and Recommendations

The Republic of Serbia has extensive unused potential for greater energy efficiency and production from RES. In particular, it could profitably develop its hydro and biofuel capacity. With relatively little adjustment to the regulatory environment, Serbia could enable private enterprise to produce enough biofuels to meet local demand and even to export, while creating up to 24,000 new jobs by 2020. Projections suggest that with minor adjustments in the regulatory system, renewable energy could easily rise to one-third of Serbia's overall primary energy consumption, which now relies on fossil fuels for 93% of its supply. Currently, hydropower comprises almost all of the renewable energy sources employed in Serbia. With relatively little effort, the country could obtain more than 18% of current fossil fuel usage from biofuels, 5% from wind power, and 1% from solar power [33].

Serbia has an extensive body of laws addressing energy issues, including the 2004 Law on Energy and the 2005 Energy Development Strategy through 2015, the 2009 regulation on acquiring the status of privileged power producers [34], and the 2010 decree on incentive measures (feed-in tariffs) regarding power generation from RES and by combined heat and power systems. All these legislative instruments are in line with the European Union Directives and supported by international agreements [35], such as the Energy Community with Southeast Europe and the European Union, Copenhagen Accord, and Kyoto Protocol. Some obstacles still remain in the area of regulatory and institutional capacity, mainly the lack of sublegal documents - regulations, ordinances, rules, etc. - that make it possible to implement the above laws.

Moving forward, Serbia's policy makers should provide training to increase the number of energy experts, put to use the necessary regulations, and ensure that adopted prices (feed-in tariffs) [36] for renewable send the right signals to investors. The Government needs to continue working on two-way education with the population on conservation and alternative energy. Serbia has a broad range of small but quite enthusiastic environmental and RES activist groups that should be engaged to help facilitate this information conveyer belt between citizens, experts, the government, and the state. Finally, it is recommended that all countries make RES and energy efficiency a key objective of their foreign direct assistance programs to Serbia in all major theme areas: including economic growth, good governance, and civil society [33].

References

1. The World Bank, Europe and Central Asia. Serbia. Country Brief 2009, September 2009. http://www.worldbank.rs/WBSITE/EXTERNAL/COUNTRIES/ECAEXT/SERBIAEXTN/0,,contentMDK:20630647 ~ menuPK:300911 ~ pagePK:141137 ~ piPK:141127 ~ theSitePK:300904,00.html
2. The World Bank, The International Bank for Reconstruction and Development. Doing Business in 2006, Creating Jobs Office of the Publisher, World Bank, Washington, D.C. (2006). ISSN 1729-2638
3. The Republic of Serbia, Ministry of Mining and Energy. The Energy Sector Development Strategy of the Republic of Serbia by 2015. Official Gazette of the RS, No. 35/05, Belgrade (2005)
4. The Republic of Serbia, Ministry of Mining and Energy. Energy Law: Official Gazette of the RS, No. 84/04, Belgrade (2004)
5. Benthem, A., Romani, M.: Fuelling growth: what drives energy demand in developing countries. Energy J. **30**, 91–114 (2009)
6. The Republic of Serbia, Ministry of Mining and Energy. Renewable Energy Sources in Serbia. http://www.mre.gov.rs/navigacija.php?IDSP=299&jezik=eng
7. E-Energy Market, The International Portal for Products and Services Related to Bioenergy. Biomass Opportunities in Serbia, 2 January 2010. http://www.e-energymarket.com/news/singleview/link//07e29ab93b/article/16/biomass-opportunities-in-serbia.html
8. Ilić, M., Grubor, B., Tesić, M.: The state of biomass energy in Serbia. Therm. Sci. J. **8**, 5–19 (2004)
9. Stević, S.: Cadastre of Small Scale Hydropower Plants in The Republic of Serbia (in Serbian). Energoprojekt-Hidroinženjering & "Jaroslav Černi" Institute, Belgrade (1987)
10. Stević, S., et al.: Improvement of Energy Efficiency for Utilization of Renewable Energy Sources as Function of Sustainable Development (in Serbian). Prepared for the Ministry of Science and Environment Protection, Belgrade (2009)
11. Reservoir Capital Corporation, Focused on Renewable Energy. Natural Resource Opportunities in Serbia and Southeast Europe. Company Overview, February 2010. http://www.reservoircapitalcorp.com/s/Home.asp
12. Milivojević, M., Martinović, M.: Utilization of geothermal energy in Serbia. In: International Geothermal Conference, Reykjavik, Iceland (2003)
13. Soleša, M., Đajić, N., Parađanin, Lj.: Production and Utilization of Geothermal Energy (in Serbian). Faculty of Mining and Geology, Belgrade (1995)
14. Reservoir Capital Corporation, Focused on Renewable Energy. Reservoir Granted Geothermal Exploration Permit in Serbia, January 2010. http://www.reservoircapitalcorp.com/s/News.asp?ReportID=378958
15. Putnik, R.: Possibilities for Utilization of Wind Energy for Electricity Production (in Serbian). Electric Power Industry of Serbia (Elektroprivreda Srbije), Belgrade (2002)
16. Hiester, T., Pennell, W.: The Meteorological Aspects of Siting Large Wind Turbines. PNL-2522, Pacific Northwest Laboratory, Richland, WA (1981)
17. Greenstar Alternative Energy, Clean Energy. Belo Blato Project Description. http://www.greenstarae.com/index.php?option=com_content&task=view&id=88&Itemid=95
18. Gburčik, P.: Study of Energy Potential for Utilization of Solar and Wind Power in Serbia (in Serbian). Center for Multidisciplinary Studies, University of Belgrade, Belgrade (2004)
19. Balkan Business Insight, Balkan Investigative Reporting Network. News. Serbia Builds First Solar Energy Plant, June 2008. http://www.balkaninsight.com/en/main/news/10804/

20. Pavlović, T., et al.: Tesla and Solar energy. In: Sixth International Symposium Nikola Tesla, Belgrade (2006)
21. Mercados, Energy Markets International. Our Projects News - Europe & CIS. Serbia: Development of the Solar Energy Market. http://www.mercadosemi.es/EuropeandCIS.htm
22. USAID, Serbia Competitiveness Project. Cleaner Heating Solutions for Belgrade Citizens, February 2010. http://compete.rs/?q=en/node/1695
23. Energy in Central & Eastern Europe, Austrian Energy Agency. Feed-in Tariffs Introduced in Serbia, February 2010. http://www.enercee.net/single/article/58/feed-in-tariffs-introduced-in-serbia.html
24. The Republic of Serbia, Ministry of Mining and Energy. Decree on the Requirements for Obtaining the Status of the Privileged Electric Power Producer and the Criteria for Assessing Fullfilment of These Requirements. Official Gazette of RS, No. 33/09, Belgrade (2009)
25. The Republic of Serbia, Ministry of Mining and Energy. Decree on Incentive Measures for Electricity Generation Using Renewable Energy Sources and For Combined Heat and Power (CHP) Generation. Official Gazette of the RS, No. 84/09, Belgrade (2009)
26. Copenhagen Accord. United Nations, Framework Convention on Climate Change (UNFCC). Conference of the Parties, Fifteenth Session, Copenhagen (2009). FCCC/CP/2009/L.7
27. International Renewable Energy Agency (IRENA). About IRENA. Signatory States (2010). http://www.irena.org/index.php?option=com_content&view=article&id=90&Itemid=93
28. Energy Community. Contracting Party Profiles. Serbia, 23 February 2010. http://www.energy-community.org/portal/page/portal/ENC_HOME/ENERGY_COMMUNITY/Profiles/Serbia/General
29. European Parliament, Council of the European Union. Directive 2001/77/EC on the Promotion of Electricity Produced from Renewable Energy Sources in the Internal Electricity Market. Brussels: Official Journal of the European Communities, 27 October 2001. L 283/33
30. European Parliament, Council of the European. Union Directive 2003/30/EC on the Promotion of the Use of Biofuels or Other Renewable Fuels for Transport. Brussels: Official Journal of the European Communities, 8 May 2003. L 123/42
31. European Parliament, Council of the European. Directive 2009/28/EC on the Promotion of the Use of Energy from Renewable Sources and Amending and Subsequently Repealing Directives 2001/77/EC and 2003/30/EC. Brussels: Official Journal of the European Communities, 23 April 2009. L 140/16
32. United Nations. Kyoto Protocol to the United Nations Framework Convention on Climate Change. Intergovernmental Panel on Climate Change, Kyoto (1998)
33. Jefferson Institute, Promoting European Energy Security. Serbia's Capacity for Renewables and Energy Efficiency. Policy Association for an Open Society (PASOS), Praha (2009). ISBN 978-86-86975-05-8
34. Stevovic, I., Jovanovic, J., Stevovic, S.: Sustainable management of Danube renewable resources in the region of Iron Gate: Djerdap 1, 2 and 3 case study. Manag. Environ. Qual.: Int. J. **28**(5), 664–680 (2017)
35. Stević, I.: Strategic orientation to solar energy production and long term financial benefits. ARHIV ZA TEHNIČKE NAUKE/ARCHIVES FOR TECHNICAL SCIENCES, **1**(17) (2017)
36. Pavlović, T.M., Milosavljević, D.D., Mirjanic, D., Pantić, L.S., Radonjić, I.S., Piršl, D.: Assessments and perspectives of PV solar power engineering in the Republic of Srpska, Bosnia and Herzegovina. Renew. Sustain. Energy Rev. **18**, 119–133 (2013)

The Impact of Electric Cars Use on the Environment

Tomislav Pavlović[1], Dragoljub Mirjanić[2], Ivana Radonjić Mitić[1], and Andjelina Marić Stanković[1](✉)

[1] Faculty of Science, University of Niš, Visegradska 33, 18000 Nis, Serbia
andjelinamaric14@gmail.com

[2] Academy of Sciences and Arts of RS, Sarajevo, Bosnia and Herzegovina

Abstract. This paper gives basic information on electric cars, types of electric cars (hybrid, chargeable hybrid, battery, electric cars with fuel cells, solar cars), on current state of the electric cars use and its impact on the environment. The highest energy saving in relation to IC engine vehicles is achieved in city driving. Electric cars are as ecologically clean and energy efficient became the subject of the energy policy of developed countries (USA, Japan, and Germany). In developed countries, regulations are adopted according to which ECs must be represented in a certain percentage in total car production. Such regulations also apply in other countries.

Keywords: Electric cars · Internal combustion engine (IC) · Environment

1 Introduction

Accelerated development of electric cars (EC) is primarily motivated by ecological reasons, because they do not emit exhaust gases and do not pollute the environment. In 1950, there were 50 million motor vehicles in the world, and 500 million in 1990. For the beginning of this century, it is projected that the number of motor vehicles will exceed one billion. Accelerated growth in the number of motor vehicles is a serious threat to the environment. In underdeveloped countries, internal combustion engines (IC engines) are inadequately maintained, posing an additional threat to the environment.

For the rise of EC on the market, it is necessary to provide the appropriate infrastructure, which primarily includes the battery charging stations (batteries) of electric cars. The world is working on the development of battery charging technology, user safety and device safety in an electric car. Within the IEC (International Electrotechnical Commission) and SAE (Society of Automobile Engineers), standards are applied for electric cars. In the United States, the National Council for the Development of Infrastructure for Electric Cars was established.

A car of 1200 kg with an IC engine consumes about 8 l of gasoline per 100 km, and an electric car of the same mass about 15 kWh of electricity. At present prices 8 l of gasoline costs about 10 euros, and 15 kWh about 1 euro. In terms of fuel costs, an electric car is about 10 times cheaper than IC-engine cars [1–4].

I. Karabegović (Ed.): NT 2019, LNNS 76, pp. 541–548, 2020.
https://doi.org/10.1007/978-3-030-18072-0_62

2 Types of Electric Cars

Nowadays there are following types of electric cars:

(1) Hybrid car (HC).
(2) Chargeable hybrid car (CHC).
(3) Battery electric car (BEC).
(4) Electric car with fuel cells.
(5) Solar car.

2.1 Hybrid Cars

A hybrid car is a car that is powered by an electric motor and an IC engine.

A hybrid car consists of an internal combustion engine, an electric generator, an energy DC/AC converter and a battery capacity of 1–2 kWh for peak power or electric drive at the vehicle start. A battery is charged using a generator that drives the IC engine or when decelerating and regenerating braking. With this car there is no possibility of charging the battery from the external power grid. An electric motor is used when starting the vehicle and speeds up to 50 km/h or when additional power or fuel savings is required (Fig. 1).

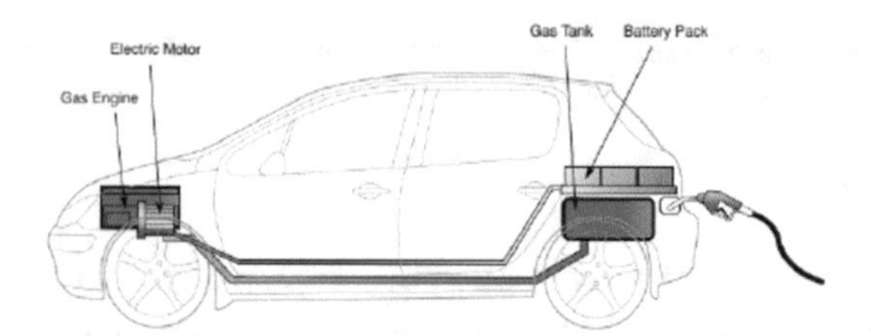

Fig. 1. Hybrid car

2.2 Chargeable Hybrid Car

The Chargeable Hybrid Car (RHC) has a battery capacity of 8–16 kWh, which is charged via an external plug or via a generator that launches an IC motor or regenerative braking while slowing down or stopping the car. With RHC, the primary drive is the IC motor, and the additional drive is a powerful electric motor.

The advantages of hybrid cars over cars with the IC engine are reflected in good performance characteristics (high torque even in low revs) and low emission of exhaust gases, and the disadvantages are high cost and weight (Fig. 2).

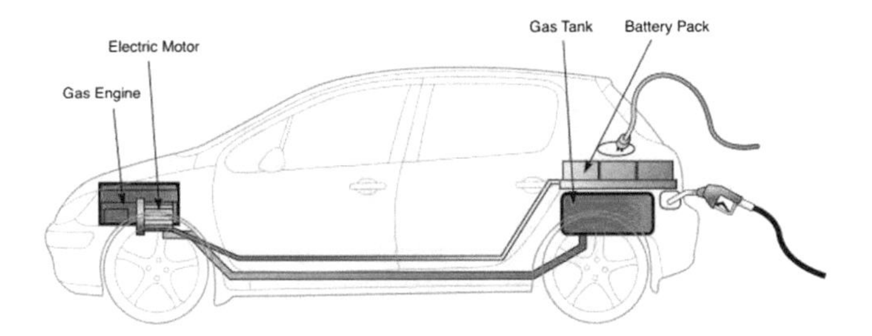

Fig. 2. Chargeable hybrid car

2.3 Battery Electric Car

The battery electric car (BA) is a fully electric car which uses only an electric drive consisting of a powerful electric motor and a high capacity battery pack. For a radius of 500 km, the capacity of the battery should be 75 kWh. The modern BA travels 120–150 km with a 30 kWh battery (Fig. 3).

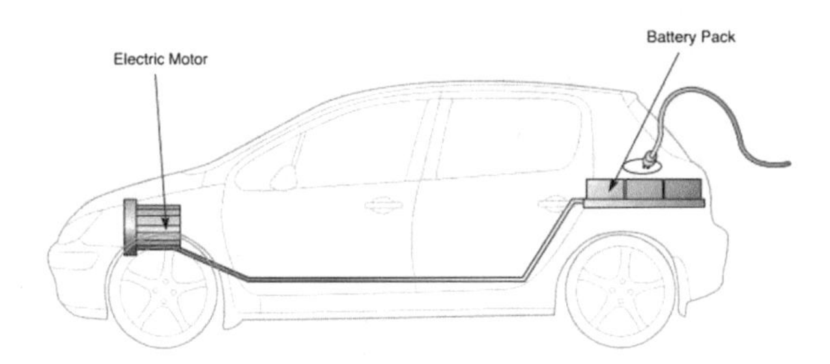

Fig. 3. Battery electric car

Disadvantages of BA are reflected in a small radius of movement, relatively low acceleration, low speeds and poor curves behavior due to the high battery weight [1–4].

2.4 Electric Car with Fuel Cells

An electric fuel cell car is supplied with electricity from fuel cells in which hydrogen is burned in the presence of air, and electricity and water are obtained. The resulting electricity is stored in *Li-ion* batteries.

A fuel cell that is installed in a Nissan X-trail electric car with fuel cells is given in Fig. 4.

Fig. 4. Fuel cell installed in Nissan X-trail electric car with fuel cells

French *Post Office* and *RENO trucks* have built a truck of 4.5 t with fuel cells. The fuel cells were manufactured by *Symbio FCell* company. The vehicle moves silently and can reach 200 km without recharging. An electric truck with fuel cells is given in Fig. 5.

Fig. 5. Electric truck with fuel cells

A hydrogen tank, which is an explosive gas, must be installed in electric vehicles with fuel cells. Electric cars with fuel cells belong to vehicles with zero emission of harmful gases [1, 2, 6, 7].

2.5 Solar Cars

A solar car denotes an electric car powered by electricity generated by solar cells (Figs. 6 and 7).

Fig. 6. Solar car – Toyota Solar Prius

Fig. 7. PV station for electric car charging

Solar cars are still not used for commercial purposes, but mostly for the racing of solar cars. The most significant solar cars race is the *World Solar Challenge* race across Australia (Fig. 8).

Fig. 8. Solar car Solar Challenge

Using solar car, the *Silent Scout* an Australian Hans Tolstrup traveled the route of 1000 km from Perth to Sydney in 1983 with an average speed of 23 km/h. Four years later, in 1987, H. Tolstrup opened the first race of solar cars in Australia. 23 solar cars from seven countries participated in the race. On the track 3010 km long, General Motors' Solar Runner won at an average speed of 67 km/h.

At the second world competition of solar cars held in Australia in 1990 on the same track as the first competition, the Swiss car Duh Bijela II won at an average speed of 62.7 km/h [8, 9].

3 Current State of Electric Cars Use

3.1 Worldwide

Worldwide, there is an increase in the use of electric cars, but still insufficient to change the negative impact of vehicles with IC engines on the environment.

The International Energy Agency *(IEA)* announced that the number of electric cars in the world at the end of 2016 exceeded two million vehicles. Electric vehicles make up only 0.2% of total vehicles for the transport of passengers.

Worldwide electric cars are mostly used in China, USA, Japan, Canada, Norway, Great Britain, France, Germany, the Netherlands and Sweden. China is the largest market for electric cars after the United States. In China, the sale of electric cars has increased due to the state's efforts to reduce air pollution. The IEA report notes that the European market for electric vehicles relies on six states that give fiscal privileges to motivate people to buy such a type of vehicle.

A comparative overview of newly registered electric cars in the world in 2014 and 2015 is given in Fig. 9.

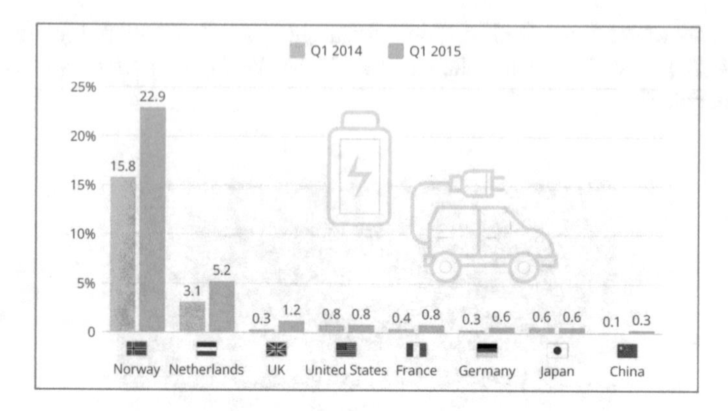

Fig. 9. A comparative overview of newly registered electric cars in the world in 2014 and 2015

In the United States, there is a very well developed network of charging stations for electric car batteries, and there are more than 25,000 in Europe. Tesla filling stations for electric car batteries in the US and Europe are shown in Fig. 10.

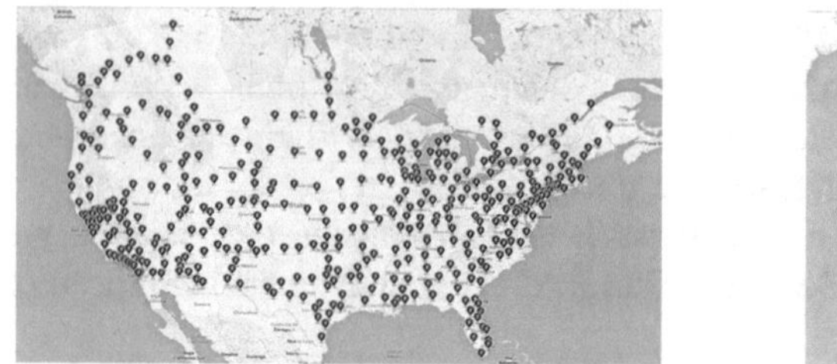

Fig. 10. Tesla supercharger stations in USA (left) and in Europe (right)

All locations for charging electric car batteries appear on the *Tesla car's* software so drivers at all times have information about the nearest stations including the power of charging. Electric cars are especially suitable as delivery vehicles in post offices around the world.

3.2 Serbia

There are few electric cars in Serbia. Hybrid electric cars in Serbia are mostly used by taxi drivers. In Serbia there are only four charging stations for electric car batteries: Bubanj potok, Šid, Subotica-south and Presevo. The development of electric car charging stations can be expected to increase the number of electric cars in Serbia. In connection with this, it is necessary to reorganize and build new SN and NN networks and corresponding TS stations in Serbia [1–4].

4 Impact of Electric Car Use on the Environment

The biggest exhaust gases from IC engines pollution is in cities. Therefore, the use of EC in cities would contribute to the reduction of air pollution by exhaust gases from the IC engine.

In addition, EC is moving silently as compared to cars with an IC engine.

Electric cars do not depend on fossil fuels whose reserves are limited. 60% of the total oil consumption in the developed countries is for transport purposes. For the production of electricity, in addition to fossil fuels (coal, oil, natural gas), nuclear energy and renewable energy sources (hydropower, wind, solar, biomass, geothermal energy) are used.

The highest energy saving in relation to IC engine vehicles is achieved in city driving.

Electric cars are as ecologically clean and energy efficient became the subject of the energy policy of developed countries (USA, Japan, Germany). In developed countries, regulations are adopted according to which ECs must be represented in a certain percentage in total car production. Such regulations also apply in other countries. This is expected to increase the number of ECs in the world [1].

5 Conclusion

The energy efficiency of the electric motor is significantly higher than the energy efficiency of the internal combustion engines. Electrically powered batteries do not consume energy as they stop in traffic jam. A part of the energy lost by braking is up to 20% returned through a regenerative braking.

Petrol engines efficiently use only 15% of the fuel energy for moving the vehicle or powering up additional equipment. The efficiency of the diesel engines is 20%, and the electric motor is 80%.

By 2020, it is estimated that from 9 to 20 million ECs will be sold, and by 2025, 40 to 70 million electric cars will be sold. In order to achieve the goal of limiting carbon dioxide emissions into the atmosphere, it is expected that by 2040, 600 million electric vehicles will be sold.

Due to the development and increasing of life standards in China and India, it is expected that by 2020, the number of cars in the world will double. Increased use of electric cars will lead to the increased consumption of electricity.

In addition to the construction of new electricity sources for the use of ECs, it is necessary to build a new medium and low voltage network, adequate transformer stations and new EC charging stations. Furthermore, an advanced electricity networks (SMARTGRIDS) and the integration of ECs and electric vehicles have become one of the most important areas of research and development in the EU and other parts of the world.

References

1. Pavlović, T.M., Mirjanić, D.L., Milosavljević, D.D.: Electric power industry. Academy of Sciences and Arts of the Republic of Srpska, Banja Luka, p. 527 (2018, in Serbian)
2. Katić, V.A., et al.: Mali električni automobili - ispitivanje osnovnih pogonskih karakteristika, Naučno-stručni simpozijum Energetska efikasnost | ENEF 2013, Banja Luka, B2-22–B2-30 (2013)
3. Katić, V.A., et al.: Electrification of the vehicle proposition system – an overview, Facta Universitatis: series Electronics and Energetics, vol. 27, no. 2, pp. 299–316 (2014)
4. Tovilović, N.: Testiranje pogona električnog automobila – prva faza, diplomski rad, Fakultet Tehničkih nauka, Novi Sad (2013)
5. Lauc, T.: Prikaz i testiranje električnog vozila, diplomski rad, Fakultet Tehničkih nauka, Novi Sad (2013)
6. Global EV Outlook 2017: International Energy Agency, OECD/IEA (2017). www.iea.org
7. Vranić, A.: Uvođenje automobila na električni pogon po ugledu na Evropsku poštu, Industrija 64, Beograd (2016)
8. Pavlović, T., Čabrić, B.: Physics and Techniques of Solar Energy, p. 342. Građevinska knjiga, Beograd (2007)
9. Roche, D.M.: Speed of light. Photovoltaic special, research centre, Sydney (1997)

Vacuum Technology of Effective Nano Suspension Fuel Manufacturing

Evgueni A. Deulin(✉) and E. I. Ikonnikova

MT-11, BMSTU, 2nd Baumanskaya st. 5, 105005 Moscow, Russia
deulin@bmstu.ru

Abstract. The effective Vacuum technology of nano-structured fuel-suspension manufacturing, based on reproducible resource is presented. The fuel-suspension heat ability is rather higher then the ability of initial components: coil and ethylene. The technology supplies the potential customers with ecology clean and profitable fuel, because it is based on the reproducible resources that may be used in regions, being isolated of petrol manufacturing companies. The fuel is ecologically clean and effective.

1 Introduction

The technology, being presented is a new technology of effective and environmentally clean fuel manufacturing is based on the reproducible vegetable resources that may be used in regions, being isolated of petrol manufacturing companies. The large number of automotive companies now work with natural sources of fuel [1]. In the role of raw stock they consider wastes of forest, peat and coil processing industries. It is wonderful, that heat ability of fuel-suspension, being manufacturing with mentioned reassures is higher, than heat ability of initial fuel components.

The base point for fuel consumers in South Europe is the fact, that presented technology provide the consumers with cheap and ecology clean fuel in conditions of consumer separation from petrol manufacturing companies, and we must know, that these companies usually are predators of oil market.

2 Theoretical Base of the Technology

For heat ability estimation of fuel-suspension it is necessary to understand the features of the fuel manufacturing, that consist in physics of nano-scale processes at milling of rough raw material (coil, turf, wood, grain) abrasion procedure till the size of the product (particles) rich the volume with the sphere radius less 100 nm [2]. This “dry friction” process (abrasion procedure) goes in vacuum with oxygen absence at water vapor pressure (about 10^3 Pa) till monolayer of water molecules forming process on the milled particles surfaces is realizes. The sorbed water molecules dissociate into OH radical and hydrogen atoms [3]. The milling time t depends on hydrogen atoms diffusion from the surfaces of the particles being milled, reach the centers of the particles.

I. Karabegović (Ed.): NT 2019, LNNS 76, pp. 549–556, 2020.
https://doi.org/10.1007/978-3-030-18072-0_63

This time is large, as we know [4], that the depth x of the atoms penetration (diffusion) depends on time and may be expressed with 2nd Fike equation:

$$C_0(x,t) = (C_{surfH} - C_i) \cdot \left(1 - erf\left(\frac{x}{2 \cdot \sqrt{D \cdot t}}\right)\right) + C_i \tag{1}$$

where:

C_{surfH} - hydrogen atoms concentration in first sorbed on the surface monolayer;
C_i - initial hydrogen atoms concentration in the volume of raw material;
D - diffusion coefficient;
x - distance from the object surface;
t - time.

The authors results [5] analysis shows that this physical process leads to high hydrogen saturation of the raw material (coil, turf, grain), that 3–4 times higher, then hydrogen saturation being rich with traditional methods [6].

The fuel-suspension technology applies that suspension is based on two component usage: 1 – nano-particles rich with hydrogen, 2 – liquid component (ethylene or methylene) being produced of vegetable resources. The mentioned 2nd component may be used as independent cheap automobile fuel instead of petrol, as they realize in Brazil and in Argentine, where the petrol international monopolies keep too high price, but the process of individual ethylene manufacturing is profitable.

The new tasks being solved in our technology is full-scale decision of mineral resources also as the vegetable resources usage in vacuum as in protected media.

The heat ability of fuel calculation may be done with the Eq. (1) usage if we may calculate the initial hydrogen atoms concentration on the particles surfaces. In this case we had to return to (1) equation, where the C_{surfH} parameter depends on media parameters and may be presented:

$$C_{surfH} = \Theta N_{1\Pi}\, d_0\, F_T \tag{2}$$

Where:

F_T - the particles surface being rubbed in vacuum,
$N_{1\Pi}$ - the "number" of free places for the hydrogen atoms being sorbed on the particles surface,
d_0 - sorbed atom diameter (d_0 = 0.3 nm),
Θ - coverage coefficient (number of sorbed atoms monolayers) for hydrogen isotopes atoms.

It may be calculated [3]:

$$\Theta = e^{\frac{E_1 - E_L}{RT}} \cdot p \cdot \left((p_s - p) \cdot \left(1 + \left(e^{\frac{E_1 - E_L}{R \cdot T}} - 1\right) \cdot \frac{p}{p_s}\right)\right)^{-1} \tag{3}$$

E_1 - physical sorption heat for 1st monolayer, J/·kilomole,
E_L - heat of evaporation for L-monolayer of sorbet gas, J/·kilomole,
p, p_s - residual pressure, saturated vapor pressure, Pa,
R - gas constant,
T - sorbed gas temperature,
C_i - initial hydrogen atoms concentration in the volume of raw material at /sm^3,
D - hydrogen diffusion coefficient in raw material,
x - root square distance of the hydrogen atom from the object surface.

3 Physical Base of Fuel-Suspension High Efficiency

The main source and reason of fuel –suspension high efficiency is large number N_T of hydrogen atoms being accumulated in every particle of fuel-suspension. This number depends on dry friction process parameters, also as on raw material parameters and may be calculated with equation:

$$N_T = F_T \bullet \int^{x} \int^{t} d\, C_0(x.t)/dx \bullet dt \tag{4}$$

The summarized hydrogen atoms number being accumulated in unit volume of solid component of fuel-suspension V_N may be calculated:
$N_{\sum} = N_T \bullet V_N$, where:

$$V_N = \frac{4}{3}\pi R^3 = \frac{4}{3} \cdot 3.14 \cdot (10^{-7})^3 = 4.18 \cdot 10^{-21} M^3$$

The heat ability of the fuel being manufacturing may be detected with equation:

$$A_T = N_{\sum 1}(N_{\sum} A_{OH} + V_P A_P) \tag{5}$$

Where:

$N_{\sum 1}$ - the solid nano-particles number in unit volume of fuel-suspension, /m^3
V_p - The volume of liquid component per./one nano-particle volume, м^3.

The average volume of one fuel nano-particle, formed with one particle size (particle sphere radius R = 100 nm = 10^{-7} m):

A_{OH} - hydrogen-oxygen atoms linking energy, J/at,
A_p - specific heat ability of fuel-suspension solid component J/m^3.

Heat ability of nano-fuel solid component may be calculated as a summarized hat ability of hydrogen atoms A_{OH} in unit volume:

$$A_H = N_{\sum 1} \cdot A_{OH} \cdot N_{\sum} \tag{6}$$

$N_{\sum}$ - summarized number of hydrogen atoms in unit volume of solid nano-particles of fuel.

Whereas: $V_p \cong V_N$, where: $V_p = 4.18 \cdot 10^{-21}$ m^3 - volume of fuel liquid component/per one solid micro particle, the figure $N_{\sum 1}$ may be calculated with equation:

$$N_{\sum 1} = \frac{0.5}{V_N} = \frac{0.5}{4.18 \cdot 10^{-21}} = 1.19 \cdot 10^{20}\ atoms \tag{7}$$

Atoms of hydrogen and oxygen linking energy for one H atom, according [2] may be calculated:

$$A_{OH} = 140.9 \cdot 10^6 \text{ J/kg} = 2.35 \cdot 10^{-19} \text{ J/}atoms \tag{8}$$

The summarized hydrogen atoms number in unit volume of fuel was detected as: $N_{\sum} = _{C_{x,t}} \cdot V_N$, where figure $C_{x,t} = 8 \cdot 10^{21} \div 2 \cdot 10^{21}\ atoms/\text{sm}^3 = 8 \cdot 10^{27} \div 2 \cdot 10^{27}\ atoms/\text{m}^3$ - is a hydrogen atoms concentration, that was experimentally detected with SIMS analysis usage [6, 7]. The experiments and calculation show, that hydrogen atoms concentration on the depth $\boldsymbol{x}$ = 0.1 μ from the micro-particle surface, i.e. in the center of particle, after 24 s of friction is equal to figure: $C_c = 4 \cdot 10^{27}\ atoms/\text{m}^3$. In this case the summarized hydrogen atoms number in unit volume is:

$$N_{\sum} = 4 \cdot 10^{27} \cdot 4.18 \cdot 10^{-21} = 16.72 \cdot 10^6\ atoms \tag{9}$$

The summarized hydrogen atoms energy for this number is:

$$A_H = 1.19 \cdot 10^{20} \cdot 2.35 \cdot 10^{-19} \cdot 16.72 \cdot 10^6 = 46.75 \cdot 10^7 \text{ J} \tag{10}$$

We must remember, that the initial heat ability of solid coil in unit volume is 69.4 J, and it increases with help of hydrogen atoms being penetrated (dissolute) into solid volume of the micro particle.

The summarized heat ability of solid particle filled with hydrogen is; $A_H + A_{\sum C} = 46.75 \cdot 10^7\,\text{J} + 69.4\,\text{J}$, i.e. we receive figure, that 10^7 times more the initial raw material heat ability. These figures contain the explanation of raw solid material heat ability large increasing after milling process. The second reason of our fuel heat ability increasing is additional heat ability of liquid component of fuel-suspension that may be calculated:

$$A_{\sum P} = N_{\sum 1} \cdot A_P \cdot V_P, \tag{11}$$

Where: $A_P = 3 \cdot 10^7$ J/kg $= 3.8 \cdot 10^4$J/m^3 - specific heat ability of fuel liquid component.

As a result, we receive: $A_{\sum p} = 1.19 \cdot 10^{20} \cdot 3.8 \cdot 10^{4} \cdot 4.18 \cdot 10^{-21} = 18.9 \cdot 10^{3}$ J.

In conclusion we may see, that the heat ability of nano-structured fuel is based on beat ability of three base components:

$$A_T = A_H + A_{\sum C} + A_{\sum p}, \quad (12)$$

where:

$N_{\sum 1}$ - The summarized number of hydrogen molecules accumulated in fuel nanoparticles per unit volume
V_p - volume of natural fuel liquid component per one nano particle
A_{OH} - energy of atom hydrogen oxidation
A_p - heat energy of nano structured fuel liquid component.

It was shown theoretically and experimentally with the Secondary Ion Mass Spectrometry analysis, that depending on the friction time (0.1–2 min) the depth x of hydrogen atoms penetration reaches a depth of 1–10 μ. [5]. and its concentration increases many times. So we can see on Figs. 1 and 2, for the coil sample, that the hydrogen and deuterium penetration on the depth 2–5 μ increases 2–5 times after 0.5 min friction process.

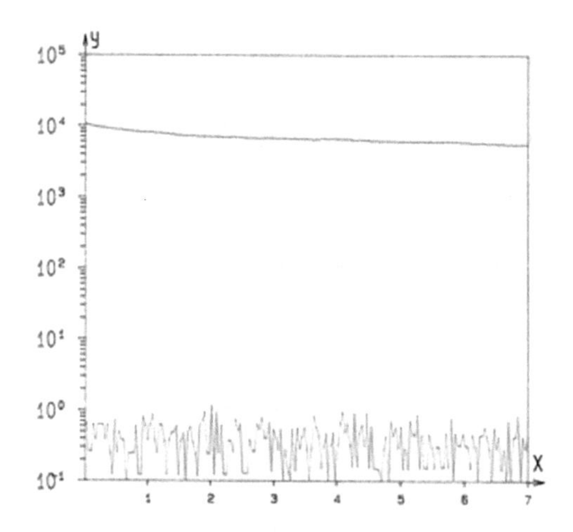

Fig. 1. AxisY-hydrogen (upper) and deuterium (low) atoms concentration (at/sm^3 multiplied by 10^{17}) initial distribution in coal as a function of X-depth, microns

It was shown experimentally many times [6, 7] that hydrogen atoms concentration increases 200–300 times in comparison with ground one. In the solid materials of friction pair.

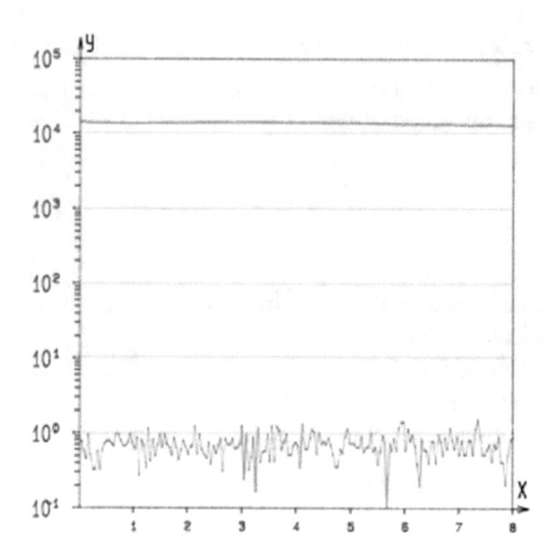

Fig. 2. Hydrogen (upper) and deuterium (low) atoms concentration (at/sm^3 multiplied by 10^{17}) distribution in coal after 30 s of friction

4 The Results Presentation

The theoretical and experimental results of the investigations that were done in BMSTU, Moscow may be illustrated with figures, below, that visually show the advantages of the presented technology, that was patented [2] in Russia.

The components of nano-structured fuel parameters are: hydrogen, coil, ethylene and comparison of their heat parameters are presented on Figs. 1 and 2. We must remember, that hydrogen atom is the smallest and the flyweight atom, that visually shown on these figures (Figs. 3 and 4).

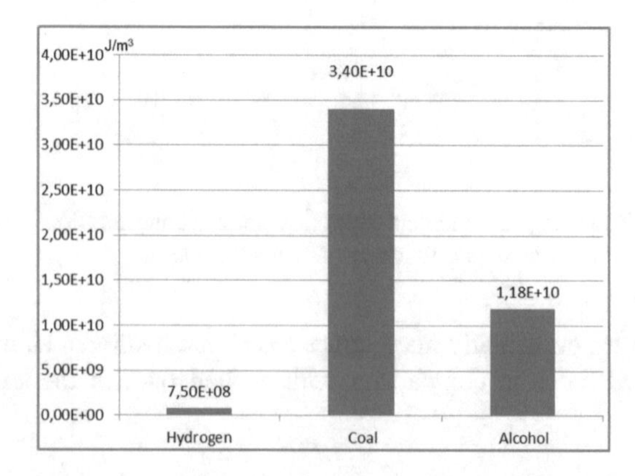

Fig. 3. "Volume" heat ability (J/m^3) of fuel-suspension base components: Hydrogen, Coil and Ethylene

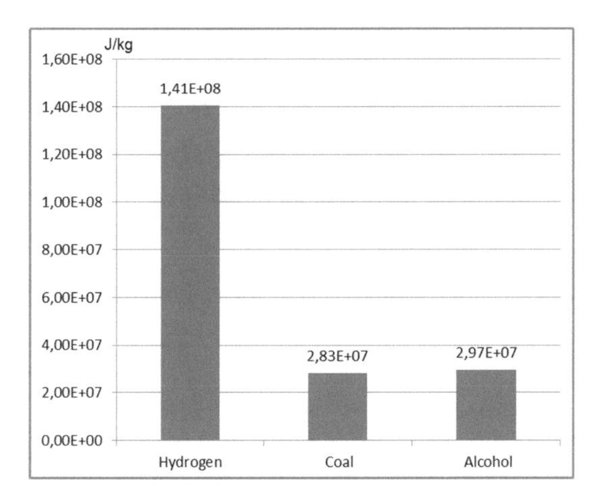

Fig. 4. "Mass" heat ability (J/kg) of fuel-suspension base components: Hydrogen, Coil, Ethylene

The Fig. 5 results comparison of different fuels show that nano-structured fuel-suspension heat ability higher, then the ones of coil and ethylene (as a base components of fuel), At that time the technology authors know, that the heat ability patented fuel may be increased with some technology parameters control, that we will not discus in the paper.

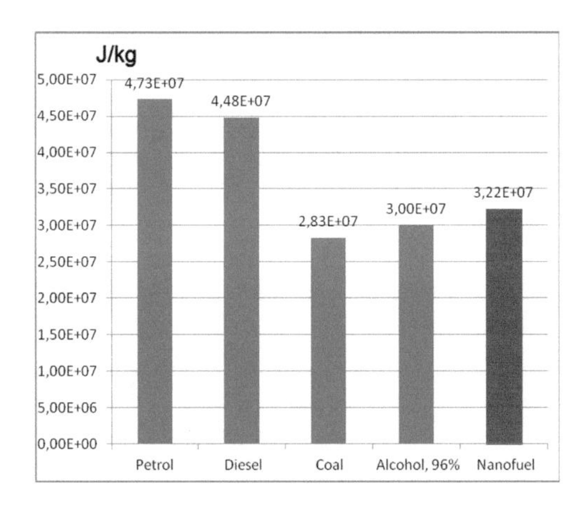

Fig. 5. "Mass" heat ability (J/kg) of nanostructured fuel variant (right) in comparison with traditional fuels

5 Conclusion

As a result of fuel technology analysis we had to show main features of this technology:

1. Heat ability of solid mineral coil peace after it milling process, till it rich's nano-scale peace's size had to be increased on 13.05%.
2. Heat ability of solid (coil) component of fuel suspension is more than heat ability of ethylene on 10%.
3. The heat ability of fuel-suspension is 3.8%. more than heat ability of ethylene – base component of fuel.
4. The main advantages of presented fuel are:

   - availability of raw resources usage in regions being separated from oil resources,
   - the renewal resources usage in fuel manufacturing technology,
   - the fuel ecology cleanliness in utilization process (in automobile),
   - the high efficiency of fuel manufacturing technology.

References

1. Bartz-Ostfildern, W.J. (Hrsg): Alkylated Naphthalenes as Performance Synthetic Fuels. Automotive and Industrial Lubrication. Techniche Academie Esslingen. Book of Syn., pp. 238–239 (2006)
2. Deulin, E.A., Moos, E.N.: Patent of Russia № 2 444 561 C1 dated 10.03.2013. The method of nano-structured fuel manufacturing. (in Russian)
3. Redhead, P.A., Hobson, J.P., Kornelson, E.V.: The Physical Basis of Ultrahigh Vacuum. Chapmen and Hall LTD., London (1968)
4. Deulin, E.A.: Exchange of gases at friction in vacuum. In: ECASIA 1997, pp. 1170–1175. Wiley, November 1997
5. Deulin, E.A., et al.: Mechanics and Physics of Precise Vacuum Mechanisms, p. 234. Springer, Heidelberg (2010)
6. Patent of RF №2373262, МПК C10 L 5/44, published 27.04.2009г and Patent of RF - №2281312, МПК C10 B
7. Deulin, E.A., Nevshupa, R.A.: Deuterium penetration into the bulk of a steel ball of a ball bearing due to it's rotation in vacuum. Appl. Surf. Sci. **144–145**, 283–286 (1999)

Photovoltaic Systems with Sun Tracking Position

Edin Šemić[1](✉) and Malik Čabaravdić[2]

[1] Faculty of Mechanical Engineering, University "Džemal Bijedić" of Mostar, University Campus, 88000 Mostar, Bosnia and Herzegovina
edin.semic@bih.net.ba
[2] University of Zenica, 72000 Zenica, Bosnia and Herzegovina

Abstract. This paper describes the work of solar photovoltaic systems and the types of photovoltaic panels. Solar energy can be transformed in many ways into electrical, and the simplest way is through photovoltaic cells. The work principles of photovoltaic cell is based on photoelectric effect. Solar panels can be fixed, or mobile panels with one or two rotation axis. Mobile systems can be optimally positioned in relation to the sun, no matter where the sun is in the sky.

Keywords: Photovoltaic cell · Solar panel · Position · Solar radiation

1 Introduction

In recent years, interest has increased for renewable energy sources by developed countries of the world, which are conducive to development and research in this direction. One of the reasons is certainly the high price of fossil fuels, climate change, and environmental pollution [2].

Renewable energy sources include energy: geothermal energy, water energy, wind, solar radiation and biomass energy.

The sun is an inexhaustible source of energy that can be exploited to produce electricity today and with the technological achievements of today without harming the environment. The advantage of renewable energy sources is, above all, the fact that they do not have any harmful effects on the environment as well as the ability to supply electricity to consumers in areas where a distribution network has not been developed or doesn't exist at all. The scientists believe that, in the future, solar and wind energy will become the main electricity producers.

The main disadvantage of this system is that the production of electricity depends on sunlight. Large panels of large power plants are also required, which makes investing in this system more cost-effective.

2 Solar Systems

2.1 Solar Radiation

Solar radiation is the basic energy source for all processes in the Earth's atmosphere. The sun radiates a huge of energy that mostly converts into heat. The surface

I. Karabegović (Ed.): NT 2019, LNNS 76, pp. 557–565, 2020.
https://doi.org/10.1007/978-3-030-18072-0_64

temperature of the Sun is about 5778 K (5505 °C), and radiation expanse (calculated energy by surface and time) on the surface of the Sun is about 63500 W/m^2 [1].

The Earth's Radiation with the solar radiation at the top of the Earth's atmosphere is called the Solar constant. The distance between Earth and Sun changes over the year, so the actual amount of Sun's constant is changed, so it can be calculated with the average S_0 = 1396 W/m^2 (Fig. 1).

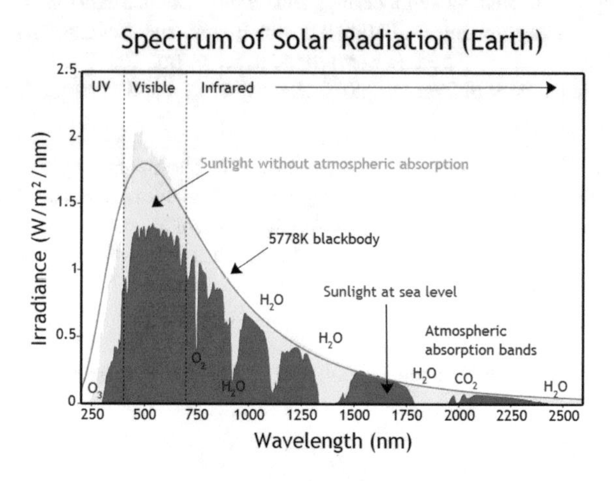

Fig. 1. Spectrum of solar radiation [6, 7]

Compared to other European countries, Bosnia and Herzegovina belongs to countries with good annual solar radiation, especially Herzegovina. On average, about 1500 kWh/m^2 of solar energy per year falls on the ground, indicating an excellent potential for producing electricity from the sun's energy (Fig. 2).

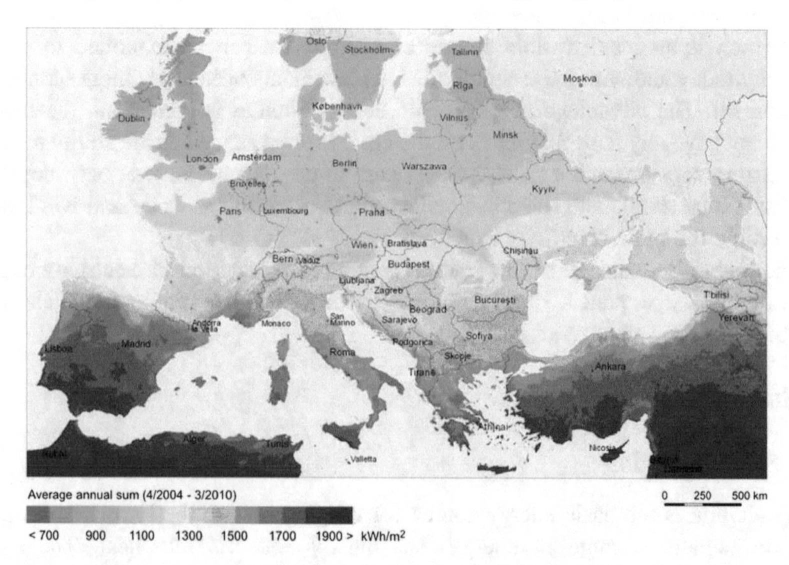

Fig. 2. Solar potential in Europe [10]

2.2 Solar Cells

Solar (photovoltaic) cell is a semiconductor element that converts solar energy into electrical photovoltaic effect. According to quantum physics, light has a dual nature, it is both particle and wave. The particles of lights called photons. When photons hit the photovoltaic cell, they can be rejected from it, passed through it or absorbed. Only absorbed photons give the energy to release electrons, i.e. photovoltaic effect. Efficiency of cells ranges from 10 to 25%, depending on the performance (Fig. 3).

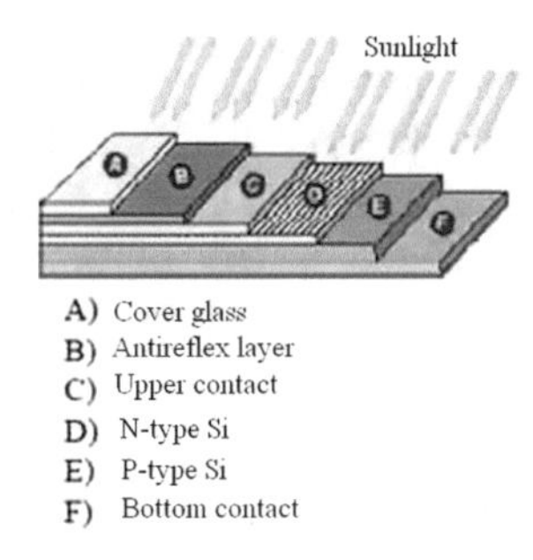

Fig. 3. Solar cell construction [8]

Figure 4 illustrates the principle of the photovoltaic effect. When the semiconductor absorbs enough light, the electrons are ejected and move to the upper layer while the cavities are moving in the opposite direction where the electrons are expected. The voltage cell will be obtained on the external contacts of the PN juncton.

The solar cells can be connected serial or parallel. By serial connection of solar cells, voltage increases, and the current strength will remain unchanged. Parallel connection of solar cells, the current gain is collected and the voltage remains unchanged.

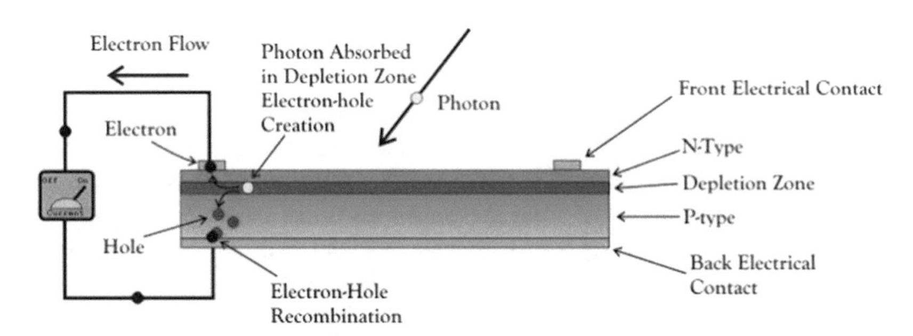

Fig. 4. Photoelectric conversion of solar energy [11]

2.3 Historical Development of Solar Cells

The first (silicon) solar cell was discovered in 1941 by American engineer Russell Ohl. Its utilization was below 1%. Subsequently, in 1954, a group of researchers at the Bell

Laboratories in New York produced silicon solar cells with a efficiency of 6%, and the first solar cells were used in space exploration.

In the 70s of the twentieth century, solar cells began to be applied to Earth during the oil crisis when began the demand for alternative energy sources.

Nowadays, especially the last ten years, the number of solar cells installed in the world is exponentially increasing, especially dominated by Japan, Germany, USA, Taiwan and China (Fig. 5).

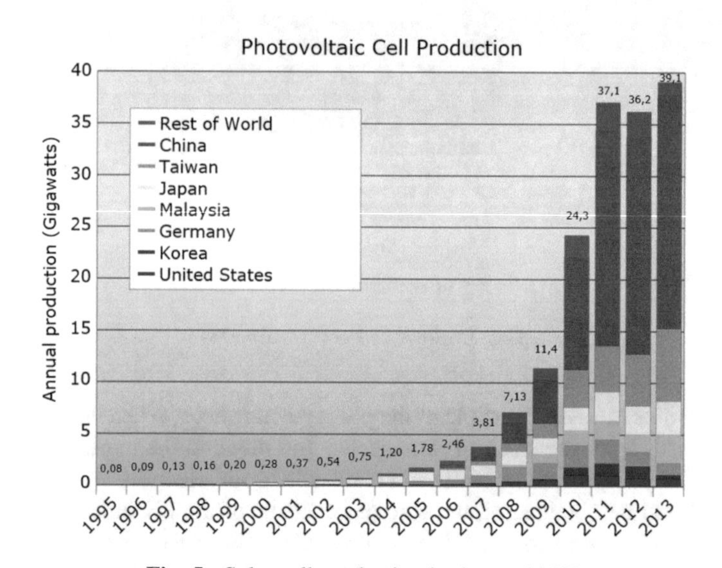

Fig. 5. Solar cell production in the world [9]

2.4 Solar Cell Production Technology

The most commonly used materials for producing solar cells are monocrystalline silicon, polycrystalline silicon and thin layer materials.

Solar cells in crystal silicon technology consist of two layers of semiconductor material, with metallic connections used to collect generating charge carriers (electricity). Generated one cell voltage is about 0.5 V, so the cells are connected serially within the protective "sandwich" from the tanned glass and plastic to the photovoltaic module. The photovoltaic modules with crystalline silicon cells are most represented on the market because they have high efficiency and longevity [4].

The first generation are made of monocrystalline silicon (c-Si) cells, and polycrystalline silicon (pc-Si). The difference between these materials is in the crystal structure, i.e. atomic arrangement in the cells. Polycrystalline silicon consists of large crystalline grains of a few hundred micrometers to a few millimeters (Fig. 6).

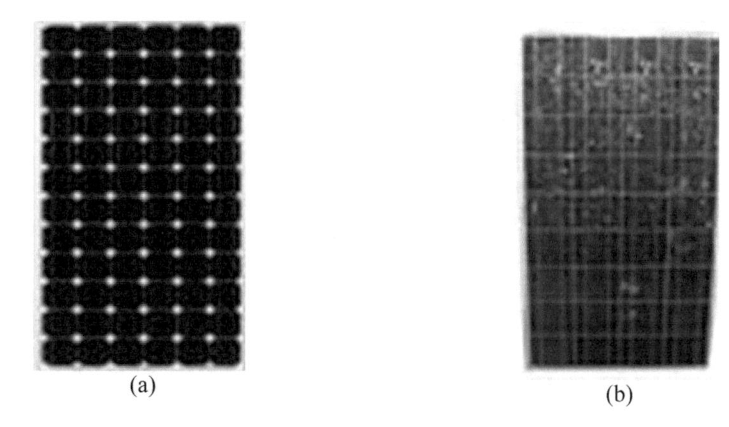

(a) (b)

Fig. 6. Monocrystalline (a) and polycrystalline (b) solar cell

The second generation are thin-cell solar cells. This group consists of: Amorphous silicon (a-Si), Cadmium teluride (CdTe), Copper-indium-galite selinide (CIGS).

The third generation is made up of: Amorphous-nanocrystalline solar cells, Photo-electrochemical (PEC) cells - Graetzel's cells, Polymers sun cells, Sunscreen cells synthesized in coating (DSSC) (Fig. 7).

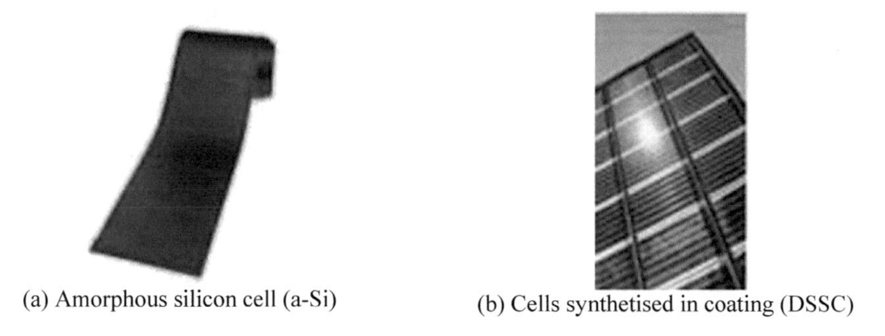

(a) Amorphous silicon cell (a-Si) (b) Cells synthetised in coating (DSSC)

Fig. 7. Second and third generations of solar cells

The fourth generation are hybrid-organic solar cells with a polymers matrix. The fourth generation still has no industrial application, as it is currently in the phase of research and technology optimization.

Figure 8 shows the power that various solar cell generations give on a sunny day.

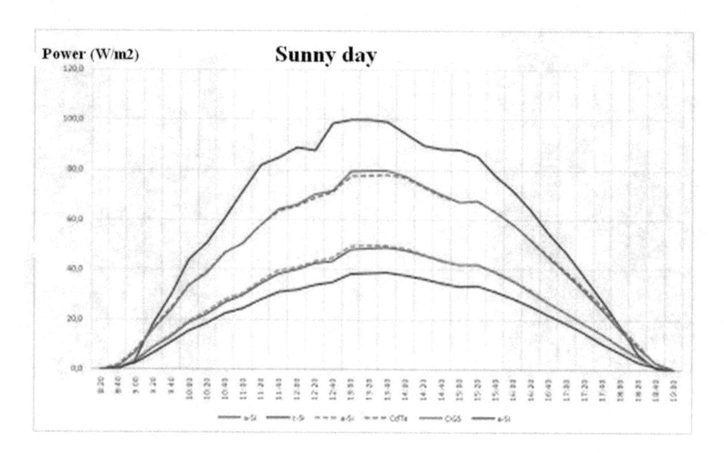

Fig. 8. The power given by different generation of solar cells

3 Solar Panel Constructions

3.1 Fixed System

The term “fixed system” refers to a system that is placed at a certain location at an optimum angle and has no ability to move the active surface. Such a panel compared to the sun represents a stationary point. Figure 10 shows one system like that [4, 5].

Fixed solar panels are usually placed under optimum inclination, i.e. to achieve maximum efficiency they need to be directed in the direction that catches the most sun. Fixed systems are simple in design, easy to set, design and maintain. Since they have no moving parts, fixed systems are flexible and do not require expensive maintenance. The disadvantage of this system is the fact that it is not optimally aligned. This means it will produce less energy than the mobile system (Fig. 9).

Fig. 9. Fixed axis solar panel

3.2 Single Axis Mobile System

Single axis mobile systems have one degree of freedom that acts as a part of rotation. The axis of rotation of single axis trackers is typically aligned along a true North meridian. It is possible to align them in any direction with advanced algorithms.

There are several types of uniaxial systems: horizontal single axis system, horizontal single axis system with tilted modules, vertical single axis system, tilted single ayis system and polar single axis system. The orientation of the module with respect to the axis is important when modeling performance.

The horizontal type is used in tropical areas where the sun becomes very high at noon and the days are short. On the other hand, the vertical type is used in areas with a higher latitude, where the sun is not very high, but summer days can be very long.

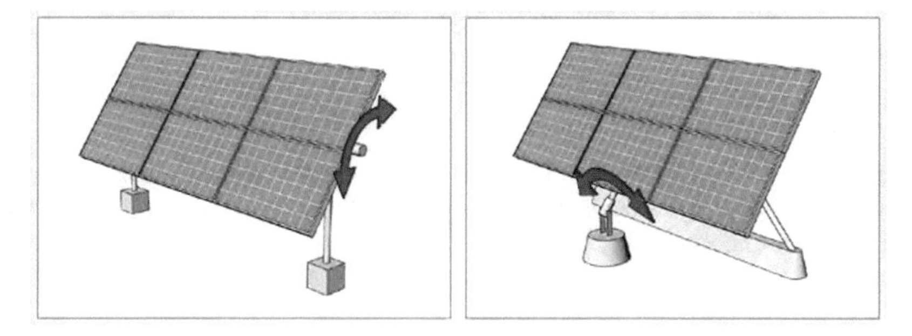

Fig. 10. Horizontal and tilted single axis solar panel

3.3 Dual Axis Mobile System

Dual axis mobile systems have two degrees of freedom that act as axes of rotation. These axes are usually perpendicular to each other. The axis that is fixed relative to the ground is considered as a primary axis. The axis that is referenced to the primary axis can be considered a secondary axis.

There are several common implementations of the dual axis mobile systems. They are classified by the orientation of their primary axes in relation to the ground. Two common implementations are tip-tilt dual axis systems and azimuth-altitude dual axis systems.

The orientation of the module relative to the axis of the system is important when modeling performance. Dual axis systems typically have modules oriented parallel to the secondary axis.

No matter where the Sun is in the sky, the dual axis systems are able to be in direct contact with the Sun at an optimum angle (Fig. 11).

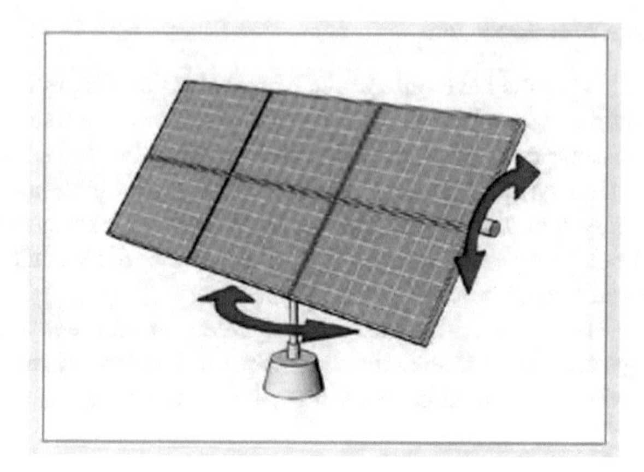

Fig. 11. Dual axis solar panel

4 Conclusion

Solar energy is one of the most reliable alternative energy sources in the modern era. Many studies have been made to improve the efficiency of photovoltaic systems. Previously used photovoltaic panels were fixed at a certain angle that collected solar energy and transformed it into electrical. These photovoltaic panels are ineffective because they are fixed only at certain angles.

This problem can be solved by using solar solar tracking system. The solar sun tracking system is one of the best approaches, as it collects more solar energy in relation to fixed panel systems. The mobile system, or "Solar Tracker", follows the position of the sun throughout the day from east to west on day and season.

This paper presents a comparison between fixed panels, single axis and dual axis solar panel systems. Based on solar radiation, the output power and the total energy provided by the solar panels depend, among other things, on which of these three systems is used.

In terms of price, single axis solar panels are cheaper, and therefore initially more desirable than the dual axis system.

Also, the energy efficiency of the single axis system compared to the dual axis would be more differentiated at different locations [3].

The dual axis sun monitoring system provides more electricity than other systems, but it is important to note that the performance of three different systems does not have to be so drastically different.

References

1. Chowdhury, K.I., Alam, M., Bakshi, P.S.: Performance comparison between fixed panel, single-axis and dual-axis sun tracking solar panel system. BRAC University, Dhaka, Bangladesh (2017)
2. Odak, T., Marković, I., Novak, T., Meštrović, Z.: The photovoltaic system with sun tracking position. Polytechnic & Design, vol. 5, no. 3 (2017)
3. Umihanić, M., Ćehajić, N., Salihović, N.: Comparative analysis of electricity energy for fixed photovoltaic systems in different parts of B&H. Tech. Bull. **9**(2), 128–135 (2015). ISSN 1846-6168
4. Basaran Filik, Ü., Filik, T., Nezih Gerek, Ö.: A hysteresis model for fixed and sun tracking solar PV power generation systems. Energies J. **11**, 603 (2018). Department of Electrical and Electronics Engineering, Anadolu University, Eskisehir, Turkey
5. Mikulović, J., Đurišić, Ž., Kostić, R.: Determination of optimum inclination angles of photovoltaic panels, Infoteh-Jahorina, vol. 12, pp. 243–248, March 2013
6. https://en.wikipedia.org/wiki/Sunlight
7. http://blog.meteo-info.hr/meteorologija/suncevo-zracenje/
8. http://www.izvorienergije.com/energija_sunca.html
9. https://hr.wikipedia.org/wiki/Solarna_fotonaponska_energija
10. http://www.solarni-kolektori.net/karta-sucevog-zracenja/
11. http://kep-power-testing-blog.blogspot.com/2017/06/photoelectric-conversion-of-solar-energy.html

New Technologies in Agriculture and Ecology, Chemical Processes

Efficiency of Precipitation and Removal of Pb(II) and Zn(II) Ions from Their Monocomponent and Two-Component Aqueous Solutions Using Na_2CO_3

Amra Selimović(✉), Halid Junuzović, Sabina Begić, and Ramzija Cvrk

Faculty of Technology Tuzla, University of Tuzla, Ul. Univerzitetska 8, 75000 Tuzla, Bosnia and Herzegovina
amra.selimovic@untz.ba

Abstract. Chemical precipitation is the most widely used method for heavy metal removal from water, and its effectiveness depends on several factors such as the type and initial concentration of heavy metals present in water, the precipitating agent used and the pH of the solution. In this paper an experiment of chemical precipitation and removal of Pb(II) and Zn(II) from their monocomponent and two-component aqueous solutions was carried out in a laboratory by batch process, using sodium carbonate as precipitating agent. The influence of initial concentrations of Pb(II) and Zn(II) and pH values of their aqueous solutions on the efficiency of precipitation and removal of lead and zinc ions was examined. By increasing the pH of the aqueous solutions, a higher efficiencies of chemical precipitation and removal of Pb(II) and Zn(II) were obtained, with higher removal efficiencies being achieved for the lead. The efficiency of removal of heavy metals was higher in solutions that had higher initial concentrations of heavy metal ions.

Keywords: Heavy metal · Water treatment · Chemical precipitation · Sodium carbonate

1 Introduction

Comprising over 70% of the Earth's surface, water is undeniably the most valuable natural resource existing on our planet [1]. However clean water supplies are under threat from urbanisation, industry and agricultural development [2]. Among various types of water pollutants, heavy metals are the largest class of contaminants and also the most difficult to treat [3]. Pollution caused by heavy metals has become one of the most serious environmental problems today [1]. Their multiple industrial, domestic, agricultural, medical and technological applications have led to their wide distribution in the environment, raising concerns over their potential effects on human health and the environment [2]. Source waters (surface water and ground water) have been increasingly contaminated due to increased industrial and agricultural activities [3]. Although some heavy metals are biological essential, they are all non-degradable and

I. Karabegović (Ed.): NT 2019, LNNS 76, pp. 569–575, 2020.
https://doi.org/10.1007/978-3-030-18072-0_65

tend to bioaccumulate in living organisms, expressing toxicity when present above threshold concentrations. The accumulation of heavy metals certainly has adverse effect on aquatic flora and fauna and may constitute a public health problem where contaminated organisms are used for food [3].

There are various methods used for water and wastewater treatment to decrease heavy metal concentrations, such as: membrane filtration, ion-exchange, adsorption, chemical precipitation, nanotechnology treatments, electrochemical and advanced oxidation processes [4]. Chemical precipitation is the most widely used for heavy metal removal from inorganic effluent [5]. This method is based on the addition of alkaline chemicals (precipitating agents) to water, to raise the pH of the solution and thereby reduce the solubility of metal ions which form insoluble compounds, i.e. precipitates that can be easily removed from the water by subsequent separation procedures. Chemical precipitation is a well-established technology with ready availability of equipment and various chemicals, which requires low maintenance since only replenishment of chemicals is needed, with no need for sophisticated operators [6]. The most common precipitation processes are three types including hydroxide precipitation, carbonate precipitation, and sulfide precipitation [7]. Although hydroxide precipitation is the most commonly used, the main drawbacks of this procedure are voluminous sludges of poor filterability and requirements for high pH values to achieve optimal precipitation of some metal ions such as lead and zinc [8]. Since metal carbonate precipitates can settle and be dewatered more easily than the metal hydroxide precipitates [9], carbonate precipitation can be an effective treatment alternative to hydroxide precipitation.

Generally, the effectiveness of a chemical precipitation process is dependent on several factors, including the type and concentration of ionic metals present in solution, the precipitating agent used, the reaction conditions (especially the pH of the solution), and the presence of other constituents that may inhibit the precipitation reaction [10]. Adjusting the pH of the aqueous solution of metal ions to alkaline conditions is a major parameter that significantly improves the removal of heavy metals by chemical precipitation [11]. Since the optimal pH for precipitation depends both on the metal to be removed and on the counter ion used (hydroxide, carbonate, or sulfide), the best treatment procedure must be determined on a case-by-case basis [12].

The purpose of this research was to study the effect of initial concentrations of Pb(II) and Zn(II) and pH of their aqueous solutions on the efficiency of their precipitation and removal from monocomponent and two-component aqueous solutions using sodium carbonate (Na_2CO_3) as precipitating agent.

2 Experimental Part

2.1 Materials

In experiments were used chemicals of analytical grade: lead(II) nitrate (Alkaloid AD Skopje, Republic of Macedonia), zinc nitrate hexahydrate (Kemika, Zagreb, Croatia), standard solution of lead 1000 mg/L Pb(II) in 0.5 M nitric acid (from $Pb(NO_3)_2$) and standard solution of zinc 1000 mg/L Zn(II) in 0.5 M nitric acid (from $Zn(NO_3)_2$) from

Merck, nitric acid, min. 65% (Lach-Ner, Czech Republic) and sodium carbonate (Sisecam Soda Lukavac, min. 99,30%).

Laboratory *glassware* was first washed with detergent, rinsed with distilled water, then soaked in aqueous nitric acid solution and finally rinsed with demineralized water and allowed to dry naturally. Aqueous solution of sodium carbonate with molar concentration of 2 g/L was used as precipitating agent. The solution was prepared by dilution with deionized water. Monocomponent aqueous solutions of Pb(II) and Zn(II) of two different initial concentrations (500 mg/L and 50 mg/L) were prepared, and a two-component aqueous solution of both metals of initial concentrations of 500 mg/L. Each artificially prepared aqueous solution was homogenized, after which its initial pH was measured.

2.2 Methods

Removal of heavy metal ions from their aqueous solutions was carried out by the chemical precipitation method. Precipitation experiments were conducted by adding various quantities of the precipitation agent to monocomponent and two-component aqueous solutions of heavy metals. Volumes of sodium carbonate used in experiments are given in Table 1.

Table 1. Volumes of Na_2CO_3 added to monocomponent and two-component aqueous solutions of Pb(II) and Zn(II)

Initial concentration of heavy metal ions	Initial pH of heavy metal aqueous solution	Volumes of 2 g/L Na_2CO_3 (mL)				
Monocomponent solutions	*Monocomponent solutions*					
500 mg/L Pb^{2+}	5.22	8.00	14.00	15.00	17.00	25.00
500 mg/L Zn^{2+}	5,17	5.00	10.00	30.00	70.00	100.00
50 mg/L Pb^{2+}	5,16	1.00	1.80	2.00	3.00	5.00
50 mg/L Zn^{2+}	6,90	5.00	7.00	8.00	10.00	15.00
Di-component solutions	*Di-component solutions*					
500 mg/L Pb^{2+} 500 mg/L Zn^{2+}	5,00	5.00	20.00	70.00	90.00	120.00

A sample of each prepared aqueous solution of metal ions in an amount of 100 ml was transferred to an individual laboratory glass of 250 ml volume. A specific volume of precipitating agent was added to each glass and mixed with the solution by a magnetic stirrer at a rate of 300 rpm and total mixing time of 5 min. Upon expiration of the set mixing time, the pH of the solution was measured in each sample, after which the sample was filtered using Whatman No. 42 filter paper to separate the precipitate.

The analysis of the concentrations of Pb(II) and Zn(II) ions in the samples before and after the treatment and filtration was performed by flame atomic absorption spectroscopy

(FAAS), with air/acetylene type of flame. The construction of the calibration curve was performed by preparing a series of standard solutions of Pb(II) concentrations 0.2, 1, 5, 7 and 10 mg/L and Zn(II) concentrations 0.2, 0.5, 1, 1.5 and 2 mg/L and by measuring their absorbances by FAAS. Obtained equations of calibration curves for lead and zinc were y = 0,0196x + 0,0035 and y = 0,4394x + 0,1256. To calculate the efficiency of removal of metal ions from their aqueous solutions, the following equation is used:

$$Er = \frac{C_0 - C_1}{C_0} \cdot 100$$

Where Er (%) is the removal efficiency, C_0 (mg/L) is the initial concentration of heavy metal in untreated sample and C_1 (mg/L) is the final concentration of the heavy metal in the filtrate of the treated sample.

3 Results and Discussion

The effect of pH value and initial concentrations of Pb(II) and Zn(II) in their aqueous solutions on the removal of lead and zinc from mono-component and two-component aqueous solutions were studied, using Na_2CO_3 as precipitating agent. The results of the effect of pH on the efficiency of removal of lead and zinc ions from their monocomponent aqueous solutions of initial concentrations of 500 mg/L are shown in Fig. 1. Based on the data in Table 1 and Fig. 1, it can be seen that increasing the volume of Na_2CO_3 added to monocomponent aqueous solutions of heavy metal ions resulted in an increase in pH of solutions, which in turn increased the precipitation of lead and zinc ions and their removal. Higher removal efficiencies at high pH values are related to higher concentration of free carbonate ions in the solution, which react with metal ions forming insoluble metal carbonates [9].

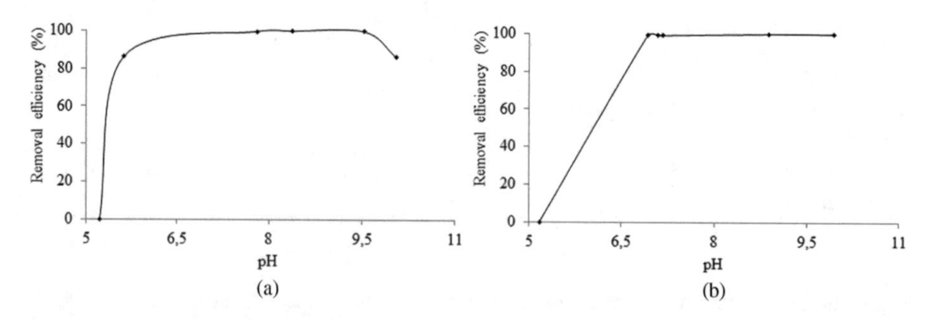

Fig. 1. Effect of pH on the efficiency of removing (a) Pb(II) and (b) Zn(II) from their monocomponent aqueous solutions of initial concentrations 500 mg/L

The efficacy of removal of Pb(II) was 99.607% at pH 7.81 and 14.00 mL of the dosed precipitating agent, and for Zn(II) it was 99.104% at pH of 7.09 and 10.00 mL of the dosed agent. Compared to hydroxide precipitation, the main advantage of using carbonate precipitation is that it can operate at a lower pH range, typically between 7 and 9 [13]. The highest removal efficiency for lead (99.814%) was obtained at a pH of 8.38 and 15.00 mL of the dosed precipitating agent, and for zinc (99.919%) at pH of 8.88 and 70.00 mL of the dosed agent. However, the data in Table 1 indicate a significantly higher consumption of sodium carbonate reagent (70.00 mL) for the precipitation of Zn(II) ions than for Pb(II) ions (15.00 mL). Peters and Shem [14] reported that some metals, such as zinc, do not precipitate easily, regardless of the amount of added carbonate, which is consistent with the results obtained in this paper.

Further increase in the dose of the precipitating agent in aqueous solutions of lead (25.00 mL) and zinc (100.00 mL) resulted in an increase in the pH of their solutions to pH 10.04 for Pb(II) and a pH 9.94 for Zn(II), which, however, reduced the precipitation and efficacy of removal of Pb(II) and Zn(II) ions (86.263% and 99.870%). The reduction of removal efficiency can be related to precipitation of metal ions as hydroxides. When pH in solution is above a certain level (e.g. 10 in some cases), the formation of metal hydroxyl complexes is enhanced, which can increase the metal solubility and reduce the precipitation effectiveness [9].

The results of the effect of pH on the efficiency of removal of lead and zinc ions from their monocomponent aqueous solutions of initial concentrations of 50 mg/L are shown in Fig. 2.

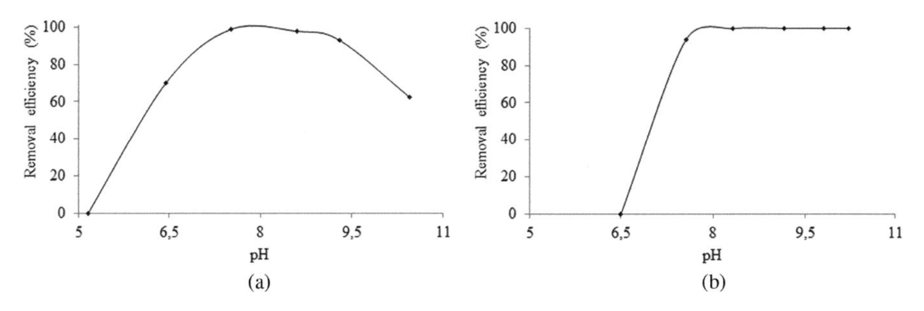

Fig. 2. Effect of pH on the efficiency of removing (a) Pb(II) and (b) Zn(II) from their monocomponent aqueous solutions of initial concentrations 50 mg/L

Removal of Pb(II) and Zn(II) ions from their mono-component aqueous solutions of initial concentrations of 50 mg/L required higher amounts of precipitating agent, compared with aqueous solutions of starting concentrations 500 mg/L. Aqueous solutions of zinc of concentration 50 mg/L had a higher initial pH (6.90) relative to the aqueous solutions of lead (5.16), but for further pH raising, they required significantly higher amounts of alkaline precipitating agent. The maximum removal efficiency for zinc (100%) was obtained at pH 9.16 and 100 mL of the dosed precipitating agent, while for the lead the maximum removal efficiency (98,722%) was obtained at pH 7.52 and precipitating agent dosage of 3.00 mL.

At pH 8.61 of the aqueous solution of Pb(II) and the pH 10,23 of the aqueous solution of Zn(II), the efficiencies of their removal were reduced, which can be explained by the re-dissolution of lead and zinc precipitates.

The results of the effect of pH on the efficiency of precipitation and removal of lead and zinc ions from their two-component aqueous solution in which the initial concentration of each heavy metal was 500 mg/L are shown in Fig. 3. At pH 5.32 the efficiency of removing Zn(II) ions was higher (98.952%) than Pb(II) ions (81.799%), while in the pH range of 6.81–9.53 the efficiency of lead removal was higher in relation to the zinc removal. This suggests a higher zinc affinity for the formation of carbonate precipitates in comparison with lead, at lower pH values. The reduced efficiency of Zn(II) removal at higher pH may be related to the formation of its hydroxides. Because metal hydroxides are increasingly soluble above or below their individual maximum precipitation point, even a slight pH adjustment to precipitate one metal may put another back into solution [6].

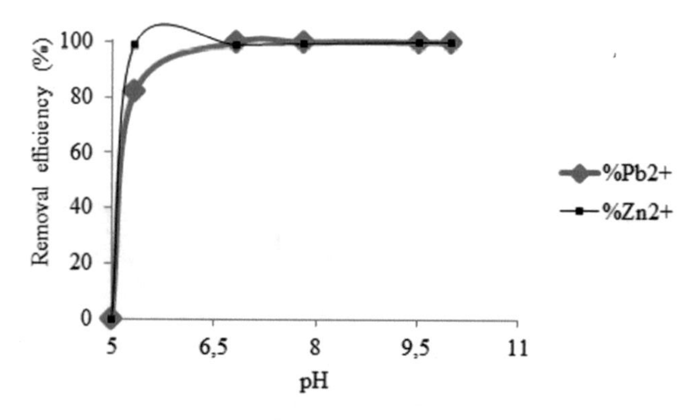

Fig. 3. Effect of pH on the efficiency of removing Pb(II) and Zn(II) of initial concentrations of 500 mg/L from a two-component aqueous solution

The maximum efficacy of removing Pb(II) ions was 100% at pH 7.81, while for Zn(II) ions was 99.991% at pH 6.81. The complete removal of lead ions was obtained in the pH interval 7.81–9.53. At pH 7, a slight decrease in the efficiency of zinc removal was observed, while at a pH of 9.53 it was again increased. This can also be explained by previously formed zinc hydroxides, which were re-dissolved at pH 7, while the zinc carbonates continued to form with a further increase in pH.

4 Conclusion

Sodium carbonate is an effective agent for the precipitation and removal of Pb(II) and Zn(II) from their mono- and two-component aqueous solutions, by increasing the pH and thereby reducing the solubility of the metal ions. Higher efficiency of removing Pb(II) was obtained by precipitation in aqueous solutions of higher initial concentrations of lead, while the efficiency of removing Zn(II) was higher in aqueous solutions of lower initial concentrations of zinc. Compared to lead ions, zinc ions require higher doses of sodium carbonate for the precipitation and removal from aqueous solutions. Increasing the pH above the optimum value can cause re-dissolution of metal precipitates and thereby reduce the efficiency of their removal from the water. Increasing the pH of the water above the optimal value for the precipitation of heavy metal can cause a dissolution of its formed precipitates, thereby reducing the efficiency of their removal from water, which should be considered in the treatment of water containing more than one heavy metal. The overall research results suggest that sodium carbonate can be efficiently used as precipitating agent for the treatment of raw water for the public supply and food industry, as well as for the treatment of waste industrial water.

REFERENCES

1. Fu, F., Wang, Q.: Removal of heavy metal ions from wastewaters: a review. J. Environ. Manag. **92**(3), 407–418 (2011)
2. Tchounwou, P.B., Yedjou, C.G., Patlolla, A.K., Sutton, D.J.: Heavy metals toxicity and the environment. EXS **2012**(101), 133–164 (2012)
3. Kundra, R., Sachdeva, R., Attar, S., Parande, M.: Studies on the removal of heavy metal ions from industrial waste water by using titanium electrodes. J. Curr. Chem. Pharm. Sci. **2**(1), 1–11 (2012)
4. Abdel-Raouf, M.S., Abdul-Raheim, A.R.M.: Removal of heavy metals from industrial waste water by biomass-based materials: a review. J. Pollut. Effects Control **5**(1), 1–13 (2017)
5. Barakat, M.A.: New trends in removing heavy metals from industrial wastewater. Arab. J. Chem. **4**(4), 361–377 (2011)
6. Akpor, O.B., Muchie, M.: Remediation of heavy metals in drinking water and wastewater treatment systems: processes and applications. Int. J. Phys. Sci. **5**(12), 1807–1817 (2010)
7. Karimi, H.: Effect of pH and Initial Pb(II) concentration on the lead removal efficiency from industrial wastewater using $Ca(OH)_2$. Int. J. Water Wastewater Treat. **3**(2), 1–4 (2017)
8. Patterson, J.W., Allen, H.E., Scala, J.J.: Carbonate precipitation for heavy metals pollutants. J. Water Pollut. Control Fed. **49**(12), 2397–2410 (1977)
9. Chen, J.P.: Decontamination of heavy metals: processes, mechanisms, and applications. CRC Press, Boca Raton (2012)
10. Dahman, Y., Deonanan, K., Dontsos, T., Iammatteo, A.: Nanopolymers. In: Dahman, Y. (ed.) Nanotechnology and Functional Materials for Engineers, pp. 121–144. Elsevier, Oxford (2017). Chapter 6
11. Kurniawan, T.A., Chan, G.Y.S., Lo, W.-H., Babel, S.: Physico–chemical treatment techniques for wastewater laden with heavy metals. Chem. Eng. J. **118**(1/2), 83–98 (2006)
12. U.S. Environmental Protection Agency: Wastewater Technology Fact Sheet: Chemical Precipitation, EPA 832-F-00-018 (2000). https://www3.epa.gov/npdes/pubs/chemical_precipitation.pdf
13. U.S. Environmental Protection Agency: Technical Resource Document: Treatment Technologies for Metal/Cyanide-Containing Wastes, EPA/600/S2-87/106 (1988). https://www.dtsc.ca.gov/HazardousWaste/Cyanide/upload/TR_summ_2000TLQY.pdf
14. Peters, R.W., Shem, L.: Separation of heavy metals: removal from industrial wastewaters and contaminated soil, International conference on emerging separation technologies for metals and fuels, Palm Coast, FL (United States), 13–28 March 1993. https://www.osti.gov/servlets/purl/6504209/

The Possibility of Improving Mineral Water Quality Using Selective Ion Exchange Column

Ramzija Cvrk[1(✉)], Mirza Softić[1], Sabina Begić[1], and Mirna Habuda Stanić[2]

[1] Faculty of Technology, University of Tuzla, Ul.Univerzitetska 8, 75 000 Tuzla, Bosnia and Herzegovina
ramzija.cvrk@untz.ba
[2] Faculty of Food Technology, University of Osijek, 31107 Osijek, Croatia

Abstract. The quality of mineral water is conditioned by a number of factors, primarily the quality of water in the water source. The aim of this study was to determinate content of Ni^{2+} and Mn^{2+} ions in the mineral water, after treatment by selective ion-exchange column. The primary objective of investigated technological process was optimal performance of the column, during the processing of water, so that the content of Ni^{2+} and Mn^{2+} ions in the mineral water, after treatment reduce to allowable concentration, and the concentration of Mg^{2+} ions in the mineral water decreases insignificantly, and thereby the cost of water treatment are not increased. Also, during the study, was determined the effect of application of ion - exchange columns in the physico-chemical water parameters, such as turbidity of the water, the water hardness and pH value.

The results showed a statistically significant difference ($p < 0{,}01$) in the content of Ni^{2+} and Mn^{2+} ions before, and after treatment, so that the mean concentration of nickel in water before processing is 0,187359 mg/l, and after the treatment, this value is 0,015600 mg/l. For the Mn^{2+} ion concentration value before processing is 0,113294 mg/l and after treatment this value was 0,008206 mg/l. It should be noted that changes in the concentration of Mg^{2+} ions there is no statistical significance, before treatment was 388,59 mg/l, and after treatment 383,80 mg/l.

Keywords: Quality of mineral water · Selective ion-exchange column

1 Introduction

Mineral water from different sources, even directly located given to flow through the different layers of the earth, and have different content. Each mineral spring has its special physical and chemical characteristics. This fact indicates the need for additional processing of mineral water, primarily as due to various natural impacts, composed of mineral water often present undesirable metal ions, and other matters that need to be reduced to the level prescribed by the Regulation, or removed completely [1].

The composition, temperature and other essential characteristics of natural mineral water must remain stable within the limits of natural changes and should not be changed during exploitation. However, natural mineral water whose content of certain

I. Karabegović (Ed.): NT 2019, LNNS 76, pp. 576–580, 2020.
https://doi.org/10.1007/978-3-030-18072-0_66

constituents exceeds the maximum limits, can be used to treat certain technological processes, to separate these ingredients, to ensure the health and safe mineral water [2].

One of the most important methods of removing some ions from the mineral water is the use of selective ion exchange column. This is especially important for water treatment processes in large industrial plants, where ion exchange resin must with stand continuous operation, large flow rate, as well as a large number of cycles of regeneration. Application of ion exchange resins in water treatment also has its economic justification because it often allows a significant reduction in operating costs of the process. Also, very often, cation exchange resins are the only way to harmonize the limit values of some ions with the legislation on drinking water [3, 4].

Ion exchange resin, precisely, their functional groups, have expressed an affinity to certain ions in solution. This property of the resin results in the appearance that the ion that is weakly bound to the functional group, can readily dissociate in solution and will be replaced with another, stronger bound ions. This phenomenon is called the principle of selectivity. In this paper described use of selective ion exchange column named Lewatit® TP 207, in the processing of mineral water.

Lewatit TP 207 is a weakly acidic, macroporous cation exchange resin with chelating imino-di-acetate (IDA) groups for the selective extraction of heavy metal cations from weakly acidic to weakly basic solutions. Divalent cations are removed from neutralized waters in the following order: Copper > Vanadium (VO^{2+}) > Uranium (UO_2^{2+}) > Lead > Nickel > Zinc > Cadmium > Iron(II) > Beryllium > Manganese > Calcium > Magnesium > Strontium > Barium >>> Sodium. Mentioned imino-di-acetate (IDA) group shows show exceptionally a high selectivity for the binding of nickel because nickel-imino-di-acetic-complex has a very high stability [5–7].

2 Materials and Methods

As the material for the research was used natural mineral water from spring water named "Kiseljak", from depth of 37, 80 meters located at Tuzla Canton area, Bosnia and Herzegovina. For the treatment to mineral water was used selective ion exchange column, newer generation, trade name Lewatit TP-207. Physico-chemical methods of analysis is carried out in this research, included the analysis of samples of mineral water before and after treatment of the selected ion exchange column. Mineral water flow during the operation of ion exchange columns was 2.5 m^3/h.

Concentration of nickel (Ni^{2+}) and manganese (Mn^{2+}) were determined by atomic absorption spectrometry, and used atomic absorption spectrometer "VARIAN 200" (with the necessary equipment, allowing corrections non specific absorbance and the nebulizer burner for acetylene/airflame). Concentration of the nickel was determined by the standard method BAS ISO 8288: 2002 [8], and concentration of the manganese by the standard method P-V-26-/B,1990 [9].

Concentration of magnesium (Mg^{2+}), calcium (Ca^{2+}), chloride (Cl^-), bicarbonate (HCO^{3-}) and total hardness were determined by standard volumetric analytical methods. Water turbidity was determined by method of turbidimetry (photometer MACHEREY-NAGEL "Nanocolor 300D"). Measurement of pH value was done by direct potentiometry, using a pH meter (Mettler Toledo Seven Go). The result for content of cations and anions is expressed as mg/l, results of total hardness is expressed in German degrees of hardness (dGH°, dH°), and results for water turbidity is expressed in *Nephelometric Turbidity Units* (NTU).

2.1 Statistical Analysis

Statistical analysis of the results was performed by SPSS statistical software (version 15). Statistical differences between investigated samples determined by *t-test* were at the level of significance of 99% ($p < 0, 01$).

3 Results and Discussion

Obtained values of measured water quality parameters indicate that are obtained concentrations of cations and anions in mineral water, after treatment by selective ion exchange column are very acceptable, and in accordance with the legislation for natural mineral water. Also, treatment of mineral water by this column had a very positive effect on some physical parameters of quality such, as turbidity, total hardness and pH of the water. Test results for the investigated parameters are shown in Tables 1 and 2.

According to the results, of the contents of Ni^{2+} ion (mg/l) in mineral water before treatment is 0.187359 (±0.0096162), while concentration same ions after treatment is 0.015600 (±0.0044919), and the obtained values clearly shows statistically significant reduction of the concentration of Ni^{2+} ions, after treatment by selected ion exchange column. Also, the results of concentration of ions Mn^{2+} (mg/l), for the samples, showed a statistically significant reduction in the concentration of these ions upon the use of selective ion exchange column, and before treatment mineral water the concentration of Mn^{2+} ions is 0.113294 (±0.0078542), while after the treatment, this value is 0.008206 (±0.0035818). Results Mg^{2+} ion concentration (mg/l) before treatment is 388.5929 (±5.86956), while the post-treatment value 383.7982 (±6.78306), as clear that there is no statistically significant difference in the concentrations of Mg^{2+} ions before and after treatment by selective ion exchange column (Table 1).

Given that after treatment of mineral water by selected ion exchange column concentrations of Ni^{2+} and Mn^{2+} significantly declined and brought to a value permitted by legislation for mineral water, and at the same time, the concentration of Mg^{2+} remains not changed statistically significant, shows to achieve the effect of selectivity of this column and the justification of its use. Results of the analysis also shows that the use of these columns in the processing of mineral water no statistically significant decrease in the concentration of Ca^{2+}, HCO^{3-} and Cl^{-} ions (Table 1). Analysis of some physical parameters, shows that application of the ion exchange column is not influenced significantly on the result of the hardness mineral water, while the pH value is significantly reduced by using this column (Table 1). The results obtained in this research show that the application of ion exchange column Lewatit TP 207 has a very high selectivity for removal of nickel (Ni^{2+}) and manganese (Mn^{2+}) from mineral water, with weakly acidic to weakly base pH value (pH = 6–9).

This is correlated with the research of authors E. Pehlivan and Turkan A. (2006) who are in their research used a columns Lewatit TP 207 and Lewatit CNP 80 to remove metal ions from waste water, and also showed high selectivity of the removal of nickel compared with other metal ions, for example Cu^{2+}, Cd^{2+}, Zn^{2+} i Pb^{2+}) [10]. This fact is very important, if we consider the toxicity of nickel, and the possibility of its negative effects on human health, whether it intakes by alimentary from water or to act through the skin. Also, the application of Lewatit TP 207 ion exchange resin showed better selectivity for the removal of nickel (Ni^{2+}) in relation to the Dowex HCR-S synthetic resin, using in the study of metal ion removal from aqueous solutions [11]. Also, other studies have shown that it is possible selectively removing manganese from water using ion exchange columns, as dissolved manganese as a divalent cation Mn^{2+}. In this process, a monovalent cation, Na^+ or H^+, is typically released from a cationic ion exchange resin (or media) as Mn^{2+} is selectively removed [12].

Table 1. Concentration of the cation and anion, total hardness and pH value of mineral water

Parameter	Before processing by ion exchange column		After processing by ion exchange column		p-value*
	Average value	Standard deviation	Average value	Standard deviation	
Ni^{2+} (mg/l)	0,187359[a]	±0,0096162	0,015600[a]	±0,0044919	0,000[a]
Mn^{2+} (mg/l)	0,113294[a]	±0,0078542	0,008206[a]	±0,0035818	0,000[a]
Mg^{2+} (mg/l)	388,5929[b]	±5,86956	383,7982[b]	±6,78306	0,044[b]
Ca^{2+} (mg/l)	67,2200[b]	±3,34976	64,8659[b]	±5,37346	0,129[b]
HCO_3^- (mg/l)	2239,3641[b]	±24,21382	2231,1018[b]	±32,14775	0,451[b]
Cl^- (mg/l)	152,3571[b]	±7,47884	146,0829[b]	±7,35095	0,071[b]
Total hardness (dGH°, dH°)	98,9300[b]	±6,36778	97,5800[b]	±6,25286	0,062[b]
pH value	7,6118[a]	±0,18205	6,5418[a]	±0,31000	0,000[a]

* p-value (t-test, $p < 0,01$): a-indicates statistical significant differences; b-indicates no statistical significant differences

In Table 2, turbidity is shown as a percentage share of samples which had the following values of turbidity in NTU units: <1 NTU, 1 NTU and 2 NTU, before and after processing by selective ion exchange column. Results shows that the application of this column very positively affect the turbidity of mineral water, because after treatment 94,118% samples had a turbidity <1 NTU, but before of treatment this value was significantly lower (47,959%).

Table 2. Turbidity of mineral water

Parameter	Before processing by ion exchange column			After processing by ion exchange column		
Turbidity unit	<1 NTU	1 NTU	2 NTU	<1 NTU	1 NTU	2 NTU
%	47,059	47,059	5,882	94,118	5,882	0

4 Conclusion

Based on the results of this research it can be concluded that using of selective ion exchange column very positive impact on the quality of the investigate mineral water. Undesirable concentration ions of nickel (Ni^{2+}) and manganese (Mn^{2+}) are removed significantly ($p < 0,01$) to a level that is below the allowable concentration.

At the same time, content of magnesium (Mg^{2+}) in the process is not significantly reduced. Also, application of this selective ion exchange column had a positive effect on the physical parameters of the water, especially on the turbidity, because after processing 94,118% samples had a turbidity <1 NTU.

References

1. Cotruvo, J., Bartram J.: Calcium and Magnesium in Drinking-water: Public health significance. World Health Organization, Geneva, pp. 43–47 (2009)
2. Dalmacija, B., Agbaba, J.: Quality control of drinking water, University of Novi Sad, Novi Sad, pp 31–68 (2006)
3. Howe, J.K., Hand, D.W., Crittenden, C.J., Trussell, R.R., Tchobanoglous, G.: Principles of Water Treatment. Wiley, New Jersey, pp. 386–387 (2009)
4. De Moel, P.J., Verberk, J.Q.J.C., van Dijk, J.C.: Drinking Water, Principles and Practices, p. 416. World Scientific Publishing, Singapore (2006)
5. Neumann, S.: Iron and Manganese Removal, pp. 2–3. Lanxess AG, Leverkusen (2007)
6. Neumann, S.: Removal of Nickel from Potable Water, pp. 1–7. Lanxsess Deutschland GmbH, Leverkusen (2008)
7. Neumann, S.: Product Information Lewatit TP 207, pp. 1–3. Lanxsess AG, Leverkusen (2012)
8. Standard methods for testing of quality drinking water, (BAS ISO 8288: 2002: Determination of cobalt, nickel, copper, zinc, cadmium and lead - Flame atomic absorption spectrometry) (2002)
9. Standard methods for testing of quality drinking water, (P-V-26-/B – Determination of manganese - Flame atomic absorption spectrometric methods) (1990)
10. Pehlivan, E., Altun, T.: Ion-exchange of Pb^{2+}, Cu^{2+}, Zn^{2+}, Cd^{2+}, and Ni^{2+} ions from aqueous solution by Lewatit CNP 80. J. Hazard. Mater. **140**, 299–307 (2007)
11. Fil, B.A., Yilmaz, A.E., Boncukcuoğlu, R., Bayar, S.: Removal of divalent heavy metal ions from aqueous solutions by Dowex HCR-S synthetic resin. Bul. Chem. Commun. **44**(3), 201–207 (2012)
12. Tobiason, J.E., Bazilio, A., Goodwill, J., Mai, X., Nguyen, C.: Manganese removal from drinking water sources. Curr. Polut. Rep. **2**, 168–177 (2016)

Biogenic Elements as Cofactors in Enzymes and Their Amount in the Chia Seed

Amra Bratovcic[1(✉)] and Edita Saric[2]

[1] Faculty of Technology, University of Tuzla, Univerzitetska 8, 75000 Tuzla, Bosnia and Herzegovina
amra.bratovcic@untz.ba
[2] Federal Institute for Agriculture in Sarajevo, 71210 Ilidza, Bosnia and Herzegovina

Abstract. In this paper, concentrations of biogenic elements such as copper, zinc, iron and magnesium in the chia seed have been determined by ICP–MS. These transition metals are important because they have the ability to bind to biomolecules and produce complex compounds (coordinating compounds or complexes). Metals which bonded to enzymes have the role of cofactors and they participate in important biological processes such as enzyme catalysis or other significant biological processes. Copper is a necessary trace element. Copper is the cofactor of important enzymes such as cytochrome oxidases and respiratory chain enzymes. It is essential for proper iron metabolism, but residual salts are toxic. It is antifungal. Zinc is one of the most important biogenic trace elements. It is necessary for preserving and transmitting genetic information. Over 300 enzymes contain zinc. It enhances immunity. It was also determined the concentration of iron which is a necessary biogenic element and in the human organism is mainly found in hemoglobin. Iron also participates in the processes of detoxification and bacterial defense, but in higher doses is toxic. Magnesium also enters into the composition of certain enzymes or is their activator. In humans, its deficiency causes muscle spasms and acts as an anti-allergic. In normal levels, they are important for stabilization of the cellular structures, but in deficiency states may stimulate alternate pathways and cause diseases. Research has shown that chia seeds are extremely rich in magnesium, iron, zinc and copper and that chia seeds can certainly be one of the ways of entering the required elements through its consumption.

Keywords: Copper · Zinc · Iron · Magnesium · Black chia

1 Introduction

Today, most people are constantly exposed to stress, and as a result, various health problems with frequent cancer diagnoses are reported and as a result they are increasingly turning to the use of natural remedies since ancient times. In fact many plants contain the necessary nutritional properties, minerals and vitamins necessary for the normal growth and development of healthy cells within the body and have a positive health effect. Trace elements are very important for cell functions at biological,

I. Karabegović (Ed.): NT 2019, LNNS 76, pp. 581–586, 2020.
https://doi.org/10.1007/978-3-030-18072-0_67

chemical and molecular levels. These elements mediate vital biochemical reactions by acting as cofactors for many enzymes, as well as act as centers for stabilizing structures of enzymes and proteins [1].

1.1 Chia Seed

Chia is rapidly growing in popularity because of its wide array of health benefits. Ancient grains are referred to as such because they have remained largely unchanged for hundreds or even thousands of years. Chia (Salvia hispanica L.) was originated from Mexico and Guatemala; it has been the part of human food for about 5500 years. The word chia is derived from a Spanish word chian which means oily, it is oilseed, with a power house of omega-3 fatty acids, superior quality protein, higher extent of dietary fibre, vitamins, minerals and wide range of polyphenolic antioxidants which act as antioxidant and safeguard the seeds from chemical and microbial breakdown [2]. The protein, fat, carbohydrate, dietary fibre, ash and dry matter contents of chia seeds ranged from 15 to 25%, 30 33%, 41%, 18 30%, 4 5% and 90 93% with a wide range of polyphenols [3]. The research which has been done by Peiretti and Gai showed that heavy metal content of seeds was within the safe limits with no potentially toxic mycotoxins and gluten [4]. The huge nutritional potential is evident from the fact that it contains 6, 11 and 4 times higher calcium, phosphorous and magnesium. The dietary fibre contained in foods and especially in whole grains is an important biocomponent due to its potential health benefit. Chia seed contains between 34 and 40 g of dietary fibre per 100 g, equivalent to 100% of the daily recommendations for the adult population; the defatted flour possesses 40% fibre, 5–10% of which is soluble and forms part of the mucilage [5]. This fibre content is higher than quinoa, flaxseed, and amaranth, even grater compared with other dried products [6].

2 The Trace Elements

There are at least 29 different types of elements including metal and nonmetals in an adult human body.

2.1 Copper (Cu)

Copper plays a very important role in our metabolism largely because it allows many critical enzymes to function properly [7]. Copper functions as a component of a number of metalloenzymes acting as oxidases to achieve the reduction of molecular oxygen. Ferroxidases are copper enzymes found in plasma, with a function in ferrous iron oxidation ($Fe^{2+} \rightarrow Fe^{3+}$) that is needed to achieve iron's binding to transferrin. Cytochrome *c* oxidase is a multisubunit enzyme in mitochondria that catalyzes reduction of O_2 to H_2O. This establishes a high energy proton gradient required for adenosine triphosphate (ATP) synthesis. This copper enzyme is particularly abundant in tissues of greatest metabolic activity including heart, brain, and liver. Dopamine β monooxygenase uses ascorbate, copper, and O_2 to convert dopamine to norepinephrine, a neurotransmitter, produced in neuronal and adrenal gland cells. Dopa, a precursor of dopamine, and metabolites used in melanin formation are oxidatively produced from tyrosine by the copper enzyme tyrosinase. Acidic conditions promotes the solubility which incorporates copper ions either in cupric form or cuprous form into the food chain. The average content of metal in the plant usually ranges from 4 to 20 mg of copper per kg of dry weight. The average adult human of 70 kg weight contains about 100 mg. The daily requirement is about 2–5 mg of which 50% is absorbed from the gastrointestinal tract (GIT) [8]. In human blood, copper is principally distributed between the erythrocytes and in the plasma. In erythrocytes, 60% of copper occurs as the copper-zinc metalloenzyme superoxide dismutase, the remaining 40% is loosely bound to other proteins and amino acids. Total erythrocytes copper in normal human is around 0.9–1.0 pg/ml of packed red cells [9]. The Copper has a selected biochemical function in hemoglobin (Hb) synthesis, connective tissue metabolism, and bone development. Synthesis of tryptophan is done in the presence of Cu. Besides these Cu as ceruloplasmin aid in the transport of iron to cells [10]. Recommended Dietary Allowance (RDA) of copper for adult men and women is 900 μg/day.

2.2 Zinc (Zn)

Zinc is a common element in human and natural environments and plays an important part in many biological processes. It plays a key role during physiological growth and fulfills an immune function. It is vital for the functionality of more than 300 enzymes, for the stabilization of DNA, and for gene expression [11]. The function of zinc in cells and tissues is dependent on metalloproteinase and these enzymes are associated with reproductive, neurological, immune and dermatological systems. They can be biochemically classified as those involved in nucleic acid and protein synthesis and degradation, alcohol metabolism, carbohydrate, lipid, and protein metabolism [12]. They include transferases, hydrases, lyses, isomerizes oxidoreductases, and transcription factors. The enzyme most essential for zinc are alkaline phosphates, alcohol dehydrogenase, carboanhydrase, glutamate and lactase dehydrogenase, and RNA polymerases. Zinc atoms have a structural role in the enzyme. The enzyme is localized in the cytosol and, along with the mitochondrial manganese-containing form, provides a defense against oxidative damage from superoxide radicals that, if uncontrolled, can lead to other damaging reactive oxygen species. RDA of Zinc for adults is 8 mg/day for women and 11 mg/day for men.

2.3 Iron (Fe)

In contrast to zinc, iron is an abundant element on earth and is a biologically essential component of every living organism. In contact with oxygen iron forms oxides, which are highly insoluble, and thus is not readily available for uptake by organisms [13]. It is well-known that deficiency or over exposure to various elements has noticeable effects on human health. Iron is an essential element for almost all living organisms as it participates in a wide variety of metabolic processes, including oxygen transport, deoxyribonucleic acid (DNA) synthesis, and electron transport. However, as iron can form free radicals, its concentration in body tissues must be tightly regulated because in excessive amounts, it can lead to tissue damage [14]. In the human body, iron mainly exists in complex forms bound to protein (hemoprotein) as heme compounds (hemoglobin or myoglobin), heme enzymes, or nonheme compounds (flavin-iron enzymes, transferring, and ferritin) [15]. The body requires iron for the synthesis of its oxygen transport proteins, in particular hemoglobin and myoglobin, and for the formation of heme enzymes and other iron-containing enzymes involved in electron transfer and oxidation-reductions [16]. Heme is the major iron containing substance. It is found in Hb, myoglobin, cytochrome while the enzymes associated with iron are cytochrome A, B, C, F 450, cytochrome C reductase, catalases, peroxidases, xanthine oxidases, tryptophan pyrrolase, succinate dehydrogenase, glucose 6 phosphate dehydrogenase, and choline dehydrogenase [17]. The RDA of iron for all age groups of men and postmenopausal women is 8 mg/day and for premenopausal women is 18 mg/day.

2.4 Magnesium (Mg)

Magnesium is one of four main electrolyte which mediate vital biochemical reactions by acting as a cofactor or catalyst in more than 300 enzyme systems that regulate diverse biochemical reactions in the body, including protein synthesis, muscle and nerve function, blood glucose control, and blood pressure regulation. Magnesium together with sodium, potassium and calcium act as centers of building stabilizing structures such as enzymes and proteins. The accumulation of metals or deficiency of these elements may stimulate an alternate pathway which might produce diseases. Although these elements account for only 0.02% of the total body weight, they play significant roles, e.g., as active centers of enzymes or as trace bioactive substances [18]. An adult body contains approximately 25 g magnesium, with 50% to 60% present in the bones and most of the rest in soft tissues. The RDA for magnesium is 310–420 mg for adults depending on age and gender.

3 Results

3.1 Instruments

In this paper have been used following instruments ICP-MS, Agilent Technologies, 7700X for determination of metals as well as microwave digestion/extraction workstation Sineo Jupiter-B for digestion for preparation of samples.

3.2 Chemicals

Nitric acid (suprapur) 67–69% produced by Carlo Erba and hydrogen peroxide (suprapur) 30% produced by Fluka. Multielement standard solution, Agilent technologies, 10 μg/ml. In Figure 1 the sample of chia seed is shown.

Fig. 1. Black chia seed

3.3 Method

The standard method for determination of trace elements in food, pressure digestion BASEN 13805:2014 was used.

In Table 1 are given the concentrations of copper, zinc, iron and magnesium in chia seed determined. The determined concentrations of trace elements in chia seed have been determined by standard BASEN 13805:2014 method.

Table 1. The amount of copper, zinc, iron and magnesium in chia seed

Element	Chia seed (mg/100 g)
Mg	408
Fe	6,93
Cu	1,61
Zn	3,94

4 Conclusion

From this research results it can be seen that chia seeds are extremely rich in magnesium, iron, zinc and copper. Chia seeds can certainly be one of the ways of entering the required elements through its consumption. It is important to note that concentrations of all tested elements do not exceed recommended daily doses. Consideration should be given to the recommended daily values of the input of the test elements because they may be toxic to human health in excessive concentrations. It should also need to consider possible interactions between elements, because large amounts of supplemental iron (greater than 25 mg) might decrease zinc absorption. Taking iron supplements between meals helps decrease its effect on zinc absorption. High zinc intakes can inhibit copper absorption, sometimes producing copper deficiency and associated anemia. Iron-deficiency anemia is a serious world-wide public health problem. Iron fortification programs have been credited with improving the iron status of millions of women, infants, and children.

References

1. Prashanth, L., Kattapagari, K.K., Chitturi, R.T., Baddam, V.R., Prasad, L.K.: A review on role of essential trace elements in health and disease. J. Dr NTR Univ. Health Sci. **4**(2), 75–85 (2015)
2. Cahill, J.P.: Ethnobotany of chia, Salvia hispanica L. (Lamiaceae). Econ. Bot. **57**, 604–618 (2003)
3. Ixtaina, V.Y., Nolasco, S.M., Tomas, M.C.: Physical properties of chia (Salvia hispanica L.) seeds. Ind. Crop Prod. **28**(3), 286–293 (2008)
4. Peiretti, P.G., Gai, F.: Fatty acid and nutritive quality of chia (Salvia hispanica L.) seeds and plant during growth. Anim. Feed Sci. Technol. **148**(2–4), 267–275 (2009)
5. Mohd Ali, N., Yeap, S.K., Ho, W.Y.: The promising future of China, Salvia Hispanica L. J. Biomed. Biotechnol. 9 (2012). https://doi.org/10.1155/2012/171956. Article ID 171956
6. Ullah, R., Nadeem, M., Khalique, A., Imran, M., Mehmood, S., Javid, A., Hussain, J.: Nutritional and therapeutic perspectives of Chia (Salvia hispanica L.): a review. J. Food. Sci. Technol. **53**(4), 1750–1758 (2016). https://doi.org/10.1007/s13197-015-1967-0
7. Harris, E.D.: Copper homeostasis: the role of cellular transporters. Nutr. Rev. **59**, 281–285 (2001)
8. Walravens, P.A.: Nutritional importance of copper and zinc in neonates and infants. Clin. Chem. **26**, 185–189 (1980)
9. Mason, K.E.: A conspectus of research on copper metabolism and requirements of man. J. Nutr. **109**, 1979–2066 (1979)
10. Turnlund, J.R.: Human whole-body copper metabolism. Am. J. Clin. Nutr. **67**(5 Suppl), 960S–964S (1998)
11. Frassinetti, S., Bronzetti, G.L., Caltavuturo, L., Croce, C.M.D.: The role of zinc in life: a review. J. Environ. Pathol. Toxicol. Oncol. **25**(3), 597–610 (2006). https://doi.org/10.1615/JEnvironPatholToxicolOncol.v25.i3.40
12. Satyanarayana, U., Chakrapani, U.: Essentials of Biochemistry, 2nd edn, pp. 210–227. Arunabha Sen Book and Allied (P) Ltd., Kolkata (2008)
13. Wood, R.J., Ronnenberg, A.: Iron. In: Shils, M.E., Shike, M., Ross, A.C., Caballero, B., Cousins, R.J. (eds.) Modern Nutrition in Health and Disease, 10th edn, pp. 248–270. Lippincott Williams & Wilkins, Baltimore (2005)
14. Abbaspour, N., Hurrell, R., Kelishadi, R.: Review on iron and its importance for human health. J. Res. Med. Sci. **19**(2), 164–174 (2014)
15. McDowell, L.R.: Minerals in Animal and Human Nutrition, 2nd edn, p. 660. Elsevier Science, Amsterdam (2003)
16. Hurrell, R.F.: Bioavailability of iron. Eur. J. Clin. Nutr. **51**, S4–S8 (1997)
17. Vasudevan, D.M., Sreekumari, S.: Text Book of Biochemistry for Medical Students, 5th edn, pp. 76–91. Jaypee Publication, New Delhi (2007)
18. Wada, O.: What are trace elements? Their deficiency and excess states. J. Japan Med. Assoc. **47**, 351–358 (2004)

Communities of Aquatic Macroinvertebrates from Konjuh Mountain Headwater Streams

Isat Skenderović(✉), Avdul Adrović, Edina Hajdarević, and Alen Bajrić

Faculty of Natural Sciences and Mathematics, University of Tuzla, Univerzitetska 4, 75000 Tuzla, Bosnia and Herzegovina
isat.skenderovic@untz.ba

Abstract. Hydrobiological studies of macrozoobenthosin the headwater streams of the protected landscape of Konjuh were carried out at five sites during the spring, summer and autumn of 2017. The invertebrate biodiversity was consisted of nine groups and 46 taxa. By analyzing the composition of macrofauna, the greatest diversity of taxa was found in the Trichoptera and Ephemeroptera group which were represented by 13 taxa each. Following them are groups with a smaller number of taxa, Plecoptera (8), Diptera (5), Coleoptera and Oligochaeta with two taxa. Other groups found in the headwater streams of Konjuh Mountain are represented by one taxon. The presence of a different number of taxa was found at the investigated sites: at the site Krabašnjica - 25, Studešnica - 15, at the Djevojačka cave and GluhaBukovica 21 taxa and at the site Tuholj - 16. Based on the analysis of the physical and chemical parameters of the water and the index of saprobity, it has been determined that the water of Konjuh Mountain headwater streams has an oligosaprobic character and is considered to be the first class of water quality.

Keywords: Aquatic invertebrates · Biodiversity · Konjuh

1 Introduction

Nowadays, the urbanization and technical-technological development has reached such a level that it directly or indirectly negatively affects biological communities. Such a negative impact is also reflected on the streams of our freshwater ecosystems, which are extremely important for humanity. Today, the most pronounced negative impact on clean and drinkable streams is uncontrolled deforestation in our mountains, which have largely escaped control. Macroinvertebrates of zoobenthos are invertebrates from the bottom of aquatic ecosystems, whose body size is over 0.5 mm. In aquatic ecosystems, they play a significant role in the matter circulation and energy flow from primary production to decomposition. The degree of sensitivity of these organisms to different physical and chemical water parameters offers the possibility to determine the degree of water saprobity. The study of aquatic organisms indicates the possibility of using them as a water quality indicator. At the beginning of the 19th century in Europe began research studies of various groups of water insects. In order to assess the degree of water pollution, statistical methods are used today. The results are presented

I. Karabegović (Ed.): NT 2019, LNNS 76, pp. 587–594, 2020.
https://doi.org/10.1007/978-3-030-18072-0_68

graphically, which provides a clearer and more precise picture of the state of waters. Mountain Konjuh is located in the southwestern part of northeastern Bosnia. Along with Ozren and Javornik, Konjuh is a part of the range of fold mountains, which together with Trebević and Majevica represent the passage of the Dinaric mountain system into the vast Pannonian Basin. There are special climatic influences in this mountainous area. Only the highest mountain parts of Konjuh, in terms of temperature and pluviometric characteristics, belong to the mountainous temperate continental climate. This climate type is characterized by short summers and longer winter periods. This area receives larger amounts of precipitation compared to the surrounding area, and due to higher absolute heights, it has lower air temperatures. With an increase in altitude, the amount of precipitation increases as well, so Konjuh peaks receive over 1200 mm of precipitation annually. Snowfall appears in the middle of autumn, and the snow cover from the peaks melts at the end of spring. The analyzed streams are in the Protected Landscape of Konjuh and some of them are streams of drinkable water for the area of town Banovići. The remaining three streams serve as sources of drinkable water for the municipalities Tuholj and Kladanj. By the 2009 Law, part of Konjuh Mountain was declared Protected Landscape "Konjuh", with total area of 8016 hectares, extending to the area of three municipalities: Banovići, Kladanj and Živinice [1]. The research has been conducted with the aim of investigating the diversity of stream entomofauna and their implementation in water quality assessment of the streams in the researched area.

2 Material and Methods

The subject of this research is the benthic macroinvertebrate community from five streams, located in the Konjuh Protected Landscape. This protected area is located in north-eastern part of Bosnia and Herzegovina, and in the south-western part of the Tuzla Canton. The protected landscape of Konjuh extends into two water catchment areas: the northern part with the basin of the Oskova River and the southern part with the basin of the river Drinjača. Of the five explored streams from Konjuh Mountain, two are located on the northern slopes of Konjuh Mountain, and these are the streams of Krabašnjica (L1) and Studešnica (L2), while the remaining three streams are on the north-eastern slopes of Konjuh Mountain, the stream of the Djevojačka cave (L3), GluhaBukovica (L4) and Tuholj (L5). Research studies of the Konjuh mountain streams were carried out during three seasons (spring, summer and autumn of 2017). Sampling of zoobenthic macroinvertebrates from the investigated streams was carried out using the "kick sampling" method. The material was processed in the Zoology Laboratory of the Faculty of Natural Sciences and Mathematics, University of Tuzla. The determination has been carried out for the majority of organisms to the species level according to the available identification keys [2–5]. In order to examine a more complete picture of the presence of macroinvertebrate communities in the research sites, a physical and chemical analysis of water was performed. Measurements of individual parameters were performed on the same day when samples were taken at the selected sites. In the terrain, the water temperature, pH value, oxygen concentration in water (mg/L), oxygen saturation (%) and conductivity (μS/cm) were measured using the multimeter (HANNA HI 9828). Water samples for laboratory analysis of physico-chemical parameters were collected in inert plastic bottles, after which they were dispatched for the analysis in the laboratory for analytical chemistry of the Faculty of Technology in Tuzla. The diversity of macroinvertebrate communities is represented by the Shannon-Weaver diversity index [6]. Cluster analysis is based on the Bray-Curtis index of similarity [7].

3 Results and Discussion

3.1 Physico-Chemical Characteristics of Water

Analysis of the physical and chemical parameters of water from the five streams of Konjuh Mountain was carried out in part at sampling points, and in part in the laboratory. Measurements of individual parameters were performed on the same day when samples were taken at the selected sites. The average values of the physico-chemical parameters of the investigated sites are given in Table 1.

Table 1. The average values of the physico-chemical parameters of the water from the investigated sites

Analyzed parameters	L1	L2	L3	L4	L5
Temperature °C	8.66	9.11	8.53	8.64	8.93
Turbidity NTU	0.98	0.34	0.23	0.33	0.07
pH	7.9	7.1	7.7	7.3	7.2
Oxygen content mg/l	12.64	10	10.3	11.3	11.61
Oxygen saturation %	116.94	88.2	105.19	109.8	103.5
Nitrates mg/l N	11.1	8.9	5.1	4.8	5.8
Nitrites mg/l N	0.008	0.008	0.005	0.01	0.001
Ammonia mg/l N	0.009	0.005	0.02	0	0
Phosphates mg/l	0.04	0.25	0.2	0.2	0.2
Sulfates mg/l	0.24	5.1	6.8	6.9	7.2
Alkalinity mg/l	28.6	36	38	31	46.5
Total hardness mg/l	166	160	178	181	214
Calcium mg/l	46	41.2	63	64.6	58.4
Magnesium mg/l	21.8	14.7	6.8	7.8	15.5
Electroconductivity $\mu S/cm^{-1}$	254	329	346	358	408

The lowest water temperature was recorded at the Krabašnjica stream in March 2017 and had a value of 6.92 °C, while the highest temperature of water was recorded at the Studešnica stream in July 2017 at 9.53 °C. The highest pH value of water was measured in the Studešnica water stream: 8.1 in October, while the lowest value of the pH of the water was measured in July at the stream Djevojačka cave 6.84. The mean value of the oxygen content in the water of the Konjuh mountain streams during the study was 10 mg/l at the Studešnica site, to 12.64 mg/l at the Krabašnjica stream.

At the same sites, the lowest and highest values of oxygen saturation were measured (Table 1). The mean nitrate values in the analyzed water samples ranged from 4.8 mg/l N at site 4 to 11.1 mg/l N at site 1. The mean phosphate values were the same 0.2 mg/l PO_4 at three investigated sites (L3, L4 and L5). The mean value of phosphate at the locality of Krabašnjica was 0.04 mg/l PO_4, while at the Studešnica site the mean phosphate value was 0.25 mg/l PO_4.

3.2 Results of Biological Analysis

The subject of this research is the zoobenthic macroinvertebrate community from five streams, located in the area of the Protected Landscape "Konjuh". The results of the analysis of the qualitative-quantitative composition of the zoobenthos of the Konjuh headwater streams are determined by the degree of diversity, especially when it comes to the fauna of water insects. The number of taxa was differentially distributed by survey sites as well as the number of individuals. During the seasonal research carried out (in spring, summer and autumn) at five streams of Konjuh Mountain, from March to November 2017, fifteen samples of zoobenthos were analyzed. In the samples, 98 taxa and 483 specimens were determined at the site. As expected, groups of macroinvertebrates sensitive to pollution (Plecoptera, Trichoptera and Ephemeroptera), as well as groups of macroinvertebrates tolerant of pollution, i.e. less sensitive (Diptera, Oligochaeta) were noticed. The most numerous are water flowers –Ephemeroptera with 180 individuals, then aquatic moths –Trichoptera (160) and two-winged flies –Diptera (64). The presence of other taxa in the total sample is lower as shown in Table 2. Only one species of snails *Ancylusfluviatilis*, with a total of 22 individuals, was detected. The largest number of individuals was found in the sample at the Studešnica site – 15 individuals, which were sampled in the spring.

Oligochaeta individuals were found at two locations, Krabašnjica and GluhaBukovica. The presence of leeches was recorded in the samples taken at the GluhaBukovica site. The presence of only one type of *Haemopissanguisuga* was found at the explored sites with a total of 3 individuals [8]. In the samples of benthos at five streams of Konjuh Mountain, 180 species of water flowers were found. The highest number of confirmed Ephemeroptera species was found at Krabašnjica (49) and Tuholj (44), while the smallest number of individuals was found in the samples which were taken at the Studešnica site, 27 individuals. Of the total of 13 taxons, which are shown in table number 2, the most common are: *Rhithrogenacarpatoalpina*– 30, *Ephemera hellenica*– 30, *Baetisrhodani*– 29. Other taxa are represented in the total sample with a small number of individuals. Five taxa of Diptera were found. The lowest number of individuals was identified in the samples at the locality of Krabašnjica– 4 individuals, while the largest number of individuals was found in the samples that were taken at the L3 site, 29 specimens. The most numerous species is Culicoides sp. with a total of 21 individuals. *Atherix ibis* is represented by 19 individuals in the total sample. Other types of Diptera are less numerous. The species of aquatic moths (Trichoptera) were found in larger numbers in the samples of the Konjuh mountain streams. A total of 160 Trichoptera species classified in 13 taxa were found in all of the explored sites in benthic samples. This is one of the biggest groups of water insects found in the samples of the Konjuh mountain streams. Most aquatic moths were found at the GluhaBukovica site,

Table 2. Qualitative-quantitative composition of zoobenthic macroinvertebrates of samples from the Konjuh mountain headwater streams

Taxons–Locality	L-1	L-2	L-3	L-4	L-5
GASTROPODA					
Ancylusfluviatilis	6	15		1	
OLIGOCHAETA					
Naispardalis	1			1	
Lumbriculidae	1				
HIRUDINEA					
Haemopissanguisuga				3	
EPHEMEROPTERA					
Rhithrogenasemicolorata	6			3	
Rhithrogenacarpatoalpina	26		4		
Cloeon simile	5		4	1	
Baetismuticus	1	3	3		1
Baetislutheri	4				
Baetisrhodani	3		7	6	13
Ephemera danica		8	1		4
Ephemera hellenica		15	4		11
Heptagenialateralis		1			4
Heptageniaquadrilineata			3	9	11
Heptageniasulphurea				11	
Habrophlebiafusca			3		
Epeorussylvicola	4			1	
DIPTERA					
Chironomusthummi	1	4	2	1	6
Atherix ibis	2		5	7	5
Simulium sp.	1		7		
Pericoma sp.		2			
Culicoides sp.		3	15	3	
TRICHOPTERA					
Athripsodesaterrimus	3				
Agapetusfuscipes	2				
Hydropsycheinstabilis	9				
Hydropsycheangustipennis	3			7	2
Hydropsyche incognita	4	1		4	11
Hydropsychesiltalai				10	4
Rhyacophiladorsalis	1		9		
Rhyacophilanubila	2		1	2	
Rhyacophilafasciata	2		1	1	
Philopotamusmontanus		33	9		
Limnephilusbipunctatus		4		23	

(*continued*)

Table 2. (*continued*)

Taxons–Locality	L-1	L-2	L-3	L-4	L-5
Limnephilusauricula				1	
Goerapilosa				11	
COLEOPTERA					
Gyrinusnatator	15	6			
Hydraenagracilis		1			
PLECOPTERA					
Capniavidua	1	2			
Perlodesmicrocephalus	1				
Dinocrascephalotes	1				
Protonemuraauberti		2	1		
Amphinemurasulcicollis			1		2
Nemouracinerea			1		3
Perlamarginata			2	2	3
Leuctra fusca					1
ODONATA					
Gomphusvulgatissimus			1		5
Individuals	105	100	84	108	86
Taxons	25	15	21	21	16

59 individuals, and the fewest at the site of the Djevojačka cave 20. The most numerous taxon *Philopotamusmontanus* is with 42 individuals, and only 33 individuals have been identified at the Studešnica site. The next species in number is *Limnephilusbipunctatus* with a total of 27 individuals, which are sampled at two sites. Other types of Trichoptera are less numerous. Coleoptera individuals found in the samples from the Konjuh Mountain sites were found in relatively small numbers. The Plecoptera individuals were identified at all streams that were explored on mountain Konjuh. A total of 23 individuals were found, which were classified into 8 taxa. The smallest number of Plecoptera are present at the site L4 (GluhaBukovica) – 2 individuals. The largest number of individuals was sampled at the site L5, 9 individuals. The most common species is the *Perlamarginata* with 7 individuals. Other species that were found at the research sites are shown in Table 2. From the representatives of dragonflies, 6 species were sampled. The species *Gomphusvulgatissimus* is present in a small number. A total of 6 individuals were found at the site of the Djevojačka cave and Tuholj.

Table 3. Values of the diversity index by locations and the research season

Location/season	Krabašnjica	Studešnica	Djevojačka cave	GluhaBukovica	Tuholj
Spring	2.19	1.77	2.31	1.72	1.34
Summer	2.06	0.56	1.56	1.87	2.18
Autumn	0.49	1.57	1.27	1.22	1.41
Average value	1.58	1.30	1.71	1.60	1.64

The results obtained by calculating the mean value of the index of diversity in relation to the site of the survey show that the communities of macroinvertebrates are best-developed at the site L3 – Djevojačka cave (H = 1.71). The lowest values of the diversity index were registered at the Studešnica site (H = 1.30). The values of the diversity index for other research sites are given in Table 3. The similarity of the macrozoobenthos communities between the research sites was verified by Bray-Curtis Cluster analysis of the similarities, or the least diversity of the investigated communities. According to the data in Table 4, it is clear that, according to the presence of the number of taxa, the most similar are GluhaBukovica and Tuholj sites (L4 and L5) with a similarity percentage of 34.02%.

Table 4. Similarities of localities of investigated sources based on the presence of macrozoobenthic communities

Step	Clusters	Distance	Similarity	Joined 1	Joined 2
1	4	65.97937775	34.02062225	4	5
2	3	67.05882263	32.94117737	3	4
3	2	75	25	2	3
4	1	78.40375519	21.59624481	1	2
Similarity matrix					
	L-1	L-2	L-3	L-4	L-5
L-1	*	15.6098	20.1058	21.5962	13.6126
L-2	*	*	25	9.6154	23.6559
L-3	*	*	*	23.9583	32.9412
L-4	*	*	*	*	34.0206
L-5	*	*	*	*	*

They are followed by sites (L3 and L5) with a similarity rate of 32.94%, while the percentage of similarities is less than 30% in all other investigated sites. The cluster analysis shows the grouping of data into four groups, with the separation of site L1 – Krabašnjica. One of these clusters is made up of two separate small clusters that are similar, and these are the L4 sites – GluhaBukovica and L5 – Tuholj. The results of macroinvertebrates a probiological analysis indicate good water quality of the researched streams. The results of the analysis of the qualitative – quantitative composition of macrobiotic zoobentos of the investigated streams of Konjuh Mountain are expected and typical for this type of habitat of Bosnia and Herzegovina and the Balkans.

4 Conclusion

By analyzing the results of the qualitative-quantitative composition of macroinvertebrate communities from five headwater streams of Konjuh Mountain, it can be concluded that the obtained values of the physico-chemical parameters of the water of the researched streams from Konjuh Mountain show that they have favorable living. The number of individuals found as well as the taxa in zoobenthos samples varied, indicating their seasonal rhythm. In an eight-month study of the Konjuh Mountain streams, 98 taxa and 483 individuals were identified. In the samples of the Konjuh mountain streams at the site 1 (Krabašnjica), 25 taxa with 105 individuals were sampled. At the GluhaBukovica site and the Djevojačka cave there was a presence of 21 taxa, while at other sites there was a smaller presence of taxa as well as individuals. In the total sample, the most numerous are water flowers – Ephemeroptera with 180 individuals, followed by aquatic moths –Trichoptera (160), while the number of other groups is significantly lower. Based on the value of the diversity index, it can be concluded that macroinvertebrate communities are best developed at the stream of the Djevojačka cave, while at the Studešnica stream the value of the diversity index is the lowest.

References

1. Zakon o proglašenju dijela područja planine Konjuh Zaštićenim pejzažom "Konjuh" ("Službene novine Tuzlanskog kantona", broj 13/09)
2. Kerovec, M.: Priručnik za upoznavanje beskralješnjaka naših potokai rijeka. SNL, Zagreb (1986)
3. Quigley, M.: Invertebrates of Streams and Rivers: A key to Identification. Edward Arnold Publishers, London (1977)
4. Sansoni, G.: Atlante per il riconoscimento dei macroinvertebrati dei corsi d'acqua italiani. Provincia autonoma di Trento. Centro italiano Studi di Biologia Ambientale (1988)
5. Waringer, J., Graf, W.: Atlas der Österreichischen Köcherfliegenlarven. Facultas Univeritätsverlag, Wien (1997)
6. Shannon, C.E., Weaver, W.: The Mathematical Theory of Communication. The University of Illinois, Illinois (1949)
7. Bray, J.R., Curtis, J.T.: An ordination of upland forest communities of southern Wisconsin. Ecol. Monogr. **27**, 325–349 (1957)
8. Skenderović, I., Adrović, A., Hajdarević, E., Bajrić, A.: Zoobenthos of macroinvertebrates in some streams of the water catchment area of lake Modrac (Bosnia and Herzegovina). In: 10th International Scientific Conference "Science and Higher Education in Function of Sustainable Development" Užice (2017)

Health Risk Assessments Based on the Contents of Heavy Metals in Sarajevo Urban Soil

Aida Šapčanin[1(✉)], Alisa Smajović[1], Ekrem Pehlić[2], Mirsada Salihović[1], and Gordan Jančan[3]

[1] Faculty of Pharmacy, University of Sarajevo, Zmaja od Bosne 8, 71 000 Sarajevo, Bosnia and Herzegovina
aidasapcanin@bih.net.ba
[2] Faculty of Health Studies, University of Bihac, Bihać, Bosnia and Herzegovina
[3] Chemilab d.o.o., Ljubljana, Slovenia

Abstract. The aim of this study was to assess risks posed to human health based on contents of heavy metals found in soil in Sarajevo urban area. The contents of Cd, Pb, Cr, Ni, Cu, Zn, Co, Se and As have been measured and the hazard coefficient (HQ), non-carcinogenic hazard index (HI) and carcinogenic risk (RI) have been calculated. Overall, HQ and HI were lower than the safe limit of 1, indicating that there is no direct risk to human health from heavy metals in the investigated area; however, these levels should be monitored in a long-term perspective. Our results suggest that children are at higher risks than the adults due to their contact with potentially polluted soil. The soil contaminated with heavy metals can be used as a diagnostic tool for health risk assessments.

Keywords: Soil · Heavy metals · Urban area · Health risk assessment · Children · Adults

1 Introduction

Health risk assessments have their scientific and social aspects. A health risk assessment is a procedure used to assess the impact of some harmful conditions on human health or on the environment in general. Scientific aspects of risk assessments refer to the preparation of the surfaces upon which such assessments will be based. The risk assessment is implemented through the risk management, namely, by taking definite measures to reduce risks, by adopting regulations and making decisions about the acceptable risks. The risk assessment itself bridges the gap between scientific knowledge and social needs. It is based on the probability of a hazardous event and on the intensity of effects that such events cause [1]. The risk assessment is a multistep procedure that comprises the following: data collection, exposure assessment, toxicity assessment and risk characterisation. A non-cancerogenic impact of particular elements is assessed by calculating the Hazard coefficient (HQ) and Hazard Index (HI). A cancerogenic impact of elements is assessed by calculating Cancerogenic Risk (RI).

I. Karabegović (Ed.): NT 2019, LNNS 76, pp. 595–603, 2020.
https://doi.org/10.1007/978-3-030-18072-0_69

HQ represents the ratio of the basic dose of a chemical element according to its Reference Dose (Rf) for the same route of exposure [2]. Rf (mg/kg day) is a maximum daily dose of the element from a particular route of exposure, and these values are established for particular ones, with different values applicable to children and adults [3]. If the basic dose is lower than Rf, then the HQ $\leq$ 1, which means that there are no harmful effects on human health. In case that the value of the basic dose is above Rf (HQ > 1), then there is a risk to human health. [2, 4]. HQ value which exceeds 10 indicates a high long-term risk of the impact of cancerogenic elements [5]. HI is the sum of all the quotients associated with risks of an impact of non-cancerogenic elements based on all three exposure routes. If the HI < 1, then there is no risk of anon-cancerogenic impact, while HI > 1 indicates the possibility of a damaging impact on human health. RI is the probability that an individual will develop a form of cancer during their lifetime due to their exposure to cancerogenic hazards.

The goal of this study was to assess health risks posed to children and adults based on the contents of heavy metals in the soil of Sarajevo urban area where Cd, Pb, Cr, Ni, Cu, Zn, Co, Fe, Sc and As were identified as potential hazardous agents in the soil, especially at playgrounds.

2 Material and Methods

Soil samples for the determination of heavy metals were prepared by microwave assisted acid digestion and determined by using an atomic absorption spectrophotometer [6]. Furthermore, the potential health risk assessment calculated for a lifetime of exposure (ingestion, inhalation and dermal), based on USEPA model [2–4, 7, 8], was determined as the cumulative carcinogenic and non-carcinogenic risk. The value of total hazard index for both children and adults was calculated for maximum, minimum and average concentration of heavy metals.

The non-carcinogenic and carcinogenic effects of heavy metals were assessed by using Eqs. (1–3):

$$ADD_{ing} = C_{soil} \times \frac{IngR \times EF \times ED}{BW \times AT} \times 10^{-6} \tag{1}$$

$$ADD_{inh} = C_{soil} \times \frac{InhR \times EF \times ED}{PEF \times BW \times AT} \tag{2}$$

$$ADD_{der} = C_{soil} \times \frac{SA \times AF \times ABS \times EF \times ED}{BW \times AT} \times 10^{-6} \tag{3}$$

ADD_{ing} - Average daily doses (mg/kg day) via ingestion
ADD_{inh} - Average daily doses (mg/kg day) via inhalation
ADD_{der} - Average daily doses (mg/kg day) via dermal contact

Reference values of some health risk assessment parameters are shown in Table 1.

Table 1. Definition and reference values of some health risk assessment parameters used for heavy metals in the soil samples

Factor	Definition	Unit	Value		Reference
			Children	Adults	
C_{soil}	Heavy metal concentration in soil	mg/kg			
IngR	Ingestion rate of soil	mg/day	200	100	[2, 9]
EF	Exposure frequency	days/year	350	350	[8, 9]
ED	Exposure duration	years	6	20	[8]
BW	Body weight of the exposed individual	kg	15	80	[8]
AT	Average time	days	365ED	365ED	[8, 9]
InhR	Inhalation rate of soil	m^3/day	7,63	12,8	[10]
PEF	Particle emission factor	m^3/day	$1{,}36 * 10^9$	$1{,}36 * 10^9$	[4]
SA	Exposed skin surface area	cm^2	2373	6032	[8]
AF	Skin adherence factor	mg/cm^2	0,2	0,07	[4]
ABF	Dermal absorption factor	none	0,001	0,001	[10–13]

2.1 A Non-carcinogenic Risk Assessment

Hazard identification basically aims to investigate chemicals that are present at any given location, their concentrations, and spatial distribution. In the study area, Cd, Pb, Cr, Ni, Cu, Zn, Co, Se and As were identified as possible hazards for the community.

Non-cancerogenic activities of elements were assessed based on calculations of the hazard coefficient (4) and the hazard index (5).

$$HQ = \frac{ADD}{RfD} \tag{4}$$

HQ – Hazard coefficient

RfD – Chronic reference dose (mg/kg day) of a specific heavy metal

$$HI = \sum_{k=1}^{n} HQ_k = \sum_{k=1}^{n} \frac{ADD_k}{RfD_k} \tag{5}$$

HI – Hazard Index

2.2 A Carcinogenic Risk (RI) Assessment

Risk characterisation predicts the potential cancerous health risk posed to children and adults in the study area by integrating all the information gathered to arrive at a quantitative assessment of cancer risk and hazard indices (Eqs. 6 and 7).

Due to the lack of accessible values for cancer slope factors (SF), we assessed only the cancerogenic risk for exposure to Pb, Cr and As, the concentrations of which were the highest in the examined samples.

$$Risk_{pathway} = \sum_{k=1}^{n} ADD_k CSF_k \tag{6}$$

Risk –unitless probability of an individual developing cancer over a lifetime
ADD_k – Average daily intake (mg/kg day) for kth heavy metal
CSF_k – Cancer slope factor $(mg/kg\ day)^{-1}$ for kth heavy metal

$$Risk_{total} = Risk_{ing} + Risk_{inh} + Risk_{der} \tag{7}$$

Reference doses and cancer slope factors for heavy metals are shown in Table 2.

Table 2. Reference doses and cancer slope factors for heavy metals

Heavy metal	RfD_{ing} (mg/kg/day)	RfD_{inh} (mg/kg/day)	RfD_{der} (mg/kg/day)	Oral CSF	Inhalation CSF	Dermal CSF
Cd	$1{,}00 * 10^{-3}$	$1{,}00 * 10^{-3}$	$1{,}00 * 10^{-5}$		6,30	
Pb	$3{,}00 * 10^{-1}$	$3{,}00 * 10^{-3}$	$5{,}00 * 10^{-4}$	$8{,}50 * 10^{-3}$	$4{,}20 * 10^{-2}$	
Cr	$3{,}00 * 10^{-3}$	$2{,}00 * 10^{-4}$	$7{,}00 * 10^{-3}$	$5{,}00 * 10^{-1}$	$4{,}10 * 10^{1}$	
Ni	$1{,}00 * 10^{-1}$	$2{,}00 * 10^{-3}$	$5{,}00 * 10^{-3}$		$8{,}40 * 10^{-1}$	
Cu	$4{,}00 * 10^{-2}$	$4{,}00 * 10^{-1}$	$1{,}00 * 10^{-2}$			
Zn	$3{,}00 * 10^{-1}$	$3{,}00 * 10^{-1}$	$6{,}00 * 10^{-2}$			
Co	$2{,}00 * 10^{-2}$	$3{,}00 * 10^{-5}$	$1{,}00 * 10^{-2}$		9,80	
Se	$5{,}00 * 10^{-3}$	$2{,}00 * 10^{-2}$	$5{,}00 * 10^{-3}$			
As	$3{,}00 * 10^{-4}$	$4{,}00 * 10^{-3}$	$3{,}00 * 10^{-4}$	1,50E	$1{,}50 * 10^{1}$	1,50

$Risk_{ing}$, $Risk_{inh}$ and $Risk_{der}$ are risk contributors through ingestion, inhalation and dermal routes

The assessment is based on the following:

- If RI < 10^{-6}, then the carcinogenic risk can be regarded as negligible.
- If RI is in the range from 1×10^{-6} to 1×10^{-4}, then the values are within the acceptable human health risk range.
- If RI > 10^{-4}, then there is a high risk for development of cancer in humans.

3 Results and Discussion

Due to the children's exposure to urban soil, non-cancerogenic index hazardous index (HI) for children was estimated using 95th percentile values of total metal concentration. The obtained results for non-carcinogenic children health risk, based on metal concentrations (minimal and maximum) in urban soils and exposure by three different routes (ingestion, inhalation and dermal), are shown in Table 3.

Table 3. The hazard coefficient and non-carcinogenic hazard index for children (95%)

Metal	HQ ingestion		HQ inhalation		HQ dermal		HI	
	Min	Max	Min	Max	Min	Max	Min	Max
Cd	0,004	0,068	$4,5 \times 10^{-9}$	$6,4 \times 10^{-7}$	3×10^{-5}	$4,2 \times 10^{-3}$	0,004	0,072
Pb	0,043	**0,54**	$1,2 \times 10^{-6}$	$1,5 \times 10^{-5}$	$1,5 \times 10^{-4}$	$\mathbf{2,8 \times 10^{-2}}$	0,043	**0,56**
Cr	0,036	**0,49**	1×10^{-4}	$1,4 \times 10^{-3}$	$9,1 \times 10^{-6}$	$1,2 \times 10^{-4}$	0,037	**0,49**
Ni	0,002	0,007	$4,2 \times 10^{-6}$	$1,3 \times 10^{-5}$	$3,3 \times 10^{-5}$	1×10^{-4}	0,002	0,007
Cu	0,003	0,034	1×10^{-9}	9×10^{-8}	$7,7 \times 10^{-5}$	$6,7 \times 10^{-4}$	0,003	0,034
Zn	0,002	0,006	$5,8 \times 10^{-8}$	$1,8 \times 10^{-7}$	$7,1 \times 10^{-5}$	$2,2 \times 10^{-4}$	0,002	0,006
Co	0,005	0,006	$1,6 \times 10^{-5}$	$1,1 \times 10^{-4}$	$6,9 \times 10^{-7}$	$4,9 \times 10^{-6}$	0,005	0,006
Se	9×10^{-4}	0,005	$6,3 \times 10^{-9}$	$3,7 \times 10^{-8}$	$5,6 \times 10^{-7}$	$3,3 \times 10^{-6}$	9×10^{-4}	0,005
As	0,06	**0,16**	$1,1 \times 10^{-7}$	$3,3 \times 10^{-7}$	$8,2 \times 10^{-5}$	$2,5 \times 10^{-4}$	0,060	**0,16**

Calculations have shown that values for HQ and HI obtained from children's playgrounds are lower than the standard values equal to 1, so that they do not pose a direct risk for children. The highest values were for Pb and Cr and the result was similar to the results obtained in Mugoša et al., study (2015). The results for HI for Pb from all investigated locations were very close to the upper limit of the safe level and it seems that they pose a long term risk. The highest values of HI are obtained if taken orally, and the lowest if inhaled. In comparison to other investigations, for example, exposure to Pb in the urban soil in Podgorica may also pose a health threat, especially to young children [14]. Because of its negative effects on the children's central nervous system, monitoring of Pb content in soil is of great importance. The hazard coefficient and non-carcinogenic hazard index for adults are shown in Table 4.

Table 4. The hazard coefficient and non-carcinogenic hazard index for adults (95%)

Metal	HQ ingestion		HQ inhalation		HQ dermal		HI	
	Min	Max	Min	Max	Min	Max	Min	Max
Cd	$5,99 \times 10^{-5}$	$8,49 \times 10^{-3}$	$5,64 \times 10^{-9}$	$7,99 \times 10^{-7}$	3×10^{-5}	$4,2 \times 10^{-3}$	0,004	0,072
Pb	$5,31 \times 10^{-5}$	$6,66 \times 10^{-4}$	$5,00 \times 10^{-7}$	$6,27 \times 10^{-6}$	$1,5 \times 10^{-4}$	$\mathbf{2,8 \times 10^{-2}}$	0,043	**0,56**
Cr	$4,55 \times 10^{-3}$	$\mathbf{6,10 \times 10^{-2}}$	$6,42 \times 10^{-6}$	$8,61 \times 10^{-5}$	$9,1 \times 10^{-6}$	$1,2 \times 10^{-4}$	0,037	**0,49**
Ni	$3,76 \times 10^{-4}$	$1,22 \times 10^{-3}$	$1,77 \times 10^{-6}$	$5,73 \times 10^{-6}$	$3,3 \times 10^{-5}$	1×10^{-4}	0,002	0,007
Cu	$4,61 \times 10^{-4}$	$4,01 \times 10^{-3}$	$4,33 \times 10^{-9}$	$3,77 \times 10^{-8}$	$7,7 \times 10^{-5}$	$6,7 \times 10^{-4}$	0,003	0,034
Zn	$2,57 \times 10^{-4}$	$7,95 \times 10^{-4}$	$2,42 \times 10^{-8}$	$7,48 \times 10^{-8}$	$7,1 \times 10^{-5}$	$2,2 \times 10^{-4}$	0,002	0,006
Co	$1,10 \times 10^{-4}$	$7,78 \times 10^{-4}$	$6,92 \times 10^{-6}$	$4,88 \times 10^{-5}$	$6,9 \times 10^{-7}$	$4,9 \times 10^{-6}$	0,005	0,006
Se	$1,13 \times 10^{-4}$	$6,69 \times 10^{-4}$	$2,65 \times 10^{-9}$	$1,57 \times 10^{-8}$	$5,6 \times 10^{-7}$	$3,3 \times 10^{-6}$	9×10^{-4}	0,005
As	$6,71 \times 10^{-3}$	$\mathbf{2,07 \times 10^{-2}}$	$4,74 \times 10^{-8}$	$1,46 \times 10^{-7}$	$8,2 \times 10^{-5}$	$2,5 \times 10^{-4}$	0,060	**0,16**

Calculations of the hazard coefficient (HQ_{min}) and non-carcinogenic hazard index (HI) are shown in Fig. 1.

Calculations of hazard coefficient (HQ_{max}) and non-carcinogenic hazard index (HI) are shown in Fig. 2.

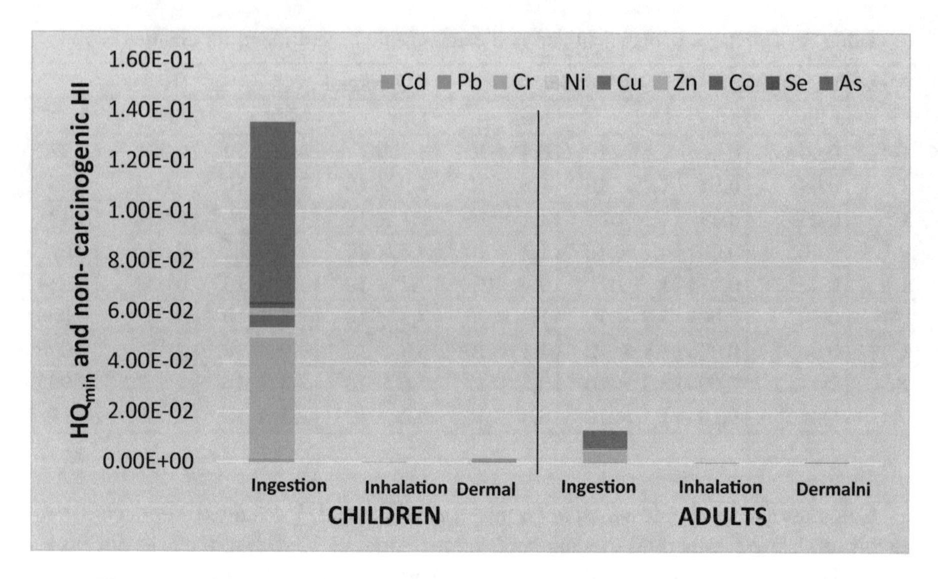

Fig. 1. The hazard coefficient (HQ_{min}) and non-carcinogenic hazard index (HI)

Comparing HI values for both children and adults, it can be concluded that children are at much higher risk of being exposed to non-cancerogenic risks caused by exposure to potentially toxic elements from the soil.

Calculations of the carcinogenic risk (for minimal heavy metal concentration) are shown in Fig. 3.

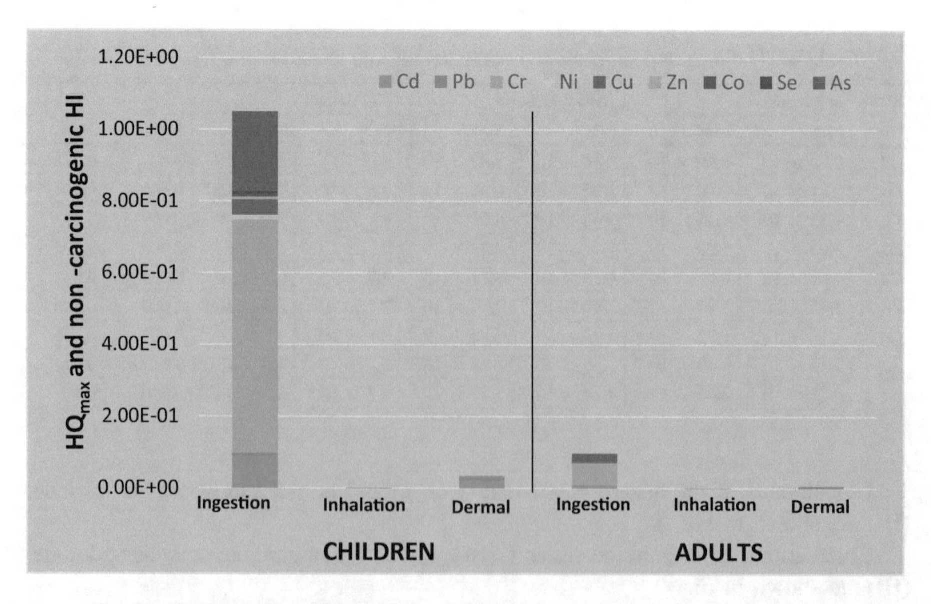

Fig. 2. The hazard coefficient (HQ_{max}) and non-carcinogenic hazard index (HI)

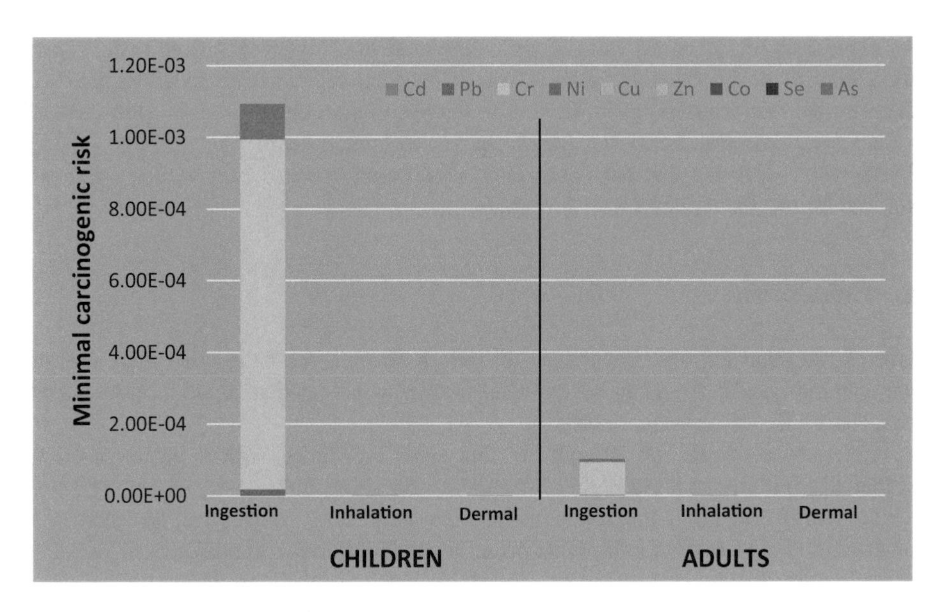

Fig. 3. The carcinogenic risk assessment (for minimal heavy metal concentration)

Calculations of the carcinogenic risk (for maximum heavy metal concentration) are shown in Fig. 4.

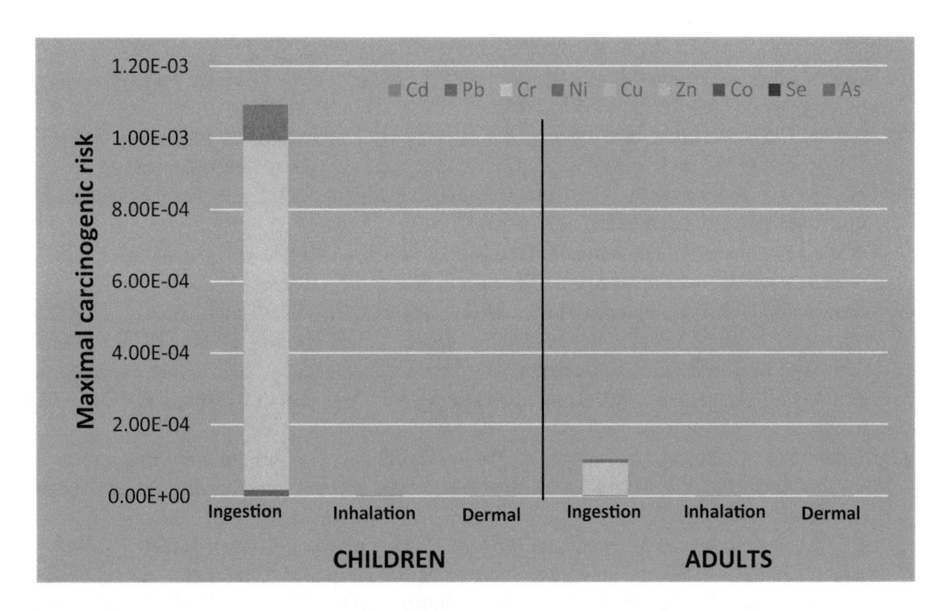

Fig. 4. The carcinogenic risk assessment (for maximum heavy metal concentration)

Similar to HI values, RI values for children were also higher compared to the values for adults. The RI value for adults was lower than 10^{-6}, which indicates that the cancerogenic risk caused by Pb, Cr and As in the soil could be negligible. However, RI values for children indicate that the risk posed to human health is acceptable [5, 15]. These results indicate that risks associated with being exposed to potentially polluted soil are higher for children than for adults.

4 Conclusion

Overall, HQ and HI were lower than the safe limit of 1, indicating that there was no direct human health risk from heavy metals in the investigated area, but for a long term perspective it should be monitored.

Calculations made for RI indicate that risks associated with being exposed to potentially polluted soil are higher for children than for adults.

Generally speaking, the soil contaminated with heavy metals can be used as a diagnostic tool to assess health risks posed to both children and adults.

Aknowledgements. This work was supported by the Federal Ministry of Education and Science in Bosnia and Herzegovina and was carried out within the framework of the project "Assessment of health risk caused by heavy metal contamination of the soil at children's playgrounds in Sarajevo" (Grant no. 0101-7552-17/15, dated 14.12.2016).

References

1. Ružić, I.: Procjena rizika po zdravlje čovjeka i kvalitetu okoliša. Hrvatske vode: časopis za vodno gospodarstvo **6**(22), 43–57 (1998)
2. USEPA: Risk Assessment Guidance for Superfund: Human Health Evaluation Manual. EPA/540/1-89/002, vol. 1. Office of Solid Waste and Emergency Response (1989)
3. USEPA: US Environmental Protection Agency, Resident Soil Table June 2017 (2017). https://semspub.epa.gov/work/03/2245085.pdf
4. USEPA: Supplemental Guidance for Developing Soil Screening Levels for Superfund Sites. OSWER 9355.4-24. Office of Solid Waste and Emergency Response (2001)
5. Salmani-Ghabeshi, S., Palomo-Marin, M.R., Bernalte, E., Rueda-Holgado, F., MiroRodriguez, C., Cereceda-Balic, F., Fadic, X., Vidal, V., Funes, M., Pinilla-Gil, E.: Spatial gradient of human health risk from exposure to trace elements and radioactive pollutants in soils at the Puchuncavi-Ventanas industrial complex, Chile. Environ. Pollut. **218**, 322–330 (2016)
6. Sapcanin, A., Cakal, M., Jacimovic, Z., Pehlic, E., Jancan, G.: Soil pollution fingerprints of children playgrounds in Sarajevo city, Bosnia and Herzegovina. Environ. Sci. Pollut. Res. **24** (12), 10949–10954 (2017)
7. USEPA: Risk Assessment Guidance for Superfund Volume I: Human Health Evaluation Manual (Part A). Office of Emergency and Remedial Response, Washington, DC (2004)
8. USEPA: Human Health Evaluation Manual, Supplemental Guidance: Update of Standard Default Exposure Factors. OSWER Directive 9200.1-120. Office of Solid Waste and Emergency Response (2014). https://rais.ornl.gov/documents/OSWER_Directive_9200.1120_ExposureFactors_corrected2.pdf

9. Department of Environmental Affairs: The Framework for the Management of Contaminated Land, South Africa (2010). http://sawic.environment.gov.za/documents/562.pdf
10. Zheng, N., Liu, J., Wang, Q., Liang, Z.: Health risk assessment of heavy metal exposure to street dust in the zinc smelting district, northeast of China. Sci. Total Environ. **408**(4), 726–733 (2010)
11. Chabukdhara, M., Nema, A.: Heavy metals assessment in urban soil around industrial clusters in Ghaziabad India: probabilistic health risk approach. Ecotoxicol. Environ. Saf. **87**, 57–64 (2013)
12. De Miguel, E., Iribarren, E., Chacon, I., Ordonez, E., Charlesworth, S.A.: Riskbased evaluation of the exposure of children to trace elements in playgrounds in Madrid (Spain). Chemosphere **66**, 505–513 (2007)
13. Shi, G., Chen, Z., Bi, C., Wang, L., Teng, J., Li, Y., Xu, S.: A comparative study of health risk of potentially toxic metals in urban and suburban road dust in the most populated city of China. Atmos. Environ. **45**, 764–771 (2011)
14. Mugoša, B., Đurović, D., Nedović-Vuković, M., Barjaktarović-Labović, S., Vrvić, M.: Assessment of ecological risk of heavy metal contamination in coastal municipalities of Montenegro. Int. J. Environ. Res. Public Health **13**(4), 393–408 (2016)
15. Qing, X., Yutong, Z., Shenggao, L.: Assessment of heavy metal pollution and human health risk in urban soils of steel industrial city (Anshan), Liaoning, Northeast China. Ecotoxicol. Environ. Saf. **120**, 377–385 (2015)

Synthesis, Characterization and Antimicrobial Activity of Silver Nanoparticles

Anera Kazlagić(✉), Enisa Omanović-Mikličanin, and Saud Hamidović

Faculty of Agriculture and Food Sciences, University of Sarajevo, Zmaja od Bosne 8, 71 000 Sarajevo, Bosnia and Herzegovina
a.kazlagic@ppf.unsa.ba

Abstract. During the past few years, metal nanoparticles received attention due to their interesting optical and electrical properties. Among them, silver nanoparticles (AgNP) showed various specific properties. In this paper, we've synthesised silver nanoparticles and tested their antimicrobial activity. As a precursor for silver, we used silver salt-$AgNO_3$. As a stabilizer and also a reducing agent, we used gallic acid monohydrate, because it is known that in strong alkaline solutions, this acid is capable of reducing silver ammonium complex, thus generating stable AgNPs. The characterization was done with UV/VIS spectrophotometer by assessment of absorption maximum λ_{max} in certain interval of time. In order to determine the inhibitory effects of silver nanoparticles, the test diffusion antibiogram method was used. Based on the obtained results, we concluded that nanoparticles synthesized in this way, show excellent antimicrobial activity and can be used as antimicrobial agent.

Keywords: Nanoparticles · Spectrophotometry · Antimicrobial activity

1 Introduction

In the past few years metal nanoparticles (MNP) have been the subject of numerous studies due to their unique electronic, mechanical, optical, magnetic and chemical properties that are different from those of bulk materials [1]. The main reason for behaviour of metallic nanoparticles can be attributed to their small sizes and very large specific surface area. Silver and other metal ions with anti-fungal, anti-bacterial and anti-viral properties have been used for a long time. The antimicrobial effects of silver (Ag) ion or salts are well known, but the effects of Ag nanoparticles (AgNPs) on microorganisms and antimicrobial mechanism have not been revealed clearly [2].

Silver nanoparticles with higher surface to volume ratio compared to common metallic silver, have shown better antimicrobial activity [3]. The investigations on the antibacterial activity of silver nanoparticles have increased due to the increasing bacterial resistance to antibiotics.

Nowadays, the attention of many scientists is focused on the development of new methods for synthesis and stabilization of metal nanoparticles. A variety of possible preparation ways have been reported for the synthesis of metallic nanoparticles, including reverse micelles process, salt reduction, ultrasonic irradiation, radiolysis,

I. Karabegović (Ed.): NT 2019, LNNS 76, pp. 604–609, 2020.
https://doi.org/10.1007/978-3-030-18072-0_70

solvothermal synthesis, electrochemical synthesis, microwave dielectric heating reduction, etc. The most commonly used bulk-solution synthetic method for metal NPs was the chemical reduction of metal salts [4–10].

However, the bactericidal property of nanoparticles depends on their stability in the growth medium. This imparts greater retention time for bacterium–nanoparticle interaction. There is a strong challenge in preparing AgNPs stable enough to significantly restrict bacterial growth [11]. Recently, biosynthetic methods employing naturally occurring reducing agents and thus promoting green chemistry principles, have emerged as a simple alternative to more complex chemical synthetic procedures to synthesize AgNPs. Those methods include three main steps: selection of solvent medium, selection of environmentally benign reducing agent, and selection of nontoxic substances for the silver nanoparticles stability [12].

In this study, we've synthesized silver nanoparticles using a gallic acid which acts as a reducing and stabilizing agent. Characterization was done using UV/VIS spectrophotometer by assessment of absorption maximum (λ_{max}). Considering a potential use of silver nanoparticles in various applications, the investigation of their antimicrobial activity was required.

In order to determine the antimicrobial activity of synthesized silver nanoparticles, in this study the disc diffusion method was used. Disc diffusion methods have always enjoyed great popularity in busy clinical microbiology laboratories because of their relative simplicity and ability to easily test multiple antimicrobial agents on each bacterial isolate. The method involves the placing of antimicrobial impregnated paper discs onto the surface of agar which has previously been seeded with the bacteria to be tested. The antimicrobial agent subsequently diffuses into the agar where it may inhibit bacterial growth in a zone surrounding the disc [13]. The antimicrobial compound diffuses from the disc into the medium. Following overnight incubation, the culture was examined for areas of no growth around the disc (zone of inhibition). The radius of the inhibition zone was measured from the edge of the disc to the edge of the zone. The end point of inhibition is where growth starts. Larger the inhibition zone diameter, greater is the antimicrobial activities. Bacteria or fungal strains sensitive to the antimicrobial are inhibited at a distance from the disc whereas resistant strains grow up to the edge of the disc [14].

2 Materials and Method

2.1 Materials

$AgNO_3$ (1.0×10^{-3}M) Centrohem p.a. >99%, $NH_3 \times H_2O$ (0.25M) Merck pro.analysi, NaOH (1.0×10^{-2}M) Centrohem p.a. >98%, gallic acid (1.0×10^{-4}M) Semikem p.a. >98%

2.2 Microorganisms

Escherichia coli and *Salmonella spp* were obtained from the river Lasva, B&H, biochemically isolated and tested (I test, MC test, VP test, C test, Urea test, H_2S test).

2.3 Synthesis of AgNPs Using Gallic Acid

Samples of AgNPs were prepared using chemical reduction method according to the slightly modified method of Wang et al. [15] In this experiment, 5.0 mL of 1.0×10^{-3}M $AgNO_3$, 160 μL of 0.25M $NH_3 \times H_2O$, and 200 μL of 10^{-2}M NaOH were mixed on magnetic stirrer at room temperature, and the mixture was stirred throughly. After that, a certain volume of 1.0×10^{-4}M gallic acid solution (200, 300, 400, 500, 600, 700, 800, 900, 1000 μL) was added into the mixture. During the process, samples were mixed vigorously. At last, the mixture was diluted to 10.0 mL with the doubly distilled water, and stirred again.

2.4 Characterization of Synthesized AgNPs

The UV-VIS absorption spectrum of the AgNPs was acquired using a UV/Visible spectrophotometer (Ultrospec 2100 pro, Amersham Biosciences). The following step was evaluation of maximum absorbance of synthesized nanoparticles.

2.5 Disc Diffusion Method

Determination of antimicrobial activity of the synthesized silver nanoparticles was achieved using the disc diffusion method, effects on particular bacteria (*Salmonella spp* and *Escherichia coli*).

By using the sterile swabs, 4–5 colonies were picked up from the clean bacterial broth. After that, preparation of the suspension consisted of the investigated bacterial culture was made (in our case those were: *Escherichia coli, Salmonella spp).* The bacterial suspension was approximately equally distributed across the whole agar area (using Mueller – Hinton nutrient agar) in order to obtain the confluent growth. That uniform layer of bacteria could be achieved by rotating the plate while continuing swabbing. After that, by using forceps, 6 discs were put approximately 15–24 cm apart from each other, and the dishes were then incubated 18 h at 37 °C.

3 Results and Discussion

The addition of gallic acid into the reaction mixture, turned the mixture into yellow, indicating the formation of silver nanoparticles. Narrow absorption peak appeared near 430 nm. As the concentration of gallic acid was higher, the color of the solution was more intensive. For the first five samples, absorption maximum went from 0,017 to 0,257 (Table 1), but for the last sample, where concentration of added gallic acid was the highest, absorption maximum was almost 0,600. Formation of the AgNPs was a result from reduction of silver ammonium complex, $[Ag(NH_3)_2]^+$, by gallic acid in the alkaline medium at room temperature.

In this study, the antimicrobial potential of AgNPs was investigated by growing *Escherichia coli* and *Salmonella spp* colonies on Mueller – Hinton nutrient agar plates supplemented with AgNPs. We investigated antimicrobial potential of synthesized AgNPs, 10 samples, for each microorganism 2 probes. Results showed in Table 2 represent average value of zone inhibition diameter.

Table 1. Absorption maximum for the first five samples.

Sample	λmax (nm)	A
1	430	0,017
2	430	0,118
3	430	0,121
4	430	0,237
5	430	0,257

Obtained results show antimicrobial potential of synthesized silver nanoparticles, although first, second and third sample didn't show measurable antimicrobial potential for *Escherichia coli*, and for the *Salmonella spp* only the first sample gave negative result (Table 2). Possible reason for negative reaction is the low concentration of synthesized AgNPs.

Table 2. Diameters of inhibition zone (Φ) for *Escherichia coli* and *Salmonella spp*

Sample no.	*Escherichia coli* Φ (mm)	*Salmonella spp* Φ (mm)
1	/	/
2	/	10
3	/	10
4	10	10
5	11	11
6	12	10
7	12	9
8	12	7
9	12	12
10	12	12

Results obtained with this method were compared with the standardized values for antibiotic Doxycycline for bacteria *Escherichia coli* and *Salmonella spp*.

Table 3. Zone of growth inhibition for both *Escherichia coli* and *Salmonella spp* (Enterobacteriaceae)

Zone of growth inhibition			
Doxycycline	R	I	S
Enterobacteriaceae	≤ 10	11–13	≥ 14

S category – states for sensitive which means that there is a great possibility that growth of particular group of microorganisms will be inhibited. The treatment with antibiotics might be very successful.

I category – is abbreviation for intermediate sensitive. It means that there is also possibility of successful inhibition of microorganisms growth, but that possibility is less when compared with the S category. Treatment is expected to show results only if maximum concentrations of antibiotics are prescribed.

R category – encompasses resistant microorganism groups. Those show no sensitivity and usage of antibiotics, regardless the dose, would probably be unsuccessful.

When compared to antibiotic Doxycyclin (Table 3), synthesized AgNPs show intermediate sensitivity to *Escherichia coli* (Fig. 1) and *Sallmonella spp* (Fig. 2) which means that there is certain possibility of inhibiting the growth of those microbes, depending upon the concentration of prescribed antibiotic.

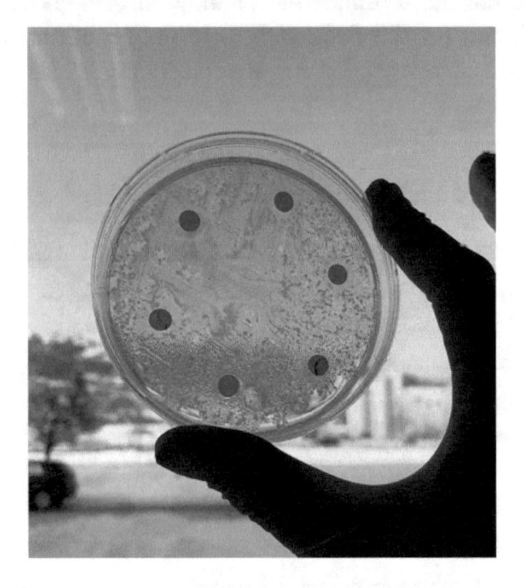

Fig. 1. Synthesized AgNPs (Sample 10) show intermediate sensitivity on *Escherichia coli*

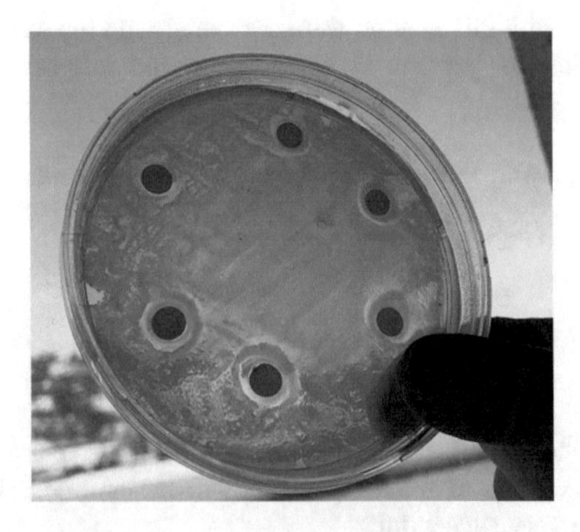

Fig. 2. Synthesized AgNPs (Sample 5) show intermediate sensitivity on *Salmonela spp*

4 Conclusion

In this paper, formation of AgNPs was secured using gallic acid which acted as a reducing and stabilizing agent simultaneously. Results obtained in the studies support the antibacterial potential of AgNPs synthesized in green manner, and it was proved by comparison with standardized antibiotic zones. These results lead us to conclusion that AgNPs synthesized this way can be used as antimicrobial agent. In order to expand and utilize the knowledge about this topic, there is a need for further investigation of the plant phenol compounds, such as gallic acid, by the means of nanotechnology techniques.

References

1. Mazur, M.: Electrochemically prepared silver nanoflakes and nanowires. Electrochem. Commun. **6**, 400–403 (2004)
2. Park, S.J., Lee, H.J., Kim, S.H., Park, Y.K., Park, Y.H., Hwang, C., Kim, Y., Lee, Y., Jeong, D.H., Cho, M.: Antimicrobial effects of silver nanoparticles (2006)
3. Kheybari, S., Samadi, N., Hosseini, S.V., Fazeli, A., Fazeli, M.R.: Synthesis and antimicrobial effects of silver nanoparticles produced by chemical reduction method DARU, vol. 18, no. 3 (2010)
4. Sergeev, G.B.: Nanochemistry. Elsevier, Amsterdam (2006)
5. Al-Dabbagh, B., Al-Shimari, H.: Preparation of nano-silver particles by chemical method for antibacteria's applications. Int. J. Nanotechnol. Appl. **3**(5), 21–26 (2013)
6. Mikac, L., Ivanda, M., Gotić, M., Mihelj, T., Horvat, L.: Synthesis and characterization of silver colloidal nanoparticles with different coatings for SERS application. J. Nanopart. Res. (2014)
7. Martinez-Castanon, G.A., Nino-Martinez, N., Martinez-Gutierrez, F., Martinez-Mendoza, J. R., Ruiz, F.: Synthesis and antibacterial activity of silver nanoparticles with different sizes. J. Nanopart. Res
8. Guzman, M.G., Dille, J., Godet, S.: Synthesis of silver nanoparticles by chemical reduction method and their antibacterial activity. Int. J. Chem. Biomol. Eng. **2**, 3 (2009)
9. Anambiga, I.V., Suganthan, V., Arunai Nambi Raj, N.: Antimicrobial and antifungal activity on glutathione stabilized silver nanoparticles – an in-vitro study. Int. J. Sci. Eng. Res. **5**(3) (2014)
10. Chaudhari, V.R., Haram, S.K., Kulshreshta, S.K., Bellare, J.R., Hassan, P.A.: Micelle assisted morphological evolution of silver nanoparticles. Colloids Surf. A: Physicochem. Eng. Aspects **301**, 475–480 (2007)
11. Shrivastava, S., Bera, T., Roy, A., Singh, G., Ramachandrarao, P., Dash, D.: Characterization of enhanced antibacterial effects of novel silver nanoparticles **Nanotechnology**, 18 (2007)
12. El-Nour, K.M.A., Eftaiha, A.A., Al-Warthan, A., Ammar, R.A.: Synthesis and applications of silver nanoparticles. Arab. J. Chem. **3**, 135–140 (2010)
13. Biemer, J.: Antimicrobial susceptibility testing by the kirby - bauer disc diffusion method. Ann. Clin. Lab. Sci. **136** (1973)
14. Jagessar, R.C., Mars, A.: Selective antimicrobial properties of phyllanthus acidus leaf extract against candida albicans, escherichia coli and staphylococcus aureus using stokes disc diffusion, well diffusion, streak plate and a dilution method. Nat. Sci. **28** (2008)
15. Wang, H., Chen, D., Wei, Y., Chang, Y., Zhao, J.: A simple and sensitive assay of gallic acid based on localized surface plasmon resonance light scattering of silver nanoparticles through modified tollens process. Anal. Sci. **27** (2011)

Some Aspects of Modern Nutrition

Jožef Božo(✉)

Technical College of Applied Sciences in Zrenjanin,
Đorđa Stratimirovića 23, 23000 Zrenjanin, Republic of Serbia
jozefbo@gmail.com

Abstract. People today are occupied by virtual reality and reality around themselves more than themselves, are becoming more and more ill and become victims of health problems, due to the inadequacy of their diet to their needs. Scientific and technological achievements have a significant impact on all aspects of today's life, but their contribution to food preparation and modern nutrition in general is not significant. In the interest of survival, with the cultural resources and scientific potential at the beginning of the 21st century, the problem of finding foods must match the contemporary needs, the taste and the spirit of the time, the conditions, the way and the style of life.

Keywords: Diet · Reduction of diversity · Molecular gastronomy

1 Introduction - The State of Nutrition Science

Scientific and technological achievements have a significant impact on all aspects of today's life, but their contribution to food preparation and modern nutrition is insignificant. This was noticed by the creators of molecular gastronomy at the end of the last century, pointing to the fact that "we know more about what is happening inside the distant stars than about what is happening in the baking steak" [1]. Molecular gastronomy has been created in a similar way by physicists and chemists, as earlier, molecular biology. However, this inclusion of science in the field of food was largely represented by the inspired application of modern scientific procedures of physics and physical chemistry in the cuisine. Symbols of modern cuisine have become foam-producing siphons, ultrasonic emulsifiers, thermostats, lyophilizers, fractional and vacuum distillers, as well as other devices, materials and processes from laboratories. Innovative dishes were prepared with current knowledge of colloid chemistry and nanotechnology in the form of polysaccharide and protein gels, emulsions. With the new textures, unusual aromas and flavors are combined. In elite restaurants around the world, distinguished chefs in this way created new specialties, gastronomic master-pieces, raising their position on the Michelin's list to an even higher level. However, after the initial run, molecular gastronomy quickly dropped out even in the field of food preparation and remained neglected as a scientific discipline. In addition to undoubtedly spectacular contributions to food preparation processes, these initial attempts of molecular gastronomy have not led to a significant, much-needed, comprehensive introduction to science in the field of nutrition. Thus, in the 21st century, at the threshold of conquest of the cosmos, the mastering of the secrets of life and artificial

I. Karabegović (Ed.): NT 2019, LNNS 76, pp. 610–616, 2020.
https://doi.org/10.1007/978-3-030-18072-0_71

intelligence, there is no proven, scientific methodology for designing a new meal or food product according to the needs of the consumer, calculation of composition and content, temperature and time of thermal treatment and other elements. The appropriate recipes, combinations and contents of the ingredients, procedures and treatment regimes for creating new products are empirically determined by means of a trial, strike or failure according to the sensory and emotional demands of the consumer. At the same time, sternly objective, growing health problems warn that the need for engaging science in the field of nutrition is more pronounced than ever [2]. The concept of "functional food" originated in Japan in the 1980s could be considered as the beginning of an effort to alleviate the problems, but the real solution could be that all food, food as a whole, becomes functional following the recommendation of the ancient Greek sages "that food be a cure".

2 Diet as a Cause of Health Problems

For us, as mammals, food has been functional since birth - mother's milk is a special body fluid that, after giving birth, supplements the interruption of contact between mother and fetus over the placenta. Unlike other beings, mammals are not immediately independent after birth but still, for some time, they depend on the mother. Mother's milk is not just food, source of material elements and energy for the growth and development of the organism, it also provides a number of vital compounds and cells necessary for the protection of health and regulation of the life processes of the newborn. Mother's milk allows the organism of the youngling and mother's organism to be connected after the birth, enabling extension of embryonic development as well as the independence of the new organism, the establishment of one's own metabolism, regulatory systems, defense systems and all other life functions of a new individual. All this takes place through eating, which points to the special biological importance of food and nutrition in the life of a mammal in relation to other beings.

It is generally accepted that the basic health problems of today, such as cardiovascular diseases, malignant diseases, diabetes, etc. beside to the increasing environmental pollution, can be attributed to the incompatibility of nutrition with the way of life and the needs of a modern man [3]. Trendy vegetarian, vitamin, mineral, basic and other restrictive diets are increasingly numerous, while health problems are spreading and intensifying [4]. The "healthy, natural diet" that is offered to solve this problem is the trend of returning with tried, traditional, ethno-food and principles, which actually represents "oil on fire" due to radical changes in the way of life and the needs of people. Sarma, Karađorđe's steak, Lenka's rolls and similar dishes, which we swear today, the favorite specialties of our grandmothers come from the time when over 80% of the population was engaged in food production, intensive physical jobs in agriculture. At the beginning of the third millennium, in modern, mechanized and computerized, "smart" economics and civilization, traditionalism just points to the lack of an appropriate approach to addressing the problem of non-harmonized nutrition to the extent that it could be expected based on the state of science and technology.

The inadequacy of traditional nutrition to the current needs of people is multiple. Traditional nutrition does not meet modern needs, nor by quantity (it is usually too large, energy-abundant), neither in quality. The World Health Organization has declared obesity as a disease, especially dangerous in children. The incidence of this disease is marked by dramatic growth. Traditional food preparation procedures, which ensure its appearance, taste, digestibility, shelf life and health safety, such as thermal treatment, salting, smoking, drying, acidification, etc. as a rule, they endanger its quality. Thermal treatment e.g. contributes to the decomposition of nutrients, changes in sensory properties and the formation of compounds in foods that endanger the health of consumers. More or less pronounced, negative consequences on the health effect of food caused also by other methods such as smoking, drying, evaporation, addition of sugar, salt, various additives and preservatives [5, 6]. Consumer awareness is becoming more and more resistant to over-processed, altered, artificial foods and it is more pronounced a growing demand for minimally processed, fresh, unladen, natural, and at the same time safe foods and meals.

3 Reduction of Food Diversity

Man's diet shows the trend of reducing the diversity of food choices and the impoverishment of the overall food composition from the beginning of civilization and the transition from the collection of food and hunts to domestication. In relation to the original diet, which lasted millions of years and formed metabolism, physiology, digestive tract, immune system and other important biological mechanisms in our ancestors, higher mammals, the hominide omnivores, civilization has for several thousand years led to the fact that the backbone of people's diet today makes only a small number of plant and animal species. The development of the food industry has made an additional contribution to reducing the number of food ingredients. Technological standards in the production of flour, vegetable oils, sugar and other foods have imposed production and, of course, the consumption of ever-cleaner, almost single-molecular products instead of former integral fruits of plants, cereals and oilseeds. Modern trends in nutrition, various restrictive diets, vegetarianism, etc. give an extra impetus to the impoverishment of foods in terms of wealth and variety of ingredients. Reducing the choice of various compounds in the diet results in a double problem - the lack of individual molecules can lead to a "hole" in metabolism, errors in the synthesis of the constituent and functional elements of the organism, while at the same time the excess of the remaining ingredients occurs. A surplus in nutrition burdens and disrupts metabolism and also impairs health. For a better understanding of this problem, the hypothesis of hormesis is indicated as fertile. Hormesis is by definition a biological phenomenon in which beneficial effects on health, higher tolerance to stress and prolongation of the life span can arise as a result of exposure to low doses of substances that are in high doses toxic and lethal.

Still dominant, the contemporary trend of eliminating fat from the diet for beauty and health, which has been going on for decades, has obviously not led to the expected alleviation of health problems, but on the contrary. It should be noted that lipid reduction in diet is carried out with the concurrent substitution of animal fat with vegetable oils and fats, hydrogenated oils. The enormous wealth of structural elements of tens of fatty acids, their countless combinations and isomers from milk fat and meat has been replaced by monotone sequences of almost one fatty acid of plant origin. This may be one of the most significant and apparently the most fateful reductions in the diversity, richness of food ingredients. Reducing fat content automatically led to an increase in the proportion of the other two essential nutrients from food, sugar and protein in the diet. We are witnesses that at the global level, the number of people with type 2 diabetes is growing like an epidemic. It is well known that the products of degradation of proteins, amino acids in excess, as energy sources, represent the burden of the organism with nitrogen compounds that can lead to a number of diseases including cancer.

Modern standards in food production, good manufacturing practice and good hygiene practice in order to ensure health safety, provide us with hygienically correct, pasteurized even sterile foods that often contain antimicrobial substances, preservatives. Civilization trends, an ever-increasing level of communal and personal hygiene, provide life in almost aseptic conditions, "under a glass bell". In this way, modern nutrition and living conditions not only do not provide positive stimulation of the creation, but directly inhibit the establishment and endanger the maintenance of the optimal symbiotic microflora of the intestines necessary for digestion and resorption of food, defense of the organism and proper functioning of the immune system without allergic and autoimmune disorders. Healthy microflora in the number of cells by the tens of times should exceed the number of cells of the organism itself, indicating its complexity and vulnerability, as well as the importance of reducing its number.

During the development of the civilization, unconsciously and unintentionally, from a health point of view, insufficiently thought out, the culture of life itself carried out the reduction of chemical and cellular components of food, the impoverishment of nutrition in terms of its biological, microbiological and chemical composition with evident health consequences. Sociological, religious, aesthetic, psychological and other reasons, taboos, customs, fashion trends, regulations and standards that have contributed to this are countless. The question that arises after all the changes that have been made towards the impoverishment of food, if in modern nutrition, as already mentioned, there is not enough science or knowledge, which is actually still available in representing the ruling conservatism in the diet?

4 Solving Problem Solution - Potential of Molecular Gastronomy

One of the creators of molecular gastronomy, Hervé This, points to the her unduly reduction in culinary art [7]. According to him, molecular gastronomy should be a science of food and nutrition at the level of molecules, much like molecular biology represents biology at the molecular level. In his paper published in 1994 in the journal *Scientific American*, he promoted *note by note cuisine* as a new technology for preparing food. The basis of this technology would be the preparation of foods using pure components or mixture of components obtained by extraction, fractionation and other processes from plant or animal tissue, instead of the tissues themselves [8]. If molecular gastronomy is initiated by molecular biology, in terms of metabolic engineering, *note by note technology* could become a kind of *gastronomic engineering*. Her author began his work with the addition of paraetylphenol in wine and whiskey, as well as the addition of limonene, ascorbic acid and other chemicals in dishes. The initial idea was to improve the taste, but soon the idea of creating new food or drink from the components was imposed. In this way, new dishes would be created not from whole meat, fish, vegetables and fruits but from their ingredients, pure compounds or mixtures, in analogy with electronic music, created without instruments (violin, trumpet), by combination of pure sounds, sound waves. In the future, note by note technology could create the composition, shape, texture, color, taste, smell and organize nutritional, health and other properties of dishes. The food industry has long been engaged in production, extracts of aromatic essential oils, mono-, di- and polysaccharides, proteins, as well as amino acids, organic acids, vitamins and others, many compounds without which modern food, drink and sweets production can not be imagined.

Significantly new in note by note technology is not only the possibility of adding useful and elimination of harmful ingredients from the food. It is of far wider importance, not only for the sensory effect, but also for the overall, nutritional, health, economic and other food effects, the ability to examine the food ingredient's interaction, and the interaction of treatments in the preparation, food production and even the interactions between treatments and ingredients [10]. Foods are multi-complex systems. Their sensory and nutritional properties, health and other effects are almost never the result of the activity of one ingredient, nor single treatment in production, but a combination of ingredients and more treatments [11]. This technology enables the examination of a combination of two or more factors of flavor, color, structure, antimicrobial, antioxidant, stimulating, inhibitory and other biological, health or aesthetic effects of food. Just like sound waves, which are getting stronger or weak when combined, create harmonic chords or dysonant tones, substances as well as treatments in combination can cause different effects (Table 1).

Table 1. Possible effects of the combination of two factors (with intensity e.g. 2 and e.g. 3, respectively)

<table>
<tr><td>Antagonistic</td><td>Summary effect is less than individual effect of a weaker factor ("2 + 3<2")</td><td>Mutually canceling</td></tr>
<tr><td>Neutral</td><td>Aggregate effect equal to the effect of more active factor ("2 + 3=3")</td><td>Treatment with a weaker factor is unessential, redundant</td></tr>
<tr><td>Additional</td><td>Summation of individual effects ("2 + 3=5")</td><td rowspan="2">Enable intensity reduction (shortening the duration of treatment, reducing the concentration of individual factors)</td></tr>
<tr><td>Synergistic</td><td>Joint effect amplification, collaboration is greater than the collection of individual effects ("2 + 3>5")</td></tr>
</table>

When choosing and combining factors, combinations with antagonistic effect can be avoided, which would eliminate individual effects and make treatments ineffective, wasting energy and materials. Antagonistic combinations burden production and can endanger the health safety of products. The effect shown in neutral action is the effect of a more active factor. Although both factors work, the weaker remains hidden, "shadowed" behind more active factor and becomes unnecessary. The combinations with additive, and especially, synergistic effects are looking for in order to mitigate the intensity of individual treatments, for example, by reducing the temperature, shortening the duration of thermal processing, reducing the concentration of components, and so on. This ensures minimal changes in food during processing, minimal consumption of ingredients, energy, treatment time and production with the optimal composition and quality of the product. Table 1 presents an overview of potential outcomes, the effects of combination only two factors or two ingredients. When dealing with a large number of factors, ingredients, which are most often the case in the real food, there are far more combinations, transient forms of these effects. According to this case, only two factors, that is, two substances of the ingredient, can be concluded about the importance of reducing the number of food ingredients. Linear change in the number of ingredients leads to an exponential change in the number of their possible combinations and actions. In this way, a small change in the presence of ingredients can drastically change the nature of their synergy and completely transform the biological effect of food. Reducing the diversity of food composition, the absence of all foods, plant and animal species from the pyramid of food could be one of the major causes of health disorders that affected us.

5 Conclusion

Nutrition innovations always ask a number of questions about taste, safety, nutrition, economy, and so on. For traditional dishes, these questions are not raised, although it is obvious that most of them would not comply with modern standards regarding nutrition and health safety. There has long been known repulsion, fear of new foods (*food neophobia*), a reflex characteristic of all primates. Food consumed in adolescence is always acceptable, while the new food and taste cause hostility. The results achieved by the molecular gastronomy for a short time of their ascent, especially in the field of aesthetics, the most insubstantial dimensions of the food, provide the basis for the expectations that in the field of physics, chemistry and biology of food, results can be even more convincing. The flavor, color, texture and biochemical effects of food are already well in the domain of contemporary scientific disciplines. It is only necessary to coordinate, deepen and intensify further, multidisciplinary research - health funds for the prevention and treatment of diseases caused by inappropriate nutrition should support them. Nutrition can not remain the "white zone" of the scientific chart just because investments in this field are not as profitable as investments in other segments of modern reality. The style of food should correspond to the spirit of the time just as style of dressing, architecture, music and painting in order to meet the needs, conditions and way of life.

References

1. Božo, J.: Molekularna gastronomija – moda, umetnost ili nauka? VII Naučno stručni skup Preduzetništvo, inženjerstvo i menadžment, Visoka tehnička škola strukovnih studija u Zrenjaninu, Zbornik radova, pp. 177–184 (2018)
2. Božo, J.: Pojedini aspekti primene začina kao antimikrobnih aditiva u proizvodnji hrane V Naučno stručni skup Preduzetništvo, inženjerstvo i menadžment, Visoka tehnička škola strukovnih studija u Zrenjaninu, Zbornik radova, pp. 207–214 (2016)
3. Božo, J.: Hrana – starenje i dugovečnost IV Naučno stručni skup Preduzetništvo, inženjerstvo i menadžment, Visoka tehnička škola strukovnih studija u Zrenjaninu, Zbornik radova, pp. 237–244 (2014)
4. Božo, J.: Prirodni biljni pesticidi u savremenoj ishrani VI Naučno stručni skup Preduzetništvo, inženjerstvo i menadžment, Visoka tehnička škola strukovnih studija u Zrenjaninu, Zbornik radova, pp. 171–179 (2017)
5. Nađalin, V., Božo, J., Đarmati, Z.: Antimicrobial action of the CO2 extract of sage (*Salvia officinalis L.).* In: 1ST Congress of Biologists of Macedonia (with international participation), Ohrid abstract, p. 105 (1996)
6. Davidson, P.M., Sofos, J.N., Branen, A.L.: Antimicrobials in Food, 3rd edn. Taylor&Francis, Boca Raton (2005)
7. This, H.: Food for tomorrow? How the scientific discipline of molecular gastronomy could change the way we eat. EMBO Rep. **7**(11), 1062–1066 (2006)
8. Đorđević, A., Đarmati, Z., Tot, N., Božo, J., Vojnović-Miloradov, M.: Supercritikal fluid extraction as a technology for production of plant extracts with high antioxidant and antibacterial activity. In: Third International Symposium and Exhibition on Environmental Contamination in Central and Eastern Europe, Warsaw, Poland abstract, p. 119 (1996)
9. Đarmati, Z., Jankov, R.M., Božo, J., Nađalin, V., Svrzić, G., Bočarov-Stančić, A.: Primena rosmanola i srodnih jedinjenja iz žalfije (Salvia officinalis L.) kao prirodnih antioksidanasa i konzervanasa u industriji mesa. Tehnologija mesa, Beograd No. 1, pp. 15–20 (1997)
10. Božo, J., Stojsavljević, T., Todorović, M.: Ispitivanje mogućnosti delimične supstitucije antibiotika antimikrobnim sredstvima neantibiotičkog porekla. Zbornik radova poljoprivrednog fakulteta Univerzitet u Beogradu, pp. 273–281 (1991)
11. Leistner, L., Gorris, L.G.M.: Food preservation by combined processes. European Commission, Final report, FLAIR Concerted Action No 7. Subgroup B (1997)

Influence of Processing on Phytonutrient Content of Cherries

Maida Djapo, Maja Kazazic(✉), and Ena Pantic

Faculty of Education, University "Dzemal Bijedic" of Mostar, University Campus bb, 88000 Mostar, Bosnia and Herzegovina
maja.kazazic@unmo.ba

Abstract. Phytochemical or bioactive compounds are secondary metabolites synthesized in plans that serve as antioxidants. Different studies have shown direct correlation between presence of bioactive compounds and health benefits of plant products. Sweet cherries contain important amount of bioactive compounds, mostly polyphenols. They are mainly used in fresh state, but also different cherry products are made to enable their consumption through the year. Aim of this study was to determine amount of polyphenols, anthocyanins and antioxidant activity in 3 different cherry cultivars and their products, jams and juice, prepared using traditional recepies. Total phenolic content was determined by the Folin-Ciocalteu method. Antioxidant activity using ABTS radical scavenging capacity assay and ferric reducing antioxidant potential (FRAP) assay. pH-differential method was used for determination of total anthocyanin content of all samples. There are significant differences in the content of all investigated bioactive compounds among selected cherry cultivars. Processing of cherries had a greater impact on the reduction of the total phenol and anthocyanin content, but did not have a significant effect on the reduction of antioxidant activity.

Keywords: Cherry · Polyphenols · Anthocyanins · Antioxidant activity · Bioactive compounds

1 Introduction

Cherries are considered as an excellent source of numerous nutrients and phytochemicals [1]. Their health benefits have been attributed to the presence of the bioactive compounds [1, 2]. Bioactive compounds in sweet cherries are mostly polyphenols: phenolic acids (hydroxycinnamic and hydroxybenzoic acids) and flavonoids (anthocyanins, flavonols and flavan-3-ols) [3]. Several studies have shown a direct correlation between the level of the phenolic compounds and the antioxidant activity in phytochemical extracts [4–7]. It is believed that one of the dominant mechanisms of the protective role of phytochemicals on human health is precisely the antioxidant activity and the capacity to capture free radicals.

Twenty two different populations of the wild cherry (*Prunus avinum*) that are naturally grown in Bosnia and Hercegovina are increasingly used in the preparation of the domestic products [8]. Although the cherry is mainly used in the fresh state, a certain part is processed into various products to ensure the possibility of its

I. Karabegović (Ed.): NT 2019, LNNS 76, pp. 617–623, 2020.
https://doi.org/10.1007/978-3-030-18072-0_72

consumption throughout the year (freezing, liqueurs, juices, jams). Processing changes the content in terms of phytochemicals and hence its bioactivity. The objective of this study was to determine whether jams and juices prepared using traditional recipes could be a good source of bioactive compounds.

In this work, we investigated total phenols, anthocyanins and antioxidant activity in fresh cherries and in 2 cherry products (jam, juice) in 3 cultivars of cherries from Hercegovina region.

2 Materials and Methods

2.1 Cherry Samples

Cherries, of the cultivars Rani hrust, Kasni hrust and Alica were hand-harvested at full maturity (full red colour) at the same locality, near Mostar, Hercegovina region, and brought to the laboratory within 4 h after harvesting. On arrival, the fruits were manually washed, after removal of the sepals and elimination of damaged fruits and analysed or processed immediately.

2.2 Preparation of Jam

Jams were prepared in laboratory conditions under atmospheric pressure. 1 kg of cherries were blended with 250 g of sugar and stirred until boiling. Mixture was boiled for 10 min. Jams were filled into hot glass jars and allowed to cool at the room temperature. Samples were kept in dark at room temperature until analysis.

2.3 Preparation of Juice

For the preparation of juice 1.5 kg of cherries, 250 g of sugar and 1 l of water was boiled for 5 min. Cherries was pressed and strained. Cherry juice was filled in hot glass bottles, cooled and stored in dark at room temperature.

2.4 Determination of Total Phenolic Content

The total phenolic content was determined with Fiolin-Ciocalteu reagent according to the method of Singleton [9] and as previously described by Kazazic [10]. Data are presented as average values of three measurements for each sample.

2.5 Determination of Antioxidant Activity

Ferric Reducing Antioxidant Potential (FRAP) Assay
The ferric reducing power of plant extract was determined using method of Benzie and Strain [11], which is based on reduction of a colorless ferric complex (Fe^{3+}-tripyridyltriazine) to a bluecolored ferrous complex (Fe^{2+}-tripyridyltriazine) by the action of electron-donating antioxidants. Procedure used was previously described by Kazazic [10]. All measurements were performed in triplicate.

Free Radical Scavenging by the Use of the ABTS Radical

The free radical scavenging capacity of plant extracts was investigated using the ABTS radical cation decolorization assay [12] as described in the work by Kazazic [10]. All solutions were prepared on the day of the experiment. All measurements were performed in triplicate.

Determination of Total Anthocyanin Content

Total anthocyanin content of all samples was determined by pH-differential method [13] and expressed as described by Mozetič [14].

3 Results and Discussion

Sweet cherries are rich source of polyphenols such as flavonoids, especially anthocyanins, to which many beneficial effects have been attributed. Three of the most common sweet cherry cultivars grown in Herzegovina region (B&H), grown at the same place and under the same conditions, were compared in relation to the contents of phenolic compounds, anthocyanins and antioxidant activity. The analyzes were performed on fresh cherries and on cherry products, juice and jam, prepared according to traditional recipes from this area. Total content of phenols, total anthocyanins and antioxidant activity in fresh sweet cherry, juices and jam are showed in Tables 1, 2 and 3.

Table 1. Total phenols (TP), anthocyanins (TA) and antioxidant capacity (TAC) using FRAP and ABTS method in sweet cherry (data are reported as mean ± standard deviations with three replication)

	c (mg GA/100 g DW)	c (CGE mg/100 g DW)	FRAP (mM Fe^{2+}/ 100 g DW)	ABTS (mg TE/100 g DW)
Alica	253.24 ± 29.70	15.66 ± 1.81	2.39 ± 0.31	6.38 ± 0.65
Kasni hrust	199.87 ± 11.28	37.95 ± 1.32	5.80 ± 0.22	9.69 ± 0.61
Rani hrust	136.31 ± 8.03	15.87 ± 1.60	2.39 ± 0.24	7.90 ± 1.98

Table 2. Total phenols (TP), anthocyanins (TA) and antioxidant capacity (TAC) using FRAP and ABTS method of the sweet cherry juices (data are reported as mean ± standard deviations with nine replication)

	c (mg GA/100 g DW)	c (CGE mg/100 g DW)	FRAP (mM Fe^{2+}/ 100 g DW)	ABTS (mg TE/100 g DW)
Alica	20.67 ± 11.77	2.12 ± 1.34	3.09 ± 0.64	7.98 ± 1.19
Kasni hrust	47.68 ± 4.68	12.79 ± 1.88	6.56 ± 0.39	10.89 ± 0.76
Rani hrust	14.10 ± 6.00	3.76 ± 0.74	12.79 ± 0.36	12.89 ± 1.12

Total phenolic content in the sweet cherries determined by Folin-Ciocalteu assay ranged from 112.25 ± 6.61 mg GA/100 g DW, which was determined for the Rani hrust cultivar, 155.58 ± 8.78 for the Kasni hrust cultivar, and 198.08 ± 23.23 mgGA/100 g DW, which was determined for Alica cultivar.

Table 3. Total phenols (TP), anthocyanins (TA) and antioxidant capacity (TAC) using FRAP and ABTS method of the sweet cherry jam (data are reported as mean ± standard deviations with nine replication)

	c (mg GA/100 g DW)	c (CGE mg/100 g DW)	FRAP (mM Fe^{2+}/ 100 g DW)	ABTS (mg TE/100 g DW)
Alica	108.6 ± 20.32	5.86 ± 1.21	1.48 ± 0.14	1.15 ± 0.37
Kasni hrust	36.88 ± 4.82	9.26 ± 2.35	1.66 ± 0.10	1.42 ± 0.34
Rani hrust	30.5 ± 19.72	7.19 ± 1.41	3.71 ± 0.12	4.74 ± 0.49

There is a statistically significant difference between the results, indicating the effect of the cultivar on the content of total phenols. This may be one of the reasons for the deviation of the results obtained in this study in relation to the results of other authors [15–17].

In addition to the influence of the cultivar, the differences in the total phenol content can be due to the different maturity of the fruits [18]. Other researched have found that year of picking of the fruit [19, 20] and the geographical location influenced the total phenol content [20].

In fresh cherries the values of total anthocyanins ranged from 12.25 to 29.54 mg CGE/100 g of dry weight. The highest content of anthocyanins was found in the cultivar Kasni hrust. This finding did not coincide with the results of [17], who analyzed 13 cultivars of cherry from Slovenia.

A higher content of anthocyanin in sweet cherries was reported previously [15, 16]. However, the various geographical locations at which plants were grown affected the composition and content of anthocyanins even though the plants were grown in the same region. The causes of variation of the anthocyanin content can also be explained by the influence of the genotype. In the work of other authors, other cultivars of cherry were analyzed. Difference in anthocyanins content for three studied varieties was statistically significant confirming the influence of the variety on the content of the anthocyanins.

The total anti-oxidant capacity (TAC) of three sweet cherry cultivars are also given in Table 1. The TAC of variety Kasni hrust was higher than in other cultivars which is not correlated with the content of total phenols and anthocyanins.

Total phenol content in processed cherry was significantly lower. In juices, the total phenol content ranged from 14.10 ± 6.0 mg/100 g DW to 47.68 ± 4.68 mg/100 g DW, and in jams from 30.5 ± 19.72 mg/100 g DW to 108.6 ± 20.32 mg/100 g DW.

The highest degree of preservation of the total phenol content was recorded in Kasni hrust and Alica varieties.

The quality of fresh cherries and cherry products is also evaluated from the aspect of the content of the anthocyanins. To determine the effect of processing on the content of anthocyanins, the content of the anthocyanins expressed as CGE mg per 100 g of dry matter in fresh samples, juices and jams was compared. The content of anthocyanins in juices ranged from 2.12 ± 1.34 to 12.79 ± 1.88 mg CGE/100 g DW, and in jams 5.86 ± 1.21 to 9.26 ± 2.35 mg/100 g DW. The content of the anthocyanins in cherry products is significantly lower. Anthocyanins have been found in large percentages in the jams. On average, the degree of degradation of anthocyanins in jams is 64%, and in juices 76%.

Degradation of anthocyanins in cherry products is caused by heat treatment. Anthocyanins are unstable components and their stability largely depends on the structure and composition of the matrix in which they are located. The loss of anthocyanins is also due to formation of complexes with other food ingredients. Loss of anthocyanins or transformations into brown components can be attributed to factors such as pH, acidity, phenolic components, sugars, oxygen, ascorbic acid and maturity of fruits [21].

In spite of antioxidant activity, compared to fresh sweet cherry cultivars, the juices and jams also represent a noticeable source of antioxidant compounds. Even considering the lower content of phenolic compounds and anthocyanins, juices have higher antioxidant capacity than fresh fruits. This fact could be at least partially explained by the formation of antioxidant Maillard products during juices processing [22]. In some cases, synergistic or antagonistic effects may occur resulting in the increase or decrease in the total antioxidant activity of the extract [23]. In jams significantly lower values of antioxidant activity in relation to fresh samples were observed which is in line with the lower content of phenol and anthocyanins.

4 Conclusion

The content of phytochemical compounds in the fruits and their stability during food processing vary depending on the geographical location, the cultivar, the food matrix and the maturity of the fruits. The results obtained in this study can be considered to be of particular interest in better defining the differences between the three different cherries cultivars that are commonly cultivated in Herzegovina. There are significant differences in the content of all investigated bioactive compounds among selected cherry cultivars.

The treatment process had a greater impact on the reduction of the total phenol and anthocyanin content, but did not have a significant effect on the reduction of antioxidant activity. Although some of the losses could have occurred, the results show that juices and jams can still be important sources of bioactive compounds in the diet with a noticeable antioxidant capacity.

References

1. McCune, L.M., Kubota, C., Stendell-Hollins, N.R., Thomson, C.A.: Cherries and health: a review. Crit. Rev. Food Sci. Nutr. **51**, 1–12 (2011)
2. Ferretti, G., Bacchetti, T., Belleggia, A., Neri, D.: Cherry antioxidants: from farm to table. Rev. Mol. **15**, 6993–7005 (2010)
3. González-Gómez, D., Lozano, M., Fernández-León, M.F., Bernalte, M.J., Ayuso, M.C., Rodríguez, A.B.: Sweet cherry phytochemicals: identification and characterization by HPLC-DAD/ESI-MS in six sweet-cherry cultivars grown in Valle del Jerte (Spain). J. Food Compos. Anal. **23**, 533–539 (2010)
4. Chaovanalikit, A., Wrolstad, R.E.: Total anthocyanins and total phenolics of fresh and processed cherries and their antioxidant properties. J. Food Sci. **69**, 67–72 (2004)
5. Tomás-Barberán, F.A., Ruiz, D., Valero, D., Rivera, D., Obón, C., Sánchez-Roca, C., Gil, M.: Health benefits from pomegranates and stone fruit, including plums, peaches, apricots and cherries. In: Skinner, M., Hunter, D. (eds.) Bioactives in Fruit: Health Benefits and Functional Foods, pp. 125–167. Wiley, New Jersey (2013)
6. Salazar, R., Pozos, M.E., Cordero, P., Perez, J., Salinas, M.C., Waksman, N.: Determination of the antioxidant activity of plants from Northeast Mexico. Pharm. Biol. **46**, 166–170 (2008)
7. Jakobek, L., Šeruga, M., Šeruga, B., Novak Jovanović, I., Medvidović-Kosanović, M.: Phenolic compound composition and antioxidant activity of fruits of rubus and prunus species from croatia. Int. J. Food Sci. Technol. **44**(4), 860–868 (2009)
8. Ballian, D., Bogunić, F., Čabaravdić, A., Pekeč, S., Franjić, J.: Population differentiation in the wild cherry (*Prunus avium* L.) in Bosnia and Herzegovina. Periodicum Biologorum **114** (1), 43–54 (2012)
9. Singleton, V.L.: Colorimetry of total phenolics with phosphomolybdic-phosphotungstic acid reagents. Amer. J. Enol. Viticult. **16**, 144–158 (1965)
10. Kazazic, M., Djapo, M., Ademovic, E.: Antioxidant activity of water extracts of some medicinal plants from Herzegovina region. Int. J. Pure App. Biosci. **4**(2), 85–90 (2016)
11. Benzie, I.F., Strain, J.J.: The ferric reducing ability of plasma (FRAP) as a measure of 'antioxidant power': the FRAP assay. Anal. Biochem. **239**, 70–76 (1996)
12. Re, R., Pellegrini, N., Proteggente, A., Pannala, A., Yang, M., Rice-Evans, C.: Antioxidant activity applying an improved ABTS radical cation decolorisation assay. Free Radic. Biol. Med. **26**, 1231–1237 (1999)
13. Wrolstad, R.E., Culbertson, J.D., Cornwell, C.J., Mattick, L.R.: J. Assoc. Off. Anal. Chem. **65**, 1417–1432 (1982)
14. Mozetič, B., Trebše, P., Hribar, J.: Anthocyanins and hydroxycinnamic acids in sweet cherries. Food Technol. Biotechnol. **40**(3), 207–212 (1982)
15. Kim, D.O., Heo, H.J., Kim, Y.J., Yang, H.S., Lee, C.Y.: Sweet and sour cherry phenolics and their protective effects on neuronal cells. J. Agric. Food Chem. 9921–9927 (2005)
16. Ballistreri, G.C., Gentile, A., Amenta, M., Fabroni, S.: Fruit quality and bioactive compounds relevant to human health of sweet cherry (Prunus avium L.) cultivars grown in Italy. Food Chem. 630–638 (2013)
17. Usenik, V., Fabčič, J., Štampar, F.: Sugars, organic acids, phenolic composition and antioxidant activity of sweet cherry (Prunus avium L.). Food Chem. 185–192 (2008)
18. Mitić, M.N., Obradović, M.V., Kostić, D.A., Micić, R.J., Pecev, E.T.: Polyphenol content and antioxidant activity of sour cherries from Serbia. Chem. Ind. Chem. Eng. Q. 53–62 (2012)

19. Alfaro, S., Mutis, A., Palma, R., Quiroz, A., Seugel, I., Scheuermann, E.: Influence of genotype and harvest year on polyphenol content and antioxidant activity in murtilla (Ugni molinae Turcz) fruit. J. Soil Sci. Plant Nutr. 67–78 (2013)
20. Viljevac Vujetić, M., Dugalić, K., Mihaljević, I., Tomaš, V., Vuković, D., Zdunić, Z., Jurković, Z.: Season, location and cultivar influence on bioactive compounds of sour cherry fruits. Plant Soil Environ. 389–395 (2017)
21. Poiana, M.A., Alexa, E., Mateescu, C.: Tracking antioxidant properties and color changes in low-sugar bilberry jam as effect of processing, storage and pectin concentration. Chem. Cent. J. **6**, 2–11 (2012)
22. Klopotek, Y., Otto, K., Bohm, V.: Processing strawberries to different products alters contents of vitamin C, total phenolics, total anthocyanins, and antioxidant capacity. J. Agric. Food Chem. **53**, 5640–5646 (2005)
23. Reber, J.D., Eggett, D.L., Parker, T.L.: Antioxidant capacity interactions and a chemical/structural model of phenolic compounds found in strawberries. Int. J. Food Sci. Nutr. **62**, 445–452 (2011)

Characterization and Investigation of Bioactivity of Copper(II) and Cobalt(II) Complexes with Imine Ligand

Emir Horozić(✉) and Jasmin Suljagić

Faculty of Technology, University of Tuzla, Univerzitetska 8, 75 000 Tuzla, Bosnia and Herzegovina
emir.horozic@untz.ba

Abstract. Imines are organic compounds which in their structure have a double bond between one carbon and one nitrogen atom. Forms very stable complexes with transition metals, which are now widely tested in order to detect a potential catalytic and pharmacological activity.

This study involves the synthesis of imines and complexes with biogenic bivalent metals (copper and cobalt) characterized by FTIR and UV/VIS spectroscopy. In order to compare color, texture and crystal size, morphological characterization was also performed. Antioxidant activity was determined using FRAP, DPPH and ABTS methods while antimicrobial activity was tested by diffusion technique on bacterial strains: Staphylococcus aureus, Enterococcus faecalis, Escherichia coli and Salmonella Enteritidis.

The results show that the structure of imine Cu(II) and Co(II) complexes differ significantly. In the formation of the Cu(II) ion, O, N and S are the donor atoms of imine, while only the S donor ligand atom is involved in the formation of a coordinate bond with Co(II). All tested methods for determining of antioxidant activity show that the antioxidative capacity of the samples decreases from the Cu(II) complex to the imine. Imin and Cu (II) complex exhibits antibacterial activity against Staphylococcus aureus and Enterococcus faecalis, while Co (II) complex has no antibacterial activity.

Keywords: Imine · Metal · Complex · Antioxidant activity

1 Introduction

Imine (Schiff base) is a nitrogen analog of an aldehyde or ketone in which the C = O group is converted to C = N − R group. It is usually formed by condensation of an aldehydes or ketones with a primary amine according to the following reaction [1]. The general scheme for obtaining imine is shown in Fig. 1. Synthesis of imine transition metal complexes by using imine as ligands appears to be fascinating in view of the possibility of obtaining coordination compounds of unusual structure and stability [2]. Imines and their metal complexes have been reported to exhibit a wide range of biological activities such as antibacterial including antimycobacterial, antifungal, antiviral, antimalarial, antiinflammatory, antioxidant, pesticidal, cytotoxic, enzyme inhibitory, and anticancer including DNA damage [3]. Imin compounds and their metal

I. Karabegović (Ed.): NT 2019, LNNS 76, pp. 624–632, 2020.
https://doi.org/10.1007/978-3-030-18072-0_73

complexes have been extensively investigated due to their wide range of applications including catalysts, crystal engineering, anti-corrosion agent [4–9].

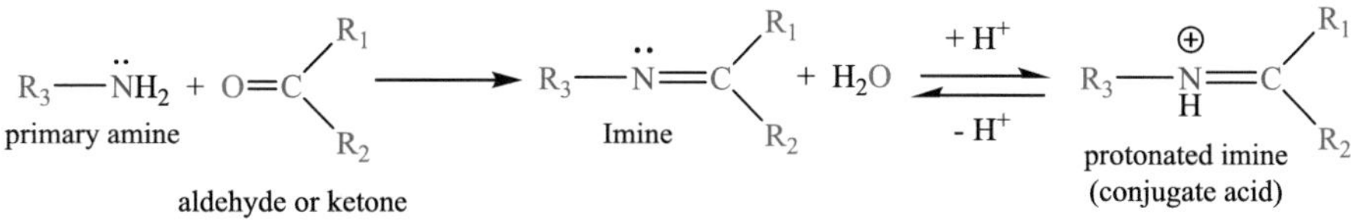

$$R_3{-}\ddot{N}H_2 + O{=}C(R_1)(R_2) \longrightarrow R_3{-}\ddot{N}{=}C(R_1)(R_2) + H_2O \underset{-H^+}{\overset{+H^+}{\rightleftharpoons}} R_3{-}\overset{\oplus}{N}H{=}C(R_1)(R_2)$$

primary amine; aldehyde or ketone; Imine; protonated imine (conjugate acid)

Fig. 1. The general reaction of forming the imine [10]

2 Experimental

All reagents used were p.a. purity and were purchased from Sinex (Bosnia and Herzegovina), EuroLab (Bosnia and Herzegovina), Sigma Aldrich (United States) and Fisher Scientific (United States).

2.1 Synthesis of Imine

The synthesis of imines was carried out according to the procedure [11]. In 60 mL of 96% ethanol, 0.91 g of thiosemicarbazide was added. The solution was transferred to a 250 mL balloon and heated to 50 °C under reflux. After this, the solution of 5-Br-salicylaldehyde, prepared by dissolving 0.201 g of 5-Br-salicylaldehyde in 30 mL of ethanol was added to the heated solution. The reaction mixture was stirred under reflux for 4 h at 60 °C and then cooled on ice, whereby the 5-Br-salicylaldehyde thiosemicarbazone crystals were precipitated.

2.2 Synthesis of M(II) Complexes

Synthesis of copper (II) and cobalt (II) complex with 5-Br-salicylaldehyde thiosemicarbazone was carried out following the published procedures [11, 12], with minor modifications. To the ethanolic solution (20 mL) of 5-Br-salicylaldehyde thiosemicarbazone (1 mmol) was added ethanolic solution (20 mL) $CuSO_4 \times 5H_2O$ (0.5 mmol), or $CoCl_2 \times 6H_2O$ (0.5 mmol). The reaction mixture was refluxed for 5 h and left overnight. The synthesized complexes were filtered through a blue strip filter paper, washed with ethanol and dried in air. The synthesized copper (II) complex was dark green, while the cobalt (II) complex was dark brown.

2.3 Spectral and Morphological Characterization

In order to determine the structure of the complex, samples were taken on the FT-IR spectrophotometer. Several milligrams of ligand, or complex, were transferred to the apparatus, pressed and recorded in the wavelength range of 4000-650 cm^{-1}. For the imaging of the UV/Vis spectra of imine and imine M (II) complexes, ligand/complex concentrations of 0.3/0.15 mg/mL in dimethyl sulfoxide (DMSO) were prepared. The absorption spectra of the prepared solutions were recorded on a double-sided UV/Vis spectrophotometer in the wavelength range of 400–800 nm. Sampling was performed at room temperature (T = 25 °C) in quartz cuvettes. The separated solid products of interaction were subjected to a microscopic analysis for the purpose of comparing the color, texture and particle size of the synthesized imine and complexes.

2.4 Determination of Antioxidant Capacity

For the in vitro antioxidative activity testing, the basic solutions of ligand and M(II) complexes of 1 mg/mL concentrations were used to test the antioxidant activity by DPPH method. For ABTS method, concentration of the ligand and the M(II) complexes was 0.166 mg/mL.

DPPH Method: To analyze the antioxidative capacity by DPPH, a 0.5 mM solution of 2,2-diphenyl-1-picyl hydrazide radical was prepared by dissolving 19.72 mg of DPPH reagent in 100 mL of methanol using an ultrasonic bath. A series of samples with different concentration of the ligand/M(II) complex was made by adding various volumes (from 10–400 μL). The tubes were then supplemented with methanol to 2 mL and then 500 μL DPPH reagent was added. Samples were incubated at room temperature in a dark space for 30 min and then recorded at 517 nm with methanol as a blank test. Inhibition of DPPH radicals was calculated according to the equation:

$$I\ (\%) = \frac{A_k - A_x}{A_k} \cdot 100 \quad (1)$$

where the Ak - absorbance of the control and Ax - absorbance of the sample. The results are expressed as an IC_{50} value.

ABTS Method: For the preparation of $ABTS^{+}$ reagents solutions of ABTS reagents (7 mM) and potassium persulphate (2.4 mM) were made. The working solution was prepared by mixing the basic solutions in equal amounts. After the solutions were mixed, they were allowed to stand for 16 h in a dark place at room temperature. The resulting working solution $ABTS^{+}$ is dark green. Due to the intensity of the green color before testing, the solution should be diluted so that its absorbance at 734 nm is about 0,700. 100 μL diluted solution of imine, or M(II) complex was transferred to the tube and then 3.8 ml of working solution ABTS+ was added. After 6 min of incubation, the absorbance is measured at 734 nm.

2.5 Determination of Antimicrobial Activity

Antimicrobial activity was studied by a diffusion method on reference bacterial strains (from ATCC collection) from gram positive group (*Staphylococcus aureus*ATCC 25923 and *Enterococcus faecalis* ATCC 51299) and gram negative bacteria (*Escherichia coli* ATCC 25922 and *Salmonella* Enteritidis ATCC 13076) by procedure form Clinical and Laboratory Standards Institute, 2009. From the bacterial strains of overnight cultures, suspensions of 0.5 McFarland turbidity were prepared (density 107–108 CFU/mL, depending on soy). The strains were then placed on the surface of the nutrient substrate-Mueller-Hinton agar (MH), dispersedin sterile Petri dishes. Substrate thickness was 4 mm. In the agar sterile drill-shaped holes were made ("wells"), into which 50 μL of imine and M(II) complexes solution in concentration of 0.6 mg/mL were added. After the plates were left at room temperature for 15 min, the substance was diffused into agar, incubated at 37 °C/24 h. After the incubation period, the size of the inhibitory zone was measured and the sensitivity of the microorganisms was expressed as follows: if the zone for inhibition of microorganism growth was greater than 20 mm, it was labeled with three pluses (+++), representing the highest sensitivity of the microorganisms. If the inhibitory zone ranged from 16 to 20 mm, it was marked with two pluses (++). Very weak sensitivity is marked with a plus (+) if the inhibitory zone is 10–15 mm in diameter. For the inhibitory zone less than 10 mm or if absent, the minus (−) has been used [13, 14]. Ciprofloxacin (CPF) was used as a control antibiotic.

3 Results and Discussion

3.1 Structure of the Ligand and M(II) Complexes and Spectral Characterization

The reaction of imine synthesis (5-Br-salicylaldehyde thiosemicarbazone) is shown in Fig. 2.

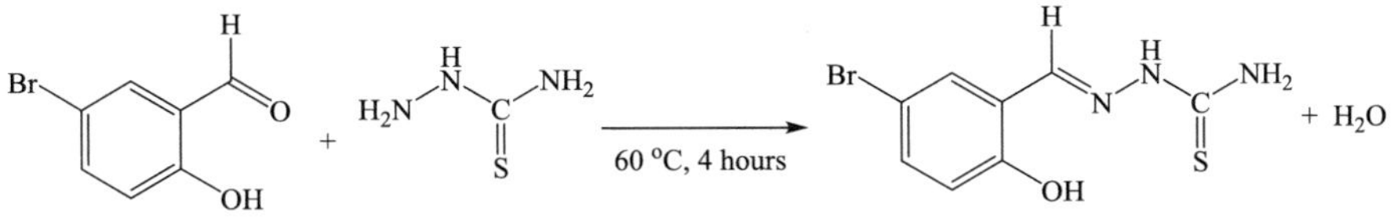

Br H O OH + H_2N H N C NH_2 S 60 °C, 4 hours Br H N H N C NH_2 S OH + H_2O

Fig. 2. Synthetic procedure of the imine

Figure 3 shows the structure of M (II) complexes. The structure of the Cu (II) complex with the synthesized imine is relatively complex. The complex consists of a positive and negative part that in 2013 was assumed by Pahont and associates [12]. In the cationic part of the complex, the imine molecule coordinates the copper ion as a tridentate donor ligand. O, N and S donor atoms participate in the formation of the bond. In addition to the three atoms belonging to the imine, the fourth O-donor atom originates from the molecule of water from the used metal salt. It is assumed that the cationic form has tetrahedral geometry (coordinate number 4). The anionic part of the molecule is more complex. In addition to the three donor atoms which are derived from imines, one molecule of water and a sulphate ion, which also lead to salt, are linked to the metal ion and are co-ordinated to Cu (II). In this case, it is a complex part of the trigonal bipyramidal geometry (coordination number 5).

The Co(II) complex, in contrast to the Cu(II) complex, has a simpler structure. The metal ion surrounds 4 atoms, two S atoms from the structure of the imine, while two chlorine atoms originate from the metal salt that was used to synthesize this complex.

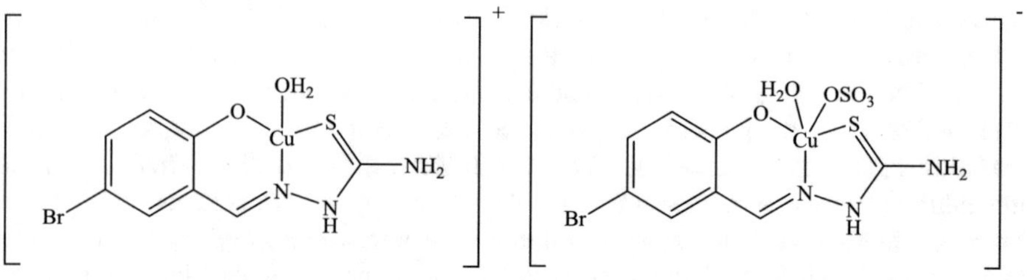

Fig. 3. The proposed structure of imines M(II) complexes: (A) complex with Cu(II) and (B) complex with Co(II)

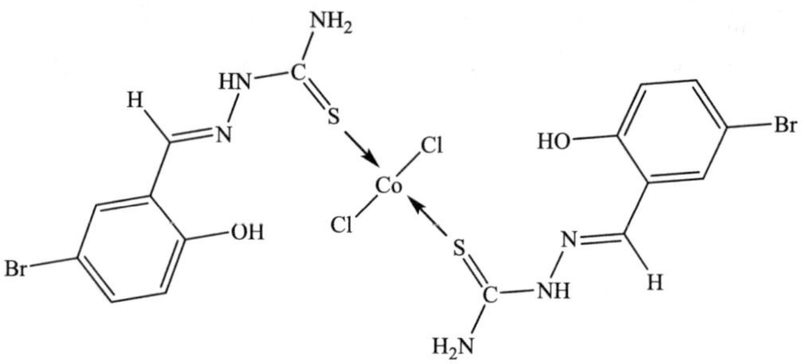

Although the imine contains a significant number of donor atoms, in this case Co(II) ion co-ordinates as monodentate ligand. At the metal center, two chlorine atoms are coordinated. On the basis of all of the above, it can be concluded that it is a complex of square-planar geometry.

In terms of color, the synthesized copper(II) complex has a dark green, complex cobalt (II) dark brown color, while the parent ligand is characterized by white color.

Table 1 shows selected infra-red frequencies and electronic spectral data of the prepared complexes.

Table 1. Selected infra-red frequencies and electronic spectral data of the prepared complexes

Sample	Infra-red spectral bands (cm^{-1})					Electronic spectral bands (nm)
	v(C = N)	v(C-O)	v(C = S)	v(O-H)	v(N-H)	
Imine	1599.14	1123.11	1541.25	3242.92	3454.04	741
Cu(II) complex	1600.32	1126.00	1531.10	–	3395.61	426
Co(II) complex	1604.82	1137.29	–	3227.52	3456.80	412

Based on the position of the absorption maximum in the UV/Vis spectrum, the value for the energy of splitting the d-subunits for the Cu(II)-imine system, or Co(II)-iminewas calculated, yielding 280 kJ/mol and 291 kJ/mol.

3.2 Morphological Characterization

The morphology of the imine crystals and the synthesized complexes is shown in Fig. 4.

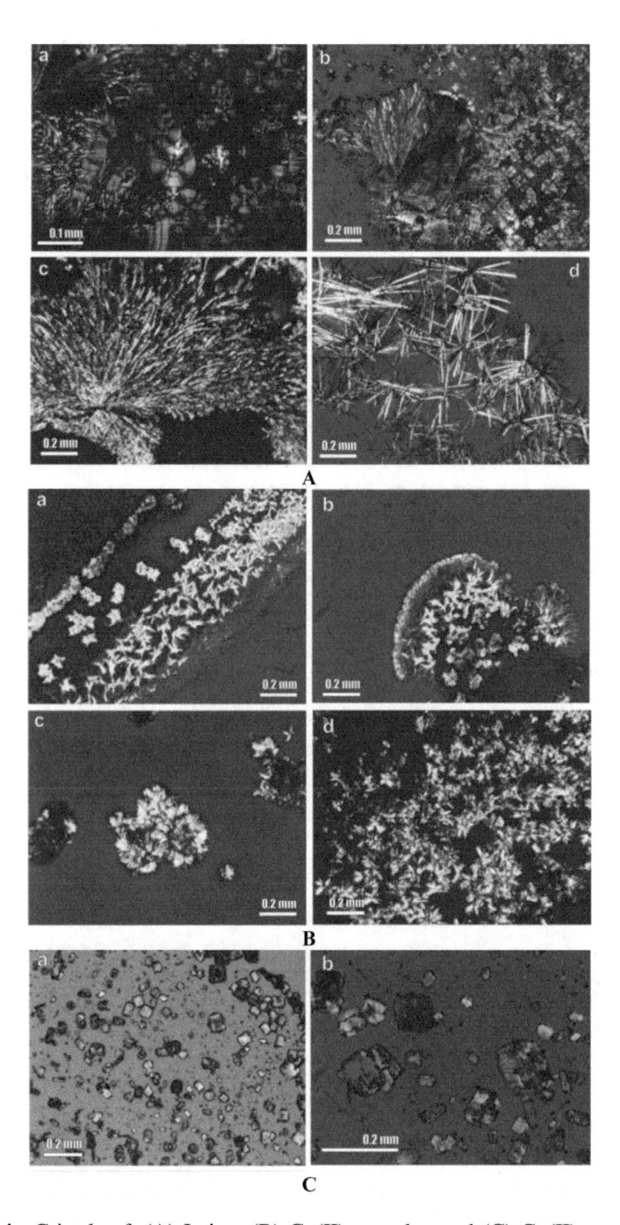

Fig. 4. Cristals of: (A) Imine; (B) Cu(II) complex and (C) Co(II) complex

In Figs. 4(a) and (b) spherical crystals with characteristic crystalline coatings are observed. Optical characteristics show undulatory darkness (half circle is light and forms "cross" and the other half is dark, turning the micro table changes (the light becomes dark and vice versa), spherical/round/crystal diameter up to 0.10 mm and the radially lightweight aggregates (they have a common crystal center from which the crystal is irregularly branched.) Interferential colors are gray and order. In terms of representation, spherical morphology is slightly represented. In Fig. 4(c) radial-light aggregates are dominant. Interfering colors are yellow first order. The crystal length reaches up to 0.5 mm. In Fig. 4(d) there are needle-like crystals for live intereferous colors (blue-purple), and order. Ignition crystals are radially-oriently oriented and their length ranges up to 0.35 mm. Crystals of Cu(II) complex developed in radial-airy forms. They belong to small crystal groups (up to 0.05 mm). Interfering colors are yellow order. The correctness of crystal growth is noted, the central parts are superficial (0.05 mm), unlike peripheral (0.01–0.02 mm). Crystals of Cu (II) complexes developed in radial-air forms. They belong to a group of small crystals (up to 0.05 mm). Interference colors are yellow and straight. The accuracy of crystal growth is noticeable, the central parts are coarse-grained (0.05 mm) as opposed to peripherals that are fine-grained (0.01–0.02 mm). Complex with Co(II) has crystallized into euhedral (right) crystals of live interfering colors first order. The crystals are fine grained (up to 0.10 mm). The crystalline forms associate with rectangular forms.

3.3 Antioxidant Capacity

The results of the antioxidant activity of imine and imine complexes, obtained by DPPH and ABTS method, are shown in Fig. 5. On the basis of DPPH results obtained, the best antioxidative capacity shows the Cu(II) complex with IC_{50} value of 0.051 g/L while the weakest antioxidant effect is recorded in the ligand (IC_{50} = 0.098 g/L). The results obtained by the ABTS method fully correspond to the results obtained by the DPPH method.

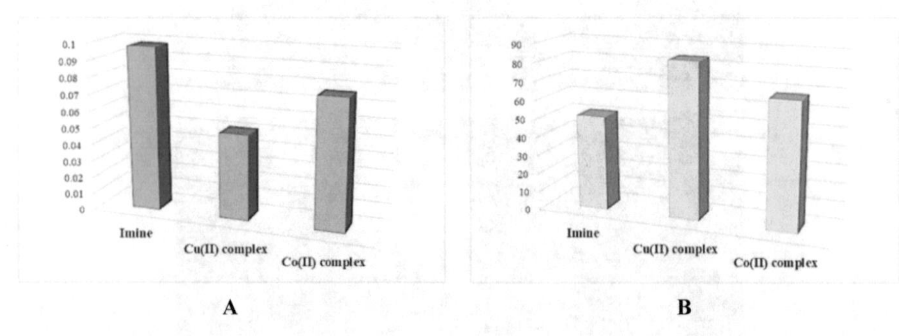

Fig. 5. Results of antioxidative activity tests: (A) DPPH method and (B) ABTS method

3.4 Antimicrobial Activity

Table 2 shows the results of antimicrobial activity. The results show that the complex with Co(II) has no antimicrobial activity. Imin shows high activity according to *E. faecalis* and *S. aureus*. Cu (II) complex exhibit significant activity towards *S. aureus*. Comparing the results of antimicrobial activity of imine, M(II) complexes and control antibiotic, it is concluded that CPF has a higher activity to all tested bacterial strains. The antimicrobial activity of the synthesized compounds and the CPF is similar in the case of *S. aureus*.

Table 2. Results of the antimicrobial activity

Sample	Inhibition zone [mm]			
	E. coli	*S. aureus*	*S.* Enteritidis	*E. faecalis*
Imine	–	29	–	23
Cu(II) complex	5	29	–	18
Co(II) complex	–	–	–	–
CFL	43	31	40	27

4 Conclusion

The synthesized Cu (II) and Co (II) complexes with 5-bromosalicylaldehyde thiosemicarbazone are structurally different. In the formation of the complex with Cu (II) ions, all donor atoms are involved, while only sulfur contributes to the formation of a bond with Co (II). Certain interactions certainly affect the complexity of structures and probably their bioactivity. The results obtained by the diffusion method indicate the existence of a certain antimicrobial activity of imine and Cu (II) complex at relatively low concentration. The highest antioxidant capacity was found in the Cu (II) complex and the smallest in the 5-bromosalicylaldehyde thiosemicarbazone.

References

1. Maihub, A.A., El-ajaily, M.M.: Synthesis, characterization and biological applications of schiff base complexes containing acetophenone or resemblance compounds. Acad. J. Chem. **3**(6), 46–59 (2018)
2. Divya, K., Pinto, G.M., Pinto, A.F.: Application of metal complexes of Schiff bases as an antimicrobial drug: a review of recent works. Int. J. Curr. Pharm. Res. **9**(3), 27–30 (2017)
3. Hameed, A., Al-Rashida, M., Uroos, M., Abid, S.A., Khan, K.M.: Schiff bases in medicinal chemistry: a patent review (2010–2015). Expert Opin. Ther. Pat. **27**(1), 63–79 (2017)
4. Zoubi, W.A.: Biological activities of schiff bases and their complexes: a review of recent works. Int. J. Org. Chem. **3**, 73–95 (2013)
5. Gupta, K.C., Sutar, A.K.: Catalytic activities of Schiff base transition metal complexes. Coord. Chem. Rev. **252**(12–14), 1420–1450 (2008)
6. Cozzi, P.G.: Metal-Salen Schiff base complexes in catalysis: practical aspects. Chem. Soc. Rev. **33**(7), 410–421 (2004)

7. Krishnamohan Sharma, C.V.: Crystal engineering - where do we go from here? Cryst. Growth Des. **2**(6), 465–474 (2002)
8. Ahamad, I., Prasad, R., Quraish, M.A.: Thermodynamic, electrochemical and quantum chemical investigation of some Schiff bases as corrosion inhibitors for mild steel in hydrochloric acid solutions. Corros. Sci. **52**(3), 933–942 (2010)
9. Antonijevic, M., Petrovic, M.: Copper corrosion inhibitors. Rev. Int. J. Electrochem. Sci. **3**(1), 1–28 (2008)
10. Horozić, E., Cipurković, A., Ljubijankić, N.: Anti-infective and anti-tumor activity of some metal complexes (M^{II}-M^{IV}) with Schiff bases. Technol. Acta **10**(1), 27–35 (2017)
11. Ljubijankić, N., Tešević, V., Grgurić-Šipka, S., Jadranin, M., Begić, S., Buljubašić, L., Markotić, E., Ljubijankić, S.: Synthesis and characterization of Ru(III) complexes with thiosemicarbazide-based ligands. Bull. Chem. Technol. Bosnia Herzeg. **46**, 1–6 (2016)
12. Pahontu, E., Fala, V., Gulea, A., Poirier, D., Tapcov, V., Rosu, T.: Synthesis and characterization of some new Cu(II), Ni(II) and Zn(II) complexes with salicylidene thiosemicarbazones: antibacterial: antifungal and in vitro antileukemia activity. Molecules **18**(8), 8812–8836 (2013)
13. Pirvu, L., Hlevca, C., Nicu, I., Bubueanu, C.: Comparative analytical, antioxidant and antimicrobial activity studies on a series of vegetalextracts prepared from eight plant species growing in Romania. J. Planar Chromatogr. **275**, 346–356 (2014)
14. Horozić, E., Cipurković, A., Ademović, Z., Kolarević, L., Bjelošević, D., Zukić, A., Hodžić, S., Husejnagić, D., Ibišević, M.: Synthesis, characterization and in vitro antimicrobial activity of the Cu(II) and Fe(III) complexes with 1-cyclopropyl-6-fluoro-4-oxo-7-(piperazin-1-yl)-1,4-dihydroquinoline-3-carboxylic acid. Bull. Chem. Technol. Bosnia Herzeg. **51**, 1–5 (2018)

Identifying Material Attributes for Designing Biodegradable Products

Indji Selim[1(✉)], Ana M. Lazarevska[2], Daniela Mladenovska[3], Tatjana Kandikjan[2], and Sofija Sidorenko[2]

[1] International School of Architecture and Design, University American College Skopje, Treta Makedonska Brigada 60, 1000 Skopje, Republic of Macedonia
indji.selim@uacs.edu.mk

[2] University Ss. Cyril and Methodius Skopje, Skopje, Republic of Macedonia

[3] University Mother Teresa, Skopje, Republic of Macedonia

Abstract. Nowadays, for a number of applications, biodegradable plastics are perceived as an environmentally friendly and often a more sustainable alternative to the commonly used materials. Having this in perspective, a relevant material selection, which simultaneously is affected by the manufacturing constrains, consumers' needs, market rules, etc., becomes one of the crucial phases in the process of product development. The process becomes even more complex, when the concept of product life cycle is considered.

This paper aims to identify relevant attributes for biodegradable materials by conducting a tailor-made survey among students enrolled in design studies. The unjudgmental and creative thinking, in addition to the benefit of not being affected and inhibited by the standard-based restrictions were the main reasons to select these students as relevant stakeholders for this survey. Therefore, utilizing the Ashby and Johnson (2002) categorization methodology which classifies the material attributes into four groups (general, technical, environmental and aesthetical), the students are inspired to propose additional ones, based on their preferences. Further, preferences-based weighting among the complete set of attributes is performed by utilizing the multi-criteria analysis (MCA), and the obtained weights are used as guidance for overall ranking of the attributes' set.

The results of this survey are a sound base towards improving the material selection process influenced both by the designer and the manufacturer perspective.

Keywords: Biodegradable materials · Attributes · Design · Multi-criteria analysis

1 Introduction

Driven by the consumer preferences toward environmental-friendly products, the usage of bio-based raw materials, as well as the governmental policies toward green procurements, the market for/of biodegradable materials is in a constant growth, and it

I. Karabegović (Ed.): NT 2019, LNNS 76, pp. 633–639, 2020.
https://doi.org/10.1007/978-3-030-18072-0_74

is expected to develop significantly during the coming years. The market growth is driven by continuous research and development (R&D) activities, increased environmental awareness, and implementation of more and more restrictive environmental regulations. In 2015, the production capacities for biodegradable plastics account for nearly 1% of the total global plastics production [1]. Nevertheless, the high cost of biodegradable plastics has been a major barrier in the growth of the market. Despite of this, it is expected that by 2020 the share of bio-based and biodegradable plastics will increase to a 2.5% share of fossil plastics production [1]. Even though in general biodegradable plastics are more expensive than their fossil-based equivalent, nowadays, several examples of biodegradable plastic products have already reached cost competitiveness. Furthermore, the price of conventional plastics is depending on oil prices, while the prices of biodegradable plastics depend on biomass prices which are prone to more stable trends compared to oil [2].

The most suitable end-of-life solution depends on the type of biodegradable plastic, the volume on the market, the application and the available collection and processing infrastructure [2]. Apart from the end-of-life options suitable for conventional plastics, certified biodegradable plastic products can be composted, digested or biodegraded on agricultural land. Substitution of fossil-based plastics by biodegradable plastics generally leads to lower carbon footprint. Moreover, the achieved reduction of the carbon footprint for biodegradable plastics is generally significantly larger than that by biofuels.

Since the biodegradable plastics belong to a large set of materials having different properties, the most suitable end-of-life solution depends on the type of biodegradable plastic and the specific application. While in most regulations mechanical recycling is the preferred end-of-life solution, energy recovery via incineration is preferred over land filling. Additionally, due to their high calorific value, plastics (both conventional and biodegradable) can be used in general waste incineration as a substitute for a certain percentage of fossil fuels [3, 4]. When comparing environmental impacts of recycling and incineration, Tyskeng and Finnveden (2010) indicate that various factors influence the solution selection, whereby one of the main principles should be the answer to the question "what replaces what?" [4].

One of the main application areas for biodegradable plastics is the food packaging industry. Therefore, the relations between biodegradable materials, resources depletion, availability and food quality must be carefully considered. The relation between food and biodegradable plastic basically is twofold: food and biodegradable plastic might compete for the same feedstock and biodegradable plastic can be used as food packaging. Although at a much smaller scale, the debate over food vs. biodegradable plastic resembles the debate over food vs. biofuel. Namely, during the 2008 strong spike in food prices, the use of edible feedstock for the production of biofuels was strongly criticized as a large driver for the rise in the food prices (Mitchell) [5].

2 Identifying Material Attributes for Designing Biodegradable Products

In general, regardless of the used material, product designers aim towards incorporating technical with aesthetic, combining practical utility with emotional delight. On the other side, consumers buy products because of their proper functionality. Additionally, the product has to be easy and convenient to use, and it must characterize with a satisfying personality. The latter strongly depends on the industrial design (ID) of the product. Therefore, in the competitive market, the market share of a certain product is won (or lost) through its visual and tactile appeal, the way it is perceived and the emotions it generates [6]. Due to its complexity and the involved diversifying features, the ID process involves both objective and subjective evaluation. Objective criteria derive from the consumer needs and the manufacturer requirements, while subjectivity is identified and introduced by the designer to cover the variety of possible solutions.

This paper deals with identifying relevant attributes for designing biodegradable products, based on the opinions and preferences of students enrolled in design studies. The idea of involving students in this survey is justified by the fact that they are still not burdened with the prejudgments and learnt experiences. The herein proposed scheme assists the designer via providing guidance throughout the decision making process and facilitates identifying, more comprehensively, materials availability and user needs [7]. Being aware that some characteristics of biodegradable plastic can be a disadvantage in one application, and an advantage in another (for example, the low water vapour barrier of the bio-based plastic polylactic acid (PLA) is a disadvantage for a water bottle but an advantage in vegetables and fruit packaging), the survey was performed for two different applications of biodegradable plastic - water bottle and salad packaging.

2.1 Description of Multi-criteria Analysis (MCA)

When it comes to new product development, among the crucial decisions is the choice of materials and manufacturing technologies. Materials selection is a multi-criteria decision making (MCDM) problem which optimally should result in finding the best compromise between material properties and design requirements [8], such as functional conditions, design limits, user behaviours and environmental conditions [9, 10].

Decision-making is the study of identifying and choosing alternatives to find the best solution based on different attributes/criteria and considering the decision-makers' preferences. In order to facilitate this type of analysis, a set of methods, referred as MCDM methods are developed, due to the need of formalizing the decision-making process in situations involving multiple criteria/attributes. This major class of methods is further divided into Multi-objective decision-making and Multi-attribute decision-making (MADM) [11]. In MADM a small number of alternatives are to be evaluated against a set of attributes which are often hard to quantify [12]. These methods can provide solutions to challenging and complex management problems. It should be pointed out that methods and results are not necessarily comparable. Every method has its restrictions, mostly due to initial model assumptions [13].

Firstly, a decision-making problem should be clearly defined, structured in a main goal, attributes and sub-attributes, together with the constraints, the degree of uncertainty and key issues. After this step, the problem can be framed indicating the evaluation criteria. The phase in which the model is built, can be described as a process where the core of the problem is extracted from the complex picture. After creating the model and identifying the relevant attributes, the next step is assessing the importance of the mapped attributes into criteria, i.e. determining weights to indicate relevant importance of each group of attributes, as well as the relative importance of the individual attributes in each group. Herein, determining the weights is performed via simple ranking. (e.g. in case of 3 attributes, the ranking is performed by numbers 1, 2, and 3, denoting "1" for the most important and "3" for the least important). Further, the best ranked attribute receives the highest ranking score (RS), as per:

$$RS = (n + 1) - r \tag{1}$$

where:

r is ranking of the attributes,

n is the number of attributes in the corresponding group.

Then, weighting factor (w_j), for each attribute i is obtained by the following formula:

$$w_j = \frac{RS_j}{\sum_{i=1}^{n} RS_i}, \mathrm{j} = 1, n \tag{2}$$

Due to the normalization of the weights (Eq. 2), the sum of weights per group of attributes is equal to 1:

$$\sum_{i=1}^{n} w_i = 1 \tag{3}$$

2.2 Problem Definition, Attributes Ranking, Results and Discussion

As previously described, the survey was performed for sustainable material selection in two different applications of biodegradable plastic - water bottle and salad packaging. Based on the sustainability paradigm and having in mind the numerous and often conflicting criteria specified by manufacturers, designers and consumers, the hierarchy structure (the model) of the analysed problem was created. The first hierarchy level has four groups of attributes (general, technical, environmental and aesthetical), categorized accordingly to Ashby and Johnson [6]. Further, in the second hierarchy level, each group consists from different number of attributes related with the specified topic.

Table 1. Pool of students included in the survey per universities, study programs and number of participants

University	Study program	Number of participants (consistent participants)
Ss. Cyril and Methodius Skopje (Faculty of Mechanical Engineering)	Industrial Design (Undergraduate studies)	16 (15)
Ss. Cyril and Methodius Skopje (Faculty of Mechanical Engineering)	Industrial Design (Post-graduate studies)	3 (2)
University American College Skopje (Faculty of Architecture)	Architecture and Design	19 (11)
University Mother Teresa Skopje (Technical Faculty)	Mechanical Engineering and Management	6 (6)
Total		**44 (34)**

All inconsistent questionnaires (missing ranking scores) were discarded from the assessment. Further, the interviewees were given the opportunity to add up to three attributes per category. However, only one interviewee added only one attribute (volume) in the aesthetical category, which is not a sufficient justification for changes in the originally identified model (hierarchy). The total number of participants with consistent answers is given in Table 1.

Figure 1 depicts the hierarchy model, the attributes and their corresponding weights for two cases of biodegradable material selection – biodegradable water bottle and biodegradable salad packaging.

The results indicate that the group of technical attributes is denoted as the most important (and equally important) for material selection of both products. The least important (also for the both products) are aesthetical attributes. In terms of the second level attributes, it should be pointed out that "recyclability" and "shape" are identified as the most important features (criteria) for both products.

As per the specific preferences of the participants related with their studies-major, it should be pointed out that there is certain correlation. For example, aesthetical attributes are preferred by architecture and design students (0.23/0.15), but in terms of technical attributes the preferences of the same groups of students aggregated separately are higher than the average weights (preferences) of the total interviewees (0.35/0.35).

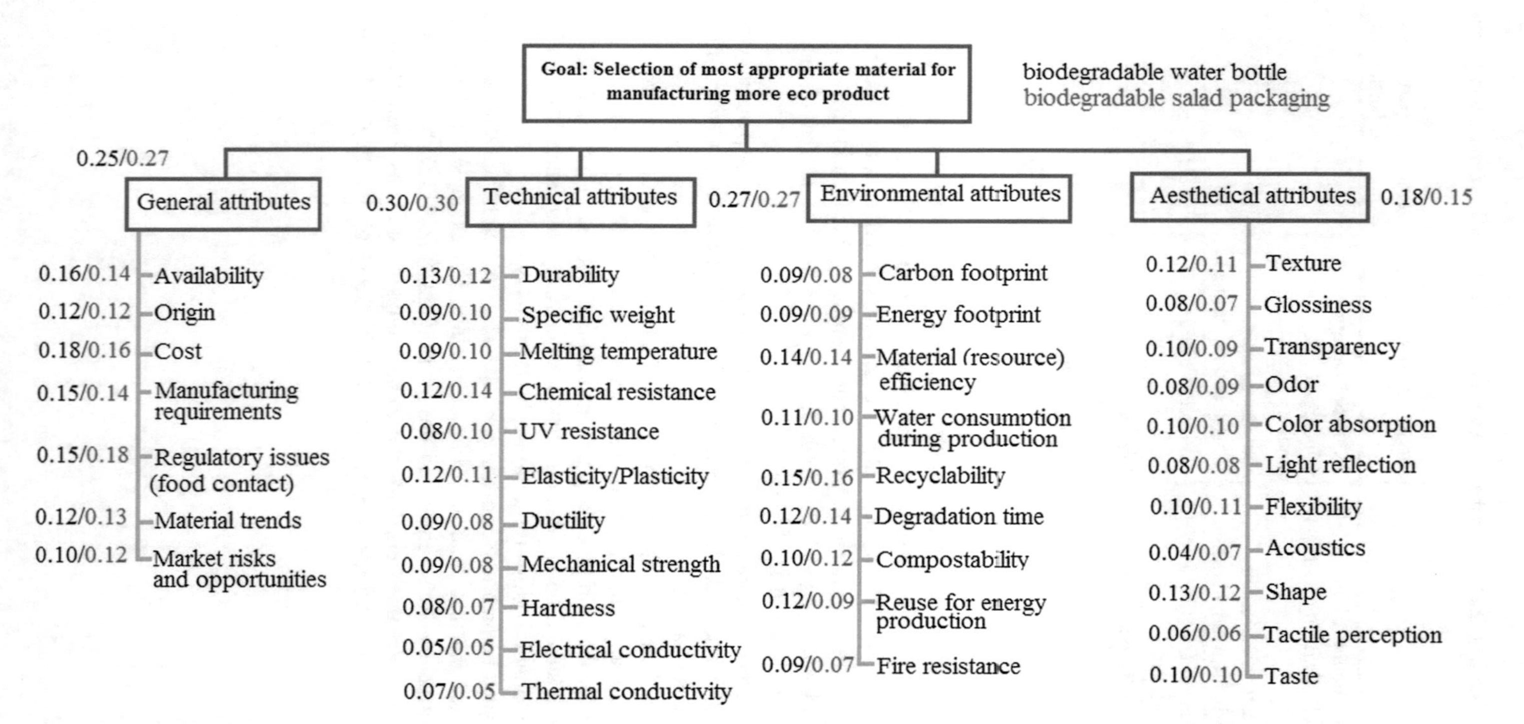

Fig. 1. Hierarchy model, identified attributes and their corresponding weighting factors, for both cases of material selection – biodegradable water bottle and biodegradable salad packaging

3 Conclusion

This paper deals with the attributes of biodegradable materials and their importance during the product design process. Multi-criteria analysis helps designers to formalize the material selection process and to reach a balance between market needs, manufacturer requirements, regulatory constraints, customer needs etc. Contemporary decision making methods usually use expert opinions and calculate importance of attributes in accordance with their preferences. However, motivated by the unjudgmental and creative thinking of the students, and the benefit of not being affected and inhibited by standard-based restrictions, in this paper, as relevant stakeholders for the survey, selected were students enrolled in design studies. Stakeholders' preferences were obtained by simple ranking scores. Next phase of the survey will be performed in standard manner by interviewing experts. Finally, the results from the both pool of participants (students vs. experts) will be compared.

References

1. Ashter, S.A.: Introduction to Bioplastics Engineering: A Volume in Plastic Design Library. Elsevier Inc. (2016)
2. van den Oever, M., Molenveld, K., van der Zee, M., Bos, H.: Bio-based and biodegradable plastics - facts and figures, Report 1722. Wageningen Food & Biobased Research Institute, Wageningen, Netherlands (2017)
3. Eriksson, O., Finnveden, G.: Plastic waste as a fuel - CO_2 neutral or not? Energy Environ. Sci. **2**, 907–914 (2009)
4. Tyskeng, S., Finnveden, G.: Comparing energy use and environmental impacts of recycling and wasteincineration. J. Environ. Eng. **136**(8), 744–748 (2010)
5. Mitchell, D.: A note on rising food crisis. The World Bank Development Prospects Group, Policy Research Working Paper 4682 (2008). http://documents.worldbank.org/curated/en/229961468140943023/pdf/WP4682.pdf
6. Ashbyand, M., Johnson, K.: Materials and Design: The Art and Science of Material Selection in Product Design. An Imprint of Elsevier Science. Butterworth-Heinemann, Oxford (2002)
7. Selim, I., Kandikjan, T., Lazarevska, A.M., Mladenovska, D.: Adaptation of the industrial design thinking process to the demands of biodegradable plastics. Poster at International Scientific Conference GREDIT 2018, Skopje, p. 161 (2002)
8. Ashby, M., Shercliff, H., Cebon, D.: Materials: Engineering, Science: Processing and Design. Butterworth-Heinemann, London (2002)
9. Ashby, M.: Material Selection in Mechanical Design. Pergamon Press, Cambridge (1992)
10. Cornish, E.H.: Materials and the Designer. Cambridge University Press, New York (1987)
11. Wang, J.J., Jing, Y.Y., Zhang, C.F., et al.: Review on multi-criteria decision analysis aid in sustainable energy decision-making. Renew. Sustain. Energy Rev. **13**, 2263–2278 (2010)
12. Hwang, C.L., Yoon, K.: Multiple Attribute Decision Making: Methods and Applications. Springer, Berlin (1981)
13. San Cristóbal, J.R.M.: Multi criteria analysis. In: San Cristóbal, J.R.M. (ed.) Multi-criteria Analysis in the Renewable Energy Industry. Green Energy and Technology. Springer, London, pp. 7–10 (2012)

New Ecological Industrial Synthesis of Alkyl Thionocarbamate from Isopropyl Dixanthogenate

Milutin M. Milosavljević(✉), Svetlana K. Belošević,
Milenko Petrović, and Milan M. Milosavljavić

Faculty of Technical Science, University of Priština,
Ul. Knjaza Miloša 7, 38220 Kosovska Mitrovica, Serbia
milutin.milosavljevic@pr.ac.rs

Abstract. A novel synthesis of N-alkyl and N,N-dialkyl-O-isopropyl thionocarbamates from isopropyl dixanthogenate and primary and secondary amines have been developed on laboratory and applied on industrial scale production. Sodium hypochlorite have been used for oxidation of amine salt of isopropyl xanthogenic acid to diisopropyl xanthogenate until all reactant have been digested. According to satisfactory yield and purity of synthesized N-alkyl and N, N-dialkyl-O-isopropyl thionocarbamates obtained by laboratory optimal synthetic procedure a satisfactory industrial adaptation on industrial scale have been done. The proposed method offer a high degree of conversion and purity of product, absence of by-products and technological applicability at industrial scale. Considering importance of the xanthogenates, application of the optimized methods of thionocarbamates synthesis would provide significant improvement in sustainable development and implementation of eco-friendly production technology. This environmentally benign process represents a suitable option to existing methods.

Keywords: Isopropyl dixanthogenate · Isopropyl thionocarbamate · Amine · Isopropyl xanthatogenic acid

1 Introduction

In this work a synthesis of *N*-alkyl and *N*,*N*-dialkyl-*O*-isopropyl thionocarbamates has been performed from isopropyl dixanthogenate (Thioperoxydicarbonic acid, bis(1-methylethyl) ester) and primary and secondary alkyl amines by oxidation with sodium-hypochlorite. The following alkyl amines have been used: methyl amine, dimethyl amine, ethyl amine, diethyl amine, n-propyl amine, di-n-propyl amine, isopropyl amine and diisopropyl amine. The mechanism of this reaction has been defined by means of isolation and identification of intermediates and products of the reaction. The identification of the reaction products have been performed using FTIR, ^{1}H and ^{13}C NMR and GC MS/MS spectroscopic methods.

Thionocarbamates are derivatives of thiocarbamic acid (I)

I. Karabegović (Ed.): NT 2019, LNNS 76, pp. 640–649, 2020.
https://doi.org/10.1007/978-3-030-18072-0_75

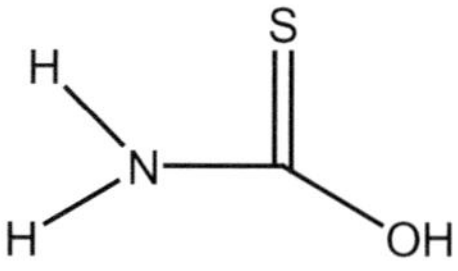

$$H_2N{-}C({=}S){-}OH \quad (1)$$

(I)

thus *N*-alkyl- and *N*, *N*-dialkyl thionocarbamates represent *N*-alkyl and *N*,*N*-dialkyl-*O*- alkyl esters of thiocarbamic acid (II)

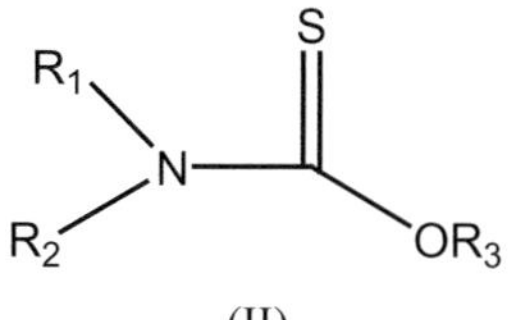

$$R_1(\mathrm{H})R_2N{-}C({=}S){-}OR_3 \quad (2)$$

(II)

Wide application of these selective collectors in flotation of copper and zinc ores is well known [1]. They are also fungicidal [2], bactericidal [3], herbicidal [4], pesticidal [5] and pharmaceutical active compounds [6].

Besides other procedures known in literature [7–11], alkyl thioncarbamates could be obtained by reaction of sodium- or potassium-xanthogenates in aqueous solution with primary or secondary aliphatic amines and elementary sulphur [12]. Thionocarbamates can also be prepared by reaction of xanthogenates and amines in the presence of nickel (II) sulphates heptahydrate as catalyst [13]. Ultrasound has significant advantages in the extraction of a great number of biologically active compounds including carotenoids. The advantages of using ultrasound for the extraction of various compounds are numerous and include effective mixing and micro-mixing, more efficient energy transfer, reduced thermal and concentration gradients, reduced temperature [14]. In the following period, we will perform the synthesis procedure intensified by ultrasound.

In this work, the investigated synthetic method present nucleophilic heterolysis of S-S bond in diisopropyl dixanthogenate (I) using amine (II) presented by reaction (3). The intermediary product is sulphenamide (*S*-(*O*-alkylcarbonothionyl)-*N*-alkylthiohydroxylamine) (III) and amine salt of isopropyl xanthogenic acid (IV). In reaction (4), sulphenamide (III) is decomposed into sulphur and the corresponding isopropyl thionocarbamate (V). Furthermore, the oxidation of amine salt (IV) is performed by addition of sodium hypochlorite giving a reactant diisopropyl dixanthogenate presented in reaction (5), and the reactions are successively repeated, i.e. the liberated amine again performs heterolysis of S-S bond of the obtained diisopropyl dixanthogenate (Fig. 1).

The assumed mechanism of this reaction is confirmed by isolation and structure determination of intermediates and products of the reaction. Namely, the intermediate in first step of the reaction of the heterolysis of S-S bond of diisopropyl xanthogenate is amine salt of isopropyl xanthogenic acid which has been isolated. The presence of the sulphur as a product of decomposition of sulphenamide in reaction mixture after filtration is nearly equal to stoichiometrically expected to the yield of reaction.

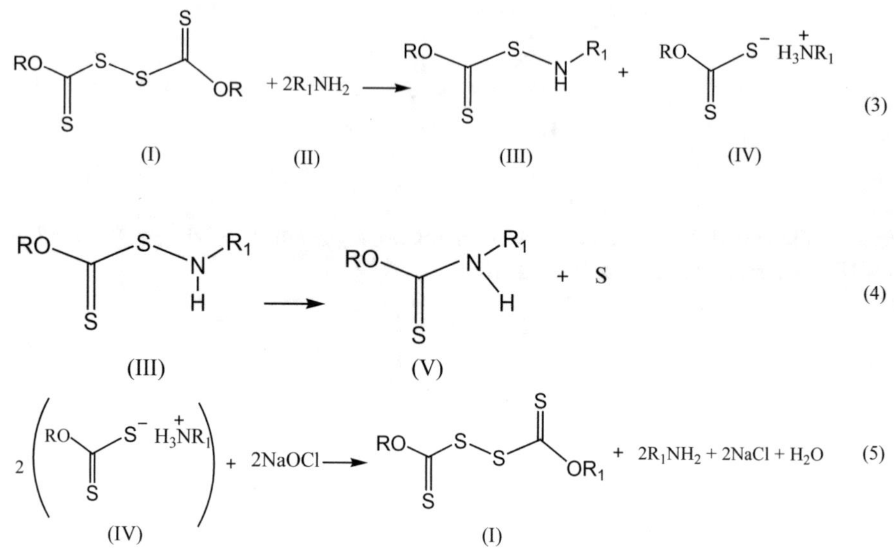

$$\underset{(I)}{RO\text{–}C(=S)\text{–}S\text{–}S\text{–}C(=S)\text{–}OR} + \underset{(II)}{2R_1NH_2} \longrightarrow \underset{(III)}{RO\text{–}C(=S)\text{–}S\text{–}NH\text{–}R_1} + \underset{(IV)}{RO\text{–}C(=S)\text{–}S^- \; H_3\overset{+}{N}R_1} \qquad (3)$$

$$\underset{(III)}{RO\text{–}C(=S)\text{–}S\text{–}NH\text{–}R_1} \longrightarrow \underset{(V)}{RO\text{–}C(=S)\text{–}NH\text{–}R_1} + S \qquad (4)$$

$$2\left(\underset{(IV)}{RO\text{–}C(=S)\text{–}S^- \; H_3\overset{+}{N}R_1}\right) + 2NaOCl \longrightarrow \underset{(I)}{RO\text{–}C(=S)\text{–}S\text{–}S\text{–}C(=S)\text{–}OR_1} + 2R_1NH_2 + 2NaCl + H_2O \qquad (5)$$

Fig. 1. Proposed reaction mechanism

In the third step, an oxidation of amine salt of isopropyl xanthogenic acid by using sodium hypochlorite, a mixed dialkyl dixanthogenate is also isolated as a crucial evidence of the reaction mechanism. Based on the identification of the isolated compounds, the assumed mechanism of the reaction on that way has been proved.

2 Experimental Part

Optimum reaction condition for synthesis of *N*-ethyl-*O*-isopropyl thioncarbamate

In a three-necked flask (250 cm^3), equipped with magnetic stirrer, dropping funnel, condenser and a thermometer, 100 cm^3 of water and 20, 65 g (0, 075 mol) of 98% diisopropyl dixanthogenate were added. During one hour, 12,25 cm^3 (0, 15 mol) of 68% ethyl amine solution is added with a vigorous stirring, on that way an increase of temperature to 30 °C occur. After that period, 20,50 g (0,075 mol) of sodium-hypochlorite (130 g of active chlorine/1000 cm^3) is added drop wise. The temperature of reaction mixture gradually rose to 45 °C during course of reaction of one hour and half, after which the reaction is completed. In the course of reaction sulphur particles precipitate in the quantity almost corresponding stoichiometrically to the reaction yield.

The reaction mixture was filtered on Buchner funnel, the precipitated sulphur is separated as filtration cake from an aqueous emulsion phase of *N*-ethyl-*O*-isopropyl thionocarbamate. The product is separated from the filtrate by two ethereal extraction, dried with sodium-sulphate and ether was distilled at atmospheric pressure. Pure product was obtained by fractional vacuum distillation at 105 °C/(6,6 mbar). 21,50 g of *N*-ethyl-*O*-isopropyl thionocarbamate is obtained, representing the yield of 85,9%. The GC purity is 97%.

All other isopropyl thioncarbamates were synthesized in an analogous manner to above procedure, using the appropriate amines under reaction conditions presented in Table 1.

Table 1. Reaction conditions for synthesis of *N*-alkyl and *N,N*-dialkyl-*O*-isopropyl thioncarbamates

Compound	Diisopropyl dixanthogenate mol	Amine mol	Sodium-hypochlorite mol	Reaction time (h)	Temperature (°C)
iPrOC(S) NHMe	0.075	0.15	0.075	2.5	30–45
iPrOC(S) NMe_2	0.075	0.15	0.075	2.5	30–45
iPrOC(S) NHEt	0.075	0.15	0.075	2.5	30–45
iPrOC(S) NEt_2	0.075	0.15	0.080	2.5	30–45
iPrOC(S) NHPr	0.075	0.15	0.075	2.5	30–45
iPrOC(S) NPr_2	0.075	0.15	0.100	3.0	40–48
iPrOC(S) NHiPr	0.075	0.15	0.100	3.0	40–53
iPrOC(S) NiPr	0.075	0.15	0.130	4.5	40–55

Me - methyl group; Et- ethyl group, Pr - propyl group, iPr - isopropyl group

The ^{1}H NMR and ^{13}C NMR spectral measurements were performed on a Bruker AC 250 spectrometer at 62.896 MHz. The spectra were recorded at room temperature in deuterated chloroform ($CDCl_3$) in 5 mm tubes. The chemical shifts are expressed in ppm (d) values referenced to the TMS (tetramethylsilane) reference standard signal.

All mass spectra were recorded on a Thermo Finnigan Polaris Q ion trap mass spectrometer, including TraceGC 2000 (Thermo Finnigan Corp., Austin, TX, USA), integrated GC-MS/MS system. DIP (direct insertion probe) mode has been used to introduce the sample and EI/MS/MS technique to acquire the spectra. Ionization conditions were: ion source temperature 200 °C, maximum energy of electron excitation 70 eV, corona current 150 μA. MS/MS conditions were: collision gas helium, collision cell pressure 1 mT, isolation time 8 ms, qz value 0.30, excitation time 15 ms, multiplier 1400 V. The precursor ions were resonantly excited by adjusting voltage in the range 1–2 V, depending on the compound investigated. The data obtained were processed using Xcalibur™ 1.2 software and analysis of experimental data as a paper [15].

Gas chromatographic analysis have been performed on Perkin-Elmer 8700 equipped with FID detector and filled column with 5% OV-210 on Gas-Chrom Q (length 2 m, diameter 0.3175 cm (1/8''). Injector temperature 250 °C; Detector temperature

270 °C; Column programme mode: 50 °C (5 min) → 10 °C/min → 130 °C (15 min); carrier gas nitrogen (purity 99,99%) flow 1 cm^3/min; air flow 250 cm^3/min (purity 99,99%); hydrogen flow 25 cm^3/min (purity 99,99%).

3 Experimental Procedures for Isolation of Intermediates and Products of the Reaction

In a 250 cm^3 three-necked flask equipped with magnetic mixer, dropping funnel, condenser and a thermometer, 100 cm^3 of water and 20.65 g (0.075 mol) of 98% diisopropyl dixanthogenate were added. Thereafter, by gentle stirring a 12.25 cm^3 (0.15 mol) of 68% ethyl amine was added during half an hour. The temperature rose to 30 °C. The reaction mixture is filtered, filtration cake consist mainly from sulphur (2.0 g, determined analytically precipitating as sulphate). The filtrate is then transferred to the separation funnel, the organic phase separate in the upper layer (*N*-ethyl-*O*-isopropyl thionocarbamate, confirmed by FTIR, ^{1}H and ^{13}C NMR, GC MS/MS spectroscopic data, 5 g of crude product). The aqueous solution is transferred into a beaker and hydrochloric acid (1:1) was added with a continuous stirring until pH of the solution become slightly acidic (pH $\approx$ 6). At the bottom of the beaker a light yellow substance, insoluble in water, isopropyl xanthogenic acid was deposited. Diethyl ether extract of isopropyl xanthogenic acid is dried with sodium-sulphate. Ether is distilled, and purity of isopropyl xanthogenic acid (7.0 g) is determined potentiometrically. The structure of isopropyl xanthogenic acid is confirmed by FTIR, ^{1}H and ^{13}C NMR and GC MS/MS data.

In second experiment, repeated as above described, the isolated amine salt of isopropyl xanthogenic acid (the lower aqueous solution) was filtered to remove sulphur, and after was treated with 20.50 g (0.075 mol) of sodium-hypochlorite (150 g of active chlorine in 1000 cm^3). In the course of time the temperature is maintained at 45 °C, during 1 h. After the reaction is completed, the reaction mixture is filtered. The filtration cake contains a diisopropyl dixanthogenate having a melting point 54 °C (lit. m.p. 55 °C), (structure confirmed by FTIR, ^{1}H and ^{13}C NMR, GC MS/MS data, Table 4).

Third experiment aqueous phase which contain amine salt of isopropyl xanthogenic acid were treated with zinc-sulphate thus corresponding salt of isopropyl xanthogenic acid precipitate. Both salts are characterized using FTIR, ^{1}H and ^{13}C NMR techniques as well as atomic absorption spectroscopy. By this way undoubtedly have been proved that ammonium salt of isopropyl xanthogenic acid is a intermediary product in a reaction mechanism presented in Scheme 1.

4 Results and Discussions

In this work a synthesis of eight isopropyl thioncarbamates has been performed, starting from diisopropyl dixanthogenates exerting nucleophilic heterolysis S-S bond using the appropriate primary and secondary alkyl amines. Intermediary product, a amine salt of isopropyl xanthogenic acid have been oxidized by use of sodium-hypochlorite. Using appropriate optimal synthetic method described in the experimental part a series of *N*-alkyl- and *N,N*-dialkyl-*O*-isopropyl thionocarbamates have been obtained with the good yields given in Table 2.

Based on the results presented in Table 2, it is shown that satisfactory yields have been achieved in synthesis of isopropyl thioncarbamates using the method used in this work and also with satisfactory purity. In Fig. 2 is given graphical presentation of the yield in [g] for different compounds.

Table 2. Yields and purities of synthesized thionocarbamates

Compound	Yield			GC purity (%)	Boiling point (°C)/ pressure (mbar)
	g	mmol	%		
iPrOC(S)NHMe	17.89	13.17	87.8	97.9	101.2/6.6
iPrOC(S)NMe$_2$	19.73	13.05	87.0	97.2	104.6/6.6
iPrOC(S)NHEt	19.54	12.89	85.9	97.0	105.0/6.6
iPrOC(S)NEt$_2$	23.34	12.99	86.6	97.4	107.5/6.6
iPrOC(S)NHPr	21.23	12.95	86.3	98.2	110.0/7.9
iPrOC(S)NPr$_2$	27.68	12.27	81.8	98.9	120.0/7.9
iPrOC(S)NHiPr	19.57	12.13	80.9	99.8	115.0/7.9
iPrOC(S)NiPr$_2$	22.79	11.00	73.27	97.9	126.0/7.9

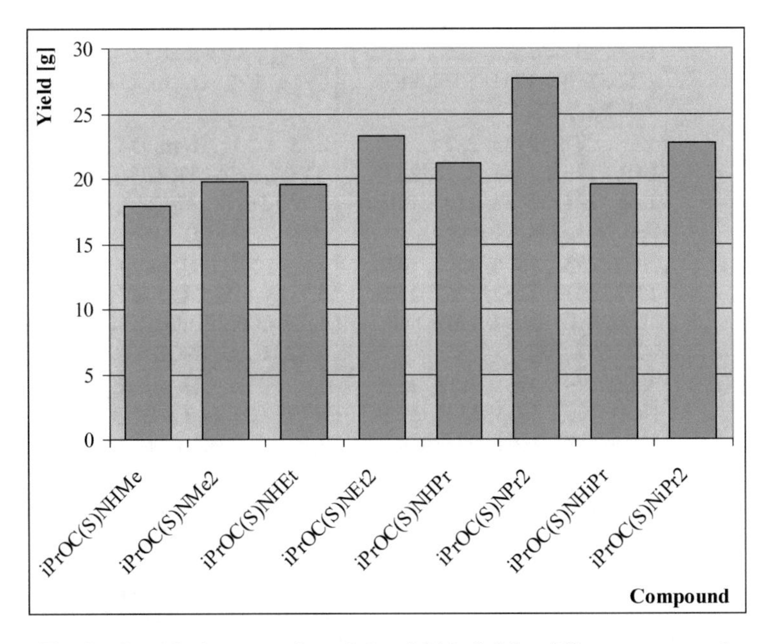

Fig. 2. Graphical presentation of the yield in [g] for different compounds

Somewhat lower yields have been achieved in synthesis of *N,N*-di-n-propyl thioncarbamates (81.8%) and *N*-isopropyl thionocarbamates (80.9%) and the lowest yield has been achieved with *N,N*-diisopropyl thioncarbamates (73.27%). The two voluminous isopropyl groups possess a significant steric hindrance effect in the first phase of the reaction of nucleophilic heterolysis of S-S bond in diisopropyl dixanthogenates. The structure of amines has vital influence on the nucleophilic reactive ability of amines.

In experimental part, dealing with determination of the reaction mechanism of the synthesis of isopropyl thioncarbamates, intermediates and products of the reaction are identified by various instrumental methods (FTIR, ^{1}H and ^{13}C NMR, GC MS/MS). The mechanism of this reaction is defined based on the confirmation of structure of isolated intermediates and products of the reaction.

The data from FTIR, ^{1}HNMR and MS spectroscopic methods of analysis of synthesized thioncarbamates are presented in Table 3, and the produced intermediaries and products in determining the reaction's mechanism are presented in Table 4.

Table 3. FTIR, ^{1}HNMR and CI MS data of synthesized isopropyl thioncarbamates

Tionkarbamat	IR, ν_{max} cm^{-1}	^{1}H NMR, δ ppm	MS, m/z
iPrOC(S) NHMe	3261, 2978, 2935, 2874, 1521, 1450, 1402, 1386, 1373, 1333, 1298, 1212, 1158, 1102, 1053, 964, 901, 834, 772	1.1–1.50 (9H, m, O-CH(C$\underline{H}_3$)$_2$, 2.94 (1H, d, N-$\underline{H}$), 3.1 (3H, d, N-C$\underline{H}_3$), 5.58 (1H, m, O-CH)	133
iPrOC(S) NMe$_2$	2977, 2935, 2874, 1502, 1426, 1378, 1317, 1282, 1252, 1182, 1150, 1104, 1094, 1079, 967, 922, 887, 774	1.04–1.55 (9H, m, O-CH (C$\underline{H}_3$)$_2$, 3.05–3.80 (2H, dq, N-C$\underline{H}_3$), 5.72 (1H, m, O-C$\underline{H}$)	147
iPrOC(S) NHEt	3261, 2978, 2935, 2874, 1521, 1450, 1402, 1386, 1373, 1333, 1298, 1212, 1158, 1102, 1053, 964, 901, 834, 772	0.95–1.50 (9H, m, O-CH (C$\underline{H}_3$)$_2$ + N-CH$_2$-C$\underline{H}_3$), 3.10–3.70 (2H, dq, N-C$\underline{H}_2$), 5.50 (1H, m, O-C$\underline{H}$)	147
iPrOC(S) NEt$_2$	2977, 2935, 2874, 1502, 1426, 1378, 1317, 1282, 1252, 1182, 1150, 1104, 1094, 1079, 967, 922, 887, 774	0.95–1.55 (12H, m, O-CH (C$\underline{H}_3$)$_2$ + 2xN-CH$_2$-C$\underline{H}_3$), 3.25–4.00(2H, dq, 2xN-CH$_2$), 5.60(1H, m, O-CH)	175
iPrOC(S)NH (nPr)	3263, 2965, 2934, 2875, 1525, 1459, 1403, 1355, 1337, 1261, 1204, 1102, 1071, 984, 929, 772	0,90 (3H, t, CH$_2$)$_2$-C$\underline{H}_3$), 1,15–1,80 (12H, m, O- CH(C$\underline{H}_3$)$_2$), 3,15-3,6 (2H, m, N-CH2)	161
iPrOC(S)N (nPr)$_2$	2967, 2934, 2875, 1497, 1465, 1421, 1372, 1318, 1285, 1248, 1229, 1181, 1152, 1098, 924, 772	0.90 (6H, t, 2x(CH$_2$)$_2$-C$\underline{H}_3$), 1.32 (6H, d, O-CH(C$\underline{H}_3$)$_2$), 1.40 (4H, m, 2xN-CH$_2$-C$\underline{H}_2$), 3.20–3.80 (4H, dt, 2xN-CH$_2$), 5.55 (1H, m, O-CH)	203

(continued)

Table 3. (*continued*)

Tionkarbamat	IR, ν_{max} cm^{-1}	^{1}H NMR, δ ppm	MS, m/z
iPrOC(S)NH (iPr)	3249, 2977, 2934, 2875, 1515, 1458, 1394, 1372, 1338, 1311, 1217, 1149, 1100, 1002, 990, 922, 814, 772, 708, 709	1.05–1.50 (12H, m, O-CH (C$\underline{H}_3$)$_2$ + N-CH(C$\underline{H}_3$)$_2$), 3.95 (1H, m, N-CH), 4.35 (1H, m, O-CH), 5.35–5.65 (1H, bm, NH)	162
iPrOC(S)N (iPr)$_2$	2977, 2932, 2873, 1461, 1396, 1375, 1341, 1313, 1221, 1167, 1151, 1098, 1004, 924, 709	0,96–1,50 (6H, t, 2 × CH (C$\underline{H}_3$)$_2$), 3,95 (2H, d, 2 × N-CH), 1,36 (6H, d, O- CH(C$\underline{H}_3$) 2), 5,60 (1H, m, O-CH)	203

Table 4. FTIR, ^{1}H NMR and MS data of intermediates and products of the reaction

Compound	FTIR	^{1}H NMR	MS
[iPrOC(S) S]$_2$	2978, 2930, 2872, 1457, 1449, 1373, 1353, 1271, 1145, 1082, 1027, 1003, 897, 795	1.42 (9H, d, OCH(C$\underline{H}_3$)$_2$), 5.65 (2H, m, OC$\underline{H}$)	270
iPrOC(S) SH	2978, 2930, 2872, 2651, 2550,1457, 1449, 1373, 1353, 1271, 1145	1.15 (6H, s, OCH(C$\underline{H}_3$)$_2$, 1.4 (1H, s, S$\underline{H}$), 3.52 (1H, s, OC$\underline{H}$)	136
(iPrOC(S) S)$_2$Zn	2972, 2936, 2877, 1456, 1343, 1262, 1135, 1062, 1003, 892,	1.38 (9H, d, OCH(C$\underline{H}_3$)$_2$), 5.32 (2H, m, OC$\underline{H}$)	–

Based on reproductive experimental results obtained in laboratory conditions of synthesis, industrial trial production was carried out. The technological scheme of the industrial production process is shown in Fig. 3.

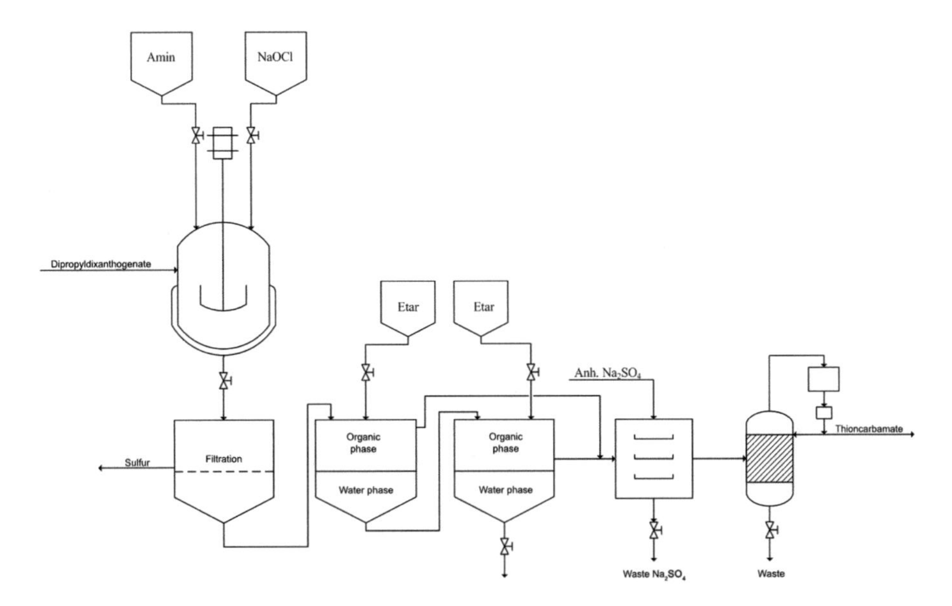

Fig. 3. Technological scheme of industrial procedure for production of thioncarbamates

The results of the industrial test production are presented in Table 5. Based on the results shown in Table 5, it can be seen that the highest conversion was achieved in the production of N-ethyl-O-isopropylthiocarbamate (95.1%), and the obtained product has high purity (99.1%). Also, the achieved conversion in all three batches is higher than the achieved yields in laboratory conditions.

Table 5. Results of the industrial process of trial production of thiocarbamate

Thioncarbamates	Reactants								Reakcioni uslovi		Sidei product	Product		
	Amin		Sulfuric acid		Na-iPrX		NaOCl		Time	Temperature	Sulfur	Yield		GC
	kg	kmol	kg	kmol	kg	kmol	m^3	kmol	h	°C	kg	kg	%	%
iPrOC(S)NHEt	35.4	0.55	28.1	0.28	158	1.0	0.66	1.5	2.5	30.0–40.0	24.8	139.8	95.1	99.1
iPrOC(S)NH(nPr)	33.2	0.55	28.1	0.28	158	1.0	0.66	1.5	2.7	35.0–45.0	24.3	146.4	92.1	92.1
iPrOC(S)N$(nPr)_2$	54.4	0.55	28.1	0.28	158	1.0	0.66	1.5	3.8	35.0–45.0	21.0	174.6	86.0	98.2

5 Conclusion

In the work a laboratory procedure of synthesis of *N*-alkyl and *N*,*N*-dialkyl-*O*-isopropyl thionocarbamates starting from diisopropyl dixanthogenates in the reaction with primary and secondary alkyl amines and by oxidation with sodium hypochlorite is shown. The obtained products are identified by instrumental methods (FTIR, ^{1}H and ^{13}C NMR and GC MSMS). Purity of the reaction's products, i.e. of isopropyl thioncarbamates is determined by GC method.

The mechanism of the reaction is determined by isolating and identifying of the intermediates and products of the reaction. It has been proved that the reaction is performed in two stages, the first one present the nucleophilic heterolysis of S-S bond of diisopropyl dixanthogenates using amines. The second stage presents a successive oxidation reactions of amine salt of isopropyl xanthogenic acid to diisopropyl dixanthogenate and the iterative reactions of heterolyses of the product obtained (diisopropyl dixanthogenate) using amines. Satisfactory reaction yield and simple work-up allow simple industrial application of the optimized laboratory synthetic procedure.

Acknowledgements. This work was supported by the Ministry of Education, Science and Technological Development of Serbia (Project Numbers 43007).

References

1. Walter, W., Bode, K.-D.: Syntheses of thiocarbamates. Angew. Chem. Int. Ed. **6**(4), 281–293 (1967). https://doi.org/10.1002/anie.196702811
2. Thind, T., Hollomon, D.W.: Thiocarbamate fungicides reliable tools in resistance management and future outlook. Pest Manag. Sci. **74**(7), 1547–1551 (2017). https://doi.org/10.1002/ps.4844

3. Chisholm, M.H., Extine, M.W.: Reactions of transition metal-nitrogen.sigma. bonds. 3. Early transition metal N, N-dimethylcarbamates. Preparation, properties and carbon dioxide exchange reactions. J. Am. Chem. Soc. (JACS) **99**(3), 782–792 (1977). https://doi.org/10.1021/ja00445a020
4. Pellegrini, G., Losco, G., Quattrini A., Arsura, E.: S-benzyl-N, N-disec-butylthiocarbamate and its use as a rice field herbicide and a rice growth stimulant, U.S. Patent no. 3930838, United States Patent and Trademark Office (USPTO), Alexandria, Virginia, USA (1976)
5. Rinehart, J.K.: S-naphthyl N-alkylthiolcarbamates, U.S. Patent no. 4059609, United States Patent and Trademark Office (USPTO), Alexandria, Virginia, USA (1977)
6. David, R.: Method and compositions for inhibiting tumor cell metabolism, U.S. Patent no. 5240914, United States Patent and Trademark Office (USPTO), Alexandria, Virginia, USA (1993)
7. Hall, V.J., Siasios, G., Tiekink, E.R.T.: Triorganophosphinegold (I) Carbonimidothioates. Aust. J. Chem. **46**(4), 561–570 (1993). https://doi.org/10.1071/CH9930561
8. Milosavljević, M.M., Marinković, A., Petrović, S.D., Sovrlić, M.: A new ecologically friendly process for the synthesis of selective flotation reagents. Chem. Ind. Chem. Eng. Q. **15**(4), 257–262 (2009). https://doi.org/10.2298/CICEQ0904257M
9. Milosavljević, M.M., Vukićević, I.M., Šerifi, V., Markovski, J.S., Stojiljković, I., Mijin, D. Ž., Marinković, A.D.: Optimization of the synthesis of N-alkyl and N, N-dialkyl thioureas from waste water containing ammonium thiocyanate. Chem. Ind. Chem. Eng. Q. **21**(4), 501–510 (2015). https://doi.org/10.2298/CICEQ141221006M
10. Friedrich, K.: Phenoxy-phenoxy-alkyl-thionocarbamate compounds, U.S. Patent no. 4060629, United States Patent and Trademark Office (USPTO), Alexandria, Virginia, USA (1977)
11. Reich, P., Martin, D.: Cyanic acid esters. IV: Molecule spectroscopic investigations of cyanic acid esters. Chem. Ber. **98**(7), 2063–2069 (1965)
12. Walter, W., Bode, K.D.: Oxidation products of thiocarboxylic acid amides. XV: Oxidation of thiocarbamic acid O-aryl esters to ortho-substituted aryloxyiminomethanesulfenic acids. Justus Liebigs Annalen der Chemie **698**(1), 122–130 (1966)
13. Johnson, G., Rafferty, M.F.: Amide, sulfonamide, urea, carbamate, thiocarbamate, and thiourea derivatives of 4'hydroxybenzylamine having anti-inflammatory and analgesic activity, U.S. Patent no. 980366, United States Patent and Trademark Office (USPTO), Alexandria, Virginia, USA (1990)
14. Avdeenko, A.P., Belova, E.A., Dašić, P.V., Konovalova, S.A., Baklanova, L.V., Krstić, S.S., Milosavljević, M.M.: Efficient two-frequency ultrasound extraction of β-carotene from the fungus Blakeslea Trispora. Hemijska industrija **71**(4), 329–336 (2017). https://doi.org/10.2298/HEMIND151110043A
15. Burya, A.I., Yeriomina, Y.A., Cui, H., Sapeshko, S.V., Dašić, P.: Investigation of the properties of carbon plastics based on polyetheretherketone. J. Res. Dev. Mech. Ind. **8**(1), 9–22 (2016)

Determination of Essential Metals in Drinking Water from Northwest Area of Bosnia and Herzegovina with AAS Method

Ekrem Pehlić[1(✉)], Aldina Aldžić-Baltić[1], Minela Žapčević[1], Aida Šapčanin[2], Kemal Salkić[3], and Huska Jukić[1]

[1] Faculty of Health Studies, University of Bihac, Ul. Nositeljahrvatskogtrolista 4, 77 000 Bihać, Bosnia and Herzegovina
pehlic_ekrem@yahoo.com
[2] Faculty of Pharmacy, University of Sarajevo, Sarajevo, Bosnia and Herzegovina
[3] Agricultural Institute of the Una-Sana Canton, Sarajevo, Bosnia and Herzegovina

Abstract. Safe and good quality drinking water is the basis for good human health. Drinking water can contain heavy metals that can cause serious health problems, but also essential elements that are very desirable and of great importance to the human organism. Therefore, the aim of this study was to determine the content of essential metals in drinking water from the northwestern Bosnia and Herzegovina. 24 samples of fresh water from eight municipalities were collected from different sources, in which nutrients were analyzed: sodium (Na), potassium (K), iron (Fe), calcium (Ca), magnesium (Mg) and selenium (Se). Measured metal concentrations (Na, Ca, Mg and Fe) were within the limits allowed by the Regulations of B&H and WHO guidelines, while the Se and K concentrations were slightly above the prescribed limits. Since these are essential heavy metals, their presence in drinking water is not a threat to human health.

Keywords: Fresh water · Essential elements · AAS

1 Introduction

Water is one of the most important substances in nature, and represents an indispensable natural resource needed for the life of living beings. It is generally obtained from two principal natural sources; surface water such as fresh water lakes, rivers, streams, etc. and ground water such as borehole water and well water. Bosnia and Herzegovina is one of the few countries in Europe and the world that has significant reserves of pure drinking water. Water has unique chemical properties due to its polarity and hydrogen bonds which means it is able to dissolve, absorb, adsorb or suspend many different compounds, thus, in nature, water is not pure as it acquires contaminants from its surrounding and those arising from humans and animals as well as other biological activities. The most common causes of chemical misalignment of water are the elevated content of heavy metals. They exist in water in colloidal,

I. Karabegović (Ed.): NT 2019, LNNS 76, pp. 650–654, 2020.
https://doi.org/10.1007/978-3-030-18072-0_76

particulate and dissolved phases with their occurrence in water bodies being either of natural origin (e.g. eroded minerals within sediments, leaching of ore deposits and volcanism extruded products) or of anthropogenic origin (i.e. solid waste disposal, industrial or domestic effluents). Heavy metals can cause serious health problems and ailments of neuropsychiatric disorders such as aggressive behavior, memory loss, depression, irritability, tiredness, infertility, gout, hypertension and headaches. Some of the metals are essential to sustain life-calcium, magnesium, potassium and sodium must be present for normal body functions. Therefore, due to the insight into the quality of spring water, the aim of the research is to determine concentrations of essential heavy metals such as Na, K, Ca, Mg, Fe and Se.

2 Materials and Methods

Water samples for analysis were collected from different distribution systems in eight municipalities: Cazin, Bihac, Buzim, Bosanski Petrovac, Bosanska Krupa, Velika Kladusa, Sanski Most and Kljuc. The survey includes sources (Table 1) that supply most of the population of north-western Bosnia and Herzegovina.

In these samples, the essential metals Na, K, Mg, Ca, Fe and Se were analyzed. Prior to analysis on the atomic absorption spectrophotometer, the samples were prepared by wet digestion in the Milestone S.r.l microwave oven-START D. After digesting, the samples were analyzed using the AAS SHIMADZU series AA - 6800. Water samples are acidified with nitric acid so that the pH is adjusted to 2. Before the measurement, the samples are filtered through a 0.45 μm filter paper (Filtres Fioroni). For the calibration of the instrument, a multistandard AAS Fluka concentration of 1000 $\mu g\ dm^{-3}$ was used, from which the series of standards required for calibration and determination of concentrations in water samples were made.

2.1 Results

The contents of the essential elements in the analyzed samples are presented in Table 2. The table also shows the values of maximum permissible concentrations (MPC) according to the Rulebook on the health of drinking water and the guidelines for water quality of World Health Organization (WHO).

The measured sodium concentrations ranged from 0.48–5.97 mg/L and all were below the maximum permissible of 200 mg/L according to B&H guidelines for water quality.

Although it is generally agreed that sodium is essential to human life, there is no agreement on the minimum daily requirement. In general, sodium salts are not acutely toxic because of the efficiency with which mature kidneys excrete sodium. However, acute effects and death have been reported following accidental overdoses of sodium chloride. Acute effects may include nausea, vomiting, convulsions, muscular twitching and rigidity, and cerebral and pulmonary oedema. Excessive salt intake seriously aggravates chronic congestive heart failure, and ill effects due to high levels of sodium in drinking-water have been documented.

Table 1. Name of source and sampling municipality

Number of samples	Municipality of sampling	Name of sources
1.	Sanski Most	Zdena
2.	Ključ	Okašnica
3.	Bosanski Petrovac	Sanica
4.	Velika Kladuša	Kvrkulja
5.	Bužim	Musići
6.	Bihać	Privilica
7.	Cazin	Stovrela
8.	Bosanska Krupa	Ada
9.	Sanski Most	Dabar
10.	Ključ	Bušotina
11.	Bosanski Petrovac	Smoljani
12.	Velika Kladuša	Šumatac
13.	Bužim	Pivnice
14.	Bihać	Klokot
15.	Cazin	Vignjevići
16.	Bosanska Krupa	Luke
17.	Sanski Most	Zdena
18.	Ključ	Okašnica
19.	Bosanski Petrovac	Sanica
20.	Velika Kladuša	Kvrkulja
21.	Bužim	Musići
22.	Bihać	Privilica
23.	Cazin	Stovrela
24.	Bosanska Krupa	Ada

Concentration of K in sample 22 (Privilica) in the town Bihac was in compliance with MPC values according to BiH regulation, while concentrations of other samples ranged from 14.20 to 22.04 mg/L and were above the permissible (12 mg/L). Although concentrations of potassium normally found in drinking-water are generally low and do not pose health concerns. Currently, there is no evidence that potassium levels in municipally treated drinking water, even water treated with potassium permanganate, are likely to pose any risk for the health of consumers.

The highest Ca content is in sample 19 (Sanica) with a concentration of 75.51 mg/L which is slightly above the prescribed upper limit of 75 mg/L, while the other samples had far less concentrations. Ca as an essential element is very important in the human body and its inadequate intakes have been associated with increased risks of osteoporosis, nephrolithiasis (kidney stones), colorectal cancer, hypertension and stroke, coronary artery disease, insulin resistance and obesity.

Mg concentrations in all samples were below MPC of 50 mg/L according to WHO, while the Rulebook in B&H do not regulate maximum concentrations of Mg in natural and spring water. Both calcium and magnesium are essential to human health.

Table 2. Values of concentrations of essential metals in analyzed drinking water samples

Number of sample	Concentrations mg/L					
	Na	K	Ca	Mg	Fe	Se
	*200	*12	**75	**50	*0,20	*0,01
1.	2,90	19,52	5,65	0,68	0,23	0,02
2.	4,13	21,23	10,45	2,10	0,22	0,02
3.	5,92	21,37	6,20	1,73	0,22	0,01
4.	2,07	21,16	7,74	0,47	0,23	0,01
5.	2,93	18,85	5,66	1,03	0,23	0,02
6.	5,92	19,74	6,21	1,85	0,22	<0,01
7.	1,88	20,01	9,86	2,42	0,23	0,03
8.	1,36	22,04	11,48	2,77	0,23	<0,01
9.	1,01	21,06	6,19	0,55	0,22	0,01
10.	1,08	18,91	5,70	2,03	0,22	0,03
11.	5,97	19,65	9,09	0,55	0,23	0,02
12.	2,41	21,29	11,20	1,03	0,24	0,04
13.	5,92	20,63	8,38	2,59	0,24	0,04
14.	3,01	21,49	5,64	2,54	0,23	0,02
15.	0,48	21,27	9,09	1,11	0,22	<0,01
16.	5,94	21,65	5,55	1,10	0,22	0,02
17.	3,95	20,95	6,20	2,07	0,22	0,05
18.	0,50	21,82	9,79	2,35	0,23	0,01
19.	5,94	20,73	75,51	1,13	0,23	0,03
20.	1,40	14,20	11,55	1,13	0,22	0,04
21.	0,72	21,10	5,64	1,79	0,24	0,02
22.	0,85	9,96	5,70	2,03	0,24	0,02
23.	5,43	21,56	6,96	1,11	0,23	0,01
24.	1,50	21,93	11,79	2,03	0,22	0,02

*Maximum permissable concentrations rulebook B&H/mg/L
** Maximum permissable concentrations WHO/mg/L

Inadequate intake of either nutrient can impair health. Low magnesium levels are associated with endothelial dysfunction, increased vascular reactions and decreased insulin sensitivity.

Iron is an essential element in human body, and it is present in the form of hemoglobin, myoglobin and enzymes that contain hem. Fe content in drinking water samples are within limits prescribed by Rulebook of water quality in B&H.

Water is not normally a major source of selenium intake, but it is important that a proper balance be achieved between recommended intakes and undesirable intakes in determining an appropriate guideline value for selenium in drinking-water. While for most parts of the world, the concentration of selenium in drinking-water will not exceed 10 µg/L, there are circumstances in which selenium may be elevated above normal concentrations. Concentrations of Se in most of the samples were above the MPC values prescribed by the Rulebook in B&H.

3 Conclusion

Of the total 144 analyzed samples, three samples of drinking water had concentrations of heavy metals below the detection limit of the method employed. The measured concentrations of Se and K were slightly above the upper limit and as such do not pose a risk to human health. The content of other metals (Na, Ca, Mg and Fe) did not exceed the permissible values regulated by the Rulebook on natural mineral and natural spring waters, the Rulebook on water quality in B&H and the World Health Organization guidelines.

References

1. Adepoju-Bello, A.A., Ojomolade, O.O., Ayoola, G.A., Coker, H.A.B.: Quantitative analysis of some toxic metals in domestic water obtained from Lagos metropolis. Nig. J. Pharm. **42** (1), 57–60 (2009)
2. Calcium and magnesium in drinking-water: public health significance. World Health Organization (2009)
3. Council Directive 98/83/EC, The quality of water intended for human consumption, L 330/32 (1998)
4. Iron in Drinking-water Background document for development of WHO Guidelines for Drinking-water Quality. World Health Organization (2009)
5. Jaishankar, M., Tseten, T., Anbalagan, N., Mathew, B.B., Beeregowda, K.N.: Toxicity, mechanism and health effects of some heavy metals. Interdiscip. Toxicol. **7**(2), 60–72 (2014)
6. Marcovecchio, J.E., Botte, S.E., Freije, R.H.: Heavy metals, major metals, trace elements. In: Nollet, L.M. (ed.) Handbook of Water Analysis, 2nd edn, pp. 275–311. CRC Press, London (2007)
7. McMurry, J., Fay, R.C.: Hydrogen, oxygen and water. In: Hamann, K.P. (ed.) McMurry Fay Chemistry, 4th edn, pp. 575–599. Pearson Education, New Jersey (2004)
8. Mendie, U.: The nature of water. In: The Theory and Practice of Clean Water Production for Domestic and Industrial Use, pp. 1–21. Lacto-Medals Publishers, Lagos (2005)
9. Potassium in Drinking-water Background document for development of WHO Guidelines for Drinking-water Quality. World Health Organization (2009)
10. Pravilnik o zdravstvenoj ispravnosti vode za piće u BiH (Službeni glasnik BiH, br. 40/10)
11. Selenium in Drinking-water Background document for development of WHO Guidelines for Drinking-water Quality. World Health Organization (2011)
12. Sodium in Drinking-water Background document for development of WHO Guidelines for Drinking-water Quality. World Health Organization (2003)
13. Sodium, chlorides and conductivity in drinking water. WHO Regional Office for Europe, Copenhagen (1979). (EURO Reports and Studies No. 2)
14. World Health Organization: Guidelines for drinking water quality, fourth edition incorporating the first addendum, vol. 394, p. 433 (2017)

Salmonellaspecies - From Production to Dining Table

Huska Jukić[1(✉)], Samira Dedić[1], Miloš Rodić[2], and Zlatko Jusufhodžić[2]

[1] Faculty of Health Studies, University of Bihać, Nositelja hrvatskog trolista 4, 77000 Bihać, Bosnia and Herzegovina
huskaj037@gmail.com

[2] Public Institution "Veterinary Institute", 77000 Bihać, Bosnia and Herzegovina

Abstract. Salmonella spp. is one of the most common bacterial causes of food poisoning. Clinical manifestation of Salmonella infections in people is often acute gastroenteritis, which does not require treatment. However, when the infection becomes invasive, antimicrobial therapy is necessary.

Salmonella spp. may be present in the intestinal tract and other tissues of poultry and animals with red meat, without showing any symptoms of the disease in the animal. This microorganism is a long-term issue in raw poultry meat and the effect was found in 70% of broilers' carcasses. This contamination agent penetrates the egg though narrow cracks and faecal contamination of the shell and through the ovary infection in chicken.

The aim of this paper was to investigate the potential presence of Salmonella spp. in fecal samples of laying hens from broiler farms, as well as table eggs, animal feed, and mortalities that sporadically occurred on the farms in the Tuzla Canton region. Continuous testing of samples for the presence of Salmonella spp. was carried out on 300 samples of faeces in the period of two years. The presence of bacteria was confirmed in 121 sample. Out of 400 samples of eggs, the presence was identified in 23 samples. Out of 40 samples of animal feed, Salmonella spp. was determined in one sample. On sporadic mortalities on farms, 50 carcasses were tested and the presence of Salmonella spp. was found in 29 samples.

Samples were tested by methods BAS EN ISO 6579-1: 2018, Horizontal method for the detection, enumeration and serotyping of Salmonella – Part 1: Detection of Salmonella spp. and Method for the detection of Salmonella spp. from animal faeces and environmental samples from primary production.

Keywords: Salmonella spp. · Faeces · Table eggs · Zoonosis

1 Introduction

Foodborne diseases pose a serious threat to public health, resulting in significant economic consequences in many parts of the world. Salmonellosis is one of the most important bacterial zoonotic foodborne diseases in the world. Human salmonellosis is classified as one of the most common and economically important zoonotic diseases. It is the second most common disease after campylobacteriosis [1, 2]. Special care is

I. Karabegović (Ed.): NT 2019, LNNS 76, pp. 655–662, 2020.
https://doi.org/10.1007/978-3-030-18072-0_77

needed in developing countries due to poor hygiene conditions that favour its spread. The most common transmission is mainly through food, such as meat, dairy products and egg by-products [3–6].

As a very nutritious food, the consumption of eggs is constantly increasing in the world, which simultaneously increases the number of cases of foodborne diseases. Eggs that have been infected with Salmonella and consumed raw or under cooked are responsible for many epidemics of salmonellosis [7, 8].

Bacteria of the genus *Salmonella* spp. are among the most important pathogens of the family *Enterobacteriaceae*. The first salmonella which causes typhoid was discovered in a patient who died of the red typhoid by Eberth in 1880. The other bacterium, which in many aspects looked like the cause of the red typhoid, was isolated from pigs suffering from hog cholera by Salmon and Smith. Suggested by French bacteriologist Liginerés, this bacterium was named salmonella in honour of Salmon [9].

Salmonella spp., like the vast majority of other bacteria of the family *Enterobacteriace*, are gram-negative facultative anaerobes capable of fermenting glucose and reducing nitrates to nitrites, catalase positive and oxidase negative [10], rod-shaped associated with red content, ranging from 0.7–1.5 to 2.0–5.0 μmin size [11]. Salmonella, in addition to glucose, ferment dulcitol, mannitol and maltose, producing gas and acid, but do not ferment lactose, sucrose, malonate, and salicin. Most salmonella produce H_2S, although there are serovarsthat do not produce it or are not very positive [12]. They do not form spores and are capsule-free; they are all motile (except *S. gallinarum* and *S. pullorum*), due to peritrichous flagella and some strains produce fimbriae [13]. A large number of salmonella strains are sensitive to chloramphenicol, ureido-penicillins and aminoglycosides [14].

Different animal species can suffer from salmonellosis and present a potential reservoir of human infection. Globally, contaminated food products of animal origin, especially eggs and egg products, are often involved in the outbreak of human salmonellosis [15]. Rats contribute to the spread of salmonella on chicken farms [16]. Eggs are sterile in the oviduct of healthy laying hens. After laying, the shell surface of the egg can be contaminated with various micro-organisms found in the environment. Microbiological contamination of eggs can lead to hygiene problems and the problem of egg deterioration. Consequently, the danger of consumer poisoning arises. *Salmonella enteritidis*, the most common isolated serotype of human poisoning in Croatia, is in more than 90% of salmonella isolates, just as in most European countries [17]. Salmonella can enter the food chain through carcasses contaminated by faeces in a slaughterhouse during production or when handling food [18]. Infected animals excrete salmonella through feces, secretion and excretion, and salmonella in infected animals are also found in their milk and meat, as well as in eggs. On the surface of egg shells, there are many species of Salmonella because they are common in the chicken digestive system. Salmonella seldom penetrate the shell, unless the eggs are handled carelessly. If shell eggs come into contact with contaminated water during cooling, bacteria can penetrate the egg shell. It is important to note that in the case of washing eggs, cold eggs must be washed in warm water, otherwise bacteria may penetrate the inside of the egg [19]. In order to prevent the possibility of laying hens being infected by salmonella from feed, which is one of the main sources of these bacteria, it is important to carry out regular analysis of raw materials and ready-made mixtures and to monitor the production in all phases. It is also important to improve the storage and transport of the mixtures and to minimize the risk of contamination [20]. De Knegt et al. (2015) examined the occurrence of salmonella serovars in animals and humans, comparing the cases of salmonellosis in broilers, turkeys, pigs, laying hens and people during travel, as well as outbreaks in 24 European Union countries. Salmonella data for animals and humans, covering the period from 2007 to 2009, were mainly obtained from studies and reports published by the European Food Safety Authority. Results showed laying hens as the most important reservoir of human salmonellosis in Europe, with 42.4% (7 903 000 cases, 95% credibility interval 4 181 000-14 510 000) of cases, 95.9% of which was caused by *S. enteritidis* [21]. In 2014, 88,715 confirmed cases of salmonellosis were reported by 28 EU member states, which is a 23.4 casesper 100,000 population. In the same year, 65 deaths were reported by 11 EU member states with an incidence rate of 0.15% per 43.955 confirmed cases [2].

According to the European Food Safety Authority (EFSA), a total of 94,530 cases of salmonellosis were reported in the EU countries in 2016, with an incidence rate of 20.4 cases per 100,000 of population [22].

2 Materials and Methods

A microbiological analysis of a total of 790 samples (Table 1) was done in a laboratory for microbiology of food, water and animal feed, which is a part of the Department of Food Hygiene and Technology of Animal Origin and Water at the Public Institution "Veterinary Institute" Bihać, accredited body (standard BAS EN ISO/IEC 6579-1: 2018) [23].

Table 1. Number and type of samples in research

Sample	Number of samples
Feces	300
Eggs	400
Animal feed	40
Laying hen carcasses	50
Total	**790**

For the isolation and identification of bacteria from the genus Salmonella, pre-enrichment in non-selective liquid medium and selective liquid and solid nutrients have been used according to the standard method instructions and ISO method (BAS EN ISO 6579-1: 2018). Standard method for the detection of Salmonella spp. in the samples includes non-selective pre-enrichment, followed by selective enrichment and selective isolation on solid media, followed by biochemical and serological confirmation of suspect colonies (ISO, 2017) [24].

Pre-warmed buffered peptone water (BWP) was used as a diluent in a quantity of 1:10 in relation to the test quantity (2550 g of sample and 225450 ml buffered peptone water). The prepared sample is then incubated at 37 °C for 18 h ± 2 h. After this incubation, 1 ml of the test sample is pipetted in 10 ml of MKTN broth and as such it is incubated at 37 °C for 24 h ± 2 h. Pipette 0.1 ml of the test sample in 10 ml of RVS broth at the same time and incubate at 41.5 °C for 24 h ± 2 h. For larger amounts (1 l or more), it is recommended to pre-warm the BWP at 34–38 °C before mixing with the test part.

Culture obtained after incubation from RVS broth or MSRV and MKTTn broth (7.3.) streak with a 10 μl loop on the first selective medium (XLD agar) to obtain well isolated colonies. In the same way, streak another selective medium (BG agar). From the positive growth on the MSRV agar, determine the furthest point of the fuzzy growth from the inoculation point and dip a 1 μl loop within the boundary of fuzzy growth. Extract the loop carefully not to have larger clumps of MSRV agar. Inoculate the surface of the XLD agar plate to obtain well-isolated colonies. In the same way, inoculate another selective medium (BG agar).

Typical *Salmonella* spp. colonies grown on the XLD agar have a black centre and a poorly transparent reddish-coloured zone due to the colour change of the indicator. Colonies of Salmonella spp. grown on BG agar change the colour of agar itself. They are grey-reddish/pink and barely convex.

Mark suspicious colonies on each plate. Choose at least one typical or suspicious colony for subculture or confirmation. If this is negative, choose 4 more suspicious colonies subcultured from different media for selective enrichment/isolation that show the growth of suspected colonies. Streak the selected colonies on the previously dried surface of the non-selective agar media (nutrient agar) in a way that will allow the growth of well-isolated colonies. Such inoculated plates are incubated at the temperature between 34 °C and 38 °C for 24 h ± 3 h.

If we have well-isolated colonies of pure culture on a selective medium, biochemical confirmation can be carried out directly on a suspicious, well-isolated colony from a selective medium. The growth on non-selective agar (nutrient agar) can be carried out in parallel with biochemical tests to check the purity of colonies taken from the selective agar plate.

In cases of disease outbreak testing, the confirmation of additional colonies (for example, 5 typical or suspected colonies) from each combination of selective enrichment/isolation medium can be useful.

The β-galactosidase test is used to distinguish the *Salmonella enterica* subspecies of arizonae and diarizonae and other members of the family *Enterobacteriaceae* (all give a positive reaction) from other subspecies of Salmonella enterica (generally they give a negative reaction). For the detection of β-galactosidase, paper disks are used according to the manufacturer's instructions ("Fluka"). The Indole test can be used when there is a need to differentiate Salmonella (generally indole negative) from E. coli and Citrobacter (both indole positive) since these organisms can give typical reactions on some Salmonella isolation plates. Pure colonies that demonstrated typical Salmonella biochemical reactions are further tested for the presence of Salmonella O-, H-antigens (and in areas where Salmonella Typhi and Vi-antigen are expected) by agglutination using polyvalent antisera. Pure colonies are grown on a non-selective agar medium (nutrient agar) and tested for autoagglutination. Auto-agglutinous strains cannot be tested for the presence of Salmonella antigen. The strains confirmed as *Salmonella spp*. can be further identified as serovars by serotyping.

3 Results and Discussion

The results of microbiological analysis of the samples indicate a large number of unsatisfactory samples on *Salmonella spp.* which is the result of insufficient implementation of hygiene measures. Out of the total number of samples tested, bacteria of the genus *Salmonella* spp. were found in 18.1% (168/928) samples, Table 2.

Table 2. The results of microbiological analysis of samples positive for *Salmonella* spp.

Sample	Number of samples	Number of positive samples
Feces	300	121
Eggs	400	23
Animal feed	40	1
Laying hen carcasses	50	29
Total	**790**	**174**

Continuous testing of samples for the presence of *Salmonella* spp. was carried out on 300 samples of faeces in the period of two years. The presence of bacteria was confirmed in 121 sample (40.33%). Out of 400 samples of eggs, the presence was identified in 23 samples (5.75%). Out of 40 samples of animal feed, *Salmonella* spp. was determined in one sample (2.5%). On sporadic mortalities on farms, 50 carcasses were tested and the presence of *Salmonella* spp. was found in 29 samples (58%).

Al-Obaidi et al. (2011) conducted a study to determine the prevalence of *Salmonella* spp. in eggs. It was found that the total percentage of *Salmonella spp.* in table eggs at retail stores in Baghdad were 20.0% and 38.7% for local and imported eggs, respectively [25].

Garcia et al. (2011) studied the *Salmonella* spp. contamination on a laying hen farm in feces samples (n = 50), cloacal swabs (n = 150), eggshells (n = 50) and egg contents (n = 50), taken each from 50 randomly selected cages. The results showed that most feces samples (92%) were positive for Salmonella spp., followed by eggshells (34%) and cloacal swabs (4%). No Salmonella spp. were detected in the egg contents [26].

Rahman et al. (2004) studied the presence of *Salmonella* spp. in healthy chicken flockson the farms in Assam, Arunachal Pradesh and Meghalaya, India. A total of 832 cloacal swabs from poultry were examined for the isolation of Salmonella. It was reported that among all the isolates obtained 35 (4.21%) samples were positive for *Salmonella* spp., of which 15 (42.9%) were *S. enteritidis*, 6 (17.1%) *S. gallinarum* and 14 (40%) *S. Typhimurium* [27]. The presence of salmonella in eggs and poultry meat samples was studied by Nagappa et al. (2007). Of 200 analysed samples, 4 samples (one meat and 3 eggs) were found contaminated with *Salmonella* enterica var Typhimurium [28].

The isolation of S*almonella* was attempted by Orji et al. (2009). They isolated Salmonella from poultry droppings and other environmental sources in the Awka region of Nigeria and different *Salmonella* serotypes were isolated from all the sources. *Salmonella paratypi A* had an isolation rate of 12.5% from poultry droppings, 4.2%

from fresh beef, and 2.1% and 4.2% from meat retailers' aprons and tables, respectively. Other serotypes isolated from the sources included *Salmonella* Typhimurium, *Salmonella* Enteritidis, *Salmonellagallinarum, Salmonella* Pullorum, *Salmonella* Typhi *and Salmonella* Agama. *Salmonella* Typhi was not isolated from poultry feces throughout the entire study [29]. The Salmonella control scheme for laying hens in a broader sense includes thorough cleansing and disinfection of the entire production chain, programs of permanent education of producers, hygiene measures and eradication, food hygiene, food supplements, competitive exclusion and vaccination. In a narrower sense, the control scheme includes bacteriological searches of feces, eggs, unhatched eggs, organs of dead day-old chickens and meconiums, raw materials, ready-made mixtures, table eggs and serum and blood analysis (Hafez, 2001, 2005) [30, 31]. Based on the results of the analysis of bacteria of the genus *Salmonella* spp. throughout a two-year period, it can be determined that *S. Enteritidis* was the dominant serotype (Table 3).

Table 3. Positive results of microbiological analysis of samples with dominant serotype

Sample	*S. enteritidis* Positive	*S. Typhimurium*	*S Infantis*	Other *Salmonella*	Total positive
Feces	66	5	32	18	121
Eggs	11	0	5	7	23
Animal feed	0	0	0	1	1
Carcasses	9	0	18	2	29
Total	**86**	**5**	**55**	**28**	**174**

Salmonella enteritidis is the primary cause of concern of egg contamination, and significant research focuses on understanding the relationship between serovars and egg production. As is known, a relationship between laying hens and *S. enteritidis* infection has been established. As a result, more effective measures to suppress the colonization of *S. enteritidis* in laying hens and to limit the contamination of eggs have been developed. However, the work should be carried out on the gastrointestinal ecology of *S. enteritidis* and its association with the autochthonous microbial population [32].

Salmonella spp. can be connected to eggs in two ways: contamination by fecal matter from poultry or trans-ovarian infection that occurs during egg development within the individual. Eggshell contamination can be controlled by paying attention to hygienic conditions, while trans-ovarian infection can be controlled only by ensuring the complete absence of Salmonella in production. Public health concern over antimicrobial resistant bacteria has recently increased. A new study could be undertaken to investigate the prevalence of enterobacteria resistant to antimicrobial antibiotics in eggs purchased on small poultry farms and farmers' markets.

4 Conclusion

In order to protect and improve health and animal welfare, to prevent the outbreak and spreading of the disease, to control and eradicate infectious animal diseases, as well as diseases that can be transmitted from animals to people, it is necessary to establish adequate veterinary control of keeping and poultry production. Therefore, the failure to properly handle eggs in a shell is a potential health risk for consumers. The presence of salmonella zoonotic isolates in table eggs is a major challenge for egg production and public health. In order to avoid an economic disaster in the poultry sector, it is essential that regulatory and public health authorities, in cooperation with egg producers, organise workshops in order to develop effective control strategies against salmonella in the poultry and egg sector.

References

1. EFSA: Analysis of the baseline survey on the prevalence of Salmonella in holdings with breeding pigs, in the EU, 2008–2011. EFSA J. **9**, 2329 (2011)
2. EFSA: The European union summary report on trends and sources of zoonoses, zoonotic agents and food-borne outbreaks in 2014. EFSA J. **13**, 4329 (2015)
3. Suzuki, A., Kawanishi, T., Konuma, H., Takayama, S., Imai, C., Saitoh, J.: Salmonella spp. and Staphylococcus aureus contamination in liquid whole eggs and studies on bacterial contamination of liquid frozen eggs. J. Food Hyg. Japan **22**, 223–232 (1981)
4. Holmberg, S.D., Wells, J.G., Cohen, M.L.: Animal to man transmission of antimicrobial resistant Salmonella spp: investigation of US outbreaks, 1971-1983. Science **225**, 883–885 (1984)
5. Gast, R.K., Beard, C.W.: Research to understand and control Salmonella Enteritidis in chicken and eggs. Poult. Sci. **72**, 1157–1163 (1993)
6. Koidis, P., Bori, A.: Isolation of Salmonella spp. from egg-laying production plants. De. Te. Elleni. Kteni. Etai. **50**, 238–243 (1999)
7. Nunes, I.A., Helmuth, R., Schroeter, A., Mead, G.C., Santos, M.A., Solari, C.A.: Phage typing of Salmonella Enteritidis from divergent sources in Brazil. J. Food Prot. **66**, 324–327 (2003)
8. Nygard, K., de Jong, B., Guerin, P.J., Andersson, Y., Olsson, A., Giesecke, J.: Emergence of new Salmonella Enteritidis phage types in Europe? Surveillance of infections in returning travelers. BMC Med. **2**, 32–33 (2004)
9. Ellermeier, C., Slauch, J.: The genus salmonella. In: Dworkin, M., Falkow, S., Rosenberg, E., Schleifer, K.-H., Stackebrandt, E. (eds.) The Prokaryotes, pp. 123–158. Springer, New York (2006)
10. Yan, S., Pendrak, M., Abela-Ridder, B.: An overview of Salmonella typing- public health perspectives. Clin. Appl. Immunol. Rev. **4**, 189–204 (2003)
11. Grimont, P.A.D., Grimont, F., Bouvet, P.: Taxonomy of the genus Salmonella (2000)
12. Janda, M.J., Abbott, S.L.: The enterobacteria, Philadelphia, Lippincott-Raven (1998)
13. Škrinjar, M.: Mikrobiološka kontrola životnih namirnica. Tehnološki fakultet, Univerzitet u Novom Sadu, Novi Sad (2001)
14. Karakašević, B.: Mikrobiologija i parazitologija.-Medicinska knjiga, str. 677–683, Beograd-Zagreb (1987)

15. Chousalkar, K., Gole, V.C.: Salmonellosis acquired from poultry. Curr. Opin. Infect Dis. **29**, 514–519 (2016)
16. Kabir, S.M.L.: Avian colibacillosis and salmonellosis: a closer look at epidemiology, pathogenesis, diagnosis, control and public health concerns. Int. J. Environ. Res. Public Health **7**, 89–114 (2010)
17. Mlinarić Galinović, G., Ramljak, M., Andreis, S., Bedenić, B., Brudnjak, Z., Horvat Krejči, D., Hunjak, B., et al.: Specijalna medicinska mikrobiologija i parasitologija. udžbenik Visoke zdravstvene škole. Merkur A.B.D. Zagreb (2003)
18. Daniels, N.A., MacKinnon, L., Rowe, S.M., Bean, N.H., Griffin, P.M., Mead, P.S.: Foodborne disease outbreaks in United States schools. Pediatr. Infect Dis. J. **21**, 623–628 (2002)
19. Baker, C.R.: Milk Food Technol. **37**(5), 265–268 (1974)
20. Mcilroy, S.G.: How do birds become infected by a Salmonella serotype? World Poult. Suppl. 15–17 (1996)
21. De Knegt, L.V., Pires, S.M., Hald, T.: Attributing foodborne salmonellosis in humans to animal reservoirs in the European Union using a multi-country stochastic model. Epidemiol. Infect. **143**(6), 1175–1186 (2015)
22. EFSA: The European Union summary report on trends and sources of zoonoses, zoonotic agents and food-borne outbreaks in 2016. EFSA J. **15**(12), 5077 (2017). https://doi.org/10.2903/j.efsa.2017.5077
23. BAS EN ISO 6579-1:2018 Mikrobiologija lanca hrane – Horizontalna metoda za detekciju, određivanje broja i serotipizaciju Salmonella- Dio 1: Detekcija Salmonella spp. (EN ISO 6579-1:2017, IDT; ISO 6579-1:2017, IDT)
24. ISO 2017 Horizontalna metoda za otkrivanje, određivanje broja i serotipizaciju Salmonella - Deo 1: Otkrivanje Salmonella spp. (SRPS EN ISO 6579-1:2017). Međunarodna organizacija za standardizaciju, Ženeva, Švajcarska
25. Al-Obaidi, F.A., Al-Shadeedi, S.M., Al-Dalawi, R.H.: Quality, chemical and microbial characteristics of table eggs at retail stores in Baghdad. Int. J. Poult. Sci. **10**(5), 381–385 (2011)
26. Garcia, C., Soriano, J.M., Benítez, V., Gregori, P.: Assessment of Salmonella spp. in feces, cloacal swabs, and eggs (eggshell and contentseparately) from a laying hen farm. Poult. Sci. **90**, 1581–1585 (2011)
27. Rahman, H., Bhattacharya, D.K., Murugkar, H.V.: Prevalence of Salmonella in poultry in Northeastern India. Ind. J. Vet. Res. **13**, 1–7 (2004)
28. Nagappa, K., Tamuly, S., Saxena, M.K., Brajmadhuri, Singh, S.P.: Isolation of Salmonella Typhimurium from poultry eggs and meat of Tarai region of Uttaranchal. Ind. J. Biotech. **6**, 407–409 (2007)
29. Orji, M.U., Onuigbo, H.C., Mbata, T.I.: Isolation of Salmonella frompoultry droppings and other environmental sources in Awka. Nigeria. Int. J. Infect. Dis. **9**, 86–89 (2009)
30. Hafez, H.M.: Salmonella infections in poultry: diagnosis and control. Periodicum Biologorum **103**, 103–113 (2001)
31. Hafez, H.M.: Governmental regulations and concept behind eradication and control of some important poultry diseases. World's Poult. Sci. J. **61**, 569–581 (2005)
32. Ricke, S.C., Dawoud, T.M., Shi, Z., Kaldhone, P., Kwon, Y.M.: Foodborne Salmonella in laying hens and egg production. Food Feed Saf. Syst. Anal. 153–171 (2018). https://doi.org/10.1016/b978-0-12-811835-1.00009-9

Non-cancerogenic Risk to Human Health with Pb, Cu, and Zn Intake from Soil in the Area of Herzegovina

Alma Mičijević[1,2](✉), Aida Šukalić[1], and Alma Leto[1]

[1] "Džemal Bijedić" University of Mostar, 88000 Mostar, Bosnia and Herzegovina
alma.micijevic@unmo.ba
[2] Faculty of Agromediteranean, University "Džemal Bijedić", USRC "Mithat Hujdur Hujka", 88104 Mostar, Bosnia and Herzegovina

Abstract. The aim of this paper was to determine the level of heavy metals (Pb, Cu, and Zn) concentrations in the total form. Following that determination, by using the Hazard Quotient Index (HQI) a non-cancerogenic risk to adult and children health by oral, digestive and dermal intake of heavy metals from the soil into the body has been calculated. The survey included three locations in the narrower part of Herzegovina in 2012, including Dubrave, Blagaj, and Mostar. Heavy metal analysis was performed according to the international standard ISO 11047 which specifies the atomic absorption spectrometry method for determining one or more elements in the soil extract. By calculating the HI (Hazard Index) at all locations, it was found that HI < 1 and therefore it does not pose a risk to adult health by heavy metals intake into body from the soil. By calculating HI at all locations for children, it was found that at the site of Mostar the value of HI for oral Pb intake from the soil is 3.03 and therefore poses a health risk.

Keywords: Heavy metals · Risk · Land · Non-cancerous index

1 Introduction

In recent years more and more attention has been paid to the soil, water and air contamination and their entry into the food chain. Contamination of agricultural soil is caused by the introduction of harmful substances through water, air or their accumulation on or in the soil at concentrations above permitted. Soil contamination with heavy elements used for agricultural purposes is the result of atmospheric precipitation, application of pesticides, artificial fertilizers and irrigation of water of poor quality [1]. Heavy metals accumulate in soil profiles over a long period of time because they are not biodegradable and cause long-term consequences in terms of fertility and soil quality, while some may cause toxicity to plants but also to humans and animals if such plants are grown and used for food production [2–5]. Exposure to metals can result in negative health effects, and for this reason, European and international agencies in the area of risk assessment define a set of health-based limit values and risk assessment methodology for the purpose of protecting human health. The basic elements of risk

I. Karabegović (Ed.): NT 2019, LNNS 76, pp. 663–671, 2020.
https://doi.org/10.1007/978-3-030-18072-0_78

assessment of chemical hazards include four steps: hazard identification, hazard characterization, exposure estimation, and risk characterization. Methods of estimation of non-cancerogenic risk are usually based on the use of the target particle hazard (HQ), the ratio between the estimated dose of the pollutant and the reference dose below which there will be no significant risk [6]. And if such a ratio exceeds the whole, there may be concern about possible health effects [7]. Identification of hazards basically aims to investigate the chemicals present on any site, their concentrations, and spatial distribution. Pb, Cu and Zn have been identified as potential hazards for the community in the field of investigation. The effects of heavy metals on living organisms can be carcinogenic and mutagenic. Heavy metals toxicity in children can result in neurological damage (decreased intelligence, memory loss, learning difficulties and coordination problems). Exposure to lead may occur in several ways, including air inhalation and food, water, soil or dust intake. Excessive exposure to lead can lead to epileptic attacks, mental retardation and behavioral disorders [8]. The danger has been increased due to the low mobility of lead in the environment [9]. Zinc in the soil comes from atmospheric precipitation, applying various measures in agriculture (fertilization, pesticide application), and discharge of sewage sludge, waste from the metal industry, as well as by the application of ashes. The fate of various types of zinc is different and depends on its chemical properties and their affinity to a particular soil, and it is, therefore, difficult to determine the overall level of zinc contamination [10]. Besides being an essential element, copper can also be toxic. Symptoms of acute poisoning include abdominal pain, nausea, vomiting, and diarrhea, and in extreme cases, death may also occur. Chronic poisoning usually occurs by inputting contaminated water and food as a result of their contact with the copper pipe or food preparation utensils.

2 Methodology

Research locations

The selection of locations was in different zones of the city and suburban area of Mostar. Three different sites are located at a distance of about 50 km so that they can provide relevant data on the presence of the investigated metals. The proposed sites are reliably assumed to belong to different soil types.

Location 1: The Dubravë plateau area, located about 40 km south of Mostar, is positioned on a private estate, away from the highway route and approximately 200 m away from the airline. This site is assumed to be unladen with metals, as it is considered to belong to unpolluted soil.

Location 2: Suburban settlement Blagaj, on the south side of Mostar, is 10 km away from the city itself. The sampling location in this area was 3 km away from the main road. It is assumed that this area should not be burdened with the heavy metals content both in the soil and in the tested fruits.

Location 3: Near the closed Zalik archery in the northern suburban part of Mostar, where significantly, increased quantities of explored metals. This site is highly polluted soil. Based on the mechanical components of the soil, a textural designation according to the USDA classification were detected. It was found that the soil at all three locations belongs to the sandy type of soil.

Soil sampling

The soil is sampled using standard methods using a chromed probe, by taking five individual samples per plot of the parcel, which were collected in one average sample, weighing about 2 kg. The soil is sampled with two different depths of 0–30 cm and 30–60 cm. A total of six soil samples were analyzed on basic soil quality indicators, especially on the metal content of the study: Pb, Cu, and Zn. Sampling was carried out at the time of vegetation, but before the ripening time of the fruit, in the period through September and October 2012. All analyzes were carried out in the laboratory of the Federal Institute of Agriculture in Sarajevo, according to methods defined by law. The Bylaw Act currently in force in the Federation of Bosnia and Herzegovina, which is directly related to the research topic, is the Guidelines on the Determination of Permissible Quantities of Hazardous and Hazardous Substances in the Land and Methods of their Examination (Official Gazette of FBiH, No. 72/09) defining the limit values of soil pollutants in their total form which refer only to agricultural soil, while the limit values for other soil types are not yet legally defined (Table 1).

Table 1. Limit values of heavy metals in the total form

Element	Limit values (mg/kg)		
	Sandy soil	Sandy-loam soil	Hard ground
Copper (Cu)	50	65	80
Lead (Pb)	50	80	100
Zink (Zn)	100	150	200

Source: Guidance on Determination of Permissible Quantities of Harmful and Hazardous Substances in the Land and methods of their testing (Official Gazette FBiH, No. 72/09)

Sample analysis

Preparation of samples for instrumental analysis of heavy metal content in soil was done by means of aqua regia, and then their content in the extract was determined by atomic absorption spectrometry (AAS). The extraction of heavy metals in gold foil was carried out according to the international standard ISO 11464. This standard specifies the method of extraction of trace elements by gold-plating using appropriate atomic spectrometric techniques. According to this standard, soil samples are reduced to particles of less than 2 mm to digestion with aqua regia. The dried sample then is extracted with a chloride/nitric acid mixture, incubated at room temperature for 16 h, followed by refluxing for two hours. The extract is evaporated - purified (filtered) and volume supplemented with nitric acid. The International Standard ISO11047 specifies an atomic absorption spectrometry method for determining one or more elements in soil extract extracted with aqua regia obtained in accordance with ISO11466 (Table 2).

Table 2. Measured average values at locations in soils

Element	Measured average values		
	Lead (Pb)	Copper (Cu)	Zink (Zn)
Dubrave	16,590	29,265	84,455
Mostar	85,610	52,245	165,865
Blagaj	20,300	22,305	59,330

Risk assessment

Given the various adverse effects that heavy metals can impose on human health, the corresponding non-cancerogenic risks were calculated for children and adults according to the USEPA risk assessment model. The purpose of exposure estimation was to measure or estimate the intensity, frequency, and duration of human exposure to the unclean environment. In the study, exposure estimation was performed by calculating the acceptable daily intake (ADI) of heavy metals due to swallowing, inhalation and dermal contact between adults and children in the study area. Adults and children were separated due to their behavioral and physiological differences. Given the various adverse effects that heavy metals may have on human health, the corresponding non-cancerogenic risks were calculated for children and adults according to the USEPA risk assessment model:

$$ADI_{ing} = \frac{CxIRxEFxEDxCF}{BWxAT}$$

$$ADI_{inh} = \frac{CxIR_{air}xEFxED}{PEFxBWxAT}$$

$$ADI_{dermally} = \frac{CxSAxFExAFxABSxEFxEDxCF}{BWxAT}$$

In which ADIing, ADIinh, ADIdermal chronic daily injections or doses were administered by the oral route (mg/kg/d), inhalation (mg/m^3 for non-cancerous and g/m^3 for carcinogenic elements) and dermal (mg/kg/d). C concentration of heavy metals in mg/kg in soil, IR in mg/d ingestion factor, IRair in m^3/d inhalation factor, EF in days/years frequency of exposure, ED duration of exposure in year, BW body weight in kg, AT period in which is the average dose expressed in days, with the skin surface exposed in cm^2, FE is the fraction of the dermal exposure ratio to the soil, the AF coagulation factor for skin in mg/cm^2, ABS is a dermal absorption factor, CF chronic conversion factor in kg/mg (Table 3).

Table 3. Exposure parameters used to assess health risks through different exposure paths to soil [5].

Parameters	Unit	Definition	Value	
			Children	Adults
ABS	–	Dermal absorption factor	0.1	0.1
AF	mg/cm^2	Soil adhesion factor for skin	0.2	0.07
BW	Kg	Average weight	15	70
ED	Year	Exposition duration	6	30
EF	d/godina	Exposition frequentation	350	350
FE	–	Dermal exposure ratio	0,61	0,61
IngR	mg/d	Soil ingestion factor	200	100
IR $_{air}$	m^3/d	Inhalation factor	10	20
SA	cm^2/event	The surface of the exposed skin	2,800	5,700
AT$_{nc}$	D	The average time for non-canc. hazards	ED × 365	
AT$_{ca}$	D	Average time for cancerous hazards	LT × 365	
CF	kg/mg	Precalculation factor	10^{-6}	
PEF	m^3/kg	Particle emission factor of soil - air	1.36×10^9	

Non-cancerogenic risk

The risk factor was calculated using the term "hazard quotient" (HQ). HQ is the ratio of two variables which expresses the likelihood of an adverse effect on an individual. It is defined as acceptable daily intake (ADI) divided by the value of the toxicity threshold referred to as the chronic reference dose (RfD) in mg/kg/day of specific heavy metals as shown in the equation:

$$HQ = \frac{ADI}{RfD}$$

For n number of heavy metals, the non-cancerogenic effect for the population is the result of the sum of all HQs for particular heavy metals. This is known under the term Hazard Index (HI) and it is described in the USEPA document. The mathematical equation is shown in the following way:

$$HI = \sum_{k=1}^{n} HQ_k = \sum_{k=1}^{n} \frac{ADI_k}{RfD_k}$$

In which HQk, ADIk, RfDk are the values of the metal. If the HI value is less than 1, the exposed population will not have harmful non-cancerogenic effects. If the value of HI is greater than 1, then there is a possible potential risk to the health of the exposed population by non-cancerogenic hazards.

3 Results and Discussion

The average concentrations of heavy metals at location Dubrave ranged from the following values in mg/kg: Pb 16,59, Cu 29,27 and Zn 84,46. At location Blagaj average values in mg/kg were Pb 20,30, Cu 22,31 and Zn 59,33. At location Mostar average values in mg/kg were Pb 85,61, Cu 52,26 and Zn 165,87. The results showed

Table 4. Reference doses (Rfd) in (mg/kg day) [5].

	Pb	Cu	Zn
oral RfD	3,60E-03	3,70E-02	3,00E-01
inhal RfD			
derm RfD		2,40E-02	7,50E-02

that at sites Dubrava and Blagaj metals are present in the soil but in quantities which do not exceed the maximum allowed values. At the site of Mostar (Zalik), all three elements showed values higher than MDK. Of all three elements, the highest values are in the presence of Zn and this is as much as 91.4% more than the permissible values. In the locality of Mostar (Zalik) the presence of Pb is greater in the surface layers, and the concentrations of Cu and Zn are somewhat greater in the deeper layers of the soil.

The non-cancerogenic risk of heavy metals for adults and children
The non-cancerogeic risk factors for children and adults were calculated on the basis of RfD as shown in Table 4 and the ADI values in Tables 5 and 6. The calculation was done for each site separately, and the HI values are shown in Tables 7 and 8.

When HI is less than 1, then there is no risk to human health, but if these values exceed 1, then there is concern about the risk of non-cancerogenic risk (USEPA, 2004). HI values for adult oral, dermal and inhalative pathways of heavy metals in organisms

Table 5. Average daily intake (ADI) values in mg/kg/d from the soil for adults at locations

Location	Exposure routes	Pb	Cu	Zn
Dubrave	Ingestion	4,55E−04	4,01E−04	1,16E−03
	Inhalation	2,34E−06	4,13E−07	1,19E−06
	Dermal	5,53E−05	9,76E−05	2,82E−03
	Total	**5,12E−04**	**4,99E−04**	**3,97E−03**
Blagaj	Ingestion	2,78E−04	3,06E−04	8,13E−04
	Inhalation	2,86E−07	3,15E−07	8,37E−07
	Dermal	6,77E−05	7,44E−05	1,98E−04
	Total	**3,46E−04**	**3,80E−04**	**1,01E−03**
Mostar	Ingestion	1,17E−03	7,16E−04	2,27E−03
	Inhalation	1,03E−07	7,40E−07	2,34E−06
	Dermal	2,85E−04	1,74E−04	5,53E−04
	Total	**1,46E−03**	**8,91E−04**	**2,83E−03**

Table 6. Average daily intake (ADI) values in mg/kg/d from the soil for children at locations

Lokacija	Exposure routes	Pb	Cu	Zn
Dubrave	Ingestion	2,12E−03	3,74E−03	1,08E−02
	Inhalation	1,17E−07	2,06E−07	5,95E−07
	Dermal	3,62E−04	6,39E−04	1,84E−03
	Total	**2,48E−03**	**4,38E−03**	**1,26E−02**
Blagaj	Ingestion	2,60E−03	2,85E−03	7,59E−03
	Inhalation	7,16E−07	7,86E−07	4,18E−07
	Dermal	4,43E−04	4,87E−04	6,48E−03
	Total	**3,04E−03**	**3,34E−03**	**1,41E−02**
Mostar	Ingestion	1,09E−02	6,68E−03	2,12E−02
	Inhalation	6,04E−07	3,68E−07	1,17E−06
	Dermal	1,87E−03	1,14E−03	3,62E−03
	Total	**1,28E−02**	**7,82E−03**	**2,48E−02**

Table 7. Hazard Index (HI) in mg/kg/d from the soil for adults at sites

Lokacija	Exposure routes	Pb	Cu	Zn
Dubrave	Ingestion	1,26E−01	1,08E−02	3,87E−03
	Inhalation			
	Dermal		4,07E−03	3,76E−02
	Total	**1,26E−01**	**1,49E−02**	**4,15E−02**
Blagaj	Ingestion	7,72E−02	8,27E−03	2,71E−03
	Inhalation			
	Dermal		3,10E−03	2,64E−03
	Total	**7,72E−02**	**1,14E−02**	**5,35E−03**
Mostar	Ingestion	3,25E−01	3,14E−03	3,03E−02
	Inhalation			
	Dermal		7,25E−03	7,37E−03
	Total	**3,25E−01**	**1,04E−02**	**3,76E−02**

Table 8. Hazard Index (HI) in mg/kg/d from the soil for children at locations

Lokacija	Exposure routes	Pb	Cu	Zn
Dubrave	Ingestion	5,89E−01	1,01E−01	3,60E−02
	Inhalation			
	Dermal		2,66E−02	2,45E−02
	Total	**5,89E−01**	**1,28E−01**	**6,05E−02**
Blagaj	Ingestion	7,22E−01	7,70E−02	2,53E−02
	Inhalation			
	Dermal		2,03E−02	8,64E−02
	Total	**7,22E−01**	**9,73E−02**	**1,12E−01**
Mostar	Ingestion	3,03E+00	1,81E−01	7,07E−02
	Inhalation			
	Dermal		4,75E−02	8,64E−02
	Total	**3,03E+00**	**2,28E−01**	**1,57E−01**

were lower than 1 at all sites, meaning that there is no risk to human health from non-cancerogenic hazards. The calculated HI values at the location of Dubrave for Lead is 1.26E−01, for copper is 1.49E−02, and for zinc is 4.15E−02; at the location of Blagaj HI values for lead 7,72E−02, for copper 1,14E−02, for zinc 5.35E−03; at the location Zalik HI value for lead 3.52E−01, for copper 1.04E−02, for zinc 3.76E−02. The calculated HI values at the locations for children for Lead are 5.89E−01, for copper 1.28E−01, for zinc 6.05E−02; at the location of Blagaj HI values for lead 7,22E −01, for copper 9,73E−02, for zinc 1,12E−1; at the location of the Zalik HI value for lead 3.03, for copper 2.28−01, for zinc 1.57E−01. At the Zalik site, the value of HI for children for lead is 3.03, meaning there is a risk of non-cancerogenic Pb intake via the oral route. At other locations, the measured values were below 1, and there is no risk of non-cancerogenic risks by heavy metal intake by any entry means into the body.

In his research, Tayel et al. investigated heavy metals in some industrial zones in Alexandria, Egypt in 2015 [11]. The maximum values of the tested heavy metals were: 1.49 mg/kg Pb, 7.30 mg/kg Cu, and 151.09 mg/kg Zn. In our research, the values for Pb and Cu are greater, while the values for Zn are in the same range as in this research. The Hazard Index (HI) values for all tested heavy metals were below 1 (1.62E-1 for adults, 2.44E-1 for children) and do not have or pose any of the non-cancerogenic risks if swallowed, dermally contacted or inhaled.

Kamunda et al. indicate the value of the hazard index for all input routes of 2.13, which makes non-cancerogenic effects significant for the adult population [12]. For children, the value of the hazard index was 43.80, which is a serious non-cancerogenic risk effect on children living in the researched area. In the Luo et al. research, the concern is expressed about the non-cancerogenic risk of oral lead intake for children, although the HI value was lower than 1 [13].

4 Conclusion

At the locations of Dubrave and Blagaj, the investigated metals Pb, Cu, and Zn are present in the soil, but in quantities which do not exceed the permissible values, while all three elements on the locality of Mostar showed values above MDK. Of all three elements of the greatest value are in the presence of Zn and this is as much as 91.4% more than the permissible values. The hazard index for adults at all sites was lower than 1, so there is no risk to adult health by heavy metals intake from the soil. The hazardous index at the Mostar site for children for oral lead intake is 3.03 and represents a potential risk of non-cancerogenic hazards. The obtained results point to the need for monitoring of agricultural soil in this and wider area in order to reduce the potential risk of non-cancerogenic hazards.

References

1. Škrbić, B., Đurišić-Mladenović, N.: Distribution of heavy elements in urban and rural surface soils: the Novi Sad city and the surrounding settlements, Serbia. Environ. Monit. Assess. **185**, 457–471 (2013)

2. Dellantonio, A., Fitz, W.J., Čustović, H., Grünewald, H., Repmann, F., Schneider, B.U., Edwards, D., Zgorelec, Ž., Marković, M., Wenzel, W.W.: Trace element uptake in agricultural and wild plants grown on alkaline coal ash disposal sites in Tuzla (Bosnia and Herzegovina) (2007)
3. Dellantonio, A., Fitz, W.J., Čustović, H., Repmann, F., Schneider, B.U., Grünewald, H., Gruberd, V., Zgorelec, Ž., Zerem, N., Carter, C., Marković, M., Puschenreiter, M., Wenzel, W.W.: Environmental risks of farmed and barren alkaline coal ash landfills in Tuzla, Bosnia and Herzegovina. Environ. Pollut. **153**(3), 677–686 (2008)
4. Mulligan, C.N., Young, R.N., Gibbs, B.F.: Remediation technologies for metalcontaminated soils and groundwater: an evaluation. Eng. Geol. **60**, 193–207 (2001). https://doi.org/10.1016/S0013-7952(00)00101-0
5. Nicholson, F., Smith, S., Alloway, B., Carlton-Smith, C., Chambers, B.: An inventory of heavy metals inputs to agricultural soils in England and Wales. Sci. Total Environ. **311**, (1–3), 205–219 (2003). https://doi.org/10.1016/S0048-9697(03)00139
6. U.S. Environmental Protection Agency: Risk Assessment Guidance for Superfund Volume I: Human Health Evaluation Manual (Part E, Supplemental Guidance for Dermal Risk Assessment); USEPA: Washington, DC, USA (2004)
7. Ruppert, D., Opsomer, J.D., Breidt, F.J., Moisen, G.G., Kauermann, G.: Comments on Model-assisted estimation of forest resources with generalized additive models. JASA **102**, 409–411 (2007)
8. Iqbal, M.P.: Lead pollution - a risk factor for cardiovascular disease in Asian developing countries. Pak. J. Pharm. Sci. **25**, 289–294 (2012)
9. Lasat, M.M.: Phytoextraction of metals from contaminated soil: a review of plant/soil/metal interaction and assessment of pertinent agronomic issues. J. Hazard. Subst. Res. **2**, 5–25 (2000)
10. Kabata-Pendias, A.: Trace Elements in Soils and Plants, 4th edn. CRC Press Taylor & Francis Group, Boca Raton (2011)
11. Tayel, M., Hosny, G., El- Darier, S.M., El-Sayed, S.: A comparative health risk assessment for exposure to heavy metals in some industrial areas. Sci-Afric J. Sci. Issues Res. Essays **5** (1), 006–014, January 2017. (ISSN 2311-6188)
12. Kamunda, C., Mathuthu, M., Madhuku, M.: Health risk assessment of heavy metals in soils from witwatersrand gold mining basin, South Africa. Int. J. Environ. Res. Public Health **13**, 663 (2016)
13. Luo, X., Ding, J., Xu, B., Wang, Y., Li, H., Yu, S.: Incorporating bio accessibility into human health risk assessments of heavy metals in urban park soils. Sci. Total Environ. **424**, 88–96 (2012)

Intense Appearance of Chestnut Gall Wasp (*Dryocosmuskuriphilus* Yasumatsu) in Bosnia and Herzegovina

Zemira Delalić(✉)

Biotechnical Faculty, University of Bihac, Luke Marjanovica bb, 77000 Bihac, Bosnia and Herzegovina
zemirabtf@gmail.com

Abstract. This paper presents the research of biological characteristics, prevalence, intensity of appearance in year 2016 and 2017 and a prognosis of further expansion of chestnut gall wasp (Dryocosmus kuriphilus Yasumatsu) in Bosnia and Herzegovina. Chestnut gall wasp is an invasive, quarantine pest which was first time registered in year 2015 on 34 locations in Bosnia and Herzegovina. In year 2016, this specific pest was found on 41 location, and in 2017 it was found on 48 locations. Research has shown that the expansion location of chestnut gall wasp has expanded in relation to localities from the first findings in 2015. Based on the research results, a hot spot map of chestnut gall wasp in Bosnia and Herzegovina has been made for 2017. First galls were found on June 2nd in 2016, and on June 5th in 2017. First adult chestnut gall wasp was found on July 3rd in 2016, and on July 8th in 2017. The study also determined the intensities of the D. kuriphilus phenomenon in all sites where this pest was found, and the intensity of occurrence (category 3) was the highest on the sites in the municipality of VelikaKladuša and Cazin. Intensive appearance of chestnut gall wasp creates a possibility of expansion of chestnut blight fungus (Cryphonectriaparasitica (Murr.) Barr. According to research in Switzerland, there is a possibility for new entry spots for the spores of chestnut peel cancer through abandoned galls of chestnut gall wasps, which fungi can saprophytically populate, which can cause spreading of the infection to neighboring branches. Due to that problem, it is necessary to conduct a research to determine if there is a risk of spreading of this disease in the areas of sweet chestnut (Castanea sativa) in Bosnia and Herzegovina.

Keywords: Chestnut gall wasp (*Dryocosmus kuriphilus*) · Sweet chestnut (*Castanea sativa*) · Intensity appearance · Invasive species

1 Introduction

Genus *Castanea* Mill. belongs to family *Fagaceae*, and it includes 7 economically and ecologically significant wood spieces which are widespread in a temperate forest belt of the northern hemisphere [25]. The European sweet chestnut (*Castanea sativa* Mill.) is found in the Mediterranean from the Caspian lake to the Atlantic Ocean, where chestnut forests occupy an area of 2.530.000 ha [4, 8]. There are three significant locations in

I. Karabegović (Ed.): NT 2019, LNNS 76, pp. 672–679, 2020.
https://doi.org/10.1007/978-3-030-18072-0_79

Bosnia and Herzegovina where European chestnuts are represented. The first Herzegovinian site is 200 hectares, the second in eastern Bosnia is 300 hectares. The third richest chestnut area in Bosnia is the area of northwest Bosnia, covering 7000 hectares [14]. In recent few decades, sweet chestnut is a highly endangered species due to the chestnut blight fungus (*Cryphonectria parasitica* (Murr.) Barr. which causes drying and decaying of trees. In Bosnia and Herzegovina, this disease is a major cause of the devastation of chestnut forests [24]. Because of its exceptional aggressiveness and pathogenicity, *C.parasitica* has been included in the IUCN (International Union for Conservation of Nature) list, among hundred world's most dangerous invasive species that are the greatest threat to biodiversity [19]. In Europe it has been proclaimed a quarantine pathogen and is included in the EPPO A2 list (European and Mediterranean Plant Protection Organization). A new invasive, quarantine pest that is threatening to the chestnut stands is the chestnut gall wasp (*Dryocosmus kuriphilus* Yasumatsu). It originates from China, and for first time, it was discovered outside of its natural area in Japan in 1941 [18] and on the Korean peninsula in 1961 [22]. Outside of Asia, this pest was first recorded in the US in 1974 [22] and Nepal in 1999 [3]. In Europe, it was first recorded in 2002, in Italy, in the region Piemont [23], 3 years later (2005) in Slovenia [13] and 2009 in Hungary [5]. Appearance of the pest has been confirmed in Netherlands (2010), Austria (2011), Czech Republic (2012), Slovakia (2012), Spain (2012) and Germany (2012) [20]. Chestnut gall wasp is a dangerous pest that causes a large loss of yield due to the appearance of the galls on leaves and on the new growth of the tree [2]. In Croatia it was first recorded in 2010 [15], and in Bosnia and Herzegovina in 2015 [6]. New researches have shown that chestnut blight fungus could appear more intensively after the attack of the chestnut gall wasp [21]. The aim of this research is to show biological characteristics in 2016 and 2017, to establish its localities of distribution, the intensity of occurrence and to predict the spread of the chestnut gall wasp in Bosnia and Herzegovina. This paper will consider the possibility of enhancing of the chestnut blight fungus (*C. parasitica*) due to the occurrence of the chestnut gall wasp.

2 Material and Methods of Work

The distribution and intensity of the infection with chestnut gall wasp were made on a wide area of Una-Sana Canton (north-western part of Bosnia and Herzegovina), where chestnut forms a stand or appears with other deciduous species. The biology of chestnut gall wasp was investigated at the sites of Pećigrad, Donja Koprivna (Municipality of Cazin) and at the locations of Johovica and Gornja Vidovska (Municipality of Velika Kladuša). Galls from chestnut plants were collected every week from May 15th till June 20th of 2016 and 2017. They were taken to entomological laboratory of the Faculty of Biotechnology for analysis. Galls were opened to track larval stages, the start and duration of the galls, and the emergence of the adults. The intensity of the appearance on chestnut trees was evaluated on the number of galls on the new branch growth. (Table 1). At each location, 10 chestnut trees were investigated by visual inspection of the new growth (as far as the view could reach from the ground level). The intensity of the attack is rated by categories: 0–3.

Table 1. Intesity of the attack of chestnut plants (*Castanea sativa*) with chestnut gall wasp (*Dryocosmus kuriphilus*) [16]

Intensity of infection (numbered)	Intensity of infection (descriptive)	Determination method
0	No galls	Visual inspection of new growth on the tree (as fas as could be seen from the ground level) has shown no galls
1	Weak attack	One gall per new growth, but only on some growth on the tree, by visual inspection they are not found easily, but they have to be looked for
2	Medium attack	1–5 galls per growth, but not on all which could be seen from the ground level
3	Strong attack	More than 5 galls per growth, all growth that could be seen from the ground level has been attacked

3 Results and Discussion

Chestnut gall wasp was found at 41 location in 2016 in Bosnia and Herzegovina, and in 2017 it has been found at 48 locations. The biggest number of locations is located in the areas of Cazin and Velika Kladusa, where chestnut is present in large areas (Tables 2 and 3). Based on the appearance locations of chestnut gall wasp, a hot spot map has been made in 2017 (Fig. 1). *D.kuriphilus* is one of 2 species from phylum *Cynipini* which causes galls on species from the *Castanea* genus [1]. Monitoring of the biology at 4 locations found that this wasp has one generation per year. During the spring, during vegetative growth on leaves and new branch growth of sweet chestnut, the insect formed 5-20 mm large, easily visible reddish or green galls. In galls, there are one or more chambers wuth white larvae. The older galls are brown and hard. Chestnut gall wasp is a monophagous species that is fed with chestnut only. In the beginning of June, white pupae are formed which turn darker in a few days. First galls were found on June 2nd in 2016 and on June 5th in 2017, which is a ecological factor. First adult chestnut gall wasp was found on July 3rd in 2016 and on July 8th in 2017. In 2015 in this area, first grown species was registered on July 10th [6]. There is one generation per year, and it reproduces parthenogenetically (no males), and embryo is produced asexually without fertilization [18]. During the research, it was found that the area of spreading of chestnut gall wasp has spread to the locations of its first findings in 2015 [6]. In 2016 and 2017 there were 14 new locations compared to 2015. Intensity of the attack (category 3) was the largest in the municipalities of Velika Kladusa and Cazin. The chestnut forests of this municipality borders with chestnut forests of the neighboring state of Croatia, from where this chestnut gall wasp has entered. However, high attack intensities have been recorded on a large number of locations in 2017, as the chestnut gall wasp has spread, and is present for 3 years. Although the wasp has determined first time in 2015, most likely it was present in the perimeter areas before. This could be concluded on the presence of a large number of old, dry galls. As a consequence of the gall wasp attack, the drying of chestnut new growth as well as the complete drying of young plants has been observed. In each subsequent year, the number of females increases, which contributes to the parthenogenetical reproduction and greater increase of the galls. The third year of occurrence can be classified in the greatest intensity of occurrence 3. These are mostly locations where the wasp appeared in 2015. The occurrence of the wasp where it first appeared is classified in category 1.

Table 2. Localities and infestation rates of *D.kuriphilus* (2016. i 2017) on municipalities Velika Kladuša and Cazin

Locality (Municipality ofVelika Kladuša)				Locality (Municipality of Cazin)			
Infestation rate				Infestation rate			
2016		2017		2016		2017	
Mala Kladuša	3	Mala Kladuša	3	Skokovi	3	Skokovi	3
Todorovo	2	Todorovo	3	Rošići	3	Rošići	3
Šumatac	2	Šumatac	3	Ponjevići	2	Ponjevići	3
Gornja Vidovska	3	Gornja Vidovska	3	Krakača	3	Krakača	3
				Gnjilavac	3	Gnjilavac	3
Drenovac	2	Drenovac	3	Pećigrad	3	Pećigrad	3
Marjanovac	3	Marjanovac	3	Šabići	3	Šabići	3
Johovica	2	Johovica	3	Mujakići	3	Mujakići	3
Fazlići	2	Fazlići	3	Krivaja	3	Krivaja	3
Zborište	3	Zborište	3	Ljubijankići	2	Ljubijankići	3
Kumarica	3	Kumarica	3	Brezova Kosa	3	Brezova Kosa	3
Stabandža	2	Stabandža	3	Bašče	2	Bašče	2
Brdo	2	Brdo	2	D.Koprivna	2	D.Koprivna	3
Crvarevac	1	Crvarevac	3	G.Koprivna	2	G.Koprivna	2
Jablan	1	Jablan	2	Karnova Glavica	2	Karnova Glavica	3
Podzvizd	1	Podzvizd	2	Lanište	1	Lanište	2
		Šiljkovača	1	Pivnice	1	Pivnice	2
						Bajramovići	1
						Stijena	1
						Šljemena	1

According to previous experiences with the expansion intensity of the chestnut gall wasp [15, 16], it can be expected that in coming years, occurrence of the gall wasp in new locations will move to a higher category of attack intensity. During next few years, intense emergence and spreading of chestnut gall wasp in other areas of Bosnia and Herzegovina can be expected. For now, the pest is not registered in the chestnut forests of the Herzegovina locality and of Eastern Bosnia. In the chestnut forests of this area, it is necessary to conduct a study of complexes of natural enemies, in particular parasites of chestnut gall wasps in Bosnia and Herzegovina. In the homeland of this pest, China, natural enemies, especially *Hymenoptera* parasites, effectively regulate its population [3]. It is also expected that indigenous parasites, very common in oak gall wasps, will adapt to the newborn pest [2]. In Italy, up until now, 16 species of indigenous parasites, has been adapted to the chestnut gall wasp, but with a low percentage of parasiticity for now [3]. *Torymus sinensis* Kamijo (*Hymenoptera:Torymidae*) is already used as a biological agent of suppression in Japan and Korea.

Table 3. Localities and infestation rates of *D.kuriphilus* (2016. i 2017) on municipalities Bužim and Bosanska Krupa

Locality (Municipality of Bužim)				Locality (Municipality of Bosanska Krupa)			
Intenzitet napada				Intenzitet napada			
2016		2017		2016		2017	
Bag	2	Bag	3	Tromeđa	2	Tromeđa	3
Vrhovska	2	Vrhovska	3	Bućevci	1	Bućevci	2
Zaradostovo	2	Zaradostovo	2	Pištaline	1	Pištaline	1
Čava	2	Čava	2			Ljusina	1
Konjodor	1	Konjodor	2				
Lubarda	2	Lubarda	2				
		Radoč	1				
		Elk.Rijeka	1				

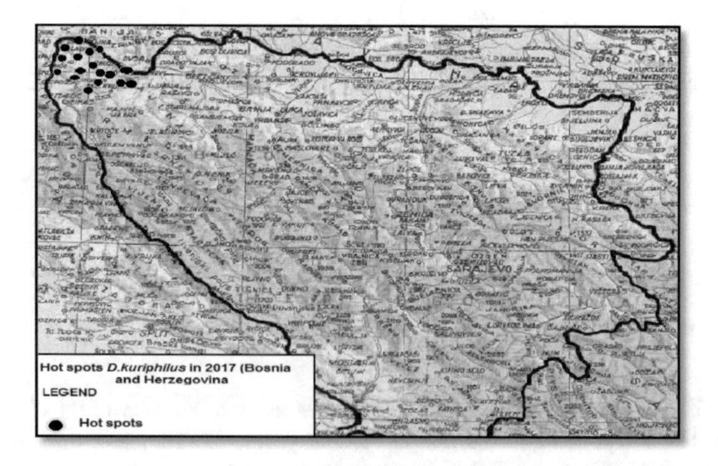

Fig. 1. Hot spots map of *Dryocosmus kuriphilus* in 2017 (Bosnia and Herzegovina)

Research shows its high efficiency [18]. In Italy, breeding experiments were carried out by releasing this parasite in nature. Genetic analysis confirmed the establishment of population in forests and plantations of sweet chestnut in Croatia and Slovenia [17].

3.1 Possibility of Strengthening of Chestnut Blight Fungus Due to Appearance of Chestnut Gall Wasp in Bosnia and Herzegovina

In Europe, the disease was first observed in 1938 in northern Italy and it spread rapidly throughout Italy and neighboring countries [11]. In Slovenia, *C.parasitica* was first recorded in 1950 near Nova Gorica [7] and in Croatia near Opatija in 1955 [9]. The disease spread very fast and aggressively, and in 1961 it was recorded in Bosnia and Herzegovina [12]. In the chestnut forests of Una-Sana Canton, three groups of symptoms of cancer like formations were spotted: active cancer (AR), carcinogenic cancer (KR) and surface necrosis (PN) [7]. Prospero and Forster showed in research conducted in Switzerland that the new point of entry of the chestnut blight spores are most likely the galls from chestnut gall wasps. Here the spores are not transmitted through the insects, but through the galls, through which the spores enter, considering that the chestnut blight is a wound parasite [9, 10]. The research was done on a sample of 24 young branches. In half of the examined branches, cancer symptoms were found, and there were no visible symptoms of the disease in the other 12 branches. It is assumed that the outflow holes of the gall wasp on the galls are the point of entry for the spores of the chestnut blight. After the grown wasp leaves, fungus can saprophytically inhabit the extinct tissue of the abandoned gall and then spread to the adjacent branches and initiate the formation of cancer substances.

4 Conclusion

Sweet chestnut (*C. sativa*) has been under a strong influence of 2 harmful biotic factors in Europe over the past 50 years. This refers to chestnut blight (*C. parasitica*), which caused great damage to chestnut forests. New pest that is threatening sweet chestnut, is the invasive, quarantine species, chestnut gall wasp. In Bosnia and Herzegovina it was first registered in 2015. The chestnut gall wasp is located at 41 location in Bosnia and Herzegovina in 2016, and in 2017 at 48 locations. During the study, it was found that the area of prevalence of chestnut gall wasp has spread to the locations of its first finding in 2015. However, the high intensity attacks of *D.kuriphilus* have been recorded on a large number of locations in 2017 as the gall wasp has spread and is present for 3 years. During next few years we can expect intensive emergence and spreading of chestnut gall wasps in other areas in Bosnia and Herzegovina (Herzegovinian locality and eastern Bosnia). In sweet chestnut forests of this area, it is necessary to conduct a study of complexes of natural enemies, in particular parasites of chestnut gall wasp in Bosnia and Herzegovina. Studies in Switzerland show that a new place for blight spores to enter could be the abandoned galls. Therefore, it is necessary to carry out studies to determine whether there is a risk for this type of infection spreading in the chestnut forests in Bosnia and Herzegovina.

References

1. Ács, Z., Melika, G., Pénzes, Z., Pujade-Villar, J., Stone, G.N.: The phylogenetic relationships between *Dryocosmus*, Chilaspis and allied genera of oak gallwasps (*Hymenoptera, Cynipidae: Cynipini*). Syst. Entomol. **32**(1), 70–80 (2007)
2. Aebi, A., Schonrogge, K., Melika, G., Alma, A., Bosio, G., Quacchia, A., Picciau, L., Abe, Y., Morya, S., Yarka, K., Seljak, G., Stone, G.N.: Parasitoid recruitment to the globally invasive chestnut gall wasp *Dryocosmus kuriphilus*. In: Ozaki, K., Yukawa, J., Ohgushi, T., Price, P.W., (eds.) Ecology and Evolution of Galling Arthropods and Their Associates, pp. 103–121. Springer, Tokyo (2006)
3. Aebi, A., Schönrogge, K., Melika, G., Quacchia, A., Alma, A., Stone, G.N.: Native and introduced parasitoids attacking the invasive chestnut gall wasp Dryocosmus kuriphilus. EPPO Bull. **37**, 166–171 (2007)
4. Conedera, M., Manetti, M.C., Giudici, F., Amorini, E.: Distribution and economic potential of the Sweet chestnut (*Castanea sativa* Mill.) in Europe. Ecol. Med. **30**, 179–193 (2004)
5. Csóka, G., Wittmann, F., Melika, G.: The oriental sweet chestnut gall wasp (*Dryocosmus kuriphilus* Yasumatsu 1951) in Hungary. Növényvédelem **45**(7), 359–360 (2009)
6. Delalić, Z.: Prvi nalaz karantenskog štetnika kestenove ose šiškarice (*Drycosmus kuriphilus*) na Unsko-sanskom kantonu (BiH), Biljni lekar, 1/58–65 (2016)
7. Delalić, Z., Kuduzović, A., Dolić, B., Rošić, A.: Pojava kestenove ose šiškarice (*Dryocosmus kuriphilus*) na Unsko-sanskom kantonu i povezanost sa rakom kore kestena (*Cryphonectria parasitica*). In: Radova, Z. (ed.) 3rd International Conference, Mostar, "NEW TECHNOLOGIES NT-2016", Development and Application, pp. 462–468 (2016)
8. Fernández-López, J., Alía, R.: Technical Guidelines for genetic conservation and use for chestnut (*Castanea sativa*) - EUFORGEN-NH (EUFORGEN- Noble Hardwoods), EUR (Regional Office for Europe) (2003)
9. Glavaš, M.: Gljivične bolesti šumskoga drveća. Zagreb, Šumarski fakultet (1999)
10. Halambek, M.: Istraživanje virulentnosti gljive *Endothia parasitica* (Murr.) And. Uzročnika raka pitomoga kestena (*Castanea sativa* Mill.), disertacija, Šumarski fakultet, Sveucilište u Zagrebu, p. 132 (1988)
11. Heiniger, U., Rigling, D.: Biological control of chestnut blight in Europe. Annu. Rev. Phytopathol. **32**, 581–599 (1994)
12. Ježić, M.: Raznolikost gljive *Cryphonectria parasitica* Murrill Barr i njezin uticaj na populacije pitomog kestena (*Castanea sativa* Mill.), Doktorski rad, Prirodoslovno-matematički fakultet, Biološki odsjek, Sveučilište u Zagrebu (2013)
13. Knapić, V., Seljak, G., Kolšek, M.: Experience with *Dryocosmus kuriphilus* Yatsumatsu erradication measures in Slovenia. OEPP/EPPO Bull. **40**, 169–175 (2010)
14. Macanović, A.: Ecological and syntaxonomic analysis of Sweet chestnut forests (*Castanea sativa* Mill.). In: BiH. ANUBIH Sarajevo. Odjeljenje prirodnih i matematičkih nauka. Zbornik radova, Posebna izdanja. vol. 343, pp. 201–220 (2012)
15. Matošević, D., Pernek, M., Hrašovec, B.: Prvi nalaz kestenove ose šiškarice (*Drycosmus kuriphilus*). Šumarski list br. **9–10**, 497–502 (2010)
16. Matošević, D.: Pojava, širenje i štetnost kestenove ose šiškarice (*Dryocosmus kuriphilus*) u Hrvatskoj, Radovi (Hrvat. šumar. inst.) **44**(2), 113–124 (2012)
17. Matošević, D., Kos, K., Rot, M., Lacković, N., Celar, F.A., Žežlina, I., Lukić, I.: Biološko suzbijanje invazivne kestenove ose šiškarice unesenom vrstom parazitoida *T. sinensis* u Hrvatskoj i Sloveniji. In: Zbornik sažetaka 61.seminara biljne zaštite, p. 31 (2017)

18. Moriya, S., Shiga, M., Adachi, I.: Classical biological control of the chestnut gall wasp in Japan. In: Van Driesche, R.G. (ed.) Proceedings of the 1st International Symposium on Biological Control of Arthropods, pp. 407–415 (2003)
19. Novak-Agbaba, S.: Monitoring raka kore pitomog kestena na trajnim plohama. Radovi Šumararski institut. Izvanredno izdanje **9**, 199–211 (2006)
20. Quacchia, A., Moriya, S., Bosio, G.: Effectiveness of *Torymus sinensis* in the biological control of *Dryocosmus kuriphilus* in Italy. Acta Horticult. **1043**, 199–204 (2014)
21. Prospero, S., Forster, B.: Chestnut gall wasp (*Dryocosmus kuriphilus*) infestations: new opportunities for the chestnut blight fungus *Cryphonectria parasitica*. New Dis. Rep. **23**, 35 (2011)
22. Rieske, L.K.: Success of an exotic gallmaker, *Dryocosmus kuriphilus*, on chestnut in the USA: a historical account. Bull. OEPP/EPPO Bull. **37**, 172–174 (2007)
23. Sartor, C., Dini, F., Marinoni, D.T., Mellano, M.G., Beccaro, G.L., Alma, A., Botta, R.: Impact of the Asian wasp *Dryocosmus kuriphilus* (Yasumatsu) on cultivated chestnut: yield loss and cultivar susceptibility. Sci. Hortic. **197**, 454–460 (2015)
24. Treštić, T.: Rak pitomog kestena u Bosni i Hercegovini s posebnim osvrtom na populacionu strukturu patogena, Magistarski rad, Šumarski fakultet, Sarajevo (2000)
25. Wang, Y., Kang, M., Huang, H.: Microsatellite loci transferability in chestnut. J. Am. Soc. Hort. Sci. **133**(5), 692–700 (2008)

Assessment of the Quality of Water of the River Una in the National Park Una on the Basis of Selected Microbiological Parameters

Aida Šahinović Alešević[1(✉)], Asmir Aldžić[2(✉)], Huska Jukić[3(✉)], and Aida Džaferović[3(✉)]

[1] Faculty of Science, University of Sarajevo, Zmaja od Bosne 33-35, 71000 Sarajevo, Bosnia and Herzegovina
aida.alesevic1@gmail.com

[2] Faculty of Health Studies, University of Bihac, Nositelja hrvatskog trolista 4, 77000 Bihać, Bosnia and Herzegovina
asmiraldzic@hotmail.com

[3] Faculty of Biotechnical, University of Bihac, Luke Marjanovića bb, 77000 Bihać, Bosnia and Herzegovina
huskaj037@gmail.com, aida.btf@gmail.com

Abstract. Una River Water Quality Assessment was conducted in the Una National Park area as field and experimental type. Due to its unique natural values such as water quality, diversity of rare flora and fauna, specific sedimentation Una River belongs to the most beautiful and most interesting rivers in Europe. Earlier researches have proved that Una River water in the Una National Park has a distinctly clean water and clear in layers of blue-green colour. The biggest threat, especially in the area of the Una National Park, is hydrological changes and anthropogenic impacts, which can impair the quality of the Una River water. To determine the microbiological quality of Una River in the Una National Park area on the basis of selected microbiological parameters, including the total coliform bacteria with the emphasis on Escherichia coli is the main task that has been successfully accomplished through this research. A total of 72 samples were sampled from six different sites (Martin Brod, Kulen Vakuf, Orasac, Strbacki Buk, Lohovo, and Ripac) within the Una National Park within six months (January–June) and they were analysed in laboratory using the membrane filtration method. By analysing the microbiological quality of water based on the selected microbiological parameters the emphasis was placed on the presence of coliform bacteria. The results of the tested samples were evaluated as positive or negative. A total of 14 samples were positive, from the Lohovo and Ripac sites. It is known that coliform bacteria are secreted by faeces which through wastewater reach into natural waters and thereby disturb their integrity. To conclude, many anthropogenic influences are found at Lohovo and Ripac sites, compared to other sites whose water samples showed negative results.

Keywords: Una River · Coliform bacteria · Escherichia coli · Membrane filtration · Microbiological quality

I. Karabegović (Ed.): NT 2019, LNNS 76, pp. 680–686, 2020.
https://doi.org/10.1007/978-3-030-18072-0_80

1 Introduction

Since the beginning of the mankind development water as the basic life necessity was and remained the extreme preoccupation of man since then. River Una, a border river between Bosnia and Herzegovina and Croatia, known for its unique natural values together with the rivers Bosna, Drina and Vrbas makes up 75.5% of the total water resources of Bosnia and Herzegovina. The surface of the Una River Basin as irregular and triangular shape is 9 368 km^2, 212 km long, with an average fall of 1.67%. It springs in the area of Zadar County in the village of Suvaja. Rijeka Una is famous for its specific flora and fauna as well as sedimentary barriers that represent the most beautiful hydrogeological phenomenon of this area which are present throughout the entire stream of the river Una [1]. The greatest threat, especially in the area of the Una National Park, is the hydrological changes and anthropogenic influences that can impair the quality of the Una water. By analysing the microbiological quality of water based on selected microbiological parameters the emphasis is placed on the presence of coliform bacteria [2]. Una River water in the Una National Park can support the reproduction of microbial communities including the pathogens. As the pathogens are responsible for the emergence of health problems in humans the microbiological characterisation of water sources is one of the most important parameters in order to achieve effective water treatment. Water can be contaminated by pathogens of various diseases such as typhoid, paratyphoid, cholera, amoebic dysentery but also dangerous carcinogens, radioactive and other chemicals [3]. Large numbers of rivers in Bosnia and Herzegovina, under the high impact of pollution through untreated wastewater of settlements and industry, have very low water quality. Only the upper streams of several rivers such as Una, Sana and Neretva have generally good and preserved water quality [4].

2 Aim of the Work

The purpose of this research was to determine the quality of the Una River water in the area of the National Park Una based on selected microbiological parameters. To achieve the stated goal the following tasks were set:

1. Field tour of the Una River watercourses in the Una National Park;
2. Water sampling;
3. Laboratory treatment of water samples;
4. Analysis of microbiological parameters;
5. Application of appropriate methods for determining the presence of microorganisms;
6. Determination of the overall quality of the Una River water in the Una National Park.

3 Material and Methods of Work

3.1 Material of Work

In order to assess the quality of surface waters Una River in the Una National Park area one sample in the amount of 500 ml is taken from different locations. Each bottle contains a label containing the following information: place, time and date of sampling as well as the name and surname of the person who sampled the material. For sub-sampling, immerse the probe (a bottle tied to a 2 m long rod) into the surface water (up to 50 cm below the surface of the water) directed to the current stream. Then the contents of the bottle (water) are poured into the appropriate sterile bottle; fill with water up to 3/4 of the bottle volume. The samples were placed in a cooling device (transport cooler) and transported to the laboratory three hours after sampling (transport should be carried out no later than 6 h after sampling). In the laboratory, processing of prepared samples with a dilution series of 0.1:100 and 1:100 is used using the selected method. Endoagar Nutritional Substances (14053 Sartorius), the necessary equipment and accessories were used.

3.2 Methods of Work

The total number of coliform bacteria can be determined in two ways, membrane filtration and most probable number method (MPN-Most probable number). The method of membrane filtration has three phases: filtering, incubation and differentiation. Membrane processes are widely used in the technology of food, drink, medicine, drinking water, ultra-pure water and wastewater treatment [5]. During the filtration process, samples of 0.1:100 ml and 1:100 ml were taken for each area of research in the membrane filtration with sterile funnels and vacuum pump. In the funnel with samples of 0.1 ml and 1 ml, distilled water was added to a mark indicating a volume of 100 ml. A membrane filter of a pore diameter of 0.45 μm was placed on the filter holder, after which the samples were filtered. After filtration the membrane was removed from the holder by the tweezers (flammered every time) and transferred to the appropriate Endo agar substrate, ensuring that there is no air between the membrane and the agar substrate. Such substrates were incubated at 37 °C. The appearance of red or metallic glow on the surface of the colony background confirms the production of indole from tryptophan or the occurrence of *Escherichia coli* in the samples which is the main task and goal of this research. The most probable number of coliform bacteria (MPN) is the statistical method based on theory of probability [6]. For the processing of data and the presentation of positive results of the number of colonies, the number of colonies obtained in 1 ml of the sample was taken and expressed as CFU/100 ml according to the formula:

$$\textit{Escherichia coli}/100\,\text{ml} = \frac{\text{number of colonies} \times 100}{\text{volume of the filtered sample}}$$

4 Results of Work

The final results show that from the 72 samples collected at the locations of Martin Brod, Kulen Vakuf, Orasac, Strbacki Buk, Lohovo and Ripac belonging to the Una National Park, the area through which the Una River flows, 14 samples are positive for the presence of faecal coliform bacteria (*Escherichia coli* in this case) while the remaining 58 is negative (Table 1). From the 14 samples, 4 samples belong to Lohovo site for the month of May and June, while the remaining 10 samples belong to the site Ripac, for five months (February–June) (Fig. 1). The reason for these negative results is primarily the absence of anthropogenic impacts in the locations of Martin Brod, Kulen Vakuf, Orasac and Strbacki Buk, while positive results are a consequence of a some minor anthropogenic impact on the Lohovo site and significantly higher in the Ripac area.

Table 1. The final results of the laboratory analysis of Una River water samples in the Una National Park area based on the selected microbiological parameters

RESULTS OF WORK			
Locations	**Number of samples**	**Positive samples**	**Negative samples**
Martin Brod	12	-	12
Kulen Vakuf	12	-	12
Orasac	12	-	12
Strbacki Buk	12	-	12
Lohovo	12	4	8
Ripac	12	10	2
TOTAL NUMBER OF SAMPLES: **72**		NUMBER OF POSITIVE SAMPLES: **14**	

The first part of the positive results refers to the Lohovo locality as presented in the table (Table 2). As such, they are not surprising because this is an area that is highly exposed to anthropogenic influences. One of the main reasons for the emergence of pathogenic bacteria is the great anthropogenic impact in the Lohovo area, especially housing, sewage and agriculture to a large extent influencing the occurrence of faecal coliform bacteria that lead to a change in water ecosystems and a negative impact on human health in this locality. After the incubation and occurrence of *Escherichia coli* specific colonies, colonies counting method and the calculation of total number *Escherichia coli* in 100 ml of sample were used. In May, one filtered sample of 1 ml volume contained 300 CFU/ml, while in June one filtered sample of 1 ml contained 500 CFU/100 ml. The average value is 400 CFU/ml which can be explained and confirmed by the significant anthropogenic effect of eco-settlements on the banks of the Una River in the Lohovo area.

Table 2. Results of laboratory analysis of samples from Lohovo locality (characterised as positive and negative) and calculation of the number of positive colonies

<table>
<tr><th colspan="2">LOHOVO LOCALITY
Period: January-June</th><th colspan="3">SAMPLES
Number of samples: 12</th></tr>
<tr><td colspan="2">January</td><td colspan="3">Negative</td></tr>
<tr><td colspan="2">February</td><td colspan="3">Negative</td></tr>
<tr><td colspan="2">March</td><td colspan="3">Negative</td></tr>
<tr><td colspan="2">April</td><td colspan="3">Negative</td></tr>
<tr><td colspan="2">May</td><td colspan="3">Positive</td></tr>
<tr><td colspan="2">June</td><td colspan="3">Positive</td></tr>
<tr><th colspan="5">COLONIES NUMBER CALCULATION</th></tr>
<tr><th>Month</th><th>Number of colonies formed</th><th>The volume of the filtered sample</th><th>Escherichia coli/100ml</th><th>Average value</th></tr>
<tr><td>May</td><td>3</td><td>1ml</td><td>300 CFU/100ml</td><td rowspan="2">400 CFU/100ml</td></tr>
<tr><td>June</td><td>5</td><td>1ml</td><td>500 CFU/100ml</td></tr>
</table>

The second part of the positive results is related to the Ripac site. Locality Ripac is the one with the largest number of positive samples. Except in January all other samples are positive for the presence of faecal coliforms, specifically *Escherichia coli* (Table 3). Columns of metal gloss formed on the basis of these samples shows the correctness of the method used and the procedure as a whole. In February and March the volume of a filtered sample of 1 ml on the endo-agar surface formed a colony for each month. One filtered sample of the month of February contained 100 CFU/ml as in March. The sample of the month of April contained 700 CFU/100 ml, the month of May 1100 CFU/100 ml and the month of June as high as 1200 CFU/ml. Going to the warmer months, the rise in temperature, the summer season of tourism, the increased activity of all residential tourist objects, more active agricultural activities as anthropogenic influence create ideal conditions for the presence of as many coliforms as possible.

Table 3. Results of laboratory analysis of samples from Ripac locality (characterised as positive and negative) and calculation of the number of positive colonies

<table>
<tr><th colspan="2">RIPAC LOCALITY
Period: January-June</th><th colspan="3">SAMPLES
Number of samples: 12</th></tr>
<tr><td colspan="2">January</td><td colspan="3">Negative</td></tr>
<tr><td colspan="2">February</td><td colspan="3">Positive</td></tr>
<tr><td colspan="2">March</td><td colspan="3">Positive</td></tr>
<tr><td colspan="2">April</td><td colspan="3">Positive</td></tr>
<tr><td colspan="2">May</td><td colspan="3">Positive</td></tr>
<tr><td colspan="2">June</td><td colspan="3">Positive</td></tr>
<tr><th colspan="5">COLONIES NUMBER CALCULATION</th></tr>
<tr><th>Month</th><th>Number of colonies formed</th><th>The volume of the filtered sample</th><th>Escherichia coli/100ml</th><th>Average value</th></tr>
<tr><td>February</td><td>1</td><td>1 ml</td><td>100 CFU/100ml</td><td rowspan="5">600 CFU/100ml</td></tr>
<tr><td>March</td><td>1</td><td>1 ml</td><td>100 CFU/100ml</td></tr>
<tr><td>April</td><td>7</td><td>1 ml</td><td>700 CFU/100ml</td></tr>
<tr><td>May</td><td>11</td><td>1 ml</td><td>1100 CFU/100ml</td></tr>
<tr><td>June</td><td>12</td><td>1 ml</td><td>1200CFU/100ml</td></tr>
</table>

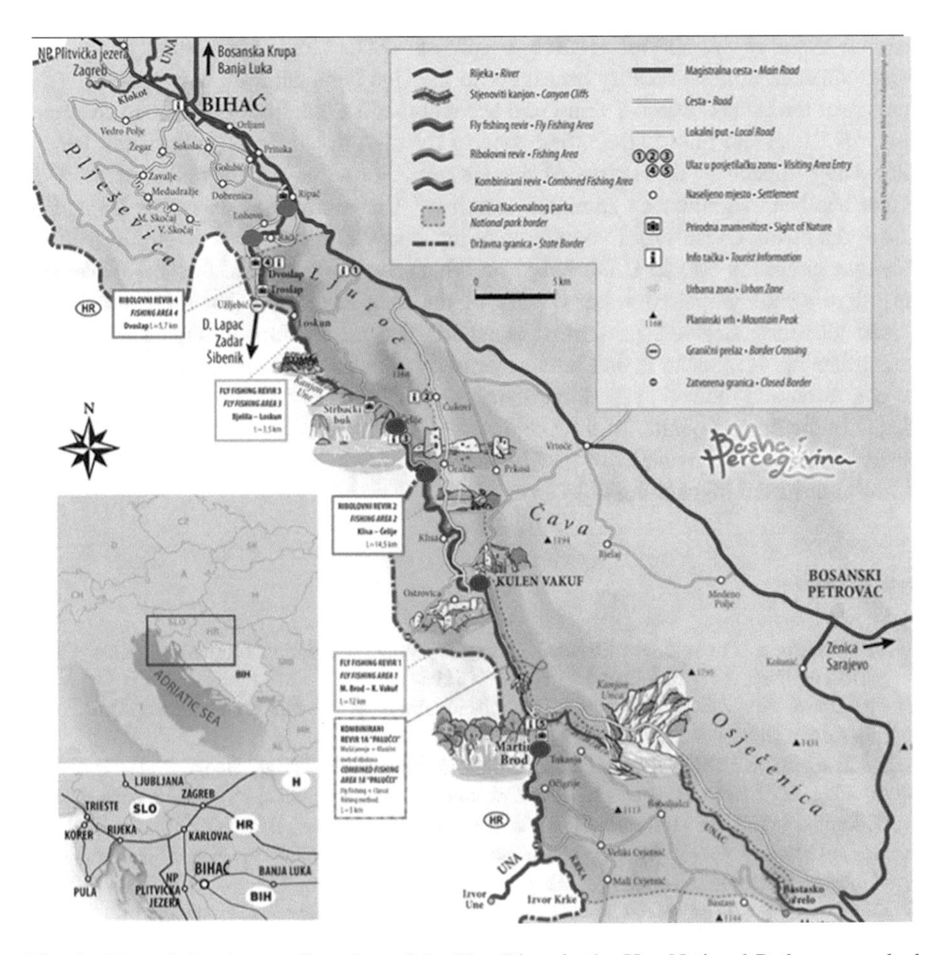

Fig. 1. Map of the six sampling sites of the Una River in the Una National Park area marked with red circles (blue circles marked locations with positive results)

5 Conclusion

Field and experimental research on the water quality of the Una River has been successfully completed for a period of six months (January–June) in which 72 samples were taken and analysed of which 58 were negative for the presence of faecal coliform bacteria and 14 were positive. Research proved that Martin Brod, Kulen Vakuf, Orasac and Strbacki Buk sites for all six months of sampling did not show any positive result regarding the presence of coliform bacteria. Based on the analysis of the microbiological parameters, it was proved that the water quality of the Una river in the area of the mentioned sites corresponds to the quality of Class I water.

The analysed Una River water samples at the Lohovo site showed a positive result for the presence of coliform bacteria in May and June, while for the remaining months the result was negative. The isolated bacteria *Escherichia coli* at the Lohovo site indicated an average contamination of water caused by wastewater and faeces with an average value in the 100 ml positive sample of 400 CFU/ml. The quality of the Una River water at Lohovo locality belonging to the Una National Park corresponds to the quality of the II class water. Taking into account the fact that from Lohovo downstream towards Ripac and then to the urban zone of the city, the anthropogenic factor has an increasing impact, in these parts of the Una river flow the water quality of the Una River has been significantly impaired. Analysed Una water samples at the Ripac site showed a positive result for five months of the study. The average value in the 100 ml positive sample is 600 CFU/ml, based on which we can conclude that the Ripac site has significantly impaired water quality of the Una River and that the Lohovo and Ripac localities meet the prescribed standards that ensure functioning. Through the strengthening of legislation and human awareness, in the long term, through all forms of education and indicating the importance and value of the Una River it is possible to maintain the unique quality of the Una River water in the area of the National Park, as well as to improve the quality of the Una water quality in the Lohovo, Ripac and Bihac bringing it to the highest possible level.

References

1. Matoničkin, I., Pavletić, Z.: Postanak i razvoj najmlađih sedrenih tvorevina u rijeci Uni s biološkog stanovišta. Krš Jugoslavije, JAZU (1964)
2. Plan upravljanja za Nacionalni park Una. Nacionalni park Una-službena stranica, April 2011. http://nationalpark-una.ba
3. Karakašević, B.: Mikrobiologija i parazitologija. Medicinska knjiga, Beograd (1987)
4. Džankić, N., Makić, H., Budimlić, A.: Prirodni i antropogeni uticaj na kvalitet voda slivnog područja rijeke Une. Univerzitet u Bihaću, Biotehnički fakultet, Bihać (2006)
5. Gesan-Guiziou, G.: Membrane processes in the dairy industry, NanoMemCourse on Nanostructured Materials and Membranes for Food Processing, Cetraro, ITM-CNR (2010)
6. Duraković, S.: Primjenjena mikrobiologija. Prehrambeno-tehnološki inženjering, Zagreb (1996)

New Technologies in Civil Engineering, Education, Control Quality

Detection of Changes in the River Bed and Identification of Boundary Changes Using Topographic Data from Different Epochs

Slobodanka Ključanin[1(✉)], Enes Hatibović[2], and Emina Hadžić[3]

[1] Technical Faculty Bihać, University of Bihać, Ul.Irfana Ljubujankića bb., 77000 Bihać, Bosnia and Herzegovina
slobodanka63@yahoo.com

[2] University of Sarajevo School Science and Technology, Sarajevo, Bosnia and Herzegovina

[3] University of Sarajevo, Sarajevo, Bosnia and Herzegovina

Abstract. The river Bosna is the right tributary of the Sava River, which has its source at the foot of Mount Igmannear Sarajevo. The entire river basin of Bosna is located in Bosnia and Herzegovina and represents the most densely populated place in the Federation of Bosnia and Herzegovina. The floods of the Bosna River occur from time to time. Some of them were disastrous, as they were in 2014. The problem of flooding in recent years is the focus of numerous research around Europe and the world due to numerous floods that have resulted in massive destruction and human casualties. The increased frequency of extreme natural disasters is associated with the climate change, which is getting more extreme year after year. When it comes to river basins, they mostly do not know the political boundaries, and problems related to the definition of borders (state, entity, cantonal, municipal, cadastre) are always in trend, especially when it comes to the lives of people who live or own real estate near the river.

This article identifies and analyzes the changes occurring within the administrative boundaries defined by the middle of the river trough. The time-flow changes (1959, 1968, 1974/1977, 2008 and 2012) of the Bosna river basin (in the section of the lower flow) were analyzed, and therefore the change of the municipal and cadastral border. As a result of the change in the flow of the river Bosna, the loss/gain of arable land is evident. It also provided an overview of procedures and legislation in Bosnia and Herzegovina dealing with the establishment of administrative boundaries, procedures for their renewal/change, and the boundaries of water resources that are directly related to monitoring the flow of watercourses.

Keywords: Bosnia river · Time-flow changes · Municipal and cadastral border · Administrative boundaries

1 Introduction

The total border length of Bosnia and Herzegovina is 1.551 km. Its land border encompasses 905 km, the river border (the rivers Sava, Drina and Una) 625 km and the maritime border approximately 21 km [1]. The length of the internal boundaries

I. Karabegović (Ed.): NT 2019, LNNS 76, pp. 689–697, 2020.
https://doi.org/10.1007/978-3-030-18072-0_81

following the rivers (the boundaries demarcation between the entities, district, cantons, cities, municipalities and cadastral municipalities) has been determined; however it has never been maintained except for some individual cases. It's somewhat natural for the rivers to, more or less, change their watercourse in time. Nevertheless, the climate changes are more frequent and more expressed; hence the river courses undergo a change. The climate change aftermath is an increase in natural disasters. One such example is the floods in the catchment area of river Bosna, which occurred in May, 2014. It has been determined that the precipitation has had one-hundred and even five hundred of return period in some part of the catchment areas [2]. The aftermath of the aforementioned floods to the environment, property and people is acutely destructive. Managing the consequences of the catastrophic floods requires exceptional engagement on the part of the country and its institutions as well as its citizens.

According to the information by the Ministry for Human Rights and Refugees, the flood and landslides aftermath (which occurred in May, 2014) is people displacement (89.981 people have been displaced), 1.943 completely demolished living units, whilst 41.306 living units have been damaged. Overall, it's 43.249 living units, i.e. families who need housing [3]. After urgent interventions in the first days of the catastrophe, things started "to normalize" and the inhabitants started coming back to their homes and properties. It's already been mentioned that natural disasters know no boundaries, however after settling, the problem of reconstructing administrative boundaries and property boundaries still remains.

2 The Problem Regarding Reconstruction of Administrative Boundaries Which Have Been Defined by the River Courses

Administrative boundaries define spatial area of legislative jurisdictions, elective, statistical and maritime areas [4]. In general, a boundary is an imaginary line designating mutual perimeter and a boundary of two adjacent parcels. The aforementioned definition clarifies the way a boundary turns into a legal line where proprietary rights of one individual meet the rights of the other [5]. Administrative boundaries are mostly fixed and often don't change. With regard to natural boundaries (sea, lakes, rivers etc.) the situation is a bit more complex since water bodies are in perpetual motion and they affect land boundaries (erosion, river course changes etc.). While defining natural boundaries one has to take into account whether the certain boundary is associated with tides, whether the boundary (i.e. the middle of the watercourse) relates to the survey of the waterbody at a low water level, high water level, or one takes into account a medium water level (tides or a medium sea level) [6]. It is expected that, due to climate change influence on the sea level and the river water level, natural boundaries become unreliable i.e. alterable in a short period of time. Consequently, analyzing the existing status of legal regulations and procedures pertaining to keeping track of the damage (disappearance of complete or part of cadastral parcels in the waterbody) is a vital to proprietary's parcels that are adjoining natural boundaries. Administrative units are territorial and they are bordered with administrative boundaries (country, entity, canton, city or municipality). One of the smallest territorial units is cadastral municipality (which is examined in this article as well as cadastral parcel), which commonly encompasses one populated area. Acts on areas and names of populated places within cadastral municipality are enacted by municipal council that beforehand gathers professional opinion by the Federal Administration for Geodetic and Property Affairs (FGA). In order to mark a boundary between cadastral municipalities which are at the same time municipality boundaries, a panel is constituted by municipal council of those municipalities [7]. Establishing and marking cadastral municipalities is administered by placing landmarks and designations along the land, and by demarcating boundaries and drafts into the register, and not later than seven days of the final boundary demarcation. If after a certain time period, a considerable disagreement appears between the cadastral data and actual condition on the terrain (which cannot be eliminated with regular maintenance), another cadastre of real estate measurement is conducted (also the territorial boundaries) [8]. On the other hand, when one speaks of rivers and streams, boundaries of water resources are also determined and they can be owned by Federation, canton, city, municipality, legal entity or natural person.[1][9]. Defining boundaries of water resource encompasses its designation in the field, methods of defining the boundaries and their entry into the water system information, that is into the water cadastre, cadastral records and cadastral mappings for a specific area. The procedure of determining affiliation of a parcel to the public water resource with regard to surface water category one is set in motion by a proposal of a competent Water Agency [10]. The boundary of water resource is defined by a study which is conducted by a company registered for land measurements. Based on the study, the Ministry of Agriculture, Water Management and Forestry issues a decision on the boundary designation of the water resource (the boundaries are displayed descriptively in the decision). The decision is forwarded to the authorized municipal department for cadastre and to the land registry of the municipal court. The boundary of the water resource is not marked in the field. The decision is delivered to the land registry because of the registration into the official copy of the land registry of the parcel which is completely or partly situated in the water resource. Water Agency has defined water resource in almost all municipalities at the watercourse of the first category. If one requires a change in the water resource boundary (for instance, creating a regulated trough), a study on the boundary modification at a level of cadastral municipality is conducted and the previous decision is modified and supplemented. Therefore, the competent institutions for Water Agency are responsible for the implementation of activities regarding protection from harmful water influence. The regulations regarding demarcation of water resource and determining cadastral plot to the public water resource are used for defining the way a land owner is compensated (Water Law, article 143, 145, 146, 147, 148 and 149) [9]. However, with the access to the topographic maps, digital orthophoto and cadastral mappings and studies, it has been determined that the watercourse update (natural boundaries) was executed periodically (depending on the year when the topographic map or orthophoto was made), whilst the watercourse data in the cadastre hasn't been

[1] If a natural person wants to sell his property which is a water resource, the right to repurchase it has the Federation, canton, city or municipality.

updated at all. This implies that neither the modifications of the administrative boundaries which follow the middle of the watercourse were made, nor the changes relating to the way one uses the land were implemented in the defined water areas. This pertains to the period of the first establishment of cadastral municipalities and the first land survey on the cadastral mapping until this very day. In this article, we are examining the watercourse change of the river Bosna from the period of 1959 to 2012 in the area of administrative/cadastral municipality Odžak/Ada, that is the parcel number 754 – the river Bosna. By analyzing the changes that have appeared, one could draw conclusions of implementation procedures in relation to the changes in cadastral studies and mappings, and at the same time about the administrative boundary changes.

3 Applied Methodology of Research and Detection of Watercourse Change of the River Bosna

For the purpose of writing this article, one has thoroughly searched for reliable data on demarcation of administrative boundaries, water areas, cadastral parcels, and official, technical registers which marked the position of the river Bosna. Diverse information has been used by interviewing geodetic, legal and construction experts. Everyone has been acquainted with the problem of watercourse change, however they haven't had a contact with the problem solving of administrative boundaries and cadastral parcel boundaries after the position change of the river and its trough. Accordingly, the authors of this article have been compelled to use the internet sources for gathering data:

- Integrated management strategy of the border in Bosnia and Herzegovina over the period of 2015 to 2018 (published in 2015)
- Regulations on the land survey SR BiH (Official gazette SR BiH, number 22/84, 12/87 and 26/90)
- Law on the land survey and cadastral register SR BiH (Official gazette SR BiH, number 14/78)
- Water Law (Official gazette of the Federation of Bosnia and Herzegovina, number 70/06)
- Regulation on methods of determining water resource boundary and on methods of determining cadastral plot to the public water resource (Official gazette of the Federation of Bosnia and Herzegovina, number 26/09)
- Topographic map at scale 1:50000 (in 1968)
- Topographic map at scale 1:25000 in 1974 and 1977
- Digital orthophoto at scale 1:5000 in 2008 and 2012
- Land lot diagram number 754 – The River Bosna, cadastral municipality Ada, Municipality Odžak in 1959.

Apart from the above mentioned resources, others were used that deal with administrative boundaries, water resources and data maintenance of the cadastre.

The data set is scanned and georeferenced based on available information on the reference system in Bosnia and Herzegovina. This enables data to be compared to the same scale and to perform analysis of the change of the position of the river. The data was georeferenced via software for image processing ERDAS 9.0. Multiple steps have been used for the extraction of the river polygon from the digital orthophoto. The first step was image classification. In this case ISO Cluster Unsupervised Classification has been utilized. The further step that was needed in order to get the river polygon was reclassification. Reclassification allows us to adjoin the river polygon with those parts that were not classified as such after the first method. The next step was to determine a certain number of midway points of the river recorded in 1959, and to track the midway fluctuations over the years.

CAD and GIS have been used for the graphic data processing relating to the watercourse change. These systems enable examination of topographic map legibility (scanned and georeferenced in space), vector data, and even various spatial measurements and analysis.

4 Watercourse Change Detection of the River Bosna over the Period 1959–2012

The river Bosna has its source at the foot of Mount Igman (Sarajevo). It's around 273 km long and it flows into the river Sava [11]. The coverage of the derived comparisons and analysis constitutes the rectangle restricted with coordinates: upper right corner Y = 6528488, X = 4986471 and lower left corner Y = 6530983, X = 4984489.

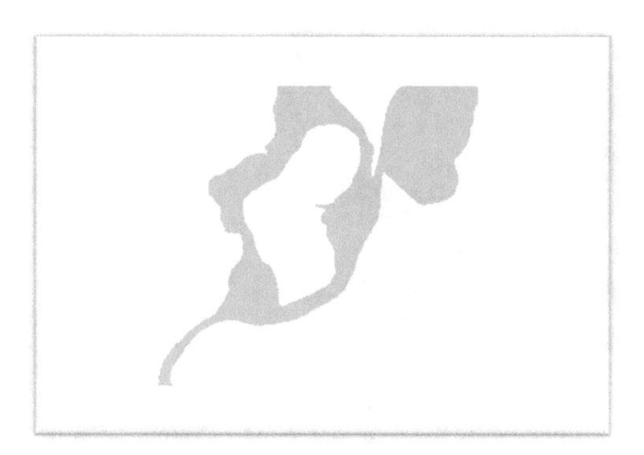

Fig. 1. Graph of the cadastral parcel (made in 1959) number 754, cadastral municipality Ada

The parcel is situated in the municipality Odžak, the cadastral municipality Ada. The river Bosna data, that is the parcel number 754 data, was gathered in 1959 and then it was registered into the cadastral record and plotted into cadastral maps. The shape of this parcel didn't change in time, and today its original form and size can be seen on Geoportal FGA (Fig. 1).

In order to analyze the change of the watercourse and the river bed of the River Bosna, its cadastral shape was compared with its subsequent cadastral-topographic entry, i.e. with the topographic maps at scale 1:50000 (made in 1968), 1:25000 (made in 1977), and the digital orthophoto at scale 1:5000 (made in 2008 and 2012) (Figs. 2a, 3a, 4a). You can see the midpoint flow deviation graphs in Figs. 2b, 3b and 4b.

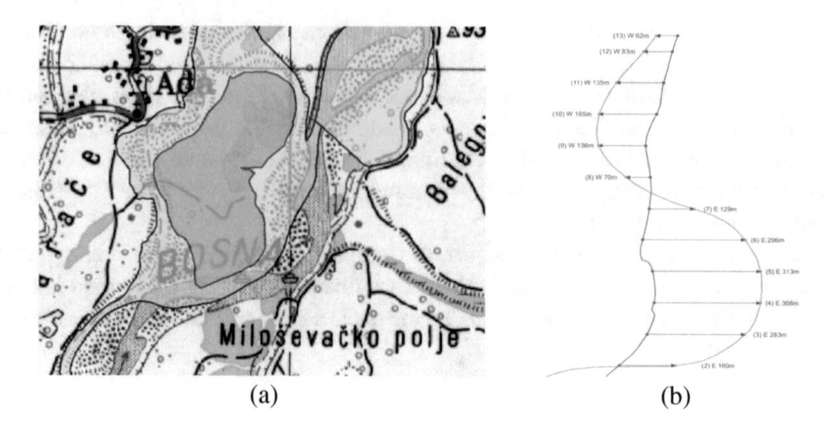

Fig. 2. (a) Map 1:50000 covered by the river parcel; (b) Graphic representation of midway point deviations (the green line is the watercourse of 1959)

One can notice a drastic change in the river watercourse within the period of 10 years by looking at the Fig. 2. Moreover, some new watercourse changes occur in the next 10 years which are recorded via topographic maps at scale 1:25000 (Fig. 3).

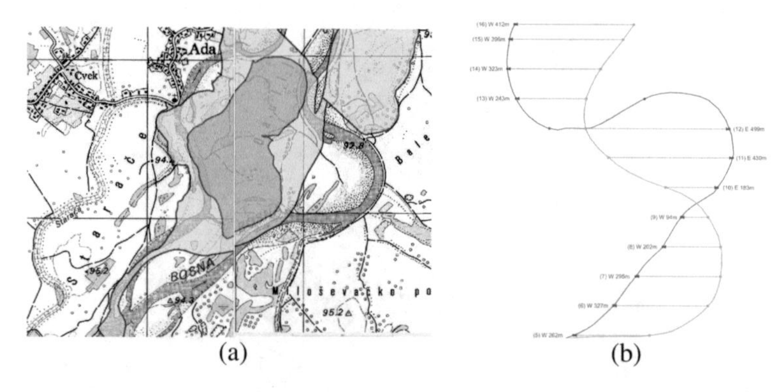

Fig. 3. (a) Map 1:25000 covered by the river parcel; (b) Graphic representation of midway point deviations (the green line is the watercourse of 1959)

Considering that in the meantime the cadastral data hasn't been updated, some differences appeared in the area of neighboring administrative boundaries by comparison with the territorial coverage surveyed in 1959.

Monitoring the production timeline of different topographic maps and digital orthophoto, it is obvious that the data in the cadastral register hasn't changed. If the cadastral municipality Ada boundary and the administrative boundary of municipality Odžakare a natural boundary, and they clearly are, why the data hasn't been updated regularly along with the change of the river Bosna watercourse. If it is necessary to make a change in the cadastral register by the law based on defining the water resources – why hasn't that been done, not even once, for almost 60 years? How were the owners compensated for the lost agricultural land by the municipality Odžak due to the river

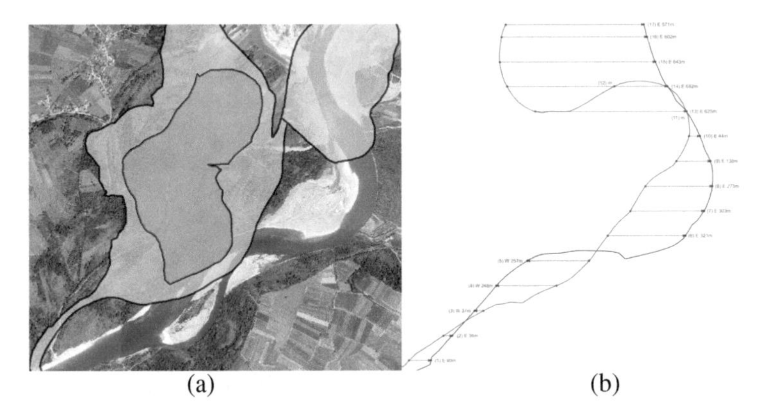

(a) (b)

Fig. 4. (a) DOF 1:5000 (made in 2008) covered by the river parcel; (b) Graphic representation of midway point deviations (the blue line is the watercourse of 1977)

Bosna watercourse change? Or were they at all? What's more, if the municipality Odžak obtained additional land surface due to the river Bosna watercourse change and the change of natural boundary position, how was the matter resolved?

Since the analysis of the position change of the river Bosna watercourse couldn't be carried out after the catastrophic changes in May, 2014 due to insufficient adequate topographic survey, it can be assumed that the river Bosna watercourse was drastically changed, and that agricultural area along the river was devastated. All things considered, the matter regarding the cadastral boundary which is the municipal boundary as well can easily be solved based on the legislation in force. Thus, the question is how to act and how to update this example of the natural boundary change and many others nowadays.

5 Conclusion and Procedure Proposal for Improvement of the Current State

Via undertaken analysis it is clear that the changes in the cadastral register haven't been carried out ex officio, i.e. when it comes to private properties they are almost never carried out. Perhaps it is considered that it is not crucial to take them into account, since no one can directly benefit from them. However, the land owners alongside rivers have the obligation to take account of their property. Nevertheless, how can one take care of his property when the property itself disappeared due to catastrophic flood? Who is to blame for the unrecorded regular changes? We cannot give an answer to this question with utter confidence. Presumably, there are numerous factors one needs to look into, but one can determine the genuine facts and propose a procedure for the improvement of the current state. While elaborating the problem, it becomes clear that there are legislations and ways to overcome the incurred problems with regard to current inconsistency between natural boundaries and administrative (territorial) boundaries which are recorded in the cadastral register and other official documents. In order to successfully manage your own territory (this applies to administrative units), it is necessary that they are activated on that account, but in what way? The best option would be to follow the current legal solutions. Hence, municipal councils should encourage solving the aforementioned problems in a way they plan the funds for financing new land survey of areas that weren't regularly registered into the municipal cadastre. According to the law on the land survey and cadastre of real estates, it is noted that demarcation is executed by organizing a board (the board is proposed by the municipal councils of the two neighboring municipalities). Accordingly, demarcations are recorded, and with the assistance of surveying experts the changes are recorded too, and they are documented into the cadastral registers. Another way of solving the problem of administrative boundaries is giving up on the natural boundary and determining the permanent physical boundary regardless of the watercourse change. Regarding the parcels owned by a natural person which are lost in the riverbed, or situations where one acquired "additional" land (the land that belonged to the neighboring municipality), it is possible to solve the problem by buying out the land so that the private ownership goes over to the state's (water resource). Thereby the natural person would be compensated by the enforcement of expropriation. Thus, the solutions are possible with the application of already existing legislations. The problem that was recorded throughout this research is the lack of the initiative (and possibly financial funds) on the part of authorized institutions.

References

1. Council of Ministers of Bosnia and Herzegovina: Integrated management strategy of the border in Bosnia and Herzegovina over the period of 2015 to 2018 (2015). http://www.vijeceministara.gov.ba/akti/prijedlozi_zakona/default.aspx?id=21132&langTag=bs-BA. Accessed 7 Nov 2018
2. Imamović, A.: The cause of the floods in the river Bosna basin with regard to the floods that occurred in May, 2014. In: Proceedings: Risk Management with Reference to the Floods and Reduction of Its Damaging Effects, p. 131. Academy of Sciences and Arts of Bosnia and Herzegovina, Sarajevo (2015)
3. Radio Sarajevo (2014). https://www.radiosarajevo.ba/vijesti/bosna-i-hercegovina/zbog-poplava-i-klizista-u-bih-raseljena-89981-osoba/164534. Accessed 8 Nov 2018
4. ANZLIC the Spatial Information Council. https://link.fsdf.org.au/fsdf-theme/administrative-boundaries. Accessed 8 Nov 2018
5. GIM International: Cadastral Boundaries or Legal Boundaries? (2017). https://www.gim-international.com/content/article/cadastral-boundaries-or-legal-boundaries. Accessed 7 Nov 2018
6. ICSM, ANZLIC Committee on Surveying & Mapping. Land Boundaries. https://www.icsm.gov.au/education/fundamentals-land-ownership-land-boundaries-and-surveying/land-boundaries. Accessed 7 Nov 2018
7. Federal Administration for Geodetic and Property Legal affairs of Federation Bosnia and Herzegovina. Regulations on land surveying SR BiH. http://www.fgu.com.ba/bs/pravilnici.html?file=files/Stranice/PDF%20files/Pravilnici/PRAVILNIK%20za%20snimanje%20detalja%20SR%20BiH.pdf. Accessed 12 Nov 2018

8. Federal Administration for Geodetic and Property Legal affairs of Federation Bosnia and Herzegovina. Law on the land survey and cadastral register SR BiH. http://www.fgu.com.ba/bs/zakoni.html?file=files/Stranice/PDF%20files/Zakoni/Zakon%20o%20premjeru%20i%20katastru%20zemljista.pdf. Accessed 12 Nov 2018
9. Water Law (Official gazette of the Federation of Bosnia and Herzegovina, number 70/06). https://fmpvs.gov.ba/wp-content/uploads/2017/Vodoprivreda/Vode-zakoni/zakon-vode-7006.pdf. Accessed 7 Nov 2018
10. Regulation on methods of determining water resource boundary and on methods of determining cadastral plot to the public water resource (Official gazette of the Federation of Bosnia and Herzegovina, number 26/09). https://fmpvs.gov.ba/wp-content/uploads/2017/Vodoprivreda/Vode-pravilnici/vode-prav2609i.pdf. Accessed 6 Nov 2018
11. Institute for Water Management, P.L.C Sarajevo, Federal Institute for Hydrometeorology: Hydrology report on surface water in Bosnia and Herzegovina (2010). ftp://ksh.fgg.uni-lj.si/exchange/BOSNA/source/hidroloska_studija_bosne/1%20vrelo%20bosne%20bosna/knjiga%201.pdf. Accessed 6 Nov 2018

Building Thermal Insulation Material Based on Sheep Wool

Sanela Klarić[1(✉)], Emir Klarić[2], Muhidin Zametica[2], and Melisa Gazdić[1,2]

[1] International Burch University, Francuskerevolucije bb, 71210 Sarajevo, Bosnia and Herzegovina
sanela.klaric@ibu.edu.ba

[2] Green Council, Brcanska 17, 71000 Sarajevo, Bosnia and Herzegovina

Abstract. The number of the research has confirmed that sheep wool is an ideal material for insulation in buildings. In BiH, there are facilities for wool washing and processing. Often times, well developed stations for purchasing wool through repurchasing stations can through extra investment revitalize the system. Production of thermal insulation panels from sheep wool in BiH, with the current infrastructure and qualified work force available, could provide a product which is of good quality and competitive price in view of the demanding EU or US markets. In addition, the production could easily be tailored to fit the existing facilities. Only the last step of the production process of the wool thermal insulation panels is different from the current one and requires specialised machines for panels.

The feasibility study identified that investment into the project of production of thermal insulation panels made out of sheep wool would be cost-effective for the investors, since the earning would be sufficient to cover the future expenditures and that such project would be profitable. As was identified and elaborated in the Financial Analysis section, for the purpose of this study it is not necessary to present additional arguments in favour of the economic and financial cost-effectiveness of launching the project by expressing (monetising) significant non-financial social benefits of this project in terms of money. The indicators of Financial Net Present Value (cost-effectiveness of the investment) and Financial Internal Return Rate (FNPV and FIRR) show that net revenues can cover overall investment and obtain the planned profit.

Among other things, the study identified several social benefits that would follow the launch of the project of production of sheep wool thermal insulation panels: (1) Sheep wool as building insulation material uses little primary energy and CO_2 for production, installation, usage and recycling; (2) Sheep wool waste can be used as natural fertiliser; (3) Launching production of sheep wool insulation materials would revive infrastructure of sheep breeding, shearing and wool buying. The sheep wool is made out of biodegradable proteins and after being used in building construction the material can be turned into natural fertiliser. Adequate support to sheep farming sector would bring about preservation of the autochthonous sheep breed Pramenka in BiH. It is important to stress that sheep wool can be easily recycled and reused.

Keywords: Sustainability · Sheep wool · Fusibility study · Financial analysis · Economic analysis

I. Karabegović (Ed.): NT 2019, LNNS 76, pp. 698–711, 2020.
https://doi.org/10.1007/978-3-030-18072-0_82

1 Introduction

Sheep wool is one of the unavoidable natural materials observed in the context of new housing requirements which, due to its natural properties, attracts the attention of a wide range of scientists and investors who study its properties as the most economical and the best models of using this material in the construction industry. This material has been used since ancient times in the textile industry, primarily because of its thermal properties, as well as in the construction of facilities, which will be in the focus of this study.

Contemporary requirements for sustainable development, as well as the construction sector's demands to increase the population's awareness regarding the need for healthy living, have recognized the characteristics of sheep wool as well as the advantages of exploiting this material. Sheep wool becomes the subject of many researches and plays an important role in the innovative development of the construction industry, which becomes a strategic industry of sustainable development in the EU and in the world. The study will show that minimal investments are required through adaptations of existing sheep wool processing plants to adapt these plants to modern processing plants for insulation panels made of this material. If all of the above stated information is viewed from the aspect of future investments in the clean technologies, which are a chance for development in BiH, we come to the conclusion that the infrastructure needed to establish this production process in BiH is still present, with the trained labour force. Furthermore, an important investment guideline is the fact that only in the final phase of the production process, already established woollen fabrication plants require additional interventions to reorganize the production of insulation panels.

2 Contextual Analysis

Health properties of wool are exceptional. It does not irritate the respiratory organs or the skin and precisely because of these characteristics it is used as a natural material in the textile industry. Artificial materials with similar isolation characteristics, mostly have a detrimental effect on health. The National Toxicology Program in the United States for the exploration of toxic materials classified the glass and mineral wool as cancer-causing materials. Sheep wool absorbs negative substances released from other building materials and thus provides healthy living conditions for humans.

In addition, sheep wool absorbs moisture up to 40% of its weight, without changing its insulation properties and returns it back to the air when conditions are created [1]. Sheep wool has good moisture absorption properties, and the risk of fungus-mold production is lower than mineral wool [2]. It is also biodegradable and does not harm environment. In case of fire, since woollen fibres contain moisture, sheep's wool does not heat up the fire, but it is extinguishing it while some constructive elements are protected by its incorporation. Other significant benefits of sheep wool are: reducing the bad influence of climate change; an excellent insulator with thermal properties similar to conventional thermal insulation materials; protects against noise and possesses exceptional acoustic properties, primary energy and CO2 emissions when processing, installing and recycling are at the lowest level compared to other insulating materials on the market; locally produced can achieve a favourable market price; does not require additional protective clothing when installed, does not damage the respiratory or other human organs; it promotes the conservation of biodiversity and animals of indigenous species of planet Earth. The following Table 1 shows the comparative values of thermal conductivity, heat resistance, density and primary energy values and CO2 emissions for seven different insulation materials. The values shown indicate benefits of using this natural material especially when it comes to primary energy.

Table 1. Comparison of insulation materials [3]

	Material	Thermal conductivity (W/m K)	Density (kg/m^3)	Thermal resistance (m K/W)	Primary energy (kWh/m^3)
1	Polystyrene board	0,035	25	28,6	1126
2	Mineral wool - roll	0,040	12	25,0	231
3	Mineral wool - board	0,035	25	28,6	231
4	Phenolic foam board	0,020	30	50,0	1126
5	Polyurethane board	0,025	30	40,0	1126
6	Cellulose fibres	0,035	25	28,6	133
7	Sheep wool	0,037		27,0	31

In the “Wool Feasibility Study” conducted by UNDP BiH in 2010, the number of sheep in BiH pre-war was estimated at about 1.3 million of heads. The majority of this population, around 80%, was the native autochthonous breed Pramenka with yields 0.75–3 kg of wool per run. The average weight of the cut fleece is 1.7 kg per head (Chart 1).

The sheep sector is gradually being rebuilt. Data on number of heads differs in different reports. According to the latest research conducted by the Food and Agriculture Organization (FAO), as part of the analysis of the agriculture, food and rural development sector in BiH for 2012 there are 1,515,000 [4] heads of sheep in BiH. At the same time, according to the annual reports of MoFTER (Ministry of Foreign Trade and Economic Relations of BiH), and its Sector for Agriculture, Food and Rural Development, responsible for these issues, some different data are mentioned (Table 2).

Table 2. Sheep population in Bosnia and Herzegovina 1990–2012 [5]

Year	1990	1996	2007	2012
Sheep population	1,318,673	787,759	1,030,654	1,515,000

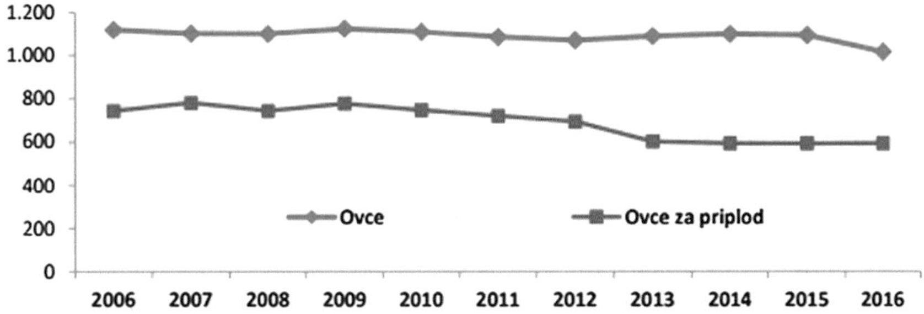

Chart 1. Sheep number in BiH from 2006 to 2016 [10]

Data obtained indicate the total amount of wool, that could be collected in BiH, varies between (1,515,000 sheep × 1,7 kg/head) or (1,200,000 sheep × 1,7 kg/head) approximately between 2,000 and 2,500 tons per year. Estimation is 80% of the sheep population in BiH belongs to Pramenka breed (fibres classified between D and E, or fibres having an average thickness in microfine between 37 and 60). Data indicate 80% of wool or about 2,000 tons is coarse wool obtained from Pramenka breed, while remaining around 20%, or 500 tons is finer wool obtained from other breeds. The quality and amount of fibre as well as the density of wool is characteristic of sheep breed. The quality and fibre density depend on the diet as well. Better nutrition of sheep provides a thicker and better fleece. Quality and density depend on the age of the sheep. There are 7.3 fibres in 1 mm at coarse fleece sheep, while in the finer flees sheep there are 29 to 88 fibres up to a maximum of 130 in 1 mm [6]. At the same time this material is available for further "growth" even though its exploitation in the traditional way decreases [7]. Certainly, issue of price is questionable, i.e. the profitability of collecting all sheared wool, so it can be assumed one part would remain outside the scope of organized collection and would remain unused. During the interview with Bahrudin Bojcic, a director of "Wool-line", the company for production and sale of wool products, interviewed by the research team, it has been stated that collection up to 1,200 tons of wool on a profitable basis in the current framework is possible. This claim implies that all collected wool, both fine and coarse, is used for the production of insulation panels.

2.1 Feasibility and Options Analysis

Research of the association Green Council has identified and considered three policy options as possible for launching a project to produce thermal insulation materials from sheep wool. The options considered are: (1) Ambitious option of producing thermal insulating panels with the maximum purchase of unwashed sheep wool in the amount of 2500 tons per year; (2) Realistic option of producing thermal insulating panels with available purchase of unwashed sheep wool in an amount of 1200 tons per year, using existing wool laundry and protecting facilities (machine laundry service in Visoko - with the required investment), (3) Realistic option of the producing of thermal insulating panels with the purchase of available unwashed sheep wool in the amount of 1200 tons per year, with the option of installing new wool laundry and protecting machines (investment in new facility and equipment).

Analysis of possible policy options, their limitations and advantages, indicated conclusion that, for the successful establishment of sheep wool thermal insulation production, only realistic and viable real option is Policy option 2. Which involves the purchase of about 1200 tons of raw unwashed wool and keeping facilities for laundry in Visoko.

3 Financial Analysis

Using invested capital or public financial resources, we can identify two very important financial indicators, namely:

- Financial Net Present Value - (FNPV)
- Financial Internal Rate of Return - (FIRR).

The net present value (NPV) is the value of the total annual net savings in the future and during the economic life, from the first to the last year, minus the investment. In order for the project to enter the sphere of financially viable, two basic conditions should be fulfilled:

- FNPV of the project should be positive or at least equal to zero (FNPV $\geq$ 0). It is necessary that the present value of the income is higher or at least the same as the present value of the project expense
- FNPV of the selected option should be higher or at least the same as with other options

The annual project's profitability indicator is the Financial Internal Rate of Return. Respectively, it reflects the readiness of the project to be sustained and this implies the revenues are higher than the expenses after the discount rate is set. The higher project's FIRR, the more sustain and better project is financially. FIRR is the decision-making indicator regarding investment option to choose. If it turns out that FIRR is negative then it is necessary to co-finance the project from other sources (grants or EU funding). This is because the banking sector will not be interested in risk financing. The financial analysis is precisely used to confirm the financial viability of the project for thermal insulation materials from sheep wool for investors, i.e. to determine whether future receipts are sufficient to cover future expenditures and to answer the question of whether the project's FIRR is higher than the discount rate in order to determine the profitability.

The financial analysis needs to determine whether the project for the production of thermal insulation materials of sheep wool is financially viable for investors, whether the future receipts are sufficient to cover future expenditures and whether the project's FIRR exceeds the discount rate, meaning that the project brings profit. If the indicators are negative, the project requires co-financing from other sources such as government grants or EU funding.

3.1 Total Investment

These investments are related to investing in fixed assets and start-up costs of the project. Costs related to employee recruitment, their training, product certification for its sale on the EU market, as well as consultancy, notary, administrative and other services are calculated.

Investments in fixed assets are calculated on the basis of an overview of the existing infrastructure, interviews and assessments of the required building facilities and equipment for the implementation of the project. Large-scale real estate investments are not calculated using information from established practices that the location of production and storage facilities can be provided in existing facilities (woollaundry in Visoko) and through the leasing of certain premises financed from the operating costs calculated in the part of the analysis treating revenues and expenditures. Due to the necessary caution in calculating cash flows, real estate investments in the amount of 300,000 km were estimated for the first two project years. Investments in the equipment have been calculated on the basis of research and interviews conducted with potential investors who have already investigated both the supplier and customer market, as well as the available supply of wool processing plants and the production of sheep woolthermal insulating panels. In total, for the existing processing option of available 1200 tons of raw wool, or 720 tons of washed and protected wool on an annual basis, three specialized machines with a capacity of 250 tons of final product are planned for two shifts, five days a week, twelve months. Following the review of available bids, the indicative price of these machines is approximately 550,000 km per a machine (including all purchases of fixed assets), therefore, a total of 1,650,000 km should be invested for three machines. Investments were carefully placed. During the first-year procurement of two machines, while during the second-year procurement of the third machine should be made. Investments in other equipment are related to the purchase of transport vehicles, warehouse equipment and office equipment (200,000 km during the first two initial years).

Table 3. Review of total project investment (KM) [8]

Investment items		Year					
		I	II	III	IV	V	VI
1.	Real estate	−200.000	−100.000				
2.	Equipment	−1.250.000	−650.000				
A	Total fixed assets (1+2)	−1.450.000	−750.000	0	0	0	0
1.	Product certification	−50.000					
2.	Training for employees	−20.000	−10.000				
3.	Consultancy, legal, notary, administrative and other services	−20.000	−10.000				
4.	Other start-up costs	−80.000	−20.000				
B	Total start-up costs (1+2+3+4)	−170.000	−40.000	0	0	0	0
	Total investment (A+B)	−1.620.000	−790.000	0	0	0	0

3.2 Business Costs and Revenues

The project's business costs included in the financial analysis cover all business expenses having the character of cash outflow. These costs do not include costs not caused by cash outflows, or billing costs that most often do not incur cash expenses such as depreciation costs. The primary cost and expense of this project are the costs of raw materials that are, for the purposes of this study, defined as the cost of washed wool protected by natural salts of Thorlan (Thorlan IW).

The price of unwashed wool in BIH is very low and ranges from 0.50 to 0.75 km for 1 kg of greasy or rough unwashed wool (35 μm to 40 μm) and 0.80–1.00 km for 1 kg of lamb wool, greater fineness (28 μm to 32 μm). In the study, Thorlan protected, washed wool is calculated as raw material at a price of 3.5 km per 1 kg of washed wool. As we have already calculated, the amount of wool that could be collected in BiH reached 2.500 tons per year (1515 sheep × 1.7 kg/head), out of which approximately 80% is coarse wool, or about 2,000 tons belonging to autochthonous sheep breed Pramenka, while the rest of about 20%, or 500 tons, is wool of greater fineness [9].

Infrastructure for the development of a new plant for the production of sheep wool insulation panels still exists. Observed from the investor's perspective, investing in clean technologies represents a strategic opportunity for BiH development. The existing laundry in Visoko can be adapted to the required ecological and production standards. Wool production facilities, only in the final stage of the production process, require special parts and machines that produce insulating elements, the rest of the process is satisfactory and should not be changed [8].

There is also an educated labour force in this sector. Based on the conducted research and interviews with an existing and potential sheep wool processing plants, it is realistic to calculate for the next few years, under conditions of inadequate and underdeveloped infrastructure for breeding, shearing and purchasing of wool, the purchase of approximately 1200 tons of unwashed wool per year. Given the 40% loss after washing, protection and drying, realistic is to expect around 720 tons of washed and protected wool [8]. The machine laundry in Visoko, with its capacities, can provide this amount of raw material for the production of thermal insulating panels with this natural material. The total annual production of the final product is realistic in the existing conditions calculated in the amount of 612 tons of sheep wool thermal insulating panels. Production would gradually increase from 200 tons of final products during the first year up to 612 tonnes in the fourth and subsequent years of the project implementation. The final product price is calculated at an average of 15 km per kilogram of the thermal insulating panel depending on its thickness and density. Total product sales are planned through wholesalers and therefore the costs of product marketing and sales are not projected. The price is projected very carefully since the EU market can achieve a significantly cheaper price (up to 30 km per kilogram). The project envisages the employment of 50 workers in the processing of washed and protected wool and the production, storage and distribution of thermal insulation panels of sheep wool. Therefore, estimates and calculations apply only to newly employed in the segment of direct production of thermal insulating panels. Employees in the segment of breeding, shearing and procurement of wool and at the stage of washing, drying and protection with Thorlan are not calculated in expenditures and cash flows in this study. Their costs are calculated in the price of washed wool being 3.5 km per kilogram.

Table 4. Review of all business expenses and revenues (KM) [8]

Items of business revenues and expenses		Year					
		I	II	III	IV	V	VI
1.	Sales income	3.000.000	6.000.000	7.500.000	9.180.000	9.180.000	9.180.000
A	Total business revenues (1+2+3)	3.000.000	6.000.000	7.500.000	9.180.000	9.180.000	9.180.000
1.	Costs of raw material	−822.500	−1.645.000	−2.065.000	−2.520.000	−2.520.000	−2.520.000
2.	Costs for employees	−600.000	−720.000	−960.000	−1.200.000	−1.200.000	−1.200.000
3.	Energy and material costs	−18.000	−60.000	−75.000	−91.800	−91.800	−91.800
4.	Expenses for production services, ongoing maintenance)	−36.000	−120.000	−150.000	−183.600	−183.600	−183.600
5.	Non-material costs	−9.000	−30.000	−37.500	−45.900	−45.900	−45.900
B	Total business expenses (1+2+3)	−1.527.500	−2.575.000	−3.287.500	−4.041.300	−4.041.300	−4.041.300
	Net business income (A+B)	1.472.500	3.425.000	4.212.500	5.138.700	5.138.700	5.138.700

3.3 Cash Flows and Net Present Value of Money

Through the next steps of financial analysis, we will present unique data from previous calculations in Tables 3 and 4, in order to provide a unique overview of cash inflows and outflows related to investments in this unique project in the region (project's Cash Flow).

The net present value of money (FNPV) is calculated using the financial parameters of the net cash flow from Table 5 using the recommended discount rate by the European Commission of 5%. In order to present the cash flows generated at different time periods at the present time, we used discounting and found that the total cash inflows of this project amount up to 18,875,788 km. With FNPV value above zero, we can conclude that the project can be accepted as financially viable by the investor. The Financial Internal Rate of Return (FIRR), which is defined as a discount rate and which calculates the net present value of financial investment (FNPV) to zero, has been calculated using the excel function for the project of production of sheep wool thermal insulating panels. Thus, we have determined its level of 6.51 and it is higher than the used discount rate of 5%, which directly tells us that the project is financially viable and that it is possible to expect a return on invested financial resources and profit. The socially valuable benefits that would be generated by the launching the project for production of sheep wool thermal insulation product would be, among other things, following: a small consumption of primary energy and CO2 in the cycle of production, installation, use and recycling of sheep wool; waste can be used as a natural fertilizer in agriculture; reviving the infrastructure for sheep breeding, shearing and procurement of wool.

Table 5. Overview of cash inflows and outflows (KM) [8]

Cash flows		Year						Total
		I	II	III	IV	V	VI	
1.	Total business income	3.000.000	6.000.000	7.500.000	9.180.000	9.180.000	9.180.000	44.040.000
A	Total cash inflow (1)	3.000.000	6.000.000	7.500.000	9.180.000	9.180.000	9.180.000	44.040.000
2.1.	Raw materials costs	−822.500	−1.645.000	−2.065.000	−2.520.000	−2.520.000	−2.520.000	−12.092.500
2.2.	Salaries and fees	−600.000	−720.000	−960.000	−1.200.000	−1.200.000	−1.200.000	−5.880.000
2.3.	Other costs	−105.000	−210.000	−262.500	−321.300	−321.300	−321.300	−1.541.400
2.	Total business costs	−1.527.500	−2.575.000	−3.287.500	−4.041.300	−4.041.300	−4.041.300	−19.513.900
3.	Total investment	−1.620.000	−790.000	0	0	0	0	−2.410.000
B	Total cash outflows (2+3)	−3.147.500	−3.365.000	−3.287.500	−4.041.300	−4.041.300	−4.041.300	−21.923.900
	Net cash flow (A+B)	−147.500	2.635.000	4.212.500	5.138.700	5.138.700	5.138.700	22.116.100
	Discount factor	1,0000	1,05000	1,10250	1,15763	1,21551	1,27628	
	Cash flow value	−147.500	2.509.524	3.820.862	4.438.983	4.227.608	4.026.311	18.875.788
	Discount rate	5%	FNPV(C)	18.875.788	FRR(C)	18,43		

4 Multi-criteria Decision Analysis

In order to make a multi-criteria analysis, primarily it is necessary to formulate the option that will be evaluated and to identify the criteria based on which evaluation of available options will be made. The identified and formulated criteria are evaluated according to their significance and the overall result of the multi-criteria analysis (Table 6).

Table 6. MKA criteria and evaluation of its significance [8]

Criterion	Significance of criterion	Ponder
Primary energy and CO2 consumption	100	0,21
Wool as waste-pollutant (elimination of ecological problems)	80	0,17
Biodegradability (wool as a natural fertilizer)	60	0,13
Possibility for providing support by existing infrastructural production and wool procurement	100	0,21
Increased employment in the sector	80	0,17
Financing and financial sustainability of the project	60	0,13
Total	480	1,00

The greatest significance is to the effects on primary energy and CO2 consumption and on the possibilities of supporting the project from the existing wool production and procurement infrastructure and in the increase of employment in this sector, given that the establishment and increase of production and exports directly affects the growth rate of the economy (Table 7).

In order to finally make a clear mathematical approach in decision making between best option, we have multiplied the rating for each option for each criterion individually with predefined criteria weights (Table 8).

Realistic option for the production of thermal insulating panels with the purchase of unwashed sheep wool in the amount of 1200 tons per year, with the use of existing laundry and wool protection machines (laundry facility at Visoko) was evaluated by the defined criteria as the most efficient, as we have already confirmed in the above presented analyses (contextual analysis, feasibility and options analysis, financial and economic analysis).

Table 7. Rating of options by criterion MKA [8]

Criterion	Option I	Option II	Option III
	Max. purchase of wool (2400 t)	Realistic purchase (1200 t) and laundry in Visoko	Realistic purchase and procurement of laundry machines
Primary energy and CO2 consumption	100	80	80
Wool as waste-pollutant (elimination of ecological problems)	100	60	80
Biodegradability (wool as a natural fertilizer)	100	80	80
Possibility for providing support by existing infrastructural production and wool procurement	10	100	100
Increased employment in the sector	100	80	40
Financing and financial sustainability of the project	60	100	80

Table 8. MKA: Weighted rating of the offered options [8]

Criterion	Option I	Option II	Option III
	Max. purchase of wool (2400 t)	Realistic purchase (1200 t) and laundry in Visoko	Realistic purchase and procurement of laundry machines
Primary energy and CO2 consumption	21	17	17
Wool as waste-pollutant (elimination of ecological problems)	17	10	13
Biodegradability (wool as a natural fertilizer)	13	10	10
Possibility for providing support by existing infrastructural production and wool procurement	2	21	21
Increased employment in the sector	17	13	7
Financing and financial sustainability of the project	8	13	10
Total	76	83	78

5 Quantitative Analysis of the Factor's Impact on the Project

By identifying the factor, a quantitative analysis of their impact on the economic net present value (ENPV) of the project was carried out in such a way that the key factors are those whose change in estimated value by 1% causes the ENPV of the project to change by more than 1%.

Table 9. Quantitative analysis of factors influence on ENPV [8]

Critical factor	Value of the factor	% of factor change	New factor value	ENPV	New value ENPV	% of change ENPV
1	2	3	4	5	6	7
Total investment	2.410.000	1%	2.434.100	18.875.788	18.899.888	0,13%
	2.410.000	−1%	2.385.900	18.875.788	18.851.688	−0,13%
Amount of business expenses	19.139.400	1%	19.330.794	18.875.788	19.067.182	1,01%
	19.139.400	−1%	18.948.006	18.875.788	18.684.394	−1,01%
Amount of business revenues	42.840.000	1%	43.268.400	18.875.788	19.289.194	2,19%
	42.840.000	−1%	42.411.600	18.875.788	18.462.382	−2,19%

Analysis presented in the Table 9 shows that total investments are not critical factors of the project. Total business revenues are a critical factor of the project because they have a significant impact on the project's effects, i.e. by changing the value of this factor by 1%, the net present value of the project is changed by more than 1% (±2.03%). Business costs are at the very limit of the project critical factors.

6 Conclusions

World researches suggests sheep wool is the perfect isolation material in the building sector as confirmed by our feasibility study. Sheep wool in BiH is not of high quality and is not suitable for the fine textile industry. However, the characteristics of the local sheep's wool meet the standards required for the production of insulation panels for the construction sector. There is a great tradition of sheep breeding in BiH. There is also a tradition and infrastructure for the processing and production of sheep wool products and qualified labour force. At the same time, sheep wool in BiH is an ecological waste because large quantities of this material are not used or treated inadequately. With additional fewer investments, once well-developed and organized infrastructure for collecting and buying wool, through planned purchasing stations, could be revitalized. A locally produced thermal insulating panel made of local sheep wool, with revitalization of existing infrastructure and the involvement of local skilled labour, may be a price-competitive product for a demanding EU or US market. Existing facilities could be easily adapted to modern standards and production requirements. The final step in the production of sheepwoolthermal insulation panels presents the only difference and requires special processing machines. All other steps in production remain the same as the existing ones.

Using sheep wool for the construction of thermal insulation panels greatly reduces CO2 emissions in the production process. CO2 emissions are also reduced in the process of transport, installation, maintenance and recycling. Sheep wool waste is used to improve the quality of agricultural land, and the by-product lanolin is used in the medical industry. Particularly important is that taking care of this material, which is ecological waste and inadequately disposed in nature, affects the reduction of harmful impacts on climate change and helps to heal the planet Earth.

The requirements, standards and new EU directives in the field of energy efficiency in the construction sector lead to the recognition of thermal properties of sheep wool. These thermal properties have been known since ancient times, and with contemporary research the value of applying this material into construction is proven. This feasibility study found that investing in a project for the production of thermal insulation materials of sheep wool is financially viable for investors. Financial income and sustainability indicators (FNPV and FIRR) show that total investment can be covered from net income and achieved projected profit.

At the end of this study, considering all the undeniable advantages and benefits of this traditional, natural, biodegradable and long-lasting material with outstanding properties of hygroscopicity, low CO2 emissions and low energy consumption, we consider it necessary to suggest that this material, in BiH and the region, should be approached on the strategic and sustainable way. It is necessary to support future research in BiH and in the region, which will be focused on other innovative possibilities of producing and processing sheep wool with the research on the benefits of combining sheep wool with other natural and recycled materials such as wood, clay, lime, and straw. It is necessary to strategically approach all the benefits of clean technologies that can lead to balanced socio-economic development.

References

1. Zach, J., Korjenić, A., Petranek, V., Hroudova, J., Bednar, T.: Performance evaluation and research of alternative thermal insulations based on sheep wool. Energy Build. **49**, 246–253 (2012)
2. Korjenić, A., Klarić, S., Hadžić, A., Korjenić, S.: Sheep wool as a construction material for energy efficiency improvement. Energies **8**, 5765–5781 (2015). https://doi.org/10.3390/en8065765. ISSN 1996-1073. file:///C:/Users/user/Downloads/energies-08-05765%20(1).pdf
3. Fordham M.: Environmental Design (Ed. by Randall Thomas) (2006)
4. FAO, Sektorskeanalize, Mesarski i mljekarskisektor u BiH (2012). http://www.fao.org/3/a-aq182e.pdf
5. Sljepčević, S.: Studijaisplativostitržištavune UNDP BiH, 2010 uključujućipodatkeiz FAO izvještaja za 2012, godinu
6. Krajinović, M.: Ovčarstvo i kozarstvo. Redakcija “Savremenepoljoprivrede” Dnevnik, Novi Sad (1992)

7. Klarić, S., Šamić, D., Katica, J., Kurtović, A., Duerod, M., RosoPopovac, M.: Vodičenergetskaefikasnost u zgradarstvukaopolaznica za ostvarivanjeodrživogdruštveno-ekonomskograzvoja u Bosni i Hercegovini. Green Council (2016)
8. Zametica, M., Klarić, E., Klarić, S.: Studijaizvodivostiprojektaproizvodnjetermoizolacionih-materijalaodovčjevune. Green Council, Sarajevo (2016)
9. Klarić, S.: Održivostanovanje-drvo, ovčjavuna, slama-izazovi i potencijalitradicionalnih-prirodnihmaterijala. International Burch University Publications, Sarajevo (2015)
10. MVTEO Godišnjiizvještajizoblastipoljoprivrede, ishrane i ruralnograzvoja za Bosnu i Hercegovinu za (2016). http://www.mvteo.gov.ba/izvjestaji_publikacije/izvjestaji/default.aspx?id=9082&langTag=bs-BA

New Trends in Engineering Wood Technologies

Sanela Klarić[1(✉)] and Murco Obučina[2]

[1] International Burch University, Francuskerevolucije bb, Ilidza, 71210 Sarajevo, Bosnia and Herzegovina
sanela.klaric@ibu.edu.ba

[2] Mechanical Faculty, University Sarajevo, 71000 Sarajevo, Bosnia and Herzegovina

Abstract. In the age of accelerated advancement of new technology, innovative and contemporary materials, the application of ecological, natural and green materials is more than urgent need for reduction of pollution and climate changes mitigation. Wood is natural material, traditionally used in building sector from the beginning. In recent years, wood has rapidly developed into a high-tech construction material. Today we are witness of the multi-storey wooden buildings all over the world. Market for the engineering wood product (EWPs) is rapidly grooving every year.

Bosnia and Herzegovina (BiH) is very rich county with forest. The wood processing industry has been traditionally a well-developed industry in BiH. Regardless that the legislation and standards are not on place in BiH the forestry sector and wood industry today do have significant export results. New technologies in EWP should be improved and developed in existing or new companies in BiH, in order to ketchup with the increasing engineering wood market abroad. This paper will present recent innovations in wood industry, applications and market trends to open discussion between high education institutions, industry and business representatives, local government and civil society organizations for future sustainable and responsible development of this important sector in BiH and region.

Keywords: Engineering wood · Environment protection · Energy efficiency · Market · Innovation

1 Introduction

Wood is in nowadays used for complex and tall constructions with respectable dimensions or for multi-storey buildings in the cities all over the world. Wood has unique ecological qualities fulfilling the growing demand for the energy efficient use of resources in the construction process. Growing environmental awareness presents one of the most important advantages of wood as a construction material, where the choice is motivated by the fact that wood is a renewable material and that its use reduces CO2 emissions – provided that the raw material is harvested in forests where sustainable forestry is practiced including replanting and adoption of management plans [1]. The introduction of new products in the construction sector is generally met with hesitation,

I. Karabegović (Ed.): NT 2019, LNNS 76, pp. 712–727, 2020.
https://doi.org/10.1007/978-3-030-18072-0_83

low awareness, and high uncertainty in the marketplace; therefore, the communication of information is vital to market success [2]. Latest technical research brought the big improvements in fire safety, load capacity and noise protection issues. The healthy and carbon neutral material with all other advances of new technologies in engineering wood products, the application of the wood increased in all aspects especially in building sector. Wood as building material today has been considered as a substitute for conventional materials such as steel or concrete. All players in building sector such as contractors, architects or construction companies today considered the new technologies of the wood as renewable, durable, energy efficient and in many cases as lower costly.

Every cubic metre of wood used as a substitute for other building materials reduces CO2 emissions to the atmosphere by an average of 1,1 t CO2. If this is added to the 0,9 t of CO2 stored in wood, each cubic metre of wood saves a total of 2 t CO2. Based on these figures, a 10% increase in the percentage of wooden houses in Europe would produce sufficient CO2 savings to account for about 25% of the reductions prescribed by the Kyoto Protocol [3] (Fig. 1).

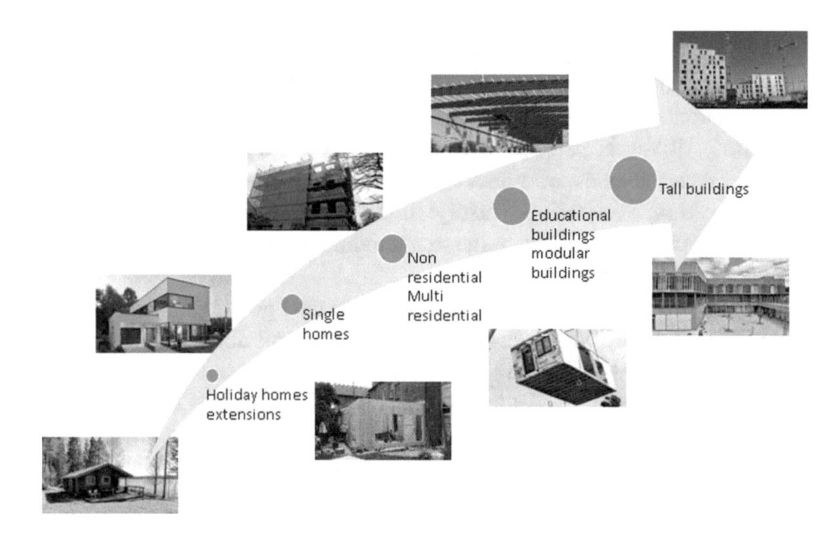

Fig. 1. Development dynamics in wood construction in last three decades

Wood on the best way fulfils all the requirements of contemporary architecture with many additional environment benefits. The building elements are prefabricated under weather-independent conditions at the production site. A high degree of prefabrication means extremely short erection times of only a few months and the construction sites are low-noise and clean [4]. Wood as constructive material is not only suitable for residential buildings but is also particularly useful for the construction public buildings such as kindergartens, schools, sport buildings, nursing homes and office buildings.

Around 56% of surface of Bosnia and Herzegovina (BiH) is covered by forest [5]. From the ancient time the forest resources have been the most important resource in

BiH with high economic value. Tradition in forest management and wood production is very long and strongly connected with the economic development. After the post war the bh forest sector has been recovering for the years to develop a modern processing wood industry and to arrange stable and sustainable management in forest. Most of the old and many new companies do see the perspective in the innovative technologies in this important sector in country.

2 Wood as an Engineering Material

With the development of technology, the use of wood has significantly improved. The development of new wood-based products has led to the elimination of strength-reducing defects in the wood structure, but also the negative effects of anisotropy and inhomogeneity of wood have been reduced. The application of "engineeringprinciples" has led to an increase in size and improved mechanical properties of wood-based products and permits new and broadened applications. The wooden "skyscrapers", i.e. the multi-storey timber buildings that have gathered momentum in European countries in recent years, is one example of such development. In order to improve in the future the physical and mechanical properties of the wood material and the products that it is used for, wood can be modified by chemical, thermo-hydro-mechanical or other modification actions. The wood modification area is undergoing huge developments at the present time, driven in part by environmental concerns regarding the use of wood treated with certain classes of preservatives. Several "new" technologies such as thermal modification, acetylation, furfurylation and different impregnation processes have been introduced successfully at the market and shows on the potential theses new technologies have.

One of the key features in wood processing is to change the dimensions of the tree to an engineered material with well-defined properties and dimensions adapted to the purpose for which it is to be used. Such technical well-defined wood products have widely been in development for the last century and nowadays they are commonly called EWPs - engineered wood products. EWPs are among the most important, innovative wood products used worldwide. EWPs, also referred to as reconstituted wood, wood-based products, or wood-based composites, are wood components in general and structural components for industrial use in the production of furniture, in interior and exterior joinery, and in construction of buildings and in infrastructural timber constructions such as bridges, towers, and pathways. The base of the EWPs is usually sawn timber, veneer, wood chips or fibres, which relate to adhesive to appropriately structures and shapes. EWPs have successfully replaced classic building materials such as metal, concrete, and bricks. Its advantage is in the high load capacity in relation to weight, proper dimensional stability and flexibility to the adaption to required shapes and dimensions of the structural element. EWPs have in general poorer mechanical properties than sawn timber in the longitudinal direction, but because of their more consistent properties in both planar directions and in the cross section of the composite, the safety margins in strength can be kept narrower than when sawn timber is used, Fig. 2.

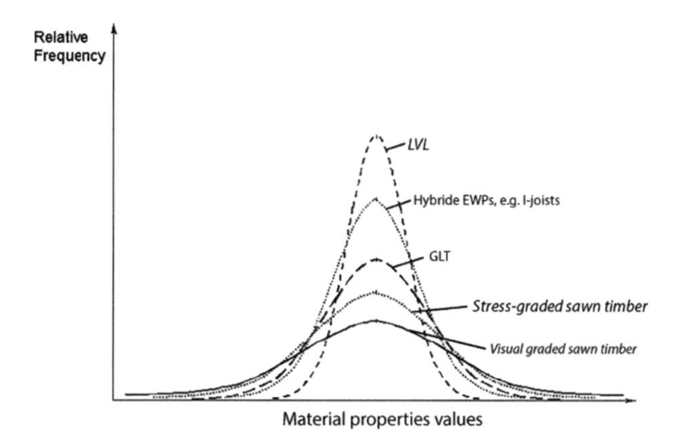

Fig. 2. Principal distribution and variability in mechanical properties of wood and different EWPs. LVL – laminated veneer lumber, GLT – glued-laminated timbe [6].

Figure 3 show the principal ways that are currently being used in industry to convert trees to wood products. After the forest is harvested, the roundwood are sorted into different classes depending on their intended use. To produce wood for construction purposes, sawmilling is today the dominant process, yielding sawn timber in well-defined dimensions as well as by-products such as bark, sawdust and chips. An alternative process is the production of veneers for subsequent use for board manufacture (plywood), elements for construction purposes (laminated veneer lumber) or the manufacture of moulded products (laminated veneer products) and high-density materials for interior and special applications (high-pressure laminated veneer). After the transformation of the roundwood in the sawmill, the sawn timber does not usually have the dimensions required in the final use, and a lot of effort is being made in the forest products industry to transform the sawn timber into dimensions and grades that suit the products requested by the consumers. Joining wood is a major step in these processes and, adhesives of different types are here the key component.

Trees or parts of trees that are not suitable for use in the sawmill or in veneer processes have, if used for industrial purposes at all, three main uses: for paper-pulp production, for the manufacture of wood-based composites or for energy conversion. These processes also use the by-products from the sawmill and veneer processes or residues from other wood-working industries, and in some cases also agricultural waste. The board industry produces a variety of wood-based panel products (oriented strand board, flake board, particleboard, hardboard, insulation board, low-, medium- and high-density fibreboards, cement-bonded board etc.) based on comminute wood in different sizes from long and thin flakes (veneer flakes) to fibre bundles that are commonly bonded together by an adhesive or by integral bonding achieved by an interfelting of fibres and in some cases by a ligneous bond. Other materials may be added to improve certain board properties. Wood plastic composites (WPC) are a rather new building material on the market, based on a thermoplastic matrix and a wood component. The matrix is usually recycled polyethylene or polypropylene, and the wood is sawdust or shaving residues from the wood industry [7].

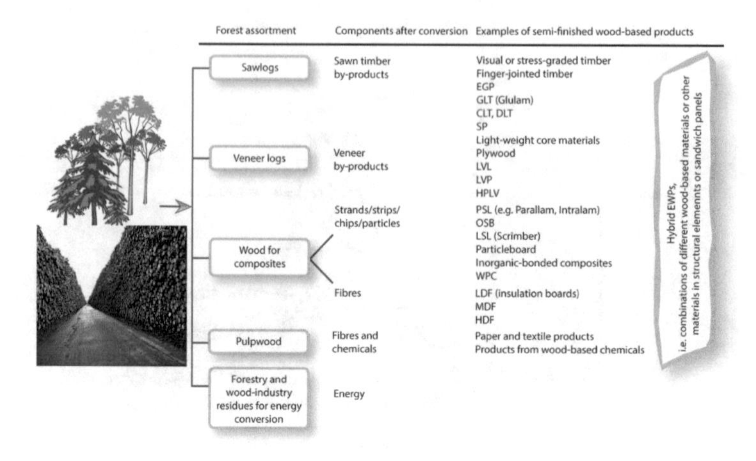

Fig. 3. The industrial use of the forest resource. Depending on species, dimensions and categories, the forest raw material is after harvesting sorted into different classes according to their industrial use. EGP – edge-glued panels, GLT – glued-laminated timber, CLT – cross-laminated timber, LVL – laminated veneer lumber, LVP – laminated veneer products (moulded products), HPLV – high-pressure laminated veneers, LDF/MDF/HDF – low/medium/high-density fibreboards, PSL – parallel strand lumber, OSB – oriented strand boards, LSL – laminated strand/scrimbed lumber, WPC – wood plastic composites [8]

The construction of environment-friendly and energy-efficient buildings, labelled with the term green building, is an on-going global trend. Different types of innovative EWPs are used in the construction of green buildings resulting in new opportunities to the forest products industry, Fig. 4.

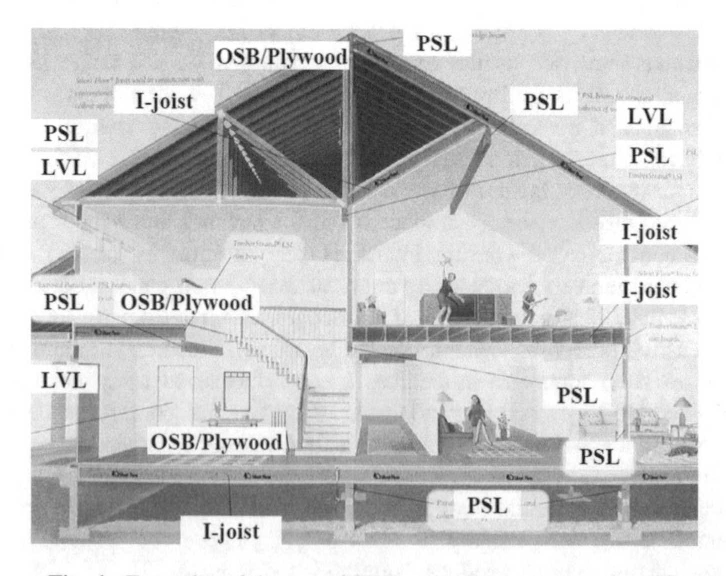

Fig. 4. Examples of the use of EWPs in building construction [8].

A constant demand for wood and wood products is, on the other hand, present on the world market, and a balance in wood supply will be difficult to hold already in 2020 on the European level. In the period from 2030 to the 2040, the missing volume of roundwood could be as large as 50 million m^3 per annum [9]. This in combination with a global increased environmental concern the development of new innovative EWPs is important. In such a context EWPs have two important advantages in construction: (1) positive aspects a structural material which include its strength, environment-friendliness, simple handling and appropriateness for industrial use, (2) a high total raw-material utilization from forest to finished products, provided that most by-products from production is utilized in new products.

There are also numerous challenges associated with the use of EWPs, especially in construction of timber buildings, and these challenges are best met through further research and more pilot projects to increase the knowledge of life cycle costs, construction costs, maintenance costs, acoustics, through the general increase in the number of timber buildings that are being erected. As a first step, this chapter will give an introduction to the most important characteristics of sawn timber and the most common EWPs in the market.

Forestry and forest-related industries have never before been as focused as they are today in discussions regarding the major challenges of the future. Instead of utilising the earth's limited resources, we have to use renewable materials, fossil fuels must be phased out and individual consumption must to a greater degree reflect the concerns for the climate and care for the environment. In this context, the emissions of carbon dioxide have been in focus for a long time. One way of reducing the emission of carbon dioxide is to use a greater proportion of wood products and to increase the life of these products so that the carbon is bound over a longer period of time or replace energy-intensive materials with wood-based products [6].

Although wood is a natural material, bonded wood products have caused some environmental concern related to formaldehyde and other volatile compounds in the adhesive formulations used in bonded wood-based products. Heating increases the problem, as it raises the vapour pressure of reactive chemicals. Isocyanates can react rapidly with compounds in the human body. Both Emulsion Polymer isocyanate adhesive (EPI) and polyurethane (PU) adhesives contain isocyanates. The preparation and processing of epoxy also poses significant health risks in the form of e.g. allergies. There are also legal requirements which aim to depress the levels of emission especially of free formaldehyde.

The foremost area of environmental concern with regard to adhesives has been formaldehyde emissions from bonded products, mainly those using urea formaldehyde (UF) adhesives. Products bonded with UF-based adhesives such as plywood, MDF and PBs are often used e.g. in kitchen joinery and furniture and they may therefore lead to an increase in the level of formaldehyde in the indoor air. Formaldehyde can also be cleaved from acid-curin ginks and varnish hes. Formaldehyde can react with biological systems in reactions similar to those that are used for the curing of adhesives. The problem can arise from both un reacted and generated formaldehyde. Un reacted formaldehyde is a problem during the manufacturing operation and in freshly produced composites, but formaldehyde emissions from composites decrease with time after production. Formaldehyde can also be generated by the decomposition of some formaldehyde copolymer adhesives, in particular the UF adhesives. These adhesive bonds are more prone to hydrolysis, generating free formaldehyde.

The greatest concern about formaldehyde is with PB, due to the large volume of indoor usage and the large amount of adhesive in the product. The particle board industry has therefore together with adhesive manufacturers over a long period focused on reducing for maldehyde emission from the boards. Figure 5 shows there duction in formaldehyde emission from particle board sin Europe from 1970 to 1990.

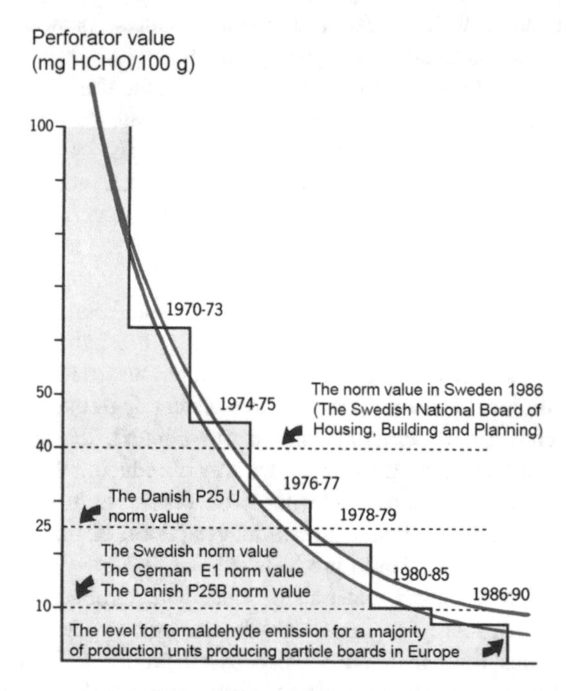

Fig. 5. Change in content of cleavable formaldehyde from PBs in Europe during 1970 to 1990, measured as perforator value, expressed as mg formaldehyde per 100 g dry material. The perforator method is described in the standard EN-120 [8].

The emission classifications for boards according to the formaldehyde emissions is:

- E1 (≠ 0.1 ppm = 0,124 mg/m ≠ air or 8 mg/100 dry plate,
- E2 (between 0.1 and 1 ppm), and
- E3 (>1 ppm)

Producers and customers should prefer boards classified E1to minimize the negative impact on human health. In 2004, IARC (International Agency for Research on Cancer) classified formaldehyde emission higher than 7 mg/m^3 as proven carcinogens.

There are many materials in the construction and housing industries that compete with wood, e.g. steel and concrete for frames and large constructions, bricks for walls and facades, and PVC and other plastics for windows, building features and furniture. For wood as a material to be competitive against other materials, the wood's environmental advantages alone are not sufficient, i.e. that wood shows lower emissions of carbon dioxide according to calculations based on LCA criteria. Wood must also be competitive for its technical qualities, show a high material utilisation during further processing, and not least, show a competitive economic yield during usage.

3 New Trends in Wooden Architecture

Wood is a resource that is locally available in sufficient quantities all over the world. It also requires the low amount of energy for its production, transport, maintenance and disposal. Many of decision makers together with architects and wooden industry start to experiment, to search and to test wooden materials within different designs and technologies. Engineering wood and prefabricated wooden construction become challenge for many stakeholders who are competing today for the best solution. It is important to mention that wood is not only optimal material for new constructions but with its small mass, timber represents an ideal building material for reconstructions and adding additional storeys because strengthening of the bearing structure is often not required. Wooden buildings are durable with a pleasant indoor climate and aesthetic qualities and many more.

Nine store building Murray Grove in London, build in 2009, is the first tall urban housing project to be constructed entirely from pre-fabricated solid timber. The nine-storey tower was built from wooden load bearing walls and floor slabs to the wooden stair and lift cores.

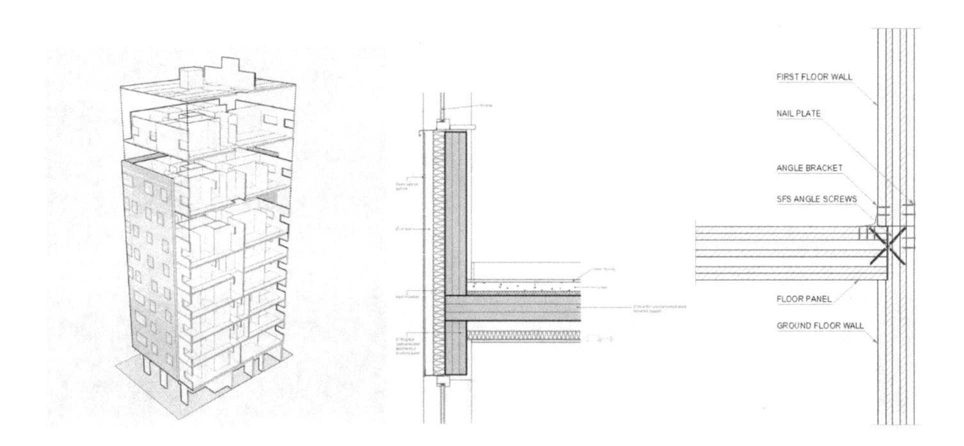

Fig. 6. Section showing honeycomb structure and details of H/V connection, [10]

As Fig. 6 shows, the nine-storey tower is built from a tight honeycomb of structural panels, with a timber core providing stability. The inset balconies with structural balustrades was designed to strengthen the outer structural wall. Cross-laminated timber product from KLH represent the system of horizontal beams and vertical structural wall boards is manufactured from spruce grown in sustainable forests. Another unanticipated benefit was that the use of timber sped up the construction process, allowing the entire structural frame to be completed by four people in 27 days, using little more than a portable crane and handheld electric screwdrivers (Fig. 7).

Fig. 7. Photos from the site, [10]

Murray Grove building in London present one of the pioneering innovative vision supported by UK government to insure the EU 202020 strategic goals especially UK government goals that was even more ambitious then EU. Today many of wooden building exist in the UK as well as the new ones are in the development or planning phase (Fig. 8).

Fig. 8. Murray Grove building in London today [10]

Today in Vienna the HoHO building is under construction. The idea for all stakeholders involved was the additional noticeable improvements and new tangible experience of the wooden EWP element in the world's tallest wood structure today. RLP Rüdiger Lainer + Partner developed a wood tower concept HoHO office building as an modular structure that allows individuality, as very flexible space that could be modified at any time by user needs. This building has 24 floors and it is cca 84 m height.

This wooden building has been designed according to the TQB (Total Quality Building) evaluation system of the ÖGNB that required efficient use, fire protection and strong structural planning standards. The deliberately simple construction system relay of four prefabricated serial components: columns, main beams, deck slabs and facade elements. The base surface of wooden composite ceilings, which are based on wooden supports in the final facade layer are attached to the core load-bearing structure of reinforced concrete (Fig. 9).

Fig. 9. HoHO building details of combination of concrete and wood prefabricated elements and final facades; [11]

The timber construction system provides high efficiency in terms of thermal insulation and serviceability. Wood-composite floors are secured to the central concrete supporting cores and extend out to the building edge. These floor panels are supported by a wooden column system around the outline of the building. This structure then supports pre-fabricated external wall modules that combine solid wood panels with an 'earthy' concrete shell to form the building's facade (Fig. 10).

Fig. 10. 3D model [12]

In the high-rise building in Aspern's Urban Lakeside of Vienna, the use of wood provides a sensual experience. The prefabricated external wall modules permit a sculptural and richly varied facade design.

The eco-school being built in the heart of Montreuil, Paris, France represent the new technology of prefabricated panels made of wood and straw. This is a new green technology that will increase energy efficiency and eco-friendly solution not only in France but all EU. Montreuil has the ambition to be one of the largest eco-neighbourhoods in Europe, an example of ecological and social policies coming together in the transformation of urban space (Fig. 11).

Fig. 11. Eco school during building time and as final, [13]

Straw is a well-known insolation natural material as a waste after harvesting of the different cereals on the agriculture fields. In line with environment responsible building innovations with natural materials as a part of prefabricated elements become very popular among scientists and entrepreneurs. Natural materials as straw, sheep wool, wood fibre and other recycling textiles and cotton show great results in regards law primary energy and the emission. Prefabricated elements made of these materials are very durable, energy efficient, healthy and economic (Fig. 12).

Fig. 12. Montage of the prefabricated elements on site, [14]

Today in Montreuil, Paris, France this is not the only one eco building. This community decide to invest more in comprehensive eco solution for water, air and soli utilization and use and many other green city solutions and measures (Fig. 13).

Fig. 13. Eco school today [14]

Those three mentioned projects are examples of innovative approach to wood and EWP in combination with some common materials as concrete or very innovative approaches in combination with natural waste such as straw. All of them are very successful pilot projects with many advantages. Innovative technologies of wood and other natural materials is present the most valuable feature of chosen project as motivation for all future innovative approaches to building sector as the most polluted one on the Earth.

4 Perception of the Engineering Wood by the Architects

All players in the building sector do have significant influence in the new development and trends of the building materials. The architects are the key decision makers in building sector especially in the selection of materials in the construction sector. In addition, some studies found architects to be environmentally conscious specifiers of construction materials, and, as such, are an important target group for research particularly as environmental and sustainability issues are becoming more salient issues generally [15].

Slovenia is in advance when we are talking about national initiatives to promote wood material in building sector. There are also initiatives in Croatia and BiH, but primarily at the local level, mostly related to wood-cluster development in Croatia, and initiatives by local architects who are willing to explore the advantages of using wood to develop unique designs for residential and tourist buildings in BiH. Wood construction in Slovenia is transitioning from a formative to a growth phase, while in Croatia, BiH and Macedonia it is still in the formative phase [16].

Conducted study of the architect's perception of EWP in Slovenia, Croatia, BiH and Sweden show that there is an general positive perception in terms of familiarity with different EWPs in all 4 countries. The most common applications are structural (GLT, SWP, PW, OSB) (Fig. 14).

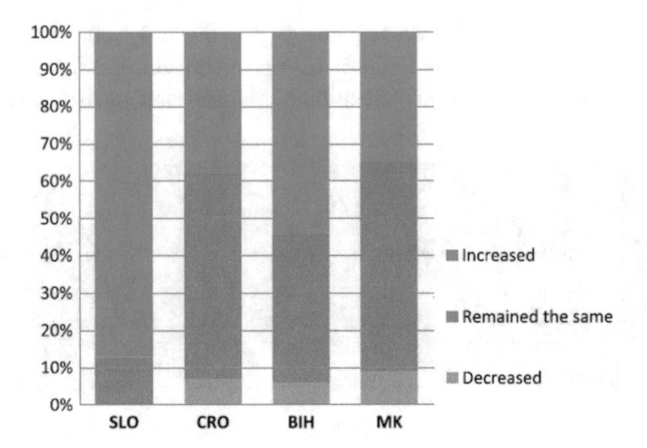

Fig. 14. Responded perception about usage of EPWs in last 5 years in Slovenia, Croatia, BiH and Macedonia, [15]

Due the strong tradition link to the wood industry in all countries when we are talking about architects' opinion on whether there has been increased use of new EWPs in the last 5 years, except Slovenia the majority agree in Croatia, BiH and Macedonia that EWP use has remained the same. The research findings show that all three countries are facing with general a lack of knowledge of the principles of design options in using timber for structural performance. Most architects do have experience with wood in residential buildings linked to interior design and flooring but no experience with tall wooden structure, prefabricated wood and EWPs in building construction.

From the other side when we check the export figures of sawn timber on the market, we could recognize a lot of space for the improvement of this picture in the future. Second bigger export in BiH after furniture is a sawn timber of 36%. Prefabricated house is 4% [5] only and should be a platform for more innovative and sophisticated technology application in BiH in order to increase added value export products such as EWP. Sawn timber has been losing its historical dominance now when EWP increasing the use and number of applications in the building sector. New figures for EWP market demand should influence all stakeholders in those countries to ketch up with innovation technologies in wooden sector and to change export figures as soon as possible.

Architects could be good advocators for an increased use of EWPs but increased efforts must be placed to increase their knowledge. Architects could be driving force but in strong collaboration among representatives of the wood industry, builders and housing associations, including contractors and civil engineers, to better understand the technical and the business potential of wood as a multi-purpose building material.

All mentioned will contribute to an understanding of the probability that innovative bio-based building materials with minimal environmental impact chosen in residential and non-residential buildings, as well as high-rise buildings, and to a greater understanding of the drivers of and barriers to increased use [15]. Increased use of EWPs and other bio based natural, renewable materials is very important element of a more sustainable future built environment. Regional cooperation is also very important to support this sustainable and responsible development on Balkan.

5 Wooden Construction Market

The global sustainability concerns along with the persistent energy crisis are among the major factors that are expected to propel market growth over the years to come. The transition from concrete-and steel-based construction to wood building materials is set to actuate demand from the market over the next eight years.

Significant attributes such as high insulation, promising thermal performance, as well as high chemical resistance, are projected to influence the market positively. In addition, the advantages such as low cost and ease of construction offered by glulam relative to other building materials have made it affordable for all consumer segments.

The demand for wood-based residential buildings, including single-family homes and multifamily apartments, is increasing due to their aesthetic appeal and design flexibility. Various benefits, such as improved fire resistance and durability, of the product are expected to augment market growth over the forecast period. Residential and commercial segments are expected to collectively account for around 80% of the global market revenue by 2025 [17]. The growing demand from these segments is attributed to the low cost of construction and easy maintenance of EWP. Innovations in building and design technologies are projected to boost the demand for EWP over the forecast period.

Easy availability of timber in forests of Central European countries and the presence of advanced timber processing industries are the prime factors responsible for market growth in Europe. Beside Europe countries like Canada, U.SA., Japan, Australia, and China are rapidly developing markets for EWP as well. Germany and Italy cumulatively accounted for around 60% of the Europe market revenue owing to the high acceptance and growing consumption of wood as a building material. Austria, being a rich source of softwood spruce and pine, is the largest production of glue laminated timber across the globe. Staretagies, EU programs and projects together with Climate changes urgent requirement all together represent huge driving force for the EWP rapidly growing market today. BiH in the EU integration process should harmonize all legislations and Strategic plans to ketch up with this economic opportunities.

6 Findings and Recommendations

Wood has been experiencing a rebirth as a contemporary construction material thanks to general public's increasing environmental awareness, Climate changes mitigation needs and great interest of energy efficiency in EU. Wood industry focused on innovative technologies in timber sector continuously increasing with new innovative products and solutions. Many countries harmonize all legislations and standards based on latest researches and improvements of environment, safety, health, energy efficiency and other important building required conditions.

Traditional wooden industry represents great assets for application of innovative technologies on the core existing infrastructure. Market demands for wood as building construction material should be driving force for countries rich with forest and tradition in wood industry. Countries in the region should develop the responsible strategies for the EWP development by ensuring economic competitiveness, exports of the added values of the products, self-sufficiency in energy, efficient management of the natural resources, investment in science and innovations, local know-how and services, branding and marketing.

With the development of the innovative technologies, the use of wood significantly improved. The application of "innovative engineering principles has led to improvement of material properties and broader application. Future improvement of physical and mechanical properties wood could be more modify to future construction and environment demands. EWPs have successfully replaced classical building materials such as concrete, metal and brick. EWP advantage is in their high load-supporting capacity in relation to their weight, their proper dimensional stability and their flexibility for adaptation to the required shape and dimensions of the structural building element.

All stakeholders in building sector should cooperate. Architects as important ones in all steps of the developments and design should be educated to be main driving force in the future development. Very important is to harmonize the curriculums in the schools and high education institutions in line with EU standards and curriculums as well as with all Climate changes mitigation needs. Innovative technologies focused on the green jobs, green economic growth, better quality of life, green jobs, green budget reform, green procurement and education for green development will ensure sustainable and responsible development of country.

7 Conclusions

Wood is usually perceived as traditional material. The properties of this materials with improvements made by innovative technologies made it possible to design free shape and highly complex and multy store tall buildings. Thank to new innovative technologies and digitalization timber construction industry is experiencing significant innovations in prefabrication and eco-friendly technologies and structural application. Long term cooperation programs and strategies including long term education and training planning and promotional campaigns are needed to make this possible to use EWP products in residential and non-residential architecture with no limitations.

There are numerus challenges associated with the construction of timber buildings which will be the best met through future innovative scientific experiments and researches as well as pilot projects and new EWP products. Improvement of those products and constructions should be made based on joint efforts and findings with focus on life cycle costs, primary energy and emission reductions. There are also the opportunities for future development and trends in high prefabrication and combination of renewable materials.

BiH as country traditionally linked to forestry and wood production facing a lot of challenges today. Position of the BiH and all existing infrastructure and efforts of all stakeholders are crucial for future innovative, sustainable and responsible development of this important sector. Additional value in value chain should be our focus in all future planning and actions. BiH could ketch up with neighbour countries in the region to collaborate in common aim to find place on the demanding and increasing EWP market in EU.

References

1. Ramage, M.H., Burridge, H., Busse-Wicher, M., Fereday, G., Reynolds, T., Shah, D.U., Wu, G., Yu, L., Fleming, P., Densley-Tingley, D.: The wood from the trees: the use of timber in construction. Renew. Sustain. Energy Rev. **68**, 333–359 (2017)
2. Robichaud, F., Kozak, R., Richelieu, A.: Wood use in nonresidential construction: a case for communication with architects. For. Prod. J. **59**(1/2), 57–65 (2009)
3. Guy-Quint, C.: Tackle Climate Change: Use wood. European Parliament, Brussels (2006)
4. ID: WOOD Clustering knowledge, Innovation and Design in the SEE Wood Sector, thematis doddier: Construction (2015). www.idwood.eu
5. Obučina, M., Kitek Kuzman, M., Sandbeg, D.: Use of Sustainable Wood Building Materials in B&H, Slovenia and Sweden, University Sarajevo, p. 15 (2017). ISBN 978-9958-601-65-1
6. Sandberg, D.: Additives in wood products – today and future development. In: Kutnar, A., Muthu, S.S. (eds.) Environmental Impact of Traditional and Innovative Forest-Based bioproducts, pp. 105–172. Springer Science + Business Media, Singapore (2016)
7. Carus, M., Gahle, C.: Injection moulding with natural fibres. Reinf. Plast. **52**, 18–25 (2008)
8. Obućina, M., Kitek, K.M., Sandberg, D.: Use of sustainable wood buildings materials in Bosnia and Herzegovina, Slovenia and Sweden, Mašinskifakultet Sarajevo, p. 216 (2017). ISBN 978-9958-601-65-1
9. Trischler, J., Sandberg, D., Thörnqvist, T.: Evaluating the competition of lignocellulose raw material for their use in particleboard production, thermal energy recovery, and pulp- and papermaking. BioResources **9**, 6591–6613 (2014)
10. Offcial page of Wood Institute Austria PROHOLZ. http://www.proholz.at/zuschnitt/33/holz-in-the-city/
11. Official page of architect office. http://www.woschitzgroup.com/en/projects/hoho-vienna-wooden-tower/
12. Official site of magazine of architecture. https://www.e-architect.co.uk/wp-content/uploads/2016/10/timber-tower-hoho-r111016-5.jpg#main
13. EU association of straw bale architecture. http://esbg2015.eu/urban-strawbale-school-in-montreuil-paris/
14. Official news site of the Municipality Montreulion France. https://montreuilonthemove.wordpress.com/2013/04/17/ecolution-montreuil-city-green/
15. Kuzman, M.K., Klarić, S., Barčić, A.P., Vlosky, R.P., Janakieska, M.M., Grošelj, P.: Architect perceptions of engineered wood products: an exploratory study of selected countries in Central and Southeast Europe. Constr. Build. Mater. **179**, 360–370 (2018)
16. Kuzman, M.K.: Wood in Contemporary Slovenian Architecture 2010–2015, University of Ljubljana, Biotechnical Faculty, Department of Wood Science and Technology, Ljubljana (2015)
17. https://www.reportlinker.com/p05291684/Glue-Laminated-Timber-Market-Analysis-By-Application-By-Region-And-Segment-Forecasts.html

The Lack of Techniques Used in Management as a Factor Causing Delays in Construction Projects in B&H and Other Delay Factors Detected

Ahmed El Sayed(✉), Emad Mamoua, and Adnan Novalić

Faculty of Engineering and Natural Sciences, International Burch University, Ul.Francuskerevolucijebb, Ilidža, 71210 Sarajevo, Bosnia and Herzegovina
ahmed.elsayed@ibu.edu.ba

Abstract. The success of any construction project is measured by covering the scope and quality and finishing on time and budget. Many factors affect the delays in construction projects and this topic is the subject of many researches worldwide. Generally, factors can be sorted into those that we can predict and avoid them happening, and others which we cannot predict and prevent. This paper investigates the factors affecting the delays in construction projects in Bosnia and Herzegovina, and to evaluate the weight of them. Also, this paper explores the usage of management software in the planning and execution phases of projects, and up to which level of precision the inputs are determined. Quantitative research method, based on collecting surveys from different type of construction projects participants, such as designers, contractors, and project managers, was used. Number of projects analysed in this research is 132 projects finished during the past 5 years. Project size varied from single family house and settlements or high-rise buildings. This paper found that inadequate use of management software, along with the issues related to contractual relations, scheduling and estimations, changes in demands by investors, lack of adequate equipment, and shortage of skilful manpower and adequate materials are main reasons causing delays in construction projects in B&H. This paper concludes that different approaches should be used when it comes to the planning and scheduling phase of projects, and proper use of software with precise determination of inputs is needed in order to reduce potential lateness of project delivery.

Keywords: Construction projects · Delay factors · Management software · MS project · Primavera · Contractual relations

1 Introduction

A successfully finished project is the one that has achieved the scope set, in accordance to the quality defined, within the time frame, and within the budget prepared [1]. However the success of projects depends also on the user's satisfaction, constructions laws and regulations of country, and other stakeholders' expectations [2]. Delays in construction, according to Karake-Shalhoub [3] is a problem that faces all countries,

I. Karabegović (Ed.): NT 2019, LNNS 76, pp. 728–735, 2020.
https://doi.org/10.1007/978-3-030-18072-0_84

and occurs generally in all construction projects, but the size of it varies from project type, country, company, and other factors. Many researchers investigated the factors affecting the delays, and stated that it is highly important to find out and determine the main reasons of delay in order to develop measures to reduce their impact. Mansfiled, Ugwu, and Doran [4] suggested that the process of construction depends on some predictable and other unpredictable factors. In general, it can be stated that cost, time, quality, and scope are affected mainly due to inefficiencies in project phases, and leads directly to the dissatisfaction of investors.

In Thailand, a study was done in 1996 that explored obstacles and challenges affecting the construction industry, and sorted them into the next 3 categories: resources and materials supplement issues, client and consultations issues, and issues related to the inadequacy of contractors and subcontractors [5].

Another study done in Hong Kong by Chan and Kumaraswamy in 1997 [6] and concluded that reasons causing delays differs depending on the point of view, as the reasons listed out by contractors didn't match those that were listed out by investors or designers and vice versa.

In a study that focused on the Nigerian construction industry projects, results proved that poor material and resources management, poor contracting skills, and disrespecting payments by investors are key factors causing delays [4]. Another study in Nigeria done by Aibinu and Jagboro [7] and concluded that both cost and time overrun of projects were the biggest delay effects in the construction industry.

Mezher and Tawil [8] run a survey to analyse factors causing delays in construction projects in Lebanon, and they found that financing issues with investors, contractual poor skills, and project management skills are among key factors.

In a study [9] made in Ghana in 2003, authors found that contractors and consultants highlighted the impact of monthly payment difficulties on delays in construction projects, while investors and users blamed the poor management skills of contractors for Delays. However, the other factors causing delays, such as lateness in material procurement and variation and escalation of materials prices during the project, were confirmed by all participants.

Battaineh [10] in 1999 and Al-Momani [11] in 2000 analysed the reasons of delays in Jordanian construction projects. They concluded that delays in Jordan occur almost in every project. For instance, Battaineh [10] found that the actual time for finishing infrastructure projects in Jordan is 160.5%, while for buildings is 120.3%. Al-Momani [11] found that the poor designs, weather and climate conditions, changes in demands made by investors, poor site conditions, disrespecting the payments and financial issues by investors, and poor planning and determination of work quantities are main reasons for delays in Jordan.

Mwandali [12] in 1996 investigated the factors that have an effect on the Kenyan Railway construction projects, and among his findings he stated that poor communication and project management skills, poor purchase planning and procurement, unskilled managers, low level of workers motivation, and weak monitoring and controlling systems are leading factors causing delays in Kenya.

In Saudia Arabia, Assaf and Al-Hejji [13] explored the causes of delay in construction industry projects that are of bigger size. They have found that the average delay is varying between 10% and 30% of the planned duration. Also, they have found that 70% of project were finished with certain time overrun. Users listed the labor-related factor and the contractor-related factors as main reasons for delays, while contractors listed the investor-related factors and consultant-related factors as the main reasons for delays. However, in a combined analysis, the Owner and contractor related factors were sorted as the biggest cause of delays, with poor designs and unskilled workers following with high impact on delays.

In order to achieve a successful construction management in site, a comprehensive planning is needed, as well as very efficient layout of facilities at the site [14]. As initial site drawings and plans made during the planning phase of the project are not displaying the construction progress, there is always a need to add manually progress to the site drawing, which cause delays or misunderstandings [14]. In 1990, Retik et al. [15] analysed the possibility and feasibility of introducing the graphics made in computers in the process of planning and scheduling the construction work, and managed to highlight some need functions to be added in newly designed software. Zhang [16] went a step forward by developing reports in a 3D graphical construction model. Later on, a 4D model that is adjustable to user's demands was developed by Willams [17] in 1996, with the purpose of planning the construction progress on the basis of simulation, communication, and visualization. In the same year, Collier and Fisher [18], Mckinney et al. [19] contributed to the same direction of development with 4D Computer Aided Design Tools. In 1998, McKinney and Fisher [20] analysed the hybrid 4 dimensional application regarding its effectiveness by using software such as Primavera, AutoCad and others. Another progress was made in 2001, when Kamat and Martinez [21] managed to present a three-dimensional model that can describe and delineate the activities of construction project process.

2 Methodology

The Aim of this paper is to explore factors causing delays in the Bosnian and Herzegovinian construction industry, and to explore the management software used in the phase of planning and determining the time needed for having the job done. This paper also aims to sort the delay factors according to their impact and frequency, and to recommend potential solution based on using advanced technologies for avoiding delay factors that are predictable, and reduce the effect of the unpredictable factors.

Authors set several research questions: How significant are delay factors, related to contractor's capabilities and skills? Are delays caused by the adequacy of equipment used for construction significant? Does the level of workers motivation affect the delays in construction? Is the shortage of skilled labourer in construction companies a delay factor in Bosnian Construction Industry? Do issues related to dishonouring the payment certificates by investors affect the delays significantly? Has the design and its quality an impact on delays, and how significant is it? Are contracts between involved parties precisely defined, or the issues related to poor contractual relations are playing big role in delays? What type of tendering is used, and what effects are related to the tendering type? What software is used for scheduling and planning of construction in B&H?

The target population selected for this research included construction companies from B&H that have been actively participating and involved in the construction process in the previous 5 years, with no less than 2 medium sized projects (with investment value between 5–15 million BAM) realized yearly, or one big size project annually. Selected companies have realized total of 132 projects in B&H in the past five years.

Quantitative research methodology was used in this paper, based on a survey divided into four sections, and delivered to selected companies. The first section is related to general information about the company, defining its size, number of projects realized in the previous 5 years and their sizes, total years of existence, types of projects realized. The second section is related to performance evaluation, where the information related to number of projects realized within the budget and time frames were given. Third part deals with the construction management software in use. Final part reports the importance of each delay factor suggest by authors and based on literature review.

The questionnaire included delay factors related to investors, contractors, design and designers, consultants, construction workers, equipment and materials, and project management tools. Two main categories related to the delay factors were set by authors: Frequency level and Severity level. A system of ranking between 1 and 4 was used for factors. In the category of Frequency, value 1 in the scale referred to *Never*, 2 meant *Occasionally*, 3 was for *Frequently*, and 4 referred to *Constantly*. In the category of Severity, number 1 in the scale meant *No Effect*, 2 referred to *Fairly Severe*, 3 stood for *Severe*, and 4 referred to *Very Severe*.

As for evaluation of results gained, two main criteria were used: Number and size of projects realized by company, and the frequency and severity of factors. In the first stage, the value on the scale for each factor was multiplied by the number of projects realized by the company filling the questionnaire in order to add the size of the company to the results. The next step considered the value of factors divided by total number of projects. If the value is between 2.5 and 3, it is considered that the delay factor has recognizable impact. Values 3 and higher refers to factors with significant impact on delays in construction industry.

3 Results and Discussion

Total of 7 contractors responded to the questionnaire, with minimum number of projects of 10, and maximum of 70. Different size of projects was included, but all projects were above the minimum acceptable criteria of this research, which is medium size. In Table 1, factors that were found as recognizable or significant are listed, with detail analysis of their impact.

The frequency of occurring was sorted out first, as all delay factors that were considered recognizable or significant were listed, and then the severity of these factors were listed to show the weight of the impact these factors have.

Some factors, with high frequency, had a severity value under the recognizable value (2.5/4). Such factors were considered as non-significant, as they happen frequently, but without any recognizable impact on the delay.

Table 1. Recognizable and significant delay factors

Delay factor	Value of frequency	Value of severity	Delay factor	Value of frequency	Value of severity
Delay in delivery of materials by supplier	2,841	3,409	Shortage of skilled workers	2,871	3,530
Worker productivity and motivation	2,879	2,894	Insufficient number of equipment	2,909	2,909
Inadequately modern equipment	2,583	1,811	Delay in honouring payment certificates	3,333	3,121
Slow decision-making by the Investor	3,053	3,826	Client initiated variations	3,174	3,667
Necessary variations	2,856	3,477	Mistakes in soil investigation	2,705	3,250
Poor design	3,689	3,614	Delays in drawings delivery	2,500	2,477
Delays in changes ordered by the Investor	3,212	3,212	Investor's interferences in the construction	3,091	2,985
Changes in the scope of the project	3,606	3,439	Delays in obtaining permits	2,826	3,432
Insufficient communication between parties	2,773	2,902	Poor professional management	2,780	2,833
Delay by subcontractors	2,811	3,303	Inaccurate time estimates	2,924	3,000
Inaccurate cost estimates	2,871	3,091	Unreliable Subcontractor	2,742	3,439

In case that a factor had a low rate of frequency and high value of severity, this factor also were excluded from the category of significant factors. Such factors occur rarely, but could have recognizable impact on the delay. However, due to the low value of frequency, it can be stated that they are mainly avoided, and there is no need for further investigation.

Only factors with high rate of frequency and severity are taken as significant factors causing delays in Bosnian and Herzegovinian Construction Industry. If majority of contractors agreed about their frequency, and the impact they have on construction whenever they happen, that refers to their importance and danger, and therefore authors highlighted them in the conclusion of this paper.

Total of 22 delay factors were found with recognizable frequency, and 7 of them were highlighted with significant frequency. However, among these 22 factors, two are excluded due to their low rate of severity, and they are "Delays in drawings delivery" and "Inadequately modern equipment".

Delays in honouring payment certificates by Investors proved to be among the significant delay factors, as it appears frequently in construction industry in B&H, and has a significant impact on delays, as Contractors stop working or reduce the effort when the payment certificates are not honoured timely. Contractors, in the planning phase, develop a financing schedule for the project, and it is generally depending on the payment certificates agreed with Investors in certain dates.

In the construction process, decisions need to be made timely by all participating sides, and if only one side is not responding on time, the process is slowed down and a delay is caused. The factor related to the quickness of Investors respond and decision-making proved to be significant, as it occurs frequently on sites, and has highly significant impact on delays.

Any kind of changes during the construction phase causes certain delays, especially when is made by client. The factor was considered as significant, according to the contractors participating in this research. In case that Investor decide to change the scope of project, that will have a huge impact on delivering works on time, and will cause a serious delay in the construction process. However, if the scope is reduced, the project could be finished earlier, and no delays are expected.

Poor designs, which are not adopted to the construction site, or with low level of details, are among the leading factors causing delays. It is clear that the Design is the main document for all further steps in construction projects, and any mistake, or incompleteness, will surely cause delays, as there will be a need for adopting new design, and agreeing new costs. Poor designs also affect the preciseness of the bill of quantity, and therefore the agreement between Investors and Contractors. With the rate of 3.689 of 4 regarding frequency, it can be stated that this problem is happening very often, and with 3.614 out of 4 regarding severity, it is clear that this issue has huge impact on delays.

When there is a need for changes, Investor is issuing a change order to designers to prepare the new solution. It is quite often that Designers at that time have moved on to other projects, and therefore they propose the changes with lateness, due to the time needed to get back into the project. It is proved that this factor is causing a significant delay and that it is quite popular issue regarding the frequency of appearance.

Factors, such as "workers motivation and productivity", "Shortage of skilled workers", and "Delays in obtaining permits" have a significant impact on delays too, but are not occurring frequently as other factors considered as significant in this research.

Contractors included in this research mainly use Microsoft Project as project management tools. 3D and 4D management tools, which can run real-time situations and follow the progress of construction are out of use, as none of contractors are using any of them. It seems that contractors are relying on basic software, and not willing to upgrade their management skills, or to offer a more comprehensive view of the construction progress. Among delay factors, the poor communication between participating parties were mentioned, but it seems that it is not caused only because of Investors or Head Engineers, but also due inability of Contractors to present the real progress of works using advanced technological models.

4 Conclusion

In this paper, authors investigated a wide range of delay factors in construction industry worldwide, and analysed the frequency and severity of them in Bosnian and Herzegovinian Construction industry. Using quantitative research methods, and with total of 132 projects analysed in the past 5 years and done by different construction companies, authors managed to find out the significant factors causing delays, and to highlight their frequency and severity.

Total of 6 factors proved to be significant, and they are: Delay in honouring payment certificates by Investors, Slow decision-making by Investors, Client Initiated variations, Poor designs, Delays in preparing changes ordered by Investors, and Changes in the scope of projects.

The lack of advances in the field of construction management is observed, as contractors tend not to use any advanced construction management tools to follow the progress of construction, and are opting to use generally MS Project or Primavera. With having 3D and 4D models for following the construction process available widely, it can be concluded that, at this stage, the construction industry in B&H is way behind the modern trends. Investors appreciate when they can follow using software and tools the daily progress of works, and to be able estimate themselves the required time to finish the job. A better communication between participating parties could be achieved by introducing new project management tools in construction industry.

Finally, authors recommend and urge running wider studies in this field, in order to approve the results of this study, and to find more factors that cause delays in B&H construction industry.

References

1. Atkinson, R.: Project management: cost, time and quality, two best guesses and a phenomenon, its time to accept other success criteria. Int. J. Project Manag. **17**(6), 337–342 (1999)
2. Ibironke, O.T., Oladinrin, T.O., Adeniyi, O., Eboreime, I.V.: Analysis of non-excusable delay factors influencing contractors' performance in Lagos state, Nigeria. J. Constr. Dev. Ctries. **18**(1), 53 (2013)
3. Karake-Shalhoub, Z.: Trust and loyalty in electronic commerce: an agency theory perspective. Greenwood Publishing Group (2002)
4. Mansfield, N.R., Ugwu, O.O., Doran, T.: Causes of delay and cost overruns in Nigerian construction projects. Int. J. Project Manag. **12**(4), 254–260 (1994)
5. Ogunlana, S.O., Promkuntong, K., Jearkjirm, V.: Construction delays in a fast-growing economy: comparing Thailand with other economies. Int. J. Project Manag. **14**(1), 37–45 (1996)
6. Chan, D.W., Kumaraswamy, M.M.: A comparative study of causes of time overruns in Hong Kong construction projects. Int. J. Project Manag. **15**(1), 55–63 (1997)
7. Aibinu, A.A., Jagboro, G.O.: The effects of construction delays on project delivery in Nigerian construction industry. Int. J. Project Manag. **20**(8), 593–599 (2002)
8. Mezher, T.M., Tawil, W.: Causes of delays in the construction industry in Lebanon. Eng. Constr. Archit. Manag. **5**(3), 252–260 (1998)

9. Frimpong, Y., Oluwoye, J., Crawford, L.: Causes of delay and cost overruns in construction of groundwater projects in a developing countries; Ghana as a case study. Int. J. Project Manag. **21**(5), 321–326 (2003)
10. Battaineh, H.T.: Information system of progress evaluation of public projects in Jordan, Irbid (1999)
11. Al-Momani, A.H.: Construction delay: a quantitative analysis. Int. J. Project Manag. **18**(1), 51–59 (2000)
12. Mwandali, D.: Analysis of Major Factors that Affect Projects Management: A Case of Kenya Railways Projects, Nairobi (1996)
13. Assaf, S.A., Al-Hejji, S.: Causes of delay in large construction projects. Int. J. Project Manag. **24**(4), 349–357 (2006)
14. Chau, K.W., Anson, M., Zhang, J.P.: Four-dimensional visualization of construction scheduling and site utilization. J. Constr. Eng. Manag. **130**(4), 598–606 (2004)
15. Retik, A., Warszawski, A., Banai, A.: The use of computer graphics as a scheduling tool. Build. Environ. **25**(2), 132–142 (1990)
16. Zhang, J.P.: A New Approach to Construction Planning and Site Space Utilization through Computer Visualization (1996)
17. Williams, M.: Graphical simulation for project planning: 4D-planner. In: Proceedings of the Third Congress on Computing in Civil Engineering, Anaheim, California (1996)
18. Collier, E., Fischer, M.: Visual-based scheduling: 4D modeling on the San Mateo County Health Center. In: Proceedings of the Third Congress on Computing in Civil Engineering, Anaheim, California (1996)
19. McKinney, K., Kim, J., Fischer, M., Howard, C.: Interactive 4D-CAD. In: Proceedings of the Third Congress on Computing in Civil Engineering, Anaheim, California (1996)
20. McKinney, K., Fischer, M.: Generating, evaluating and visualizing construction schedules with CAD tools. Autom. Constr. **7**(6), 433–447 (1998)
21. Kamat, V.R., Martinez, J.C.: Visualizing simulated construction operations in 3D. J. Comput. Civil Eng. **15**(4), 329–337 (2001)

New Approaches and Techniques of Motivation for Construction Industry Engineers in B&H

Ahmed El Sayed[1(✉)], Suad Špago[2], Fuad Ćatović[2], and Adnan Novalić[1]

[1] Faculty of Engineering and Natural Sciences, International Burch University, Ul.Francuske revolucije bb, 71210 Ilidža, Sarajevo, Bosnia and Herzegovina
ahmed.elsayed@ibu.edu.ba
[2] University "DžemalBijedić", 88104 Mostar, Bosnia and Herzegovina

Abstract. Motivation of engineers in the construction industry is the subject of many researches, due to the effect of lack of motivation on the work delays, and the tendency to not remain in one company longer than few years. This paper investigates what motivation factors are significant for engineers in construction Industry in Bosnia and Hercegovina. Quantitative research methods based on a sample of 252 engineers filling the Multidimensional Work Motivation Scale (MWMS) were used. Results are analysed using IBM SPSS Statistics 21 and IBM SPSS Amos 22. Selected target population included different profiles of Engineers regarding the age, gender, family status, level of education, professional experience, residential status, and bank credit status. Results proved that identified regulations has significant impact on the motivation of engineers with higher level of education. This paper found also that external regulations are significantly affecting the motivation of engineers with unsolved residential status. Lastly, results showed that number of family members has significant impact on the introjected and identified regulations. This paper concluded that new motivation model that will include dynamic following of engineers working in B&H construction industry regarding their private, financial, and professional development should be developed and used in order to ensure higher level of job satisfaction of engineers and increase the willingness to remain longer at their workplaces.

Keywords: Motivation theories · Construction industry · Human behavior · Engineers · Intrinsic motivation · Identified regulations · Introjected regulations · Productivity · Job satisfaction

1 Introduction

Construction industry differs from other industries due to the rigorous deadlines, common conflicts between different participants, work intensity, work site and its conditions, security in wok, and evaluation of costs [1]. The motivation of humans is a subject of investigation since the earliest days. Many theories emerged during the 19th

I. Karabegović (Ed.): NT 2019, LNNS 76, pp. 736–745, 2020.
https://doi.org/10.1007/978-3-030-18072-0_85

century, such as the social man by Mayo (1945), economic man by Taylor (1947), and complex man by Shein (1980) [2].

Maslow based his motivation theory on the hierarchy of needs, and made a contribution to the investigations related to motivation of workers. His theory stated that human have five universal needs ranked as in pyramid, with the lowest part of it is physiological needs, and the highest is self-actualization needs. In his theory, only by satisfying one need, human start to look for satisfying the upper need in the pyramid [3].

Other motivation theories were developed later. Alderfer's theory classified needs into only three main groups, and explained that it is not required to achieve one need in order to seek for the higher category [4].

In the modern world, civil engineers are responsible for conducting wide range of various jobs, which sometimes not related to the civil engineering profession. In general, managers of construction industry companies believe that motivation is mainly gained by external regulations, such as payment increasing, or bonuses [5].

The effect of the payment increasing on the productivity level was tested by Maloney [5] in 1986. He stated that job satisfaction has direct impact on the workers motivation in construction industry, and that the productivity will increase if the job of workers is well designed to suits their profile, and when they are given the chance to grow and advance in their careers.

Giritli, Setyesilisk, and Horman [6] in 2013 analysed the relationship between organization commitment and job satisfaction of engineers. They surveyed using the 5-point Likert-type scale 350 engineers in construction industry in Turkey. With 65% response, they states that the relationship between the job satisfaction and organizational commitment is significant, and the level of job satisfaction affects the performance level of engineer work.

According to Tinbergen [7], Organizational commitment has a major impact on the organizational productivity. In the research done by Savery and Syme [8], the relationship between organizational commitment and country productivity, the willingness to assist the company in problems solving, and workers' willingness to fight and hold on their work position was found. Another aspect that is directly affected by organizational commitment is the company's effectiveness, and the decreasing of the rate of turnover per working personal [9].

According to Cong and Van [10], only in cases when the companies or organizational top management understand the impact of construction workers job satisfaction on the productivity of workers and the on the organizational improvement, they start employing different methods of motivation to start investigate the job satisfaction of their employee. In their study, they aimed to explore factors that are effecting the motivation of PVNC workers. They have found that financial gain, along with promotion are the most significant motivation factors for workers.

According to [11] and [12] two motivation theories has been used commonly for the purpose of exploring the construction industry's engineer level of motivation: Maslow's and Herzberg's ones.

In a research done by Ruthankoon and OluOgunlana [13] in which they tested the Herzberg's theory in construction industry. The target population analysed in their research was made of 125 engineers and foremen. They found that the Herzberg's theory is not applicable entirely to construction sites in Thailand. Several other researchers found that Herzberg's theory is not applicable, such as Park et al. (1988), Willams (1992), and Jensen (1993) [13].

Maloney and Mcfillen [14] concluded that Expectancy theory can be used for construction workers. However, they believed that there is a need for further tests related to the validity of the results gained by this theory.

In 1997 in Iran, a study was made by Zakeri, Olomolaiye, Holt and Harris [15] concluded that main factors of motivation in Iranian construction industry are: Fairness of salary determination, rewarding the well-done job and praising it, and timely receiving monthly payments.

Shoura and Singh [16] used the Maslow's motivation theory as basis for their research, and found that an emphasis should be put on the importance of fulfilling the basic survival needs before looking for the higher need in the hierarchy. However, according to [12], Maslow's theory doesn't fit for testing the motivation of construction industry workers.

Levinson [17] concluded that there is a significant impact of the company's management on the motivation of construction workers, due to their role in setting the conditions of work and the salaries.

According to Skitmore et al. [18] found that lack of motivators, stress at work, and inappropriate motivation methods lead to demotivation of construction workers, which causes delays at works and other negative consequences. According to Hewage and Ruwanpura [19], poor communication also leads to demotivation of construction workers.

Smithers [20] found that gender, as demographic factor, has no significant impact on motivation, while the non-recognition at work done, long working hours, poor planning and unrealistic distribution of resources, chaotic situation at site, and aggressive management style are main demotivating factors in construction industry. According to [21] and [22], the correlation between gender and motivation is not significant.

2 Methodology

The Aim of this paper is to explore motivation factors for engineers in Bosnia and Herzegovina construction industry. Also, this paper aims to develop a new model of motivation, and to recommend set of actions for companies' managers to improve the motivation level of their employed engineers, and their satisfaction with job. Different profiles were included in this research.

Authors set several research questions: Are young engineers in B&H motivated more by external regulations than by any other motivational factors? Does gender have any significant impact on the level of job satisfaction of engineers in B&H? Are married engineers motivated more by external regulations, compared to single engineers? Does the number of family members cause the demotivation of engineers? What is the effect of educational level on job satisfaction and motivation? Is there an effect of having bank credit on the motivation method needed to be used? Are older engineers more satisfied with their job than younger ones?

As starting model for the research, authors proposed a model that connecting three main group of factors: Demographic factors as the independent variables, Job satisfaction factors, and work motivation factors. Beside the effect of demographic factors on the motivation and job satisfaction factors, another relationship was established between work motivation factors and job satisfaction [23] (Fig. 1).

Fig. 1. Basic proposed model [23]

As for population selected for this study, architects, civil engineers, and survey engineers working in construction industry in B&H, Serbia, and Montenegro were surveyed. Companies involved in the research included all sizes, and types of work performed. From total of 400 engineers selected for the research, 252 responded and filled the questionnaire, which is 64%. The sample size is considered to be good, according to [24].

Quantitative research method was used in this paper, based on using the Multidimensional Work Motivation Scale (MWMS) proposed by [25], in combination with the Minnesota Satisfaction Questionnaire (MSQ) proposed by [26]. Both instruments are set with the 5-point Likert Scale, where 1 refers to Very Dissatisfied, 2 refers to Dissatisfied, 3 refers to Neither, 4 refers to Satisfied, and 5 refers to very satisfied [26].

The questionnaire included demographic factors with 9 items, MSQ with 20 items, and MWMS with 19 items. Demographic factors included in the research are: age, gender, family status and size, educational level, professional experience, residential and bank credit status [23].

According to [27], MSQ includes in its three dimensions the next factors: the use of own abilities, chances for making achievements, ability of being active at work, ability for promotion, having authority at work, the policies and practices of firm, fairness of payment and salary per work done, relationship among workers, ability to be creative at work, chance to work individually, moral values respect, recognition received per done work, growth of social status, the relationship between supervision and workers, the technical competencies of top management, ability to work on various works, job security, working conditions. The reliability of intrinsic satisfaction dimension varied in the range between 0.84 and 0.91, while the reliability of extrinsic satisfaction dimension varied within the range of 0.77 and 0.82 [26]. Engineers scored generally the highest results for reliability [26].

Gange et al. [25] in 2015 proposed the Multidimensional Work Motivation Scale and tested it in total of 9 countries with 7 different languages. Total of 3 435 workers were survey for the purpose of determining the validity and reliability of the scale. The MWMS included 19 items that are sorted in 6 different dimensions: amotivation, external material regulations, external social regulations, introjected regulations, identified regulations, and intrinsic motivation [25].

In this paper, for the purpose of analysing results of both scales, Confirmatory Factor Analysis (CFA) and Modal Fit Analysis (MFA) were used. Correlation and regression were used to determine the significance of the relationship between dimensions and factors.

3 Results and Discussion

The sample had the next characteristics [23]:

- 64.8% are younger than 39 years,
- 62.3% are male engineers and 37.7% female engineers,
- 65.5% are married engineers,
- 46% have at least one child,
- 43.7% hold Bachelor degree, 50.8% hold master degree, and 5.5% hold PhD degree,
- 55.9% have less than 10 years of professional experience,
- 40.5% didn't solve the residential status yet,
- 29.8% have a bank credit to repay.

The Cronbach's alpha coefficient was used to test the reliability of all dimensions of both scales used. For determining the loading of items in both scales, Principal Component Analysis was used in SPSS 21. It has been suggested in [28] and [29] that if majority of model fit indicators are within the range that is acceptable, it can be stated that the model is acceptable for further process of analysis.

Loadings of items, according to the Principal Component Analysis, ranged between 0.514 and 0.874 for Intrinsic Job Satisfaction Dimension, and ranged between 0.770 and 0.855 for Extrinsic Job Satisfaction Dimension [23]. According to [30], the minimum acceptable loading per item should be 0.4. The Cronbach's alpha coefficients for intrinsic and extrinsic dimensions of the MSQ were 0.918 and 0.907, and it can be concluded that both satisfies for further testing [23].

Loadings of items for the MWMS were also determined using the Principal Component Analysis, and ranged all above 0.4, excluding two items (MWMS 6 and MWMS 10), that authors excluded from further analysis [23]. Cornbach's alpha coefficients for all dimensions of the MWMS varied between 0.605 and 0.945, and according to [31] and [32], values above 0.6 for Cronbach's alpha coefficient are considered as satisfying.

Other model fitness tests that were used are: The ratio of the minimum discrepancy and its degree of freedom (CMIN/DF), the comparative Fit Index (CFI), The Normed Fit Index (NFI), the Tucker-Lewis Coefficient (TLI), and the Root mean square error of approximation (RMSEA). All the mentioned model fitness tests had values within the acceptable range. However, the only value that was not in the acceptable range is the RMSEA for MWMS, but it was suggested that for bigger samples, and when other tests are satisfying, the value of RMSEA may exceed the value of 0.1. In Table 1, results of tests and criteria for evaluation are previewed.

Table 1. Model-fitting indexes for both tested models in this paper [23]

Fit index	Criteria	Scale 1 Job satisfaction	Scale 2 Work motivation
CMIN/DF	Between 2 and 5	2.256	4.568
CFI	Close to 1	0.954	0.870
NFI	Close to 1	0.921	0.841
TLI	Close to 1	0.944	0.832
RMSEA	Close or less to 0.1	0.071	0.119

To answer the research questions set in the start of this research, Linear Regression, which is a model-based technique that can be used as Pearson Correlation extension, was used. With this technique, the significant of the relationship between two variables, in which one is independent and other is dependent, is determined.

First research question set by authors investigated the relationship between age, as an independent variable, and motivation dimensions (External regulation, Introjected regulation, Identified regulation, and Intrinsic motivation). Results gained using SPSS 21 proved that the relation between age and identified regulations is not significant ($p = 0.868 > 0.05$), while the relationship between age and other motivation dimensions is significant. As the value of unstandardized beta coefficient was negative for them (external regulation -0.144, introjected regulation -0.180, intrinsic motivation -0.222), it can be stated that younger engineers are motivated by these dimensions compared to older engineers. Finally, according to the values of standardized beta coefficient (external regulation -0.572, introjected regulation -0.604, intrinsic motivation -0.526), it can be concluded that young engineers are motivated more by introjected motivation dimension than by any other motivation dimension. However, as the difference in the Beta value is small, it can also concluded that they are highly motivated by the other dimensions [23].

Second research question was related to the gender and the job satisfaction. The relationship between the gender as independent variable, and job satisfaction dimensions as dependent variable was test using linear regression in SPSS 21, and results proved that there is a significant (for intrinsic job satisfaction dimension $p = 0.004 < 0.05$, while for extrinsic job satisfaction dimension $p = 0.01 < 0.05$) relationship between gender and both dimensions of MSQ (intrinsic and extrinsic job satisfaction dimensions). The value of the unstandardized Beta coefficient for the relationship between intrinsic job satisfaction dimension and gender (unstandardized Beta = 0.325), and negative for the relationship between extrinsic job satisfaction dimension and gender (unstandardized Beta = -0.319), it can be concluded that female engineers are more satisfied with intrinsic job satisfaction dimension, while male engineers are more satisfied with extrinsic job satisfaction dimension [23].

Third research question tested the relationship between family status as independent variable, and external regulations as motivation dimension. The linear regression gained by SPSS 21 proved that there is significant relationship between the two variables ($p = 0.000 < 0.05$), and due to the negative value of unstandardized Beta coefficient (unstandardized Beta = −*0.402*) it can be stated that married engineers are less motivated by external regulations than single engineers [23].

Fourth research question explored the effect of the number of family members as independent variable, on the amotivation of engineers as dependent variable. Results of linear regression proved that there is a significant relationship between the two variables ($p = 0.000 < 0.05$), and the positive value of unstandardized Beta coefficient (unstandardized Beta = *0.155*) shows that by increasing the number of family members, the amotivation of engineers increases [23].

Fifth research question looks for the effect of educational level as independent variable, on job satisfaction and motivation dimensions as depended variables. Linear regression results showed that there is a positive (unstandardized Beta for extrinsic job satisfaction dimension = *0.271*, for intrinsic job satisfaction dimension = *0.151*, and for identified regulation = *0.178*) significant relationship between the level of education and both dimensions of job satisfaction and identified regulation (for extrinsic job satisfaction dimension $p = 0.000 < 0.05$, for intrinsic job satisfaction dimension $p = 0.008 < 0.05$, and for identified regulation $p = 0.000 < 0.05$). Results proved also that there is a significant negative relationship ($p = 0.003 < 0.05$, unstandardized Beta = −*0.135*) between educational level and introjected regulation, while the relationship between educational level and external regulation is not significant ($p = 0.461 > 0.05$). These results leads to the conclusion that engineers with higher educational level are more satisfied with both job satisfaction dimensions than engineers with lower educational level, and engineers with lower educational level are more motivated by introjected regulations than engineers with higher educational level [23].

Sixth research question is investigating the relationship between the Bank credit debt as independent variable, and external regulation as dependent variable. Using linear regression, results gained show that there is a significant ($p = 0.001 < 0.05$) negative (unstandardized Beta = −*0.257*) relationship between the two variables. It can be concluded that indebted engineers are more motivated by external regulation if compared to engineers without Bank debt [23].

Final research question is exploring the relationship between age as independent variable, and intrinsic job satisfaction dimension as dependent variable. Using the linear regression, results show that there is a significant ($p = 0.000 < 0.05$) positive (unstandardized Beta = *0.251*) relationship between the two variables, and older engineers are more satisfied by intrinsic job satisfaction dimension than their younger colleagues [23].

4 Conclusion

In this paper, results proved that different profile of engineers, based on different demographic factors, require different motivation systems. Profile of engineers changes as a consequence of human life, and therefore there is no standard motivation system that is applicable for all engineers. Also, with aging, and starting of family, other obligations and responsibilities start to occupy engineers, and therefore they tend to look for different dimensions of life. This paper aimed to show for the main elements of engineers' profile the motivation dimensions that were approved to have impact on them and their level of job satisfaction.

Several conclusions can be made in accordance to the results and discussion made in this paper, and they are [23]:

- Younger engineers, in the start of their careers, are tending to try proving to others and themselves that they are adequate for the working position, hard-workers, and potentials for future, and therefore they are more motivated by introjected motivation factors compared to other factors. However, due the needs they have, according to Maslow's theory of motivation, the are also motivated by external motivation factors. Older engineers are more satisfied with the intrinsic factors, as the have already overcomed the first parts of Maslow's hierarchy pyramid.
- Female and male engineers are not satisfied by same factors of job satisfaction. Male, due to their natural role in life in B&H, are more looking for extrinsic job satisfaction factors, while female engineers are looking for intrinsic job satisfaction factors.
- Married engineers are in stage when they tend to build their home with other person, and therefore they ae nor only motivated with external regulations, while single engineers are looking to build themselves and ensure the required needs for development in life. With the development of marriage life, the number of family members increases, and engineers, due to the new responsibilities as humans, start to be demotivated for work. In this period, other motivation factors, such as identified and introjected regulation are decreasing the amotivation, and improving the satisfaction of engineer.
- It is important to follow the educational development and level of each employee, as those with higher educational level are differently motivated than those with lower educational leve. Also, both categories are satisfies with different job satisfaction factors.
- Engineers with bank debts are more motivated by job safety and payment increasment, as they are under pressure due to the fear of losing job or inability to repay the credit.

This paper recommends to managers of construction and design companies in B&H and surrounding region to pay more attention to motivation of workers and their level of job satisfaction. Unfortuantely, in B&H, it is still believed that only external motivation factors, such as salary, is the only way to motivate workers, regardless their stage of professional and private life.

References

1. Nave, H.J.: Construction personnel management. J. Constr. Div. **94**(1), 95–108 (1968)
2. Reddy, R.J.: Organisational Behaviour. APH Publishing, New Delhi (2004)
3. Cherrington, D.J.: Organizational Behaviour: The Management of Individual and Organizational Performance. Allyn and Bacon, Boston (1994)
4. Alderfer, C.P.: Existance, Relatedness, and Growth: Human Needs in Organizational Settings. Free Press, Calefornia (1972)

5. Maloney, W.F.: Understanding motivation. J. Manag. Eng. **2**(4), 231–245 (1986)
6. Giritli, H., Sertyesilisik, B., Horman, B.: An investigation into job satisfaction and organizational commitment of construction personnel. Glob. Adv. Res. J. Soc. Sci. **2**(1), 1–11 (2013)
7. Tinbergen, J.: The influence of productivity on economic welfare. Econ. J. **62**(245), 68–86 (1952)
8. Savery, L.K., Syme, P.D.: Organizational commitment and hospital pharmacists. J. Manag. Dev. **15**(1), 14–22 (1996)
9. Bayram, L.: Yönetimde Yeni Bir Paradigma: Örgütsel Bağlilik. Sayiştay Dergisi **59**, 125–139 (2005)
10. Cong, N.N., Van, D.N.: Effects of motivation and job satisfaction on employees' performance at Petrovietnam Nghe An Construction Joints Stock Corporation (PVNC). Int. J. Bus. Soc. Sci. **4**(6), 212–217 (2013)
11. Kazaz, A., Manisali, E., Ulubeyli, S.: Effect of basic motivational factors on construction workforce productivity in Turkey. J. Civil Eng. Manag. **14**(2), 95–106 (2008)
12. Ogunlana, S.O., Pien Chang, W.E.: Worker motivation on selected construction sites in Bangkok. Thailand. Eng. Constr. Archit. Manag. **5**(1), 61–81 (1998)
13. Ruthankoon, R., Olu Ogunlana, S.: Testing Herzberg's two-factor theory in the Thai construction industry. Eng. Constr. Archit. Manag. **10**(5), 333–341 (2003)
14. Maloney, W.F., McFillen, J.M.: Motivation in unionized construction. J. Constr. Eng. Manag. **112**(1), 122–136 (1986)
15. Zakeri, M., Olomolaiye, P., Holt, G.D., Harris, F.C.: Factors affecting the motivation of Iranian construction operatives. Build. Environ. **32**(2), 161–166 (1997)
16. Shoura, M.M., Singh, A.: Motivation parameters for engineering managers using Maslow's theory. J. Manag. Eng. **15**(5), 44–55 (1999)
17. Levinson, H.: Management by whose objectives? Harvard Bus. Rev. **81**(1), 107–116 (2003)
18. Ng, S.T., Skitmore, R.M., Lam, K.C., Poon, A.W.: Demotivating factors influencing the productivity of civil engineering projects. Int. J. Project Manag. **22**(2), 139–146 (2004)
19. Hewage, K.N., Ruwanpura, J.Y.: Carpentry workers issues and efficiencies related to construction productivity in commercial construction projects in Alberta. Can. J. Civil Eng. **33**(8), 1075–1089 (2006)
20. Smithers, G.: The effect of the site environment on motivation and demotivation of construction professionals. In: Proceedings: 16th Annual ARCOM Conference of ARCOM, Glasgow (2000)
21. Gilbert, G.L., Walker, D.H.: Motivation of Australian white-collar construction employees: a gender issue? Eng. Constr. Archit. Manag. **8**(1), 59–66 (2001)
22. Eskildsen, J.K., Kristensen, K., Westlund, A.H.: Work motivation and job satisfaction in the Nordic countries. Empl. Relat. **26**(2), 122–136 (2004)
23. El Sayed, A.: The effect of motivation on the productivity of construction engineers - case study: Bosnia and Herzegovina and surrounding region, Beau Bassin: AV AkademikerVerlag (2017)
24. Comrey, A.L., Lee, H.B.: A First Course in Factor Analysis. Psychology Press (2013)
25. Gagné, M., Forest, J., Vansteenkiste, M., Crevier-Braud, L., Van den Broeck, A., Aspeli, A., Bellerose, J., Benabou, C., Chemolli, E., Güntert, S.T., Halvari, H.: The multidimensional work motivation scale: validation evidence in seven languages and nine countries. Eur. J. Work Organ. Psychol. **24**(2), 178–196 (2015)
26. Weiss, D.J., Dawis, R.V., England, G.W.: Manual for the Minnesota Satisfaction Questionnaire. Minnesota studies in vocational rehabilitation (1967)

27. Gillet, B., Schwab, D.P.: Convergent and discriminant validities of corresponding job descriptive index and minnesota satisfaction questionnaire scales. J. Appl. Psychol. **60**(3), 313 (1975)
28. Hu, L., Bentler, P.: Evaluating model fit. In: Hoyle, R. (ed.) Structural Equation Modelling: Concepts, Issues and Applications, pp. 76–99). Sage Publications, Thousand Oaks (1995)
29. Bollen, K.A.: Structural Equations with Latent Variables. John Wiley & Sons, New York (1989)
30. Hair, J., Black, W., Babin, B., Anderson, R.: Multivariate Data Analysis, 7th edn. Prentice-Hall Inc, Upper Saddle River (2010)
31. Churchill Jr, G.A., Peter, J.P.: Research design effects on the reliability of rating scales: a meta-analysis. J. Mark. Res. **21**, 360–375 (1984)
32. Nunnally, J.C.: Psychometric Theory. McGraw Hill, Englewood Cliffs, New Jersey (1988)

Sustainable Innovation in Architectural Design

Jovana Jovanovic[1(✉)], Xiaoqin Sun[2], Zoran Cekic[1], and Suzana Koprivica[1]

[1] Faculty of Civil Engineering and Management, University Union Nikola Tesla, 11000 Belgrade, Serbia
jocka747@gmail.com
[2] University of Science and Technology, Changsha, China

Abstract. Architecture always tries to find a far-sighted redefinition and revival through various concepts. The design is there to support innovative and random functional volumes, which appear in the form of bizarre, scattered constellations of new housing estates. By itself, the design is inwrought in a metropolitan glossary and it becomes almost a routine. The ongoing technologies and machinery transform the view on urban design. Eco-towns made up of living modules that adapt bioclimatic techniques emerge all over the world and serve as pores to the world. The constantly new scenarios of sustainable urban design impose the tumultuous growth of the cities. Poor designs of dwellings nowadays can have underlying flaws as they do not contribute to the energy saving and reduce energy consumption inside them. Sustainable urban design brings a new horizon into focus driven by economic, energetic, technological and cultural changes.

Keywords: Design · Metropolitan glossary · Eco-towns · Sustainable urban design · Horizon

1 Introduction

Sustainable design is the multicellular transition towards overwhelmingly invasive design thinking. As a design thinking, it stretches over the concept of sustainability, which is still unclear and ambiguous. In its grip, it has innovative concepts and contents that color the minds of consumers. It always questions how well and how much something can be restructured with the new pattern of existence. The sustainable design seems to be a much popularized term of everyday life. It can be quoted as, brainstorming design ideas through a series of posters to investigate communicational capabilities among different entities and sustainability [1]. On the other side, sustainability metrics and models were used to rate sustainable design within the automotive industry [2].

Sometimes, sustainable design is in a close correlation with solar energy implementation [3]. Healthy urban development can be reached only if it is both resilient and sustainable, which is further founded on vulnerability and pertinacity of cities [4]. Even [5] subjected the Portuguese building model to different sustainability assessment systems trying to measure the sustainability of made adaptations of that model to the context of old urban centers. Articles [6, 7] discussed the urban sustainability

I. Karabegović (Ed.): NT 2019, LNNS 76, pp. 746–757, 2020.
https://doi.org/10.1007/978-3-030-18072-0_86

indicators, benchmarks, and scores that are crucial to guarantee urban sustainability and resilience. It was shown that local authorities and their actions have a great impact on the choice of the most compatible sustainability indicators. It was also investigated to what extent occur the stakeholder's involvements in the urban sustainability [8]. Article [9] proved that vernacular building practices in the Himalayan area of north India are sufficient for bringing new building legislation of hillside sustainable settlements. In [10] even indicated the list of sustainability indicators in urban mobility and demonstrated the urban mobility as a reflection and bond to sustainable development.

The aim of this manuscript is to point out the dominant role of sustainable innovation in urban architectural design as an administrative element of emerging and newborn cities. Such a manuscript is a path towards new building designs, urban systematizations by finding the new, generic and defining formula of a triplet: design, sustainable design, and sustainable urban design. Except for the stated, the modern shifts of urban building style are present in this manuscript. The 3D Builder Program is used as a drawing programme for buildings of specific and sustainable design. Efforts for greening and reinventing the existing urban design of Paris are contextualized. Overall significance is mirrored in drawing a line what meanings have the design, sustainable design and sustainable urban design for vibrant and contemporary cities, building units and surroundings.

2 Modern Architecture

The core and noteworthy traits of modern architecture should be the simplicity and minimalism hidden in complex architectural forms, which are founded on the revival concept of landscapes. Systems design, which was rooted around the mid-twentieth century, remained as a key tool for prospering building innovations [11]. During construction of a certain typology of object, serial, accompanying steps are appraisals of system performance, materials durability and sustainability (matching properties of utilized thermal and acoustic insulations), testing of structural framing. Modern architecture encompasses a variety of modern building structures such as dumbwaiters, solar walls, escalators, flat, green roofs, different designs of attics and partition walls which enable the external and internal flow of a building. This type of architecture is also often a fast-track construction of low-rise and high-rise buildings, where life-cycle cost is above construction cost of building, because of built-in design. Various designs and volumes of attic spaces are accomplished by setting up different forms of roofs such as hipped roofs, roofs with a skylight and roofs with dormers [11].

The concept of modern architecture is often vernacular architecture, preserved in locally available, natural renewable materials which often leads to the self-sustainability and self-sufficiency of object. Vernacular architecture, which is equated to modern architecture has genesis in the term, genius loci", which means being a reflection of the place [12].

Uninhabited, savage areas with very little surrounding infrastructure request a higher load of adaptability to the area. This concept comprises a certain adaptation to the harsh area conditions and surroundings, by even so-called, site-craft" and usage of a few modern wall-finishing techniques, like a technique of a dry stone wall, without binders [13].

Even the Swiss architect Le Corbusier, an advocate of modern architecture, implemented the bioclimatic and ecological design of dwellings, in the first mid of 20th century, by his principles of the free facade and "mur neutralisant" which served as a precocious form of a dual facade. His architectural control of microclimate knocked down the visual and ambient frontiers among interior and exterior. Le Corbusier fiercely pointed out that the healthy living area is merely the one, which is naturally lighted and air-permeable [14].

Many Swedish architects, in the 20th century, experimented with socially inclusive designs, which were adapted for wider social masses and leaned on Le Corbusier's design philosophy. At that time housing was more accessible to all. Building blocks were plain, with unadorned facades, some of them with timber frames covered by asbestos-cement sheets while others were with steel frames. Other implications of Le Corbusier's design in Swedish modernistic architecture was the usage of rough concrete [15].

Modern architecture also encourages asymmetrical architectural forms and compliant relationships among the detail and entirety. Exponents of modernism in architecture are container constructions, which can be seen as component-based systems. Influential nature of modern architecture is perceived today, by shaping, (greening) urban tissues and mitigating urban heat islands.

Urban heat islands are determined by a specific urban morphology. In such urban configuration, characterized by a high percentage of new-built, enclosed surfaces with neglecting thermal insulation properties, and high thermal absorption of solar radiation, urban heat islands are frequent. Higher thermal absorption coefficients have bricks, concrete, asphalt, while in contrast wood, straw and sand have less high thermal absorption coefficients [16]. Therefore national targets for planting greenery in city centers, are five percent of augmentation in planting until the 2030s and another five percent of augmentation until 2050s.

Architectural communities are at the same time conservation habitats. The mainstream idea in the continuum of conservation is materiality and the aim is to protect the materials' authenticity by not ruining the preexisting design. Dozens of iconic architectural works are the heritage of Modern epoch and tend to be preserved in a more superior manner, as opposed to 70s and 80s of the last century. Remarkable exemplars of modern architecture stretch from 1956. As a town, Brasilia was projected, up to some super modern of Zaha Hadid Architects which all have elitistic spirit. It can be observed that merely architectural works which exude with luxury, richness, variety in forms, innovations, and comfort get appreciation, acceptance of public and certain conservation. Other architectural works of modernism, if they are not marked as, publicly accepted" are often turned down and judged as architectural rubbish [17].

With the modern architecture, the burning discourse that comes in the focus is the energy consumption of existing and new building stock. According to a few statistics, in the USA are still under construction 22 million buildings, which will not use only the current produced in electric power plants, but also directed the fuel, natural gas or propane combusted in their cauldrons or furnaces. It is prognosticated that by 2040, in the USA, three-quarters of buildings will be new-built or refurbished, which creates huge interventions in buildings against the possible effects on climate changes, (global warming) and emissions of toxic gases. Although there are cuttings in the consumption of fossil fuels worldwide, buildings still get 57% of electricity from fossil fuels combustion [18].

3 Sustainable Innovation as a Keypoint of Durable Design

Sustainable practices in structural, building statics as well as in building design cherish sustainable innovations as their benchmarks. Recently, the scientists from the UK University in Lancaster found out that the nanoparticles from carrot increase the strength and durability of cement, up to 80%. Fibers of cellulose from carrot, change the properties and manner of water binding in cement, and later in a fresh, concrete mixture. It was shown that half a kilo of nanoparticles from carrot, saves up around ten kilos of cement per cubic meter of concrete. This method requires less quantity of cement and radically reduces CO2 emissions, which is important because the cement is accountable for 8% of global CO2 emissions.

The agro-concrete mixed with the Oakwood and layers of rice is also applicable as some kind of eco-concrete, which would be characterized as some sustainable innovation of standard concrete. Concrete can also be mixed with wood pellets and polystyrene.

From the viewpoint of materials, durable design is not only definitive by proper materials' choice, but also monitoring materials' specifications in advance and adequate sequence of materials in building elements. That means, when it comes to the sequence of materials, a builder should assess does glass wool match better with the normal brick or hollowed brick; should that joint be filled with another thinner insulation; what is the effect of matching anhydride slab with a concrete slab; should exterior layers be less porous… Deciding on the role of building elements is also significant, whether a wall should be curtain wall, load-bearing wall, a decorative panel or board. Contributing to the energy-efficient profile of the building, facades can be dual with integrated rotating panels with solar obscuration which move towards the sunlight. Suitable from the energy point is energy plus house which creates more energy than is needed and stores them as back-ups.

On the general level, sustainable innovations would be replenishing or delivering the functionality to certain matters. In architectural sense, it could be adapting of void and deserted spaces to alternative uses, reconstruction for a higher compatibility with surroundings, incorporating renewable energy strategies and devices in building envelopes, narrowing the thickness of walls by putting insulation materials with better performances (higher thermal inertia, better thermal conductivity, better transparency, and bigger airtightness.).

4 Materials for Green Architecture

A lately set up challenge in architectural practice is organic architecture. Materials' fabrication requests a huge energy expenditure and therefore building materials set in construction have their impact indicators such as carbon footprint [19].

The spectrum of good thermal insulation materials begins with biopolymer silica aerogels, which are lightweight and have a lower thermal conductivity coefficient than most common insulation materials-(synthetic resins, mineral and glass wool, styrofoam, polystyrene). In that manner, thermal insulation layers with aerogel incorporated can be thinner. Silica aerogels can be used as composites, whose additives rise the mechanical strength, thermal mass, density and other properties of these materials [20].

In the green architecture are also used terracotta bricks, bricks from raw soil, cane and bamboo. Phase change materials, known as PCM, susceptible to phase transitions, during which absorb and release large amounts of latent heat, are also exquisite thermal insulators and provide interior thermal comfort. They can be found in the form of sealed aluminum pouches gathered in sheets, as PCM beads with enclosed PCM liquids, and also in other micro-encapsulating and macro-encapsulating forms.

In one case-study new, cool-colored, cement-based materials were created to improve the thermal insulation properties of the external envelope of historical buildings in the Mediterranean region [21]. In the study, the traditional colored mortar was compared with cool-colored mortars.

There are many other materials of entangled and cellular fibers which have high tensile strength and are good for acoustic insulation. Plywood sheets are good thermal and acoustic insulation materials, used also for cladding the roof's horizontal or tapered deck. They can be proposed for floorings, walls, and as adhesive to vapor control layers. Rubber membranes are envisaged for flat roof's practical solutions and they can also serve as insulation layers.

5 The Formula of Sustainable Architectural Design

It is said that the urbanizations' surges change the world. Sustainability is not only an issue of environmental fit and effectiveness, but it also contains the concepts of Liveable Cities-cities that are pleasurable to live and work in [22]. According to the worldwide statistics, 75% of Europeans live in cities and urban spaces and by 2020s this percentage is going to be elevated to 80%. The high quality of life in cities is pursued by launching green initiatives, planting vegetation in building envelopes and roof ridges, cutting environmental costs for waste disposal, sanitation of sewage sludges, specific building designs, creating an atmosphere for resilient and sustainable cities. Despite more and more present urban sprawl, the Lisbon Treaty plan specified the safeguarding of European cultural heritage as one of the top priorities for Europe. Hamburg was a winner, city carrier of the European Green Capital Award for the year of 2011 [23]. In the UK, the mainstream practices are oriented towards public investments and value significance of good urban design. On behalf of doing research about design, in manuscript [24] was analyzed proactively value sensitive design through three types of investigations: conceptual, empirical and technical. Cradle-to-cradle design philosophy has been recommended as an ultimate milestone for reaching ecologically intelligent and efficient dwellings [25].

The nake design of the buildings had been insufficiently processed.

Referring on [26], the conceptual, generic defining formula of a triplet (design, sustainable design, and sustainable urban design) was created. In this article, the design, sustainable design, and sustainable urban design were fragmented in underlying elements and some elements were presented as tree crochets.

Observing design in first hand, as a strategic multi-factor process, the general value of design and its manipulating space-time definition can be stated in the following manner:

$$S = f(X.L) \quad (1)$$

$$D = f(S) = f(f(X.L)) \quad (2)$$

D-general value of design
S-shape
L-linearity
X-multiple factor

According to the made-up, model equations, the shape is a function of multiplied linearity and the general value of design is the function of shape.

Sustainable design is much more complex than a general value of design. The manipulating space-time definition of sustainable design roots from unifying general value of design and sustainable innovation.

Sustainable innovation, as an underlying element of sustainable design, is in this article considered as a multi-determinant net. Sustainable design can be defined as:

$$SD = D + SI \quad (3)$$

D-general value of design
SI-sustainable innovation
SD-sustainable design

$$D = f(S) = f(f(X.L)$$

$$SI = S/n = (f(X.L)/n \quad (4)$$

$$SD = f(f(X.L)) + f(X.L)/n \quad [(2),(4)] \text{ in } (3)$$

S-shape
L-linearity
X-multiple factor
n-number of varieties

Graphically, sustainable innovation can be fragmented into the following parts (Fig. 1):

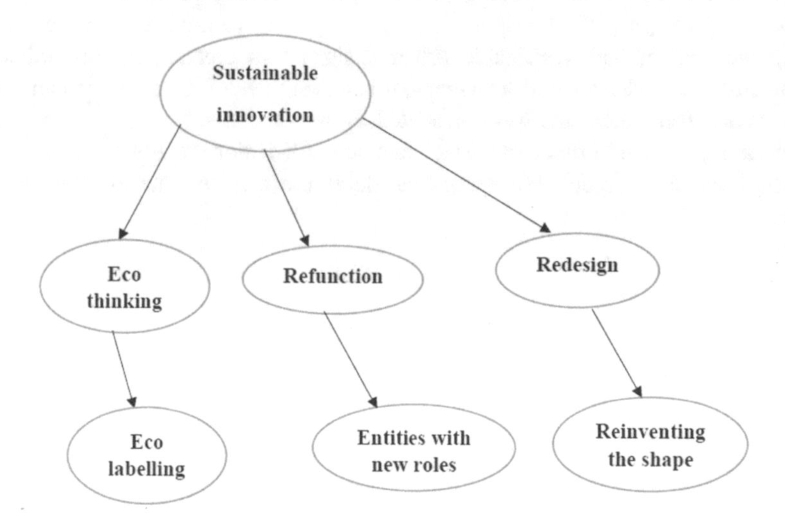

Fig. 1. The skeleton of sustainable innovation

Each applied sort of design is a step further in the practice. According to the referred, sustainable design comprises the general value of design and sustainable urban design is a sustainable design outspread by the urban background (Fig. 2).

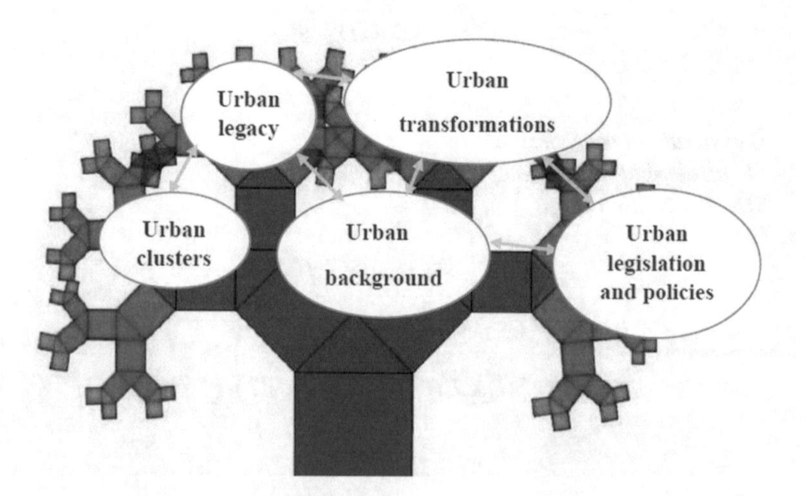

Fig. 2. The urban background (U) tree crochet structure (Available at http://www.google.com)

$$SUD = SD + U \tag{5}$$

$$SUD = D + SI + U \quad (3)\,\text{in}\,(5)$$

D-general value of design
SI-sustainable innovation
U-urban background

Further, it is equaled to:

$$SUD = f(f(X.L)) + S/n + U \quad [(2),(4)]\,\text{in}\,(5)$$

$$SUD = f(f(X.L)) + f(X.L)/n + U \quad [(2),(4)]\,\text{in}\,(5)$$

6 Modern Shifts Towards Sustainable Urban Background

Building technologies and its design record a vigorous and prompt evolution. Sustainability legislation substantially modified the construction schemes and approaches. German Passive House Standard certified specific building fabrics and lightening that reduce absolutely heat losses bringing the maximum of thermal comfort to the occupants of dwellings [27]. Low-energy dwellings with a space heating demand lower than 40 kWh/m^{2}a and passive households with a space heating demand lower than 15 kWh/m^{2}a were substituted with Zero Carbon Homes by 2016, deployed in the UK. There were 250 built households in the UK, entitled as Zero Carbon homes, by the end of 2013, from whom 16 were in Scotland [28]. The thermal comfort indicators of five certified passive households in Scotland [29] were compared by using the Building Performance Evaluation study. Modernities towards sustainable urban background can be also perceived alongside with the existence of "Trombe wall" in the building constructions. Thermal performances of Trombe wall system in the Mediterranean zone by using TRNSYS software were analyzed [30] as its optimal size by using Life Cycle Cost criteria. Energy efficiency gaining by a combination of PV modules and Trombe wall was shown [31] using simulations carried out by SAM (System Advisor Model) software. Administering the questionnaire in certain housing estates in Kwara, Nigeria [32] defined how mass housing can be created as sustainable and its linkage to the owners.

Sustainable nature of mass housing estates depends on several key variables: mass housing design, implemented construction materials, construction system and engaged technologies, consumption chain and costs. There is a constant tendency to build uniform housing units of sustainable design which is a mistake. It should exist much bigger variety in sustainable designs, forms, unities, combinations of materials. Sustainability of housing estates should be the most determined by occupants as end-of-users. The ratio of openings and occupied space, size of windows, natural ventilation or ventilation by exhausting the air from devices (mechanical), occupants'activities etc.

determine the rating and intensity of present sustainability. Sustainability of some building unit is also affected by changing the purpose of certain spatial part of building the unit. Sustainability must be valued according to tougher systems, scales, and marks. Sometimes, in order to procure sustainability of building sites is necessary to relocate fuel and oil refineries, excavating points of ores, modify the types of soil. Sustainable design forms do not need to be inevitably unusual, asymmetrical or complex forms, but rather compact and plain forms. Each city that transforms into a sustainable one, bares contextual changes implying new infrastructure schemes.

Building design confronts a lot of challenges. For instance, biomimicry is mimicking the nature and natural cycles which are managed and operated by building construction skeletons. Besides that, biomimicry could be recognized in the production of new materials (bioproduction), configurations and systems which is founded on biological principles. Biomimicry is quite new in the urban design world. There is a very profound theory underlying the principles of biomimicry which must be fully accepted in the practice. Biomimicry practical specimens can be seen in the existence of carbon-sequestering cement which is inspired by coral reefs [33]. Other specimens of applied biomimicry are ship hulls and swimming suits based on the structure of shark's skin. Movable and rotating facades on some buildings of the new era are inspired by sunflowers and their movements. In order to apply a biomimicry design, the various span of data needs to be collected from different areas: green architecture, ecology, marine biology, branches of engineering. In an advanced building practice, it happens that some other models of green urbanism are joined with biomimicry.

On Fig. 3 is exposed the pilot sketch for an unusual and sustainable design of a construction model, from 3D Builder drawing program, [34]. Figure 4 presents a far-seeing sustainable design projection of the city of Paris, where many facades are renovated [35].

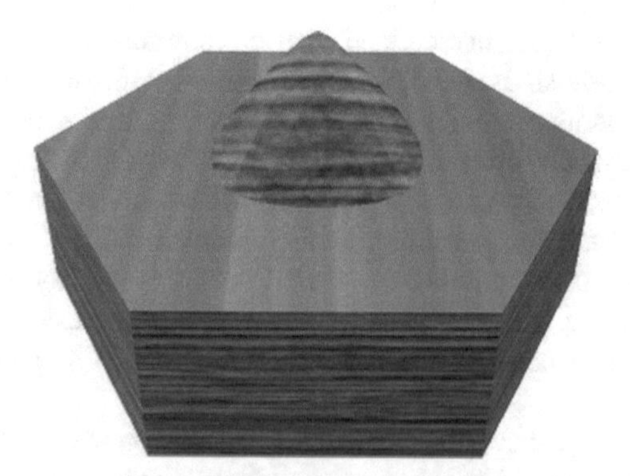

Fig. 3. The pilot sketch for an unusual (sustainable) design of a construction model from the 3D Builder drawing program

Fig. 4. Bringing it more sustainable by reinventing and greening the infrastructure of Paris (projections for the 2050s)

7 Conclusion

The notion and meaning of the sustainable urban design are still blurry and fruitage of that commitment is viewed distant, as through binoculars. Flaws of sustainable urban design lie in insufficient knowledge which exists in practice. There is just a handful of normatives assumed by European Commission which oblige stakeholders to act and construct in a sustainable manner. Sustainability and sustainable urban design must be driven by more stable and secure factors and less rely on assessments and questionnaires. Questionable is what are the thresholds for sustainable urban design and where does sustainable urban design end. Each city has so-called, metabolism" which is and can be modified by sustainability framework. Therefore, local communities, facility owners and other practitioners must be involved in each stage of a project and each part of a design stage. From top to the bottom, mindsets about sustainability criteria, sustainability valorization, sustainability tracking need to be changed and entirely improved. Liveable and sustainable cities must be a synthesis of all coherent design processes which are based on systems thinking. Design requirements of the new era and for several future decades of time horizon are sustainability, resilience, flexibility, and liveability.

References

1. Storer, I., Badni, K., Bhamra, T., Farmer, S.: Communicating sustainable design. Des. Technol. Educ.: Int. J. **10**(2), 44–57 (2005)
2. Mayyas, A., Qattawi, A., Omar, M., Shan, D.: Design for sustainability in the automotive industry: a comprehensive review. Renew. Sustain. Energy Rev. **16**(4), 1845–1862 (2012)
3. Stević, I.: Strategic orientation to solar energy production and long term financial benefits. Arch. Tech. Sci. 1(17) (2017)

4. Zhang, X., Li, H.: Urban resilience and urban sustainability: what we know and what do not know? Cities **72**, 141–148 (2018)
5. Almeida, P.C., Ramos, F.A., Silva, M.J.: Sustainability assessment of building rehabilitation actions in old urban centers. Sustain. Cities Soc. **36**, 378–385 (2018)
6. Ghellere, M., Devitofrancesco, A., Meroni, I.: Urban sustainability assessment of neighborhoods in Lombardy. Energy Proc. **122**, 44–49 (2017)
7. Lutzkendorf, T., Balouktsi, M.: Assessing a sustainable urban development: typology of indicators and sources of information. Proc. Environ. Sci. **38**, 546–553 (2017)
8. Soma, K., Dijkshoorn-Dekker, M.W.C., Polman, N.B.P.: Stakeholder contributions through transitions towards urban sustainability. Sustain. Cities Soc. **37**, 438–450 (2018)
9. Kumar, A., Lata, P.: Vernacular practices: as a basis for formulating building regulations for hilly areas. Int. J. Sustain. Built Environ. **2**, 183–192 (2013)
10. Macedo, J., Rodrigues, F., Tavares, F.: Urban sustainability mobility assessment: indicators proposal. Energy Proc. **134**, 731–740 (2017)
11. Merritt, F.S., Ricketts, J.T.: Building Design and Construction Handbook, 6th edn. (2000). ISBN 0-07-041999-X
12. Achenza, M.: Architectural sustainability - a new inspiration. COBISS.SR-ID 227459596, pp. 167–178 (2016)
13. Ragheb, A., El-Shimy, H., Ragheb, G.: Green architecture: a concept of sustainability. Proc.-Soc. Behav. Sci. **216**, 778–787 (2016)
14. Porteous, C.: The New Eco-Architecture. Alternatives from the Modern Movement. Spon Press, New York (2002)
15. Campo-Ruiz, I.: Experimenting with prototypes: architectural research in Sweden after Le Corbusier's projects. International Congress paper (2015). http://dx.doi.org/10.4995/LC2015.2015.893
16. Kazmierczak, A., Carter, J.: Adaptation to Climate Change Using Green and Blue Infrastructure. A database of case studies (2010). http://www.grabs-eu.org/membersarea/files/berlin.pdf
17. The Getty Conservation Institute: Conservation Perspectives. The Getty Conservation Institute Newsletter, vol. 28, no. 1 (2013)
18. Omar, O.: Towards eco-neighborhoods, solutions for sustainable development, construction, and energy-saving technologies. J. Architect. Urbanism **42**, 95–102 (2018)
19. Blanc, I., Peuportier, B.: Eco-design of buildings and comparison of materials. In: Proceedings of the 1st International Seminar on Society & Materials, SAM1 (2007)
20. Malfait, W., Zhao, S., Brunner, S., Huber, L., Koebel, M.: Biopolymer-Silica Aerogel. Lab. J.-Bus. Web Users Sci. Ind. (2017). https://www.laboratory-journal.com
21. Rosso, F., Pisello, A.L., Castaldo, V.L., Ferrero, M., Cotana, F.: On innovative cool-colored materials for building envelopes: balancing the architectural appearance and the thermal-energy performance in historic districts. Sustainability **9**, 2319 (2017)
22. Rogers, C.D.F.: Engineering future liveable, resilient, sustainable cities using foresight. In: Proceedings of the Institution of Civil Engineers, (ICE), Paper no. 1700031 (2017)
23. Commission, E.: Making our Cities Attractive and Sustainable, pp. 1–36. Publications Office of the European Union, Luxembourg (2011)
24. Mok, L., Hyysalo, S.: Designing for energy transition through value sensitive design. Des. Stud. **54**, 162–183 (2018)
25. Ankrah, A.N., Manu, E., Hammond, N.F., Awuah, K., Tannahill, K.: Beyond sustainable buildings: eco-efficiency to eco-effectiveness through cradle-to-cradle design. In: Sustainable Building Conference, pp. 47–56 (2013)

26. Iwaro, J., Mwasha, A.: The impact of sustainable building envelope design on building sustainability using integrated performance model. Int. J. Sustain. Built Environ. **2**, 153–171 (2013)
27. Bell, S.: The Passive House Standard: An Introduction. Passive House, pp. 14–17 (2010)
28. Passivhaus Trust: (2016). http://www.passivhaustrust.org.uk
29. Foster, J., Sharpe, T., Poston, A., Morgan, C., Musau, F.: Scottish passive house: insights into environmental conditions in monitored passive houses. Sustainability **8**(412), 1–24 (2016)
30. Jaber, S., Ajib, S.: Optimum design of the Trombe wall system in the Mediterranean region. Sol. Energy **85**(9), 1891–1898 (2011)
31. Jovanovic, J., Sun, X., Stevovic, S., Chen, J.: Energy-efficiency gain by the combination of PV modules and Trombe wall in the low-energy building design. Energy Buildings **152**, 568–576 (2017)
32. Folaranmi, O.A.: User participation in housing unit provision in Kwara State Nigeria: a basis for sustainable design in mass housing design. Interdisc. J. Contemp. Res. Bus. **4**(2), 723–732 (2012)
33. Maccowan, J.R.: Biomimicry + Urban Design, Master Program Book, pp. 1–194 (2012)
34. 3D Builder Drawing Software: Microsoft Store application
35. Ferrier, J.: Resilient cities for the 21st century, changing the infrastructure of Paris (2016)

Assessment of Activated Charcoal Efficiency with Filter Paper in the Water Purification Process

Nudžejma Jamaković(✉), Enver Karahmet, and Branka Varešić

Faculty of Agriculture and Food Science, University of Sarajevo, Zmaja od Bosne 8, 71000 Sarajevo, Bosnia and Herzegovina
nudzejma.sahmatra@gmail.com

Abstract. The water purification is process of removing unwanted chemicals, biological contaminants, suspended solids and gases from the contaminated water with a specific purpose. Activated charcoal is often used as part of filters for water purification and on the pharmacological market are available activated charcoal capsule that can consume a certain period of time for cleaning the body. The aim of this paper is to check the efficiency of activated charcoal as a water purifier, and assess whether it can really purify drinking water until adequate measures. The results showed that activated charcoal with filter paper reduces calcium, iron, chloride, lead and cadmium concentrations, the concentration of potassium, magnesium and sodium after passing through active charcoal with filter paper depends on the starting sample, potassium monitors the movement of sodium, and magnesium is disproportionate to sodium. Activated charcoal with filter paper removes carbonates very well, it doesn't remove efficiently chloride and we can say that is not a good copper purifier, but that conclusion is not safe and ultimate.

Keywords: Water · Purification · Activated charcoal · Metals · Chlorides

1 Introduction

Water is essential for life, which is not surprising existence of the term "safe water" and "water treatment" since the beginning of humanity. At the beginning of civilization man had no measuring instruments, but he used his senses to distinguish correct from incorrect water [8, 19]. Author Rajković [15] state that the World Health Organization (WHO) ranked the quality of drinking water in twelve basic indicators of the health status of a population, which confirms its important role in protecting and improving health. The importance of water is growing more and its value is compared to the black gold of the future. Forecasts go so far that the third world war will be driven by drinking water in the world. Albert Einstein said: "Take a deep look, deep in nature, and then you will understand better. There is more perfection in a drop of water than in any machine that man has made" [17, 18]. Water is a complex system, the basis of which is the chemical compound that has a unique formula of H_2O and thus the water belongs to the group of the simplest chemical compounds [7, 6]. Author Velagić-Habul [20] argues that each water molecule can generate 4 hydrogen bonds, two of which give their own hydrogen

I. Karabegović (Ed.): NT 2019, LNNS 76, pp. 758–771, 2020.
https://doi.org/10.1007/978-3-030-18072-0_87

atoms in the connection with the oxygen of the other molecule and the two adjuvants, when it accepts two atoms of hydrogen of adjacent molecules to its oxygen (solitary electronic pairs) (the only exceptions are molecules on the surface).In this way, each water molecule is surrounded by tetrahedron with four other molecules of water [6, 9, 20]. All chemical reactions in the body take place in an aqueous medium under the laws of dilute solutions, indicating that it is impossible to imagine no metabolism without water to keep the leading percentage body. In addition, the water carries with it a series of anions and cations that are needed or not needed by the body, as well as some types of bacteria, it all depends on the degree of purity of water [5, 8, 13]. Until the dissolution of a variety of substances in water comes by the circulation of water in nature [7, 11]. The circulation of water in nature is created under the influence of the sun's rays, part of the water evaporates from the sea, soil and plants. Water vapor rises in the atmosphere, it cools and condenses there and returns to Earth in the form of precipitation. Precipitation is recycled in lakes, streams, rivers or swamps to the ground, where they collect as groundwater [6, 7, 12]. Water impurities can be of different origin (Scheme 1).

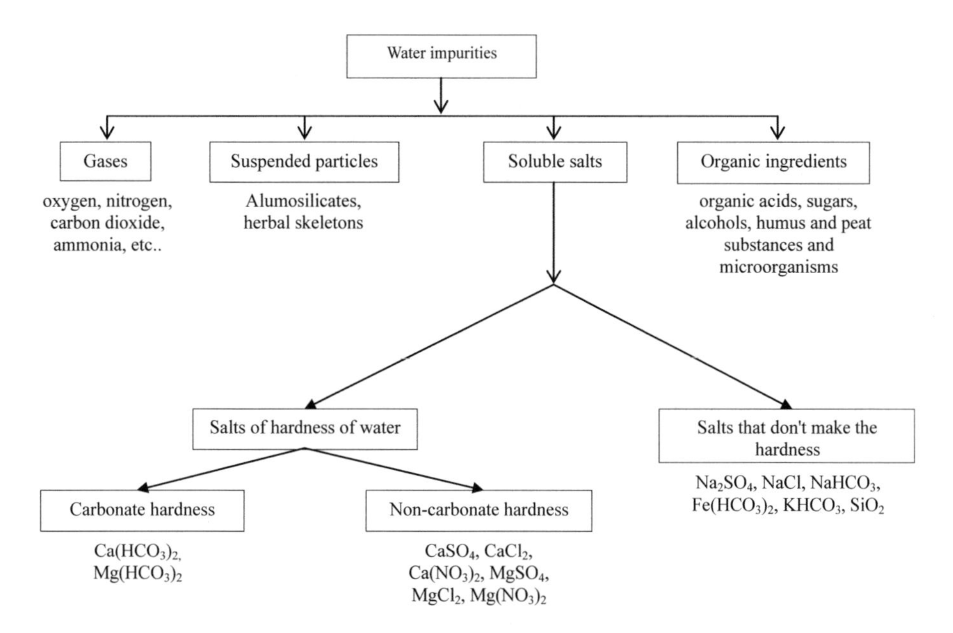

Scheme 1. Most common ingredients of natural water [6]

In Food Law [2] states that water belongs to the category of food and its composition must comply with certain regulations set out in other regulations in accordance with what the water is used for, whether it is used for drinking, is installed in food, used to produce beer, juice, and the like. The content of heavy metals and hardness of water are important in the food industry in the production of juices and beers [3, 4, 14]. Drinking water placed on the market must satisfy the basic criteria in accordance with "on the health safety of drinking water". Regulations are certain maximum permitted concentrations of certain substances in foods that will not endanger the health of the consumer. Authors Akagić and Spaho [3] state that toxic substances are substances that in small quantities endanger health or lead to death. In drinking water are: lead, zinc, copper, fluoride, arsenic, phenols, cyanide, mercury and others. Authors Đukić and Ristanović [6] reported that some natural water occurs increased content of iron and manganese (II) with insignificant toxicity, or deteriorate the quality of water, wherein the concentrations above 0.1–0.3 mg/dm^3 give the iron taste. Water like that has a metalic taste, it is painted with rust paint (which could also be visualized by color of stone), so it is unsuitable for drinking [6, 15]. In the work of Rajković and associates [15] it is said a higher amount of iron in water promotes the development of aeruginous bacteria (Clamydotrix ochracea, Clamidotrix ferrginea and Clonotrix) that develop between the iron hydroxide flakes, creating a residue of unpleasant odor and flavor, and if they accumulate in larger quantities, they can cause clogging and spraying of the pipes, which leads to to the penetration of pathogenic microorganisms and secondary water pollution. The presence of iron in the pipes makes it difficult to maintain the free chlorine in water due to absorption of chlorine. Authors Đukić and Ristanović [6] reported that the sulphate and chloride components as permanent water at high concentration negative influence on the quality of water. At a concentration above 500 mg/dm^3 in chloride, there is a fairly salty taste, and in the sulphate, some functions of the organism are suppressed and have a purgative effect. Calcium and magnesium in drinking water are not considered harmful, but desirable in a moderate amount. In larger quantities they are harmful because they carry carbonates with them. For sodium, the maximum allowable concentration in drinking water is given for its effect on human pressure. Because of this, water treatment is often done in various ways. Water purification is the process of removing undesirable chemicals, biological contaminants, suspended matter and gases from contaminated water in order to obtain water suitable for a particular purpose (drinking water, technological water, washing and cleaning water, etc.) [6, 7]. Some of the methods of water treatment are: softening (kalcijizacijski method, kalcijizacijsko-soda method, a phosphate method), permutitni method (the method of ion exchange), and coagulation and flocculation, removing odors and flavors (most efficient over a bed of activated carbon), membrane filtration, microbiological method (reagent/oxidative and without reagents), removal of iron, defluorisation, fluoridation, desilifikation, manganese removal, desalting, removal of water soluble gases, electrodialysis, hyperfiltration (reverse osmosis), etc. [3, 4, 6, 10, 15]. In addition, various filters and glasses for water purification are available on the market, and the cooking of water and activatedcharcoal are often applied in housework. Based on the review of previous studies researchers have mostly examined the effect of the cooking water to the content of the pathogen, and then to the activated carbon affects the same. It has been shown that activated charcoal effectively removes the petroleum hydrocarbons, including BTEX (benzene, toluene, ethylbenzene and xylene), followed by trihalomethanes, organic elements and some pesticides, humic substances, hydrogensulfide, chlorides, ozone and radon. It does not effectively remove bacteria, viruses, nitrates and some metals, and carbons saturated with contaminants, and is an excellent medium for the development of bacteria, which is not safe for use on the bacteriological side [19].

As enough attention is paid to the problem of bacteriological correctness of water, the question arises what happens to other water components after some form of purification. From this is derived and objective work, and that is to understand the effect of activated carbon on some of the common ingredients of water.

2 Material and Working Methods

In this research work, four water samples were used:

(1) water from the fountain (tap water),
(2) bottled water available on the market,
(3) a sample of water from the polluted river,
(4) the modal system (deionised water obtained on the deionisation apparatus) and the following salts at a concentration of 100 ppm are dissolved: $CuSO_4 \times 5H_2O$, $ZnCl_2$, $Pb(NO_3)_2$, $CdCl_2 \times H_2O$ and $FeCl_3$.

Activated charcoal with filter paper was used as a purifier. Add 1 g of active charcoal to 1 L of water sample and, after shaking, filtered through filter paper. First analyzes were done on the sample before purification (original sample), and then analyzed on the same sample after treatment.

The following analyzes were made:

(1) the content of chloride
(2) the content of the selected metals (Ca, Mg, K, Na, Fe, Cu, Pb and Cd)

The chloride content is determined by the automatic titration method.

The Na and K metals were determined on the flame spectrophotometer, and the metals Fe, Cd, Pb, Cu, Mg and Ca were determined on the atomic absorption spectrophotometer.

The analyzes were done at the Faculty of agriculture and food sciences in Sarajevo (Fig. 1).

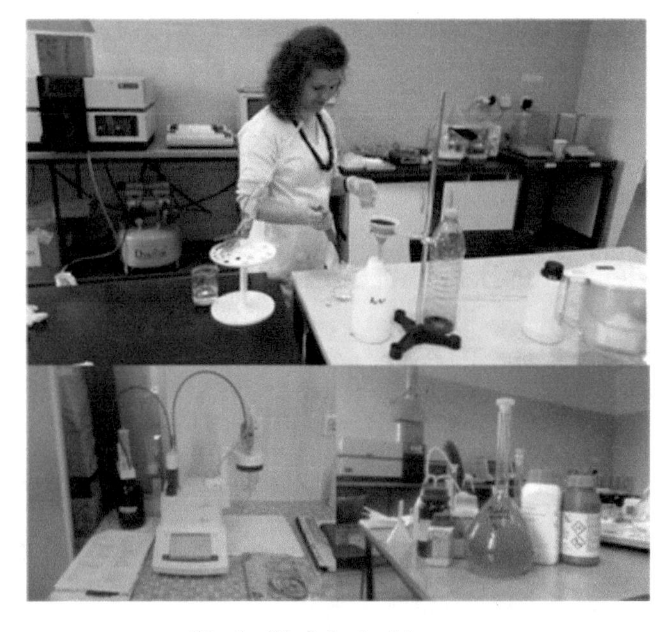

Fig. 1. Work in the laboratory

3 Results of Work and Discussion

In the continuation of the work, laboratory results are presented through diagrams with LSD test results. Each diagram shows one tested parameter for all four samples before and after the treatment, and the LSD test results were done individually for each sample within a single parameter.

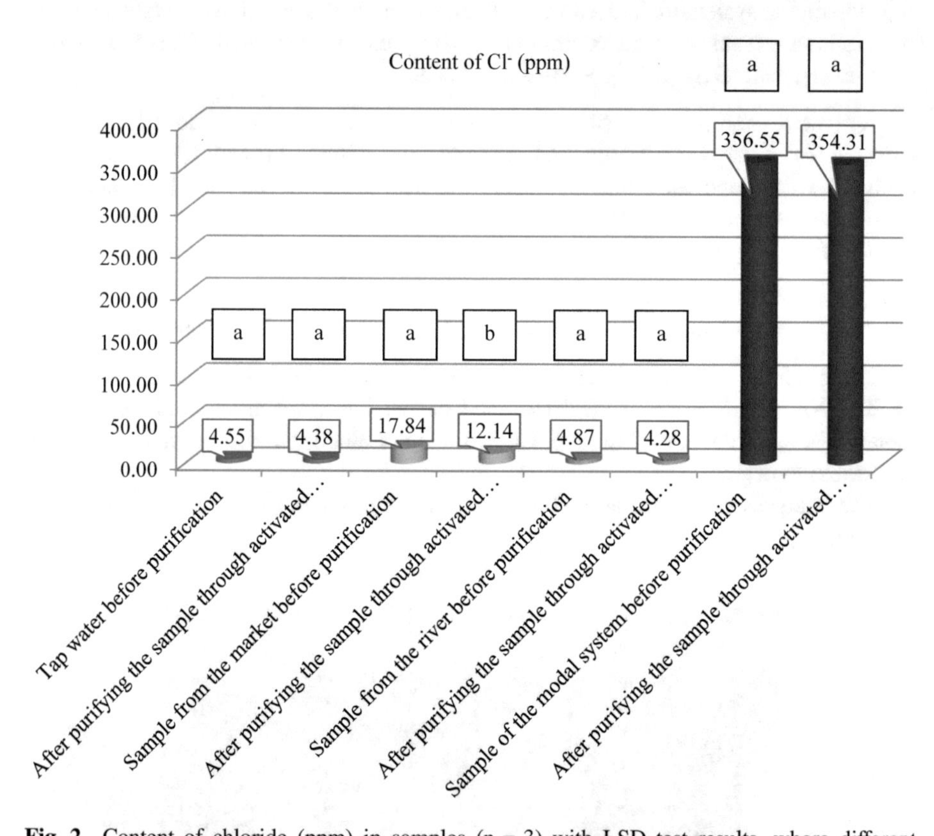

Fig. 2. Content of chloride (ppm) in samples (n = 3) with LSD test results, where different letters on the column mean the existence of statistically significant differences between samples

Activated charcoal with filter paper reduces chloride content (from 4.55 to 4.38, from 17.84 to 12.14, from 4.87 to 4.28 and from 356.55 to 354.31), which is in line with the SHL research (2012), but in this paper it is significant only for the sample from the market.

In Accordance with "on the health safety of drinking water" [1] maximum permissible concentrations in drinking water for chlorides is 250 ppm, so we can say, based on a sample of the modal system and other samples also, that the activated-charcoal with filter paper is not efficient enough in reducing chloride (Figs. 3 and 4).

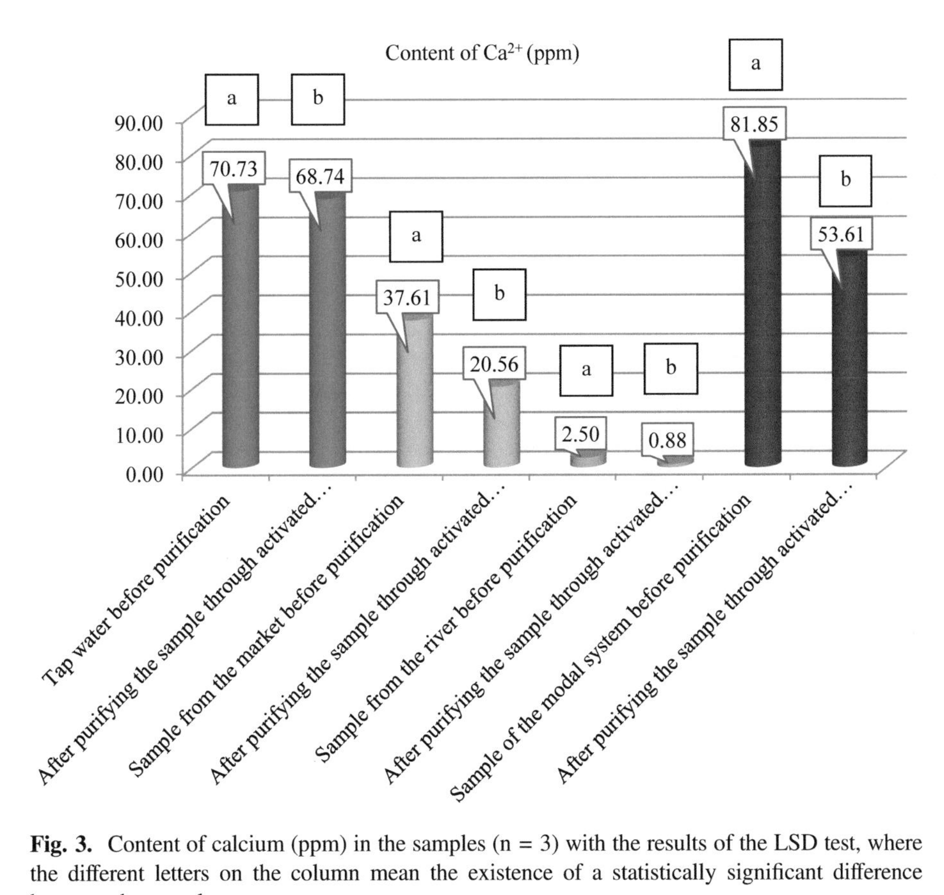

Fig. 3. Content of calcium (ppm) in the samples (n = 3) with the results of the LSD test, where the different letters on the column mean the existence of a statistically significant difference between the samples

Calcium and magnesium, and potassium and sodium are elements that are connected and difficult to detach and must be tracked together. Activated charcoal with filter paper significantly reduces calcium concentration in all samples compared to the starting sample (from 70.73 to 68.74, from 37.61 to 20.56, from 2.50 to 0.88 and from 81.85 to 53.61), while magnesium significantly decreases in tap water and from the market compared to the starting sample (from 10.65 to 9.93 and from 21.90 to 12.38), and significantly increases in the sample from the river and modal system (more organic and inorganic matter, from 1.68 to 5.26 and from 5.26 to 5.58).

In water calcium is present mainly as $Ca(HCO_3)_2$, $Ca(OH)_2$ and $CaSO_4$ and precipitates as $CaCO_3$. Magnesium is present mainly as $Mg(HCO_3)_2$ and $MgSO_4$ and precipitated as $Mg(OH)_2$ and $Mg(OH)_2CO_3$. [6] When passing through activated charcoal with filter paper, calcium is released and thus releases OH^- will be able to bind to Mg. This precipitates Mg as $Mg(OH)_2$ and $Mg(OH)_2CO_3$ (the concentration of Mg decreases). When passing through activated charcoal, magnesium is released and thus releases CO_3^{2-}, which will be able to bind to calcium. This causes calcium to precipitate as $CaCO_3$. In the case of the sample from the river and the modal system,

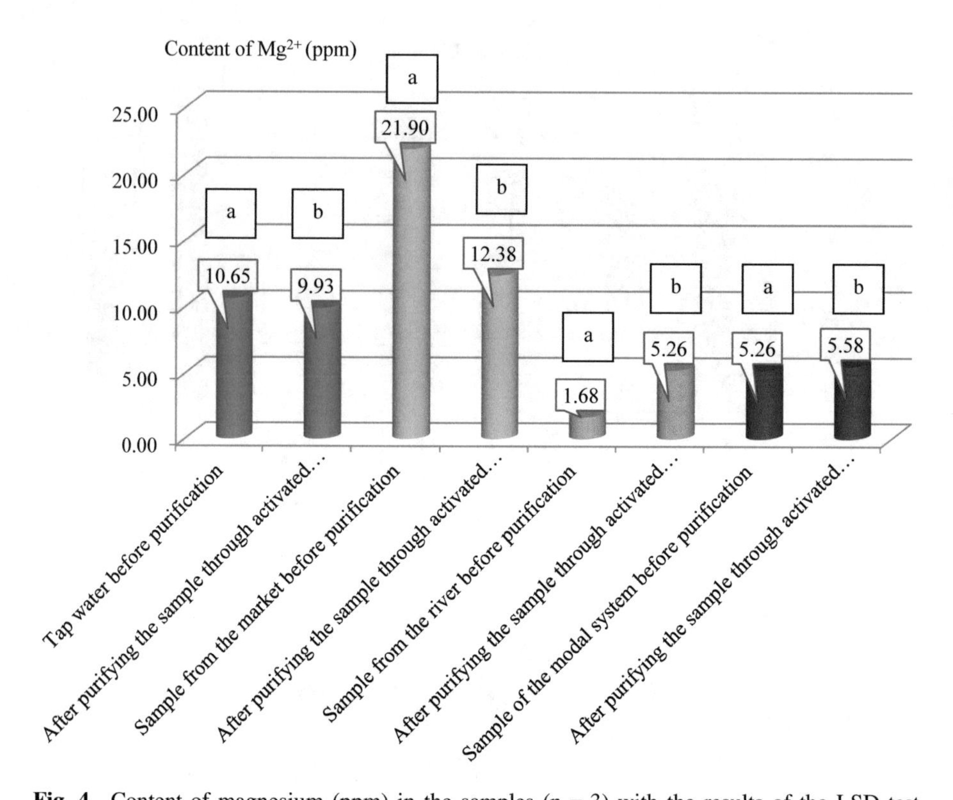

Fig. 4. Content of magnesium (ppm) in the samples (n = 3) with the results of the LSD test, where the different letters on the column mean the existence of a statistically significant difference between the samples

the magnesium is released from the carbonate and the calcium deposits as $CaCO_3$, but Mg doesn't bind to the liberated OH^-. The cause can be the reactions of other elements with the liberated OH^- (in the sample from the river there are many organic elements, in the modal system are many other metals). We can conclude that activated charcoal with filter paper removes carbonates very well, because carbonates are removed at the same time with magnesium and calcium.

In Accordance with "on the health safety of drinking water" [1] maximum permissible concentrations in drinking water for these metals are not given because they are not considered harmful, but desirable in a moderate amount. In larger quantities they are harmful because they carry carbonates with them.

Activated charcoal with filter paper increases the potassium concentration in tap water and in the sample from the market (from 0.003 to 0.237 and from 0.517 to 1.037), and in the sample from the river decreases (from 0.303 to 0.177). In all three cases, this increase/decrease is considered significant. We can assume that the activated charcoal with filter paper has little bound potassium on its particles, which has led to an increase in potassium in the samples, and that potassium bonded to organic components from the river, which caused its decrease, and in the modal system potassium was reacted with other components and that resulted in its reduction.

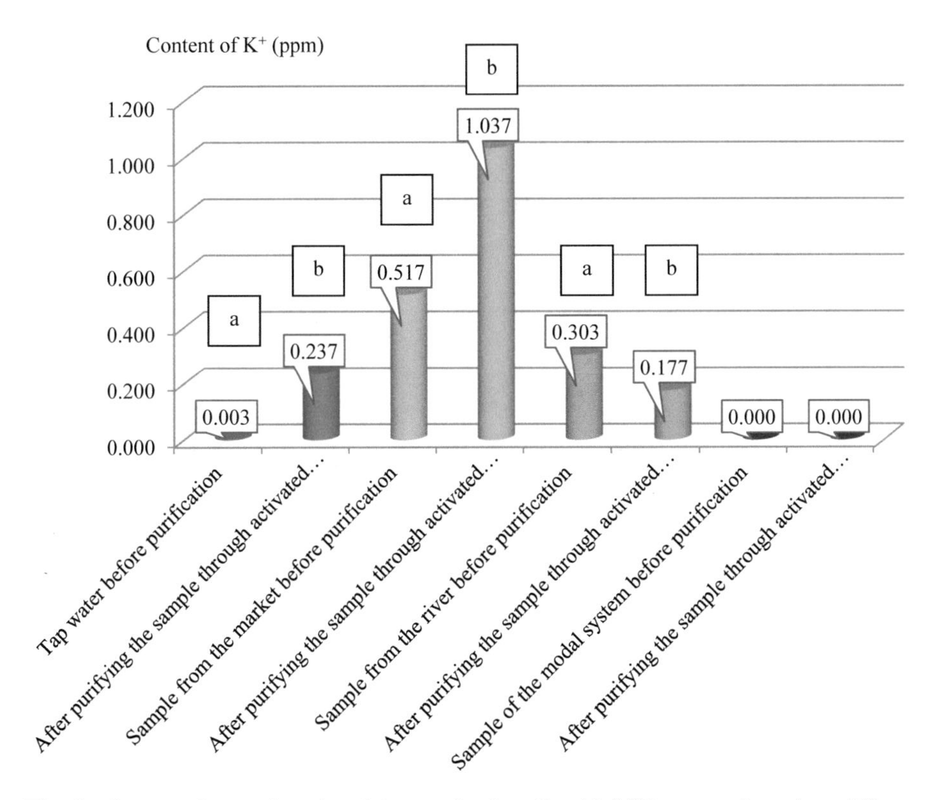

Fig. 5. Content of potassium (ppm) in samples (n = 3) with LSD test results, where different letters on the column mean the existence of statistically significant differences between samples

In Accordance with "on the health safety of drinking water" [1] the maximum permissible concentrations in drinking water for this metal are not given because it is considered to be desirable for health.

Activated charcoal with filter paper significantly increases sodium concentration (from 1.81 to 2.10 and from 42.90 to 98.70) and potassium in tap water and in the sample from the market (from 0.003 to 0.237 and with 0.517 to 1.037), and significantly reduces sodium concentration in samples with more organic and inorganic matter (river and modal system, from 4.60 to 4.20 and from 0.52 to 0.48) and significantly reduces potassium concentration in sample from the river (from 0.303 to 0.177). As previously concluded, activated charcoal with filter paper significantly reduces calcium concentration in all four samples in comparison with the starting sample, while magnesium significantly decreases in tap water and in the sample from the market compared with the starting sample, and significantly increases in the sample of the river and of the modal system (more organic and inorganic matter), we conclude that the concentration of potassium, magnesium and sodium after passing through activated charcoal with filter paper depends on the starting sample and that potassium monitors the movement of sodium and the magnesium is disproportionate to sodium.

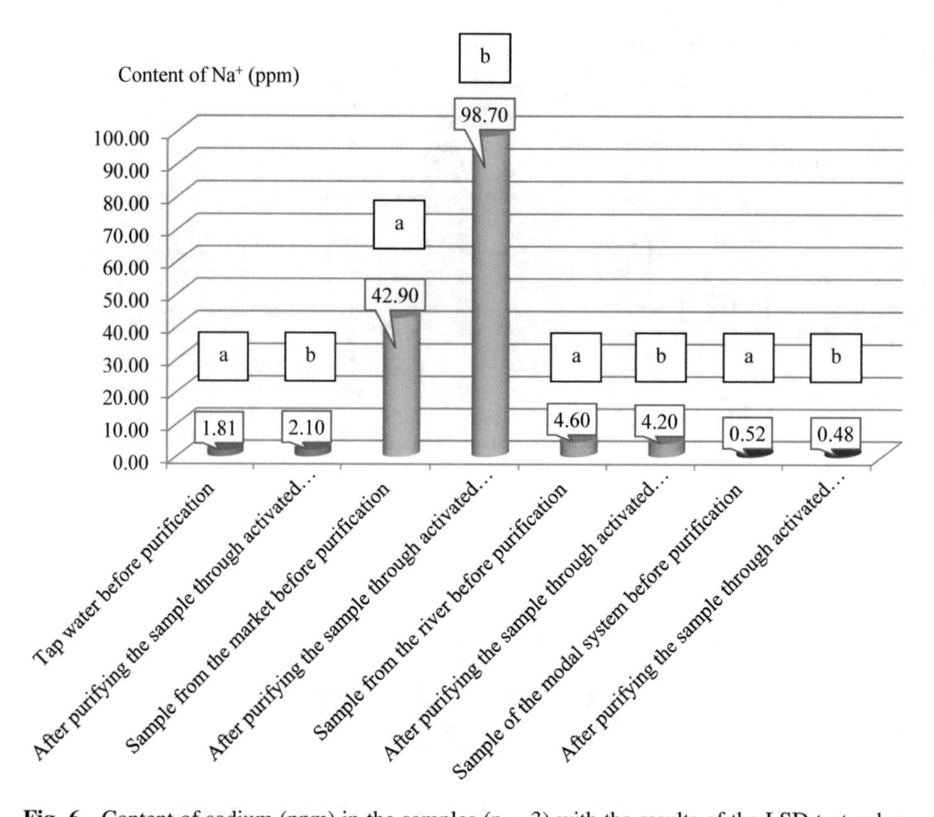

Fig. 6. Content of sodium (ppm) in the samples (n = 3) with the results of the LSD test, where the different letters on the column mean the existence of a statistically significant difference between the samples

In Accordance with "on the health safety of drinking water" [1] the maximum permitted concentration of sodium in drinking water is 200 ppm, and neither in the initial samples before purification the concentration is not exceeded.

Activated charcoal with filter paper significantly reduces the iron concentration in the sample from the river (from 0.25 to 0.00) and in the sample of the modal system (from 73.50 to 18.00, what is 75.51%).

In Accordance with "on the health safety of drinking water" [1] the maximum permitted concentration of iron in drinking water is 0.2 ppm. These purifiers are thought to be capable of lowering the iron to the recommended value, since tap water and sample from the market contains little or no. Of course, the exception is the water naturally occurring iron, such as Guber water in Srebrenica (BiH), which is used in the case of anemia. Such water is recommended to drink with straw (in order to avoid tooth decay) with the addition of a little lemon or orange (vitamin C and acids stimulate the absorption of iron).

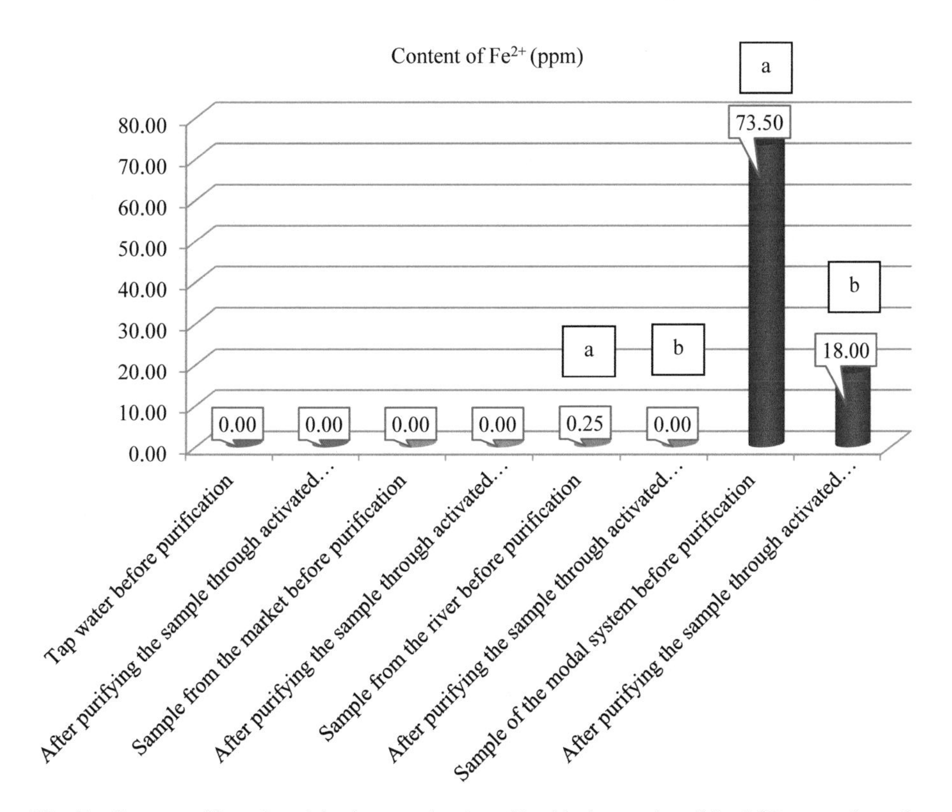

Fig. 7. Content of iron (ppm) in the samples (n = 3) with the results of the LSD test, where the different letters on the column mean the existence of a statistically significant difference between the samples

Activated charcoal with filter paper significantly increases the content of copper (from 10.60 to 12.42). As the modal system is made, among other things, with the $CuSO_4 \times 5H_2O$ salt, and the SHL research (2012)proved that activated charcoal with filter paper reduces sulphides, we can assume that there has been a decrease in SO_4^{2-} in relation to the starting sample of the modal system during purification with activated charcoal with filter paper, and that release of as many copper moles. This liberated copper with activated charcoal with filter paper hasn't been removed, which means that activated charcoal with filter paper doesn't remove copper, either increases or decreases it, as we don't know how much copper has been released completely. In this study the content of the sulphate is not monitored, but we can assume what the results would be in some subsequent researches and investigate it.

In Accordance with "on the health safety of drinking water" [1] the maximum permitted concentration of copper is 2 ppm. This toxic metal was not found in the tap water and in the sample from the market. We can say that activated charcoal isn't a good copper purifier, but that conclusion is not safe and ultimate.

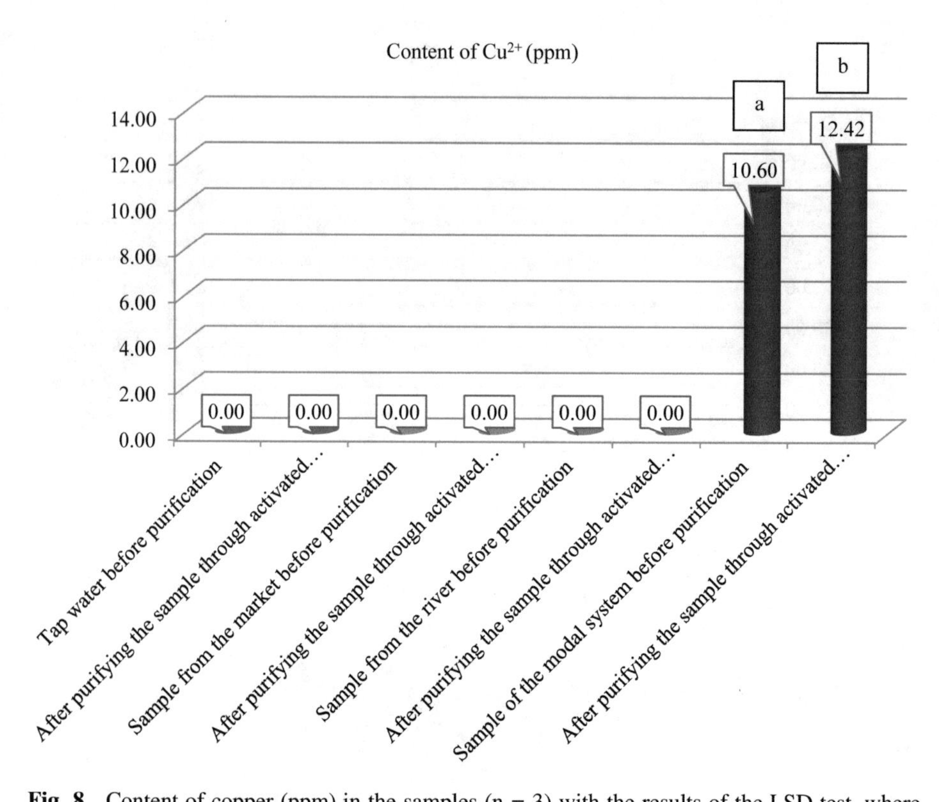

Fig. 8. Content of copper (ppm) in the samples (n = 3) with the results of the LSD test, where the different letters on the column mean the existence of a statistically significant difference between the sample

In tap water, in the sample of market and the modal system (where there was lead), significant lead reduction was demonstrated (tap water and the sample from the market from 0.017 to 0.00 and for the modal system from 3.264 to 2.952). As the modal system was made, among other things, with $Pb(NO_3)_2$ salt, and the SHL research(2012) proved that activated charcoal with filter paper reduces nitrates, we can assume that there has been a reduction in nitrate in relation to the initial sample of the modal system during purification, and thus to the release of as many lead moles as the number of NO_3^- mols was removed, and it is to be expected, as in the previous example, that the lead content will increase after the activated charcoal with the filter paper, but this doesn't happen. It doesn't happen because there is obviously activated charcoal with filter paper good lead purifier.

In Accordance with "on the health safety of drinking water" [1] the maximum permitted concentration of lead is 0.01 ppm. We can say that activated charcoal with filter paper can reduce the lead concentration below this maximum allowable value for samples tap water and from the market, and in the modal system it is quite well removed (9.55%). Although the subject matter wasn't the content of the metal in the initial samples, we can notice that in the tap water and in the sample from the market lead traces of lead which reached the maximum allowed value. From this aspect, it is

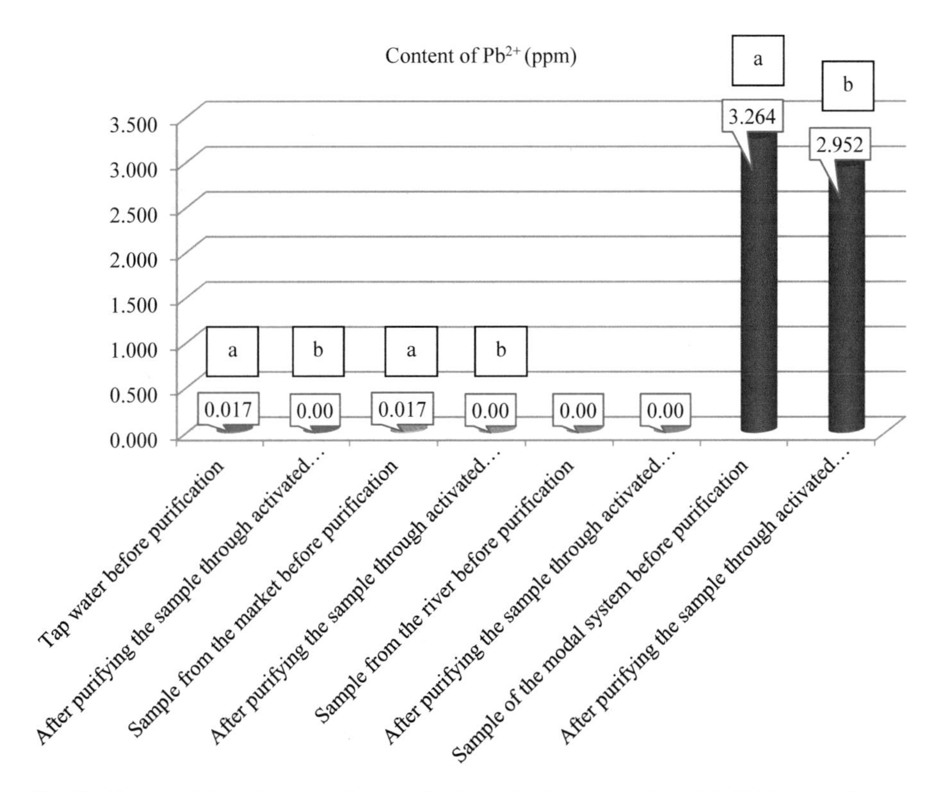

Fig. 9. Content of lead (ppm) in the samples (n = 3) with the results of the LSD test, where the different letters on the column mean the existence of a statistically significant difference between the samples

important to take into account the water we drink and purify if we are able, because neither waters from the market (at least this one being examined) are not safe enough from this aspect. Lead is a metal that can't be removed from the body and its accumulated for years in our bones. It should, of course, be examined and where everything is getting lead in the water and preventing its further maturity.

How only cadmium was found in the initial sample of the modal system, we conclude the conclusion only on the basis of that sample. It has been shown that activated charcoal with filter paper (from 68.493 to 62.086) significantly reduces cadmium content (by 9.34%).

In Accordance with "on the health safety of drinking water" [1] the maximum permitted concentration of cadmium is 0.005 ppm. The mitigating circumstance is that in tested tap water and in the water from the market the cadmium isn't found.

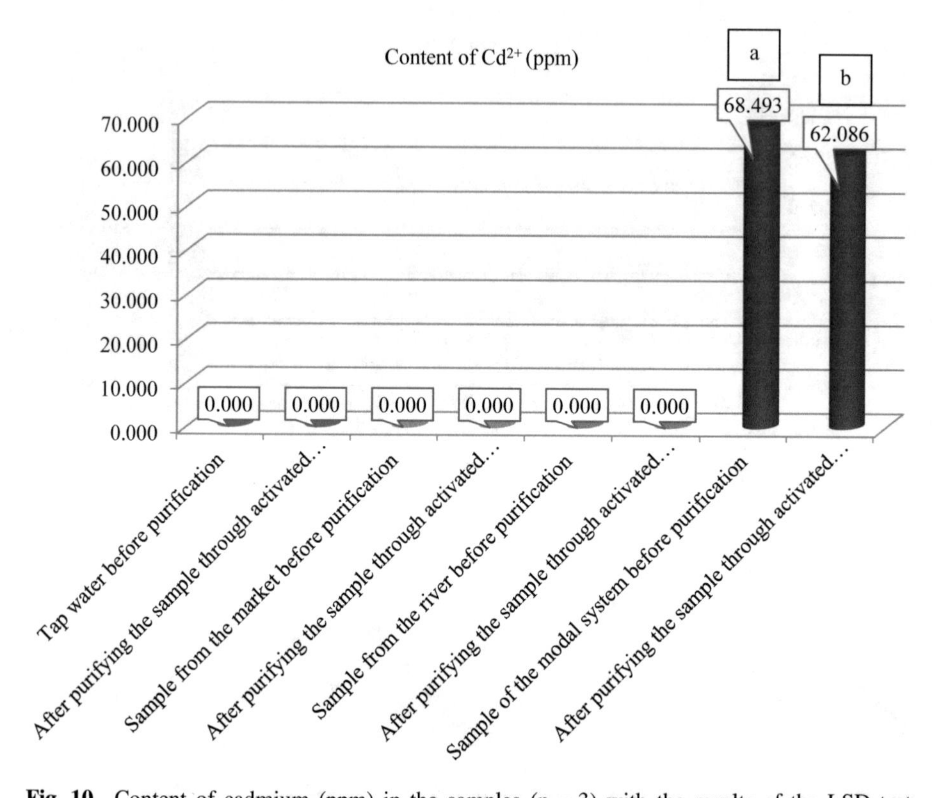

Fig. 10. Content of cadmium (ppm) in the samples (n = 3) with the results of the LSD test, where the different letters on the column mean the existence of a statistically significant difference between the samples

4 Conclusion

The results showed that activated charcoal with filter paper reduces calcium, magnesium, lead and chloride concentrations, and increases potassium and sodium concentration in tap water and in the sample from the market. In addition, reduced the concentration of calcium, potassium, sodium, iron and chloride, and increases the concentration of magnesium in the sample from the river. Activated charcoal with filter paper reduces calcium, sodium, iron, lead, cadmium, and chloride concentrations, and increases magnesium and copper (or it only doesn' t remove it, or it may even decrease) concentrations in a sample of the modal system. The concentration of potassium, magnesium and sodium after passing through active charcoal with filter paper depends on the starting sample, potassium monitors the movement of sodium, and magnesium is disproportionate to sodium. Activated charcoal with filter paper removes carbonates very well, it doesn't remove efficiently chloride and we can say that is not a good copper purifier, but that conclusion is not safe and ultimate.

References

1. Food Safety Agency of BiH: Ordinance on the healthfulness of drinking water, Official Gazette, no. 40, p. 30 (2010)
2. Food Safety Agency of BiH: Food Law, Official Gazette, no. 50/04, p. 1 (2004)
3. Akagić, A., Spaho, N.: Juice and Nectar Technology, Sarajevo, Faculty of Agriculture and Food Science, pp. 13–43 (2017)
4. Ashurst, P.R.: Production and packaging of non-carbonated fruit juices and fruit beverages, Maryland, pp. 386–420 (1999)
5. Borić, N., Ivankić, D.: Lexicon of nutrients with detailed value tables, Rijeka, Leo Commerce d. o. o., pp. 19–21 (2015)
6. Đukić, D.A., Ristanović, V.M.: Chemistry and microbiology of water, Serbia, Stylos, pp. 15–184 (2005)
7. Gržetić, I., Brčeski, I.: Water, Quality and Health, Mol d. d., Belgrade, pp. 7–95 (1999)
8. Haman, D.Z., Bottcher, D.B.: Home Water Quality and Safety, Florida, University of Florida, pp. 1, 2 and 4 (1986)
9. Kemer, F.N.: Water guide of Nalkov, Novi Sad, Serbia,, 2nd edn. pp. 11, 13, 114, 118–131, 138, 197–204, 350, 427–428 (2005)
10. Laurent, P.: Household drinking water systems and their impact on people with weakened immunity, MSF-Holland public health department, Holland, p. 1 (2005)
11. Lekić, M., Korać, F.: Repeat chemistry with tests and tasks for preparing qualification exams at faculties, Publishing, Sarajevo, p. 103 (2001)
12. Mijačević, M.: Determination of drinking water quality in Požega, Croatia, p. 5 (2016). https://repozitorij.vup.hr/islandora/object/vup%3A256/datastream/PDF/view. Accessed 20 May 2018. godine
13. Milenković, M.: General Biochemistry, Sarajevo, Svjetlost, 2nd edn. pp. 17–20, 341–342 (1991)
14. Niketić-Aleksić, G.: Technology of fruit and vegetable, Belgrade, Faculty of Agriculture, pp. 28–33 (1982)
15. Rajković, B.M., Stojanović, M., Milojković, S.: Determination of the crystalline calcium carbonate structure obtained from drinking water, Original scientific paper, University of Belgrade, Serbia (2008)
16. Rajković, B.M., Stojanović, M., Milojković, S.: Examination of drinking water quality from individual wells in Dubravica village in Branicevo district, Original scientific paper, University of Belgrade, Serbia (2015)
17. Ružinski, N.: Water in Croatia and approximation to Europe. In: Four Conference of Water in Croatia, Croatia, p. 2 (2007)
18. Slavulj, M.: Water-strategic resource of 21 century, the University of Juraj Dobrila in Pula, Croatia, pp. 1, 17 i 18 (2016)
19. State Hygienic Laboratory (SHL) at The University of Iowa: Well Water Quality and Home Treatment Systems, Coralville, Iowa, pp. 2, 4 and 15 (2012)
20. Velagić-Habul, E.: Chemistry of Food, pp. 1–12. University of Sarajevo, Sarajevo (2010)

Map of Characteristic Snow Loads on the Ground of Bosnia and Herzegovina

Rašid Hadžović[1(✉)] and Bakir Krajinović[2]

[1] Faculty of Civil Engineering, "DzemalBijedic" University of Mostar, USRC "MithatHujdurHujka", 88 104 Mostar, Bosnia and Herzegovina
rasid.hadzovic@unmo.ba

[2] Federal Hydrometeorological Institute, Sarajevo, Bosnia and Herzegovina
bakir.krajinovic@fhmzbih.gov.ba

Abstract. Every country of European Union is using own map of snow loads on the ground based on real values of snow height and snow density in accordance to Eurocode 1. Values of snow loads are given by different zones with boundaries. Maps are similar but not the same. In Czech Republic map is created in digital form and usable by geographic coordinates of the object place. In Bosnia and Herzegovina map is made in digital form together with experts from Czech Republic. Values are taken from main meteorological stations, calculated and used for map snow load. In this paper is represented 3 years work on map of snow loads on the ground of Bosnia and Herzegovina based on real values of snow height and density.

Keywords: Snow load · Snow height and density · Eurocode 1

1 Introduction

Snow load on the ground in Europe is defined as a map of load according to Eurocode 1 (EC 1). There is a global map for the whole of Europe, and each European country has its own snow map. Depending on the team of experts, maps are made in different ways, so we have divisions in zones with exact values; others are related to equations in zones that follow altitudes or with precisely defined values in places. All maps were made in print, except for the map of experts from the Czech Republic, which was done also in digital form. Bosnia and Herzegovina participated in the development of a map together with experts from the Czech Republic, so that the map of the snow load on the territory of Bosnia and Herzegovina was done in print and digital form. Bosnia and Herzegovinian Standard BAS EN 1991-1-3: 2018, Eurocode 1 - Building design - Part 1–3: General operations - Snow load, defines the snow load on the structures and the map shows the characteristic snow load on the ground.

I. Karabegović (Ed.): NT 2019, LNNS 76, pp. 772–781, 2020.
https://doi.org/10.1007/978-3-030-18072-0_88

2 Snow and Snow Data

2.1 Snow and Nature of Snow Load

Snow is a type of precipitation in a solid state, a natural phenomenon, and as any other precipitation is independent of the previous one. It can be arranged on board, in very different ways, depending on the shape of the roof, its thermal properties, the roughness of its surface, the amount of heat generated under the roof, the vicinity of adjacent buildings, the local terrain and the local meteorological climate, especially the wind, temperature variation and probability of precipitation (whether in the form of rain or snow) [1].

Large snowfall makes it difficult for people and goods to move on roads, make additional loads on structures, and thus reduce the reliability of structures. If the snow falls at the same time as the wind blows, there is an accumulation of winds on certain roof elements, which makes additional load on the roof and roof structure and therefore endangers it. Buckets of snow on one part of the roof structure are the most dangerous loads that lead to demolition of buildings and must be taken into account when designing the building. The snow load belongs to geophysical variable loads and is therefore subject to the probability theory, the so-called statistical legality, which focuses on the random process that occur in time, but also the place at which it is measured [1].

2.2 Snow Data

Research, collection and recording of data in Bosnia and Herzegovina has been carried out in the past 66 years in accordance with the Guidelines for Monitoring and Measurement at the Main Weather Stations at the Main and Secondary Measures in the Network of Weather Stations of the Republic Hydrometeorological Institute of Bosnia and Herzegovina and continued in the entity hydrometeorological institutions. During winter and snow days the Institutes make readings of the snow height (cm) on a daily basis, and every fifth day the content of the water in the snow is measured and expressed as the height of the water in the snow (VSV) in (mm/cm) [1].

In accordance with the recommendations of the World Metrology Organization (WMO) for statistical processing and obtaining quality budget results, the minimum period must be 30 years. The data provided for the analysis from the Institute are: snow depth, sampling height and water content for winter periods 1961–1991 for all major measuring points in Bosnia and Herzegovina Fig. 1. The main measuring points are shown on the map of Bosnia and Herzegovina with altitudes [1].

Fig. 1. Major metering points in Bosnia and Herzegovina with altitudes [1]

2.2.1 Heights of Snow

On the basis of daily records, the monthly maximum measured height of snow and height of snow of the sample are determined, together with the corresponding data on the amount of water in the snow in order to determine the density of the snow. As a rule, the height of the snow increases with the rise in altitude, so that the height of the snow should be higher in the mountain parts than the lowlands.

2.2.2 Snow Density Calculation

Snow density is a key element in determining the snow load and is expressed in [kN/m^3]. The density of snow is determined based on the existing data from the amount of water in the snow. In the literature there are different empirical formulas for determining the density of snow, but it is questionable to use them if the origin and location of the formulation of this formula is unknown. The snow density increases with the duration of the snow cover, and depends on the location of the building, the climate and the altitude.

According to the measured heights of the content VSV u (*mm/cm*), the density of snow is counted in (kN/m^3) [1]:

$$\rho_{snijeg} = VSV\frac{mm}{cm} \times \rho_{voda}\frac{g}{cm^3} \times 10^{-2}\left[\frac{kN}{m^3}\right] \tag{1}$$

3 Characteristic Snow Load on the Ground

According to JUS regulations we have the appearance of uneven security of the building, because the snow load, as a natural phenomenon, cannot be determined, since the snow is defined as a random variable. The snow load depends on a number of different regular variations and irregular conditions related to the climate (moisture, altitude, climate zone, proximity to neighbouring buildings, amount of heat generated under the roof…).

For the calculation of the characteristic load of snow on the ground, it is necessary to determine the snow meteorological load [kN/m^2] as a product of the height of the snow sample [m] and the calculated snow density [kN/m^3] for each month individually. After that, the maximum values for each winter are determined, which are the basic parameters for defining the characteristic snow load as a random variable. Each random variable is described by the mean value, the standard deviation, and the corresponding distribution.

For each measurement point, the calculated meteorological snow load tested on possible distribution was done individually using the software STATREL (part of the software STRUREL) as shown in Fig. 2. Measuring point Sarajevo, where it is visible that Gumbel's distribution Extreme Type I best describes the given data.

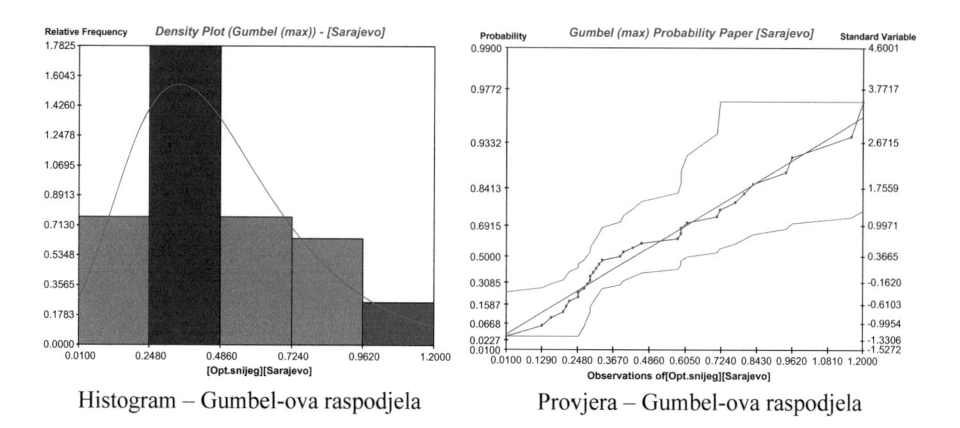

Fig. 2. Results of meteorological snow load test for Gumbel distribution for Sarajevo

In order to obtain a characteristic snow load it is necessary to be calculated for a 50-year period in accordance with the return period of the facility, for each measuring point individually according to Gumbel's distribution, and the following formula is applied [1]:

$$s_K = \widehat{X} + \frac{1}{a}[\ln n - \ln(-\ln p)] \quad (2)$$

- Characteristic snow load on the ground s_k
- modus

$$\widehat{X} = \bar{X} - \frac{c}{a} \quad (3)$$

- coefficient of dropout

$$a = \frac{1}{\sigma} \cdot \frac{\pi}{\sqrt{6}} \quad (4)$$

- number of analyzed years $n = 50$
- Fractional value of the action $p = 0{,}95$
- the mean value of the meteorological snow load $\overline{X}$
- standard deviation of meteorological snow load σ
- Euler's constant c = 0,577

Calculation of the characteristic snow load on the ground, for each measurement point, was performed and the data were entered into the numerical model for the development of map of snow load in Bosnia and Herzegovina.

The following segments of this paper describe the development of a map from the aspect of the meteorological profession in order to create a digital map.

4 Map of Characteristic Snow Load on the Ground of Bosnia and Herzegovina

The data to be considered for the purpose of producing a national snow load additive is: a characteristic snow load on the ground calculated for the main measurement sites, the water content in the snow cover or the density of the snow cover, the maximum height of the snow cover and the index for the climate classification of the climate. After controlling all data and statistical analysis, data on snow cover density were taken from 17 meteorological stations; data on the maximum snow cover height from 77 meteorological stations, while 53 stations were considered for the Kepen climate classification index. In addition to the data from the measuring points in BiH, the measuring points in the border areas from Serbia, Montenegro and Croatia were analysed [6, 7] (Figs. 3, 4 and 5).

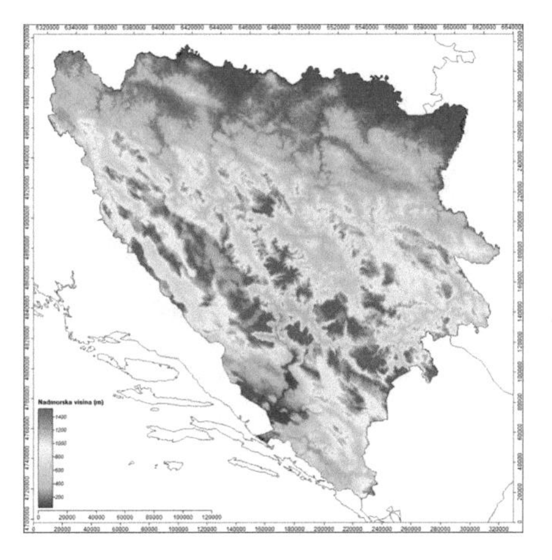

Fig. 3. Digital model of terrain resolution 200 m

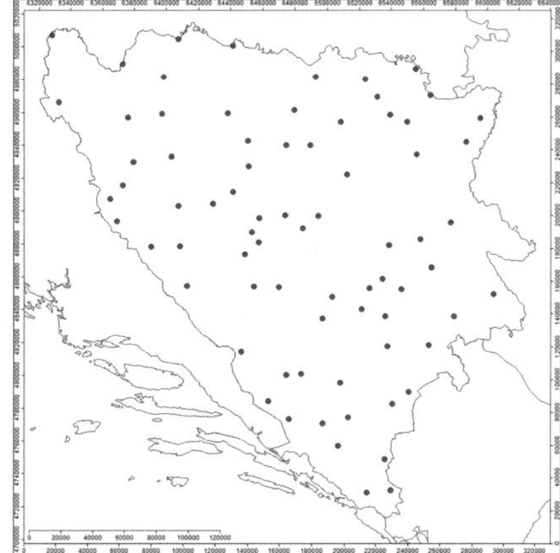

Fig. 4. All stations used for maping

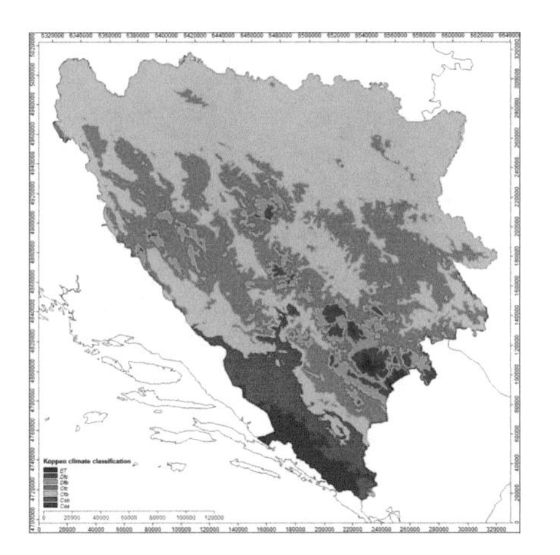

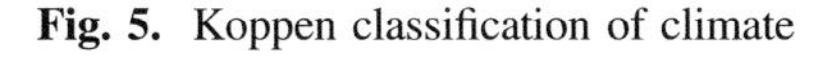

Fig. 5. Koppen classification of climate

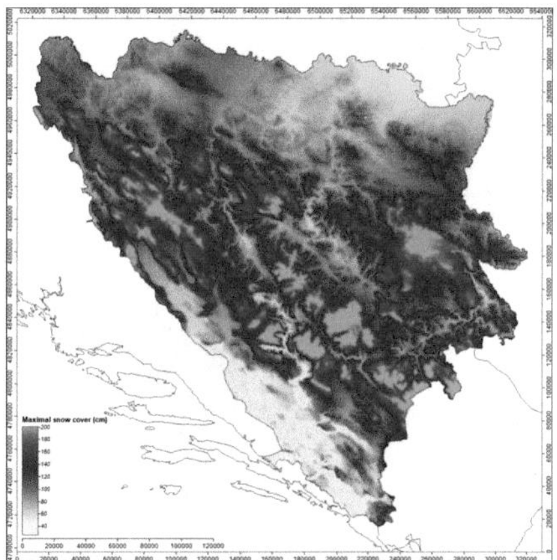

Fig. 6. Maximal snow cover

Based on data from meteorological stations where measured the water content in the snow, the load correlation and the maximum snow height were tested. On the basis of the correlation coefficient, the basis of the maximum snow height was used to improve the spatial distribution of the snow load. As an additional basis for a credible spatial image of the load we also took the index of the Kepen Climate Classification. A large number of stations (77 measuring points) that measured the height of the snow cover and 53 measuring points of the climatological stations compensated for the large deficiency of meteorological stations that performed water content in the snow cover of only 17. The Digital Land Model (DMT) Korsten as the first substrate for all mapping has a very high resolution of 200 ms. This resolution has been selected because of the

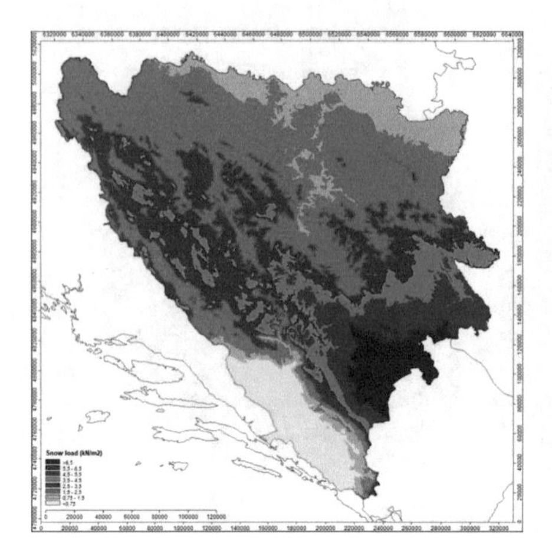

Fig. 7. Snow load and DMT

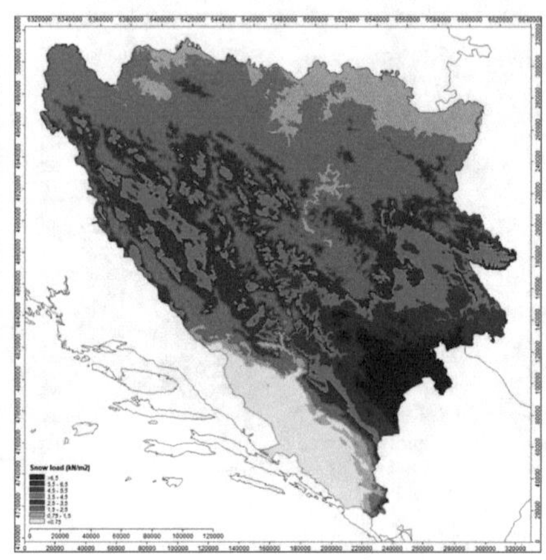

Fig. 8. Snow load with DMT and maximal snow cover

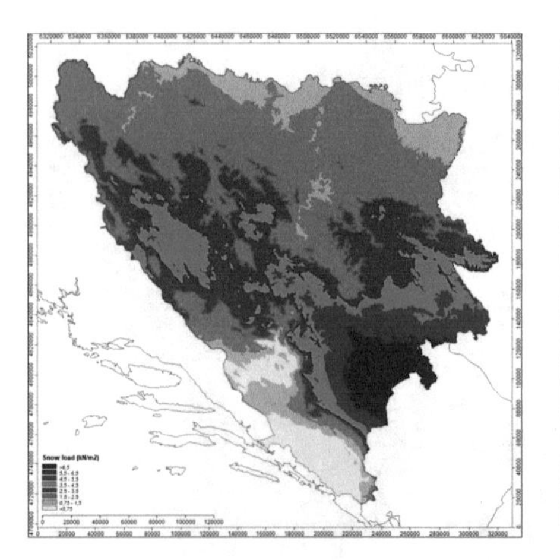

Fig. 9. Snow load with DMT and Koppenclas. of climate

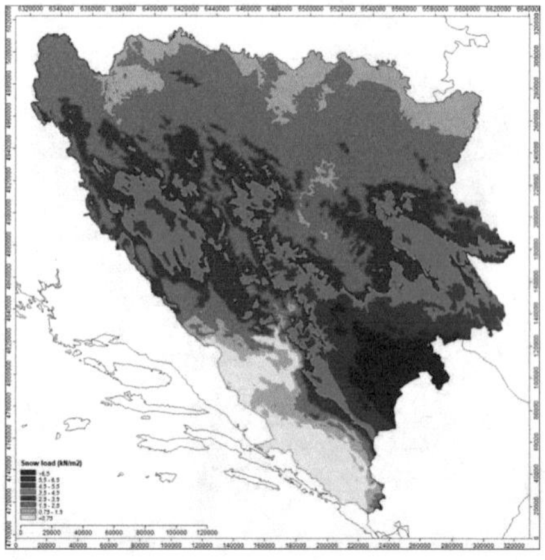

Fig. 10. Snow load with DMT and maximal snow cover and Koppenclas. of climate

project's requirements, because all displays go on a digital map whose image goes to the level of the building itself. The method used for spatial distribution is Universal Kriging, and the System for Automated Geoscientific Analysis, Version: 2.1.1.

On the map Fig. 11 is given the result of the Kriging method, which has the basis of DMT and the calculated snow load. This combination gives a 52.26% reliable spatial distribution, which is certainly not satisfactory (Fig. 9). The map of the maximum height of the snow cover is characterized by reliability of 76.07%, which is a significant improvement compared to the previous case (Fig. 10). After taking into account the Kepen climate classification index, the spatial reliability of the map is 81.08%. The final result is shown on Fig. 12 (Figs. 7 and 8).

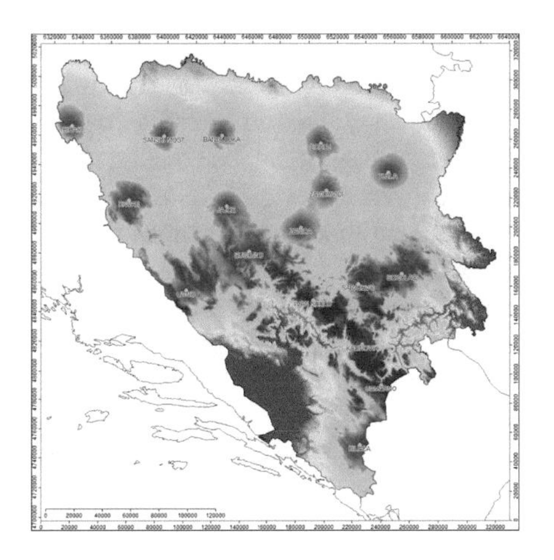

Fig. 11. Variance of snow load and DMT

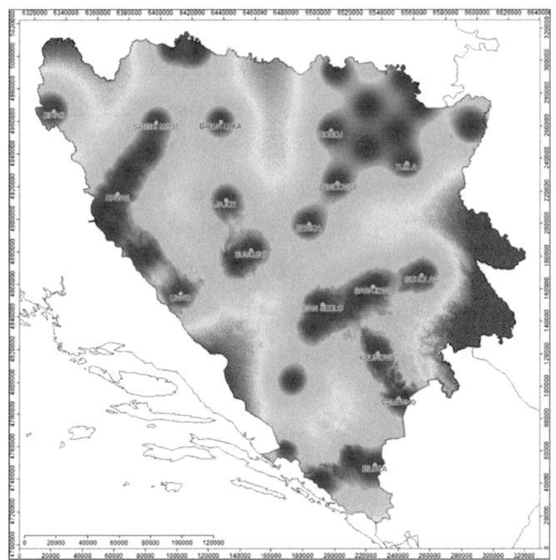

Fig. 12. Variance of maximum snow cover and DMT

Based on the complete analysis, a map of the characteristic snow load on the ground in Bosnia and Herzegovina is represented at Fig. 13.

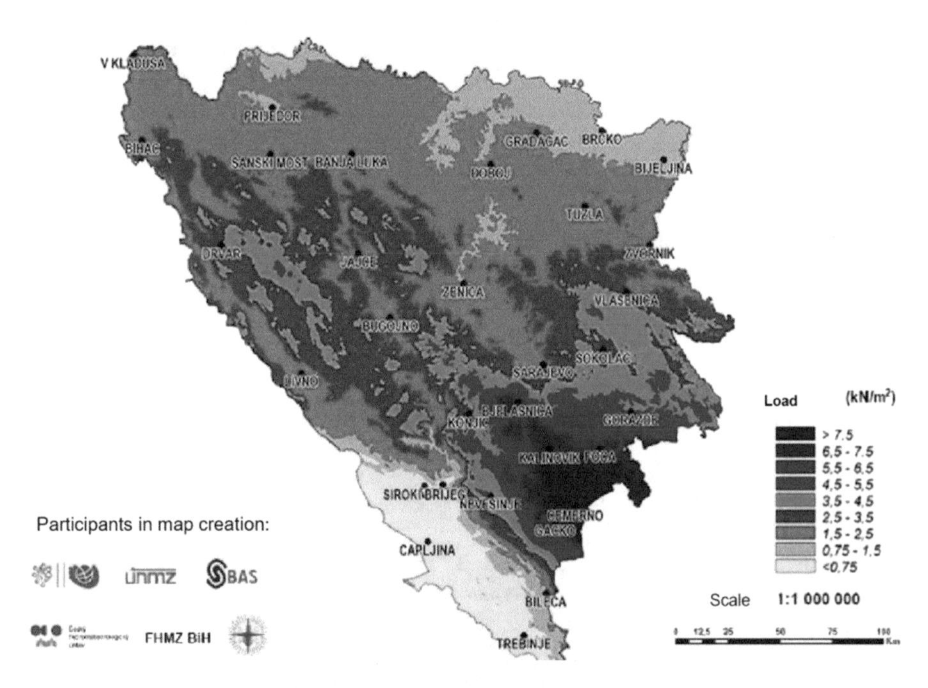

Fig. 13. Printed map of characteristic snow load on the ground of Bosnia and Herzegovina

Figure 13 is a map of the snow load on the ground in Bosnia and Herzegovina, and the digital version is available on the web: www.eurokodovi.ba/snijeg. By simply clicking or by entering the coordinates of the site, the snow load is spread across the whole of Bosnia and Herzegovina. If the printed form is used, it can be concluded that each colour represents one zone, so the minimum load value is <0.75 kN/m^2, and the maximum >7.50 kN/m^2. Due to the large differences in altitude, the snow load is given in 9 different zones with a value difference of 1 kN/m^2 in order to minimize dispersion of the result. The biggest problem was the 9th zone due to the large amounts of snow and mountain massifs in the east of Bosnia and Herzegovina. If a digital map is not available, the map thus prepared can be used, with exception of zone 9, where there is a possibility of a phenomenon of larger snow than 7.5 kN/m^2. It is recommended that the upper values are taken in the zones, and that additional effort is being made in the zone 9 to open the web site since the values can be higher than 10 kN/m^2.

5 Conclusion

Based on the available data from the Hydrometeorological Institute of FBiH for the period 1961–1991, the values of snow load for the return period of the facility of 50 years in Bosnia and Herzegovina were calculated and estimated for the main metering points and included in the program for Automated Geoscientific Analyses, Version: 2.1.1. Given the very small number of main measuring points, secondary snow gauges were also used in order to obtain a quality map of the characteristic snow load on the soil of Bosnia and Herzegovina in accordance with Eurocode 1. The map is available on the website www.eurokodovi.ba, and is owned by the Institute for standardization of Bosnia and Herzegovina www.bas.gov.ba.

References

1. Hadžović, R., Peroš, B.: Pouzdanostkonstrukcijadominantnoopterećenihsnijegom u BosniiHercegovini, GrađevinskifakultetUniverziteta "DžemalBijedić" uMostaru, Mostar (2016)
2. Eurocode 1: Actions on structures - Part 1–3: General Actions – Snow loads - National Annex
3. Tadić, M.P.: Karta karakterističnogopterećenjasnijegom, Kartografijaieurokodovi (2012). http://www.kartografija.hr/tl_files/Hkd/dogadjaji/Svjetski%20dan%20GISa/prezentcije/05_2012_Svjetski%20dan%20GIS_snijeg_MPTadic_web.pdf
4. Holický, M., Marková, J., Sýkora M.: ZatíženíStavebníchkonstrukcíPříručka K ČSN EN (PriručnikizRepublikeČeške) (1991)
5. Zaninović, K., Gajić-Čapka, M., Andrić, B., Džeba, I., Dujmović, D.: Određivanjekarakterističnogopterećenjasnijegom. Građevinar **53**(6), 363–378 (2001)
6. Jonas, T., Marty, C., Magnusson, J.: Estimating the snow water equivalent from snow depth measurements in the Swiss Alps. J. Hydrol. **378**, 161–167 (2009)
7. Krajinovic, B., et al.: Atlas klimeFederacijeBosneiHercegovine, Federalnihidrometeorološkizavod, BiH, Sarajevo (2018)

8. Luna, M.J., Morata, A., Chazarra, A., Almarza, A.: Mapping of snow loads on the ground in Spain. Geographical information systems and remote sensing: environmental applications. In: Proceedings of the International Symposium held at Volos, Grece, 7–9 November 2003
9. Sturm, M., Taras, B., Liston, G.E., Derksen, C., Jonas, T., Lea, J.: Estimating snow water equivalent using snow depth data and climate classes. J. Hydrol. **11**, 1380–1394 (2010)
10. Hadžović, R.: "Određivanjesigurnostinosivihkonstrukcijazakarakterističnoopterećenjesnijegom u BosniiHercegovini", magistarski rad odbranjen. GrađevinskifakultetUniverziteta u Sarajevu, 28 June 2004

Advantages and Deficiencies of "In-Situ" Explosive Materials Production ("Nalim" Technology)

Senaid Bajrić[1(✉)], Azra Špago[2], Merima Šahinagić-Isović[2], Fuad Ćatović[2], and Amira Kasumović[2]

[1] Zagrebinspekt d.o.o., Rudarska 247, 88000 Mostar, Bosnia and Herzegovina
pctuzla@zgi.eu
[2] Faculty of Civil Engineering, "Džemal Bijedić" University of Mostar, Mostar, Bosnia and Herzegovina

Abstract. The technology of explosive materials production by "NALIM" ("in-situ") technology represents the production of explosive materials at the point of their application, respectively production of explosive materials "in-situ". It is used for the surface exploitation of useful mineral raw material deposits both in Bosnia and Herzegovina and around the world. This production technology has a significant number of advantages over conventional stationary technology for the production of explosive materials, as well as a certain number, mostly minor, of deficiencies for which it can be claimed as insignificant in relation to the advantages of this technology.

Keywords: "NALIM" technology · BULK vehicles · Explosive materials · Mineral row materials

1 Introduction

The deposits of useful mineral raw materials in Bosnia and Herzegovina (BA), on which there are technical and economic justifications for the application of surface exploitation, are the work environments for whose demolition applies the energy of explosive materials decomposition and thus the exploitation of useful mineral raw materials.

This paper aims to evaluate the advantages and justification of the introduction and use of equipment for the application of mechanized production of explosive materials at the place of use and mechanized filling of mine drillhole at the point of production of explosive materials under the conditions of the work environments of the Banovići - Đurđevik coal basin, BA [6].

Technology of explosive material production by "NALIM" technology, in essence, has a meaning of producing explosive materials at the point of their application, respectively production of explosive materials "in-situ". This production technology has a significant number of advantages over conventional stationary technology of explosive materials production, as well as a number of, mostly minor, deficiencies that can be claimed to be not significant in relation to the benefits of this technology.

I. Karabegović (Ed.): NT 2019, LNNS 76, pp. 782–789, 2020.
https://doi.org/10.1007/978-3-030-18072-0_89

The application of explosive materials production at the point of their use ("NALIM" technology) implies the production of modern explosive materials such as ANFO, Heavy ANFO, explosive emulsions and numerous of combinations of these modern explosive materials for the general purpose [2, 4, 6, 7, 9].

Complex work environments where the drilling and blasting technologies perform, their very heterogeneous compositions, particularly in terms of geological, geomechanical, hydrogeological and other conditions, make it conditional that for their demolition are applied flexible technologies, respectively modern explosive materials whose characteristics, according to their resistance and other work environment properties that are demolished by the energy of explosive decomposition of these materias, can be changed continuously in the process of production at the place of use. These conditions cann't be achieved by the use of simple plants for the production of explosive materials on site, respectively they cann't be achieved under the conditions of the simple batch production of explosive materials.

2 Advantages of Explosive Materials Production at the Place of Consumption

Advantages of the explosive materials production at the place of consumption, compared to stationary production technologies in production plants are as follows [1, 3, 6]:

1. The complete process of explosive materials production is carried out on mobile plants, out of any pyrotechnical building facilities, excluding the construction of demanding and expensive production infrastructure. There is no need to use construction land for the production by "NALIM" technology and therefore no obligation to obtain various approvals and consents related to the construction of buildings.
2. Installations for the production of explosive materials by "NALIM" technology can be very simple, small dimensions and weights so that the same, together with the corresponding components, transports from one destination to another, whereby all manipulations, especially transport, are carried out with less dangerous, or harmless materials.
3. For the production of explosive materials which have better thermodynamic and blasting-technical characteristics, for whose production a larger number of components are used as sensitizers and/or phlegmatizers and other emulsifiers, as well as for the production of larger quantities and more types of explosive materials it is necessary in the application to introduce a special BULK vehicles that have special superstructures and other equipment for the production of explosive materials in highly precision programmed and controlled quantities of final products, capacity 5 to 12,5 t.
4. Warehouse facilities of raw material that enter the technological process of production, especially ammonium nitrate as a component with a significant share, are built on one, in accordance to all pyrotechnic and other regulations, defined location, while the warehouses of other components are mainly built as multipurpose warehouses.

5. Technology is based on modern, mostly stable components (especially ammonium nitrate), which can be provided in sufficient quantities even in conditions of major market disturbances.
6. If there is a failure on the mechanisms and installations of the mobile plant during the production of explosive materials, manual operation can complete the initiated process of production and thus make the bridging in consumer supply until the plant is brought to the proper and functional state.
7. Mechanized equipment for the production of explosive materials by the NALIM technology has great potential for modernization, computerization and high production effects in almost all conditions of work environments. It can be said that technical and technological solutions on BULK vehicles almost every 2 to 3 years offer innovation that contribute to the development in technology of explosive materials production, safety and security of their application, cost effectiveness of blasting and environment protection.
8. When applying this production methodology the blasting field, where blasting shot holes are drilled, becomes a place of production, where in a much shorter time and with significantly less manipulations with explosive materials the exposure time of employees to the potential impact of explosive materials is reduced to a relatively small extent, because components for the production of explosive materials, mechanically by means of modern plant for the production of explosive materials are dosed from a specialized vehicle directly into mine shot holes and only after about 10 min they create explosive matter.
9. There is never a need or justification for the quantities of explosive materials, produced in certain blasting fields, to exceed the requirements of the planned filling in mine shot holes on a given blasting field, so there is no a need to carry out any storaging of explosive materials or their transportation to other destinations, as well as no other manipulations with these dangerous substances.
10. The mobile plant for the production of explosive materials on the site is mainly independent of the continuous sources of energy, especially electricity. Most often these plants are completely autonomous, convenient and easy to use, which points to the fact that the training of operators for the work with the same is done in a very simple and fast manner.
11. When mass blastings by using explosive materials produced at the place of use are carried out in demanding conditions, especially in the case when the facilities exist in a gravitating environment, plants, installations, etc. which are of particular importance or are particularly sensitive to the seismic and aerial effects of blasting, all manipulations with explosive substances should be carried out by operators with a high degree of expertise and extensive professional experience, such as geological, geophysical, geomechanical, pedological, geodetic and other professions.
12. For the application of mechanized equipment for the production and use of explosive materials on site, there is no any transportation of explosive substances on public roads. In this way, various procedures for obtaining approvals for the transport of dangerous goods on public roads are avoided, eliminating restrictions on the movement of vehicles with dangerous substances due to some safety factors, and eliminating all the dangers in terms of general safety of people and property in all environments.

13. On the prepared blasting fields where mine shot holes have been drilled, the process of establishing production, production and installation of explosive substances in mine shot holes can be carried out without any danger of degradation of the working and live environment.
14. The mechanism of explosive decomposition of explosive matter in mine shot holes can be managed. Blasting can be performed fully controlled with a high degree of reliability, where the potential energy level losses are minimized or completely eliminated, and the seismic and other undesirable effects of explosive decomposition are also eliminated or reduced to the permissible limits.
15. By production of explosive materials out of the site of drilling-blasting works establishes an incessant chain of potential dangers to the ecosystems of the production facility, i.e. from the production site under the classical manufacturing conditions and factory storaging of explosive materials, through their transport to the place of use, i.e. the place of storage at the user up to the descending of explosive materials in mine shot holes and their activation. By production of explosive materials at the site of use, all of the potential hazards characteristic for the classical technology of explosive materials production are localized only to the site of application of explosive substances and a narrow gravitating area in which the seismic effects of mass blasting can be manifested.
16. In the application of "NALIM" technology, there is no need to build facilities for storing and keeping explosive materials because there is no need to create and store reserve quantities even on objects that are significant consumers of explosive materials.
17. The introduction and application of "NALIM" technology is in the function of a multidisciplinary professional or scientific team who will be able to provide a high degree of application of this technology.
18. All explosive materials produced at the site of application are much safer for manipulations than classic explosives, because as raw materials are used mainly non-explosive substances, what makes the final products to be characterized by low sensitivity.
19. The production of explosive materials at the place of use almost completely excludes the dissipation of raw materials and/or produced explosive materials to the blasting fields as well as the dissipation of different packaging used for the supply of raw materials, respectively explosive substances to blasting fields. This reduces losses in mining technology, and the environment is protected from any damage.

2.1 Deficiencies of Explosive Materials Production at the Place of Consumption

As the harmful effects of the production and application of explosive materials by NALIM technology, on the ecosystems of production area and use, the following can be pointed out [5–8]:

1. All explosive substances produced by this system are predominantly mixtures of explosive or non-explosive aggressive or non-aggressive and by other criteria

various substances which, by their separation from explosive materials mixture, due to poor homogeneity, incomplete explosive combustion, storage and improper keeping, can be released into working environments, through cracks, faults and other openings, carried by capillary waters and other natural activities of working environments, flow into working environments, sink into water streams or evaporate and go to the atmosphere.

2. Explosive substances that are produced at a place of application in a significant extent use different oils, paraffins, paraffinic resins etc. as components. These substances present particularly great potential hazards to soil, water and air because they tend to be easily ignited with the consequences of creating gaseous products with toxic and non-toxic components of smoke and soot as well as very rapid movement through all openings in working environments to very significant depths, which are very often converted into various cracks whose life is almost impossible to estimate. These substances are very efficiently and quickly transported in all water streams because they are easier than water and can be maintained on its surface for very long periods. Due to this fact, the use of modern mechanized equipment to produce explosive substances at the place of application, which has completely closed systems of raw materials and final products, should be used.
3. Under the conditions of the great fragmentation of mining technology (industry) in BH, the primary tendency of the owners of mining facilities to make production with as few investments as possible, it is resorting to use the "NALIM" technology with the simplest equipment, unprofessional personnel and raw materials of unknown and inadequate quality. In this way, produced explosive materials are cheap, but often very unstable, subject to breakdowns and interruptions of established detonation chains, therefore reducing the blasting effects, releasing harmful components that remain in the working environment or are transported above the working environment with the already mentioned consequences to the ecosystem, generating uncontrolled large seismic effects from blasting and obtained material from blasting contain a very significant number of out-of-dimension units, which request additional blasting with new potential hazards to ecosystems.

3 Application of Modern Mechanized Equipment for the Production of Explosive Materials at the Place of Use

For a longer period, technologically advanced countries produce various types of mobile mechanized equipment for the production of modern explosive materials. The upgraded bulk cargo for the filling with raw materials and production of modern explosive materials, the BULK vehicle [1, 3, 5, 6], is equipped for very effective production of ANFO, Heavy ANFO, emulsion explosives and a large number of combinations of these modern types of explosive materials, Fig. 1.

Fig. 1. Preparation for sampling of explosive material heavy ANFO on BULK vehicle

The main characteristitcs of BULK vehicle are as follows:

1. Nowadays, a dozen types of very modernly equipped BULK vehicles are produced in the world that carry raw materials for production of 5 to 12.5 t of modern explosive materials. This means that in one cycle of operations the raw materials can be delivered, produced explosive material and filled the blasting field where 12 t of explosive material will be installed.
2. A bulk vehicle is supplied with raw materials in the owner's warehouse, which by itself does not represent a dangerous facility. Charging the BULK vehicle tanks with individual components is carred out mechanically without the need for any physical strain of the operator. When filling tanks and reservoir system of the vehicle, there is no any waste of raw materials or energy sources, as everything is done in closed systems. The production capacity of the BULK vehicle is different and depends on a number of factors, but it can be assumed that some average production capacity of these vehicles is about 200 kg/min or 12 t/h, produced and in mine shot holes installed modern explosive materials. This means that a BULK vehicle with capacity of 12 t for one hour produces and fills a blasting field with 12 t of built-in explosive material, which is more than four times faster than when applying plant which works by batch production system.
3. The bulk vehicle is served by only two workers, one of which is a driver and a plant operator and the other one is carring out the filling of mine shot holes with explosive material through a dozing flexible tubes connected to the BULK vehicle and pulled into the top of the mine shot hole, what exclude any waste of the raw material or explosive material.
4. During the production of modern explosive materials on the BULK vehicle in the mine shot holes, in exactly reciped and controlled proportions, the components for the production of explosive materials are dosed and only after the time of 10 min after dosing, in the mine shot hole by itself they adopt all the characteristics and properties of the explosive materials.
5. If there are extraordinary circumstances at the blasting field that cause a work suspension at the blasting field, the work of the BULK vehicle stops and all the remaining raw materials preserves in it. Filled mine shot holes are kept as well as the classically filled blasting field until conditions are created for their activation.

This excludes any need for the explosive materials produced to be transported to and stored in the warehouse. Such work excludes any need and the possibility to transport the produced explosive materials on public roads, and to store and preserve it. This means that when using a BULK vehicle, a mining company or other owner of this vehicle may not own any pyrotechnic storage facilities, which excludes the engagement of the guard service, the warehouseman and other persons whose jobs are related to the maintenance and use of these facilities. It can be said that the use of BULK vehicles for the production of explosive materials, on the jobs for production and other manipulations with explosive materials and blasting involve about 10 workers less than in the application of classical methods of production.

6. Modern computer equipment of BULK vehicles create the conditions to quickly and efficiently change the recipes for the production of explosive materials. In this way, even in one shot hole with medium depth (up to 25 m), it is possible to change the properties of the explosive material, respectively the type of explosive material that fills the mine shot hole. The effects of these possibilities are very numerous, especially when working in complex working environments such as the working environment of the Banovići - Đurđevik coal basin.
7. Modernly equipped BULK vehicles justify the engagement of highly skilled personnel, various related professions, for drilling and blasting operations. In this way, it significantly contributes to the development of the exploatation technic and technology of useful mineral raw materials, in particulary the increase in the economy of blasting and protection of people, goods and all ecosystems in gravitating and wider areas.
8. Controlled production of explosive materials through the management of their physico - chemical and blasting - technical properties under the conditions of modern BULK vehicles application creates all the preconditions for controlling seismic and other blasting effects and to reliably manage them. On this way naturaly and by human work created goods are protected in the near and wider environment of mining facilities.

4 Conclusion

The application of flexible production of explosive materials by "NALIM" technology, which quickly adapts to all physical-mechanical, structural, geological, geomechanical, hydrological and other working environment characteristics, and which provides the two most important conditions for the application of this technology, which is safety of work and environment, people and property in these areas as well as a high degree of blasting economy, can be achieved with the use of modern BULK vehicles. The negative consequences of rocks destruction with explosive energy are necessary both in the application of classical explosive materials and classic blasting technology, as well as in the application of modern methods of production of explosive materials at the site of application and modern blasting methods. However, it is not questionable at all which technology of explosive materials production and which technology of filling the mine shot holes and blasting will have less harmful effects on ecosystems in gravitating environments. It is sufficient to point out that, in the classical methods of production and use of explosive materials, these processes take place on more locations than one of which one is a production facility and completely different location of blasting sites. There are at least two potential sources of ecosystem pollution.

References

1. Australian Explosives Industry Safety Group Inc: Code of Practice, Mobile Processing Units, 4 edn. (2018)
2. Božić, B.: Miniranje u rudarstvu, graditeljstvu i geotehnici (Blasting in mining, civil engineering and geotechnics) Geotehnički fakultet Varaždin (1998)
3. Guidelines for Bulk Explosives Facilities, Minimum Requirements, Explosives Regulatory Division, Explosives Safety and Security Branch, Minerals and Metals Sector. Natural Resources Canada (2014)
4. Persson, P.A., Holmberg, R., Lee, J.: Rock Blasting and Explosives Engineering. CRC Press, USA (1994)
5. Report, Explosion Accident during Mobile Production of Bulk Explosives, Norwegian Directorate for Civil Protection (DSB) (2015)
6. Bajric, S., et al.: Study on indicators of application of equipment for the production of explosive materials according to Nalim technology, Mining Institute Tuzla (2015)
7. Bajric, S., Catovic, F.: Nalim production technology of commercial explosives, potential hazards and prophylactics. In: International Conference – SONT, Sibenik, pp. 51–53 (2016)
8. Bajric, S., Isabegovic, J.: Research of potential hazard from the atmosphere in the production of explosives according to Nalim technology and prophylactics. In: International Conference – Air Quality, Monitoring, Legal Legislation, Zrenjanin (2011)
9. Singh, S.P.: New trends in drilling and blasting technology. Int. J. Surf. Min. Reclam. Environ. **14**(4), 305–315 (2000). https://doi.org/10.1080/13895260008953338

Urban Stormwater Management – New Technologies

Suvada Jusić(✉), Emina Hadžić, and Hata Milišić

Faculty of Civil Engineering, University of Sarajevo, Ul. Patriotskelige 30, 71000 Sarajevo, Bosnia and Herzegovina
suvada_jusic@gf.unsa.ba

Abstract. Urbanization and climate change have negative effects on the changes of natural hydrological regime (precipitation and runoff regime), which results with more frequent floods and landslides. Besides of the increased quantity of the stormwater that drainage channels have to collect, negative consequences of urbanization are also evident through increased pollution of runoff stormwater. Therefore, the drainage system and stormwater management should be adapted to these changes so that the negative effects of new hydrological conditions of precipitation and runoff and stormwater pollution in the urban areas are mitigated. This article aims at presenting challenges and opportunities for the advancement of stormwater management practices in urban areas. Some of urban stormwater management technologies are presented. These technologies are more used in urban areas of some countries of Europe.

Keywords: Urban stormwater · Stormwater management · Decentralized technologies · Runoff hydrograph · Green roof · Porous pavements · Bioretention

1 Introduction

Half of the world's population lives in urban areas and it is expected that, by 2050, that figure will rise to above two-thirds. Cities have paved over natural green spaces to make way for streets, homes, and commercial developments. Urban stormwater can be defined as the extreme runoff from pervious and impervious surfaces that include roofs, driveways, pavements, footpaths, and roads infrastructure characteristic of urban areas. Actually when it rains, urban stormwater no longer has an opportunity to sink into the land and recharge groundwater basins. Also climate change has caused more frequent and intense storms. Increased urbanization and climate changes have a direct impact on the local hydrologic cycle. Due to climate change and urbanization stormwater volumes and pollution are getting more and more important [1]. If the effects of urbanization and climate changes are not appropriately managed, channel geomorphology and aquatic ecology will degrade, stream base flow will decrease, water quality will diminish, and flooding frequency will increase.

Managing stormwater runoff is a particular challenge in urban areas. The design of conventional stormwater management systems is currently undergoing scrutiny and

I. Karabegović (Ed.): NT 2019, LNNS 76, pp. 790–797, 2020.
https://doi.org/10.1007/978-3-030-18072-0_90

revision [2]. The result of that revision is modern sustainable urbane drainage methods and measures available for the management of stormwater.

2 Urban Stormwater Management

Extreme urban stormwater events are considered a threat to urban infrastructure, urban economy and ecosystem, when not planned and managed properly. Urban stormwater management (USWM) is tool to provide flow rate control and also adequate control of runoff quality [3]. Managing stormwater needs to start with an understanding of quantity and quality aspects of the water that will be anticipated for an urban area. Generally there are two approaches of stormwater management described below.

2.1 Conventional Approach

Current conventional urban drainage systems were built to manage stormwater for purpose of flood control. The networks, called municipal combined or separate storm sewer systems, generally contain numerous elements, including open channels, catch basins, road-drainage systems, curbs, gutters, ditches, and underground storm drains. The main idea of drainage design that stormwater is a waste product and must be removed as quickly as possible away from the source and into adjacent rivers and lakes [3, 4]. These systems were usually installed without considering water quality aspects. As this conventional approach focuses primarily on managing stormwater for flood protection it will not lead to a path for sustainable development.

Generally impacts of urbanization and climate change also, on hydrograph are increased peak discharge and volume of runoff, reduce time of concentration, reduced base flow from the catchment and increased drainage flow [2]. When comparing the hydrographs of natural surface and after installation of (conventional) storm sewers, the latter is characterized by a higher total runoff volume, a higher peak flow and a shorter time of concentration (Fig. 1).

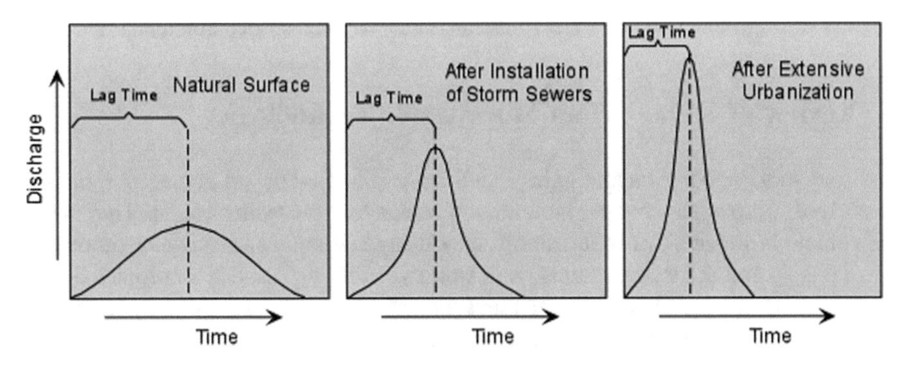

Fig. 1. Effect of urbanization on runoff hydrograph [2]

In fact, it is taking less time for surface flow to reach streams (time of concentration) causing reduced stream flows to extremely low levels for extended periods simply because the watershed is draining so quickly. Finally, there are increased surface flow velocities due to a combination of higher peak volumes, reduced time of concentration and smoother hydraulic surfaces of drainage pipes.

Very often, conventional drainage system don't have enough design capacity for reception all increased stormwater volumes caused by (after) extensive urbanization and climate change impacts (Fig. 1). In that situation more water is leaving the conventional system causing more frequent and severe flooding.

2.2 New Urban StormWater Management Approach

Modern (integrated and sustainable) stormwater management should aim at both flood control and pollution control especially because of the EU Water Framework Directive (WFD) which emphases the control of diffuse pollution as a key factor in enabling good ecological status. This new urban stormwater management (USWM) approach provides methods that allow source control to handle the quality and quantity of the runoff at local level - at or nearby the source (decentralized technologies). These methods are named differently: 'Sustainable (Urban) Drainage Systems' (SUDS), 'Low Impact Development' (LID) or 'Best Management Practices' (BMPs) [4, 5]. These methods to stormwater management represent a diverse range of control procedures, which integrate stormwater quality and quantity control as well as enabling social and amenity perspectives to be incorporated into stormwater management approaches at the source.

USWM techniques are designed to maximize stormwater reduction and provide flow rate control [1]. Also, the goals of modern USWM design, is typically to preserve groundwater, prevent geomorphic changes in waterways, prevent flooding risks, protect water quality, and maintain aquatic life. A common thread found in all those sustainable concepts relates to three core benefits when a shift is made away from traditional to more sustainable approaches [6]: - a more 'natural' water cycle (hydrograph is more similar to first graph of Fig. 1 (natural surface)); - enhancement of water security through local source diversification; and - water resource efficiency and reuse.

3 Review of Some Urban Stormwater Technologies

The four most popular technologies identified in the Europe (in some countries) are green roof, porous pavements, bioretention basins and bioswales [6, 7]. They reduce the volume of urban stormwater runoff, they delay and reduce stormwater runoff peak flow (Fig. 1) and reduce pollutants from stormwater flow. These technologies offer an option for decentralized stormwater management (source control system) from urban areas and they are explained briefly in this chapter.

3.1 Green Roofs

A green roof system is a vegetative layer grown as an extension of an existing roof. It is built on new and existing roof structures which need to be prepared to fit this special purpose [8, 9]. For example, it needs to have a good waterproofing and root repellent system; it needs to include a drainage layer and a filter cloth, a mulch layer and lightweight growing medium and plants (Fig. 2) [6].

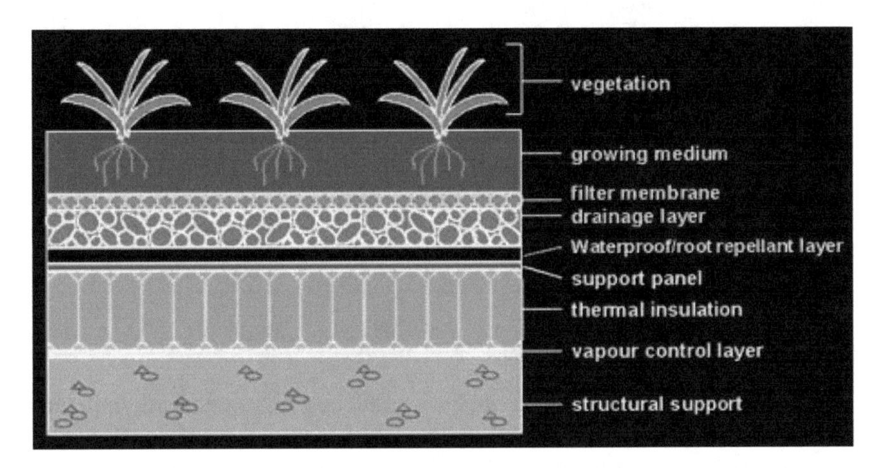

Fig. 2. Schematic representation of principal green roof components [6]

Although the initial investment cost of the green roof technology is higher than standard roofing, the costs can be recovered by different benefits/advantages (aesthetically pleasing and recreational area, increased durability of flat roofs, reduction of building energy consumption, improvement in air quality etc.). Three factors have a major influence on the hydraulic as well as pollutant removal performance of green roofs: i. Precipitation (duration, intensity); ii. Substrate layers (type, thicknesses); and iii. The types of vegetation used in the green roofs. Using an appropriate design which integrates different technical options can prove useful to increase the efficiency of green roofs in different regions and under different climatic conditions. The technique of applying green roof systems is well developed in many European countries (for example, green roofs are commonly used in Germany and Sweden, also in Turkey, Italy etc.).

3.2 Porous Pavements

Pervious pavements are permeable surfaces where the runoff can pass and infiltrate into the ground. The system allows the majority of water to be stored and infiltrated into the sub-grade soil and the excess water flows through an underdrain system (Fig. 3) [2, 10]. Types of porous pavements are porous asphalt pavements, porous concrete pavements and garden blocks. Porous asphalt pavements are popularly used in urban areas in roads and parking lots. Porous concrete pavements are used in open walkways parking areas. Garden blocks are used in pavements in gardens that are only used for walking.

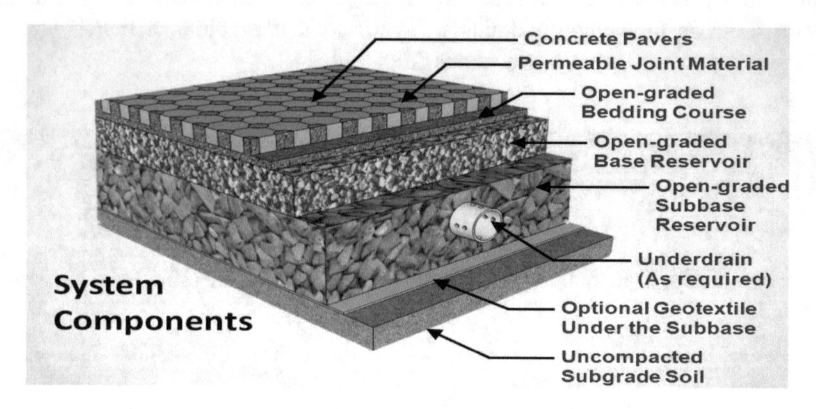

Fig. 3. Permeable pavements system components – typical elements [11]

Porous pavements have higher initial capital cost than standard impermeable pavement however its overall cost can be lower considering the additional cost of the associated drainage infrastructure (curb, catch basins, piping, and ponds) for standard impervious pavements [6].

Application of porous pavement requires the knowledge of the local soil and groundwater condition. Permeable concrete segmental paving emerged in Germany, Sweden and Austria in the 1980s as a means of flood mitigation. In Germany, porous pavement has been installed with increasing frequency, more than any other country in the world.

3.3 Bioretention Basins

It is a shallow excavated surface depression containing mulch and a prepared soil mix and planted with specially selected native vegetation that captures and treats runoff (Fig. 4). During storms, runoffs collect in the depression and gradually filter through the mulch and prepared soil mix and root zone [11, 12]. They are commonly located in parking lot islands or within small pockets in residential land uses.

Biorentention systems have relatively low construction and maintenance costs as well as low maintenance requirements. Like other infiltration based stormwater management systems, their design and performance depends on the availability of permeable soil layer. Bioretention systems help to recharge groundwater, however they also pose a threat to pollute groundwater and soil if design guidelines are not followed properly. The filtered runoff can either infiltrate into the native soil or be collected in a perforated underdrain and returned to the storm sewer system.

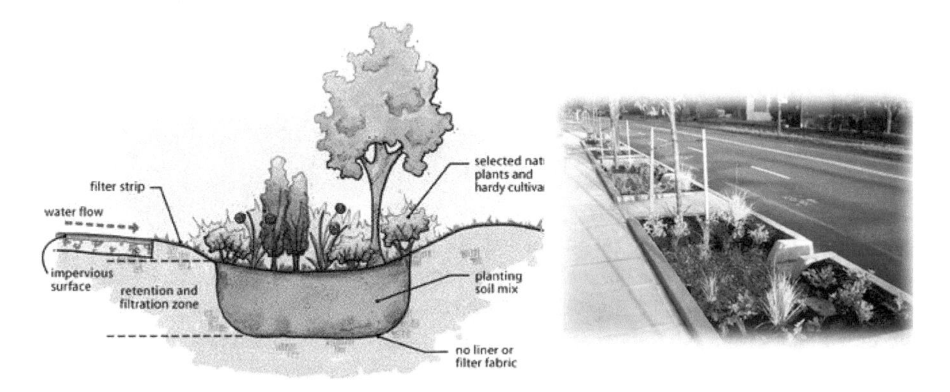

Fig. 4. Typical design of a bioretention area with underdrain [2]

3.4 Bioswales

Bioswales are landscaping features that slow and collect polluted stormwater runoff where it will infiltrate soils and be treated by natural elements [10]. Above ground, bioswales can be subtle and feature typical turf grass or designed as an attractive flower garden with native plants and grasses. Whatever vegetation is used, it is important that it protects against soil erosion. Bioswales are similar to rain gardens but whereas rain gardens are typically smaller and used for residential purposes, bioswales are designed to handle larger quantities of water generated from impervious surfaces like parking lots and city streets [12]. They tend to be long and narrow, often require engineered soils for adequate drainage, and are much deeper than a typical rain garden. Bioswales are often parabolic or trapezoidal in shape and should be large enough to handle the amount of rainfall received in 24-h during a 10-year storm event at your location. In order to facilitate stormwater infiltration, bioswale soils may need to be amended with compost and sand. In some cases, rock trenches or perforated underdrains can be installed down the center of the swale (Fig. 5).

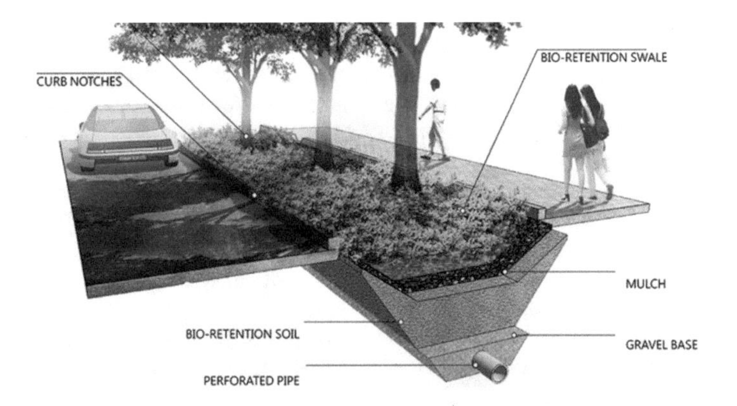

Fig. 5. Typical design of a bioswale of city street [12]

When stormwater infiltrates a bioswale, the purified water slowly recharges groundwater and prevents contamination of our waterways with polluted, unfiltered stormwater runoff.

4 Conclusion

In contrast to the conventional urban stormwater management (USWM) approach, a source control approach (by new USWM decentralized technologies), can be integrated into a catchment-wide planning framework of community development, with a focus on green space, native landscaping, natural hydrologic functions (hydrographs of natural surface) and various other techniques to generate less runoff. Decentralized technologies design shows that it's possible to effectively manage runoff while also creating inviting, attractive landscapes.

Decentralized solutions may also slow down the augmentation of existing infrastructure. For example, conventional water and wastewater pipe network design has to cater for peak demands, i.e. infrastructure needs to be large enough to handle peak loads. If decentralized solutions could mitigate these peak loads, i.e. provide additional local capacitance, investments could be deferred, resulting in a substantial reduction in capital costs. Which type of USWM decentralized technology is appropriate for application in a particular location depends on a set of different factors, including infiltration capacity, groundwater level, soil permeability and contamination, surface runoff characteristics, local climate, land availability and ground slope.

References

1. Andrew, W.S.: Stormwater management performance of green roofs, Thesis and Dissertation Repository, The University of Western Ontario, p. 121 (2015)
2. Jotte, L., Raspati, G., Azrague, K.: Review of Storm Water Management practices - Raport. SINTEF Building and Infrastructure, Trondheim (2017)
3. Zhang, D., Gersberg, R.M., Jern Ng, W., Tan, S.K.: Conventional and decentralized urban stormwater management: a comparison through case studies of Singapore and Berlin. Germany Urban Water J. **14**(2), 113–124 (2017)
4. Cone, W.C.: Stormwater management trends: a review of tools, techniques and methods for design and development of theland with implications for sustainable design. Landscape Archit. Reg. Plan. Masters Projects (2005). https://scholarworks.umass.edu/larp_ms_projects/11
5. Hoang, L., Fenner, R.A.: System interactions of stormwater management using sustainable urban drainage systems and green infrastructure. Urban Water J. **13**(7), 739–758 (2016)
6. Trincheria, J., Yemaneh, A.: New Knowledge on Urban Stormwater Management Final Report of the BalticFlows Project. Hamburg University of Applied Sciences and Technical University of Hamburg-Harburg (2017)
7. Maeda, P.K., Chanse, V., Rockler, A., Montas, H., Shirmohammadi, A., Wilson, S.: Linking stormwater best management practices to social factors in two suburban watersheds. PLoS ONE **13**(8), e0202638 (2018)

8. Kok, K.H., Sidek, L.M., Abidin, M.R.Z., Basri, H., Muda, Z.C., Beddu, S.: Evaluation of green roof as green technology for urban stormwater quantity and quality controls. In: IOP Conference Series: Earth and Environmental Science 4th International Conference on Energy and Environment (2013)
9. Shafique, M., Kim, R., Kyung-Ho, K.: Green roof for stormwater management in a highly urbanized area: the case of Seoul, Korea. Sustainability **10**, 584 (2018). https://doi.org/10.3390/su1003058. https://www.mdpi.com/journal/sustainability
10. Askarizadeh, A., Rippy, M.A., Fletcher, T.D., Feldman, D.L., et al.: From rain tanks to catchments: use of low-impact development to address hydrologic symptoms of the urban stream syndrome. Environ. Sci. Technol. **49**(19), 11264–11280 (2015)
11. Hinman, C.: Rain Garden Handbook for Western Washington, A Guide for Design, Maintenance, and Installation. Washington State University, Department of Ecology State of Washington, (2013). https://fortress.wa.gov/ecy/publications/documents/1310027.pdf
12. Echols, S., Pennypacker, E.: Artful Rainwater Design: Creative Ways to Manage Stormwater. Island Press, Washington (2015)

Forecasting Accessories Forming on the Basis of Historical Prototyping

M. V. Taube, N. V. Bekk, and Irina Andreevna Boychenko(✉)

Department of Industrial Design, Novosibirsk State University of Architecture, Design and Arts, 630000 Novosibirsk, Russia
ir.boychenko@mail.ru

Abstract. The basis of modern design for women's accessories is a statistical information obtained through questionnaires distributed among the population. However, this method cannot meet the requirements of medium-term forecasting, which would be more beneficial in modern conditions. The medium-term forecast is based on historical databases. It is reasonable to start collecting data for the medium-term forecast, referring to the beginning of the XIX century. The era of Empire style was the founder for the development of women's and men's accessories thanks to the changed way of life and the way of life people got used to. The French revolution changed people's attitude to life, thus becoming a trigger for the formation of modern society in our understanding. A special aspect in the formation of society was the emergence of a dress code the phenomenon we understand now, but which was completely innovative then. Dress code soaked through all social classes, dictating to them how to look for each and every moment in their life. Partially because of its emergence a need appeared for sophisticated system not only in costumes but also for accessories, especially for women. Also, The French Revolution led people to push their own horizons, relocate more often and get out of their habitat, which undoubtedly led to the need to carry things, and therefore dictated the shape and volume of leather goods. The most distinguished female images at that time were: home, ball room, outdoor, social. Home image did not require special devices for carrying things. In contrast to the home, ball room, outdoor and social required some accessories, which formed the basis for the first three types of design of women's leather accessories. The same types of design are easy to detect through 20th century, no less relevant they are now. Their design and design criteria, no ware the basis for building medium-term forecasts.

Keywords: Assortment · Leather goods · Historical prototypes · Forecasting · Constructions

1 Introduction

The problems of modern shaping of leather goods, which are often interpreted as accessories, include a wide range of components of women's wardrobe. It is believed that men's wardrobe provides less options for the number of products.

The basis for research is mostly a statistical analysis of sociological data, to which research on contemporary gender relations has recently been attributed. However,

I. Karabegović (Ed.): NT 2019, LNNS 76, pp. 798–802, 2020.
https://doi.org/10.1007/978-3-030-18072-0_91

analysis of only the investigation, unsupported material about the causes, cannot provide a complete picture of the forecast of the planned product range. This means that it will not give a likely answer, how to create a positive dynamic in the development of one or another range.

Prediction based on historical databases can be considered more objective. As early as the 19th century, psychology and biology specialists proposed the concept of genetic memory, the essence of which is that many behavioral lines are laid down in people of previous generations of ancestors. This is one of the reasons explaining the concept of cyclical fashion, as well as the possible possibility of predicting new products from historical prototypes.

2 Other Topics

Analysis of the design characteristics and the shaping of the costume [1–3] showed that research can be considered reliable, taking the 19th century as a starting point. From the point of view of the genetic code, this time interval is the most significant for the analysis of the suit, both in duration and transformation. Information on the genetic code, referring even to the mid-end of the 18th century, will be far for us and hardly adaptive due to the mismatch of the mentality of the average person then and now. The system for interpreting data from the 19th century is amenable to modern understanding. After the French Revolution, it is customary to take the countdown of a new era in the life of mankind. People have a different attitude to life, their position in society has changed, and this has resulted in education and new needs. One of the needs was related to moving around the city or between cities - carrying things. Previously, products for carrying items and things were quite simple in form of bales, baskets, chests, boxes. Their dimensions, in principle, are not very different. They were bulky, and not always one person could transfer them. For free movement around the city, such things are not convenient for ordinary citizens, especially women and girls. Therefore, it is possible to call the revolutionary appearance of accessories, with other functions and designs. Prototypes of models of accessories models existing in this period in men's and women's wardrobe were considered. After the French Revolution, a radically new costume emerged in society. Now it is called one term "Empire Style." However, the costume of this period will be correctly designated as the costume of the Directory and Consulate due to the features of its cut and the materials used. This costume marked not just the birth of a new style of clothing. With it starts the development of two important phenomena in modern society - accessories and dress code. Today, the dress code phenomenon pervades the whole society, a business dress code, official, evening, corporate. And it's hard for us to imagine what was once wrong. The appearance of the dress code at the beginning of the 19th century was a reaction of European society to the process of the abolition of class inequality that began after the French Revolution. The dress code with its principle and rules for aristocrats and ordinary citizens exacerbated the difference in the wealth of the rich and the poor, singled out people with capabilities and taste, discarding those who did not have such opportunities.

One of the rules of good tone was not to appear in the same dress at the social events twice. Poor families of aristocrats had to look for ways to update their once-worn dress in order to refresh the image. The dress code regulated what time of day the material is preferred and what kind of cut the dress or costume should have, be it a home image or a ball one. The problem of dress code was easier to resolve with a help of accessories. And in modern reality, we also note that accessories are status products and clearly meet the requirements of the dress code. If we talk about accessories, the costume of the Directory and the Consulate was originally exclusively white or monotonous light shades. To breathe life into the "white marble statues", the girls used jewelry, whether it be bracelets on their wrists, ankles and forearms, necklaces, fieronnieres, hair bands, earrings, pendants, brooches, etc. Also stoles were widely used, which were distinguished from modern ones by a larger size, 2–2.5 human height (Fig. 1). In addition, traditional ribbons, shawls and scarves were used.

Fig. 1. Costume with attributes in the Empire style

The silhouette of Empire style, after 5 years, gradually changed and approached a new image of the Romantic era. The colors and the concept of female and male images, their fullness began to change.

It is undeniable that the culture of women's handbags was born during the Empire, together with the dress code for clothes, they also became extremely important.

If women's clothing carried more information in a hierarchically-social order, pointing to the social status of the girl, the status of the event itself, the wealth of the family and much more, the accessories were a responsive way to express themselves. Attempting to convey to those around you "I", to reveal your identity. Handicraft has become an extremely popular tool for self-expression. Girls embroidered, weaved from beads, worked with applique.

3 Conclusion

Thus, there are three types of designs of primordially feminine accessories (Figs. 2 and 3).

Fig. 2. Type of construction "pouch"

The simplest design was a "pouch" (Fig. 2), a handbag made round with a neck, tied with cords, at the same time serving asa handle. In the era of the Empire style, it was customary to fasten such handbags to the belt of a dress and wear it rather low, at the very hem, therefore the handles at that time were extremely long. Such handbags were used everywhere and with any costume, traditionally they belong to ballroom costumes. Complicated design is a bag with a handle-support (Fig. 3).

Another design was the type of handbag with a frame lock (Fig. 3). The side details were made by hand, the dimensions of the bags allowed to decorate them not only with flower compositions, but also with real images of people and everyday life.

A new type of design has appeared - a bag with a valve. Girls could not make such bags from the skin of their own; they had to be ordered, which meant the high cost of the product. Also, with leather goods, it was not customary to appear at social events. Such handbags were, however, convenient in any weather, which determined their initial use as "street" products, visit stores with purchases, or survive a long journey.

Thus, at the beginning of the 19th century, three main types of female products were laid, which with minimal transformations reached the 21st century.

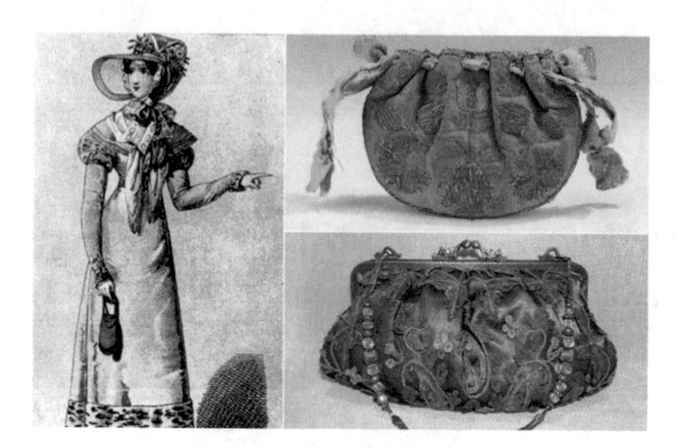

Fig. 3. Types of designs of women's handbags of the Empire era: handle-holding, bag with frame lock

References

1. Parmon, F.M.: Costume Composition: A Textbook for Universities, 318 p. Legprombytizdat, Moscow (1997). (In Russian)
2. Cherenkova, S.S., Markelova, A.A., Beck, M.V.: The development of leather goods in the XXI century. On Sat Technical regulation: the basic basis for the quality of materials, goods and services: international. Collection of scientific papers. IPS & P (branch) DGTU, Shakhty, pp. 59–60 (2013). (In Russian)
3. Beck, M.V., Fedorov, S.S., Kozlova, V.V., Klyuev, I.V.: Dressing Women Bags: A Tutorial. Publishing House - South, Krasnodar, 48 p (2015). (In Russian)

Customizing of the Techniques Used for Designing of the Orthopedic Footwear

N. V. Bekk[1], V. V. Kostyleva[2], and T. S. Lapina[3](✉)

[1] Novosibirsk State University of Architecture, Design, Arts, 630000 Novosibirsk, Russia
[2] The Kosygin State University, 117997 Moscow, Russia
[3] Novosibirsk Institute of Technology of The Kosygin State University, 630000 Novosibirsk, Russia
tatianaana@rambler.ru

Abstract. The search of scientific information was carried out in order to expand opportunities for rehabilitation of people having low limbs (legs and feet) diseases. The techniques were developed for customizing of the orthopedic footwear intended for different abnormalities of low limbs. The offered techniques were tested in the industrial conditions having proved that their application decreases the costs of designing process and improves the quality of ready-made footwear.

Keywords: Footwear · Customizing · Designing · Orthopedic boots · Correction · Feet abnormalities

1 Introduction

The development of means of technical rehabilitation and creation of the barrier free environment cover a lot of spheres including shoe-making industry. Mass-produced footwear cannot meet the needs of people having low limbs deformations. In spite of numerous developments of the Russian and foreign authors in the field of designing of mass-produced footwear [1–3], the orthopedic footwear is paid far less attention. The conventional type classifications of orthopedic footwear and recommendations of colour solutions of the potential products have not been developed, as well as there are no available techniques for designing of customized orthopedic footwear. Thus, one of the goals of the improvement of the rehabilitation means is a creation of the methodical approach to the development of the orthopedic footwear designing.

2 The Techniques of Designing of the Orthopedic Footwear

Having studied the abnormalities in movements and having carried out the anthropometric examinations of feet, the classifications of corrective elements of an orthopedic insole, hard details and typical constructions of orthopedic boots were developed, as well as the classification of boots was created taking into account the rehabilitation effect. All the researches mentioned above formed the basis for development of the

I. Karabegović (Ed.): NT 2019, LNNS 76, pp. 803–807, 2020.
https://doi.org/10.1007/978-3-030-18072-0_92

techniques used for designing of the constructions of orthopedic boots taking into account various feet deformations.

The customized footwear as the mean of rehabilitation for the unable people is made in two ways – as complementation of the orthopedic footwear intended for an impersonal customer with an individual insole and designing of footwear by individual parameters of customer's feet with an individual orthopedic insole.

In manufacture of the orthopedic footwear by individual parameters the cases occur when the left and right feet differs from each other greatly. Where in a designer must create the anatomically correct footwear and achieve the best possible esthetic appearance of the designed product. In manual designing of orthopedic footwear (without usage of the computer technologies) it is necessary to consider the peculiarities of the most common variants of abnormalities of low limbs.

2.1 The Technique for Designing of Footwear with Different Dimensions

The designing of footwear for different feet dimensions worth to consider firstly. In case that one of the feet matches the average parameters and there is a ready pattern for making the boots, it is necessary to modify the ready-made base pattern model for the foot with increased girth by fixation in the area of the most bulging heel and those of balls. It is important to note that in a number of cases the quarter lasts are used. Hence, to create the construction with high rehabilitation effect it is necessary to ensure close fitting of the upper last to the ankle part of the last. After trying of the base pattern model on the last it is necessary to measure the value of increasing in the area of a quarter paying attention to the distribution over the ankle part of the last.

In case of increasing of the amount of the balls area of a foot the base pattern model must be fixed regarding the curve of the vamp and then the adjustments are to be made as it will be described by the method of designing of footwear intended for a short leg.

In terms of visual perception the proper construction of a top line is very important. Based on the analysis of the difference of dimensions in a single semi-pair from a visual point of view the conclusion was drawn that top lines are recommended to design in one size relating to the back seam in case that the dimensions difference is not more than 21%. If the dimensions difference exceeds the given value it is recommended to increase the top line leaving the equal distance to the quarters edges.

With increase of dimensions the front arch of a quarter is shortened. Wherein the change of the appearance of means used for fixation of the boot on a foot should be taken into account. For the laced boots the decrease of the number of eyelets in a larger semi-pair is appropriate. Wherein the distance from the side eyelets to the quarters edges must remain equal (Fig. 1). In case of the fixation means such as Velcro tapes or straps with buckles the width of straps should be considered beforehand in the relation to the both semi-pairs. When the dimensions difference is more than 45% the number of straps of the semi-pairs can be different according to the taste of a customer.

The very special case from the point of view of designing is the creation of orthopedic boots for lower limbs having significant swellings (more than 42% of average foot fitting). In this case various constructions of boots can be offered with possibility of adjustment of fitting parameters of inner space of the boot (laces, Velcro tapes, inserting of elastic materials) and easy entry for the foot. The special attention

Fig. 1. The boots designed by the offered technique.

should be paid to designing of warm boots with height covering the swellings of an ankle or a calf when the sharp contrast between the fitting parameters occurs.

In the considered case two additional measures of a foot are needed (Fig. 2). The first one is the height of the area without a swelling in the heel part (L1) used for the calculation of the counter height and building of the bootleg angle. The second measure is the length of the area without a swelling in the toe part (L2) used for the calculation of the vamp-quarter attachment point location, as well as the location of a cut tongue which is necessary for escaping the folds in the area of the foot bending. The area where the tongue overlaps the vamp must be shifted for 0.7–1 cm from the place of the swelling towards the toe.

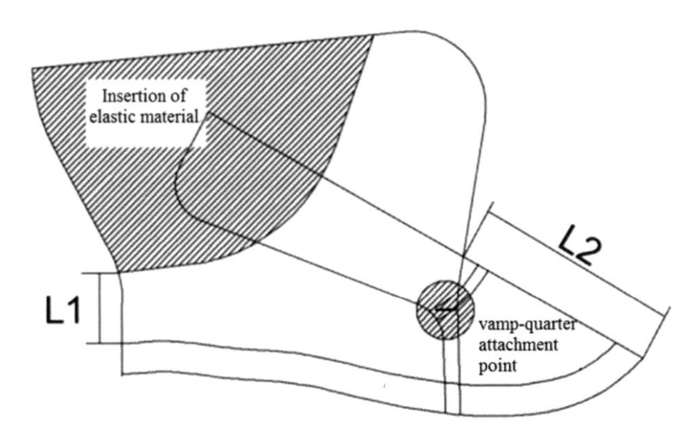

Fig. 2. The key moments of designing of the high boots used for significant swellings of legs

To achieve the best fitting of the model on a foot it is recommended to use the inserting from elastic material providing a close fitting of a leg and adjustable variability in dependence on the swelling amount. The construction of a laced boot is recommended to use for the best fitting parameters adjustment. To make the boot wearing more comfortable a zip can be used on the inner side.

2.2 The Technique for Designing of Footwear with Different Heights

The purpose of different heights of boots belonging to the single pair can be both a short leg and significant deformations of one foot. The boots of different heights in one pair is prescribed by a doctor and agreed with a customer beforehand. Designing the given constructions needs strict following the height and width parameters of the boot quarters.

It worth to take into account that in the height difference of not more than 3 cm the number of such fixing elements as buckles and Velcro tapes are recommended to remain equal. In case of lacing the number of eyelets can differ. In the height difference of more than 3 cm the number of buckles and Velcro tapes must be enough to fix the foot safely. From esthetic point of view it is advised to remain the heights of counters and a top line equal. If the midsole filler is made from cork, the decorative elements of an instep girth area must be placed on the anatomically correct level by stretching the edges of the details upward and downward the vertical axis of the footwear.

2.3 The Technique for Designing of Footwear with Different Sole Lengths

In the made-to-measure orthopedic boots the difference of sole lengths of 10 mm is common and does not present any difficulties for designing. In the given case the length addition can be carried out evenly by lengthening the vamp and a back part of the quarter.

The technique for designing of boots with different sole lengths is applicable when the difference between the right and left feet is more than 10 mm. The designing of top lines is necessary to make by the technique described above. Designing the vamps it is necessary to calculate the lengths of the both semi-pairs of the footwear. combining the different sole length with midsole filler in the toe part compensating the shortness of a leg, the height of the filler must be considered increasing the length of the vamp. The whole quarter or the one consisting of details is expanded evenly along the horizontal axis. Where in the in step girth curve must be built for every semi-pair separately.

The lengths and heights of details which the quarter is to consist of are drawn by a designer based on the most harmonic visual perception of the projected construction. Distributing the fixing elements (buckles and Velcro tapes) over a foot it is necessary to consider the disease of a customer and availability of the hard details in the boot. Thus, when the shorter leg or both feet don't need the additional fixation, the number of fixing elements can be equal. In case that both feet need the additional fixation in the inner space of a boot, the number of fixing elements must differ.

2.4 The Technique for Designing of Footwear for a Short Leg

The process of designing of the orthopedic boots having the midsole filler intended for compensation of legs shortness begins as in the cases previously described with selection or construction of the base pattern model for a normal leg (not shortened) leg, fixation of it and measurement of the potential corrections. Besides measurement of the quarter amount it is necessary to make adds in the area of a feather edge.

After making the corrections the refinement of the base pattern model is carried out relating to the coordinate axes. To do this, having put the corrected base pattern in the coordinate axes it is necessary to make changes in the inclination of the quarter top relatively to the base pattern model intended for a normal leg by putting it in the area of balls and a heel. Then the shape of the ready-made model is corrected. The last stage of the process presents the transferring of the drawing lines from the initial base pattern to the designed model making the necessary corrections. The designing can be carried out both manually and using a computer.

3 Conclusion

The given techniques have been introduced at the Russian prosthetic and orthopedic enterprises. As the experiment eighty-four pairs of boots intended for customers having various foot abnormalities were designed. The examples of boots made by the developed designing techniques are shown in the Fig. 3 to demonstrate the results of the done work.

Fig. 3. The examples of boots made by the developed designing techniques.

With probability of 87% (seventy-three models) the ready-made models didn't need any corrections. The rest 13% needed to make not more than two corrections which were connected with the complicity of the designed constructions and individual making of them.

References

1. Maximova, I.A.: Development and justification of the technology of special footwear manufacture in the conditions of mass production, Moscow State University of Design and Technology, Moscow (2009)
2. Bekk, N.V.: Modelling, designing and quality control of orthopedic footwear for children and adults, Infra-M, Moscow (2016)
3. Kostyleva, V.V.: Development of shoe constructions by orthopedic parameters, Moscow State University of Design and Technology Moscow (2016)

Author Index

I. Karabegović (Ed.): NT 2019, LNNS 76, pp. 809–812, 2020.
https://doi.org/10.1007/978-3-030-18072-0

Printed in the United States
By Bookmasters